Calculus

Ideas and Applications

Calculus

Ideas and Applications

Alex Himonas

Alan Howard

University of Notre Dame

John Wiley & Sons, Inc.

ASSOCIATE PUBLISHER	Laurie Rosatone
ACQUISITIONS EDITOR	Michael Boezi
MARKETING MANAGER	Julie Z. Lindstrom
FREELANCE DEVELOPMENTAL EDITOR	Anne Scanlan-Rohrer
SENIOR PRODUCTION EDITOR	Ken Santor
SENIOR DESIGNER	Kevin Murphy
ILLUSTRATION EDITOR	Sigmund Malinowski
ILLUSTRATION STUDIO	Techsetters, Inc.
PHOTO EDITOR	Hilary Newman
PHOTO RESEARCHER	Elyse Rieder
PROGRAM ASSISTANT	Heather R. Olszyk
COVER PHOTO	© James Randklev/The Image Bank/Getty Images

About the cover: Calculus can be applied to many fields, whether you are a wildlife biologist tracking the population growth of a fish species for research purposes, or the owner of a fishing company tracking the same species to determine your company's future profit or loss. The same calculus techniques can help sociologists predict human population growth and its impact on resources.

This book was set in Times Roman by Techsetters, Inc. and printed and bound by the Von Hoffmann Corporation. The cover was printed by the Von Hoffman Corporation.

This book is printed on acid free paper. ♾

ISBN 0-471-40145-5

Printed in the United States of America

10 9 8 7 6 5 4 3 2 1

Preface

This book is intended for students in business and economics, the life sciences, the social sciences, and the liberal arts. There are two versions: a brief version that is more than sufficient for a one-semester or two-quarter course covering material that includes functions, limits, derivatives, integrals, an introduction to multivariable calculus, and trigonometric functions, with interesting real-life applications throughout; and this full, one-year version consisting of all of the above, plus differential equations and applications, higher order approximations, and probability and statistics.

Approach

In this book we tell the calculus story directly to the student in a conversational style, and we focus on the main characters of the story, suppressing peripheral information that may distract the student from the central themes. At the same time, we make this story interesting and relevant to the student by using well-chosen "real-world" examples close to their interests, both to motivate and apply the mathematics they are learning. We have tried to write a text that is less rambling and more structured than the experimental ones and at the same time more interesting, clear, informal, and lively than the traditional ones. We have tried to write a book that speaks directly to the student, so that the student is able to read it, understand it, learn from it, and gain an appreciation of the subject's enormous power and beauty.

The major differences between our approach and that of conventional texts are the following:

Concepts introduced when most needed. In general, we have taken great pains to find the appropriate place in the text to introduce each new concept. Asymptotes, for instance, are encountered in the first chapter immediately after discussing rational functions, where they seem most simple and natural, as opposed to many traditional texts, which bring them in as an afterthought in the section on using calculus to graph functions.

Early approach to limits. For us, the concept of the limit is the key idea in calculus, leading to both the derivative and the integral and to the asymptotic behavior of functions and sequences. Thus, it is important to focus on the limit before introducing the derivative. We motivate the study of limits by providing a number of applications that do not depend on the derivative, such as the behavior of a rational function near a zero of its denominator. Using limits, we can introduce the exponential and logarithmic functions before the derivative, which is a major feature of our book. Having introduced all the major characters (the functions) of calculus, and armed with the limit, we take the student on an exciting voyage through the ideas of calculus, in which all functions make their contributions both to the theory and applications. Other books may arrive at the derivative earlier, but we cover all the basic material of a one-semester calculus course (by the end of our fifth chapter) in a way that is at least as fast and is certainly more efficient and rich with examples and applications. It also prepares the student better for the later chapters, which deal with more interesting and sophisticated ideas and applications, such as the qualitative behavior of differential equations.

Placement of exponentials and logarithms. We introduce exponentials and logarithms before discussing derivatives, and we stress their application to questions of compound interest and population growth. In our opinion, these functions and their uses are particularly important for this group of students, and it is essential that they have the maximum possible practice in working with them. In addition, having these functions at our disposal makes it much easier to give interesting applications and instructive examples and homework exercises when we discuss the rules of differentiation, particularly the chain rule. In our opinion, the latter is much easier to grasp if the student can see examples involving logarithms and exponentials.

Differential equations. We introduce differential equations early and use them to motivate and practice many of the calculus techniques. In addition to their historical importance in the development of calculus, differential equations appear everywhere in applications of mathematics to engineering, business, and science, including life and social sciences. Thus, we consider it an important goal that our students learn early on how to use differential equations to model real-life situations and how to study their behavior by analytic, geometric, and numerical methods.

Linear approximation. We use linear approximation as the basic tool in explaining a number of facts and techniques, including the Fundamental Theorem of Calculus and Euler's Method.

Informality with integrity. We keep our explanations as clear and informal as possible without sacrificing mathematical integrity. In the same spirit, we try to talk directly to the student and to keep our style from being too stiff or didactic. All theorems, definitions, and rules are clearly highlighted in the text.

Features of this Book

Applications. One of our major goals is to present the student with real and interesting examples and exercises. These examples and exercises come from business and economics, life sciences, and social sciences. They range from simple but engaging human-interest problems (where to place the appliances in a kitchen to minimize walking distance) to important global issues (how the International Whaling Commission can ensure the survival of gray whales by restricting the allowable size of the annual catch). Many of our applications are constructed using real data and include a reference for the source of the data. The Applications Index lists all applications in the text. For examples of real and interesting applications, see pages 143–144, 232, 279–280, 338–339, and 441–442.

Exercises. We provide a large and varied collection of exercises, at all levels of difficulty from drill to challenging conceptual problems, many of which lead the student in exciting and interesting directions. For examples of interesting exercises, see pages 303–304, 434–435, 444–445, and 511–512.

Technology. We make a thorough and thoughtful incorporation of technology, featuring graphing calculators, spreadsheets, and computer algebra systems. The computer is a very powerful tool, but it is more expensive and less portable than a graphing calculator. The use of spreadsheets is prevalent in business, and they can be very effective in a variety of mathematical applications. We stress the use of these devices to enhance understanding not to replace it, with warnings of the pitfalls involved in mindless use of technology. And, we have done it in a non-intrusive way that puts the instructor in charge of how much or little to use, and even allows him/her to avoid it completely without disrupting the flow of material.

Many of the exercises incorporate the use of technology and are indicated by one of three icons: Graphing Utility ⌇ (for using graphing calculators or graphing software), Computer Algebra System 🄲, or Spreadsheet ▦. In addition, Applying Technology headings indicate subsections that cover the use of technology in detail. All of these uses of technology are optional and can be covered at the discretion of the instructor. For examples of our use of technology, see pages 105–106, 185–186, and 285–286.

Problem-Solving Tactics. These suggestions for approaching problems appear in the margins "just in time" where it is anticipated that students might have problems with concepts or techniques. They provide concise problem-solving strategies without interrupting the flow of text. See pages 210, 280, 380, 426, and 517.

Practice Exercises. These short sets of "warm-up" exercises appear just before each problem set. Each set of 2–4 basic exercises helps students get ready for working the exercise set by giving them simple problems with solutions available at the end of the corresponding set of exercises. See pages 186, 302, 361, and 523.

Historical Profiles. These are short essays about the individuals who either created the mathematics discussed or used it in influential ways. They provide students with the human connection to the mathematics. In addition to mathematicians, such as Newton, Lagrange, David Blackwell, and Cathleen Morawetz, we include scientists in other fields, such as Marie Curie and Robert Solow. For examples, see pages 112, 151, 177, and 492.

End-of-chapter Summary, Review Questions, Review Exercises, and Practice Exam. These end of chapter review materials help students test their understanding. The summary lists all major concepts and terms from each chapter. Review questions are four to eight conceptual questions that often require students to explain concepts in their own words. Review exercises provide about 35–50 exercises that cover the skills learned in the chapter. The Practice Exam is 12–25 questions designed to simulate a typical chapter test. See pages 248–252, 310–315, and 457–461.

Projects. Each chapter ends with one or two projects that extend the applications and the concepts learned in the chapter. These can be used for group activities. The main goal of these projects is to encourage active learning and further readings and connections. See pages 253–255, 461–462, and 530–531.

Supplements

The following supplements are available with the text.

Instructor's Solutions Manual. (ISBN 0-471-26640-X) Solutions to all exercises in the text, for instructors only.

Student Solutions Manual. (ISBN 0-471-26639-6) Solutions to every odd-numbered exercise in the text. Available for sale to students.

Instructor's Manual with Test Bank. (ISBN 0-471-26642-6) Contains sample exams for each chapter.

Instructor's Resource CD-ROM. (ISBN 0-471-26641-8) This CD-ROM contains all materials in the Instructor's Manual, as well as the complete solutions from the Instructor's Solutions Manual.

Activites and Technology Manual. (ISBN 0-471-43192-3) This manual consists of class activities and technology features, including *Mathematica* notebooks and Maple worksheets, as well as basic instructions for calculators and Excel worksheets.

eGrade. *eGrade* is an on-line assessment system that contains a large bank of skill-building problems, homework problems, and solutions. Instructors can now automate the process of assigning, delivering, grading, and routing all kinds of homework, quizzes, and tests while providing students with immediate scoring and feedback on their work. Wiley *eGrade* "does the math"... and much more. For more information, visit www.wiley.com/college/egrade or speak with your Wiley representative.

Calculus Machina for Applied Calculus. *Calculus Machina for Applied Calculus* is a Web-based, intelligent software package that solves and documents calculus problems in real time. For students, *Calculus Machina* is a step-by-step electronic tutor. As the student works through a particular problem online, *Calculus Machina* provides customized feedback from the text, allowing students to identify and learn from their mistakes more efficiently. Instructors can use *Calculus Machina* to preview problems and explore functions graphically and analytically. For more information, visit www.wiley.com/college/machina or speak with your Wiley representative.

Supplements for the instructor can be obtained by sending a request on your institutional letterhead to Mathematics Marketing Manager, John Wiley & Sons Inc., 111 River Street, Hoboken, NJ, 07030, or by contacting your local Wiley representative.

Acknowledgements

We would like to thank the following reviewers: Margaret Ehrlich, Georgia State University; Klara Grodzinsky, Georgia Institute of Technology; Tim Howard, Columbus State University; Christine Keller, Southwest Texas State University; Donna Lee McCracken, University of Florida; Michael McJilton, California Polytechnic State University, San Luis Obispo; Dot Miners, Brock University; Steven Plotnick, The State University of New York at Albany; Joseph W. Rody, Arizona State University; Donald R. Sherbert, University of Illinois at Urbana-Champaign; Rienk Venema, Boise State University. In addition, we would like to thank Tracy Johnston for her excellent applications suggestions and her accuracy check of the text and solutions manual, and Neil Wigley and Ed Lodi for doing a wonderful job accuracy-checking the entire text.

This book in different forms has been used at the University of Notre Dame since 1996 for Mathematics 105 and 108. In this final form it incorporates most of the changes suggested by the instructors of those courses, to whom we wish to express our thanks. They are: Andrew Arana, Rebekah Arana-Jager, Katrina Barron, Chris Bendel, Mario Borelli, Mark Buckles, Peter Byers, Amarjit Budhiraja, Ovidiu Calin, Jianguo Cao, Yu C Chen, Peter Cholak, Dan Coman, Michael Dekker, Matt Dyer, Laszlo Feher, Michael Gekhtman, Jennifer Gorsky, Phillip Harrington, Jacob Heidenreich, Jeff Hirst, Dan Isaksen, Mark Johnson, Audi Kirkowski, Julia Knight, Greg Lysik, Charles McCoy, George McNinch, Juan Migliore, Tang Mouktonglan, Chris Monico, Liviu Nicolaescu, Viorel Nitica, John Palmieri, Annette Pilkington, Claudia Polini, Joachim Rosenthal, Alexander Samuel, Brian Smyth, Stefan Stolz, Evgeny Vassiliev, Steve Walk, Michael Weiss, Warren Wong, Jay Wood, and Fred Xavier. In addition, the following members of the Business College faculty provided many helpful suggestions: Thomas Cosimano, David Hartvigsen, Carl Mela, Ram Ramanan, and Ed Trubac. The authors are also grateful to John Affleck-Graves, Steven Buechler, John Derwent, Thomas Frecka, Alex

Hahn, and Laurence Taylor for their involvement in the creation and evolution of Math 108, and to Cheryl Huff for typing the lecture notes of the pilot version of Math 108 in the spring of 1996.

We are especially indebted to Ovidiu Calin and Yu Chen for their invaluable assistance at all stages of the preparation of the manuscript. We express our gratitude to Chris Bendel and John Palmieri for many helpful comments and suggestions, and to John Derwent and Larry Taylor for their help with computer graphics. Also, we would like to thank Aris and Eirene Alexandrou and Nahid Erfan for their assistance during the preparation of the manuscript.

We thank the students of Math 105 and 108 at Notre Dame for their patience and invaluable feedback over the last seven years. In particular, we want to thank Jason Perkins for his assistance.

We also want to acknowledge the great support we received from the secretarial and administrative staff of the Department of Mathematics at Notre Dame: Patti Strauch, Theresa Bishop, Judy Hygema, Robin Lockhart, and Carole Martin.

We would like to thank the wonderful people at Wiley, especially our executive editor, Ruth Baruth, and our associate publisher, Laurie Rosatone; our development editor, Anne Scanlan-Rohrer; our marketing manager, Julie Lindstrom; our program assistant, Heather Olszyk; our production editor, Ken Santor; our illustration editor, Sigmund Malinowsky; our photo editor, Hilary Newman; and our designers, Kevin Murphy and Nancy Field. We would also like to thank Jay Beck of Wiley Custom Publishing for all the help he gave us with the custom versions of this book during the last six years.

For their valuable suggestions and information, we would like to thank Charlie Dresser, Tracy Johnston, Neil Wigley, and Ed Lodi.

Finally, we would like to thank our families for their great patience and unfailing support.

Alex Himonas
Alan Howard

October, 2002

About the Authors

Alex Himonas received his B.S. from Patras University in Greece, and his M.S. and Ph.D. from Purdue University. He is currently professor of mathematics at University of Notre Dame, where he has won the Kaneb Teaching Award and received numerous National Science Foundation grants. His research interests include regularity of solutions to partial differential equations and the Cauchy problem for non-linear evolution equations. Alex grew up in rural Greece, surrounded by five siblings and ninety-nine sheep. His mathematical talent first surfaced when his father asked him to calculate the volume of their wine barrels in order to determine their holding capacity. His first encounter with Riemann sums came when he was invited to help the local surveyor measure the neighboring farms. As a youth, Alex was an avid beekeeper and also had a passion for astronomy. He spent many nights staring at the sky and studying the constellations. These days, Alex returns to Greece every few years. Outside of mathematics, he enjoys swimming, gardening, and family time with his wife and two children.

Alan Howard received his B.A. from Rutgers University and his Ph.D. from Brown University. He is currently professor of mathematics and associate chair at University of Notre Dame, where he has won numerous teaching awards, including the Madden Award for excellence in teaching freshmen, the Shilts/Leonard Teaching Award, the Notre Dame President's Award, and the Kaneb Teaching Award. He has published research papers in complex manifolds and complex analysis. Alan was born and raised in Brooklyn, New York, where the only stars he ever noticed were those who played baseball for his beloved Brooklyn Dodgers. Alan's undergraduate major was English literature, and he has maintained a lifelong interest in reading and writing fiction. In the mid-1980s he was fortunate to have two mystery novels published under the pseudonym of N. J. McIver: *Come Back, Alice Smythereene* by St. Martin's Press and *An Assassin Prepares* by Doubleday. Alan is also an enthusiastic amateur musician, whose taste runs the gamut from Claudio Monteverde to Bill Monroe. His most persistent daydream these days is of sitting in as guest mandolinist with the Nashville Bluegrass Band. While waiting for that to happen, he enjoys music sessions with his wife, daughters, and granddaughters, singing such timeless classics as *There Ain't No Flies on Me* and *The Teddy Bears' Picnic*.

Contents

*Optional

Chapter 10   Higher-Order Approximations   **614**

Chapter 11   Probability and Statistics   **664**

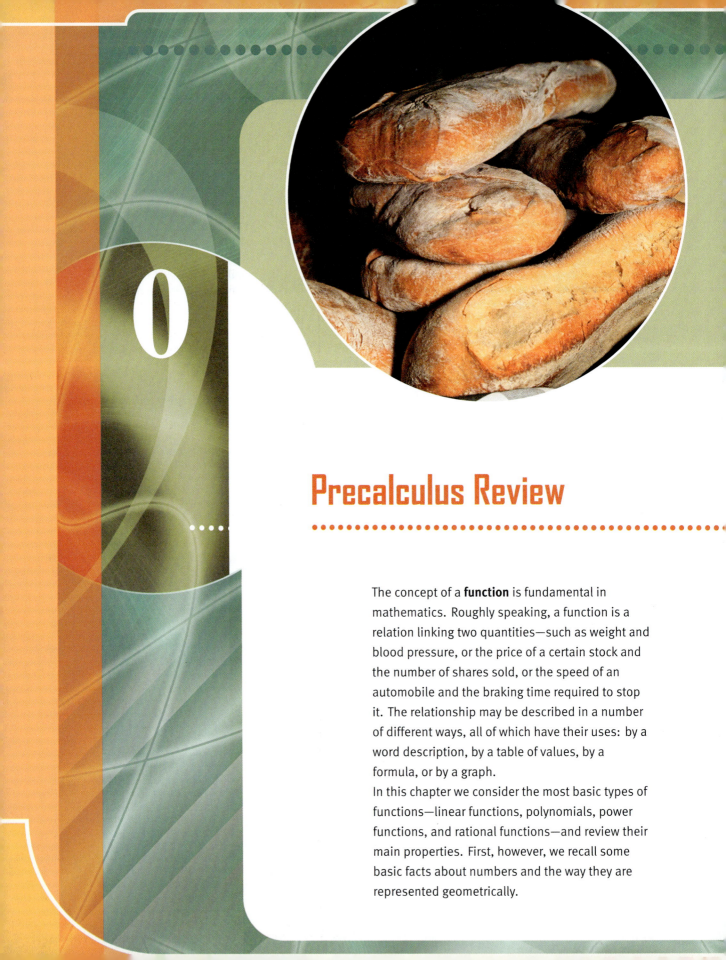

Precalculus Review

The concept of a **function** is fundamental in mathematics. Roughly speaking, a function is a relation linking two quantities—such as weight and blood pressure, or the price of a certain stock and the number of shares sold, or the speed of an automobile and the braking time required to stop it. The relationship may be described in a number of different ways, all of which have their uses: by a word description, by a table of values, by a formula, or by a graph.

In this chapter we consider the most basic types of functions—linear functions, polynomials, power functions, and rational functions—and review their main properties. First, however, we recall some basic facts about numbers and the way they are represented geometrically.

The number system we generally use in both scientific and everyday work is called the **real number system**. The real numbers include the familiar **integers** (or whole numbers), both positive and negative, such as -1, 0, 1, 2, etc., and **fractions**—that is, quotients of integers, such as $\frac{2}{3}$, $-\frac{1}{10}$, $\frac{21}{5}$. (Of course, we can think of an integer as a fraction with denominator 1.) The collection of integers and fractions is called the set of **rational numbers**.

Every rational number can be represented in decimal form. For instance, we all know that $\frac{1}{5} = 0.2$ and $\frac{7}{200} = 0.035$. Some fractions are represented by never-ending decimals with infinitely many places, such as the familiar $\frac{1}{3} = 0.3333\ldots$, where the dots at the end indicate that the pattern continues indefinitely. A less familiar example is $\frac{199}{110} = 1.8090909\ldots$, where the 09 pattern continues indefinitely. (To check the decimal expansion, divide the denominator into the numerator and keep working out decimal places until you see the pattern.) We abbreviate these repeating patterns in the following way, where the overbar indicates indefinite repetition:

$$\frac{1}{3} = 0.\overline{3} \quad \text{and} \quad \frac{199}{110} = 1.8\overline{09}.$$

In fact, every rational number can be represented by a decimal with either a finite number of digits or infinitely many that eventually form a repeating pattern. And vice versa! *Every such decimal is a rational number.*

However, the rational numbers do not make up the entire real number system. There are also **irrational** numbers, such as $\sqrt{2}$ (which is the length of the diagonal of a square whose side has length 1), and π (which is both the ratio of the circumference of a circle to its diameter and the area of a circle of radius 1). It can be proved that neither of these numbers is equal to the ratio of two integers.[1]

Irrational numbers may also be represented by decimals—those that have infinitely many digits and no repeating pattern. For instance,

$$\sqrt{2} = 1.1414213562373095\ldots \quad \text{and} \quad \pi = 3.141592653589793\ldots.$$

Conversely, *every* decimal with infinitely many digits and no repeating pattern represents an irrational number.

The set of real numbers can be visualized geometrically as a horizontal line, with every point of the line representing a real number and vice versa. To do that, we first choose a point O on the line, called the **origin**, to represent the number zero. Then we choose

- a positive direction, which is usually taken to the right, and
- a point on the positive side of the origin to represent the number 1.

That gives us a **unit** to measure distances on the line. Then we place a given positive number x at a point P on the positive ray and at distance x from the origin, as shown in

---

[1] The fact that $\sqrt{2}$ is irrational was known over 2500 years ago and is attributed to Pythagoras. A proof can be found in the famous *Elements* of Euclid.

$$-x \quad -5/2 \quad -\sqrt{2} \quad -1 \quad 0 \quad 1 \quad \sqrt{2} \quad 5/2 \quad x$$
$$P' \qquad\qquad\qquad\qquad O \qquad\qquad\qquad\qquad P$$

Figure 0.1.1

Figure 0.1.1. And we place the number $-x$ on the negative ray at a point P', which is symmetric to P with respect to the origin. Conversely, to any given point P we associate a real number x, which is its distance from the origin if P is on the right, and the negative of that distance if P is on the left.

Thus we have a one-to-one correspondence between the points of a given line and the real numbers. The number associated with a point is called its **coordinate**, and a line with coordinates is called an **axis** or **real line**.

Notation   The set of the real numbers is usually denoted by the symbol $\mathbb{R}$.

Terminology   In this book the word *number* with no qualifying adjective will always mean *real number*.

Intervals

The set of numbers defined by the inequalities $a < x < b$, where a and b are given numbers, is called an **open interval**. It consists of all numbers between a and b, not including the endpoints a and b. A standard way of denoting this open interval is by writing (a, b).

An interval of the form $a \leq x \leq b$, in which the endpoints are included, is called a **closed** interval. It is denoted by $[a, b]$.

We will also need to consider half-open intervals, described by inequalities such as $a < x \leq b$ or $a \leq x < b$. In those cases we use a parenthesis for the open side and a square bracket for the closed side. In summary:

- (a, b) stands for the set of x satisfying $a < x < b$:
- $[a, b]$ stands for the set of x satisfying $a \leq x \leq b$:

Euclid (c. 325 B.C.–265 B.C.) is most famous for his great book *The Elements*, comprising 13 volumes and containing virtually all known mathematics up to that time. For over 2,000 years it has played a central and unique role in mathematical teaching, and was a major influence upon Isaac Newton in his development of calculus and the theory of gravitation. Books one to six deal with plane geometry, books seven to nine with number theory—including proofs that $\sqrt{2}$ is irrational and that the set of prime numbers is infinite, as well as a method for finding the greatest common divisor of two integers, known as the **Euclidean algorithm**. Books ten through thirteen deal with three-dimensional geometry.

- $(a, b]$ stands for the set of x satisfying $a < x \leq b$:
- $[a, b)$ stands for the set of x satisfying $a \leq x < b$:

In addition to representing numbers, the symbols b and a may also denote the symbols ∞ (infinity) and $-\infty$ (minus infinity), respectively, to indicate that the interval extends infinitely far to the right or left. In other words,

- (a, ∞) stands for the set of x satisfying $a < x$:
- $[a, \infty)$ stands for the set of x satisfying $a \leq x$:
- $(-\infty, b)$ stands for the set of x satisfying $x < b$:
- $(-\infty, b]$ stands for the set of x satisfying $x \leq b$:

Absolute value

The distance of any number x on the real line from the origin is called its **absolute value** and denoted by $|x|$. For example, $|8| = 8$ and $|-8| = 8$. The following is a more precise definition.

Definition 0.1.1

The *absolute value* of a real number x, denoted by $|x|$, is defined as follows:

$$|x| = \begin{cases} x & \text{if } x \geq 0 \\ -x & \text{if } x < 0. \end{cases}$$

The **distance** between any two real numbers x_1 and x_2 is equal to $|x_2 - x_1|$. For example, the distance between 3 and -2 is $|-2 - 3| = |-5| = 5$.

The Cartesian plane

In studying relations between quantities, we deal not just with numbers but *pairs* of numbers of the form (x, y), where x and y are both real numbers. We call such a pair an **ordered pair**, which means that the order in which the numbers are listed counts—for instance, $(3, 1)$ is not the same as $(1, 3)$. Just as the set of real numbers can be represented by a line, there is a correspondence between the ordered pairs of real numbers and the points of a plane. In other words, each pair (x, y) corresponds to a unique point of the plane.

To set up the correspondence, we take two axes that meet perpendicularly at their origins. We usually take one axis to be horizontal with the positive direction to the right and call it the **x-axis**. The other axis, called the **y-axis**, is taken to be vertical with the upward direction as positive. We call the point where they meet the **origin** of the plane.

Now, to any ordered pair of numbers (x, y) we can associate a unique point P in the plane by starting at the origin and moving a distance of $|x|$ units along the x-axis—to

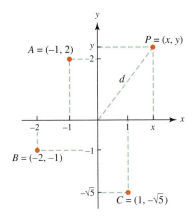

Figure 0.1.2

the right if $x > 0$, and to the left if $x < 0$. From there we next move parallel to the y-axis for a distance of $|y|$ units—going up if $y > 0$ and down if $y < 0$. (If $y = 0$, we simply stay where we are on the x-axis.) The point P at which we arrive corresponds to the pair (x, y). The first number of the pair is called the **x-coordinate** of P, and the second is called its **y-coordinate**. A few examples are shown in Figure 0.1.2.

Conversely, starting with a point P, we can determine its coordinates in the following way. Starting at P, we move parallel to the y-axis until we meet the x-axis. The coordinate x of the point at which we meet the x-axis is the x-coordinate of P. Then, starting again from P, we move parallel to the x-axis till we meet the y-axis. The coordinate y of the meeting point is the y-coordinate of P.

The set of ordered pairs of real numbers is usually denoted by the symbol $\mathbb{R}^2$. A plane with coordinates is known as a **Cartesian plane** and the coordinates are sometimes called **Cartesian coordinates**, in honor of the French mathematician René Descartes (1596–1650).

Imagine a flat area—a park, for example—in which you are jogging. If it has been assigned a system of Cartesian coordinates, you can report your location by giving your coordinates. In fact, there are cities whose streets form a rectangular grid, which is essentially a system of Cartesian coordinates.

Distance in the plane and the equation of a circle

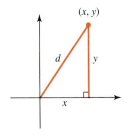

Figure 0.1.3

By using the Pythagorean theorem (see Figure 0.1.3), we see that the **distance** d from the origin $(0, 0)$ to a point $P = (x, y)$ of the plane is given by the formula

$$d = \sqrt{x^2 + y^2}.$$

More generally, the distance between two points $A = (x_1, y_1)$ and $B = (x_2, y_2)$ is given by the formula

$$d = \sqrt{(x_2 - x_1)^2 + (y_2 - y_1)^2}.$$

For example, if $A = (-1, 2)$ and $B = (-2, -1)$, the distance between them is

$$d = \sqrt{(-2 - [-1])^2 + (-1 - 2)^2} = \sqrt{10}.$$

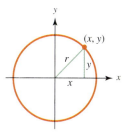

Figure 0.1.4

Equation of a circle For any positive number r, the equation $x^2 + y^2 = r^2$ represents a circle of radius r centered at $(0, 0)$. That follows from the distance formula, since the circle consists of all points of distance r from the origin (see Figure 0.1.4).

More generally, the equation of the circle with center at a point (x_0, y_0) and radius r is given by the formula

$$(x - x_0)^2 + (y - y_0)^2 = r^2, \tag{1}$$

which says that a point (x, y) is on the circle if and only if its distance from (x_0, y_0) equals r.

EXAMPLE 0.1.1

Find the equation of the circle with center at $(7, 5)$ and radius 3.

SOLUTION Using (1), we see that the equation is $(x - 7)^2 + (y - 5)^2 = 9$.

0.2 Functions

One of the fundamental concepts of mathematics is that of a **function**. Roughly speaking, a function is a rule (or operation) that assigns a unique value (the *output*) to a given number (the *input*). There are four standard ways of describing a function:

1. by a word description,
2. by a formula,
3. by a table, and
4. by a graph.

HISTORICAL PROFILE

Pythagoras (c. 569 B.C.–475 B.C.) is often described as the first pure mathematician. He traveled in Egypt around 535 B.C., which may have influenced his later mathematical work. He was the leader of a community that combined religious with scientific pursuits and was perhaps the first on record to give equal status to women. In addition to the celebrated Pythagorean theorem, other discoveries attributed to Pythagoras and his followers are the irrationality of $\sqrt{2}$ and the fact that the strings of a musical instrument produce harmonious intervals when the ratios of their lengths are integers. In particular, a string that is half the length of another will produce a tone an octave away. Aristotle wrote that "the Pythagorean, having been brought up in the study of mathematics, thought that things are numbers and the whole cosmos is a scale and a number."

These are not mutually exclusive, however. A word description can usually be expressed more precisely by a formula, which in turn may lead to a table or a graph. Conversely, a graph or table may suggest a formula. Here are examples of each way of describing a function.

Converting from Celsius to Fahrenheit: a formula and a graph

As you probably know, there are two temperature scales used in the United States—the Celsius scale for scientific work and the Fahrenheit scale for everyday use. The formula converting Celsius to Fahrenheit is

$$F = \frac{9}{5}C + 32. \tag{2}$$

In this context we think of C as the input in the formula and F as the output, and we say that F is a *function* of C. We can assign any value we choose to C, and that choice determines F. With that in mind, we call C the **independent variable** and F the **dependent variable**. For example, if $C = 100$ (the boiling temperature of water), then $F = 212$.

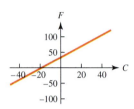

Figure 0.2.1

Figure 0.2.1 shows the graph of F as a function of C, drawn for C between -50 and 50. It consists of all ordered pairs (C, F) that satisfy formula (2), and it continues indefinitely to both the left and right. Following the standard custom in mathematics, the independent variable (in this case, C) is represented along the horizontal axis. As you can see, the graph is a straight line. In Section 0.4 we will see the reason for that.

U.S. population: a table and a graph

A table of the population of the United States (in millions) from 1910 to 2000 is shown in Figure 0.2.2. In this case, the year is the independent variable, the population is the dependent variable, and the table represents the population as a function of the year. If t denotes an entry in the top row (the year), we will write $p(t)$ to represent the corresponding entry in the second row. For example, $p(1940) = 132$.

A graph of the table is shown in Figure 0.2.3. Each data pair in the table is marked on the graph, and the points are connected by straight-line segments. Both the table and

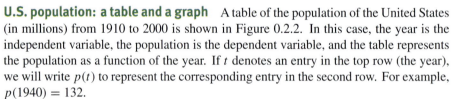
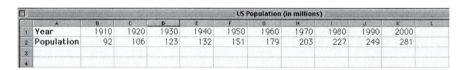

	A	B	C	D	E	F	G	H	I	J	K	L
1	Year	1910	1920	1930	1940	1950	1960	1970	1980	1990	2000	
2	Population	92	106	123	132	151	179	203	227	249	281	
3												
4												

Figure 0.2.2

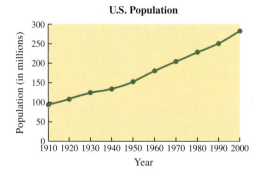

Figure 0.2.3

TABLE 0.2.1

Time (in hours)	0	1	2	3	4	5	6	7	8	9
Average cell count	9.6	18.3	29.0	47.2	71.1	119.1	174.6	257.3	350.7	441.0

Time (in hours)	10	11	12	13	14	15	16	17	18
Average cell count	513.3	559.7	594.8	629.4	640.8	651.1	655.9	659.6	661.8

the graph were made using an Excel spreadsheet, which we will discuss further at the end of this section.

Logistic population growth: a table, a formula, and two graphs   In the early part of the twentieth century the American biologist Raymond Pearl studied a model of population growth known as the **logistic model**, which had been introduced almost 100 years earlier by the Belgian mathematician P. F. Verhulst but remained neglected until Pearl reintroduced it. Pearl tested the model against various sets of experimental data. One of them, a table giving the growth of a culture of yeast cells, is shown in Table 0.2.1 (based on data taken from *The Quarterly Review of Biology*, Vol. 2, 1927, p. 533).

The top row is the time t (in hours); the second row is the average number of yeast cells observed over a number of repetitions of the experiment. A graph of Table 0.2.1 (also created using Excel) is shown in Figure 0.2.4 on page 8.

In addition to a table of experimental data, Pearl had an equation, called a **logistic equation**, that was intended to model the growth of the yeast cell population. In fact, one of the purposes of his experiment was to determine how well the actual data fit the model. The graph of the logistic equation is shown in Figure 0.2.5 on page 8. At this stage we need not be concerned with the actual equation. We will return to it in Chapter 6, after we have developed the calculus needed to understand it.

A demand curve: a graph   A bakery has determined that the quantity of bread sold when the unit price ranges from \$3 to \$1 is modeled by the graph shown in Figure 0.2.6. It is called a **demand curve** for bread, which relates the price to the quantity that consumers demand. At a given price p (the independent variable) the diagram assigns a unique quantity q (the dependent variable) that the consumers demand. For example, when $p = 3$, then $q = 50$. Here we have a situation where the demand (the quantity sold) is a function of the price, and this function is represented by a graph. In this example, the dependent variable q is represented on the horizontal axis and the independent variable p on the vertical, following the convention in economics, but opposite to the custom in mathematics. We will return to this example in Chapter 6 when we see how calculus can be applied to economic models.

Following the common thread of these examples leads to:

Definition 0.2.1

A function f is a rule that assigns a unique value $f(x)$ to every number x in a given set D.

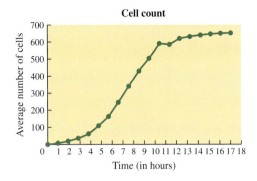

Figure 0.2.4

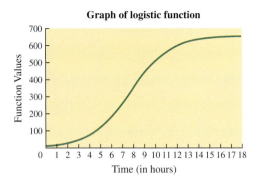

Figure 0.2.5

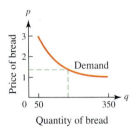

Figure 0.2.6

The number x, which we can think of as the *input*, is called the **independent variable**. The set D, which consists of all possible input numbers, is called the **domain** of the function. In the formula converting Celsius to Fahrenheit, for instance, the domain consists of all real numbers. In the table of U.S. population, on the other hand, the domain consists of the ten integers 1910, 1920, ..., 2000.

The *output* number that f assigns to x is denoted by $f(x)$ (read as "f of x") and is called the **value** of the function at x. Following customary usage, we often refer to "the function $f(x)$" even though $f(x)$ is, strictly speaking, the output value. We may also describe a function by writing an equation of the form $y = f(x)$, and we refer to y as the **dependent variable** to indicate that its value depends on the choice of x.

Substituting a number for the independent variable yields a value of the function. If $f(x) = x^2$, for example, then

$$f(3) = 3^2 = 9 \quad \text{and} \quad f(-1) = (-1)^2 = 1.$$

We may also substitute a letter (representing a number) or an algebraic expression (representing a combination of numbers). For instance, again using $f(x) = x^2$, we have

$$f(a + h) = (a + h)^2 = a^2 + 2ah + h^2 \quad \text{and} \quad f\left(\frac{1}{c}\right) = \left(\frac{1}{c}\right)^2 = \frac{1}{c^2}.$$

EXAMPLE 0.2.1

Let f be the function defined by the formula

$$f(x) = 3x^2 - 15x + 18,$$

with the set of all real numbers $\mathbb{R}$ as domain. Find the values obtained by setting

(i) $x = -1$,   (ii) $x = 2$,   (iii) $x = t + 1$.

SOLUTION   (i) $f(-1) = 3(-1)^2 - 15(-1) + 18 = 36$.
     (ii) $f(2) = 3 \cdot (2^2) - 15 \cdot 2 + 18 = 0$.
     (iii) $f(t + 1) = 3(t + 1)^2 - 15(t + 1) + 18$
                  $= 3(t^2 + 2t + 1) - 15t - 15 + 18 = 3t^2 - 9t + 6$.

The domain of a function (the set D) may consist of *all* real numbers, or it may be restricted to some subset of numbers. There are two reasons for such a restriction:

- first, because the laws of mathematics—such as no division by zero and no square root of a negative number—may not allow certain numbers to be used as input; and
- second, because of the context in which the function appears.

The domain may not always be specified, in which case it is understood to be the largest domain compatible with the laws of mathematics and the context in which the function appears. If the domain is not restricted by the context, we refer to it as the **natural domain** of the function.

For example, let's consider the well-known formula for the area of a square, $A = s^2$. This defines the dependent variable A (the area) as a function of the independent variable s (the length of side).

What is the domain of this function? There are two answers to that question. If we simply think of the function defined by the formula $f(s) = s^2$, its natural domain is the entire set of real numbers $\mathbb{R}$. In other words, *any* number can be used as input without violating any mathematical rules. On the other hand, in the geometric context in which we are considering this function, it would not make sense to assign a negative number as input, because a square cannot have a side of negative length. In this case, then, we specify the function *and* its domain by writing $f(s) = s^2$, $s \geq 0$.

EXAMPLE 0.2.2

Suppose $f(t) = t - (1/t)$. Describe the natural domain of this function and evaluate each of the following:

$$f(3), \quad f(-1), \quad f(c^2), \quad f(2a), \quad f(z-1), \quad f(x+h).$$

SOLUTION Zero is the only number that cannot be used as input, so that the natural domain consists of all $t \neq 0$. By replacing t wherever it appears, we get the following evaluations:

$$f(3) = 3 - \frac{1}{3} = \frac{8}{3}.$$

$$f(-1) = (-1) - \frac{1}{(-1)} = 0.$$

$$f(c^2) = c^2 - \frac{1}{c^2} = \frac{c^4 - 1}{c^2}.$$

$$f(2a) = 2a - \frac{1}{2a} = \frac{4a^2 - 1}{2a}.$$

$$f(z-1) = (z-1) - \frac{1}{z-1} = \frac{(z-1)^2 - 1}{z-1} = \frac{z^2 - 2z}{z-1}.$$

$$f(x+h) = (x+h) - \frac{1}{x+h} = \frac{(x+h)^2 - 1}{x+h}.$$

The set of values taken by the dependent variable—the set of all the numbers $f(x)$, in other words—is called the **range** of the function.

EXAMPLE 0.2.3

Determine both the natural domain and the range of each of the following functions:

$$\text{(i)}\ f(x) = x^2, \quad \text{(ii)}\ f(x) = x^3, \quad \text{(iii)}\ f(x) = \frac{1}{x^2}, \quad \text{(iv)}\ f(x) = \frac{1}{1+x^2}.$$

SOLUTION (i) Since any number can be used as input, the natural domain is the set of all real numbers $\mathbb{R}$, which we can also denote in interval notation by $(-\infty, \infty)$. To determine the range, we recall that the square of a number cannot be negative, so that $f(x) \geq 0$. Furthermore, any $y \geq 0$ can be written as the square of some number x (take x to be the square root of y). Therefore, the range is the set of all nonnegative real numbers, which we can denote in interval notation by $[0, \infty)$.

(ii) Once again, the interval $(-\infty, \infty)$ is the natural domain because any number can be cubed. In this case the output can be positive, negative, or zero, and every number y is the cube of some number x. Therefore, the range is $(-\infty, \infty)$.

(iii) In this case, we must have $x \neq 0$, since the denominator cannot be zero. Any other number can be substituted for x, which means the domain is the set of all $x \neq 0$. The output must be positive, and any positive number can occur. (If $y > 0$, then letting $x = 1/\sqrt{y}$ gives $y = 1/x^2$.) Therefore, the range is $(0, \infty)$.

(iv) Since $x^2 \geq 0$ for any x, the denominator satisfies

$$1 + x^2 \geq 1. \tag{3}$$

Therefore, any number can be used as input without the risk of division by zero, and the natural domain is $(-\infty, \infty)$. To determine the range, we note first that $f(x) > 0$ for all x and, second, that dividing both sides of Eq. (3) by $(1 + x^2)$ gives

$$1 \geq \frac{1}{1+x^2},$$

which means that $f(x) \leq 1$. Moreover, it is not hard to show that every number in the interval $(0, 1]$ can be written in the form $1/(1 + x^2)$ for some x (see the exercises). Therefore, the range of f is the interval $(0, 1]$.

The graph of a function

Suppose we start with a function defined by an equation of the form $y = f(x)$. We can get a very useful geometric picture of the function by drawing its **graph**, which is defined as the set of all points in the Cartesian plane whose coordinates (x, y) satisfy the equation.

By plotting all those pairs on a set of coordinate axes, we get a curve representing the function. Each point on the curve corresponds to one input-output pair (x, y). In practice, of course, we can't plot *all* the pairs, but we can get a rough sketch of the graph by plotting a number of points and interpolating a simple smooth curve between them. (In doing that, we are assuming that the graph is a continuous, smooth curve. Both of these terms, in fact, have precise meanings that we will discuss later.)

> ### EXAMPLE 0.2.4

〜 The length of a square's side is given as a function of its area by the formula

$$s = \sqrt{A}, \quad A \geq 0.$$

Sketch the graph of this function by constructing a table of values for A from 0 to 2.5 in steps of 0.25, plotting the points, and connecting them with a smooth curve.

SOLUTION   Using a calculator and rounding the answers to two decimal places, we get the following table:

A	0	0.25	0.5	0.75	1	1.25	1.5	1.75	2	2.25	2.5
s	0	0.5	0.71	0.87	1	1.12	1.22	1.32	1.41	1.5	1.58

By plotting these pairs, we obtain the picture on the left in Figure 0.2.7. Interpolating a smooth curve between the points gives the graph shown on the right.

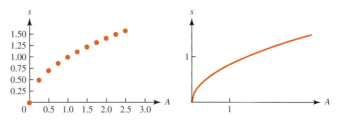

Figure 0.2.7

Plotting points and interpolating between them, as we have just done, can only give us an approximate picture of the graph. To get an accurate representation, we may need to plot many points at relatively short distances from one another. Computers and graphing calculators do essentially the same thing—compute a large table of values, plot the corresponding points, and interpolate between them—and they are invaluable tools that can save us a lot of tedious work. For instance, Figure 0.2.8 shows two versions of the graph of $y = x^3 - 3x$, one created with a TI-83 Plus graphing calculator, the other with a computer program called Mathematica.

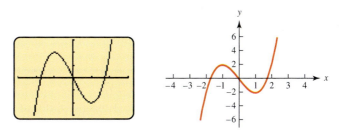

Figure 0.2.8

We will discuss the use of calculators in Section 0.7. Even with a calculator, however, we may get a misleading picture of the graph unless we first analyze its key features. In Chapter 4 you will learn how to use calculus to do that.

When is a curve the graph of a function?

It is important to note that not every curve in the plane is the graph of a function. A curve may be given by an equation in x and y, but it is not the graph of a function unless we can put the equation in the form $y = f(x)$.

In principle, there is a simple criterion for when we can do that: The curve is the graph of a function if no two of its points have the same x coordinate. That is because a function gives a *unique* output value y for any choice of x in its domain. Conversely, any curve with this property is indeed the graph of a function. In geometric terms, we can state it as follows.

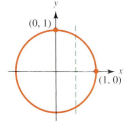

Figure 0.2.9

The Vertical Line Test

A curve in the plane is the graph of a function of the form $y = f(x)$ if and only if no vertical line meets the curve more than once.

For example, the circle given by the equation $x^2 + y^2 = 1$ is not the graph of a function, as you can see by looking at Figure 0.2.9, in which the dashed line meets the circle twice. To understand what happens when we try to solve this equation for y in terms of x, let's first isolate the term involving y on one side of the equation,

$$y^2 = 1 - x^2,$$

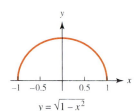

$y = \sqrt{1 - x^2}$

Figure 0.2.10

and then take the square root of both sides. Of course, we need to restrict x to the interval between -1 and 1; otherwise, the right-hand side is negative and the square root does not exist. A more serious difficulty, however, is that any positive number has both a positive and a negative square root. Thus, if we set $x = 0$, we get both 1 and -1 as possible values for y. In general, $y = \pm\sqrt{1 - x^2}$, which means that we do not get a unique output value for each x. In fact, we get *two* solutions

$$y = \sqrt{1 - x^2} \quad \text{and} \quad y = -\sqrt{1 - x^2},$$

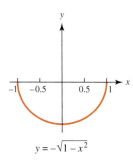

$y = -\sqrt{1 - x^2}$

Figure 0.2.11

each of which describes y as a function of x. Each of these functions has the closed interval $[-1, 1]$ as its domain. The first, with the positive root, has the upper semicircle as its graph, and the second (with the negative root) has the lower semicircle. As you can see in Figures 0.2.10 and 0.2.11, each passes the vertical line test.

EXAMPLE 0.2.5

The curve shown in Figure 0.2.12 has equation $y^2 = x$. Check that it fails the vertical line test and then use it to define two different functions.

SOLUTION Any vertical line that passes through the positive x-axis meets the curve at two points. Thus, it fails the vertical line test and is not the graph of a function.

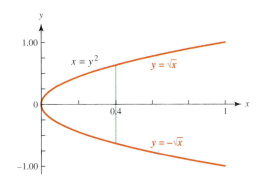

Figure 0.2.12

Solving for y gives two values, $y = \pm\sqrt{x}$, which defines two functions, $f(x) = \sqrt{x}$ and $f(x) = -\sqrt{x}$, each with domain $[0, \infty)$.

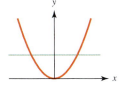

Figure 0.2.13

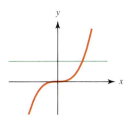

Figure 0.2.14

The horizontal line test

To repeat an important point: If $y = f(x)$, there is only one y value for each x in the domain. That is the basis of the vertical line test. However, several different choices of x may lead to the same value for y. If $y = x^2$, for instance, there are two x's for every positive y. In geometric terms, every horizontal line in the upper half-plane meets the graph twice, as shown in Figure 0.2.13.

If a function does have the property that distinct choices of x yield distinct values of y, we say it is **one-to-one**. More formally, $f(x)$ is one-to-one if $f(x_1) \neq f(x_2)$ for any distinct pair of numbers x_1 and x_2 in the domain of f. The function $f(x) = x^2$, defined over its natural domain, is not one-to-one, as we saw above. On the other hand, $f(x) = x^3$ *is* one-to-one, because any two distinct numbers have different cubes. The graph is shown in Figure 0.2.14. As you can see, no horizontal line meets it more than once. That is because no y value corresponds to two different x's. This geometric interpretation of a function being one-to-one leads to the following criterion.

The Horizontal Line Test

A function $f(x)$ is one-to-one if and only if no horizontal line meets its graph more than once.

EXAMPLE 0.2.6

For each of the three plane curves shown below, state whether it is the graph of a function, and, if so, whether the function is one-to-one.

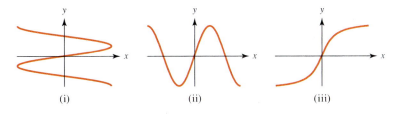

(i)   (ii)   (iii)

SOLUTION   Graph (i) fails the vertical line test and is therefore not the graph of a function. Graph (ii) passes the vertical line test and is the graph of a function, but it is not one-to-one because it fails the horizontal line test. Graph (iii) passes both tests and is the graph of a one-to-one function.

In applying the horizontal line test, it is important to consider the domain of the function.

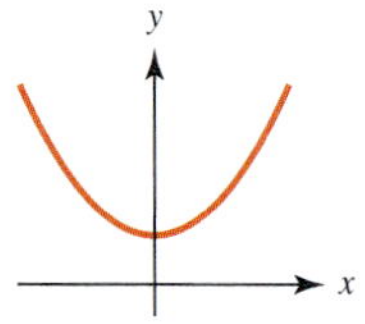

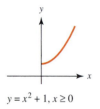

$y = x^2 + 1$

Figure 0.2.15

EXAMPLE 0.2.7

Determine whether each of the following functions is one-to-one:

$$\text{(i) } f(x) = x^2 + 1 \quad \text{and} \quad \text{(ii) } f(x) = x^2 + 1, \quad x \geq 0.$$

SOLUTION   In (i) the domain is not specified, so we assume it is the natural domain, which consists of all real numbers. As you can see in Figure 0.2.15, the graph of $y = x^2 + 1$ fails the horizontal line test, and the function is not one-to-one. We can also reach the same conclusion by noting that $f(-x) = f(x)$ for every x in the domain.

In (ii), however, we have restricted the domain to nonnegative x, so that no two x values give the same y value. Therefore, the function is one-to-one. The graph (shown in Figure 0.2.16) is simply the right-hand branch of the previous one. It clearly passes the horizontal line test.

$y = x^2 + 1, x \geq 0$

Figure 0.2.16

Inverse functions

Here is an important property of one-to-one functions: If $f(x)$ is one-to-one, we can solve the equation $y = f(x)$ for x as a function of y.

For example, the equations $y = x^3$ and $x = \sqrt[3]{y}$ are equivalent, by which we mean that any (x, y) pair satisfying one of them must also satisfy the other. The first equation is of the form $y = f(x)$. The second is in the form $x = g(y)$ and is the result of solving the first for x in terms of y.

A concise way of stating that the two equations are equivalent is by writing

$$y = x^3 \iff x = \sqrt[3]{y},$$

where the symbol $\iff$ is used to indicate that both equations say the same thing. Both express the same relationship between x and y, but *the roles of the two variables have been reversed*. In the equation on the left, x is the input, y is the output, and y is a function of x. In the equation on the right, y is the input, x is the output, and x is a function of y.

In general, if two functions f and g are related in that way—that is, if

$$y = f(x) \iff x = g(y) \tag{4}$$

we say that they are **inverses** or that each is the **inverse** of the other. Once again, the symbol $\iff$ means that the two equations express the same relationship. Any pair (x, y)

satisfying one of them must also satisfy the other. The input of one equation is the output of the other and vice versa.

EXAMPLE 0.2.8

Find the inverse of the function $f(x) = 2x - 1$. Express it as a function of x.

SOLUTION We start with the equation $y = 2x - 1$ and solve for x as a function of y. In this case, we have

$$y = 2x - 1 \iff x = \frac{1}{2}(y + 1). \tag{5}$$

The equation on the right defines the inverse function.

To express the inverse function as a function of x, we simply permute x and y in the right-hand equation of (5) to get $y = \frac{1}{2}(x + 1)$. Thus, the inverse function of $f(x) = 2x - 1$ is given by $g(x) = \frac{1}{2}(x + 1)$. These functions satisfy the inverse relationship stated in (4).

The reason for permuting x and y is to display the inverse function with x as the independent variable. Figure 0.2.17 shows the graphs of the inverse functions $f(x) = 2x - 1$ and $g(x) = (x + 1)/2$ of Example 0.2.8, drawn on the same set of axes. The illustration reveals a relationship between the two graphs, which is more clearly displayed in Figure 0.2.18. Each of the graphs is the reflection of the other in the 45-degree line (the line making a 45-degree angle with each coordinate axis). That line, shown as a dashed line, is the graph of $y = x$, and reflecting across it is the geometric equivalent of permuting x and y. For instance, the points $(3, 5)$ and $(5, 3)$ are mirror images of one another across the 45-degree line.

One other point worth noting about inverse functions, f and g, is this:

- the domain of f is the range of g, and
- the domain of g is the range of f.

In other words, the input of one is the output of the other, and vice versa.

<aside>
PROBLEM-SOLVING TACTIC

To find the inverse of a one-to-one function,

1. Solve the equation $y = f(x)$ for x in terms of y, so that

    $y = f(x) \iff x = g(y)$

2. Permute x and y in the right-hand equation to get $y = g(x)$.

Then $g(x)$ is the inverse of $f(x)$.
</aside>

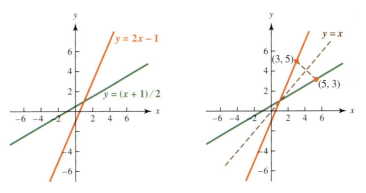

Figure 0.2.17   Figure 0.2.18

> **EXAMPLE 0.2.9**

Let $f(x) = x^2 + 1$, $x \geq 0$. Find and graph the inverse, and determine its domain and range.

SOLUTION   We have already seen (in Example 0.2.7) that f is one-to-one. To find the inverse, we observe that

$$y = x^2 + 1, \quad x \geq 0 \iff x = \sqrt{y - 1}, \quad y \geq 1.$$

Next, we permute x and y in the right-hand equation to get $y = \sqrt{x - 1}$. Thus, the inverse function is given by $g(x) = \sqrt{x - 1}$.

The domain of f is $[0, \infty)$, and its range is $[1, \infty)$. Therefore, the domain of g is $[1, \infty)$ and its range is $[0, \infty)$.

The graph of $y = x^2 + 1$, $x \geq 0$, was displayed in Figure 0.2.16. It is shown again in Figure 0.2.19, where it has been labeled both with the equation $y = x^2 + 1$ (with $x \geq 0$ omitted but understood) and the equation $x = \sqrt{y - 1}$. Remember: These two equations are equivalent, which means that they have the same graph.

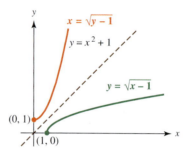

Figure 0.2.19

Permuting x and y gives the inverse, expressed by $y = \sqrt{x - 1}$. Its graph is obtained by reflecting the graph of $y = x^2 + 1$ in the 45-degree line, which is drawn as a dashed line.

Practice Exercises 0.2

1. If $f(x) = x^2 + 8$, find $(f(3 + h) - f(3))/h$ for $h \neq 0$. Simplify as much as possible.

2. Determine the natural domain of the function $f(x) = \sqrt{x - 5}$.

3. Figure 0.2.20 shows the graphs of two functions. Determine whether either of them has an inverse, and, if so, sketch the graph of the inverse function.

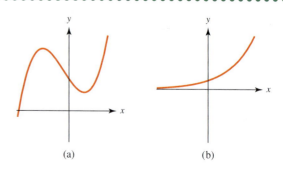

(a)   (b)

Figure 0.2.20

Exercises 0.2

1. If $f(x) = x^3$, find $f(2)$, $f(-1)$, $f(1/3)$, $f(a)$, $f(-a)$, $f(a+1)$, $\dfrac{f(x+h) - f(x)}{h}$.

2. If $f(x) = x^2 - 3x + 2$, find $f(0)$, $f(3)$, $f(-3)$, $f(1/2)$, $f(2t)$, $f(t-2)$.

3. If $f(x) = 1/x$, find $f(2)$, $f(1/2)$, $f(-1)$, $f(a/3)$, $f(3/a)$, $\dfrac{f(a+h) - f(a)}{h}$.

4. If $f(x) = \dfrac{x}{x+1}$, find $f(0)$, $f(2)$, $f(-2)$, $f(-1/2)$, and $f(x-1)$.

5. If $g(x) = x^3 - 2x + 1$, find $g(1)$, $g(-2)$, $g(-c)$, $g(a-1)$, and $g(a+h) - g(a)$.

6. If $f(t) = t + (1/t)$, show that $f(a) = f(1/a)$ for any $a \neq 0$.

7. If $h(x) = x^2 - x + 1$, show that $h(a) = h(1-a)$ for any a.

In Exercises 8–16, find the natural domain of the function.

8. $f(x) = \sqrt{x-1}$

9. $f(x) = \sqrt{1-x}$

10. $f(x) = \sqrt{1-x^2}$

11. $g(x) = \dfrac{1}{\sqrt{x-1}}$

12. $f(t) = \dfrac{t}{t-1}$

13. $h(t) = \dfrac{t}{t^2 - 1}$

14. $g(x) = x^3 - 3x^2$

15. $h(u) = u^3 - \dfrac{3}{u^2}$

16. $f(x) = \dfrac{x}{x^2 + 1}$

17. Show that if $0 < y \leq 1$ and $x \geq 0$, the equation $y = 1/(1 + x^2)$ can be solved for x as a function of y.

In Exercises 18–23, describe the range of the given function.

18. $f(x) = x^4$, $-\infty < x < \infty$

19. $f(x) = 1 + x^2$, $-\infty < x < \infty$

20. $f(x) = \sqrt{x-1}$, $x \geq 1$

21. $f(x) = \dfrac{1}{x}$, $x \neq 0$

22. $f(x) = \dfrac{1}{\sqrt{x}}$, $x > 0$

23. $f(x) = \dfrac{2}{3 + x^2}$

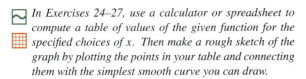 *In Exercises 24–27, use a calculator or spreadsheet to compute a table of values of the given function for the specified choices of x. Then make a rough sketch of the graph by plotting the points in your table and connecting them with the simplest smooth curve you can draw.*

24. $f(x) = x^2$ for x from -2 to 2 in steps of 0.25

25. $f(x) = 2x - x^2$ for x from -1 to 3 in steps of 0.5

26. $f(x) = 1/x$ for x from $\frac{1}{5}$ to 2 in steps of $\frac{1}{5}$

27. $f(x) = \sqrt{1-x}$ for x from -1 to 1 in steps of 0.2

In Exercises 28–33, state whether or not the given curve is the graph of a function of the form $y = f(x)$. Justify your answer.

28.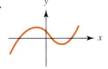

29.

30.

31.

32.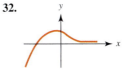

33.

In Exercises 34–39, determine whether the curve shown is the graph of a one-to-one function. Justify your answer.

34.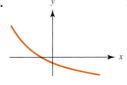

35.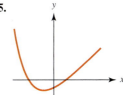

36.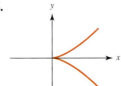

37.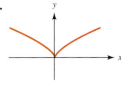

38.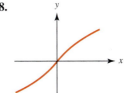

39.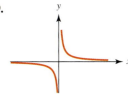

In Exercises 40–45, determine whether $f(x)$ has an inverse. If it does, find an explicit formula for the inverse and specify its domain and range.

40. $f(x) = 1/x^2$, $x \neq 0$ **41.** $f(x) = 1/x^2$, $x > 0$

42. $f(x) = 1/x^2$, $x < 0$ **43.** $f(x) = 1/x^3$, $x \neq 0$

44. $f(x) = 1/(x - 2)$, $x \neq 2$

45. $f(x) = \sqrt{x - 2}$, $x \geq 2$

46. Let $f(x) = 1/(x - 1)^2$, $x > 1$. Find the inverse function and specify its domain and range.

47. Let $f(x) = -1/x$, $x > 0$. The graph of this function is shown in Figure 0.2.21. Find the inverse function, specify its domain and range, and draw its graph.

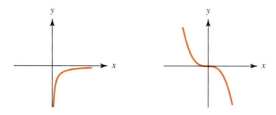

Figure 0.2.21   **Figure 0.2.22**

48. Let $f(x) = -x^3$. The graph of this function is shown in Figure 0.2.22. Find the inverse function, specify its domain and range, and draw its graph.

Solutions to practice exercises 0.2

1. We have

$$\frac{f(3 + h) - f(3)}{h} = \frac{(3 + h)^2 + 8 - (3^2 + 8)}{h}$$

$$= \frac{3^2 + 2 \cdot 3 \cdot h + h^2 + 8 - 3^2 - 8}{h}$$

$$= \frac{6h + h^2}{h} = 6 + h.$$

2. For the square root to make sense as a real number, we must have $x - 5 \geq 0$. Therefore, the natural domain of the function $f(x) = \sqrt{x - 5}$ is the interval $[5, \infty)$.

3. The graph in (a) does not pass the horizontal line test, and therefore the function does not have an inverse. The graph in (b) does pass the horizontal line test, which means the

function has an inverse. To find its graph, we reflect the given graph in the 45-degree line. The result is shown in Figure 0.2.23.

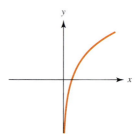

Figure 0.2.23

0.3   Geometric Properties of Functions

Increasing and decreasing functions

We say that $f(x)$ is **increasing** if its graph rises to the right (in other words, the height of the graph increases as x increases). More precisely, we have the following.

> **Definition 0.3.1**
>
> A function $f(x)$ is *increasing* on an interval (a, b) if
>
> $$f(x_1) < f(x_2) \text{ whenever } a < x_1 < x_2 < b.$$
>
> In other words, the value of $f(x)$ becomes greater as x increases over the interval (a, b).

A simple example of an increasing function is given by the Celsius-to-Fahrenheit conversion formula

$$y = \frac{9}{5}x + 32.$$

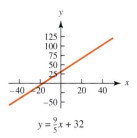

$y = \frac{9}{5}x + 32$

Figure 0.3.1

As we have already seen, its graph (reproduced in Figure 0.3.1) is a straight line that rises to the right.

We similarly say that a function $f(x)$ is **decreasing** if its graph falls as it goes to the right (in other words, the height of the graph decreases as x increases). More precisely, we have the following.

Definition 0.3.2

A function $f(x)$ is *decreasing* on an interval (a, b) if

$$f(x_1) > f(x_2) \text{ whenever } a < x_1 < x_2 < b.$$

In other words, the value of $f(x)$ becomes smaller as x increases over the interval (a, b).

A function may be increasing over one part of its domain and decreasing over another. In Figure 0.3.2, for instance, we have the graph of a function that is increasing on the intervals $(-1, 6)$ and $(9, 10)$ and decreasing on the interval $(6, 9)$.

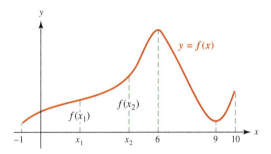

Figure 0.3.2

EXAMPLE 0.3.1

Where is the function $f(x) = |x|$ increasing and where is it decreasing?

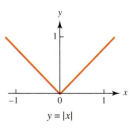

$y = |x|$

Figure 0.3.3

SOLUTION The graph of $y = |x|$ is shown in Figure 0.3.3. On the interval $(0, \infty)$ the graph rises as it moves to the right, and the function is increasing. On the interval $(-\infty, 0)$ the graph falls to the right, and the function is decreasing.

For a somewhat more precise analysis, we first observe that $|x| = x$ if $x > 0$. Therefore, $|x_1| < |x_2|$ whenever $0 < x_1 < x_2$, which means that $|x|$ is increasing on the interval $(0, \infty)$.

On the other hand, $|x| = -x$ if $x < 0$. And, if $x_1 < x_2 < 0$, we have $-x_1 > -x_2$. (Multiplying both sides of an inequality by -1 *reverses* the inequality.) Therefore, $|x_1| > |x_2|$ whenever $x_1 < x_2 < 0$, which means that $|x|$ is decreasing on $(-\infty, 0)$.

Symmetries in graphs

Observe that the graph of the function $f(x) = |x|$ in the last example is symmetrical with respect to the y-axis. That is because replacing x by $-x$ does not change the value of the function—that is, $|-x| = |x|$ for all numbers x. It follows that for any point on the right side of the graph there is a symmetrical point of the same height on the left side of the graph.

In general, the graph of a function $f(x)$ is **symmetric about the y-axis** if the function satisfies the condition

$$f(-x) = f(x).$$

A function with this property is also called an **even** function. As shown in Figure 0.3.4, every point on the graph has a "mirror image" directly across the y-axis—that is, a corresponding point on the graph with the same vertical coordinate but whose horizontal coordinate is the negative of the given one.

Another important type of symmetry is shown in Figure 0.3.5. For any point on the graph, say with coordinates (a, b), the point with coordinates $(-a, -b)$ is also on the graph. We say that a graph with that property is **symmetric about the origin**. A function whose graph has that kind of symmetry satisfies the condition

$$f(-x) = -f(x)$$

and is called an **odd** function.

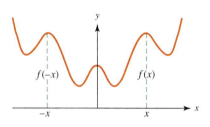

Figure 0.3.4
Symmetry about y-axis.

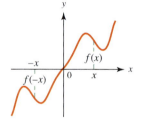

Figure 0.3.5
Symmetry about the origin.

EXAMPLE 0.3.2

Check each of the following functions for symmetry, either about the y-axis or about the origin and determine where each is increasing or decreasing. Use that information to sketch the graph.

(i) $f(x) = x^2$,   (ii) $f(x) = x^3$,   (iii) $f(x) = 1 - x$.

SOLUTION   (i) Since $(-x)^2 = x^2$, the function $f(x) = x^2$ is even, and therefore its graph is symmetric about the y-axis. The graph of $y = x^2$ is shown in Figure 0.3.6. The figure shows it is increasing on the interval $(0, \infty)$ and decreasing on $(-\infty, 0)$. We can also determine that without seeing the graph, by reasoning as follows.

Suppose x_1 and x_2 are two numbers satisfying $x_1 < x_2$. If both numbers are positive, then we also have $x_1^2 < x_2^2$. Therefore, $f(x)$ is increasing on the interval $(0, \infty)$. On

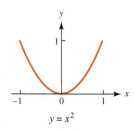

$y = x^2$

Figure 0.3.6

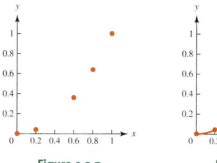

Figure 0.3.7   **Figure 0.3.8**

the other hand, if x_1 and x_2 are both negative, squaring *reverses* the inequality, so that $x_1^2 > x_2^2$. (For instance, $-3 < -2$, but $9 > 4$.) Therefore, $f(x)$ is decreasing on $(-\infty, 0)$.

In fact, we can use this information, together with the symmetry, to draw a rough sketch of the graph. Because of the symmetry, we need only draw the part of the curve in the right half of the plane and then reflect it in the y-axis to get the rest of the graph. To sketch the right-hand branch, we merely compute and plot a number of points on the graph, such as $(0, 0)$, $(0.2, 0.04)$, $(0.6, 0.36)$, and so forth, as shown in Figure 0.3.7. Next, we connect the points with an increasing curve, as shown in Figure 0.3.8. Then, we reflect in the y-axis to get a sketch of the graph, as shown in Figure 0.3.6.

(ii) Since $(-x)^3 = -x^3$, the function $f(x) = x^3$ is odd, and therefore its graph is symmetric about the origin. The graph of $y = x^3$ is shown in Figure 0.3.9. Observe that this function is increasing over its entire domain—that is, on the interval $(-\infty, \infty)$. That's because $x_1^3 < x_2^3$ whenever $x_1 < x_2$, no matter whether the numbers are positive or negative. (For instance, $-3 < -2$, and $-27 < -8$.)

Once again, we can get a rough sketch by first plotting a limited number of points on the right-hand side and interpolating between them with an increasing curve, as shown in Figure 0.3.10. Then we exploit the symmetry by reflecting about the origin to obtain the graph on both sides.

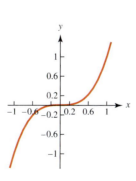

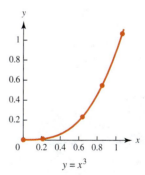

$y = x^3$

Figure 0.3.9   **Figure 0.3.10**

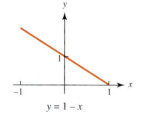

$y = 1 - x$

Figure 0.3.11

(iii) The function $f(x) = 1 - x$ is neither even nor odd, so its graph is not symmetric about the y-axis or the origin. The function is decreasing on $\mathbb{R}$, for, if $x_1 < x_2$, then $1 - x_1 > 1 - x_2$. (Recall once again that multiplying an equality by -1 reverses the inequality. Adding 1 to both sides preserves it.) The graph, shown in Figure 0.3.11, is a line passing through the points $(1, 0)$ and $(0, 1)$.

Intercepts

In sketching a graph, it is helpful to find the points where it crosses the coordinate axes. The graph of $y = f(x)$ crosses the y-axis if and only if zero is in the domain of f. If it is, we call the number $f(0)$ the **y-intercept** of the graph. It is the y-coordinate of the point where the graph crosses the y-axis. The graph cannot have more than one y-intercept because of the vertical line test.

To find the points where the graph crosses the x-axis, we need to solve the equation $f(x) = 0$. Each solution is called an **x-intercept** of the graph (and also a **zero** of the function). It is the x-coordinate of an intersection point of the graph and the x-axis.

EXAMPLE 0.3.3

Find the x- and y-intercepts, if any, of each of the following graphs:

$$\text{(i)} \ \ y = 3x - 2, \quad \text{(ii)} \ \ y = x^2 - 1, \quad \text{(iii)} \ \ y = x^2 + 1, \quad \text{(iv)} \ \ y = \sqrt{x - 1}.$$

SOLUTION All four graphs are shown in Figure 0.3.12.

(i) Setting $x = 0$ gives $y = -2$, which is the y-intercept. Solving $3x - 2 = 0$ gives $x = \frac{2}{3}$, which is the x-intercept.

(ii) Setting $x = 0$ gives $y = -1$, which is the y-intercept. There are two x-intercepts, $x = 1$ and $x = -1$, obtained by solving $x^2 - 1 = 0$.

(iii) Setting $x = 0$ gives $y = 1$, which is the y-intercept. The equation $x^2 + 1$ has no solution (in real numbers), and there is no x-intercept.

(iv) The natural domain of this function is the interval $[1, \infty)$. Therefore, the graph never crosses the y-axis, and there is no y-intercept. There is an x-intercept, $x = 1$, obtained by solving $\sqrt{x - 1} = 0$.

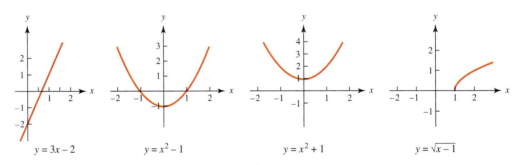

$$y = 3x - 2 \qquad y = x^2 - 1 \qquad y = x^2 + 1 \qquad y = \sqrt{x - 1}$$

Figure 0.3.12

Vertical and horizontal translation

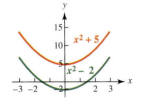

Figure 0.3.13

Given a function $f(x)$ and a constant c, we can add them together to form a new function $f(x) + c$. For instance, if $f(x) = x^2 - 2$ and $c = 7$, we get $f(x) + c = x^2 + 5$. Figure 0.3.13 shows the graphs of both functions, drawn on the same set of axes. As you can see, the two graphs run parallel to one another, with the graph of $y = x^2 + 5$ remaining seven units above the graph of $y = x^2 - 2$.

The reason is not hard to see: For every point on the graph of $y = x^2 - 2$, such as $(1, -1)$, we get a corresponding point on the other graph, in this case $(1, 6)$, by using the same x-coordinate but adding 7 to the y-coordinate.

This same type of relation holds between the graphs of $y = f(x)$ and $y = f(x) + c$ for any function, and for the same reason. The graphs run parallel, with the second remaining c units above the first if $c > 0$ and below if $c < 0$. We say that the graph of $y = f(x) + c$ is obtained from that of $y = f(x)$ by a vertical shift—also called a **vertical translation**.

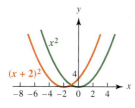

We can also shift the graph horizontally. Suppose $c > 0$. Figure 0.3.14 shows the graphs of $y = x^2$ and $y = (x + 2)^2$, drawn on the same set of axes. As you can see, for every point on the graph of $y = x^2$ there is a corresponding point two units to the left on the graph of $y = (x + 2)^2$. For instance, $(0, 0)$ on the right-hand graph corresponds to $(-2, 0)$ on the left-hand one. In other words, the graph of $y = (x + 2)^2$ is a copy of that of $y = x^2$ shifted two units to the left.

Figure 0.3.14

To see why the graph shifts to the left, think of a point (a, a^2) on the graph of $y = x^2$. The point $(a - 2, a^2)$ is on the graph of $y = (x + 2)^2$, and it is at the same height as (a, a^2) but shifted two units to the left.

For the same reason, we can conclude that if $f(x)$ is any function and c is a positive constant, the graph of $y = f(x + c)$ is a copy of that of $y = f(x)$ shifted c units to the left. Similarly, the graph of $y = f(x - c)$ is obtained from that of $y = f(x)$ by a horizontal shift of c units to the right. In both cases, we say that one graph has been obtained from the other by a **horizontal translation**.

To summarize: Suppose $f(x)$ is a function and c is a positive constant.

> - The graph of $y = f(x) + c$ is obtained by translating the graph of $y = f(x)$ upward by c units.
>
> - The graph of $y = f(x) - c$ is obtained by translating the graph of $y = f(x)$ downward by c units.
>
> - The graph of $y = f(x + c)$ is obtained by translating the graph of $y = f(x)$ to the left by c units.
>
> - The graph of $y = f(x - c)$ is obtained by translating the graph of $y = f(x)$ to the right by c units.

EXAMPLE 0.3.4

The graph of $y = x^3 - 3x$ is drawn in Figure 0.3.15. Use it to find the graphs of

$$\text{(i) } y = x^3 - 3x - 2 \quad \text{and} \quad \text{(ii) } y = (x - 1)^3 - 3x + 3.$$

SOLUTION   Let $f(x) = x^3 - 3x$.

(i) We can write this equation as $y = f(x) - 2$. Its graph is obtained from the graph in Figure 0.3.15 by shifting downward by two units. The result is shown in Figure 0.3.16.

(ii) Since $-3x + 3 = -3(x - 1)$, we can write the given equation as $y = f(x - 1)$. Its graph is obtained from the graph in Figure 0.3.15 by shifting one unit to the right. The result is shown in Figure 0.3.17.

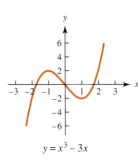

$$y = x^3 - 3x$$

Figure 0.3.15

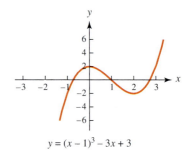

$$y = x^3 - 3x - 2$$

Figure 0.3.16

$$y = (x-1)^3 - 3x + 3$$

Figure 0.3.17

Practice Exercises 0.3

1. Let $f(x) = 1/x^2$, $x \neq 0$. Where is f increasing and where is it decreasing?

2. Determine the symmetry, if any, in the graph of

  (a) $f(x) = \dfrac{1}{1+x^2}$,   **(b)** $f(x) = \sqrt[3]{x}$.

3. Using the graph of $y = 4 - x^2$ (shown in Figure 0.3.18), obtain the graph of $y = 6 - (x-1)^2$ by applying translations.

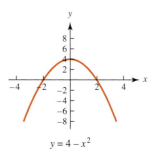

$$y = 4 - x^2$$

Figure 0.3.18

Exercises 0.3

In Exercises 1–6, find where the function with the given graph is increasing and where it is decreasing.

1.

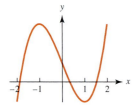

2.

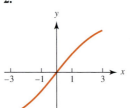

4.

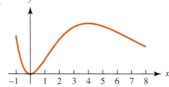

5.

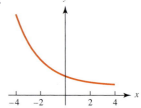

3.

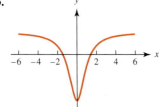

6.

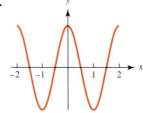

In Exercises 7–14, determine where the function is increasing and where it is decreasing.

7. $f(x) = -2x + 7$ **8.** $f(x) = \dfrac{x}{3} - 1$

9. $f(x) = (x + 1)^2$ **10.** $f(x) = x^3 + 1$

11. $f(x) = \dfrac{1}{x}, \; x \neq 0$

12. $f(x) = \sqrt{x - 1}, \; x \geq 1$

13. $f(x) = \dfrac{1}{(x - 1)^2}, \; x \neq 1$

14. $f(x) = \dfrac{1}{1 + x^2}$

In Exercises 15–23, state whether the function is odd, even, or neither and what kind of symmetry, if any, its graph has.

15. $f(x) = 2x$ **16.** $f(x) = 2x - 1$

17. $f(x) = x^4 - 3x^2 + 1$

18. $f(x) = x^3 - 3x + 1$

19. $f(x) = x^3 + 5x$ **20.** $f(x) = \dfrac{x}{x^2 + 1}$

21. $f(x) = \dfrac{x^2}{x^2 + 1}$ **22.** $f(x) = 1 - |x|$

23. $f(x) = \dfrac{x}{|x| + 1}$

In Exercises 24–26 use a calculator or spreadsheet to compute a table of values of the given function for the specified choices of x. Next, plot the points in your table and use them to make a rough sketch of the graph for $x \geq 0$. Finally, use symmetry to complete the sketch of both branches of the graph.

24. $f(x) = x^3 - x$, x from 0 to 1.5 in steps of 0.25

25. $f(x) = \dfrac{10}{x^2 + 1}$, x from 0 to 2 in steps of 0.4

26. $f(x) = \dfrac{2x}{4x^2 + 1}$, x from 0 to 2 in steps of 0.5

In Exercises 27–34, find the x- and y-intercepts, if any, of the graph.

27. $y = 3x + 4$ **28.** $y = x^2 + 3$

29. $y = x^2 - x$ **30.** $y = \sqrt{x + 1}$

31. $y = x^3 + 1$ **32.** $y = \sqrt[3]{x - 8}$

33. $y = \dfrac{1}{x}$ **34.** $y = \dfrac{1}{(1 + x^2)}$

35. Figure 0.3.19 shows the graph of $y = x^3$. Match the following equations with the graphs below it:

(i) $y = (x - 2)^3$ (ii) $y = (x + 1)^3$ (iii) $y = x^3 - 2$

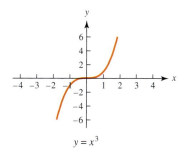

$y = x^3$

Figure 0.3.19

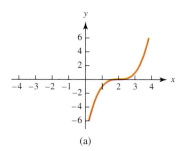

(a)

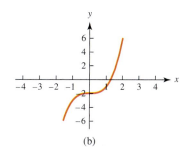

(b)

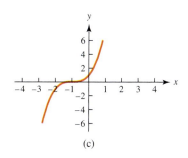

(c)

36. Figure 0.3.20 shows the graph of $y = |x|$. Match the following equations with the graphs below it:

(i) $y = |x| - 1$ (ii) $y = |x - 1|$ (iii) $y = |x + 1|$

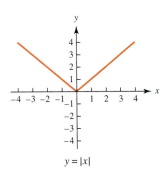

$$y = |x|$$

Figure 0.3.20

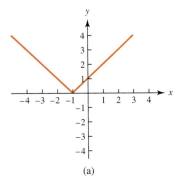

(a)

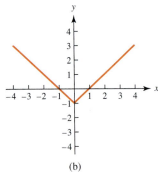

(b)

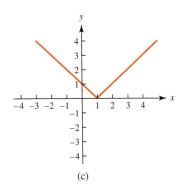

(c)

In Exercises 37–40, apply translations to the graph of $y = x^2$ (Figure 0.3.21) to obtain the graph of the given equation.

37. $y = x^2 - 2$ **38.** $y = (x - 2)^2$

39. $y = (x - 1)^2 + 2$ **40.** $y = (x + 2)^2 - 1$

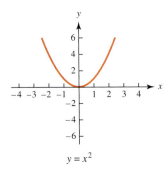

$$y = x^2$$

Figure 0.3.21

In Exercises 41–44, apply translations to the graph of $y = \sqrt[3]{x}$ (Figure 0.3.22) to obtain the graph of the given equation.

41. $y = \sqrt[3]{x + 2}$ **42.** $y = \sqrt[3]{x} + 2$

43. $y = 1 + \sqrt[3]{x - 2}$ **44.** $y = \sqrt[3]{x + 2} - 1$

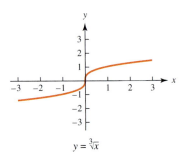

$$y = \sqrt[3]{x}$$

Figure 0.3.22

45. Given a function $f(x)$, let $g(x) = f(-x)$. The graph of $y = g(x)$ is the reflection (or mirror image) in the y-axis of the graph of $y = f(x)$. That is because for any point (x, y) on the graph of f, the point $(-x, y)$ is on the graph of g. Use the graph of $y = 1 + x + x^3$ (Figure 0.3.23) to obtain the graph of $y = 1 - x - x^3$.

46. By applying a reflection and a translation to Figure 0.3.23, draw the graph of $y = 2 - x - x^3$.

47. Given a function $f(x)$, let $g(x) = -f(x)$. The graph of $y = g(x)$ is the reflection (or mirror image) in the x-axis of the graph of $y = f(x)$. That is because for any point (x, y) on the graph of f, the point $(x, -y)$ is on the graph of g. Using the graph of $y = x^3 - 3x$ (Figure 0.3.24), draw the graph of $y = 3x - x^3$.

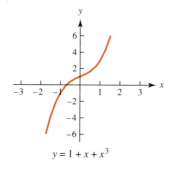

$y = 1 + x + x^3$

Figure 0.3.23

48. By applying a reflection and a translation to Figure 0.3.24, draw the graph of $y = 3x - 3 - (x - 1)^3$.

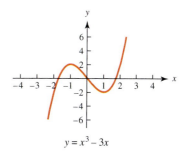

$y = x^3 - 3x$

Figure 0.3.24

Solutions to practice exercises 0.3

1. If $0 < x_1 < x_2$, then $x_1^2 < x_2^2$ and $1/x_1^2 > 1/x_2^2$. Therefore, $f(x)$ is decreasing on $(0, \infty)$. On the other hand, if $x_1 < x_2 < 0$, then $x_1^2 > x_2^2$ and $1/x_1^2 < 1/x_2^2$. Therefore, $f(x)$ is increasing on $(-\infty, 0)$.

2. (a) Since $1/(1 + (-x)^2) = 1/(1 + x^2)$, the function is even, and its graph is symmetric about the y-axis.
(b) Since $\sqrt[3]{-x} = -\sqrt[3]{x}$, the function is odd, and its graph is symmetric about the origin.

3. Replacing x by $(x - 1)$ causes a shift of one unit to the right, and replacing the constant 4 by 6 (i.e., adding 2) causes a shift of two units upward. Therefore, the graph of $y = 6 - (x - 1)^2$ (shown in Figure 0.3.25) is obtained from that of $y = 4 - x^2$ by translating one unit to the right and two units up.

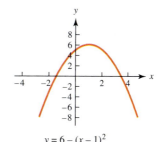

$y = 6 - (x - 1)^2$

Figure 0.3.25

0.4   Linear Functions

The formula for converting Celsius to Fahrenheit

$$F = \frac{9}{5}C + 32 \tag{6}$$

that we encountered in the last section is an example of what is called a **linear function**.

Definition 0.4.1

A linear function is a function of the form $f(x) = mx + b$, where m and b are given numbers.

We refer to m and b as **constants**. Their values are fixed, as opposed to the variable x, which can be assigned any value in the domain of the function.

The constant b is called the **y-intercept**. It is the height at which the graph crosses the vertical axis. That is simply because substituting $x = 0$ into the equation $y = mx + b$ gives $y = b$, which means the graph goes through the point $(0, b)$, as shown in Figure 0.4.1. If b is negative, the graph crosses the y-axis *below* the origin, and we interpret the height as being negative in that case.

The graph of a linear function is a straight line. To understand why, we need to discuss the **slope** of a line.

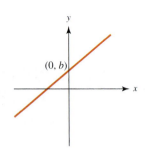

Figure 0.4.1

The slope

Given a nonvertical line in the plane, choose any two points on the line—say, (x_1, y_1) and (x_2, y_2), as shown in the Figure 0.4.2. The difference in height between the first and second points is $y_2 - y_1$, sometimes referred to as the **rise**. The horizontal change from one point to the other is $x_2 - x_1$, sometimes called the **run**. The quotient of these two differences—the rise over the run—is called a **difference quotient**. It does *not* depend on the particular pair of points chosen. If we chose two other points on the line, say, (x_3, y_3) and (x_4, y_4), we would obtain the same quotient. The explanation comes from geometry: The triangles shown in Figure 0.4.2 are similar, so that

$$\frac{y_4 - y_3}{x_4 - x_3} = \frac{y_2 - y_1}{x_2 - x_1}.$$

We sometimes write the difference quotient in the form

$$\frac{\Delta y}{\Delta x}, \tag{7}$$

where the symbols Δy and Δx denote the rise and run. (The Greek letter Δ, called "delta," stands for "difference," from the Greek word $\Delta\iota\alpha\varphi o\rho\acute{\alpha}$, pronounced "diafora.")

The difference quotient given by (7) is called the **slope** of the line, usually denoted by m. To summarize:

$$\boxed{\text{Slope} = m = \frac{y_2 - y_1}{x_2 - x_1} = \frac{\Delta y}{\Delta x}} \tag{8}$$

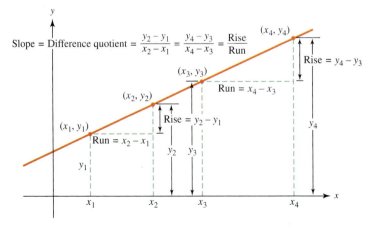

Figure 0.4.2

> ### EXAMPLE 0.4.1

Find the slope of the line connecting the points $(-1, -2)$ and $(2, 3)$.

SOLUTION Putting these numbers into formula (8) gives

$$\frac{3 - (-2)}{2 - (-1)} = \frac{5}{3}.$$

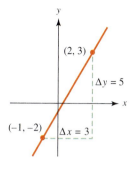

Figure 0.4.3

The line is shown in Figure 0.4.3.

It does not matter which of the two points we take first in Eq. (8), as long as we make the same choice in both numerator and denominator. For instance, the slope we just computed, for the line through $(-1, -2)$ and $(2, 3)$, can also be written as

$$\frac{-2 - 3}{-1 - 2} = \frac{-5}{-3} = \frac{5}{3}.$$

To repeat: Given a nonvertical line, the difference quotient for any pair of its points is the same. *Conversely: Given a point (x_1, y_1) and a number m, the points in the plane whose difference quotient with (x_1, y_1) equals m coincide with those of the straight line of slope m passing through (x_1, y_1).*

Thus, to show that the graph of a linear equation

$$y = mx + b \tag{9}$$

is indeed a straight line, we first choose any input x_1 and compute the output $y_1 = mx_1 + b$, thereby obtaining a point (x_1, y_1) on the graph. Any other point (x, y) on the graph also satisfies $y = mx + b$ and its difference quotient with (x_1, y_1) is equal to

$$\frac{y - y_1}{x - x_1} = \frac{mx + b - mx_1 - b}{x - x_1} = \frac{m(x - x_1)}{x - x_1} = m.$$

Thus, the graph of $y = mx + b$ coincides with the straight line passing through (x_1, y_1) and having slope m.

The slope measures the steepness of the line—how much the graph rises (or falls) vertically for each unit of change in the horizontal direction. If the slope is positive, the line goes upward to the right, and if it is negative, the line goes downward. If $m = 0$, the line is horizontal and has equation $y = b$. A vertical line is given by an equation of the form $x = a$ and is not the graph of a function. Its slope is undefined. Figure 0.4.4 shows the four possiblities.

Finding the equation of a line from geometric data

As we have just seen, if (x_1, y_1) is any point on a line of slope m, then any other point (x, y) on the line must satisfy the equation

$$\frac{y - y_1}{x - x_1} = m.$$

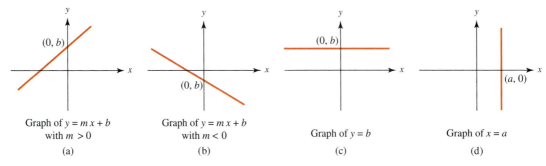

Graph of $y = mx + b$
with $m > 0$

(a)

Graph of $y = mx + b$
with $m < 0$

(b)

Graph of $y = b$

(c)

Graph of $x = a$

(d)

Figure 0.4.4

Multiplying both sides by $(x - x_1)$ gives the equation of the line in a form called the **point-slope form**:

$$y - y_1 = m(x - x_1). \tag{10}$$

By expanding and collecting terms, we can rewrite this equation in a form called the **slope-intercept form**:

$$y = mx + b, \tag{11}$$

where $b = y_1 - mx_1$. This is the form we started with in Eq. (9).

EXAMPLE 0.4.2

Find the slope-intercept form of the equation of the line connecting the points $(-1, 1)$ and $(2, -3)$ as shown in Figure 0.4.5.

SOLUTION   We first find the slope:

$$m = \frac{-3 - 1}{2 - (-1)} = -\frac{4}{3}.$$

Then we put the slope and one of the given points into formula (10). Using $(-1, 1)$, we obtain the point-slope form:

$$y - 1 = -\frac{4}{3}(x + 1). \tag{12}$$

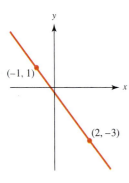

Figure 0.4.5

By expanding and collecting terms, we can put this in the slope-intercept form:

$$y = -\frac{4}{3}x - \frac{1}{3}. \tag{13}$$

If we use the point $(2, -3)$ instead, we get the point-slope form

$$y + 3 = -\frac{4}{3}(x - 2).$$

Although this looks different from Eq. (12), it is not hard to check that expanding the right-hand side and collecting terms lead to the same slope-intercept form.

If we multiply both sides of Eq. (13) by 3 and rearrange terms, it becomes

$$4x + 3y = -1.$$

By a similar manipulation, we can write the equation of any line in a form called the **standard form**:

$$Ax + By = C, \tag{14}$$

where A, B, and C are constants. Conversely, Eq. (14) can be converted into the slope-intercept form, provided $B \neq 0$. We simply solve (14) for y to obtain

$$y = -\frac{A}{B}x + \frac{C}{B},$$

which is the slope-intercept form $y = mx + b$ with $m = -A/B$ and $b = C/B$.

If $B = 0$, Eq. (14) becomes $x = C/A$, which is the equation of a vertical line, for which the slope is undefined.

A linear function and its graph are completely determined once we know its slope and a single point on the line. The slope by itself determines a family of parallel lines. In fact, **parallel lines** have the *same slope* and any two lines that have the same slope must be parallel. (These statements can be proved by means of similar triangles. See the exercises that follow.)

> ## EXAMPLE 0.4.3

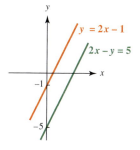

Figure 0.4.6

Find the equation of the line through the point $(2, 3)$ parallel to the line

$$2x - y = 5.$$

SOLUTION By rewriting the given equation in slope-intercept form, $y = 2x - 5$, we see that it has slope 2. To solve the problem then, we need to find the equation of the line through $(2, 3)$ with slope 2. The lines are shown in Figure 0.4.6. Using Eq. (10) and simplifying, we get

$$y = 2x - 1.$$

Linear models

Linear functions are the simplest functions, but they play an essential role in calculus. In fact, one of the underlying ideas of calculus is the use of linear functions to analyze the behavior of more complicated ones. In addition, linear functions can be used to model certain phenomena, as in the following example taken from economics.

EXAMPLE 0.4.4

Modeling cost, revenue, and profit A small audio company produces x compact disc players that sell for $180 each. There are two types of costs involved: the **cost per unit item** (for material, labor, etc.) and a **fixed cost** (for rent on the company's building, its accountant's fees, etc.). Assuming that the fixed cost amounts to $6,600 a week and the cost per unit item is $70, find

(i) the cost of producing x items,

(ii) the revenue earned from the sale of x items, and

(iii) the profit from producing and selling x items.

SOLUTION (i) Producing x units per week will cost $70x$ plus the fixed cost, for a total of $70x + 6,600$. Writing $C(x)$ for the cost of producing x items, we have

$$C(x) = 70x + 6,600, \quad \text{for } x \geq 0. \tag{15}$$

We say that Eq. (15) **models** the cost.

(ii) Since each disc player sells for $180, selling x players will earn $180x$ dollars in revenue. If we write $R(x)$ for the revenue earned from selling x players, we have an equation that models the revenue:

$$R(x) = 180x.$$

(iii) The profit is simply the difference between the revenue and the cost. Writing $P(x)$ for the profit from producing and selling x items, we have

$$P(x) = R(x) - C(x) = 180x - (70x + 6,600) = 110x - 6,600.$$

The three functions we considered in the last example appear frequently in mathematical economics.

- The **cost function** $C(x)$ is the cost of producing x items.
- The **revenue function** $R(x)$ is the income earned from the sale of x items.
- The **profit function** $P(x)$ is the difference between the revenue and cost; that is, $P(x) = R(x) - C(x)$.

The variable x, which represents the number of units of a certain item produced and sold, is called the **production level**.

The graphs of the cost and revenue functions of the last example are shown in Figure 0.4.7, drawn on the same set of axes. Their intersection point is $(60, 10,800)$, whose coordinates are found by solving the equation $R(x) = C(x)$ for x, then substituting that x into either the cost or revenue function to get the y-coordinate.

The x-coordinate of the intersection point—the solution of the equation $R(x) = C(x)$, in other words—is called the **break-even point**. Since the profit function $P(x)$ is the revenue minus the cost, we can also describe the break-even point as the number x for which $P(x) = 0$. In the last example, for instance, we can find the break-even point

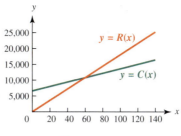

Figure 0.4.7

by solving $110x - 6,600 = 0$, which gives $x = 60$. Notice that if $x < 60$, the value of $P(x)$ is negative, which means the company loses money. Thus, in order to show a profit, the company must produce (and sell) more than 60 players a week. At exactly 60 they break even.

In Example 0.4.4 the cost function was linear, but that may not always be the case. In the next section we will consider an example of a nonlinear cost function. Similarly, the revenue and profit functions may or may not be linear.

Demand and supply curves

In economics, the quantity of a certain item produced and sold is modeled by two curves, called the demand and supply curves. The **demand curve** gives the quantity q that consumers will demand at each given price p, and the **supply curve** gives the quantity q that producers will supply at each given price p. In a free-market economy, the demand curve is decreasing and the supply curve is increasing. Their intersection point is called the **equilibrium point** and is denoted by (p_e, q_e). The equilibrium price p_e is the one toward which a free-market economy drives the price of the item.

EXAMPLE 0.4.5

When the price of a medium pizza was \$12, a restaurant sold about 300 of them per week. When the price was reduced to \$8, the sales climbed to 500 per week.

  (i) Find the demand function, assuming that it is linear.
  (ii) If the supply curve is $q = 100p - 600$, then find the equilibrium point.

SOLUTION   (i) We assume that the quantity q is a linear function of the price p, that is, $q = mp + b$. Then

$$\Delta q = 500 - 300 = 200,$$
$$\Delta p = 8 - 12 = -4,$$

and $m = \Delta q / \Delta p = -50$. Therefore, the demand curve is $q - 300 = -50(p - 12)$, or $q = -50p + 900$.

  (ii) To find the equilibrium point, we set $-50p + 900 = 100p - 600$, and obtain $p = 10$. Substituting it into either the demand or supply function, we obtain $q = -50 \times 10 + 900 = 400$. Thus, the equilibrium point is $(p_e, q_e) = (10, 400)$. The supply and demand curves are shown in Figure 0.4.8.

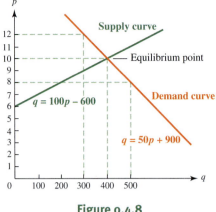

Figure 0.4.8

Remark   In economics it is customary to use the vertical axis for the independent variable p (price) and the horizontal axis for the dependent variable q (quantity).

Slope as a rate

In addition to measuring the steepness of the graph, the slope of a linear function also measures the **rate** at which the dependent variable y changes with respect to the independent variable x. What that means is this: If the slope is m, then a one-unit change in x will cause a change of m units in y. To illustrate that, consider the equation that converts Celsius to Fahrenheit

$$y = \frac{9}{5}x + 32,$$

where x and y are the number of degrees Celsius and Fahrenheit, respectively. The slope of this function is $\frac{9}{5}$, which means that the Fahrenheit reading changes by $\frac{9}{5}$ of a degree for each 1-degree change in Celsius. As another example, consider the cost function of Example 0.4.4

$$C(x) = 70x + 6,600.$$

The slope is 70, which reflects the fact that raising the production level by one unit increases the cost by $70.

That is an important point—that slope measures both the *steepness* of the graph and the *rate* at which the dependent variable changes. We will come back to it in Chapter 3.

Linear regression: the least-squares line

As we have already seen, there is a line passing through any two given points, (x_1, y_1) and (x_2, y_2), and we can write a linear equation $y = mx + b$ that describes it (provided that $x_1 \neq x_2$.) However, in many situations, especially in statistical analysis, we have more than two points—usually coming from some table of data—that do not lie along the same straight line but may be clustered *close* to a line. In such cases we often try to find the line that "best fits" all the given points. To put it another way, we want to find the linear relationship, $y = mx + b$, that comes closest to modeling the data.

For example, suppose a consumer agency is testing a certain model of truck to see how many miles per gallon of gasoline it uses at different speeds of highway driving, and it comes up with the following table of data:

speed (in miles per hour)	50	55	60	65
gasoline use (in miles per gallon)	11.1	9.7	8.9	7.3

(16)

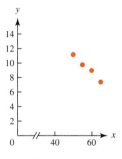

Figure 0.4.9

By plotting each data pair as a point in the plane, with x being the speed and y the miles per gallon, we get the diagram shown in Figure 0.4.9. This type of diagram, in which pairs of related numbers are plotted as points in the plane, is called a **scatter diagram**.

Although the points do not lie along a straight line, they seem to cluster around a straight line. The agency's analysts may suspect there is actually an approximate linear relationship between the two variables, and they want to determine the linear function that fits the data best. In fact, there is a standard method in statistics for determining that function. Its graph is called the **regression line** and also the **least-squares line**, and it is determined in the following way:

Definition 0.4.2

Given a set of data points—say, (x_1, y_1), (x_2, y_2), and so forth, up to (x_n, y_n), where n is some positive integer—the regression line is given by the equation $y = mx + b$, where

$$m = \frac{n(x_1 y_1 + \cdots + x_n y_n) - (x_1 + \cdots + x_n)(y_1 + \cdots + y_n)}{n(x_1^2 + \cdots + x_n^2) - (x_1 + \cdots + x_n)^2} \tag{17}$$

and

$$b = \frac{(y_1 + \cdots + y_n) - m(x_1 + \cdots + x_n)}{n}. \tag{18}$$

The rationale behind these formulas will be explained in Chapter 7. For now, we will simply accept and use them. They may seem imposing, but they are easy to apply, particularly with a calculator to perform the computations.

EXAMPLE 0.4.6

Find the regression line for the data points given by table (16) and use it to estimate the number of miles per gallon the truck will get at 70 miles per hour.

SOLUTION In this case, $n = 4$, and the four data points are $(50, 11.1)$, $(55, 9.7)$, $(60, 8.9)$, and $(65, 7.3)$. Noticing that several expressions occur more than once in formulas (17) and (18), let's begin by computing them with the help of a calculator.

$$x_1 + x_2 + x_3 + x_4 = 50 + 55 + 60 + 65 = 230$$
$$y_1 + y_2 + y_3 + y_4 = 11.1 + 9.7 + 8.9 + 7.3 = 37$$
$$x_1^2 + x_2^2 + x_3^2 + x_4^2 = 50^2 + 55^2 + 60^2 + 65^2 = 13,350$$
$$x_1 y_1 + x_2 y_2 + x_3 y_3 + x_4 y_4 = 50 \cdot (11.1) + 55 \cdot (9.7) + 60 \cdot (8.9) + 65 \cdot (7.3) = 2,097.$$

Entering these numbers into formula (17) and using a calculator give

$$m = \frac{4 \cdot (2,097) - (230) \cdot (37)}{4 \cdot (13,350) - (230)^2} = -0.244.$$

Next, doing the same with formula (18) gives

$$b = \frac{37 + (0.244) \cdot (230)}{4} = 23.28.$$

Therefore, the regression line has the equation

$$y = -0.244x + 23.28.$$

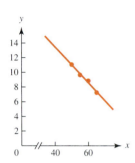

Figure 0.4.10

Figure 0.4.10 shows the scatter diagram and the regression line on the same set of axes.

To use the regression line to estimate the number of miles per gallon at 70 miles per hour, we simply substitute $x = 70$ into the line's equation, which gives

$$-0.244(70) + 23.28 = 6.2.$$

Both graphing calculators and spreadsheet programs will compute the least-squares coefficients, which is of inestimable help in dealing with large sets of data points. We will discuss their use, with examples, in Section 0.7.

Practice Exercises 0.4

1. Find the equation of the line through the point $(-2, 3)$ with slope 8.

2. A software company is currently producing an average of 500 copies of a certain computer game per week at a cost of $42,500. It assumes the cost is a linear function of the quantity produced, and it estimates that the cost will increase by $1,875 per week for every additional 25 copies. Find the cost at the production level of 550 copies per week.

Exercises 0.4

In Exercises 1–8, determine the slope and y-intercept and sketch the line.

1. $y = \frac{1}{2}x - 1$ **2.** $y = -x + 2$ **3.** $y = 4$

4. $y = -2x$ **5.** $x - 2y = 5$ **6.** $3x + y = 7$

7. $y = -3$ **8.** $2x + 5y = 1$

In Exercises 9–22, find the equation of the given line.

9. The line with slope -3 and y-intercept 2

10. The line with slope $\frac{1}{2}$ and y-intercept -1

11. The line through $(0, 0)$ with slope $-\frac{1}{2}$

12. The horizontal line through the point $(-3, 5)$

13. The horizontal line with y-intercept $-\frac{5}{4}$

14. The vertical line through the point $(-2, \frac{1}{3})$

15. The line through the point $(0, 3)$ with slope -2

16. The line through the point $(-1, 4)$ with slope 3

17. The line through the points $(2, 1)$ and $(-1, 5)$

18. The line through the points $(\frac{1}{2}, 0)$ and $(0, -\frac{3}{2})$

19. The line through the points $(\frac{2}{3}, 1)$ and $(-\frac{4}{7}, 1)$

20. The line through the points $(-2, \frac{1}{2})$ and $(-2, -1)$

21. The line through the point $(\frac{3}{2}, -\frac{1}{2})$ and parallel to the line $5x - 3y = 1$

22. The line through the point $(0, 4)$ and parallel to the line $x + 2y = 3$

23. Write a linear function that gives the Celsius reading C in terms of the Fahrenheit F [see Eq. (6)] and find the Celsius readings corresponding to $F = 60,\ 70,\ 80$, and 90. What is the rate at which the Celsius reading changes with respect to the Fahrenheit?

24. A long-distance phone service charges a fee of 9 cents a minute for every call and a service fee of $6.50 per month. Write a linear function that gives the monthly charge (in dollars) as a function of the number of minutes used.

25. A salesman is paid a fixed weekly salary of $400 plus a commission of 8% of his total sales revenue. Write a linear function expressing the salesman's weekly income as a function of the amount of sales revenue. At what rate does the salesman's income change with respect to the sales revenue?

26. A small shop making classical guitars has fixed expenses of $2,100 per month. Each guitar costs $450 to make and sells for $1,050.
(a) Write the monthly cost, revenue, and profit as functions of the number of guitars made.
(b) Find the break-even point.
(c) How much profit will the shop make if it produces and sells nine guitars a month?
(d) Suppose the shop lowers the price to $950. How many does it have to produce and sell a month in order to make a profit?

27. The daily profit function (in dollars) for a small bookstore is given by $P(x) = 1.5x - 300$, where x is the number of books sold.
(a) What is the break-even point?
(b) How many books must the store sell in a given day to earn a profit of $750?
(c) At what rate does the profit change with respect to the number of books sold?

28. An artist wants to earn money by making hand-painted neckties and selling them at craft fairs. She estimates that each tie costs $12 to produce, and she will also need to budget $200 per month for transportation and entry fees to the fairs.
(a) Write the cost (in dollars) as a function of the number of ties produced.
(b) She estimates she can make and sell 50 ties a month. What price does she have to charge to break even? What

price does she have to charge to make a profit of $300 per month?
(c) Suppose she charges $20 per tie. Write the profit function and determine how many ties she has to produce and sell to make a profit of $300 per month.

29. A car is traveling due east on the Indiana toll road at a steady speed of 1.2 miles per minute. At noon it is exactly 12 miles east of the Chicago Skyway. Letting t be the number of minutes past noon and s the number of miles east of the Chicago Skyway, write an equation expressing s as a linear function of t. Where is the car at 20 minutes past noon? At 1 P.M.?

In Exercises 30–32, a pair of demand, $D(p)$, and supply, $S(p)$, curves is given. In each case, sketch the curves and find the equilibrium price p_e and quantity q_e.

30. $D(p) = -40p + 470$ and $S(p) = 50p - 250$

31. $D(p) = -20p + 2{,}400$ and $S(p) = 10p - 600$

32. $D(p) = -12p + 1{,}200$ and $S(p) = 8p - 400$

33. A bus company sold about 1,000 tickets per week in its airport line when the price for each ticket was $30. After it raised the price to $40, it sold 100 fewer tickets.
(a) Find the demand curve, assuming that it is linear.
(b) If the supply curve is $q = 45p - 675$, then find the equilibrium point.

34. A bakery has determined that it cannot produce any bread at a price of 75 cents or less per loaf, but at a price of $1.25 per loaf it can produce 80 loaves per week. On the demand side, it has estimated that at a price of $1 per loaf the demand is 450 loaves per week, and at $3 per loaf the demand is 150 loaves per week. Find the demand and supply curves, assuming that they are linear, as well as the equilibrium point.

In Exercises 35–37, plot the given table of data as a scatter diagram and find the equation of the regression line using a calculator or spreadsheet for computations.

35.

x	1	2	3	4	5
y	3.5	5.25	9.25	9.75	14

36.

x	-2	-1	0	1	2
y	2.2	1.2	1.4	1	0.6

37.

x	0	2	4	6	8	10	12
y	-0.8	-1	-0.2	0.2	-2.0	0.8	-0.6

38. A company's sales revenue in its first five years of operation are given in the following table:

year	1	2	3	4	5
sales revenue	3.5	5.25	9.25	9.75	14

Find the equation of the regression line for this data set and use it to predict the revenue in the sixth year of operation, assuming an approximately linear relationship between years of operation and sales revenue.

39. An education study compared the amount spent per pupil in six school districts with the average SAT scores of graduating seniors, with the following result (with spending per pupil in thousands of dollars):

district	1	2	3	4	5	6
spending per pupil	3.9	5.8	4.4	3.6	4.1	4.3
average SAT score	880	974	887	865	890	856

Find the equation of the regression line for this data set and use it to predict the average SAT score in a district spending $4,000 per pupil, assuming an approximately linear relationship between the variables.

40. A study compared the age of seven chestnut trees with their diameters (measured at a height of 5 feet), resulting in the following table:

age (in years)	4	5	7	7	10	10	12
diameter (in feet)	0.7	0.8	1.0	2.1	2.0	3.4	4.6

Plot a scatter diagram for this data set and find the equation of the regression line. (Notice that in this case there are points with the same x-coordinate, but that makes no difference either in plotting the diagram or using the regression formulas.)

41. Show that *two lines, neither horizontal nor vertical, are perpendicular if and only if the product of their slopes is* -1 in the following sequence of steps:
(a) Assuming the lines are not parallel, let (x_0, y_0) be their point of intersection. Choose another point on each line and connect them as shown in Figure 0.4.11.
(b) Use the following fact from geometry: θ is a right angle if and only if the sides of the triangle satisfy the Pythagorean theorem. Therefore, L_1 and L_2 are perpen-

dicular if and only if
$$(x_1 - x_0)^2 + (y_1 - y_0)^2 + (x_2 - x_0)^2 + (y_2 - y_0)^2$$
$$= (x_2 - x_1)^2 + (y_2 - y_1)^2.$$

(c) By expanding and collecting terms, show that this equation is equivalent to
$$(y_1 - y_0)(y_0 - y_2) = (x_1 - x_0)(x_2 - x_0).$$

(d) Show that the last equation is equivalent to the statement that the product of the slopes of L_1 and L_2 equals -1.
(e) Finally, show that if the lines are parallel, the product of their slopes cannot equal -1. Use the fact that parallel lines have the same slope (see Exercise 46).

In Exercises 42–45, use the result stated in Exercise 41.

42. Find the intersection point of the lines $x + 2y = 2$ and $2x - y = 1$ and show that they meet at right angles.

43. Two lines meet at the point $(-1, 0)$, with one passing through $(2, -1)$ and the other through $(0, 3)$. Are they perpendicular?

44. Find the line through the point $(-1, 2)$ and perpendicular to the line $y = 3x - 5$.

45. State whether each of the following pairs of lines is parallel, perpendicular, or neither:
(a) $3x + 2y = 1$ and $3x - 2y = -1$
(b) $3x + 2y = 1$ and $2x - 3y = -1$
(c) $3x + 2y = 1$ and $3x + 2y = -1$
(d) $3x + 2y = 1$ and $2x + 3y = -1$

46. Show that two nonvertical lines are parallel if and only if they have the same slope by referring to Figure 0.4.12 and proceeding as follows:
(a) Let L_1 and L_2 be two nonvertical lines, and let P_1 and P_2 be their points of intersection with a fixed vertical line.
(b) Recall from plane geometry that L_1 and L_2 are parallel if and only if angles θ_1 and θ_2 are equal.
(c) Show that the angles θ_1 and θ_2 are equal if and only if the right triangles $P_1 A_1 B_1$ and $P_2 A_2 B_2$ are similar.
(d) Show that the triangles are similar if and only if L_1 and L_2 have the same slope.

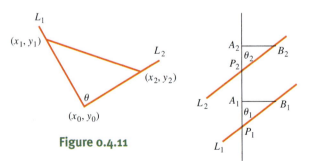

Figure 0.4.11

Figure 0.4.12

Solutions to practice exercises 0.4

1. Using the point-slope formula with $(x_1, y_1) = (-2, 3)$ and $m = 8$, we find $y - 3 = 8[x - (-2)]$ or $y = 8x + 19$.

2. We assume that the cost function is of the form $C(x) = mx + b$, where x denotes the number of units produced. To find the slope m, we use the information that $\Delta C = 1,875$ when $\Delta x = 25$. Therefore,

$$m = \frac{\Delta C}{\Delta x} = \frac{1,875}{25} = 75.$$

Using the point-slope formula with this m and with $(x_1, y_1) = (500, 42,500)$, we find that $C - 42,500 = 75(x - 500)$ or $C(x) = 75x + 5,000$. Therefore, the cost at the production level of 550 units per week is $C(550) = 46,250$.

0.5 Quadratic Functions

After linear functions, the next simplest class is that of **quadratic functions**.

Definition 0.5.1

A quadratic function is a function of the form

$$f(x) = ax^2 + bx + c,$$

where a, b, and c are constants with $a \neq 0$.

To find the roots of such a function, that is, to solve the equation

$$ax^2 + bx + c = 0,$$

we have the **quadratic formula**:

$$x = \frac{-b \pm \sqrt{b^2 - 4ac}}{2a}. \tag{19}$$

The quantity $b^2 - 4ac$ is called the **discriminant**.
There are

- two distinct solutions if $b^2 - 4ac > 0$,
- a single solution if $b^2 - 4ac = 0$, and
- no solutions if $b^2 - 4ac < 0$.

(If you have learned something about imaginary or complex numbers, then you know that the last statement is not strictly true. There are solutions, but they are not real numbers, which are the only ones we are concerned with in this book.) Later in this section, we will explain how to derive the quadratic formula by completing the square.

Factoring a quadratic and determining its sign

If a quadratic function, $f(x) = ax^2 + bx + c$, has the real numbers r and s as its roots—if $f(r) = 0$ and $f(s) = 0$, in other words—then we can write the quadratic in factored form, as follows:

$$ax^2 + bx + c = a(x - r)(x - s). \tag{20}$$

This fact can be proved in a number of ways. One of them, by completing the square, is taken up in the exercises that follow.

If $a = 1$, then (20) reduces to

$$x^2 + bx + c = (x - r)(x - s) = x^2 - (r + s)x + rs,$$

so that $b = -(r + s)$ and $c = rs$. By using these formulas and a bit of trial and error, you can sometimes obtain a quick factorization. If $a \neq 1$, you can rewrite the quadratic as

$$ax^2 + bx + c = a\left(x^2 + \frac{b}{a}x + \frac{c}{a}\right)$$

and try to factor the expression in parentheses. If all else fails, you can fall back on the quadratic formula to find the roots, provided that $b^2 - 4ac \geq 0$.

EXAMPLE 0.5.1

Factor each of the following quadratics:

$$\text{(i) } x^2 - 5x + 6, \quad \text{(ii) } 4x^2 - 8x + 3, \quad \text{(iii) } x^2 - x - 1.$$

SOLUTION (i) We look for numbers r and s with $r + s = 5$ and $rs = 6$ and easily come up with $r = 3$ and $s = 2$. Therefore,

$$x^2 - 5x + 6 = (x - 2)(x - 3).$$

(ii) Writing $4x^2 - 8x + 3 = 4(x^2 - 2x + \frac{3}{4})$, we need to find r and s with $r + s = 2$ and $rs = \frac{3}{4}$. A bit of trial and error gives $r = \frac{1}{2}$ and $s = \frac{3}{2}$, so that

$$4x^2 - 8x + 3 = 4\left(x - \frac{1}{2}\right)\left(x - \frac{3}{2}\right).$$

(iii) In this case we need $r + s = 1$ and $rs = -1$. Trial and error is not helpful, and any attempt to find r and s leads back to the very quadratic we are trying to factor. Therefore, we fall back on the quadratic formula, which gives the roots

$$\frac{1 + \sqrt{5}}{2} \quad \text{and} \quad \frac{1 - \sqrt{5}}{2}.$$

Thus, we obtain the factorization

$$x^2 - x - 1 = \left(x - \frac{1 + \sqrt{5}}{2}\right)\left(x - \frac{1 - \sqrt{5}}{2}\right).$$

Factoring is useful for determining the sign of a quadratic function, as we see in the following example.

EXAMPLE 0.5.2

Determine where the quadratic function $f(x) = x^2 - 2x - 3$ is positive and where it is negative.

SOLUTION   We start by factoring the quadratic:

$$f(x) = x^2 - 2x - 3 = (x + 1)(x - 3).$$

If both factors on the right have the same sign, their product is positive. How can that occur? First, if $x > 3$, both factors are positive. Second, if $x < -1$, both are negative. In each case, $f(x) > 0$.

On the other hand, if $-1 < x < 3$, then $x + 1 > 0$ and $x - 3 < 0$. Therefore, their product is negative and $f(x) < 0$. We conclude:

- $f(x) < 0$ on the interval $(-1, 3)$, and
- $f(x) > 0$ on the intervals $(-\infty, -1)$ and $(3, \infty)$.

Figure 0.5.1

There is another way to arrive at that same conclusion that will prove useful when we work with more complicated functions. It is based on the fact that a quadratic can only change sign at the points where its graph crosses the x-axis—that is, at the x-intercepts, which are the roots $x = -1$ and $x = 3$. They divide the x-axis into three intervals, $(-\infty, -1)$, $(-1, 3)$, and $(3, \infty)$, as shown in Figure 0.5.1. Since $f(x)$ can only change sign at -1 and 3, it must have the same sign throughout each of these intervals. Therefore, we only need to test one point in each interval. Whatever the sign is at that point, it stays that way throughout the interval. For instance,

- in $(-1, 3)$ we can test $x = 0$, and since $f(0) = -3 < 0$, we conclude that $f(x) < 0$ throughout $(-1, 3)$;
- in $(-\infty, -1)$ we can test $x = -2$, and since $f(-2) = 5 > 0$, we conclude that $f(x) > 0$ throughout $(-\infty, -1)$; and
- in $(3, \infty)$ we can test $x = 4$. Since $f(4) = 5 > 0$, we conclude that $f(x) > 0$ throughout $(3, \infty)$.

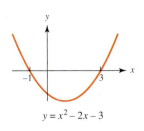

$y = x^2 - 2x - 3$

Figure 0.5.2

The changes in sign are illustrated in Figure 0.5.1 and the graph of the function in Figure 0.5.2. Although this sign-checking procedure may seem longer than necessary for dealing with a quadratic, it will be very helpful later. The principle on which it is based—that the function can only change sign at an x-intercept—is true for a large and important class of functions known as **continuous** functions, which we will meet in Chapter 1.

Completing the square

A quadratic function, $f(x) = ax^2 + bx + c$, with $a \neq 0$, can be rewritten in the following useful form:

$$f(x) = a(x - h)^2 + k,$$

where h and k are constants. To find them, we equate the two forms of the function

$$ax^2 + bx + c = a(x - h)^2 + k,$$

then expand the right-hand side and collect terms:

$$ax^2 + bx + c = a(x^2 - 2hx + h^2) + k$$
$$= ax^2 - 2ahx + (ah^2 + k).$$

Equating the coefficients of like powers of x gives

$$b = -2ah \quad \text{and} \quad c = ah^2 + k.$$

Solving the first of these equations for h gives

$$\boxed{h = -\frac{b}{2a}.} \tag{21}$$

Substituting that into the second equation gives

$$c = \frac{b^2}{4a} + k.$$

Therefore,

$$\boxed{k = c - \frac{b^2}{4a} = \frac{4ac - b^2}{4a}.} \tag{22}$$

Putting it together gives

$$\boxed{ax^2 + bx + c = a\left(x + \frac{b}{2a}\right)^2 + \left(\frac{4ac - b^2}{4a}\right).} \tag{23}$$

It is not necessary to memorize this formula; in fact, we strongly suggest that you do not. All you need to do is to follow the steps we took to find h and k, a procedure known as **completing the square**.

EXAMPLE 0.5.3

Complete the square of the quadratic

$$3x^2 - 5x + 4.$$

SOLUTION   We want to find h and k so that

$$3x^2 - 5x + 4 = 3(x - h)^2 + k.$$

Expanding the right-hand side and collecting terms give

$$3x^2 - 5x + 4 = 3(x^2 - 2hx + h^2) + k$$
$$= 3x^2 - 6hx + (3h^2 + k).$$

Equating the coefficients of like powers of x, we get

$$-5 = -6h \quad \text{and} \quad 4 = 3h^2 + k.$$

Solving the first equation for h gives $h = \frac{5}{6}$. Substituting this into the second gives $4 = \frac{25}{12} + k$, and, therefore,

$$k = 4 - \frac{25}{12} = \frac{23}{12}.$$

Putting it together, we have

$$3x^2 - 5x + 4 = 3\left(x - \frac{5}{6}\right)^2 + \frac{23}{12}.$$

The quadratic formula can be derived by completing the square, as follows. Given the equation $ax^2 + bx + c = 0$, completing the square puts it in the form

$$a\left(x + \frac{b}{2a}\right)^2 + \frac{4ac - b^2}{4a} = 0. \tag{24}$$

This equation is not hard to solve for x, and, after simplifying, it reduces to the quadratic formula. The details are left as an exercise.

The graph of a quadratic function

The graph of a quadratic function is called a **parabola**. In Figure 0.5.3 we see six simple examples, all drawn on a computer using the Mathematica program. (You should try creating your own graphs on a computer or graphing calculator.) The equations corresponding to these six graphs are as follows:

$$y = \frac{1}{2}x^2, \quad y = x^2, \quad y = 2x^2, \quad y = -\frac{1}{2}x^2, \quad y = -x^2, \quad y = -2x^2.$$

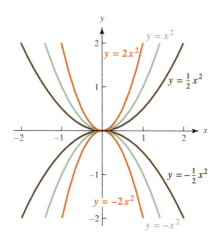

Figure 0.5.3

Because they are all even functions, their graphs are symmetric about the y-axis. Therefore, to draw each of these graphs, it suffices to draw only the right-hand branch, which we can do by plotting enough points and connecting them by a curve. Then we simply reflect the right-hand branch across the y-axis to get the left-hand branch.

These functions are all of the form $y = ax^2$, and any such function is symmetric about the y-axis and its graph goes through the point $(0, 0)$. We call that point the **vertex** of the parabola, and we call the y-axis the **axis of symmetry**. Moreover, the graph of any such function

- opens upward if $a > 0$, with $(0, 0)$ the lowest point of the graph and
- opens downward if $a < 0$, with $(0, 0)$ the highest point of the graph.

How wide or narrow the graph is depends on the size of $|a|$. The larger $|a|$ is, the narrower and steeper the graph is, as you can see in Figure 0.5.3.

For a general quadratic, $y = ax^2 + bx + c$, completing the square changes it to the form $y = a(x - h)^2 + k$. Therefore, the graph is obtained from that of $y = ax^2$ by a horizontal and vertical translation (see Section 0.3). The vertex is moved to the point (h, k) and the axis of symmetry to the line $x = h$.

For instance, the graph of $y = 2x^2 + 5$, shown in Figure 0.5.4, is obtained from that of $y = 2x^2$ by a shift of five units upward. The vertex is at $(0, 5)$, and the y-axis is the axis of symmetry.

If we replace x by $(x + 2)$, we get $y = 2(x + 2)^2 + 5$, whose graph is shown in Figure 0.5.5. It is obtained from the graph of $y = 2x^2 + 5$ by a horizontal shift of two units to the left. Its vertex is at $(-2, 5)$, and the line $x = -2$ is its axis of symmetry.

In general, by completing the square, we get

$$ax^2 + bx + c = a(x - h)^2 + k, \text{ with } h = -\frac{b}{2a} \text{ and } k = \frac{4ac - b^2}{4a}.$$

The graph is obtained from $y = ax^2$ by a vertical shift of $|k|$ units (upward if $k > 0$, downward if $k < 0$) and a horizontal shift of $|h|$ units (to the right if $h > 0$, to the left if $h < 0$). Its vertex is at the point (h, k), and the line $x = h$ is its axis of symmetry.

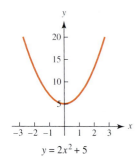

$y = 2x^2 + 5$

Figure 0.5.4

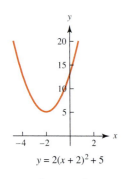

$y = 2(x + 2)^2 + 5$

Figure 0.5.5

Once again, we strongly suggest that you do not memorize this information, but simply follow the procedure for completing the square.

If $a > 0$, the vertex is a minimum point of the graph, and if $a < 0$, it is a maximum point.

EXAMPLE 0.5.4

Draw the graphs of the quadratic functions

$$\text{(i)} \quad y = x^2 - 2x - 3 \quad \text{and} \quad \text{(ii)} \quad y = -x^2 + 2x.$$

For each, find the vertex, the axis of symmetry, the x- and y-intercepts, and the maximum or minimum value.

SOLUTION (i) We begin by completing the square:

$$x^2 - 2x - 3 = (x - h)^2 + k$$
$$= x^2 - 2hx + (h^2 + k).$$

Equating coefficients of like powers of x and solving give $h = 1$ and $k = -4$. Thus, we can rewrite the equation of the graph in the form

$$y = (x - 1)^2 - 4.$$

This shows that the parabola is obtained from $y = x^2$ by a vertical shift of four units downward and one unit to the right. It opens upward, with the line $x = 1$ as its axis of symmetry and its vertex at $(1, -4)$. It follows that -4 is the minimum value of the function, occurring at $x = 1$.

To find the y-intercept, we set $x = 0$, which gives $y = -3$. Therefore, the graph crosses the y-axis at $(0, -3)$. To find the x-intercepts, we solve $x^2 - 2x - 3 = 0$, which gives $x = -1$ and $x = 3$. The graph is shown in Figure 0.5.6.

(ii) As before, we begin by completing the square:

$$-x^2 + 2x = -(x - h)^2 + k$$
$$= -x^2 + 2hx + (k - h^2),$$

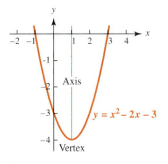

Figure 0.5.6

from which we get $h = 1$ and $k = 1$. The equation of the graph then becomes

$$y = -(x - 1)^2 + 1,$$

and it is obtained from the graph of $y = -x^2$ by a shift of one unit upward and one unit to the right. The parabola opens downward with vertex at $(1, 1)$ and the function takes the maximum value of 1 at $x = 1$. The axis of symmetry is again given by the line $x = 1$. Setting $x = 0$ gives $y = 0$, which is the y-intercept. Solving $-x^2 + 2x = 0$ shows that the x-intercepts are at $(0, 0)$ and $(2, 0)$. The graph is shown in Figure 0.5.7.

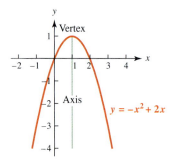

Figure 0.5.7

It will be useful in later applications to summarize what we now know about the maxima and minima of quadratic functions.

Optimization of quadratic functions

> If $f(x) = ax^2 + bx + c$, then by completing the square, we see that:
>
> 1. $f(x)$ has a **minimum** value if $a > 0$ and a **maximum** value if $a < 0$;
> 2. the minimum or maximum occurs at $x = -b/2a$;
> 3. the value of $f(x)$ at $x = -b/2a$ is $(4ac - b^2)/4a$.

In Chapter 4 we will use the methods of calculus to reach these same conclusions in a totally different way. What is more, we will be able to apply those methods to a much larger class of functions to determine their minimum and maximum points and to get information about their graphs. The quadratics provide the simplest examples of how to find maximum and minimum values, and they give us important insight into the more general case.

A quadratic model of revenue

In introducing the cost and revenue functions in Section 0.4, we assumed they were linear, but that may not be realistic. For instance, the price of an item may fall as the supply grows more plentiful. The following is a modification of Example 0.4.4 taking that into account.

EXAMPLE 0.5.5

A company producing compact disc players has a fixed cost of $6,600 a week and a cost per item of $70. The company estimates that, within the range of 0 to 400 players per week, the price of a player starts at $180 and goes down by 25 cents for each one available. Find and graph the cost and revenue functions for producing and selling up to 400 players per week. Then find the profit function and the break-even points. Finally, determine how many players the company should produce per week to maximize its profit.

SOLUTION   As in Example 0.4.4, the cost for producing x units is given by the cost function

$$C(x) = 6,600 + 70x. \tag{25}$$

To find the revenue function, we first derive a formula for the price p of each disc player in terms of x, the number produced and sold. According to our assumptions, the price, which is initially $180, loses $\frac{1}{4}$ dollar per player sold. That leads to the formula

$$p = 180 - 0.25x, \quad 0 \le x \le 400.$$

If the company sells x players at a price of p dollars each, its revenue is xp, and so we get the revenue function:

$$R(x) = xp = -0.25x^2 + 180x, \quad 0 \le x \le 400. \tag{26}$$

As you can see, this function is quadratic with $a = -0.25$, $b = 180$, and $c = 0$.

The profit function is the revenue minus the cost:

$$P(x) = -0.25x^2 + 110x - 6,600. \tag{27}$$

A graph of $R(x)$ over the domain [0, 400] is shown in Figure 0.5.8. In Figure 0.5.9 the same graph is drawn, together with the cost function given by formula (25). As you can see, the graphs $R(x)$ and $C(x)$ intersect in two points, which we can find by solving the equation $R(x) = C(x)$ or, equivalently, $P(x) = 0$. Using the quadratic formula, we get two solutions:

$$x = 2(110 - \sqrt{110^2 - 6,600}) \approx 71.7 \quad \text{and} \quad x = 2(110 + \sqrt{110^2 - 6,600}) \approx 368.3.$$

The symbol $\approx$ means "is approximately equal to." In this case the solutions have been rounded off to one decimal place.

Thus, there are *two* break-even points. If the company makes and sells less than 71.7 or more than 368.3 players per week, they will lose money. Between those numbers they make a profit. Of course, they can't make exactly 71.7 or 368.3 players. They have to round these numbers up or down to the nearest whole number.

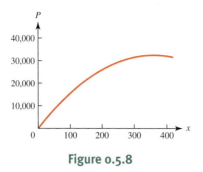

Figure 0.5.8

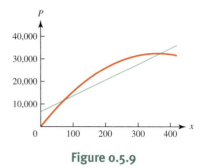

Figure 0.5.9

Finally, by completing the square, we can rewrite the profit function in the form

$$P(x) = -0.25\,(x - 220)^2 + 5{,}500.$$

(You should work that out as an exercise.) Therefore, $P(x)$ has a maximum value of 5,500, attained at $x = 220$.

A law of Galileo

An important law of nature, first discovered by Galileo, states that if an object is moving under the influence of the Earth's gravity, with no other forces acting on it, its height until it hits the ground is given as a function of time by the equation

$$h = -gt^2 + v_0 t + h_0. \tag{28}$$

In this formula, t denotes the time (measured in seconds) and h denotes the height of the object. The letter g represents a physical constant, whose value depends on the unit used to measure the height. If h is measured in feet, then $g = 16$, and if it is measured in meters, then $g = 4.9$. (These numbers are rounded-off versions of more accurate decimal expansions that have been determined experimentally.)

The constants h_0 and v_0 stand for the object's height and velocity, respectively, at time zero—called the **initial height** and **initial velocity**. The initial velocity may be negative. That will be the case, for instance, if you throw a ball downward from the top of a building. In other words, the sign of v_0 is positive if the initial velocity is in the direction of increasing height and negative in the direction of decreasing height.

EXAMPLE 0.5.6

Suppose you throw a baseball from the top of a 100-foot building. Express its height (in feet) as a function of the time (in seconds) in each of the following cases, and determine when the object hits the ground.

(i) You release it from a position of rest.
(ii) You throw it downward with a velocity of 40 feet per second.
(iii) You throw it upward with a velocity of 40 feet per second.

In the last case, find the maximum height of the ball.

SOLUTION (i) In this case, $h_0 = 100$ and $v_0 = 0$, and Eq. (28) takes the form

$$h = -16t^2 + 100.$$

(We set $g = -16$ because h is measured in feet.) To find when the ball hits the ground, we set $h = 0$ and solve for t, taking the positive solution as our answer:

$$t = \sqrt{\frac{100}{16}} = \frac{10}{4} = 2.5 \text{ seconds.}$$

(ii) In this case, $h_0 = 100$ and $v_0 = -40$, and Eq. (28) becomes

$$h = -16t^2 - 40t + 100.$$

To find when the ball hits the ground, we set $h = 0$ and solve for t. The quadratic equation gives two solutions (as you should verify): $t = 5(\sqrt{5} - 1)/4$ and $t = -5(\sqrt{5} + 1)/4$. Only the first of these is positive, and we conclude the ball hits the ground at the end of $5(\sqrt{5} - 1)/4 \approx 1.545$ seconds.

(iii) In this case, Eq. (28) takes the form

$$h = -16t^2 + 40t + 100. \tag{29}$$

As before, we set $h = 0$ and solve for t to find the time the ball hits the ground. Applying the quadratic formula to the equation

$$-16t^2 + 40t + 100 = 0$$

gives $t = 5(1 \pm \sqrt{5})/4$. Choosing the positive solution, we conclude that the ball hits the ground at $t = 5(1 + \sqrt{5})/4 \approx 4.045$ seconds.

To find the maximum height that the ball reaches, we must find the maximum value of the quadratic given by (29). Completing the square changes this to

$$h = -16\left(t - \frac{5}{4}\right)^2 + 125$$

(as you should verify), from which we see that the maximum height is 125 feet, occurring at $t = \frac{5}{4}$ seconds.

Practice Exercises 0.5

1. Complete the square of the quadratic $2x^2 - x - 1$ and then factor it. Find the vertex, axis of symmetry, and x- and y-intercepts of the graph, and determine where the function is positive and where it is negative.

2. A bike company estimates that its cost function is given by the formula $C(x) = 6{,}000 + 40x$ and its revenue by the formula $R(x) = (150 - 0.5x)x$, where x denotes the units of bikes sold per month. Find the number of bikes that the company should sell to maximize its profit.

Exercises 0.5

In Exercises 1–8, complete the square of the quadratic. Then state whether the graph opens upward or downward, find the vertex and axis of symmetry, and sketch the graph.

1. $y = -x^2 + 6x + 11$ **2.** $y = 3x^2 - 6x + 7$

3. $y = 2x^2 + 5x$ **4.** $y = -2x^2 + x + 1$

5. $y = -0.1x^2 - 1.2x + 3.6$

6. $y = \frac{1}{2}x^2 - 6x + 1$

7. $y = 3x^2 - 2x + 1$ **8.** $y = 3x^2 + 2x + 1$

In Exercises 9–14, factor the quadratic without using the quadratic formula to find the roots. Then determine where the function is positive and where it is negative.

9. $f(x) = x^2 - 5x - 14$ **10.** $f(x) = 2x^2 - x - 1$

11. $f(x) = x^2 - \frac{5}{6}x + \frac{1}{6}$ **12.** $f(x) = 4x^2 - 9$

13. $f(x) = \frac{1}{4}x^2 - 2x + 3$

14. $f(x) = 3x^2 + 5x - 2$

In Exercises 15–18, factor the quadratic by any means and then determine where the function is positive and where it is negative.

15. $f(x) = x^2 - 2x - 1$ **16.** $f(x) = x^2 - \frac{3}{4}$

17. $f(x) = x^2 + \frac{1}{10}x - \frac{1}{5}$ **18.** $f(x) = 4x^2 - 8x + 1$

19. At a daily production level of x units, with $0 < x < 240$, a company has cost and revenue functions

$$C(x) = 5,000 + 1,800x \quad \text{and} \quad R(x) = -10x^2 + 2,400x.$$

Find the break-even points. At what production levels does the company make a profit?

20. The price of a certain product varies according to how much is available. At a production level of x units the price per unit is $3.40 - 0.01x$.
(a) What is the revenue if x units are produced and sold?
(b) If the cost of producing x units is $100 + 1.20x$, find the profit function and the break-even points.

21. A theater company charges the same price for all tickets, regardless of day of the week or seat location. It has found that at a price of $16 it sells 900 tickets a week, but at a price of $20 per ticket it only sells 780. Assume that q, the number of tickets sold, is a linear function of the price p.
(a) Write an equation expressing q as a function of p (demand curve).
(b) Write the revenue as a function of the price.

22. A restaurant has determined that its monthly cost function is given by $C = 3,000 + 16q$, where q is the number of dinners sold per month. Moreover, $q = 1,200 - 20p$, where p is the average price of a dinner.
(a) Solve for p in terms of q and write the revenue and profit as functions of q.
(b) Find the break-even points in terms of q. In other words, find the minimum and maximum number of dinners per month that the restaurant can serve without losing money.
(c) Economists usually take the price as the independent variable because it can be controlled by the company. Write the cost, revenue, and profit as functions of p and find the break-even points in terms of p.
(d) Determine the range of prices for which the restaurant makes a profit.

23. In Exercise 22, what price should the restaurant charge to maximize
(a) its revenue?   **(b)** its profit?

24. A software company is currently selling an average of 400 copies of a certain computer game per week at a price of $120. It assumes the demand is a linear function of the price, and it estimates that every $5 decrease in price will add an average of 50 sales per week.
(a) Write the demand q as a linear function of the price p.
(b) Write the revenue as a function of the price.

25. In Exercise 24, what price should the software company charge to maximize its revenue from the sale of the computer game?

26. A rock is thrown straight up with an initial velocity of 48 feet per second from an initial height of 56 feet. Its height h at time t is given by the following formula, which is valid until the rock hits the ground:

$$h = -16t^2 + 48t + 56.$$

(a) At what time does the rock hit the ground?
(b) At what time does it reach its maximum height?
(c) What is the maximum height it reaches?

In Exercises 27 and 28, find the equilibrium quantity and price for each of the following demand and supply curves.

27. $D(q) = 0.3(q - 20)^2$, $S(q) = 2q + 10$, $0 \leq q \leq 18$

28. $D(q) = 0.005(q - 100)^2$, $S(q) = 0.1q + 2$, $0 \leq q \leq 100$

29. Show that the solution of Eq. (24) is given by

$$x = -\frac{b}{2a} \pm \sqrt{\frac{b^2}{4a^2} - \frac{c}{a}},$$

and that it can be simplified to the form

$$x = \frac{-b \pm \sqrt{b^2 - 4ac}}{2a}.$$

30. In this exercise you will verify the factorization given in formula (20) for the quadratic $ax^2 + bx + c$ in the following sequence of steps. We assume that $b^2 - 4ac \geq 0$, for otherwise there are no real roots.

(a) By the quadratic formula, the roots are

$$r = -\frac{b}{2a} + \frac{\sqrt{b^2 - 4ac}}{2a} \quad \text{and} \quad s = -\frac{b}{2a} - \frac{\sqrt{b^2 - 4ac}}{2a}.$$

Verify that $r + s = -b/a$.

(b) Using the identity

$$u^2 - v^2 = (u - v)(u + v),$$

show that $rs = c/a$.

(c) Now expand $a(x - r)(x - s)$ and show that it equals $ax^2 + b + c$, which verifies formula (20).

Solutions to practice exercises 0.5

1. We write $2x^2 - x - 1 = 2(x - h)^2 + k$. Expanding the right-hand side and collecting terms give $2x^2 - x - 1 = 2x^2 - 4hx + (2h^2 + k)$. Equating coefficients of like powers of x and solving give $h = \frac{1}{4}$ and $k = -\frac{9}{8}$. Therefore,

$$2x^2 - x - 1 = 2\left(x - \frac{1}{4}\right)^2 - \frac{9}{8}.$$

The line $x = \frac{1}{4}$ is an axis of symmetry of the graph, and the vertex is at $(\frac{1}{4}, -\frac{9}{8})$. To find the roots, we use the quadratic formula with $a = 2$, $b = -1$, and $c = -1$, to get

$$x = \frac{1 \pm \sqrt{1 + 8}}{4} = 1, \; -\frac{1}{2}.$$

These are the x-intercepts, and they are the only places where the function can change sign. Testing a point in each of the intervals marked off by the x-intercepts, we see that the function is positive on $(1, \infty)$ and $(-\infty, -\frac{1}{2})$ and neg-ative on $(-\frac{1}{2}, 1)$. Or, we can use the roots to write the factorization:

$$2x^2 - x - 1 = 2(x - 1)\left(x + \frac{1}{2}\right)$$

and use it to reach the same conclusion. The y-intercept is found by setting $x = 0$ in the quadratic, which gives $y = -1$.

2. The profit function is

$$P(x) = R(x) - C(x) = (150 - 0.5x)x - (6{,}000 + 40x)$$
$$= -0.5x^2 + 110x - 6{,}000$$

The graph is a parabola opening downward. The maximum point will be at

$$-\frac{b}{2a} = -\frac{110}{2 \cdot (-0.5)} = 110.$$

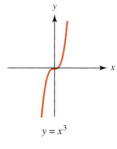

Figure 0.6.1

0.6 Polynomials, Rational Functions, and Power Functions

After linear and quadratic functions, the next type to consider is that of third-degree, or **cubic**, functions—those of the form

$$f(x) = ax^3 + bx^2 + cx + d.$$

Once we get to degree three, the graphs become more varied. For instance, two graphs of cubic functions are pictured in Figures 0.6.1 and 0.6.2. As you can see, they are quite different from one another. The graph in Figure 0.6.1 increases throughout, whereas the one in Figure 0.6.2 turns around twice—increasing to decreasing to increasing. One of the applications of calculus you will learn is how to predict the shape of a graph from its formula.

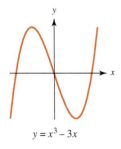

$y = x^3 - 3x$

Figure 0.6.2

Of course, there are also fourth-, fifth-, and higher-degree functions, made up of powers of x with constant coefficients. A general fourth-degree function looks like this:

$$f(x) = a_4 x^4 + a_3 x^3 + a_2 x^2 + a_1 x + a_0.$$

Notice that this time instead of using different letters for the coefficients, we used the same letter a, but with different subscripts. That is a more efficient way of writing the coefficients for high-degree functions.

Functions of this type—whether quadratic, cubic, fourth-degree, or higher—are called **polynomials**. To write a typical polynomial, we first fix a positive integer, say, n, as the degree. Then we write a_0, a_1, a_2, and so on, up to a_n to represent the coefficients. In the following definition, the symbol . . . is simply a mathematical way of writing "etc."

Definition 0.6.1

A polynomial of degree n is a function of the form

$$f(x) = a_n x^n + \cdots + a_2 x^2 + a_1 x + a_0,$$

where $a_0, a_1, \ldots, a_n$ are constants, and $a_n \neq 0$.

The number a_n is called the **leading coefficient**.

Figures 0.6.3 and 0.6.4 show the graphs of two fourth-degree polynomials. As you can see, their shapes are quite different.

One common feature of the graphs in Figures 0.6.1 through 0.6.4 is that they increase unboundedly (climb higher and higher without bound) as they move further and further toward the right. That is a property of all polynomials with a positive leading coefficient, one that we already observed in quadratics—the graph climbs higher and higher without bound as it moves far to the right. If the leading coefficient is negative, the behavior is exactly the opposite—the graph decreases unboundedly (falls lower and lower without bound) as it moves farther and farther toward the right.

How does the graph behave as it moves far toward the left? The answer to that depends on two things—the degree n of the polynomial and the sign of the leading coefficient a_n. Here is a summary of the various possibilities. Given $f(x) = a_n x^n + \cdots + a_2 x^2 + a_1 x + a_0$,

- if n is even and $a_n > 0$, the graph rises to the right and left;
- if n is even and $a_n < 0$, the graph falls to the right and left;

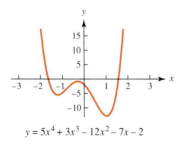

$y = 5x^4 + 3x^3 - 12x^2 - 7x - 2$

Figure 0.6.3

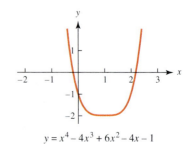

$y = x^4 - 4x^3 + 6x^2 - 4x - 1$

Figure 0.6.4

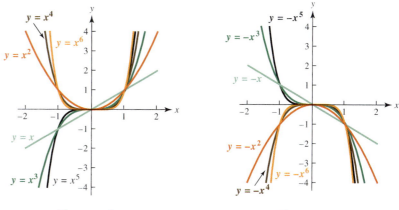

Figure 0.6.5   **Figure 0.6.6**

- if n is odd and $a_n > 0$, the graph rises to the right and falls to the left;
- if n is odd and $a_n < 0$, the graph falls to the right and rises to the left.

A simple way to remember how the graph behaves as x goes further and further to the right (or left) is the following principle: *A polynomial behaves more and more like its highest-degree term as the absolute value of x gets larger and larger.* In symbols, using $\approx$ to denote approximation, we can summarize this principle as follows:

> If x is very large in absolute value, $a_n x^n + \cdots + a_2 x^2 + a_1 x + a_0 \approx a_n x^n$.

We will have more to say about this approximation in the next section when we discuss limits. What makes it useful is that the right-hand side is much easier to analyze. The graphs of $y = x^n$ for n from 1 to 6 are drawn in Figure 0.6.5, all on the same set of axes. Figure 0.6.6 shows the graphs of $y = -x^n$ for n from 1 to 6. These illustrations will help you to visualize the behavior of $y = a_n x^n$ as x moves farther and farther to the left or right.

EXAMPLE 0.6.1

Let $f(x) = -3x^5 - 2x^4 + x^3 - x^2 + 8x + 1$. How does the graph of this polynomial behave as it moves farther and farther to the right or left?

SOLUTION   To answer this question, we analyze the behavior of the leading term, $-3x^5$. The sign of $-3x^5$ is opposite that of x, which means that the graph falls lower and lower as it moves to the right, and it climbs higher and higher as it moves to the left. Since we know that

$$-3x^5 - 2x^4 + x^3 - x^2 + 8x + 1 \approx -3x^5$$

when x has a large absolute value, we can conclude that the graph of $f(x)$ behaves the same way.

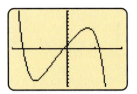

Figure 0.6.7

To help you visualize the behavior of $f(x)$, its graph is shown in Figure 0.6.7. It was created using a graphing calculator, setting $-2 \leq x \leq 2$ and $-10 \leq y \leq 10$. At present, we do not have the tools to analyze and sketch such a graph by hand, but Chapter 4 will provide the appropriate techniques.

Rational functions

If we add, subtract, or multiply two polynomials, we get another polynomial. For instance, from algebra we know that

$$(x^3 - 5x^2 + 2x - 4) + (3x^2 + x - 7) = x^3 - 2x^2 + 3x - 11$$

and

$$(x^3 - 5x^2 + 2x - 4) \cdot (3x^2 + x - 7) = 3x^5 - 14x^4 - 6x^3 + 25x^2 - 18x + 28.$$

But if we take a quotient of two polynomials, we get a new type of function called a **rational function**.

Definition 0.6.2

A rational function is a quotient of two polynomials.

An important difference between rational functions and polynomials is that polynomials are defined for *all* numbers x, but rational functions may not be. Those x which are *zeros of the denominator are not in the natural domain*.

EXAMPLE 0.6.2

Find the natural domain of the function $f(x) = 1/x$. Then discuss its geometric properties and sketch its graph.

SOLUTION   The natural domain of this function consists of all $x \neq 0$. We cannot use $x = 0$ as input because division by zero is not a legitimate algebraic operation. It follows that the graph consists of two disjoint branches, one in the right half-plane (where $x > 0$) and the other in the left (where $x < 0$).

Observe that $f(-x) = -f(x)$, which means that the function is odd and the graph is symmetric about the origin. Because of that, we can restrict our attention to the right-hand branch and then use symmetry to sketch the left-hand branch. Here are a few simple conclusions:

- If $x > 0$, then $1/x > 0$. Therefore, the right-hand branch of the graph lies entirely above the y-axis.
- The function is decreasing for $x > 0$. For if x_1 and x_2 are two positive numbers with $x_1 < x_2$, then $1/x_2 < 1/x_1$.

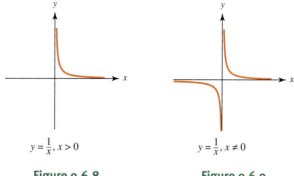

$$y = \frac{1}{x}, x > 0 \qquad\qquad y = \frac{1}{x}, x \neq 0$$

Figure 0.6.8   **Figure 0.6.9**

- The graph approaches the x-axis as it goes further and further to the right. That is because $1/x$ becomes very small when x gets very large.
- The right-hand branch climbs higher and higher without bound as it approaches the y-axis. That is because $1/x$ becomes larger and larger as x gets closer and closer to zero.

This information is a sufficient basis for a rough sketch of the right-hand branch. A more precise version, created with the Mathematica program, is shown in Figure 0.6.8. Using symmetry, we can then sketch the left-hand branch as well to get the graph shown in Figure 0.6.9.

When a graph approaches a vertical line in the way that the graph of $y = 1/x$ approaches the y-axis—either climbing or falling unboundedly—we call that line a **vertical asymptote** of the graph. And when a graph approaches a horizontal line as it moves farther and farther to the right or left—as the graph of $y = 1/x$ approaches the x-axis—we call that line a **horizontal asymptote** of the graph. We will study asymptotes more fully in Chapter 1 when we discuss the important concept of **limits**. For now, we will confine ourselves to a few informal observations.

In general, a rational function has a vertical asymptote at each point where its denominator is zero—provided that it is written as a **quotient in lowest terms**. In other words, if we write the function in the form

$$f(x) = \frac{p(x)}{q(x)},$$

where $p(x)$ and $q(x)$ are polynomials with no common factor, then there is a vertical asymptote at each point where $q(x) = 0$. That is because the denominator is close to zero near such a point, whereas the numerator is not, which means that the quotient is very large in absolute value.

EXAMPLE 0.6.3

Let $f(x) = 1/(x - 1)^2$. Find the natural domain and the vertical asymptotes of this function. Then discuss its geometric properties and sketch its graph.

SOLUTION   This function is a quotient in lowest terms, and the denominator is zero when $x = 1$. The natural domain of this function consists of all $x \neq 1$, and the line $x = 1$ is a vertical asymptote. Thus, the graph consists of two disjoint branches, one to the right of the asymptote, the other to the left.

The function is positive for all x in its domain. Its values get unboundedly large as x gets closer and closer to 1 (from either side), because the numerator is fixed at 1 and the denominator gets very close to zero. For example, if we evaluate the function at $x = 1 \pm 10^{-6}$, we obtain the output

$$f(1 \pm 10^{-6}) = \frac{1}{(\pm 10^{-6})^2} = 10^{12} = 1{,}000{,}000{,}000{,}000.$$

Three further observations:

- The function is decreasing on the interval $(1, \infty)$. For if $x_2 > x_1 > 1$, then $(x_2 - 1) > (x_1 - 1) > 0$, which means that $(x_2 - 1)^2 > (x_1 - 1)^2$ and, therefore, $1/(x_2 - 1)^2 < 1/(x_1 - 1)^2$.
- The function is increasing on the interval $(-\infty, 1)$. For if $x_1 < x_2 < 1$, then $(x_1 - 1) < (x_2 - 1) < 0$, which means that $(x_1 - 1)^2 > (x_2 - 1)^2$ and, therefore, $1/(x_1 - 1)^2 < 1/(x_2 - 1)^2$.
- Both branches of the graph approach the x-axis as they move further and further away from the line $x = 1$. (In other words, the x-axis is a horizontal asymptote.) That is because $1/(x - 1)^2$ becomes very small when x gets large in absolute value.

By using this information and plotting some points obtained by constructing a table of values, we can make a rough sketch of the graph. A more accurate one is shown in Figure 0.6.10.

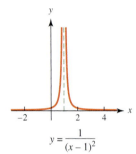

$$y = \frac{1}{(x - 1)^2}$$

Figure 0.6.10

In both of the previous examples the rational function was in lowest terms. If it is not—that is, if $p(x)$ and $q(x)$ have a common factor—there may not be a vertical asymptote, even though the denominator is zero at a certain point.

EXAMPLE 0.6.4

Let $f(x) = (x^2 + x - 2)/(x - 1)$. Determine the natural domain of this function and find its vertical asymptotes, if any.

SOLUTION   The natural domain is the set of all $x \neq 1$. We cannot use $x = 1$ as input because it would involve division by zero. At first glance, you might suspect that the line $x = 1$ is a vertical asymptote, but a closer examination shows that the quotient is not in lowest terms. In fact, the numerator can be factored as follows:

$$x^2 + x - 2 = (x + 2)(x - 1).$$

Therefore, if $x \neq 1$,

$$f(x) = \frac{(x + 2)(x - 1)}{x - 1} = x + 2.$$

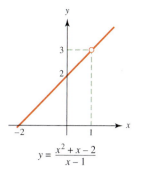

$$y = \frac{x^2 + x - 2}{x - 1}$$

Figure 0.6.11

Thus, the graph of $f(x)$ coincides with the graph of the linear equation $y = x + 2$ for all $x \neq 1$. The only difference between them is that the line includes the point $(1, 3)$, but the graph of $f(x)$ does not.

As a consequence, the graph of $f(x)$ neither rises nor falls unboundedly as x approaches 1, and there is no vertical asymptote. Instead, the graph approaches a height of 3, just as the linear graph does. Figure 0.6.11 shows the graph of $f(x)$ as a straight line of slope 1, with a "hole" at $(1, 3)$ to indicate that the point is not on the graph.

We will return to this example when we take up limits in Chapter 1.

Power functions

A power function has the form

$$y = Cx^m, \tag{30}$$

where C and m are fixed numbers. Notice that

- if m is a positive integer, it is a polynomial (with only one nonzero term), and
- if m is a negative integer, it is a rational function.

For instance, $3x^{-2}$ is the same as $3/x^2$.

Equation (30) makes sense even when m is not an integer, and in that case the function it defines is neither a polynomial nor a rational function.

Suppose m is a rational number, say, $m = p/q$, where p and q are positive integers. In that case we define

$$x^{p/q} = \sqrt[q]{x^p}. \tag{31}$$

For instance, $8^{2/3} = \sqrt[3]{8^2} = \sqrt[3]{64} = 4$. An equivalent definition is

$$x^{p/q} = \left(\sqrt[q]{x}\right)^p. \tag{32}$$

For instance, $8^{2/3} = (\sqrt[3]{8})^2 = 2^2 = 4$.

If m is negative, say, $m = -p/q$ for positive integers p and q, then we use the formula

$$x^{-p/q} = \frac{1}{x^{p/q}}.$$

For instance,

$$8^{-2/3} = \frac{1}{8^{2/3}} = \frac{1}{4}.$$

In general, there are two restrictions on the natural domain of a power function with exponent $m = p/q$, namely:

- x cannot be negative if q is even (because negative numbers do not have even roots), and

- x cannot be zero if m is negative (to avoid division by zero).

For instance,

- the domain of $x^{3/2}$ consists of all $x \geq 0$;
- the domain of $x^{-1/3}$ consists of all $x \neq 0$;
- the domain of $x^{-3/4}$ consists of all $x > 0$;
- the domain of $x^{1/3}$ consists of all the real numbers.

We close this section by describing a class of power functions that is important in economics.

The Cobb-Douglas production function

The quantity of goods produced in an economic system such as the U.S. economy depends on the amounts of capital and labor put into the system. Paul Douglas, who was a professor of economics and a U.S. senator from 1949 to 1966, studied the relation between these quantities. Working with Charles Cobb, a mathematician, he proposed the following formula, now called the **Cobb-Douglas production function**.

Let Q stand for the quantity (in dollars, say) of goods produced, and let K and L stand for the number of units of capital and labor, respectively, that are used in production. The Cobb-Douglas formula is an equation of the form

$$Q = AK^{\alpha}L^{1-\alpha}. \tag{33}$$

In this formula the symbol α (the Greek letter alpha) stands for a fixed number between 0 and 1, and the letter A denotes a positive constant. The exact values of these constants have to be determined by analyzing data and studying the particular system.

If we divide both sides of Eq. (33) by L, using the fact that

$$L = L^{\alpha}L^{1-\alpha},$$

we get

$$\frac{Q}{L} = \frac{AK^{\alpha}L^{1-\alpha}}{L^{\alpha}L^{1-\alpha}} = A\left(\frac{K}{L}\right)^{\alpha}.$$

Next, if we set $k = K/L$ and $q = Q/L$, this equation becomes

$$q = Ak^{\alpha}. \tag{34}$$

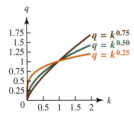

Figure 0.6.12

In this equation the independent variable is k, and it represents the amount of capital per unit of labor. The dependent variable is q, representing the amount of goods produced per unit of labor. The coefficient α is a constant between 0 and 1, and the letter A is a positive constant. As you can see, Eq. (34) defines q in terms of k as a power function. Figure 0.6.12 shows the graphs of $q = Ak^{\alpha}$ for $A = 1$ and $\alpha = 0.25, 0.50, 0.75$.

We will come back to the Cobb-Douglas production function in a later chapter when we study a mathematical model of economic growth.

Practice Exercises o.6

1. How does the graph of the polynomial function

$$f(x) = -7x^8 + 2x^5 + x^3 - 5x^2 + 9x + 2$$

behave as it moves further and further to the right or left?

2. Determine the natural domain of the function $f(x) = (x^2 - 7x + 12)/(5x^2 - 45)$ and find the vertical asymptotes, if any.

Exercises o.6

In Exercises 1–4, determine how the graph of the polynomial behaves as x moves further and further toward the right and left—specifically whether it increases without bound or decreases without bound.

1. $f(x) = x^5 - 3x^4 + 7x^3 + 6x^2 - x - 8$

2. $f(x) = -3x^4 + x^2 - 9$

3. $f(x) = -0.01x^3 + x^2 + 0.1x - 10$

4. $f(x) = 100x^6 - 82x^4 + 31x^2 - 144$

In Exercises 5–8, you are shown the graph of a polynomial of degree n with the leading coefficient a_n. Determine whether n is even or odd and whether a_n is positive or negative.

5.
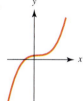

6.
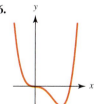

7.
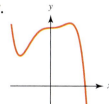

8.

In Exercises 9–20, find the vertical asymptotes, if any. If there is a vertical asymptote, determine how the graph behaves (i.e., climbs or falls) as it approaches the asymptote from either side by checking whether the function is positive or negative close to the asymptote on either side.

9. $f(x) = \dfrac{1}{(x+1)^2}$

10. $f(x) = \dfrac{1}{(x-1)^3}$

11. $f(x) = \dfrac{2x-1}{x+4}$

12. $f(x) = \dfrac{x^2-3}{3x^2+1}$

13. $f(x) = \dfrac{x}{x^2-4}$

14. $f(x) = \dfrac{x^4}{x^2-4}$

15. $f(x) = \dfrac{x^2-5x+6}{x-3}$

16. $f(x) = \dfrac{x^2-5x+6}{x+1}$

17. $f(x) = \dfrac{x-1}{x^2-1}$

18. $f(x) = \dfrac{x+1}{x^2+2x+1}$

19. $f(x) = \dfrac{x+1}{x^3-x^2-2x}$

20. $f(x) = \dfrac{x+4}{x^2-4x}$

In Exercises 21–30, compute the number without using a calculator.

21. 2^4

22. $(-4)^3$

23. $(0.3)^3$

24. $4^{1/2}$

25. 4^{-2}

26. $(0.008)^{1/3}$

27. $\left(\dfrac{1}{3}\right)^{-3}$

28. $\left(\dfrac{1}{27}\right)^{1/3}$

29. $8^{2/3}$

30. $\left(\dfrac{1}{8}\right)^{-1/3}$

In Exercises 31–36, evaluate $f(9)$ without using a calculator.

31. $f(x) = (x + 7)^{1/2}$ **32.** $f(x) = x^{1/2} + 3$

33. $f(x) = 4x^{1/2}$ **34.** $f(x) = (4x)^{-1/2}$

35. $f(x) = \left(\dfrac{1}{x}\right)^{3/2}$ **36.** $f(x) = \left(\dfrac{1}{x}\right)^{-3/2}$

In Exercises 37–42, find the natural domain of the given power function and determine where it is increasing, where it is decreasing, where it is positive, and where it is negative. Find the vertical asymptotes, if any, and how the graph approaches the asymptote. Finally, determine any symmetry of the graph. Use the information to sketch a graph of the function.

37. $f(x) = x^{1/3}$ **38.** $f(x) = x^{-1/3}$

39. $f(x) = x^{2/3}$ **40.** $f(x) = x^{-2/3}$

41. $f(x) = x^{3/2}$ **42.** $f(x) = x^{-3/2}$

43. When a bird hovers in the air by flapping its wings, it uses energy. The power it expends is a function of its length. If we let x denote the length of the bird and P the power expended, the formula defining P as a function of x can be derived from simple principles of geometry and physics. It has the form $P = Cx^{7/2}$, where C is a positive constant. (We say that P is **directly proportional** to $x^{7/2}$ and call C the **constant of proportionality**.)

Source: From *Mathematical Ideas in Biology*, J. Maynard Smith, Cambridge University Press, 1968-9, pp. 13–14.

If the bird's length doubles, by what factor does the power it expends increase?

44. If a certain type of car is going at 50 miles per hour, it requires 130 feet to stop when the brakes are applied. Assuming that the stopping distance is directly proportional to the square of the velocity, write a formula expressing the stopping distance as a function of the velocity.

45. According to Newton's law of gravitation, the force F that the Earth exerts on an object depends on the object's mass and its distance from the center of the Earth. It is given by the formula

$$F = \frac{Gm}{r^2}, \qquad (35)$$

where F is the gravitational force, m is the mass of the object, and r is its distance from the Earth's center. The letter G stands for a certain fixed number, called the **gravitational constant**, that has been determined by experiment. (If mass is measured in kilograms and distance in meters, then $G \approx 6.37 \times 10^{-11}$.) If we assume that m is fixed, formula (35) defines F as a function of r.

(a) What is the natural domain of this function? In applying it to gravitation, would you use the natural domain? If not, how would you restrict it?

(b) Where is the function increasing and where is decreasing? Does it have a vertical asymptote, and, if so, how does it approach it?

(c) If an object's distance from the center of the earth increases by 0.2%, by what factor does the gravitational force on it decrease?

Solutions to practice exercises 0.6

1. This polynomial function has the leading term $-7x^8$. Since we know that $-7x^8 + 2x^5 + x^3 - 5x^2 + 9x + 2 \approx -7x^8$ when x has a large absolute value, we can conclude that the graph of $f(x)$ behaves the same way as the graph of $y = -7x^8$, which falls lower and lower as it moves both to the right and left.

2. We obtain $5x^2 - 45 = 5(x - 3)(x + 3)$ by factoring the denominator. Thus, the natural domain of this function is the set of all numbers except ± 3. Factoring the numerator gives $x^2 - 7x + 12 = (x - 3)(x - 4)$, so that

$$f(x) = \frac{(x - 3)(x - 4)}{5(x - 3)(x + 3)} = \frac{x - 4}{5(x + 3)} \quad \text{for all } x \neq \pm 3.$$

Therefore, $f(x)$ has a vertical asymptote at $x = -3$.

0.7 Applying Technology

Computers and graphing calculators can be of great help in analyzing functions. In addition to sparing us from long, tedious calculations, they will solve equations, draw graphs, find their maximum and minimum points, and perform many other challenging tasks. They also enable us to experiment with formulas and with data tables that may give valuable insight into mathematical rules and procedures.

Three main devices available are

- graphing calculators, such as the TI-82, TI-83, TI-86, and TI-89,
- computer spreadsheet programs, such as Microsoft Excel, and
- mathematical software.

Each has its uses and limitations. Several mathematical software packages—two of the most widely used being Mathematica and Maple—are extremely powerful and will perform symbolic operations (such as factoring and simplifying) as well as numerical calculations. However, such packages may be expensive and not readily available outside an academic environment. In addition, they require you to be at a computer that can access the package.

A spreadsheet program also requires you to be at a computer, but this type of software is very widely used and easily available. Spreadsheets are particularly useful in analyzing, manipulating, and displaying large sets of data.

Graphing calculators are the cheapest and most portable of these three types of devices, and they can perform a remarkable range of applications, including those mentioned in the opening paragraph. On the other hand, using them for sophisticated operations may prove cumbersome and inefficient, and the results are generally displayed in a less clear and pleasing way than on a desktop or laptop computer.

Warning: Whatever electronic device is used must be used with care and thought or the results may sometimes be misleading. The ability to operate a calculator or run a computer program will not compensate for a lack of understanding of basic concepts.

Graphing with a calculator

To graph a function given by a formula, we enter the formula, choose certain options, and press a key to display the graph. Exactly how those steps are done varies according to the calculator model. For instance, in several of the most commonly used calculators, including the TI-82, TI-83, and TI-83 Plus, the function editor is reached through the $\boxed{\text{Y=}}$ key.

Figure 0.7.1 shows a typical calculator screen (in this case, a TI-83 Plus), in which several functions have been entered. The highlighted equals sign indicates a function that has been selected to be graphed or evaluated. We assume that you are familiar with some of the standard symbols such as ^ for exponentiation and * for multiplication (which can often be omitted). As usual, care must be taken in using parentheses. For instance, $(1 - x)^3$ (which is the third function shown) is not the same $1 - x^3$, and $1/(x^2 + 1)$ (the fourth function shown) is different from $1/x^2 + 1$.

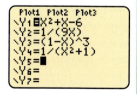

Figure 0.7.1

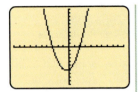

Figure 0.7.2

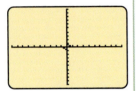

Figure 0.7.3

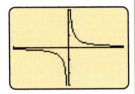

Figure 0.7.4

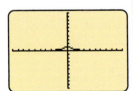

Figure 0.7.5

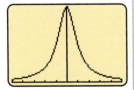

Figure 0.7.6

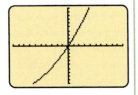

Figure 0.7.7

The most important option to be chosen is called the **viewing window**, which consists of a rectangle in the x, y-plane formed by the product of the intervals $a \leq x \leq b$ and $c \leq y \leq d$, which we will denote by $[a, b] \times [c, d]$. It designates the portion of the plane in which the graph will be drawn. The default window in most calculators is preset at $[-10, 10] \times [-10, 10]$, which works fairly well for the graph of $y = x^2 + x - 6$ (see Figure 0.7.2), but very badly for $y = 1/(9x)$ (see Figure 0.7.3). In that case, as you can see, it is almost impossible to detect the graph in the window. One reason is that the function gets relatively close to zero as x becomes large. At $x = 5$, for instance, $y = \frac{1}{45}$, which is small enough (on a scale of -10 to 10) to cause the graph to almost disappear into the x-axis. A remedy is to shrink the viewing window in both directions, so that the values of y do not get too small (in absolute value) relative to the size of the window. Figure 0.7.4 shows the graph of $y = 1/(9x)$ in the viewing window $[-1, 1] \times [-3, 3]$. Notice that the graph consists of two branches, one in the right half of the plane, the other in the left. The right-hand branch approaches the x-axis as it moves to the right.

At present you may have little recourse other than trial and error to find an appropriate viewing window, but even at this stage some knowledge of the domain and range can help you.

EXAMPLE 0.7.1

Use a calculator to draw the graph of $y = 1/(1 + x^2)$. First use the default settings and then try to determine a better viewing window.

SOLUTION   The graph obtained with the default settings is shown in Figure 0.7.5. As you can see, it is not very revealing. To find a better window, we observe two things. First,

$$0 < \frac{1}{1 + x^2} \leq 1$$

(recall Example 0.2.3), which suggests that we shrink the window in the y direction, say, by setting $0 \leq y \leq 1$. Second, the function gets small as x goes to the right or left, which suggests we might obtain a better graph by shrinking the window in the x direction as well. Figure 0.7.6 shows the graph in the viewing window $[-5, 5] \times [0, 1]$.

Of course, there are no clear "right" and "wrong" answers concerning the choice of viewing window, but some advance planning will help you get a clearer picture. Calculus will furnish you with a number of important tools for doing that.

We have now seen several examples in which the default settings are too large, but there are also cases in which they are too small to give an accurate picture of the graph. Figure 0.7.7 shows the graph of $y = 0.1x^2 + 2.2x$ in the preset viewing window. The picture suggests that the function is increasing, but that is simply because the graph does not extend over a large enough domain. Figure 0.7.8 displays it in a $[-30, 20] \times [-20, 30]$ viewing window, and we can see that the curve has a minimum point where it changes from decreasing to increasing. One of the important uses of calculus, which you will learn in Chapter 4, is in finding minimum and maximum points.

In addition to graphing functions, we can use the calculator to evaluate them, as shown in Figure 0.7.9, in which the function $f(x) = x^3 - 4x^2 + 4$ has been graphed (in

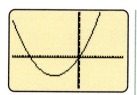

Figure 0.7.8

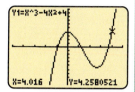

Figure 0.7.9

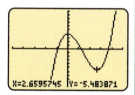

Figure 0.7.10

Figure 0.7.11

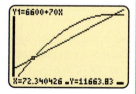

Figure 0.7.12

Figure 0.7.13

a $[-5, 5] \times [-10, 10]$ viewing window) and evaluated at $x = 4.016$. Notice the cursor marking the graph at that point.

We can also trace the curve, with the cursor's coordinates displayed as it moves. In Figure 0.7.10, for instance, it has been moved as close as possible to the point where the graph changes from decreasing to increasing. The bottom of the sreeen shows the coordinates $(2.6595745, -5.483871)$. In fact, this is only an approximation to the actual point where the graph changes. The exact coordinates, which can be found using calculus (see Chapter 4), are $(\frac{8}{3}, -\frac{148}{27}) \approx (2.\overline{6}, -5.\overline{481})$.

We can also graph several functions simultaneously. In Figure 0.7.11 we see the functions

$$C(x) = 6{,}600 + 70x \quad \text{and} \quad R(x) = -\frac{x^2}{4} + 180x,$$

both highlighted on the function editor screen of a TI-83 Plus. The graphs, created in a viewing window of $[0, 400] \times [0, 35{,}000]$, are shown in Figure 0.7.12. In this instance we have also used the TRACE function (found on all standard graphing calculators) to display a cursor that can be moved along the graph. In Figure 0.7.12 it has been placed as close as possible to the first break-even point, and its coordinates are shown at the bottom of the screen. They are only approximations to the actual coordinates of the break-even point and not as accurate as those we found in Example 0.5.5.

Tables of values: calculators and spreadsheets

All standard graphing calculators can create a table of values for a given function by means of a TABLE key or menu. The exact procedure varies according to the calculator model. To create a table of values for the cost and revenue functions defined by the function editor shown in Figure 0.7.11, we first set an initial value (in this case 0) and stepsize (in this case 0.1). The settings (using a TI-83 Plus) are displayed in Figure 0.7.13. The resulting table—as much of it as can be fit in the window—is shown in Figure 0.7.14. The other values can be obtained by scrolling down.

Spreadsheet programs can also be used to create tables of values and graphs. The spreadsheet pictured in Figure 0.7.15 contains both a graph and a table of values giving the height of a baseball thrown upward from a height of 100 feet with an initial velocity of 40 feet per second.

The times are listed in column A, from 0 to 4 in increments of 0.1. The heights are given in column B. They were obtained by substituting the corresponding times into the formula for the height of the ball:

$$h = -16t^2 + 40t + 100 \tag{36}$$

(see Example 0.5.6). Observe, for instance, that cell B4 is highlighted. The formula pertaining to it is shown at the top of the spreadsheet. It is preceded by an equals ($=$) sign, which is used in Excel to indicate that a function is to be applied. The information following "$=$" instructs the program to apply formula (36) to the entry in cell A4. If we had highlighted a different cell in column B, we would have obtained the same formula on top, with A4 replaced by the appropriate cell from column A.

It is not necessary to enter all the data in column A. After entering 0 in cell A3, we enter "$=A3+0.1$" in A4 and then use a command to repeat the operation in each successive cell. The result is that the entry in each cell is 0.1 more than the previous one.

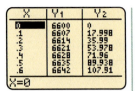

Figure 0.7.14

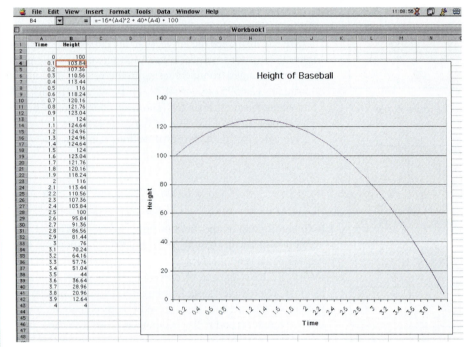

Figure 0.7.15

Similarly, it is not necessary to enter the formula in each cell of column B. Once it is entered in cell B3, we use a command to repeat it in each successive cell with an appropriate change of variable (from A3 to A4 to A5, and so forth).

To create an Excel chart from a table of data, we select **Chart** from the **Insert** menu (which you can see listed in the heading of Figure 0.7.15). A dialogue box then appears on the screen (as shown in Figure 0.7.16), from which you can choose the type of chart (the one in Figure 0.7.15 being a line chart) and decide on a number of display options, including the scale to display along the horizontal axis.

Linear regression with a spreadsheet and calculator

Both spreadsheet programs and graphing calculators are useful tools in studying regression (introduced in Section 0.4). Both devices will draw scatter diagrams, compute the coefficients of the least-squares line, and graph the line on the same set of axes as the data points.

Figure 0.7.16 shows an Excel spreadsheet containing data gathered by the World Health Organization.[2] Column B lists the per capita gross domestic product (the amount of goods and services produced per person) for 17 Latin American countries. Column C shows the life expectancy for an average male in each of those countries. To study the relation, if one exists, between these data sets, we want to plot a scatter diagram—a point for each GDP/life expectancy pair. Selecting **Chart** from the **Insert** menu at the top of the spreadsheet brings up the dialogue box labeled "Chart Wizard." As you can see, there are many types of charts to choose from, and we have selected a scatter diagram.

[2]Taken from the World Health Organization Web site http://www-nt.who.int.

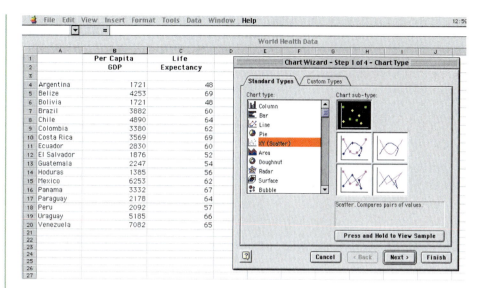

Figure 0.7.16

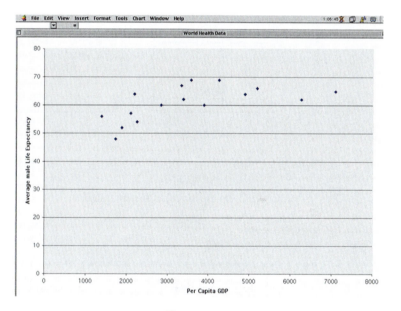

Figure 0.7.17

After deciding on a number of options (such as how to label the axes), we obtain the scatter diagram shown in Figure 0.7.17. By selecting **Add Trendline** from the **Chart** menu, we can display the least-squares line and its equation, $y = 0.0026x + 51.399$ as shown in Figure 0.7.18. (The screen appearance may vary slightly, depending on the computer, but the essentials are the same.)

Graphing calculators can also be used to compute the slope and y-intercept of the regression line and plot both the line and scatter diagram. Figure 0.7.19 shows the same data table as Figure 0.7.16, this time on a TI-83 Plus screen.

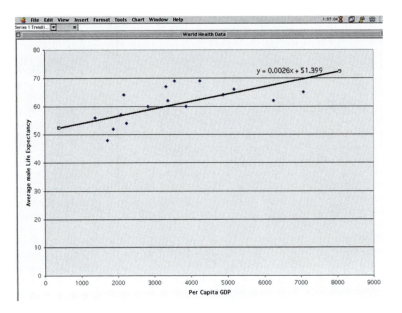

Figure 0.7.18

Figure 0.7.19   **Figure 0.7.20**   **Figure 0.7.21**

The equation of the regression line is shown in Figure 0.7.20, and the line and scatter diagram are displayed in Figure 0.7.21. The exact method for obtaining them varies according to the model of the calculator.

Exercises 0.7

Exercises 1–3 refer to the calculator screen shown in Figure 0.7.22.

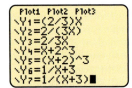

Figure 0.7.22

1. What value will be returned for each of the following?
   (a) $Y_1(3)$ (b) $Y_2(3)$ (c) $Y_3(3)$

2. What value will be returned for each of the following?
   (a) $Y_4(1)$ (b) $Y_5(1)$

3. What value will be returned for each of the following?
   (a) $Y_6(2)$ (b) $Y_7(2)$

Exercises 4–5 refer to the calculator screen shown in Figure 0.7.23.

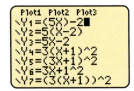

Figure 0.7.23

4. What value will be returned for each of the following?
   (a) $Y_1(-1)$ (b) $Y_2(-1)$ (c) $Y_3(-1)$

5. What value will be returned for each of the following?
(a) $Y_4(1)$ (b) $Y_5(1)$ (c) $Y_6(1)$ (d) $Y_7(1)$

In Exercises 6–10, graph the given function with a graphing calculator, using the calculator's default settings for the viewing window.

6. $f(x) = 1 - 2x + x^2$

7. $f(x) = x^3 + x^2 + x - 14$

8. $f(x) = \sqrt{x^2 + x + 1}$

9. $f(x) = \dfrac{8}{\sqrt{x^2 + 1}}$

10. $f(x) = x + \dfrac{4}{x}$

In Exercises 11–14, graph the given functions in two ways with a graphing calculator: first, using the default window and, second, using the window settings given in the exercise. Which method gives a more informative picture of the graph in each case?

11. $f(x) = 0.1x^2 + 2.4x - 10$; $[-40, 40] \times [-30, 30]$

12. $f(x) = \dfrac{2x}{x^2 + 4}$; $[-15, 15] \times [-1, 1]$

13. $f(x) = 0.003x^3 - x$; $[-30, 30] \times [-15, 15]$

14. $f(x) = 10x + \dfrac{10}{x}$; $[-5, 5] \times [-75, 75]$

In Exercises 15–17, graph the given functions using the default window. Then, try to set a better viewing window by considering the domain and range of the function, finding its intercepts (if possible), and evaluating it at a few points to get an idea how it behaves. Experiment with a number of choices until you obtain a graph that is similar to the one shown in Figure 0.7.24.

15. $f(x) = x^2 - 12x$ **16.** $f(x) = \dfrac{1}{x^2 + 5}$

17. $f(x) = \sqrt{x - 10}$

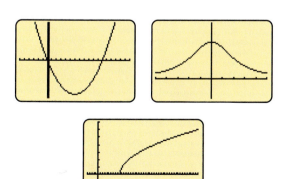

Figure 0.7.24

In Exercises 18–19, use a calculator to graph the supply and demand functions in the same window, using the specified viewing window. Then use the TRACE cursor to find the approximate coordinates of the equilibrium point.

18. $D(q) = 0.2(q - 30)^2$, $S(q) = 1.5q + 12$; viewing window $[0, 30] \times [0, 100]$

19. $D(q) = 0.05(q - 30)^2$, $S(q) = 0.2q^2 + 10$; viewing window $[0, 20] \times [0, 50]$

20. Use a calculator to graph the following on a single screen, using a viewing window of $[-1, 1] \times [0, 1]$. What conclusions can you draw about the functions?
(a) $y = x^{1/2}$ (b) $y = x^{1/3}$ (c) $y = x^{1/4}$ (d) $x^{1/5}$
(*Note:* You can use the TRACE cursor to identify each graph by changing it from graph to graph and reading the corresponding equation in the window.)

21. (Continuing the previous exercise) How would you expect the picture to change if you reset the viewing window to $[-1, 1] \times [-1, 1]$? Do it and see if you were right.

Every standard graphing calculator has an equation solver. Use it to solve the following equations (or, more accurately, to find approximate solutions). Try to find as many distinct solutions as specified. Consult your calculator manual for details.

22. $x^2 + x - 7 = 0$ (2 solutions)

23. $x^3 + x^2 + 3x - 1 = 0$ (1 solution)

24. $x^3 - 5x + 1 = 0$ (3 solutions)

25. $x - \sqrt{x} = 1$ (1 solution)

In Exercises 26–28, try to guess how many roots the function has by graphing it in an appropriate viewing window. Use the TRACE cursor to get a rough approximation to each root and then use the equation solver to obtain a better approximation.

26. $f(x) = x^3 + 7x + 2$ **27.** $f(x) = x^3 - 7x + 2$

28. $f(x) = x^4 + x^3 - 5x^2 + 3x - 24$

In Exercises 29–32, use a graphing calculator to construct a table of values for the given function with the initial value and step size as specified.

29. $f(x) = \sqrt{x}$, start at $x = 0$, step size = 0.2

30. $f(x) = -16x^2 - 40x + 100$, start at $x = 0$, step size = 0.05

31. $f(x) = 1/x$, start at $x = 1$, step size = -0.1

32. $f(x) = (1 + x^2)^{-1}$, start at -2, step size = 0.25

In Exercises 33–36, use an Excel spreadsheet to generate a table of values for the given function with the initial value, final value, and step size as specified. Then make a line graph.

33. $f(x) = \frac{9}{5}x + 32$, start at $x = -40$, end at $x = 40$, step size $= 4$

34. $f(x) = -16x^2 + 100$, start at $x = 0$, end at $x = 2.5$, step size $= 0.25$

35. $f(x) = \sqrt[3]{x}$, start at $x = -8$, end at $x = 8$, step size $= 1$

36. $f(x) = x^3 - x$, start at $x = -2$, end at $x = 2$, step size $= 0.2$

The Excel spreadsheet shown in Figure 0.7.25 lists economic indicators for a 12-state midwest region for the first quarter of each year from 1991 to 2001 (taken from the U.S. Bureau of Labor Statistics Web page, http://stats.bls.gov). Column A contains the Consumer Price Index (CPI) and column B a Wage and Salary Index. Enter the data into an Excel spreadsheet and use it for Exercises 37–39.

37. Make a line chart of the CPI, with the year as the independent variable.

38. Make a line chart of the Wage and Salary Index, with the year as the independent variable.

39. Make a scatter diagram of the CPI/Wage and Salary pairs. Then find the equation of the regression line and graph it on the same chart.

A	B	C	D
Year	1st Quarter CPI	1st Quarter Wages & Salaries	
1991	130.9	106.9	
1992	134.4	110.7	
1993	138.6	113.5	
1994	142.1	117.5	
1995	146.7	120.9	
1996	150.9	125.1	
1997	155.8	129.0	
1998	158.0	134.7	
1999	160.6	138.9	
2000	166.0	145.3	
2001	171.9	150.9	

Figure 0.7.25

40. The Excel spreadsheet shown in Figure 0.7.26 relates public sector spending on health care in 20 western hemisphere countries to infant mortality rates (taken from the World Health Organization Web site, http://www.nt.who.int). The expenditure is given as a percentage of the gross domestic product (GDP). Use an Excel

A	B	C	D
	Public Sector Expenditue on Health Care (Percentage of GDP)	Infant Mortality per 1000	
Argentina	4.3	22	
Belize	6	29	
Bolivia	4.1	66	
Brazil	1.8	42	
Canada	6.6	6	
Chile	2.5	13	
Colombia	2.9	30	
Costa Rica	6.3	12	
Ecuador	2	46	
El Salvador	2.4	32	
Guatemala	1.7	46	
Haiti	1.3	68	
Hoduras	2.8	35	
Mexico	2.4	31	
Nicaragua	5.3	43	
Panama	4.7	21	
Paraguay	1.8	39	
Peru	2.2	45	
Uraguay	7	18	
Venezuela	3	21	

Figure 0.7.26

spreadsheet or graphing calculator to plot a scatter diagram of the data pairs, find the equation of the regression line, and plot it in the same diagram.

41. Figure 0.7.27 shows an Excel spreadsheet containing CPI data for the United States during 12 quarters covering the years 1988 through 2000 (taken from the U.S. Health Care Financing Administration Web site, http://www.hcfa.gov). Column A shows the CPI for all items other than medical care, and column B shows the CPI for medical care. Use an Excel spreadsheet or a graphing calculator to plot a scatter diagram of the data pairs, find the equation of the regression line, and graph it in the same diagram.

A	B	C
Quarter	CPI less Med care	CPI Med care
1998, Q1	157.5	239.1
1998, Q2	158.4	241.4
1998, Q3	159	243.4
1998, Q4	159.5	244.7
1999, Q1	160.1	247.5
1999, Q2	161.6	249.6
1999, Q3	162.6	251.8
1999, Q4	163.6	253.4
2000, Q1	165.2	256.9
2000, Q2	166.9	259.6
2000, Q3	168.2	262.4
2000, Q4	169.1	264.2

Figure 0.7.27

42. Figure 0.7.28 shows (hypothetical) data for 15 employees of a company, relating their scores on a preliminary screening exam to a performance rating given after 6 months on the job. Use a graphing calculator or Excel spreadsheet to plot a scatter diagram of the data pairs, find the equation of the regression line, and graph it on the same diagram.

Employee	Screening Exam	Performance Rating
1	72	80
2	64	72
3	81	78
4	86	92
5	72	71
6	48	55
7	63	68
8	90	82
9	91	95
10	76	71
11	53	58
12	97	92
13	65	73
14	88	85
15	37	62

Figure 0.7.28

Chapter 0 Summary

- A **function** assigns to each input in its **domain** a unique output in its **range**. If both input and output increase together then the function is **increasing**; if the output decreases when the input increases then it is **decreasing**. The graph of a function $f(x)$ is **symmetric about the y-axis** if $f(-x) = f(x)$ (**even function**), and is **symmetric about the origin** if $f(-x) = -f(x)$ (**odd function**).

- A **linear function** is of the form $f(x) = mx + b$. Its graph is a line with **slope** $m = \Delta y/\Delta x = $ rise/run. The slope is also the **rate** at which the dependent variable changes with respect to the independent variable.

- A **quadratic function** is of the form $f(x) = ax^2 + bx + c$, with $a \neq 0$. Its graph is a parabola with vertex at $(-b/2a, (4ac - b^2)/4a)$. If $a > 0$ it opens upward and the quadratic has **minimum** value at $-b/(2a)$. If $a < 0$ it opens downward and the quadratic has **maximum** value at $-b/(2a)$.

- A function of the form $f(x) = a_n x^n + \cdots + a_1 x + a_0$ with $a_n \neq 0$ is called a **polynomial function of degree** n. For $|x|$ large it behaves like its top degree term $a_n x^n$.

- A **rational function** is of the form $r(x) = f(x)/g(x)$, where both $f(x)$ and $g(x)$ are polynomials. When written in lowest terms it has **vertical asymptotes** at the zeros of the denominator.

- A **power function** is of the form $f(x) = Cx^m$, with $C \neq 0$. A class of examples that are important in economics is that of the **Cobb-Douglas** production functions, for which $C > 0$ and $0 < m < 1$.

Chapter **0** **Review Questions**

• State a criterion for a curve to be the graph of a function. Give an example of a curve that is not the graph of a function.

• Explain why the demand curve is decreasing and the supply is increasing.

• What is the horizontal line test? Give an example of a curve that fails it.

• Explain the meaning of two functions $f(x)$ and $g(x)$ being inverses of one another. How are their graphs related?

• Given a set of n points in the plane, what criterion do we use for determining the line that fits them best?

• What form do we want to put a quadratic in by completing the square, and what information can we obtain from it?

• How does the graph of a rational function behave as it approaches a vertical asymptote? Give examples of all the possibilities.

Chapter **0** **Review Exercises**

1. Let $f(x) = x^2$. Evaluate $f(-5)$, $f(-\sqrt{2})$, $f(-1)$, $f(-3/5)$, $f(0)$, $f(3\sqrt{5})$, $f(t+3) - f(t-3)$, and $[f(x+h) - f(x)]/h$.

2. Let $f(t) = t^3$. Evaluate $f(-2)$, $f(-5^{1/3})$, $f(-1)$, $f(0)$, $f(2/3)$, $f(1)$, $f(2)$, $f(t+1) - f(t-1) - 2f(t)$, and $[f(t+h) - f(t)]/h$.

3. Let $f(x) = 1/(x-4)$. Describe the natural domain of this function and evaluate $f(-6)$, $f(-1)$, $f(0)$, $f(2)$, $f(7/2)$, and $[f(5+h) - f(5)]/h$.

4. Let $f(t) = 3t^2 - (1/\sqrt{25 - t^2})$. Describe the natural domain of this function and evaluate $f(-4)$, $f(-3)$, $f(-\sqrt{21})$, $f(0)$, $f(3)$, $f(4)$, and $f(\sqrt{21})$.

5. Let f be the function whose graph is given in Figure 0.R.1.

(a) Estimate $f(-\frac{1}{2})$, $f(0)$, $f(1)$, and $f(\frac{7}{4})$.
(b) Find all x with $f(x) = \frac{7}{4}$.
(c) Estimate the x-intercept of f.
(d) What is the domain of f?
(e) Find the intervals where f is increasing.
(f) Does f satisfy the horizontal test?
(g) Is f one-to-one in the interval $[\frac{1}{2}, \frac{3}{2}]$?
(h) Does f have an inverse in the interval $[-\frac{1}{2}, \frac{1}{2}]$?

6. Let f be the function whose graph is given in Figure 0.R.1.
(a) Estimate $f(-\frac{1}{4})$, $f(\frac{1}{4})$, $f(\frac{1}{2})$, $f(\frac{5}{4})$, and $f(\frac{3}{2})$.
(b) Find all x such that $f(x) = \frac{5}{4}$.
(c) Estimate the y-intercept of f.
(d) What is the range of f?
(e) Find the intervals where f is decreasing.
(f) Is f one-to-one in the interval $[-\frac{1}{2}, 1]$?
(g) Does f satisfy the horizontal test in the interval $[0, 2]$?
(h) Does f have an inverse in the interval $[\frac{1}{2}, \frac{3}{2}]$?

In Exercises 7–8, we refer to the function g whose graph is given in Figure 0.R.2.

7. Draw the graphs of the following functions.
(a) $y = g(x - 1)$ (b) $y = g(x + \frac{1}{2})$ (c) $y = g(x) - 1$
(d) $y = g(x) + \frac{1}{2}$ (e) $y = 2g(x)$ (f) $y = 2g(x - 1) - 1$

8. Draw the graphs of the following functions:
(a) $y = g(x + 1)$ (b) $y = g(x - \frac{3}{2})$ (c) $y = g(x) + 1$
(d) $y = g(x) - \frac{3}{2}$ (e) $y = -g(x)$ (f) $y = 2 - g(x - 1)$

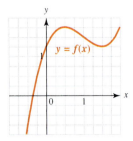

Figure 0.R.1

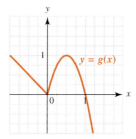

Figure o.R.2

Complete the squares for each of the following quadratic functions. Then find their maximum or minimum values, their x- and y-intercepts, and sketch their graphs by hand.

22. $f(x) = -2x^2 + 12x + 13$

23. $f(x) = -3x^2 + 10x - 8$

Sketch by hand the graphs of the following rational functions. Find their vertical asymptotes, if they have any, and state if they are even, odd, or neither.

24. $f(x) = \dfrac{1}{x}$ **25.** $f(x) = \dfrac{-1}{1+x^2}$

26. $f(x) = \dfrac{x}{x^2 - 4}$

Sketch by hand the graphs of the following functions and describe their natural domains:

27. $f(x) = x^{1/3}$ **28.** $f(x) = (x-1)^{1/3} + 1$

29. $f(x) = (x+1)^{1/4}$

30. A ball is thrown upward from a height of 15 feet with an initial speed of 8 feet per second. The height of the ball is given by the formula $h = -16t^2 + 8t + 15$ where h is the height (in feet) and t is the time (in seconds).
   (a) When does the ball reach its highest point?
   (b) How high does it go?
   (c) When does it hit the ground?

31. A company that makes office chairs has fixed costs of $5,000 per month plus a cost of $30 per chair. The company plans to sell the chairs for $50 each. Suppose it produces q chairs a month and assume that it sells every chair it produces.
   (a) Find the monthly profit function.
   (b) Find the break-even point.

32. A small company manufacturing tennis rackets has fixed costs of $700 per week. For the first 50 rackets it makes, each one costs $32 to produce. Because of overtime wages, every racket over 50 the company makes costs $38. If x is the number of rackets it produces in a given week, find the cost function.

33. A private health club has determined that the number of members depends on the price of a membership, and they are related by an equation of the form $q = 3,000 - 20p$, where q is the number of members and p is the annual price of a membership. The club has fixed costs of $20,000 per year plus an average annual cost of $40 per member.
   (a) Write the club's revenue R as a function of the price p.
   (b) Write the club's profit P as a function of the price p.
   (c) What membership price should the club set to maximize its profit?

34. A company offers dinner cruises on the Chicago River. The company has found that the average number of passengers per night is 80 if the price is $45 per person. At a

9. Sketch by hand, on the same axes, the graphs of the following functions. Also, state if they are even, odd, or neither.
   (a) $f(x) = 1$ **(b)** $f(x) = x$ **(c)** $f(x) = x^2$
   (d) $f(x) = x^3$

10. Sketch by hand, on the same axes, the graphs of the following functions. Also, state if they are even, odd, or neither.
   (a) $f(x) = -1$ **(b)** $f(x) = -x$ **(c)** $f(x) = -x^2$
   (d) $f(x) = -x^3$

11. Find the natural domain of each of the following functions.
   (a) $f(x) = \dfrac{\sqrt{x}}{x-2}$ **(b)** $f(x) = \dfrac{(x-9)^{1/3}}{(x-1)(x-2)}$

12. Find the inverse of the function $f(x) = x^3 + 1$. Then graph $f(x)$ and its inverse on the same axes.

13. Let $f(x) = 1 - x^4$, $x \geq 0$. Find and graph the inverse, and determine its domain and range.

In Exercises 14–16, find the slope of each of the given lines:

14. Through $(1, -1)$ and $(3, 2)$

15. Given by the equation $2x + 3y = 5$

16. Through $(1, -2)$, parallel to the line with equation $x - 2y = 1$

In Exercises 17–19, find the equation of each of the following lines:

17. Through $(0, -1)$ and $(-2, 5)$

18. Through the point $(4, -1)$ and parallel to the line $2x + 5y = 6$

19. Through the point $(4, -1)$ and perpendicular to the line $2x + 5y = 6$.

Determine where each of the following quadratic functions is positive and where it is negative:

20. $f(x) = x^2 - 7x + 14$

21. $f(x) = -4x^2 + 8x + 1$

price of $30 per person, the average number of passengers per night is 125. Let p denote the price and q denote the demand (the average number of passengers).

(a) Assuming that the demand is a linear function of the price, write the demand as a function of the price.

(b) Write the nightly revenue R as a function of the price.

(c) What price should the company charge to maximize its revenue?

Chapter 0 Practice Exam

1. The function $f(x)$ is defined by the following table:

x	1	2	3	4	5	6	7
$f(x)$	1.2	2.3	6.4	11.2	9.6	8.3	7.9

If $g(x) = f(x-1) + 3$, compute the value $g(4)$.

2. Find the natural domain of the function
$$f(x) = \frac{\sqrt{3x - x^2}}{x}.$$

3. If $f(x) = \begin{cases} \sqrt{x} & \text{if } 0 \le x < 2, \\ 1 + x & \text{if } 2 \le x < \infty, \end{cases}$ find the value of the product $f(1) \cdot f(2)$.

4. Choose the curves in Figure 0.E.1 that fit the following descriptions:
(i) *Not* the graph of a function
(ii) The graph of a one-to-one function
(iii) The graph of an odd function
(iv) The graph of an even function

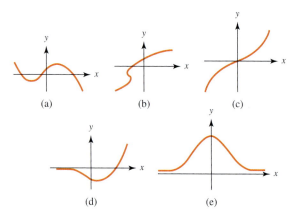

Figure 0.E.1

5. Let $f(x) = x^2 + 2x - 1$. Find the following and simplify them as much as possible.
(a) $f(-2)$ (b) $f(1-t)$ (c) $\dfrac{f(1+h) - f(1)}{h}$

6. Find the inverse of the function $f(x) = 2x + 1$. Then graph $f(x)$ and its inverse on the same axes.

7. Figure 0.E.2 shows the graph of a function. Draw the graph of the inverse function.

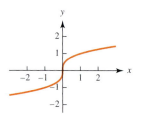

Figure 0.E.2

8. The graph of $y = 2 + 3x - x^3$ is shown in Figure 0.E.3. Use it to make a sketch of the graph of $y = 3 + 3(x + 1) - (x + 1)^3$.

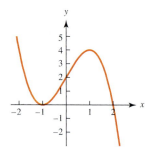

Figure 0.E.3

9. Find the slope-intercept form of the equation of the line passing through the points in the plane with coordinates $(1, -1)$ and $(-7, 3)$.

10. Sketch by hand the graphs of the following functions. Also, state if they are even, odd, or neither.
(a) $f(x) = 4 - x^2$ (b) $f(x) = 2/x$ (c) $f(x) = \sqrt{x - 1}$

11. A company making bird feeders has fixed costs of $12,000 per month. Each bird feeder costs $10 to make and sells for $40. How many does it have to make per month to break even?

12. Given the equation $y = -x^2 + 3x - 5$, complete the square of the quadratic. Then state whether the graph opens upward or downward, find the vertex and axis of symmetry, and sketch the graph.

13. A pizza restaurant sells its large pizza for $10, and at that price it is currently selling 40 of them a night. The restaurant is thinking of reducing the price, and it estimates they will gain eight sales for each dollar decrease in price.

(a) Find a formula for the number of pizzas sold q as a function of the price p.

(b) Find a formula for the revenue R as a function of the price p.

(c) What price should the restaurant charge to maximize revenue?

14. Suppose a ball is thrown upward from a height of 48 feet with an initial velocity of 32 feet per second.

(a) Write a formula giving the height (in feet) of the ball as a function of the time (in seconds) from the time it is thrown until it hits the ground.

(b) When will it hit the ground?

(c) How high does it go?

····· **Chapter o Projects** ·····································

1. Quadratic and polynomial regression In Section 0.4 we introduced the **regression** or **least squares** line, whose slope and y-intercept were given by Eqs. (17) and (18). It is the line that best fits a given set of data. Some data, however, are clearly nonlinear, and it is misleading to use the regression line based on those data to predict future results. For instance, the points in a scatter diagram, such as the one shown here, may fit a parabola better than a line: In that case, we try to find the quadratic function, given by $y = ax^2 + bx + c$, that best fits the data points. This process is called **quadratic regression**.

There are formulas for the coefficients, a, b, and c, analogous to formulas (17) and (18), but they are cumbersome. Graphing calculators, such as the TI-83 Plus, have built-in functions to calculate the quadratic regression coefficients.

(a) Use a graphing calculator to find the quadratic regression formula that gives the best fit for the following table of data for the **demand curve** of a certain item:

p	3	6	9	12	15	18	21	24	27	30
q	66	58	50	43	37	31	25	20	16	12

(b) Use a graphing calculator to find the quadratic regression formula that gives the best fit for the following table of data for the **supply curve** of the same item:

p	3	6	9	12	15	18	21	24	27	30
q	6	10	14	19	25	30	37	44	52	60

(c) Find the equilibrium price and demand.
Consult your calculator manual for specific instructions concerning quadratic regression.

(d) Some data, when plotted in a scatter diagram, seem better fitted to a cubic or higher-degree polynomial, and there are formulas and programs for **polynomial regression**. Use a graphing calculator to find the cubic equation, of the form $y = ax^3 + bx^2 + cx + d$, whose graph best fits the data in the following table:

x	1	1.5	2	2.5	3	3.5	4
y	1.2	2.2	7.1	13.8	25.2	41.6	58.3

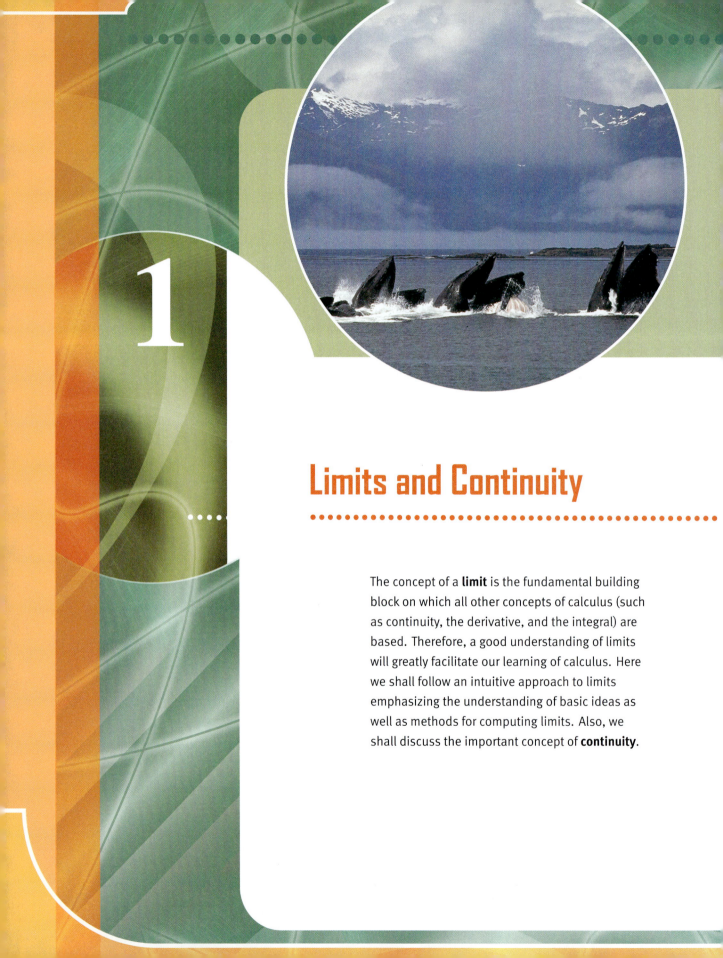

1

Limits and Continuity

The concept of a **limit** is the fundamental building block on which all other concepts of calculus (such as continuity, the derivative, and the integral) are based. Therefore, a good understanding of limits will greatly facilitate our learning of calculus. Here we shall follow an intuitive approach to limits emphasizing the understanding of basic ideas as well as methods for computing limits. Also, we shall discuss the important concept of **continuity**.

Let us begin by trying to sketch the graph of the rational function

$$f(x) = \frac{x^2 - x - 2}{x - 2}. \tag{1}$$

The domain consists of all $x \neq 2$. We cannot use $x = 2$ as input because that would involve dividing by zero. However, it is instructive to study the behavior of this function *near* the excluded point. To do that, we make a table of values, such as the following:

x	1.95	1.96	1.97	1.98	1.99	2	2.01	2.02	2.03	2.04	2.05
$f(x)$	2.95	2.96	2.97	2.98	2.99	?	3.01	3.02	3.03	3.04	3.05

Examining this table, we observe a certain pattern. First, $f(x)$ gets close to 3 as x gets close to 2. Using standard mathematical terminology, we say that $f(x)$ *approaches* 3 as *x approaches* 2. Moreover, the values of $f(x)$ approach 3 in a regular way. Each change of ± 0.01 in x results in a ± 0.01 change in $f(x)$, which suggests that $f(x)$ behaves like a linear function of slope 1. That is indeed the case, as we see by factoring the numerator

$$x^2 - x - 2 = (x - 2)(x + 1),$$

and canceling one of the factors against the denominator. Thus, for all $x \neq 2$, we have

$$f(x) = \frac{x^2 - x - 2}{x - 2} = \frac{(x - 2)(x + 1)}{x - 2} = x + 1.$$

The result is a linear function, whose domain includes $x = 2$ (which the original function did not). The graph of $y = x + 1$, shown in Figure 1.1.1, is a straight line passing through the point $(2, 3)$. The graph of $f(x)$ coincides with that line for all $x \neq 2$, which explains the regularity we observed in the table above.

To summarize: Even though $f(x)$ is not defined for $x = 2$, we know something about its behavior *near* $x = 2$, specifically that

- its values—by which we mean its output values—are close to 3 when x is close to 2, and
- we can make them as close as we want to 3 by taking x close enough to 2.

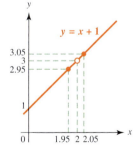

Figure 1.1.1

We symbolize this behavior by writing the following standard notation:

$$\lim_{x \to 2} \frac{x^2 - x - 2}{x - 2} = 3. \tag{2}$$

In words, we say *the limit of $\frac{x^2 - x - 2}{x - 2}$ as x approaches 2 equals 3.*

In general, we write

$$\lim_{x \to a} f(x) = L \tag{3}$$

to indicate that the values of the function $f(x)$ approach the number L as x approaches the number a, and we say that *the limit of $f(x)$ as x approaches a equals L.* In discussing the limit, we assume that $f(x)$ is defined for all x near a, but not necessarily for $x = a$, which is exactly the situation we just encountered in arriving at Eq. (2).

An exact definition of the limit of a function is given in the last section of this chapter. It is rather technical, however, and involves a higher degree of mathematical precision than we generally use in this book. The following informal definition will suffice for our purposes. As before, the symbol $\approx$ stands for "is approximately equal to."

Definition 1.1.1

Informal Definition of Limit

$$\lim_{x \to a} f(x) = L \text{ means that}$$

1. $f(x) \approx L$ for all x close (but not equal) to a, and
2. we can make the approximation as close as we want by taking x close enough to a.

As you can see, the definition avoids any discussion of what happens when $x = a$. We have already noted that the function may not even be defined for $x = a$. And even if it is, its value there, $f(a)$, may or may not equal the limit as $x \to a$. For emphasis, let's repeat that: *The limit $\lim_{x \to a} f(x)$ does not depend on the value the function takes when $x = a$ or even on whether the function is defined there.* On the other hand, we assume that the function is defined for all other x near a.

EXAMPLE 1.1.1

Determine

$$\lim_{x \to 0} \frac{(1 + x)^2 - 1}{x}$$

in two ways. First, construct a table of values for x near zero and make an educated guess at the limiting value. Second, simplify the quotient until it becomes easy to analyze.

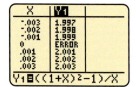

Figure 1.1.2

SOLUTION   In this case

$$f(x) = \frac{(1 + x)^2 - 1}{x}$$

and $a = 0$. This function is defined for all x around zero but not at zero. Figure 1.1.2 shows a table of values created on a calculator, using an initial value of -0.003 and step

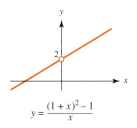

$$y = \frac{(1+x)^2 - 1}{x}$$

Figure 1.1.3

size 0.001. (Notice that ERROR corresponds to $x = 0$, in recognition that the function is undefined there.) The table suggests that $f(x) \approx 2$ when x is close to zero. To verify that, we expand the numerator of $f(x)$ to obtain

$$f(x) = \frac{(1+x)^2 - 1}{x} = \frac{1 + 2x + x^2 - 1}{x} = \frac{2x + x^2}{x} = 2 + x \quad \text{if } x \neq 0.$$

The graph of $f(x)$ is shown in Figure 1.1.3. It is identical to that of $y = 2 + x$ except at $x = 0$, where $f(x)$ is not defined. Since $2 + x \approx 2$ for all x sufficiently close to zero, we conclude that $\lim_{x \to 0} f(x) = 2$.

In the simplest cases, the function *is* defined at $x = a$ and $f(a)$ is equal to $\lim_{x \to a} f(x)$. That is, the limiting value as $x \to a$ is the same as the value when $x = a$. An example of that is the function $f(x) = x^2$. If $x = 3$, for instance, $f(3) = 9$, and

$$\lim_{x \to 3} x^2 = 9.$$

We will give a proof of this formula as part of a more precise discussion of limits in Section 1.4. However, the proof is rather technical, and for now we will simply accept the following statement as reasonable: *If a number is very close to 3, such as 2.9999 or 3.0001, its square is very close to 9.*

A similar conclusion holds if we replace 3 by *any* fixed number a, that is,

$$\lim_{x \to a} x^2 = a^2.$$

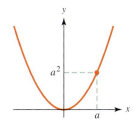

Figure 1.1.4

(See Figure 1.1.4). What is more, we can extend this formula to other power functions. For instance, if n is any fixed positive integer and a is any fixed number, we can conclude that

$$\lim_{x \to a} x^n = a^n. \tag{4}$$

By way of example, the following table of values of x^3 near $x = 2$ indicates that x^3 approaches 8 as x approaches 2:

x	1.996	1.997	1.998	1.999	2	2.001	2.002	2.003	2.004
$f(x)$	7.952	7.964	7.976	7.988	?	8.012	8.024	8.036	8.048

Since polynomials are simply sums of power functions multiplied by numerical coefficients, we can derive similar limit equations for polynomials and rational functions by combining formula (4) with the following theorem, which tells us how to combine the limit operation with sums, products, and quotients.

Theorem 1.1.1

Suppose $\lim_{x \to a} f(x) = L$ and $\lim_{x \to a} g(x) = M$.

1. For any constant c, $\lim_{x \to a} cf(x) = c \lim_{x \to a} f(x) = cL$.

2. $\lim_{x \to a} [f(x) + g(x)] = \lim_{x \to a} f(x) + \lim_{x \to a} g(x) = L + M$.

3. $\lim_{x \to a} f(x) \cdot g(x) = \left(\lim_{x \to a} f(x) \right) \left(\lim_{x \to a} g(x) \right) = L \cdot M$.

4. If $M \neq 0$, $\lim_{x \to a} \dfrac{f(x)}{g(x)} = \dfrac{\lim_{x \to a} f(x)}{\lim_{x \to a} g(x)} = \dfrac{L}{M}$.

5. $\lim_{x \to a} [f(x)]^{p/q} = \left[\lim_{x \to a} f(x) \right]^{p/q} = L^{p/q}$.

We can also apply these rules to sums and products of three or more functions by "peeling off" one term at a time, as shown in the next example.

EXAMPLE 1.1.2

Find each of the following limits:

$$\text{(i)} \ \lim_{x \to 2} 3x^2, \qquad \text{(ii)} \ \lim_{x \to 2} (3x^2 - x^3),$$
$$\text{(iii)} \ \lim_{x \to 2} \frac{x}{3x^2 - x^3}, \qquad \text{(iv)} \ \lim_{x \to 2} (x^4 + 2x^3 - 5x^2).$$

SOLUTION   (i) By formula (4), $\lim_{x \to 2} x^2 = 4$. Combining that with the first rule of Theorem 1.1.1 gives

$$\lim_{x \to 2} 3x^2 = 3 \lim_{x \to 2} x^2 = 3 \cdot 4 = 12.$$

(ii) Since $\lim_{x \to 2} 3x^2 = 12$ and $\lim_{x \to 2} x^3 = 8$, applying Theorem 1.1.1 yields

$$\lim_{x \to 2} (3x^2 - x^3) = \lim_{x \to 2} 3x^2 - \lim_{x \to 2} x^3 = 12 - 8 = 4.$$

(iii) Combining this last result with rule 4 of Theorem 1.1.1, we get

$$\lim_{x \to 2} \frac{x}{3x^2 - x^3} = \frac{\lim_{x \to 2} x}{\lim_{x \to 2} (3x^2 - x^3)} = \frac{2}{4} = \frac{1}{2}.$$

(iv) In this case, repeated application of Theorem 1.1.1 results in

$$\lim_{x \to 2} (x^4 + 2x^3 - 5x^2) = \lim_{x \to 2} (x^4 + 2x^3) + \lim_{x \to 2} (-5x^2)$$
$$= \lim_{x \to 2} x^4 + 2 \lim_{x \to 2} x^3 - 5 \lim_{x \to 2} x^2 = 16 + 2 \cdot 8 - 5 \cdot 4 = 12.$$

By repeatedly combining Theorem 1.1.1 with formula (4), we can find the limit of a rational function at any point where the denominator is not zero.

Theorem 1.1.2

If $f(x)$ and $g(x)$ are polynomials with $g(a) \neq 0$,

$$\lim_{x \to a} \frac{f(x)}{g(x)} = \frac{f(a)}{g(a)}.$$

Since a polynomial is a rational function whose denominator is the constant 1, the formula applies to polynomials.

EXAMPLE 1.1.3

Let $f(x) = x^3 - 4x^2 + 2x + 5$. Find $\lim_{x \to 3} f(x)$ and verify that it equals $f(3)$.

SOLUTION   First, we compute $f(3) = 3^3 - 4 \cdot 3^2 + 2 \cdot 3 + 5 = 2$. Next, combining formula (4) and Theorem 1.1.1, we get

$$\lim_{x \to 3}(x^3 - 4x^2 + 2x + 5) = \lim_{x \to 3} x^3 + \lim_{x \to 3}(-4x^2) + \lim_{x \to 3}(2x) + \lim_{x \to 3} 5$$
$$= \lim_{x \to 3} x^3 - 4 \lim_{x \to 3} x^2 + 2 \lim_{x \to 3} x + 5$$
$$= 3^3 - 4 \cdot 3^2 + 2 \cdot 3 + 5 = 2.$$

On the other hand, if $g(a) = 0$, the rational function $f(x)/g(x)$ may or may not have a limit as $x \to a$.

EXAMPLE 1.1.4

Suppose $f(x) = \dfrac{x^2 - 5x + 6}{x^2 - 4}$. Find

(i) $\lim_{x \to 1} f(x)$,   (ii) $\lim_{x \to 2} f(x)$,   (iii) $\lim_{x \to -2} f(x)$.

SOLUTION   (i) The domain of $f(x)$ consists of all $x \neq \pm 2$. In particular, the denominator is not zero at $x = 1$, and

$$\lim_{x \to 1}(x^2 - 5x + 6) = 2 \quad \text{and} \quad \lim_{x \to 1}(x^2 - 4) = -3.$$

Applying Theorem 1.1.1, we get

$$\lim_{x \to 1} f(x) = -\frac{2}{3}.$$

(ii) Because $\lim_{x \to 2}(x^2 - 4) = 0$, we cannot apply Theorem 1.1.1. However, by factoring both the numerator and denominator, we get

$$\frac{x^2 - 5x + 6}{x^2 - 4} = \frac{(x-3)(x-2)}{(x+2)(x-2)}. \tag{5}$$

For any x in the domain of the function f, we can cancel the factor $(x-2)$ from the numerator and denominator and get

$$\frac{x^2 - 5x + 6}{x^2 - 4} = \frac{x-3}{x+2} \quad \text{for all } x \neq 2. \tag{6}$$

But that means that both sides of this equation have the same output values for all x close to (but not equal to) 2, and we can conclude that

$$\lim_{x \to 2} f(x) = \lim_{x \to 2} \frac{x-3}{x+2} = -\frac{1}{4}.$$

(iii) Once again, the denominator approaches zero—that is,

$$\lim_{x \to -2}(x^2 - 4) = 0.$$

The factorization given in (5) does not help this time because the troublesome factor in the denominator is $(x+2)$, and we cannot get rid of it by canceling. As $x \to -2$, the denominator approaches zero but the numerator does not. In fact,

$$\lim_{x \to -2}(x^2 - 5x + 6) = (-2)^2 - 5 \cdot (-2) + 6 = 20.$$

In such cases—when the denominator has a limit of zero but the numerator does not—the quotient becomes unboundedly large in absolute value, and there is no finite limit. The following table of values for x near -2 provides numerical evidence of this fact:

x	-2.01	-2.0001	-2.000001	-2	-1.999999	-1.9999	-1.99
$f(x)$	501	50,001	$5 \cdot 10^6$	?	$-5 \cdot 10^6$	$-49,999$	-499

In this case, we say that *the limit does not exist.*

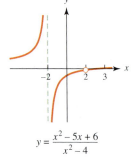

$$y = \frac{x^2 - 5x + 6}{x^2 - 4}$$

Figure 1.1.5

The graph of $f(x)$ is shown in Figure 1.1.5. It consists of two branches, one to the right of $x = -2$, the other to the left. The vertical line separating the branches (which is not actually part of the graph) is a **vertical asymptote**—a vertical line that the graph approaches, rising or falling without bound as it does. (Vertical asymptotes were introduced in Section 0.6.) Any time the denominator of a rational function approaches zero and the numerator does not, we can expect a vertical asymptote because the function grows unboundedly large in absolute value.

Notice that there is a small gap in the graph at $x = 2$, reflecting the fact that the function is not defined there. However, the graph shows what we already verified in part (ii) of Example 1.1.4—that the limit exists as $x \to 2$.

Indeterminate forms

Quotients in which both the numerator and denominator go to zero as $x \to a$, as in part (ii) of the last example, are called **indeterminate forms**. In such cases the limit of the quotient may or may not exist, and, if it does, its value cannot be predicted without simplifying the quotient in some way. The following examples illustrate some of the possibilities.

EXAMPLE 1.1.5

Find $\lim_{x \to 1} \dfrac{x^3 - 1}{x - 1}$.

PROBLEM-SOLVING TACTIC

To find $\lim_{x \to a} \dfrac{f(x)}{g(x)}$, first check $\lim_{x \to a} g(x)$.

- If $\lim_{x \to a} g(x) \neq 0$, you can find the limit by applying Theorem 1.1.1.
- If $\lim_{x \to a} g(x) = 0$ and $\lim_{x \to a} f(x) \neq 0$, the quotient has no (finite) limit.
- If $\lim_{x \to a} g(x) = 0$ and $\lim_{x \to a} f(x) = 0$, simplify the quotient.

SOLUTION In this case $\lim_{x \to 1}(x^3 - 1) = 0$ and $\lim_{x \to 1}(x - 1) = 0$, so we are dealing with an indeterminate form. To get an idea of the behavior of the quotient near $x = 1$, we first calculate a table of values, such as the one in Figure 1.1.6, created with an Excel spreadsheet.

Notice that the spreadsheet returned an error message at $x = 1$, chastising us for attempting to divide by zero. Studying the values close to 1, however, suggests that our function does have a limit of 3 (or very close to it) as $x \to 1$.

For a more precise analysis, we turn to algebra for help in simplifying the quotient. First, recall that for any two numbers, u and v,

$$u^3 - v^3 = (u - v)(u^2 + uv + v^2). \tag{7}$$

(If you have forgotten this formula, you can verify it by expanding the right-hand side and simplifying. Most of the terms cancel.) Dividing both sides of Eq. (7) by $(u - v)$ gives

$$\frac{u^3 - v^3}{u - v} = u^2 + uv + v^2, \quad \text{for } u \neq v. \tag{8}$$

By taking $u = x$ and $v = 1$, we get

$$\frac{x^3 - 1}{x - 1} = x^2 + x + 1, \quad \text{for } x \neq 1,$$

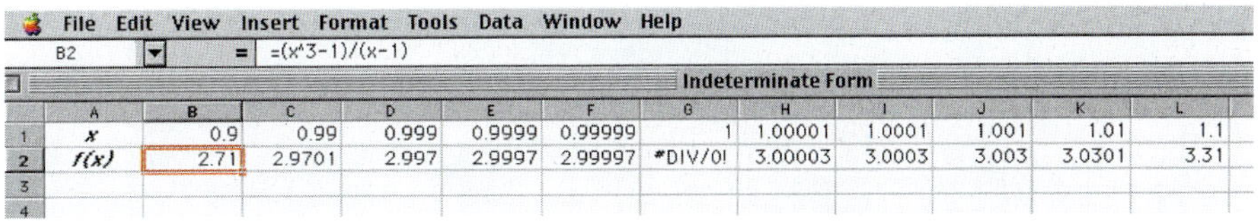

File Edit View Insert Format Tools Data Window Help
B2 = =(x^3-1)/(x-1)
Indeterminate Form

	A	B	C	D	E	F	G	H	I	J	K	L
1	x	0.9	0.99	0.999	0.9999	0.99999	1	1.00001	1.0001	1.001	1.01	1.1
2	f(x)	2.71	2.9701	2.997	2.9997	2.99997	#DIV/0!	3.00003	3.0003	3.003	3.0301	3.31
3												
4												

Figure 1.1.6

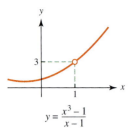

$$y = \frac{x^3 - 1}{x - 1}$$

Figure 1.1.7

and letting $x \to 1$ gives

$$\lim_{x \to 1} \frac{x^3 - 1}{x - 1} = \lim_{x \to 1}(x^2 + x + 1) = 3.$$

The graph of $(x^3 - 1)/(x - 1)$ is shown in Figure 1.1.7. It agrees with that of $x^2 + x + 1$ except at $x = 1$, where it is not defined.

The next example shows a technique for handling limits involving square roots.

EXAMPLE 1.1.6

Find $\displaystyle\lim_{h \to 0} \frac{\sqrt{4 + h} - 2}{h}$.

SOLUTION   We first observe that both the numerator and denominator go to zero as $h \to 0$. That is obvious in the case of the denominator, which is simply h itself. For the numerator, we have $\sqrt{4 + h} \approx 2$ when $h \approx 0$, and we can make the approximation as good as we want by taking h close enough to zero. For instance, if $h = 0.001$ we have $\sqrt{4.001} \approx 2.00025$, and if $h = -0.000001$, we have $\sqrt{3.999999} \approx 1.9999998$.

We conclude that the quotient is an indeterminate form. To get an idea of what happens to $f(h)$ as $h \to 0$, we start by constructing a table of values for h near zero, as shown in Figure 1.1.8

Once again, there is an error message instead of a numerical value at $h = 0$, indicating that zero cannot be used as input. Checking the values of $f(h)$ at the other points near $h = 0$ suggests that $f(h) \to 0.25$ as $h \to 0$. To verify it, we need to simplify the quotient. By recalling the formula

$$(u - v)(u + v) = u^2 - v^2, \tag{9}$$

for any two numbers u and v, and substituting $u = \sqrt{4 + h}$ and $v = 2$, we get

$$\left(\sqrt{4 + h} - 2\right)\left(\sqrt{4 + h} + 2\right) = (4 + h) - 4 = h.$$

That suggests we can simplify the indeterminate form by multiplying both numerator and denominator by $(\sqrt{4 + h} + 2)$. Multiplying *and* dividing by the same quantity does

File   Edit   View   Insert   Format   Tools   Data   Window   Help

B2   =   =(SQRT(4+h)-2)/h

Indeterminate2

	A	B	C	D	E	F	G	H	I	J
1	*h*	−0.01	−0.001	−0.0001	−0.00001	0	0.00001	0.0001	0.001	0.01
2	*f(h)*	0.2501564	0.2500156	0.2500016	0.2500002	#DIV/0!	0.2499998	0.2499984	0.2499844	0.2498439
3										
4										

Figure 1.1.8

not change the quotient, and we get

$$\frac{\sqrt{4+h}-2}{h} = \frac{\sqrt{4+h}-2}{h} \cdot \frac{\sqrt{4+h}+2}{\sqrt{4+h}+2}$$

$$= \frac{(4+h)-4}{h\left(\sqrt{4+h}+2\right)}$$

$$= \frac{h}{h\left(\sqrt{4+h}+2\right)} = \frac{1}{\sqrt{4+h}+2},$$

valid for any h near zero but $\neq 0$. The final term in this string of equations is *not* an indeterminate form and has a well-defined limit as $h \to 0$. In fact, the denominator satisfies

$$\lim_{h\to0}\left(\sqrt{4+h}+2\right) = \sqrt{4}+2 = 4,$$

and therefore

$$\lim_{h\to0}\frac{\sqrt{4+h}-2}{h} = \frac{1}{4}.$$

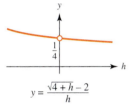

$$y = \frac{\sqrt{4+h}-2}{h}$$

Figure 1.1.9

The graph of $(\sqrt{4+h}-2)/h$ is shown in Figure 1.1.9.

One-sided limits

When a function $f(x)$ is only defined on one side of the point a, or when a function behaves differently on each side of a, it is more natural in defining the limit to require that x approaches a only from the side under consideration. Thus, we obtain

- the **right-hand limit**, denoted by $\lim_{x\to a^+} f(x)$, if x approaches a from the right, and
- the **left-hand limit**, denoted by $\lim_{x\to a^-} f(x)$, if x approaches a from the left.

If both one-sided limits exist and are equal as $x \to a^-$ and $x \to a^+$, then the two-sided limit $\lim_{x\to a} f(x)$ exists. However, if $\lim_{x\to a^-} f(x) \neq \lim_{x\to a^+} f(x)$, then $\lim_{x\to a} f(x)$ does not exist.

The rules stated in Theorem 1.1.1 apply to one-sided as well as two-sided limits.

EXAMPLE 1.1.7

Find $\lim_{x\to1^-} \dfrac{x^2-1}{\sqrt{1-x}}$.

SOLUTION In this case, the domain consists of all $x < 1$, and we can only take a one-sided limit.

As $x \to 1$ from the left, both the numerator and denominator approach zero, so we have an indeterminate form. A little numerical experimentation results in the following table:

x	0.9	0.99	0.999	0.9999	0.99999	0.999999	0.9999999	1
$f(x)$	-0.601	-0.199	-0.063	-0.02	-0.006	-0.002	-0.0006	?

It suggests that the function values approach zero as a limit.

To verify that, we once again factor the numerator:

$$(x^2 - 1) = (x + 1)(x - 1) = -(1 + x)(1 - x).$$

Moreover, $(1 - x)/\sqrt{1 - x} = \sqrt{1 - x}$ if $x < 1$. Therefore,

$$\frac{x^2 - 1}{\sqrt{1 - x}} = \frac{-(1 + x)(1 - x)}{\sqrt{1 - x}}$$
$$= -(1 + x)\sqrt{1 - x}.$$

Letting $x \to 1^-$, we get

$$\lim_{x \to 1^-} \frac{x^2 - 1}{\sqrt{1 - x}} = \lim_{x \to 1^-} -(1 + x)\sqrt{1 - x} = -2\sqrt{0} = 0.$$

The graph is shown in Figure 1.1.10.

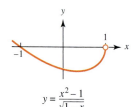

$y = \dfrac{x^2 - 1}{\sqrt{1 - x}}$

Figure 1.1.10

EXAMPLE 1.1.8

The First National Bank charges a monthly fee for a standard checking account. The size of the fee depends on the average balance in the account in any given month. For an average balance up to and including \$1,000, the charge is \$10; for an average balance greater than \$1,000 but less than or equal to \$5,000, the fee is \$5; and for an average balance greater than \$5,000, the fee is \$2. Letting $f(s)$ be the fee on an average balance of s dollars,

    (i) write a formula for $f(s)$ and draw the graph;

    (ii) find the one-sided limits in both directions at $s = 1,000$ and $s = 5,000$ and evaluate f at each of those points.

SOLUTION (i) Translating the word description into a formula leads to

$$f(s) = \begin{cases} 10 & \text{if } 0 \le s \le 1,000 \\ 5 & \text{if } 1,000 < s \le 5,000 \\ 2 & \text{if } 5,000 < s. \end{cases}$$

The graph is shown in Figure 1.1.11. Each small circle at a point where the graph jumps indicates a point that is not on the graph.

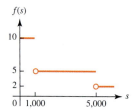

Figure 1.1.11

(ii) From either the formula or the graph we see that $f(1,000) = 10$ and $f(5,000) = 5$, and

$$\lim_{s \to 1,000^-} f(s) = 10, \qquad \lim_{s \to 1,000^+} f(s) = 5, \qquad \lim_{s \to 5,000^-} f(s) = 5, \qquad \lim_{s \to 5,000^+} f(s) = 2.$$

APPLYING TECHNOLOGY

Exploring limits with a calculator or spreadsheet   As we have seen in several examples, graphing calculators and spreadsheets can be helpful in making an educated guess about whether a limit exists and for estimating its value.

EXAMPLE 1.1.9

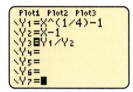

Figure 1.1.12

Figure 1.1.13

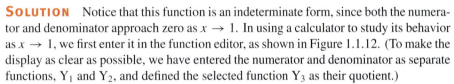

Figure 1.1.14

Use a graphing calculator or spreadsheet to explore the behavior of the function

$$f(x) = \frac{x^{1/4} - 1}{x - 1}$$

near $x = 1$. In particular, estimate $\lim_{x \to 1} f(x)$.

SOLUTION   Notice that this function is an indeterminate form, since both the numerator and denominator approach zero as $x \to 1$. In using a calculator to study its behavior as $x \to 1$, we first enter it in the function editor, as shown in Figure 1.1.12. (To make the display as clear as possible, we have entered the numerator and denominator as separate functions, Y_1 and Y_2, and defined the selected function Y_3 as their quotient.)

Next, we create a table of values in one of two ways: either by choosing an initial x-value and step size and letting the calculator generate the table, or by manually entering whatever x-values we choose. The latter method is necessary if we want to use unevenly spaced x-values. Figure 1.1.13 shows a typical table setup editor. In it we have chosen the manual entry option by selecting **Ask** for the independent variable. Finally, we create the table by entering values of x that approach 1 from both sides. The result is shown in Figure 1.1.14. It suggests that $f(x) \to 0.25$ as $x \to 1$.

A calculator is limited both in the size of the table it can display and the number of decimal places in each entry. A more convincing estimate can be made using an Excel spreadsheet, as shown in Figure 1.1.15. The x-values were entered in two rows, those in row 1 approaching 1 from the left and those in row 5 from the right. The function was entered in cell B2, as shown, and then repeated across row 2 using the **FillRight** command from the **Edit** menu. A similar procedure was used in row 6.

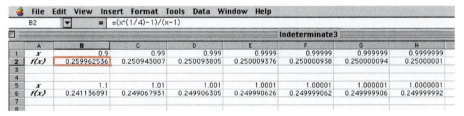

Figure 1.1.15

The data strongly suggest that

$$\lim_{x \to 1} \frac{x^{1/4} - 1}{x - 1} = 0.25,$$

which is indeed true in this case. But experimental data, no matter how convincing, may sometimes be misleading and should not be considered conclusive. It should be verified by deduction based on the properties of limits, as we did in Examples 1.1.5 and 1.1.6. In the exercises you will be asked to apply similar reasoning to the present example.

Practice Exercises 1.1

1. Find $\lim\limits_{x \to 4} \dfrac{x^2 - 6x + 8}{x^2 - 16}$.

2. Let $f(x) = \dfrac{\sqrt{x}}{x + 2}$. Find $\lim\limits_{x \to 0^+} f(x)$.

Exercises 1.1

Exercises 1–4 refer to the function f, whose graph is shown in Figure 1.1.16.

1. Compute the limits:
   (a) $\lim\limits_{x \to -1} f(x)$ **(b)** $\lim\limits_{x \to 1^+} f(x)$ **(c)** $\lim\limits_{x \to 1^-} f(x)$

2. Compute the limits:
   (a) $\lim\limits_{x \to 3^-} f(x)$ **(b)** $\lim\limits_{x \to 3^+} f(x)$ **(c)** $\lim\limits_{x \to 2} f(x)$

3. Find all x in the interval $(-2, 4)$ at which the function f has different one-sided limits.

4. Find all x in the interval $(-2, 4)$ at which the function f has no limit. Explain.

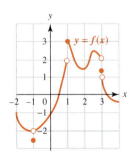

Figure 1.1.16

Exercises 5–8 refer to the function g, whose graph is shown in Figure 1.1.17

5. Compute the following limits and values:
   (a) $\lim\limits_{x \to -1} g(x)$ **(b)** $\lim\limits_{x \to 2^+} g(x)$ **(c)** $\lim\limits_{x \to 2^-} g(x)$
   (d) $g(-1)$ and $g(1)$

6. Compute the following limits and values:
   (a) $\lim\limits_{x \to 1} g(x)$ **(b)** $\lim\limits_{x \to 3^+} g(x)$ **(c)** $\lim\limits_{x \to 3^-} g(x)$
   (d) $g(2)$ and $g(3)$

7. Find all x in the interval $(-2, 4)$ at which the function g has different one-sided limits.

8. Find all x in the interval $(-2, 4)$ at which the function g has no limit. Explain.

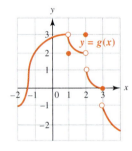

Figure 1.1.17

9. The cost of producing q units of a certain quantity is $C(q) = 10,000 + 8q$ dollars. Thus, the average cost per unit is

$$A(q) = \frac{10,000 + 8q}{q}.$$

   (a) Find the limit of $A(q)$ as q approaches 10^7.
   (b) Does $A(q)$ have a limit as q approaches zero?

10. A telephone company charges the rate of 10 cents per minute between 8 A.M. and 4 P.M., and 5 cents per minute the rest of the day.

(a) Sketch the graph of the rate $r = r(t)$ as a function of time t during the 24 hours of a day.
(b) Find the limit of r as t approaches 8 from the left and from the right.
(c) Find all points t at which the function has a limit.

11. In the first 4 years of its existence, a preschool with an excellent reputation had space for 25 students, and it was filled to capacity each year. By expanding its facility, it has been able to increase its enrollment by five students a year in each of the following 3 years.
(a) Write a function that represents the number of students versus time in years and sketch its graph.
(b) Find all points where this function has no limit and in each case compute the one-sided limits, if they exist.

In Exercises 12–31, construct a small table of values and use it to determine whether the limit exists and, if it does, to find it. Then determine it more precisely by using the properties of limits and simplifying the expression, if necessary.

12. $\lim\limits_{x \to 2}(x^2 + x - 3)$

13. $\lim\limits_{x \to -1}(x^3 - 2x^2 + 5x + 1)$

14. $\lim\limits_{x \to 1}\dfrac{1}{x^2 - 1}$

15. $\lim\limits_{x \to 1}\dfrac{x - 1}{x^2 - 1}$

16. $\lim\limits_{x \to 1}\dfrac{x + 1}{x^2 - 1}$

17. $\lim\limits_{x \to 1}\dfrac{x + 1}{x^2 + 1}$

18. $\lim\limits_{x \to -1}\dfrac{x + 1}{x^2 - 1}$

19. $\lim\limits_{x \to -1}\dfrac{x + 1}{x^2 + 1}$

20. $\lim\limits_{x \to 3}\dfrac{x^2 - 4x + 3}{x - 3}$

21. $\lim\limits_{x \to 3}\dfrac{x^2 - 4x + 3}{x^2 - x - 6}$

22. $\lim\limits_{x \to 3}\dfrac{x^2 - 4x + 3}{x^2 - 6x + 9}$

23. $\lim\limits_{x \to -1}\dfrac{x + 1}{x^3 - 1}$

24. $\lim\limits_{x \to -1}\dfrac{x^3 - 1}{x + 1}$

25. $\lim\limits_{x \to -1}\dfrac{x^3 + 1}{x + 1}$

26. $\lim\limits_{t \to 0}\dfrac{t}{t^2 + 2t}$

27. $\lim\limits_{t \to 0}\dfrac{t}{t^2 + 2t + 1}$

28. $\lim\limits_{x \to 0}\dfrac{x^4 - 16}{x^3 - 8}$

29. $\lim\limits_{h \to 0}\dfrac{(h + 3)^2 - 9}{h}$

30. $\lim\limits_{h \to 0}\dfrac{(h - 3)^2 - 9}{h}$

31. $\lim\limits_{h \to 0}\dfrac{(h - 3)^2 - 9}{(h - 2)^2 - 4}$

In Exercises 32–37 use a method similar to the one used in Example 1.1.6 to simplify the quotient and then find the limit.

32. $\lim\limits_{h \to 0}\dfrac{\sqrt{9 + h} - 3}{h}$

33. $\lim\limits_{x \to 1}\dfrac{\sqrt{x} - 1}{x - 1}$

34. $\lim\limits_{h \to 0}\dfrac{h}{1 - \sqrt{1 + h}}$

35. $\lim\limits_{h \to 0}\dfrac{1 - \sqrt{1 + h}}{h^2 + 3h}$

36. $\lim\limits_{x \to 1}\dfrac{2 - \sqrt{x^2 + 3}}{x - 1}$

37. $\lim\limits_{x \to 0}\dfrac{\sqrt{1 + x} - \sqrt{1 - x}}{x}$

In Exercises 38–43, assume that $\lim\limits_{x \to 0} f(x) = 3$ and $\lim\limits_{x \to 0} g(x) = -2$ and find the given limit.

38. $\lim\limits_{x \to 0}[2f(x) - g(x)]$

39. $\lim\limits_{x \to 0}\left[f(x)g(x)^2\right]$

40. $\lim\limits_{x \to 0}\dfrac{f(x) - 3}{f(x) + 1}$

41. $\lim\limits_{x \to 0}[f(x)g(x) + 5]$

42. $\lim\limits_{x \to 0}\dfrac{g(x)}{x^2 - 4}$

43. $\lim\limits_{x \to 0}\dfrac{xf(x) - g(x)}{f(x) - g(x)}$

In Exercises 44–52, find the one-sided limit, if it exists.

44. $\lim\limits_{x \to 4^+}\sqrt{x - 4}$

45. $\lim\limits_{x \to 0^-}\dfrac{|x|}{x}$

46. $\lim\limits_{x \to 0^+}\dfrac{|x|}{x}$

47. $\lim\limits_{t \to 2^+}\dfrac{\sqrt{t - 2}}{t^2}$

48. $\lim\limits_{t \to 2^+}\dfrac{\sqrt{t - 2}}{t - 2}$

49. $\lim\limits_{t \to 2^+}\dfrac{t^2 - 4}{\sqrt{t - 2}}$

50. $\lim\limits_{t \to 0^+}\dfrac{\sqrt{4 + t}}{2 + \sqrt{t}}$

51. $\lim\limits_{t \to 0^+}\dfrac{\sqrt{4 + t} - 2}{\sqrt{t}}$

52. $\lim\limits_{x \to 3^+}\dfrac{x^2 - 4x + 6}{\sqrt{x - 3}}$

53. According to Einstein's theory of relativity, if a rigid 1-meter rod is moving in the direction of its length with velocity v, a stationary observer will observe its length as being equal to

$$\ell(v) = \sqrt{1 - \dfrac{v^2}{c^2}},$$

where c is the speed of light, which is 300,000 kilometers per second.
(a) Find the limit of ℓ as v approaches c from the left.
(b) Find the limit of ℓ as v approaches zero from the the right.

In Exercises 54–59, assume that $\lim_{x \to 0^+} f(x) = 1/2$ and $\lim_{x \to 0^+} g(x) = -1/3$ and find the given limit.

54. $\displaystyle\lim_{x \to 0^+} \sqrt{x}\, f(x)$

55. $\displaystyle\lim_{x \to 0^+} [f(x) - g(x)]$

56. $\displaystyle\lim_{x \to 0^+} \frac{f(x) + \sqrt{x}}{g(x) + 1}$

57. $\displaystyle\lim_{x \to 0^+} \frac{|f(x)|}{f(x)}$

58. $\displaystyle\lim_{x \to 1^-} g(1 - x)$

59. $\displaystyle\lim_{x \to 1^+} f(x - 1)$

60. (This exercise refers to Example 1.1.9.)   Find $\lim_{x \to 1} (x^{1/4} - 1)/(x - 1)$ by using the following identity from algebra:

$$u^4 - v^4 = (u - v)(u^3 + u^2 v + u v^2 + v^3)$$

for any two numbers u and v. (*Hint:* Take $u = x^{1/4}$.)

In Exercises 61–66, use a calculator or spreadsheet to construct an appropriate table of values, and then use it to judge whether the limit exists and, if it does, to estimate it.

61. $\displaystyle\lim_{t \to 1^-} \frac{1 - \sqrt{t}}{\sqrt{1 - t}}$

62. $\displaystyle\lim_{x \to -1} \frac{\sqrt[3]{x} + 1}{x + 1}$

63. $\displaystyle\lim_{x \to 2^+} \frac{\sqrt{x^2 - 4}}{x - 2}$

64. $\displaystyle\lim_{h \to 0} \frac{\sqrt{9 + 4h + h^2} - 3}{h}$

65. $\displaystyle\lim_{t \to 3} \frac{\sqrt{2t + 3} - 3}{\sqrt{t + 1} - 2}$

66. $\displaystyle\lim_{x \to 16} \frac{\sqrt{x} - 4}{\sqrt[4]{x} - 2}$

Solutions to practice exercises 1.1

1. Factoring the denominator gives $x^2 - 16 = (x - 4)(x + 4)$, and factoring the numerator gives $x^2 - 6x + 8 = (x - 4)(x - 2)$. Simplifying and applying Theorem 1.1.1 give

$$\lim_{x \to 4} \frac{x^2 - 6x + 8}{x^2 - 16} = \lim_{x \to 4} \frac{x - 2}{x + 4} = \frac{4 - 2}{4 + 4} = \frac{1}{4}.$$

2. We have $\lim_{x \to 0^+} \sqrt{x} = 0$. Moreover, since $\lim_{x \to 0}(x + 2) = 2$, we also have $\lim_{x \to 0^+}(x + 2) = 2$. (If the two-sided limit exists, each of the one-sided limits exists and equals it.) Therefore, by the quotient rule for limits we have

$$\lim_{x \to 0^+} \frac{\sqrt{x}}{x + 2} = \frac{0}{2} = 0.$$

1.2   More About Limits and Asymptotes

If a rational function has a vertical asymptote—if, after putting it into lowest terms, we find a point where the denominator is zero—then the function does not have a finite limit at that point. In fact, the output grows unboundedly large in absolute value.

Definition 1.2.1

Infinite limits

1. We say that $\lim_{x \to a} f(x) = \infty$ if $f(x)$ is positive and very large for all x close to (but not necessarily equal to) a, and we can make $f(x)$ as large as we want by taking x close enough to a.

2. We say that $\lim_{x \to a} f(x) = -\infty$ if $f(x)$ is negative and very large (in absolute value) for all x close to (but not necessarily equal to) a, and we can make $|f(x)|$ as large as we want by taking x close enough to a.

The one-sided limits are defined similarly, but with x approaching a from one side only.

EXAMPLE 1.2.1

Find $\displaystyle\lim_{x\to 1}\frac{1}{(x-1)^2}$.

SOLUTION This function is a rational function in lowest terms, and it has a vertical asymptote at $x = 1$. If x is very close to 1, the denominator is very small, which means that the quotient is very large. The following table shows how large $f(x)$ is for various choices of x near 1:

x	0.9	0.999	0.999999	1	1.000001	1.001	1.1
$f(x)$	100	10^6	10^{12}	**?**	10^{12}	10^6	100

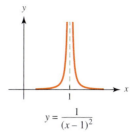

$y = \dfrac{1}{(x-1)^2}$

Figure 1.2.1

The graph is shown in Figure 1.2.1. We conclude that

$$\lim_{x\to 1}\frac{1}{(x-1)^2} = \infty.$$

EXAMPLE 1.2.2

Find $\displaystyle\lim_{x\to -3}\frac{x}{(x+3)^2}$.

SOLUTION Once again, the function is a rational function in lowest terms. In this case the vertical asymptote is at $x = -3$, and the function values are negative and large for all x near -3. The following table shows the values of $f(x)$ for selected x near -3:

x	-3.1	-3.001	-3.000001	-3	-2.999999	-2.999	-2.9
$f(x)$	-310	$-3.001\cdot 10^6$	$-3\cdot 10^{12}$	**?**	$-3\cdot 10^{12}$	$-2.999\cdot 10^6$	-290

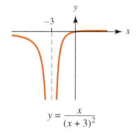

$y = \dfrac{x}{(x+3)^2}$

Figure 1.2.2

The graph is shown in Figure 1.2.2, with the dashed line representing the vertical asymptote $x = -3$. We conclude that

$$\lim_{x\to -3}\frac{x}{(x+3)^2} = -\infty.$$

EXAMPLE 1.2.3

Let $f(x) = \dfrac{1}{x}$. Determine the limiting behavior of this function as $x \to 0$.

SOLUTION This function has a vertical asymptote at $x = 0$, and its absolute value grows unboundedly large as we take x closer and closer to zero. However, it is positive

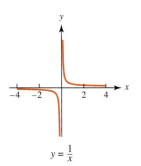

Figure 1.2.3

for all x to the right of zero and negative for all x to the left as the following table of values indicates:

x	-0.1	-0.001	-0.000001	0	0.000001	0.001	0.1
$f(x)$	-10	$-1,000$	-10^6	$?$	10^6	$1,000$	10

As the table suggests,

$$\lim_{x \to 0^-} \frac{1}{x} = -\infty \quad \text{and} \quad \lim_{x \to 0^+} \frac{1}{x} = \infty,$$

but $\lim_{x \to 0} (1/x)$ does not exist. The graph is shown in Figure 1.2.3.

EXAMPLE 1.2.4

If $f(x) = (x - 1)^{-2/3}$, find $\lim_{x \to 1} f(x)$.

SOLUTION   Rewriting the function in the form $\dfrac{1}{(x - 1)^{2/3}}$, we see that

- it is defined for all $x \neq 1$, and
- it is positive for all x near 1.

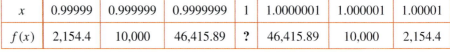

(In fact, it is positive for all $x \neq 1$.) When x is very close to 1, the denominator is very small, which means that the quotient is very large. We see that in the following table of values:

x	0.99999	0.999999	0.9999999	1	1.0000001	1.000001	1.00001
$f(x)$	$2,154.4$	$10,000$	$46,415.89$	$?$	$46,415.89$	$10,000$	$2,154.4$

Figure 1.2.4

We conclude that $\lim_{x \to 1} (x - 1)^{-2/3} = \infty$. The graph is shown in Figure 1.2.4.

Horizontal asymptotes and limits at infinity

Let's reconsider the function $f(x) = 1/x$ that we studied in Example 1.2.3. It has the y-axis as a vertical asymptote. But Figure 1.2.3 shows another line that also acts as an attractor for the graph—namely, the x-axis. As you can see, the graph gets closer and closer to the x-axis as it moves further and further to the right or left.

A horizontal line that the graph approaches in this way is called a **horizontal asymptote**. Another way of putting it is this: A line with equation $y = b$ is a horizontal asymptote if the graph tends to level off toward height b as it moves further and further to the left or right.

We can give a better definition of a horizontal asymptote by using the language of limits. To do that, we have to introduce a new type of limit—**the limit of** $f(x)$ **as** x **goes to infinity**—which we symbolize by writing $\lim_{x \to \infty} f(x)$.

Definition 1.2.2

We say that $\lim_{x \to \infty} f(x) = L$ if $f(x) \approx L$ for all x positive and very large, and we can make the approximation as close as we want by taking x large enough.

For instance, as we have just observed, $\lim_{x \to \infty}(1/x) = 0$. Here is another example of a horizontal asymptote.

EXAMPLE 1.2.5

Let $f(x) = \dfrac{x}{x+1}$. Find $\lim_{x \to \infty} f(x)$.

SOLUTION   Figure 1.2.5 shows an Excel spreadsheet with values of $f(x)$ calculated for larger and larger values of x. (The x-values are written in scientific notation—for instance, 1.00E+03 means 1.00×10^3, which equals 1,000, and 1.00E+15 means 1.00×10^{15}, which equals the number represented by a 1 followed by 15 zeros.) The table suggests that $\lim_{x \to \infty} f(x) = 1$.

	File	Edit	View	Insert	Format	Tools	Data	Windo
	B2	▼		**=**	=x/(x+1)			

	A	B	C
1	*x*	*f(x)*	
2	1.00E+01	0.909090909090909000	
3	1.00E+03	0.999000999000999000	
4	1.00E+05	0.999990000099999000	
5	1.00E+07	0.999999900000010000	
6	1.00E+09	0.999999999000000000	
7	1.00E+11	0.999999999990000000	
8	1.00E+13	0.999999999999900000	
9	1.00E+15	0.999999999999999000	
10			
11			
12			

Figure 1.2.5

To verify this, we first divide both the numerator and denominator by x. (Recall that dividing both the numerator and denominator by the same quantity does not change the quotient.) We get

$$f(x) = \frac{x}{x+1} = \frac{\dfrac{x}{x}}{\dfrac{x+1}{x}} = \frac{1}{1+\dfrac{1}{x}},$$

valid for all $x \neq 0$. As we already observed, $\lim_{x \to \infty}(1/x) = 0$. Therefore, $\lim_{x \to \infty}(1+(1/x)) = 1$, and

$$\lim_{x \to \infty} \frac{x}{x+1} = \lim_{x \to \infty} \frac{1}{1+\dfrac{1}{x}} = \frac{1}{1} = 1.$$

The graph is shown in Figure 1.2.6.

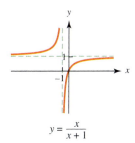

$y = \dfrac{x}{x+1}$

Figure 1.2.6

It is easy to modify the definition of the limit of $f(x)$ as $x \to \infty$ to allow for infinite (positive or negative) limits. Two simple but important examples are

$$\lim_{x \to \infty} x^k = \infty \quad \text{and} \quad \lim_{x \to \infty} (-x^k) = -\infty,$$

for any positive integer k.

For a rational function, the limit as $x \to \infty$ depends on the degrees of the numerator and denominator and, in some cases, on their leading coefficients. Here is a precise statement of what happens.

Theorem 1.2.1

Suppose $f(x)$ is a rational function whose numerator has degree n and whose denominator has degree d; that is,

$$f(x) = \frac{a_n x^n + \text{lower-degree terms}}{b_d x^d + \text{lower-degree terms}}, \quad a_n \neq 0, \ b_d \neq 0.$$

1. If $n < d$, then $\lim_{x \to \infty} f(x) = 0$, and the x-axis is a horizontal asymptote.

2. If $n = d$, then $\lim_{x \to \infty} f(x) = a_n/b_d$, and the line $y = a_n/b_d$ is a horizontal asymptote.

3. If $n > d$, then there is no horizontal asymptote, and

$$\lim_{x \to \infty} f(x) = \begin{cases} \infty & \text{if } a_n/b_d > 0 \\ -\infty & \text{if } a_n/b_d < 0. \end{cases}$$

We can prove this theorem by a procedure similar to the one we used in Example 1.2.5—divide the numerator and denominator by the highest power of x appearing in the quotient and use the fact that $\lim_{x \to \infty}(1/x^k) = 0$ if k is a positive integer. Rather than doing that, however, we prefer to discuss an easy way to understand and remember these rules by recalling the basic principle about polynomials introduced in Section 0.6. *As x goes further and further to the right (or left), any polynomial behaves more and more like its highest-degree term.* So, to study the behavior of a rational function for very large positive or negative x, *replace the numerator and denominator by their highest-degree terms.* In other words (with $\approx$ standing for approximation),

$$\begin{aligned} f(x) &= \frac{a_n x^n + \text{lower-degree terms}}{b_d x^d + \text{lower-degree terms}} \\ &\approx \frac{a_n x^n}{b_d x^d} \\ &= \frac{a_n}{b_d} x^{n-d}. \end{aligned}$$

From this we see that for large (positive or negative) x, the function $f(x)$ behaves like (a_n/b_d) times x to the power $n - d$. There are three possibilities:

- If $d > n$, then $\lim_{x \to \infty} f(x) = \lim_{x \to \infty} a_n/(b_d x^{d-n}) = 0$.
- If $n = d$, then $f(x) \approx (a_n/b_d)$ for large x, and $\lim_{x \to \infty} f(x) = a_n/b_d$.
- If $n > d$, then $\lim_{x \to \infty} f(x) = \lim_{x \to \infty}(a_n x^{n-d}/b_d) = \begin{cases} \infty & \text{if } a_n/b_d > 0 \\ -\infty & \text{if } a_n/b_d < 0. \end{cases}$

We can similarly define and study the limit as $x \to -\infty$.

Definition 1.2.3

We say that $\lim_{x \to -\infty} f(x) = L$ if $f(x) \approx L$ for all x negative and very large (in absolute value), and we can make the approximation as close as we want by taking $|x|$ large enough.

For rational functions, we can apply the same principle of replacing the numerator and denominator by their highest-order terms.

EXAMPLE 1.2.6

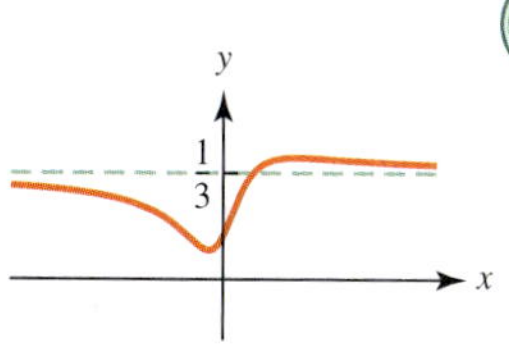

Find $\lim\limits_{x \to -\infty} f(x)$, where

(i) $f(x) = \dfrac{x^2 + x + 1}{3x^2 + 7}$,   (ii) $f(x) = \dfrac{4x^3 + 5x^2 + 7}{x^4 + 2}$,   (iii) $f(x) = \dfrac{x^3 - 2}{x^2 + 1}$.

SOLUTION   (i) If $|x|$ is very large,

$$x^2 + x + 1 \approx x^2 \quad \text{and} \quad 3x^2 + 7 \approx 3x^2.$$

Therefore,

$$\lim_{x \to -\infty} \frac{x^2 + x + 1}{3x^2 + 7} = \lim_{x \to -\infty} \frac{x^2}{3x^2} = \frac{1}{3}.$$

The graph is shown in Figure 1.2.7.

(ii) Using the same principle, we have

$$\lim_{x \to -\infty} \frac{4x^3 + 5x^2 + 7}{x^4 + 2} = \lim_{x \to -\infty} \frac{4x^3}{x^4} = \lim_{x \to -\infty} \frac{4}{x} = 0.$$

The graph is shown in Figure 1.2.8.

(iii) In this case,

$$\lim_{x \to -\infty} \frac{x^3 - 2}{x^2 + 1} = \lim_{x \to -\infty} \frac{x^3}{x^2} = \lim_{x \to -\infty} x = -\infty.$$

The graph is shown in Figure 1.2.9.

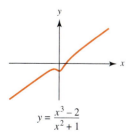

$y = \dfrac{x^2 + x + 1}{3x^2 + 7}$

Figure 1.2.7

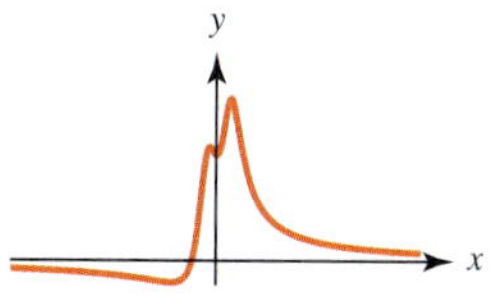

$y = \dfrac{4x^3 + 5x^2 + 7}{x^4 + 2}$

Figure 1.2.8

$y = \dfrac{x^3 - 2}{x^2 + 1}$

Figure 1.2.9

We can use limits as $x \to \pm\infty$ to arrive at a precise definition of a horizontal asymptote.

Definition 1.2.4

The line $y = L$ is a **horizontal asymptote** for the graph of a function f if either or both of the following limits is valid:

$$\lim_{x \to -\infty} f(x) = L \quad \text{or} \quad \lim_{x \to \infty} f(x) = L$$

EXAMPLE 1.2.7

Find the limits of $f(x)$ as $x \to \pm\infty$ and determine whether there are any horizontal or vertical asymptotes for each of the following:

$$\text{(i)} \;\; f(x) = \frac{4x^3 + 3x^2 + 8x}{x^3 - 8} \quad \text{and} \quad \text{(ii)} \;\; f(x) = \frac{2x^3 - 3x^2 - 5x + 1}{x^2 + 1}.$$

SOLUTION (i) Replacing the numerator and denominator by their highest-degree terms gives

$$\lim_{x \to \pm\infty} \frac{4x^3 + 3x^2 + 8x}{x^3 - 8} = \lim_{x \to \pm\infty} \frac{4x^3}{x^3} = 4.$$

It follows that the line $y = 4$ is a horizontal asymptote.

Looking for a vertical asymptote, we observe that the domain of $f(x)$ consists of all $x \neq 2$, the quotient is in lowest terms, the denominator is zero at $x = 2$, and the numerator is not. Therefore, the line $x = 2$ is a vertical asymptote. The graph is shown in Figure 1.2.10, with the asymptotes drawn as dashed lines.

(ii) The degree of the numerator is greater than the degree of the denominator, and so there is no horizontal asymptote. In fact, if x is very large (positive or negative), we get the following approximation by replacing each polynomial by its highest-degree term:

$$\frac{2x^3 - 3x^2 - 5x + 1}{x^2 + 1} \approx \frac{2x^3}{x^2} = 2x.$$

This means that the graph behaves approximately like the graph of $y = 2x$ when x is very large. In particular,

$$\lim_{x \to \infty} f(x) = \lim_{x \to \infty} 2x = \infty \quad \text{and} \quad \lim_{x \to -\infty} f(x) = \lim_{x \to -\infty} 2x = -\infty,$$

which verifies that there is no horizontal asymptote. There is no vertical asymptote either, for the denominator of this quotient is never zero, and the function is defined for all x. The graph is shown in Figure 1.2.11.

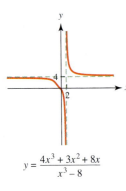

$y = \dfrac{4x^3 + 3x^2 + 8x}{x^3 - 8}$

Figure 1.2.10

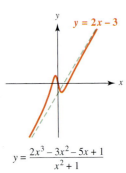

$y = \dfrac{2x^3 - 3x^2 - 5x + 1}{x^2 + 1}$

Figure 1.2.11

Remark   If the numerator is one degree greater than the denominator, as in case (ii) of the last example, then the graph $y = f(x)$ has a **slant asymptote**, which is a line $y = mx + b$ that is neither vertical nor horizontal and such that $\lim_{x \to \infty}[f(x) - (mx + b)] = 0$. In case (ii) of the last example, the slant asymptote is the line $y = 2x - 3$, shown as a dashed line in Figure 1.2.11, since $\lim_{x \to \infty}[f(x) - (2x - 3)] = 0$. To see why, divide the denominator into the numerator, to obtain a quotient and remainder, as follows:

$$f(x) = \frac{2x^3 - 3x^2 - 5x + 1}{x^2 + 1} = 2x - 3 + \frac{-7x + 4}{x^2 + 1}.$$

Therefore,

$$f(x) - (2x - 3) = \frac{-7x + 4}{x^2 + 1} \to 0$$

as $|x| \to \infty$, which means that the graph of $f(x)$ gets arbitrarily close to the line as $x \to \pm\infty$.

Sketching the graphs of rational functions

We can get a fair amount of information about a rational function and a rough idea of its graph by using its asymptotes together with any information about its domain, range, sign, and symmetry, as the next examples show.

EXAMPLE 1.2.8

Sketch the graph of the rational function $f(x) = \dfrac{x}{x^2 + 1}$.

SOLUTION   In this case the numerator has degree 1 and the denominator has degree 2. Therefore, we can conclude that

$$\lim_{x \to \pm\infty} f(x) = 0,$$

and the graph has the x-axis as a horizontal asymptote in both directions.

   To get an idea of what the graph looks like, we observe that $f(x) > 0$ for $x > 0$. That means that the graph is above the x-axis in the right half of the plane and sinks toward height zero as it moves further and further to the right. Similarly, $f(x) < 0$ for $x < 0$, so the graph lies below the x-axis in the left half of the plane and rises to height zero as it moves further and further to the left. Since $f(0) = 0$, the point $(0, 0)$ is on the graph.

   Finally, we observe that $f(x)$ is an odd function, which means that the graph is symmetric about the origin. Putting this information together, we can get a rough sketch of the graph. By way of comparison, Figure 1.2.12 shows one that was created with a TI-83 Plus calculator.

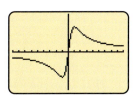

Figure 1.2.12

EXAMPLE 1.2.9

Sketch the graph of the function $f(x) = \dfrac{2x^2 - 2}{x^2 + 1}$.

SOLUTION   As x goes further and further to the right or left,

$$\frac{2x^2 - 2}{x^2 + 1} \approx \frac{2x^2}{x^2} = 2,$$

which means $\lim_{x \to \pm\infty} f(x) = 2$, and the line $y = 2$ is a horizontal asymptote.

In this case the graph always lies below the line $y = 2$. To see why, we observe that for any x,

$$2x^2 - 2 < 2x^2 + 2 = 2(x^2 + 1),$$

which means that $(2x^2 - 2)/(x^2 + 1) < 2$. By checking the sign of the function, which is the same as the sign of $2x^2 - 2 = 2(x - 1)(x + 1)$, we see that the graph

- lies below the x-axis for $-1 < x < 1$,
- lies above the x-axis for $x > 1$ and for $x < -1$, and
- crosses the x-axis at $x = 1$ and $x = -1$.

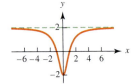

Figure 1.2.13

Also, we note that the denominator is never zero, which means that the function is defined for all x and there are no vertical asymptotes. Finally, the function is even, and therefore its graph is symmetric with respect to the y-axis. Putting together all of that information, we can infer that the graph looks like the one drawn in Figure 1.2.13, with the dashed line representing the horizontal asymptote.

EXAMPLE 1.2.10

Sketch the graph of the rational function $f(x) = \dfrac{x}{x - 1}$.

SOLUTION   Since the numerator and denominator have the same degree and the same leading coefficient,

$$\lim_{x \to \infty} f(x) = 1 \quad \text{and} \quad \lim_{x \to -\infty} f(x) = 1,$$

and the line $y = 1$ is a horizontal asymptote. Moreover, the numerator is larger than the denominator if $x > 1$, which means that the graph approaches the asymptote from above as $x \to \infty$. On the other hand, as $x \to -\infty$, the graph approaches the asymptote from below. That is because dividing both sides of the inequality $x - 1 < x$ by $x - 1$ when x is negative leads to the inequality

$$1 > \frac{x}{x - 1}.$$

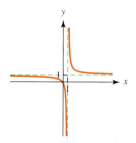

Figure 1.2.14

(*Reminder:* Dividing or multiplying an inequality by a negative number *reverses* the inequality.)

The function has a vertical asymptote at $x = 1$, and a simple sign check shows that

- $f(x) > 0$ for $x > 1$,
- $f(x) < 0$ for $0 < x < 1$, and
- $f(x) > 0$ for $x < 0$.

The graph crosses the x-axis, passing from positive to negative, at $x = 0$.
Putting these observations together leads to the graph shown in Figure 1.2.14.

The next example illustrates the economic principle of diminishing returns.

EXAMPLE 1.2.11

A company estimates that when it spends a million dollars to advertise its product, its annual revenue R, in millions of dollars, is modeled by the function $R(a) = 500 - \frac{1,000}{a+4}$.

(i) Compute $\lim_{a \to 0} R(a)$ and $\lim_{a \to \infty} R(a)$, and draw the graph of $R(a)$.

(ii) If the company is currently spending \$30 million on advertising, would you recommend increasing it to \$40 million?

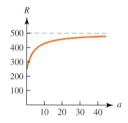

Figure 1.2.15

SOLUTION (i) We have

$$\lim_{a \to 0} R(a) = 500 - \frac{1,000}{0+4} = 250 \quad \text{and} \quad \lim_{a \to \infty} R(a) = 500 - 0 = 500.$$

Observing that $R(a)$ is increasing, $R < 500$, and $R = 500$ is a horizontal asymptote, we draw the graph of $R(a)$ as shown in Figure 1.2.15.

(ii) Since $R(30) \approx 470.59$ and $R(40) \approx 477.27$, we conclude that it is not profitable for the company to increase its advertising budget to \$40 million since $R(40) - R(30) \approx 6.69$, which is less than the additional \$10 million spent on advertising.

Practice Exercises 1.2

1. Find $\lim_{x \to 0+} \dfrac{x-1}{\sqrt{x}}$.

2. Find the limits of $f(x)$ as $x \to \pm\infty$ and determine whether there are any horizontal or vertical asymptotes for the rational function

$$f(x) = \frac{-9x^5 + 5x^2 + 6x - 1}{(x-1)(3x^4 + 1)}.$$

Exercises 1.2

Exercises 1–3 refer to the function f, whose graph is shown in Figure 1.2.16.

1. Compute the following limits and values:
   (a) $\lim_{x \to -1^-} f(x)$ (b) $\lim_{x \to -1^+} f(x)$ (c) $\lim_{x \to 0^-} f(x)$
   (d) $\lim_{x \to 0^+} f(x)$ (e) $f(-1)$

2. Compute the the following limits and values:
   (a) $\lim_{x \to 1} f(x)$ (b) $\lim_{x \to 2^-} f(x)$ (c) $\lim_{x \to 2^+} f(x)$
   (d) $\lim_{x \to \infty} f(x)$ (e) $f(0)$ and $f(1)$

3. Find all x at which the function f has no limit. Explain.

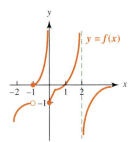

Figure 1.2.16

Exercises 4–6 refer to the function g, whose graph is shown in Figure 1.2.17.

4. Compute the following limits and values:

   (a) $\lim_{x \to -\infty} g(x)$ **(b)** $\lim_{x \to -1^-} g(x)$ **(c)** $\lim_{x \to -1^+} g(x)$

   (d) $\lim_{x \to 0} g(x)$ **(e)** $g(-1)$

5. Compute the the following limits and values:

   (a) $\lim_{x \to 1^-} g(x)$ **(b)** $\lim_{x \to 1^+} g(x)$ **(c)** $\lim_{x \to 2} g(x)$

   (d) $\lim_{x \to \infty} g(x)$ **(e)** $g(2)$

6. Find all x at which the function g has no limit. Explain.

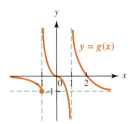

Figure 1.2.17

In Exercises 7–18, find the one-sided limit, which may be ∞ or $-\infty$ or a real number.

7. $\lim_{x \to 5^+} \dfrac{1}{(x-5)^2}$ **8.** $\lim_{x \to 5^-} \dfrac{1}{(x-5)^2}$

9. $\lim_{x \to 5^+} \dfrac{1}{x-5}$ **10.** $\lim_{x \to 5^-} \dfrac{1}{x-5}$

11. $\lim_{x \to 5^+} \dfrac{x}{x-5}$ **12.** $\lim_{x \to 5^-} \dfrac{x}{x-5}$

13. $\lim_{x \to 2^-} \dfrac{x}{x^2-4}$ **14.** $\lim_{x \to 2^+} \dfrac{x}{x^2-4}$

15. $\lim_{x \to -2^-} \dfrac{x}{x^2-4}$ **16.** $\lim_{x \to -2^+} \dfrac{x}{x^2-4}$

17. $\lim_{t \to 2^+} \dfrac{\sqrt{t-2}}{t^2-4}$ **18.** $\lim_{t \to -2^+} \dfrac{t^2+3t+2}{\sqrt{t+2}}$

In Exercises 19–30, find the given limit.

19. $\lim_{x \to \infty} \dfrac{2x-1}{5x+1}$ **20.** $\lim_{x \to -\infty} \dfrac{x+3}{4x-1}$

21. $\lim_{x \to \infty} \dfrac{x^2}{x^2+1}$ **22.** $\lim_{x \to \infty} \dfrac{x^3}{x^2+1}$

23. $\lim_{x \to -\infty} \dfrac{x^3}{x^2+1}$ **24.** $\lim_{x \to -\infty} \dfrac{x^2}{x^3+1}$

25. $\lim_{x \to \infty} \dfrac{5x^2+1}{x^3+x+2}$ **26.** $\lim_{x \to \infty} \dfrac{x^4}{10x^3+7x^2+5}$

27. $\lim_{x \to -\infty} \dfrac{9x+10}{x^2+1}$ **28.** $\lim_{x \to \infty} \dfrac{x^2+2}{2x^2+1}$

29. $\lim_{x \to -\infty} \dfrac{x^2+2}{2x^2+1}$ **30.** $\lim_{x \to \infty} \dfrac{x^4}{x^4+x^2+1}$

31. The cost of manufacturing x number of music CDs is $C(x) = 0.17x + 100$ dollars and the average cost per CD is $C(x)/x$.

   (a) What happens to the cost as x becomes very large?

   (b) Compare that to the average cost as x becomes very large.

32. Biologists studying whale migration patterns calculate that for the last 25 years, the number of whales that swim past a certain peninsula each year is approximately modeled by

$$W(y) = 200 + \frac{1{,}000}{y+1},$$

where y is the time in years. Use limits to predict the number of whales that will swim past the specified peninsula many years from now.

33. A group of 21 rabbits migrate to a large meadow. Assume that the function that best approximates their numbers over time is

$$R(t) = \frac{210}{10-t},$$

where t is in years. Graph the function $R(t)$ to study the rabbit population as time goes on. How large can this population become as $t \to 10$?

34. Suppose that the model

$$p(t) = \frac{1,200t}{t + 6}$$

describes the size of the population of a new fishery at any time t.

(a) Compute the population at $t = 0, 5$, and 100.

(b) What happens to the population as t becomes very large?

(c) Use a graphing utility to sketch the graph of $p(t)$ and the line $p = 1,200$ together.

35. Suppose that the demand curve for a certain commodity is modeled by

$$p = \frac{2,400}{q + 2},$$

where p is the price and q is the quantity.

(a) Compute the revenue R.

(b) Compute $\lim_{q \to 0^+} R(q)$, $\lim_{q \to 10} R(q)$, and $\lim_{q \to \infty} R(q)$.

(c) How large can the revenue become?

In Exercises 36–47, find any vertical and horizontal asymptotes and try to determine where the function is negative or positive and how the graph approaches the asymptote (rising or falling for vertical, from above or below for horizontal). Make a rough sketch of the graph using that information.

36. $y = \dfrac{1}{x + 3}$ **37.** $y = \dfrac{2x + 1}{x - 2}$

38. $y = \dfrac{4}{x^2 - 4}$ **39.** $y = \dfrac{3x}{x^2 - 4}$

40. $y = \dfrac{1}{x^2 + 1}$ **41.** $y = \dfrac{x^2}{x^2 + 1}$

42. $y = \dfrac{2x^2}{x - 2}$ **43.** $y = \dfrac{x^2 - 9}{x^2 - 1}$

44. $y = \dfrac{x^4}{x^4 + x^2 + 1}$ **45.** $y = \dfrac{1}{x^2 - 2x - 3}$

46. $y = \dfrac{x^3}{x^2 - 1}$ **47.** $y = \dfrac{x^3}{x^3 - 1}$

In Exercises 48–51, find the equation of the slant asymptote to the graph of the given function.

48. $y = \dfrac{x^2 - 2x + 1}{x + 5}$ **49.** $y = \dfrac{x^3 - 5x^2 + 1}{3x^2 + 7}$

50. $y = \dfrac{3x^5 + 5x^3 + 3x - 2}{x^4 + x^2 + 1}$

51. $y = 2x + \dfrac{1}{x}$

In Exercises 52–60, use a calculator or spreadsheet to construct an appropriate table of values and then use it to determine if the limit exists (either as a finite or infinite limit) and to estimate it.

52. $\lim\limits_{t \to \infty} \dfrac{\sqrt{t}}{\sqrt{1 + 4t}}$ **53.** $\lim\limits_{x \to \infty} \dfrac{\sqrt{x^2 + 9}}{x + 3}$

54. $\lim\limits_{x \to -\infty} \dfrac{\sqrt{x^2 + 9}}{x + 3}$ **55.** $\lim\limits_{t \to \infty} t - \sqrt{t}$

56. $\lim\limits_{t \to \infty} \dfrac{t - \sqrt{t}}{t + \sqrt{t}}$ **57.** $\lim\limits_{x \to \infty} \dfrac{x\sqrt{x} + 1}{x + \sqrt{x} + 1}$

58. $\lim\limits_{x \to \infty} \dfrac{\sqrt[3]{x} + 1}{\sqrt{x}}$ **59.** $\lim\limits_{x \to \infty} \dfrac{\sqrt{x^4 + 1}}{1 - x}$

60. $\lim\limits_{x \to -\infty} \dfrac{\sqrt{x^4 + 1}}{1 - x}$

Use a calculator to graph each of the following. Choose an appropriate viewing window, making sure to include any vertical asymptotes and to indicate the limits at infinity.

61. $y = \dfrac{x + 1}{x + 2}$ **62.** $y = \dfrac{3x}{x^2 + 4}$

63. $y = \dfrac{2x}{x^2 - 4}$ **64.** $y = \dfrac{x^2 - 1}{x^2 + 1}$

65. $y = \dfrac{x^2}{x + 2}$ **66.** $y = \dfrac{x^3}{x^2 - 1}$

67. Neglecting air resistance, the initial velocity v required for a rocket projected upward from the earth's surface to reach a given maximum height h is given by

$$v = \sqrt{2gR\,\frac{h}{h + R}},$$

where g is the gravitational constant (which is 9.8 meters per second squared) and R is the radius of the earth (which is 6.38×10^6 meters at the equator). Compute the velocity required so that the rocket does not return to the earth, which is called the escape velocity.

Solutions to practice exercises 1.2

1. Because of the square root in the denominator, the function is only defined for $x > 0$. The numerator is negative for x near zero, and the denominator is positive, so that the quotient is negative. Moreover, $\lim_{x \to 0^+} (x - 1) = -1$ and $\lim_{x \to 0^+} \sqrt{x} = 0$. Therefore, $\lim_{x \to 0^+} (x - 1)/\sqrt{x} = -\infty$.

2. Replacing the numerator and denominator by their highest-degree terms gives

$$\lim_{x\to\pm\infty} \frac{-9x^5 + 5x^2 + 6x - 1}{(x - 1)(3x^4 + 1)} = \lim_{x\to\pm\infty} \frac{-9x^5}{3x^5} = -3.$$

Thus, the line $y = -3$ is a horizontal asymptote. Looking for a vertical asymptote, we observe that the domain of $f(x)$ consists of all $x \neq 1$, the quotient is in lowest terms, the denominator is zero only at $x = 1$, and the numerator is not. Therefore, the line $x = 1$ is a vertical asymptote.

1.3   Continuity

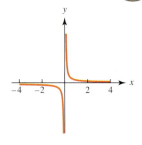

Figure 1.3.1

As we have already seen, the graph of a rational function has a "break" or "jump" at a vertical asymptote. We say that the function is **discontinuous** or has a **discontinuity** wherever such a break occurs. For instance, $f(x) = 1/x$ has a discontinuity at $x = 0$, as seen in Figure 1.3.1.

Roughly speaking, a function is **continuous** if you can draw its graph without lifting your pencil from the paper, and it is discontinuous otherwise. The discontinuities of the function occur where we have to lift our pencil. As we have just noted, that happens when a rational function has a vertical asymptote, but it can occur in other ways as well. Figure 1.3.2 shows the graph of a function on the interval $[0, 9]$ with discontinuities at the points 0, 1, 3, 5, and 7 and continuous everywhere else in the interval.

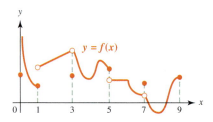

Figure 1.3.2

We can give a precise definition of continuity using the notion of a limit.

Definition 1.3.1

Definition of Continuity

A function $f(x)$ is said to be continuous at $x = a$ if

1. the number a is in the domain of the function, and
2. $\lim_{x\to a} f(x) = f(a)$.

Among other things, this says that if $f(x)$ is continuous at $x = a$, then $f(x) \approx f(a)$ for all x near a.

EXAMPLE 1.3.1

Determine where the following functions are continuous:

(i) $f(x) = x^2$,   (ii) $f(x) = |x|$,   (iii) $f(x) = \begin{cases} \dfrac{|x|}{x} & \text{if } x \neq 0 \\ 0 & \text{if } x = 0. \end{cases}$

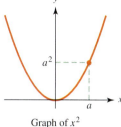

Graph of x^2

Figure 1.3.3

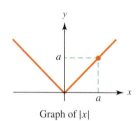

Graph of $|x|$

Figure 1.3.4

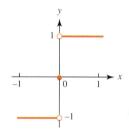

Figure 1.3.5

SOLUTION (i) The domain of $f(x)$ consists of all real numbers, and, as we saw in Section 1.1, $\lim_{x \to a} x^2 = a^2$ for every a (see Figure 1.3.3). Therefore, the function is continuous throughout the x-axis.

(ii) By the definition of $|x|$, we have $f(x) = x$ if $x > 0$, and $f(x) = -x$ if $x < 0$. If $a \neq 0$, $\lim_{x \to a} |x| = |a|$. Moreover, $|0| = 0$ and $\lim_{x \to 0} |x| = |0| = 0$. Therefore, $f(x) = |x|$ is continuous everywhere. (See Figure 1.3.4.)

(iii) Again using the definition of $|x|$, we see that

$$f(x) = \frac{x}{x} = 1 \quad \text{if } x > 0 \quad \text{and} \quad f(x) = \frac{-x}{x} = -1 \quad \text{if } x < 0.$$

In addition, $f(0) = 0$. It follows that the graph consists of two horizontal half-lines—one at height 1 in the right half of the plane, the other at height -1 on the left—and an isolated point at the origin. Since

$$\lim_{x \to 0^+} f(x) = 1 \quad \text{and} \quad \lim_{x \to 0^-} f(x) = -1,$$

we see that $\lim_{x \to 0} f(x) \neq 0$. In fact, the limit does not exist, since the two one-sided limits are unequal. Therefore, the function has a discontinuity at $x = 0$. As Figure 1.3.5 shows, it is continuous everywhere else.

A function may be discontinuous at $x = a$ even though $\lim_{x \to a}$ exists, as the next example shows.

EXAMPLE 1.3.2

Find the points, if any, where each of the following functions is discontinuous:

$$\text{(i)} \quad f(x) = \begin{cases} \dfrac{x^2 - 1}{x - 1} & \text{if } x \neq 1 \\ 1 & \text{if } x = 1 \end{cases} \quad \text{and} \quad \text{(ii)} \quad f(x) = \begin{cases} \dfrac{x^2 - 1}{x - 1} & \text{if } x \neq 1 \\ 2 & \text{if } x = 1. \end{cases}$$

SOLUTION Both functions are defined for every x, and the limit as $x \to 1$ is the same in both cases:

$$\lim_{x \to 1} f(x) = \lim_{x \to 1} \frac{x^2 - 1}{x - 1} = \lim_{x \to 1} \frac{(x - 1)(x + 1)}{x - 1} = \lim_{x \to 1} (x + 1) = 2.$$

(i) In this case, $f(1) = 1 \neq \lim_{x \to 1} f(x)$. Therefore, the function is discontinuous at $x = 1$. At every other point the function agrees with the rational function $(x^2 - 1)/(x - 1)$ and is, therefore, continuous (see Theorem 1.3.1 below).

The graph of this function is shown in Figure 1.3.6. It consists of the line $y = x + 1$ for $x \neq 1$ (the "hole" in the line indicates the missing point) and the isolated point $(1, 1)$.

(ii) In this case, $f(1) = 2 = \lim_{x \to 1} f(x)$. Therefore, the function is continuous at $x = 1$, as well as at every other x. Its graph is shown in Figure 1.3.7. The hole at $(1, 2)$ has been filled in.

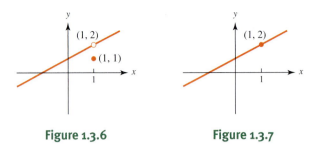

Figure 1.3.6   **Figure 1.3.7**

Functions that involve more than a single-line formula, such as those in the previous example, sometimes arise in applications.

EXAMPLE 1.3.3

The cost of sending an airmail letter from the United States to Canada in 2001 was 60 cents for the first ounce and 25 cents for each additional ounce (or fraction thereof), up to 5 ounces. If we let x stand for the weight (in ounces) and $f(x)$ stand for the cost (in dollars), then the cost is given as a function of the weight by the following multiline formula for weights up to 5 ounces:

$$f(x) = \begin{cases} 0.60 & \text{if } 0 < x \le 1 \\ 0.85 & \text{if } 1 < x \le 2 \\ 1.10 & \text{if } 2 < x \le 3 \\ 1.35 & \text{if } 3 < x \le 4 \\ 1.60 & \text{if } 4 < x \le 5. \end{cases} \tag{10}$$

Draw the graph of this function and find its discontinuities.

SOLUTION   The graph for $x > 0$ is shown in Figure 1.3.8. As you can see, it is composed of five disconnected horizontal line segments. The function is discontinuous where x equals 1, 2, 3, and 4. At $x = 1$, for instance, we have $f(1) = 0.60$ and

$$\lim_{x \to 1^-} f(x) = 0.60 \quad \text{and} \quad \lim_{x \to 1^+} f(x) = 0.85.$$

You can see that either by looking at the graph or reading the formulas defining the function. We conclude that there is no limit as $x \to 1$, only two different *one-sided*

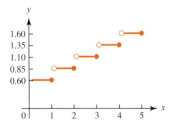

Figure 1.3.8

limits. Therefore, $f(x)$ is discontinuous at $x = 1$. Similarly, there is no limit at $x = 2$, 3, or 4. At all other points, $f(x)$ is continuous. For example, $f(1.25) = 0.85$ and $\lim_{x \to 1.25} f(x) = 0.85$. Thus, f is continuous at $x = 1.25$.

Next, we state several continuity properties that follow from the corresponding limit properties of the last section.

Theorem 1.3.1

If f and g are two functions, both continuous at a point a, and if c is a constant, then the following functions are also continuous at a:

1. the constant multiple: $cf(x)$
2. the sum of the functions: $f(x) + g(x)$
3. the product of the functions: $f(x) \cdot g(x)$
4. the quotient of the functions: $f(x)/g(x)$, provided $g(a) \neq 0$.

The following statements are consequences of this theorem:

*   Polynomial functions are continuous everywhere.
*   A rational function is continuous at every point in its domain.

The intermediate value theorem and zeros of functions

Because a continuous function has no breaks or jumps, it takes on every value between any two of its values. This important property is known as the **intermediate value property**. To be more precise, let's consider a function that is continuous on a closed interval $[a, b]$, by which we mean that it is continuous at every point inside the interval (a, b), and at the endpoints it has one-sided limits with

$$\lim_{x \to a^+} f(x) = f(a) \quad \text{and} \quad \lim_{x \to b^-} f(x) = f(b).$$

The following theorem is an exact statement of the intermediate value property.

Theorem 1.3.2

The intermediate value theorem

If $f(x)$ is a continuous function on a closed interval, $[a, b]$, and if k is any number between $f(a)$ and $f(b)$, then there is at least one number c in $[a, b]$ such that $f(c) = k$.

The theorem is illustrated in Figure 1.3.9. A precise proof is rather technical, but an intuitive geometric explanation is this: If you can draw the graph without lifting your pencil from the paper (our informal criterion for continuity), then the graph must cross every horizontal line whose height is between $f(a)$ and $f(b)$.

The following result is an important application of the intermediate value theorem.

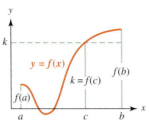

Figure 1.3.9

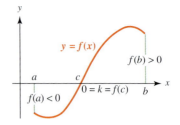

Figure 1.3.10

Theorem 1.3.3

Existence of zeros

If $f(x)$ is a continuous function on a closed interval $[a, b]$, and if $f(a)$ and $f(b)$ have opposite signs, that is, $f(a) \cdot f(b) < 0$, then there is at least one number c in (a, b) such that $f(c) = 0$.

This is a special case of the intermediate value theorem, for if $f(a)$ and $f(b)$ have opposite signs, then zero is between $f(a)$ and $f(b)$. Therefore, by the intermediate value theorem there is a c with $f(c) = 0$. This is illustrated in Figure 1.3.10.

The bisection method

The existence of zeros theorem gives us a simple method for determining whether a continuous function is zero inside an interval and for approximating the root. First, if $f(a) \cdot f(b) < 0$ for two numbers a and b, then there is a root inside the interval $[a, b]$. To approximate it, we first bisect the interval into two equal segments and keep the segment at whose endpoints $f(x)$ has opposite signs. Using that segment as our new interval, we repeat the process and continue bisecting until we have reached a satisfactory approximation.

This method, known as the **bisection method**, is illustrated in the following example.

EXAMPLE 1.3.4

Show that the function $f(x) = x^5 - 2x^4 + x^3 - 4x^2 + 2x + 1$ has a zero in the interval $(0, 1)$ and approximate this zero by the bisection method.

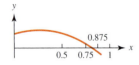

Figure 1.3.11

SOLUTION   Since $f(0) = 1$ and $f(1) = -1$, the existence of zeros theorem says there must be a point c in the interval $(0, 1)$ for which $f(c) = 0$. The graph is shown in Figure 1.3.11.

To use the bisection method, we proceed as follows.

• The midpoint of the interval is at $x = 0.5$, and we can check that $f(0.5) = 1.03125 > 0$. Therefore, $f(0.5) \cdot f(1) < 0$, so we replace the original interval by the segment $[0.5, 1]$.

• The midpoint of this new interval is at $x = 0.75$, and we can check that $f(0.75) = 0.276367 > 0$. Therefore, $f(0.75) \cdot f(1) < 0$, and we take $[0.75, 1]$ as our new interval.

- The midpoint of this new interval is at $x = 0.875$, and we can check that $f(0.875) = -0.302032 < 0$. Therefore, $f(0.875) \cdot f(0.75) < 0$, and we take $[0.75, 0.875]$ as our new interval.

The next step, as you should check, leads to the interval $[0.8125, 0.875]$, and the one after that to $[0.8125, 0.84375]$. In general, at each step we have an interval, say, I, and we replace it by a new interval as follows:

- Find the midpoint m of I.
- Compare the sign of $f(m)$ with the signs of $f(x)$ at the endpoints of I.
- Keep the endpoint of I at which $f(x)$ has a different sign from $f(m)$.
- Form a new interval bounded by that endpoint and m.

By choosing the successive intervals this way, we ensure that each of them contains a root of the function. In the present case, then, we can conclude that there is a root in the interval $[0.8125, 0.84375]$. The simplest approximation is to use the midpoint of that interval, $c \approx 0.828125$.

How close is that approximation? The distance between c and the midpoint is no bigger than half the size of the interval, which in this case amounts to $(0.84375 - 0.81250)/2 = 0.015625$. Of course, we can keep bisecting if we want a closer approximation.

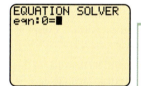

Figure 1.3.12

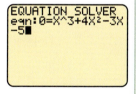

Figure 1.3.13

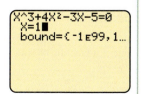

Figure 1.3.14

APPLYING TECHNOLOGY

Finding roots of functions The bisection method is a simple way to approximate the solution of an equation of the form $f(x) = 0$, provided we know an interval $[a, b]$ on which the function is continuous and with $f(a) \cdot f(b) < 0$. In addition, as we saw in Example 1.3.4, the method also gives a bound on how close our estimate is to the actual root.

However, a defect of the bisection method is that it requires many calculations. To put it another way, the estimates converge very slowly toward the root, so that the process must be repeated many times in order to get a high degree of accuracy. For that reason, more sophisticated methods, using calculus techniques, have been devised, and the better ones have been incorporated into computer packages and also into hand calculator routines. (All of these methods are variations on a procedure known as the *Newton–Raphson method*, which is included in the projects following Chapter 3 of this book.)

A standard graphing calculator has an equation solver, such as the one shown in Figure 1.3.12. The exact procedure for using it depends on the calculator model, but the basic features are the same. The equation to be solved must be put in the form $0 = f(x)$, as shown in Figure 1.3.13, in which $f(x) = x^3 + 4x^2 - 3x - 5$.

Every standard method for finding (actually, approximating) a solution requires an initial guess, either supplied by the user or by default. The guess does not have to be very accurate, but if it is too far away from the actual solution, the process may be very slow or it may even come up with a different root from the one we are looking for. Figure 1.3.14 shows a typical solver equation screen after the function and an initial guess have been entered. The choice of $x = 1$ was based on the observation that $f(0) < 0$ and $f(2) > 0$,

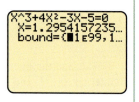

Figure 1.3.15

which means that there is a solution in the interval [0, 2], and the midpoint of that interval seemed a reasonable place to start.

In addition to an initial guess, the program must be provided with an interval in which to search for a root. The default settings are -10^{99} and 10^{99} (in scientific notation, $-1E99$ and $1E99$, as partially shown in Figure 1.3.14). Resetting them, say, to 0 and 2, would speed the process in this case, but not significantly, and we have therefore left them at the default settings.

The solution, $x = 1.2954157235\ldots$, is shown in Figure 1.3.15. A different initial guess may lead to a different solution. In this case, the function has three distinct roots, and you can find all of them by suitable choices of the initial guess (see the exercises below).

Practice Exercises 1.3

● ●

1. Let $f(x) = \begin{cases} 3 + 2x & \text{if } x < 0 \\ 3 - x^2 & \text{if } 0 \le x \le 1 \\ x & \text{if } 1 < x < 2 \\ x^3 - 6 & \text{if } 2 \le x. \end{cases}$

Where does $f(x)$ have discontinuities, if any?

2. Compute the values of $f(x) = x^3 + 2x^2 - 3x - 1$ for x from -3 to 3 in steps of size 1. How many solutions can you be sure the equation $f(x) = 0$ has in the interval $(-3, 3)$? Explain.

Exercises 1.3

● ●

Exercises 1–8 refer to the function f, whose graph is shown in Figure 1.3.16. Explain each of your answers.

1. Find all the discontinuities of the function f in the interval $[-2, 9]$.

2. Find all the discontinuities of the function f in the interval $[0, 7]$.

3. Find all the discontinuities of the function f in the interval $[2, 3]$.

4. Find all x in $[-2, 9]$ at which the function f has no limit.

5. Find all x in $[0, 5]$ at which the function f has no limit.

6. Find all x in $[-2, 9]$ at which the function f has a limit, but it is different from its value there.

7. Find all x in $[-2, 9]$ at which the limit of f is smaller than its value.

8. Find all x in $[-2, 9]$ at which the limit of f is bigger than its value.

9. Draw the graph of a function defined on the interval $[0, 4]$ that is discontinuous only at three points and has no limit only at two points.

10. Draw the graph of a function $f(x)$, which is defined on the interval $[-1, 5]$, is continuous everywhere except at $x = 0, 1, 2, 3, 4$, and has no limit at $x = 0, 3$.

11. The "discount rate" is the rate charged by branches of the Federal Reserve Bank when they extend credit to depository institutions, like banks. The following table lists the discount rates from April 18 to October 10, 2001:

Period	4/18–5/16	5/17–6/27	6/28– 8/21
Days	(0, 29]	(29, 71]	(71, 126]
Rate	4.0	3.5	3.25
Period	8/22–9/16	9/17–10/02	10/03–10/10
Days	(126, 152]	(152, 168]	(168, 176]
Rate	3.0	2.5	2.0

Sketch the graph of the discount rate as a function of time and find the points of discontinuity.

In Exercises 12–21, determine where the function is continuous and where it is discontinuous.

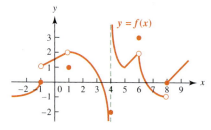

Figure 1.3.16

12. $f(x) = x^4 - \frac{1}{2}x^3 + 3x^2 - \frac{5}{7}x - \frac{3}{8}$

13. $f(x) = \dfrac{x+1}{x-1}$ **14.** $f(x) = \dfrac{x}{x^2-1}$

15. $f(x) = \dfrac{x}{x^2+1}$

16. $f(x) = \begin{cases} \dfrac{x^2-4}{x+2} & \text{if } x \neq -2 \\ 0 & \text{if } x = -2 \end{cases}$

17. $f(x) = \begin{cases} \dfrac{x^2-4}{x+2} & \text{if } x \neq -2 \\ -4 & \text{if } x = -2 \end{cases}$

18. $f(x) = \begin{cases} \dfrac{x}{|x|} & \text{if } x \neq 0 \\ 1 & \text{if } x = 0 \end{cases}$

19. $f(x) = \begin{cases} \dfrac{x}{|x|} & \text{if } x \neq 0 \\ -1 & \text{if } x = 0 \end{cases}$

20. $f(x) = \begin{cases} \dfrac{x}{|x|} & \text{if } x \neq 0 \\ 0 & \text{if } x = 0 \end{cases}$

21. $f(x) = \begin{cases} \dfrac{x^2}{|x|} & \text{if } x \neq 0 \\ 0 & \text{if } x = 0 \end{cases}$

22. Let $f(x) = \begin{cases} x^2 & \text{if } x \leq 2 \\ 2x & \text{if } 2 < x < 4 \\ \sqrt{x} & \text{if } x \geq 4. \end{cases}$

   (a) Compute $f(1)$, $f(2)$, $f(3)$, $f(4)$, and $f(5)$.
   (b) At what points, if any, is $f(x)$ discontinuous?

23. Let $f(x) = \begin{cases} x+2 & \text{if } x \leq 1 \\ \dfrac{1}{x} & \text{if } 1 < x \leq 2 \\ \sqrt{x} & \text{if } x > 2. \end{cases}$

   (a) Compute $f(0)$, $f(1)$, $f(2)$, and $f(4)$.
   (b) At what points, if any, is $f(x)$ discontinuous?

24. Let $f(x) = \begin{cases} x+1 & \text{if } x < 0 \\ x-1 & \text{if } 0 \leq x < 2 \\ \dfrac{x}{2} & \text{if } x \geq 2. \end{cases}$

   (a) Compute $f(-2)$, $f(1)$, and $f(3)$.
   (b) At what points, if any, is $f(x)$ discontinuous?

25. Let $f(x) = \begin{cases} \sqrt{x} & \text{if } 0 \leq x \leq 4 \\ x-2 & \text{if } 4 < x \leq 8 \\ x^{3/2} & \text{if } x > 8. \end{cases}$

   (a) Compute $f(1)$, $f(4)$, $f(5)$, $f(8)$, and $f(9)$.
   (b) At what points, if any, is $f(x)$ discontinuous?

26. Let $f(x) = \begin{cases} \dfrac{x}{x-1} & \text{if } x \neq 1 \\ 0 & \text{if } x = 1. \end{cases}$

   (a) Compute $f(2)$, $f(0)$, $f(-1)$, and $f(1)$.
   (b) At what points, if any, is $f(x)$ discontinuous?

27. Let $f(x) = \begin{cases} \dfrac{x^2-3x+2}{x-2} & \text{if } x \neq 2 \\ 0 & \text{if } x = 2. \end{cases}$

   At what points, if any, is $f(x)$ discontinuous?

28. Let $f(x) = \begin{cases} \dfrac{x^2+x-2}{x-1} & \text{if } x \neq 1 \\ 3 & \text{if } x = 1. \end{cases}$

   At what points, if any, is $f(x)$ discontinuous?

In Exercises 29–34, find the number c that makes $f(x)$ continuous for every x.

29. $f(x) = \begin{cases} \dfrac{x^2-3x}{x-3} & \text{if } x \neq 3 \\ c & \text{if } x = 3 \end{cases}$

30. $f(x) = \begin{cases} \dfrac{x^3-8}{x-2} & \text{if } x \neq 2 \\ c & \text{if } x = 2 \end{cases}$

31. $f(x) = \begin{cases} x^3 - 4x^2 + 3x - 1 & \text{if } x \neq 2 \\ c & \text{if } x = 2 \end{cases}$

32. $f(x) = \begin{cases} \dfrac{(x-1)^2}{x^2+1} & \text{if } x \neq 1 \\ c & \text{if } x = 1 \end{cases}$

33. $f(x) = \begin{cases} \dfrac{x^4-1}{x^3-1} & \text{if } x \neq 1 \\ c & \text{if } x = 1 \end{cases}$

34. $f(x) = \begin{cases} \dfrac{x^2+x-2}{x+2} & \text{if } x \neq -2 \\ c & \text{if } x = -2 \end{cases}$

In Exercises 35–38, extend the definition of the function to make it continuous at $h = 0$.

35. $f(h) = \dfrac{(1-h)^2 - 1}{h}$, $h \neq 0$

36. $g(h) = \dfrac{(1+h)^3 - 1}{h}$, $h \neq 0$

37. $f(h) = \dfrac{h^2}{(1+h)^2 - 1}$, $h \neq 0$

38. $g(h) = \dfrac{\dfrac{1}{2+h} - \dfrac{1}{2}}{h}$, $h \neq 0$

39. A salesman is paid a fixed weekly salary of $500. If his weekly sales exceed $5,000 he is paid a bonus of $250. If they exceed $7,500, the bonus is raised to $500. However, once they reach $10,000, the salary and bonus are replaced by a commission of 10% of the weekly sales. Sketch a graph of the salesman's income and find the points of discontinuity.

Figure 1.3.17 shows the graph of a continuous function $f(x)$ on the interval $[-3, 5]$. Use it to answer the questions of Exercises 40–45.

40. Does the function f take the value 1? How many times?

41. Does the function f take the value -1? How many times?

42. Which one of the numbers $-9, -2, 2, 3, 4$, and 5 does the function f take as values? How many times?

43. How many zeros does the function f have in the interval $[-3, 5]$?

44. Does the function $f(x) + 1$ have any zeros in the interval $[-3, 5]$? Explain.

45. Does the function $f(x) + 2$ have any zeros in the interval $[-3, 5]$? Explain.

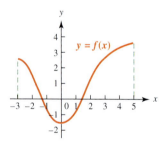

Figure 1.3.17

46. Does the function g in Figure 1.3.18 take the value 1 in the interval $[0, 4]$? Does the intermediate value theorem apply? Explain.

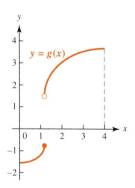

Figure 1.3.18

47. Does the function g in Figure 1.3.18 have any zeros in the interval $[0, 4]$? Does the existence of zeros theorem apply? Explain.

48. Suppose a continuous function $f(x)$ satisfies the following table of values:

x	0	0.5	1	1.5	2	2.5	3	3.5	4
$f(x)$	3.1	1.8	-0.4	0.6	1.2	0.2	-1.1	-0.9	0.1

How many roots can you be sure of $f(x)$ having on the interval $(0, 4)$, and where are they located?

49. Suppose a continuous function $f(x)$ satisfies the following table of values:

x	-4	-3	-2	-1	0	1	2	3	4
$f(x)$	-2	-3	-2	-1	1	2	1	-1	-2

How many roots can you be sure of $f(x)$ having on the interval $(-4, 4)$, and where are they located?

50. In which of the following intervals can you be sure that the equation $6x^4 + x^3 - 4x^2 - 5 = 0$ has a solution? (There may be more than one.)
(a) $[-2, -1]$ **(b)** $[-1, 0]$ **(c)** $[0, 1]$ **(d)** $[1, 2]$

51. In which of the following intervals can you be sure that the function

$$f(x) = x^4 + 2x^3 - 3x^2 - 2x + 1$$

has a root? (There may be more than one.)
(a) $[-3, -2]$ **(b)** $[-2, -1]$ **(c)** $[-1, 0]$ **(d)** $[0, 1]$
(e) $[1, 2]$ **(f)** $[2, 3]$

52. The demand and supply curves for a commodity are

$$D(q) = 0.0001q^4 + 0.01q^2 + q + 2$$
$$\text{and} \quad S(q) = -q + 10.$$

Explain why there must be an equilibrium point in the interval $0 \le q \le 8$ and then estimate the equilibrium price and quantity.

53. Estimate the equilibrium price and quantity of an item with demand and supply curves given by

$$D(q) = \frac{10}{\sqrt{q^2 + 1}} \quad \text{and} \quad S(q) = 0.04q^2 + 1.$$

In Exercises 54–57, approximate the root of $f(x)$ in the given interval by using the bisection method with a calculator. Stop after n bisections and estimate how close your approximation is to the actual root.

54. $f(x) = x^2 - 7$ in $[2, 3]$ with $n = 4$

55. $f(x) = x^3 - 5$ in $[1, 2]$ with $n = 5$

56. $f(x) = x^3 + x - 3$ in $[0, 2]$ with $n = 5$

57. $f(x) = x^3 - 3x^2 + 2x - 3$ in $[0, 3]$ with $n = 5$

In Exercises 58–61, use the equation solver on a graphing calculator to find a root of the given function in the given interval.

58. $f(x) = x^3 - 3x^2 - 2x + 3$ in $[-2, 0]$

59. $f(x) = x^3 - 3x^2 - 2x + 3$ in $[0, 2]$

60. $f(x) = x^5 - 3x^4 + 4x^2 - 5x + 2$ in $[0, 1]$

61. $f(x) = x^5 - 3x^4 + 4x^2 - 5x + 2$ in $[-2, 0]$

62. Use a graphing calculator to find all three roots of the function $f(x) = x^3 + 4x^2 - 3x - 5$. (*Hint:* One solution, obtained using the initial guess $x = 1$, was displayed in Figure 1.3.15 of this section. In setting the other initial guesses, use the following observations: $f(0) = -5$, $f(-3) = 13$, and $f(-5) = -15$.)

63. Use a graphing calculator to find all three roots of the function $f(x) = x^3 - x^2 - 4x + 1$.

Solutions to practice exercises 1.3

1. As $3 + 2x$, $3 - x^2$, x, $x^3 - 6$ are continuous functions, $f(x)$ *might* have discontinuities only at the matching points $x = 0$, $x = 1$, and $x = 2$. Since

$$\lim_{x \to 0^-} f(x) = \lim_{x \to 0} 3 + 2x = 3 \quad \text{and}$$

$$\lim_{x \to 0^+} f(x) = \lim_{x \to 0}(3 - x^2) = 3$$

and $f(0) = 3 - 0^2 = 3$, it follows that $f(x)$ is continuous at $x = 0$. Since

$$\lim_{x \to 1^-} f(x) = \lim_{x \to 1}(3 - x^2) = 3 - 1^2 = 2 \quad \text{and}$$

$$\lim_{x \to 1^+} f(x) = \lim_{x \to 1} x = 1,$$

$\lim_{x \to 1^-} f(x) \neq \lim_{x \to 1^+} f(x)$ and $f(x)$ is discontinuous at $x = 1$. Since

$$\lim_{x \to 2^-} f(x) = \lim_{x \to 2} x = 2 \quad \text{and}$$

$$\lim_{x \to 2^+} f(x) = \lim_{x \to 2}(x^3 - 6) = 8 - 6 = 2$$

and $f(2) = 8 - 6 = 2$, it follows that $f(x)$ is continuous at $x = 2$. Therefore, $f(x)$ has only one discontinuity at $x = 1$.

2. The table of values is as follows:

x	-3	-2	-1	0	1	2	3
$f(x)$	-1	5	3	-1	-1	9	35

According to the intermediate value theorem, $f(x)$ must have a root on each of the intervals $(-3, -2)$, $(-1, 0)$, and $(1, 2)$. Therefore, we can be sure there are three solutions of $f(x) = 0$ on the interval $(-3, 3)$.

1.4 The Precise Definition of Limit

Part of the power of mathematics comes from the precision of its language. In this section we will give a precise definition of **limit**, first introduced informally in Section 1.1. Once again, let's consider a function $f(x)$ defined for all x in an interval around some given point a, except possibly at a itself, and try to give the clearest possible meaning to the expression

$$\lim_{x \to a} f(x) = L, \tag{11}$$

where L is a number.

To do that, we first introduce the notion of the **approximation error**, which is defined to be $|f(x) - L|$. It measures, for any x, how far $f(x)$ is from the limit value

L. Clearly, if $f(x)$ has L as its limit as $x \to a$, the approximation error becomes very small as x gets very close to a.

How small? The answer is: *smaller than any preassigned number.* The traditional symbol in mathematics for a small preassigned error bound is the Greek letter ε (epsilon). Using this symbol, we can explain Eq. (11) by saying this: For any positive number ε, no matter how small, we can make $|f(x) - L| < \varepsilon$ for all x that are very close to a.

How close to a? That depends on ε. In general, the smaller we choose ε, the closer to a we have to take x. The distance between x and a is measured by the absolute value $|x - a|$, and the Greek letter δ (delta) is traditionally used to symbolize the maximum size of $|x - a|$.

Using these symbols, we obtain the following definition:

Definition 1.4.1

Precise Definition of Limit

$$\lim_{x \to a} f(x) = L \text{ means that}$$

given any $\varepsilon > 0$, we can find a number $\delta > 0$ such that

$$|f(x) - L| < \varepsilon \quad \text{if} \quad 0 < |x - a| < \delta.$$

Another way to state it is this: Given any ε, we can find a δ so that

$$L - \varepsilon < f(x) < L + \varepsilon \quad \text{if} \quad a - \delta < x < a + \delta, \quad x \neq a.$$

In other words, as shown in Figure 1.4.1, we can sandwich $f(x)$ between the numbers $L - \varepsilon$ and $L + \varepsilon$ by sandwiching x between the numbers $a - \delta$ and $a + \delta$, where δ is some bound that depends on ε.

To emphasize: Given ε (the bound on the approximation error) we must find a suitable δ (the maximum distance between x and a). The determination of δ depends on the choice of ε, as the following example illustrates.

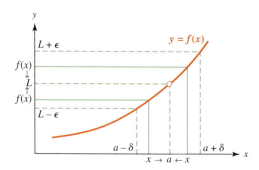

Figure 1.4.1

EXAMPLE 1.4.1

Suppose $f(x) = 2x$ and ε is a small, positive number. Find a positive number δ such that $|f(x) - 2| < \varepsilon$ if $0 < |x - 1| < \delta$.

SOLUTION We begin by observing that

$$|f(x) - 2| = |2x - 2| = |2(x - 1)| = 2|x - 1|.$$

It follows that the inequality $|f(x) - 2| < \varepsilon$ is the same as the inequality $2|x - 1| < \varepsilon$. Therefore, by setting $\delta = \varepsilon/2$, we see that $|f(x) - 2| < \varepsilon$ if $|x - 1| < \delta$.

A few remarks about the last example are in order. First, the example shows that $\lim_{x \to 1} 2x = 2$, a fact that is certainly not surprising in view of our earlier work with limits. However, the ε-δ method is more precise and gives a more accurate determination of how small $|x - 1|$ has to be in order to get a preassigned approximation error. A second remark is that in this example, we did not need to assume that $x \ne 1$. That is because the given function $f(x) = 2x$ is continuous at $x = 1$.

In general, the determination of δ may require some manipulation of inequalities.

EXAMPLE 1.4.2

Show that $\lim\limits_{x \to 3} x^2 = 9$.

SOLUTION Here $f(x) = x^2$, $a = 3$, and $L = 9$. Observe that our function is defined for all x, including 3. We must show that given any positive ε (no matter how small), we can find a positive δ so that

$$|x^2 - 9| < \varepsilon \quad \text{if} \quad 0 < |x - 3| < \delta.$$

To find an appropriate δ (given an ε), we start with the factorization

$$|x^2 - 9| = |(x + 3)(x - 3)| = |x + 3| \cdot |x - 3|.$$

Next, we observe that $(x + 3) > 0$ if x is close enough to 3, which means that $|x + 3| = x + 3$. In particular, if $2 < x < 4$, which is the same as $|x - 3| < 1$, we have $|x + 3| = x + 3 < 7$. Therefore, if $|x - 3| < 1$,

$$|x^2 - 9| = |x + 3| \cdot |x - 3| < 7|x - 3|.$$

To make $|x^2 - 9| < \varepsilon$, then, we simply have to set $\delta \le \varepsilon/7$.

Thus, by choosing δ to be any number smaller than both $\varepsilon/7$ and 1, we obtain

$$|x^2 - 9| < \varepsilon \quad \text{if} \quad |x - 3| < \delta.$$

This proves that $\lim_{x \to 3} x^2 = 9$.

As another example of the power of the ε-δ definition of the limit, we will use it to prove that if $\lim_{x \to a} f(x) = L$ and $\lim_{x \to a} g(x) = M$, then

$$\lim_{x \to a}[f(x) + g(x)] = L + M. \tag{12}$$

(See Theorem 1.1.1.)

PROOF   Let ε be a given positive number. Using the definition of limit, but with $\varepsilon/2$ as the preassigned number rather than ε, we can choose a positive number δ_1 such that

$$|f(x) - L| < \frac{\varepsilon}{2} \quad \text{if} \quad 0 < |x - a| < \delta_1. \tag{13}$$

Similarly, we can find a positive number δ_2 such that

$$|g(x) - M| < \frac{\varepsilon}{2} \quad \text{if} \quad 0 < |x - a| < \delta_2. \tag{14}$$

Now, if we choose δ to be the smaller of the numbers δ_1 and δ_2, then both statements (13) and (14) are true, with both δ_1 and δ_2 replaced by δ. That is, we have

$$|f(x) - L| < \frac{\varepsilon}{2} \quad \text{and} \quad |g(x) - L| < \frac{\varepsilon}{2} \quad \text{if} \quad 0 < |x - a| < \delta. \tag{15}$$

Using the inequality $|A + B| \leq |A| + |B|$ and (15), we conclude that if $0 < |x - a| < \delta$, then

$$\begin{aligned}
|[f(x) + g(x)] - (L + M)| &= \big|(f(x) - L) + (g(x) - M)\big| \\
&\leq |f(x) - L| + |g(x) - M| \\
&< \frac{\varepsilon}{2} + \frac{\varepsilon}{2} = \varepsilon,
\end{aligned}$$

which proves (12).

HISTORICAL PROFILE

Augustin-Louis Cauchy (1789–1857) made important contributions to the rigorous development of limits, and he played a major part in establishing the theorems of calculus on a sound, precise basis. His first job after leaving the university was in Cherbourg, France, working on port facilities for Napoleon's English invasion fleet and doing mathematical research in his spare time. Of that period, he wrote, "I get up at four o'clock each morning and I am busy from then on. I do not get tired of working, on the contrary it invigorates me." Cauchy was a staunchly religious person, whose rigid, conservative views brought him into conflict with other scientists. He was highly principled, and in 1830 his refusal to swear an oath of allegiance to the new French government caused him to lose his academic position in Paris, despite his having won the Grand Prix of the French Academy in 1816. He was denied a university position until 1848, when a new government took over. Despite these difficulties, Cauchy continued to make significant mathematical discoveries.

Exercises 1.4

In Exercises 1–7, find a δ (in terms of ε) that makes $|f(x) - L| < \varepsilon$ if $|x - a| < \delta$.

1. $f(x) = 3x$, $L = 6$, $a = 2$

2. $f(x) = 2x + 5$, $L = 3$, $a = -1$

3. $f(x) = \dfrac{x+1}{2}$, $L = 1$, $a = 1$

4. $f(x) = x^2$, $L = 4$, $a = 2$

5. $f(x) = x^2 + 1$, $L = 5$, $a = 2$

6. $f(x) = 4 - x^2$, $L = 0$, $a = 2$

7. $f(x) = x^3$, $L = 8$, $a = 2$ (*Hint:* $x^3 - 8 = (x - 2)(x^2 + 2x + 4)$.)

8. Use the ε-δ definition of the limit to prove that if $\lim_{x \to a} f(x) = L$ and c is any constant, then $\lim_{x \to a} cf(x) = cL$.

Chapter 1 Summary

- **Limit:** For a function defined near a point a (but not necessarily defined at a), we say that $\lim_{x \to a} f(x) = L$ if $f(x) \approx L$ for all x close (but not equal) to a, and we can make the approximation as close as we want by taking x close enough to a. We have many variations of this definition: when x approaches a from left or right, when $L = \pm\infty$, and when $a = \pm\infty$.

- An **indeterminate form** is a quotient of functions $f(x)/g(x)$ with $\lim_{x \to a} f(x) = 0$ and $\lim_{x \to a} g(x) = 0$. The quotient may have a finite limit, an infinite limit, or no limit. To determine its behavior we must simplify the quotient until it is no longer indeterminate.

- A rational function $p(x)/q(x)$ has $p(a)/q(a)$ as its limit as $x \to a$ if $q(a) \neq 0$. It has a vertical asymptote if $q(a) = 0$ and $p(a) \neq 0$, and it is indeterminate if $p(a) = q(a) = 0$.

- A rational function $p(x)/q(x)$ has no **horizontal asymptote** if the degree of $p(x)$ is greater than that of $q(x)$, and it has the x-axis as a horizontal asymptote if the degree of $p(x)$ is smaller than that of $q(x)$. If the degrees are equal there is a horizontal asymptote of the form $x = a_n/b_n$, where a_n and b_n are the highest-degree coefficients of the numerator and denominator.

- A function is **continuous** when its graph has no gaps, or more precisely when it has a limit at each point and that limit is equal to the value of the function there. If two numbers are in the range of a continuous function, then every number between them is also in the range (**Intermediate Value Theorem**).

Chapter 1 Review Questions

- Give examples of rational functions with none, one, two, and three vertical asymptotes.

- Give an example of a rational function having $y = 4$ as a horizontal asymptote and $x = 1$ as a vertical asymptote.

- Give two examples of a rational function whose denominator is zero at $x = 0$, one with a finite limit as $x \to 0$ and the other with a vertical asymptote.

- Write a function that has both a left and right limit at a point, both finite but not equal.

- Give an example of a function that has ∞ as a limit as $x \to 0$. Give an example of one that has $x = 0$ as vertical asymptote, but no limit, either finite or infinite as $x \to 0$.

- Is there a continuous function with outputs 1 and 3 but not 2?

- Assume that you know that a continuous function has a zero in an interval. How can you go about computing it?

Chapter 1 Review Exercises

Exercises 1–8 refer to Figure 1.R.1.

1. Compute the following limits and values:
   (a) $\lim\limits_{x \to -6} f(x)$ (b) $\lim\limits_{x \to -2^+} f(x)$
   (c) $\lim\limits_{x \to -2^-} f(x)$ (d) $f(-2)$

2. Compute the following limits and values:
   (a) $\lim\limits_{x \to -5} f(x)$ (b) $f(-5)$
   (c) $\lim\limits_{x \to 0^-} f(x)$ (d) $\lim\limits_{x \to 0^+} f(x)$

3. Compute the following limits and values:
   (a) $\lim\limits_{x \to 1^-} f(x)$ (b) $\lim\limits_{x \to 1^+} f(x)$ (c) $f(1)$

4. Compute the following limits and values:
   (a) $\lim\limits_{x \to 5^-} f(x)$ (b) $\lim\limits_{x \to 5^+} f(x)$
   (c) $f(5)$ (d) $\lim\limits_{x \to \infty} f(x)$

5. Find all x in the interval $(-7, 9)$ at which the function f has no limit. Explain.

6. Find all x in the interval $(-7, 9)$ at which the function f is not continuous. Explain.

7. Find all x in the interval $(-7, 9)$ at which the function f has a limit but it is not continuous. Explain.

8. Find all x in the interval $(-7, 9)$ at which the function f has different one-sided limits.

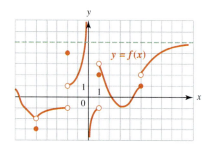

Figure 1.R.1

In Exercises 9–23, find the limits.

9. $\lim\limits_{x \to 3} \dfrac{x^2 - 9}{x - 3}$

10. $\lim\limits_{x \to -1} (x^5 - 3x^2 + 2x + 1)$

11. $\lim\limits_{x \to -1} \dfrac{x + 1}{x^2 - 3x - 4}$

12. $\lim\limits_{h \to 0} \dfrac{\sqrt{25 + h} - 5}{h}$

13. $\lim\limits_{h \to 0} \dfrac{(h - 4)^2 - 16}{h}$

14. $\lim\limits_{h \to 0} \dfrac{3 - \sqrt{9 + h}}{h\sqrt{1 + h}}$

15. $\lim\limits_{x \to 2^+} \sqrt{x - 2}$

16. $\lim\limits_{x \to 7^+} \dfrac{x}{x - 7}$

17. $\lim\limits_{x \to 7^-} \dfrac{x}{x - 7}$

18. $\lim\limits_{x \to 7^-} \dfrac{1}{(x - 7)^2}$

19. $\lim\limits_{x \to 5^-} \dfrac{1}{(x - 5)^3}$

20. $\lim\limits_{t \to 0^+} \dfrac{9 + \sqrt{t}}{\sqrt{t}}$

21. $\lim\limits_{x \to \infty} \dfrac{8x^4 + 2x}{2x^4 + 1}$

22. $\lim\limits_{x \to -\infty} \dfrac{-6x^4 + 2}{2x^2 + 1}$

23. $\lim\limits_{x \to \infty} \dfrac{x^6 - x^2}{x^8 + x^4 + 1}$

In Exercises 24–27, determine whether each of the given functions is continuous. If not, find the discontinuities. Sketch the graph of each function by hand.

24. $f(x) = \begin{cases} 2x - 1 & \text{if } x < 0 \\ 1 - x & \text{if } 0 \le x \le 1 \\ x - 1 & \text{if } x > 1 \end{cases}$

25. $f(x) = \begin{cases} 1 - x^2 & \text{if } x < 1 \\ x - 1 & \text{if } x \ge 1 \end{cases}$

26. $f(x) = \begin{cases} x^2 + 1 & \text{if } x \le 0 \\ 1 & \text{if } 0 < x < 1 \\ x^2 & \text{if } x \ge 1 \end{cases}$

27. $f(x) = \begin{cases} x^2 & \text{if } x \le 0 \\ 1 & \text{if } 0 < x < 1 \\ x^2 + 1 & \text{if } x \ge 1 \end{cases}$

28. Where, if anywhere, is the following function *not* continuous? Sketch its graph.

$$f(x) = \begin{cases} x - 2 & \text{if } x \le 0 \\ 2x - 2 & \text{if } 0 < x \le 1 \\ 1 - x & \text{if } 1 < x < 2 \\ x & \text{if } x \ge 2 \end{cases}$$

Exercises 29–32 refer to the graphs of Figure 1.R.2.

29. Which graph most closely resembles the graph of $y = 3x/(x^2 - 1)$?

30. Which graph most closely resembles the graph of $y = (x^2 + 3)/(x^2 + 1)$?

31. Which graph most closely resembles the graph of $y = x^2/(x^2 - 1)$?

32. Which graph most closely resembles the graph of $y = x^3/(x^2 - 1)$?

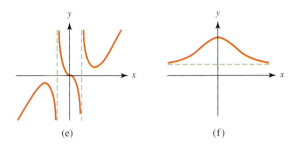

(e)   (f)

Figure 1.R.2 (continued)

In Exercises 33–37, find the vertical and horizontal asymptotes, if any, for the graph of the given function. If there are none, write "none."

33. $f(x) = \dfrac{x^2 - 1}{x^2 + 1}$ **34.** $f(x) = \dfrac{x}{x^2 - 4}$

35. $f(x) = \dfrac{x^4}{x^2 - 100}$ **36.** $f(x) = \dfrac{x + 2}{3x - 4}$

37. $f(x) = \dfrac{3x^4 + 15x^3 + 10x^2 + 5x + 20}{x^4 + 1}$

In Exercises 38–40, find any vertical and horizontal asymptotes and determine where the function is negative or positive and how the graph approaches the asymptote (rising or falling for vertical, from above or below for horizontal). Sketch the graph using that information.

38. $y = \dfrac{x}{x + 2}$ **39.** $y = \dfrac{3x^2}{x^2 + 1}$

40. $y = \dfrac{x}{x^2 - 9}$

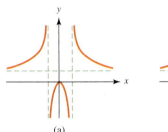

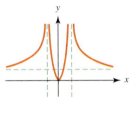

(a)   (b)

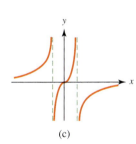

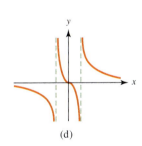

(c)   (d)

Figure 1.R.2

Chapter 1 Practice Exam

In Questions 1–4, determine whether the limit exists and, if so, find it.

1. $\displaystyle\lim_{x \to 1} \frac{x^2 - 1}{x - 1}$ **2.** $\displaystyle\lim_{x \to 1} \frac{x^2 + 1}{x - 1}$

3. $\displaystyle\lim_{x \to 0} \frac{x^2 + 1}{x - 1}$ **4.** $\displaystyle\lim_{x \to 2} \frac{x - 2}{x^2 - 7x + 10}$

5. Find all the vertical asymptotes of

$$f(x) = \frac{x^3 + 3x^2 + 2x}{x^2 + x - 6}.$$

6. Suppose $\lim_{x \to 2} f(x) = -1$ and $\lim_{x \to 2} g(x) = 4$. Find

(a) $\displaystyle\lim_{x \to 2} \frac{g(x)}{f(x) + g(x)}$ **(b)** $\displaystyle\lim_{x \to 2} \frac{1}{2} f(x) [g(x)]^2$

7. Find $\lim\limits_{h \to 0} \dfrac{h}{\sqrt{4+h}-2}$.

8. Find

  (a) $\lim\limits_{x \to 1^+} \dfrac{x-1}{|x-1|}$

  (b) $\lim\limits_{x \to 1^-} \dfrac{x-1}{|x-1|}$

9. Find

  (a) $\lim\limits_{x \to 2^-} \dfrac{2x+3}{x-2}$

  (b) $\lim\limits_{x \to 2^+} \dfrac{2x+3}{x-2}$

  (c) $\lim\limits_{x \to \infty} \dfrac{2x+3}{x-2}$

10. The graph of f is given in Figure 1.E.1.
  (a) Compute the one-sided limits of f at $x = -2$, 2, 5, and 7.
  (b) Find all x at which f has no limit.
  (c) Find all x at which f is discontinuous. Explain.

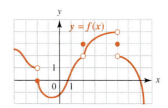

Figure 1.E.1

In Questions 11–13, list all vertical and horizontal asymptotes for the given rational function. Write "none" if there are none.

11. $f(x) = \dfrac{15x^2 + 2x + 1}{3x^2 - 27}$

12. $f(x) = \dfrac{-3x^3 + 3x + 1}{x^4 + 1}$

13. $f(x) = \dfrac{x^2}{x+1}$

14. Let $f(x) = x/(x^2 - 1)$. Find the natural domain of the function and its range, determine where it is positive and where it is negative, and find any horizontal or vertical asymptotes. Use that information to draw a rough sketch of the graph.

15. Let $f(x) = \begin{cases} x - 1 & \text{if } x < 0 \\ x + 1 & \text{if } 0 \le x \le 1 \\ 2x & \text{if } 1 < x \le 2 \\ x^2 & \text{if } 2 < x. \end{cases}$

Where does $f(x)$ have discontinuities, if any?

16. Find the number c that makes the function

$$f(x) = \begin{cases} \dfrac{x^2 - 3x + 2}{x^2 - 4} & \text{if } x \ne 2 \\ c & \text{if } x = 2 \end{cases}$$

continuous for all x.

17. Suppose a continuous function $f(x)$ satisfies the following table of values:

x	-2	-1.5	-1	-0.5	0	0.5	1
$f(x)$	-0.6	0.1	0.8	0.6	-0.2	-1.4	0.1

How many roots can you be sure of $f(x)$ having on the interval $(-2, 1)$, and where are they located?

18. If $f(x) = x^2 - 5x$, then compute

$$\lim_{h \to 0} \frac{f(3+h) - f(3)}{h}.$$

19. Show that

$$\lim_{x \to a} \frac{x^4 - a^4}{x^3 - a^3} = \frac{4}{3} a.$$

••••• Chapter ① Project ••••••••••••••••••••••••••••

An important limit   When we discuss compound interest in the next chapter we will encounter the following limit:

$$\lim_{x \to \infty} \left(1 + \frac{1}{x} \right)^x.$$

Do you think you can guess what this limit is? The answer may surprise you. It is neither 1 nor 0 nor ∞. Instead of just guessing, try experimenting with a calculator. Make a table of values of the function, much as we did in Example 1.1.9, except that in this case we want to use large values of x. Try the following settings:

(a) Start at 10 with step size 1.

(b) Start at 100 with step size 10.

(c) Start at 1,000 with step size 100.

(d) Start at 10,000 with step size 1000.

Now, after looking at the tables, try to guess the limit.

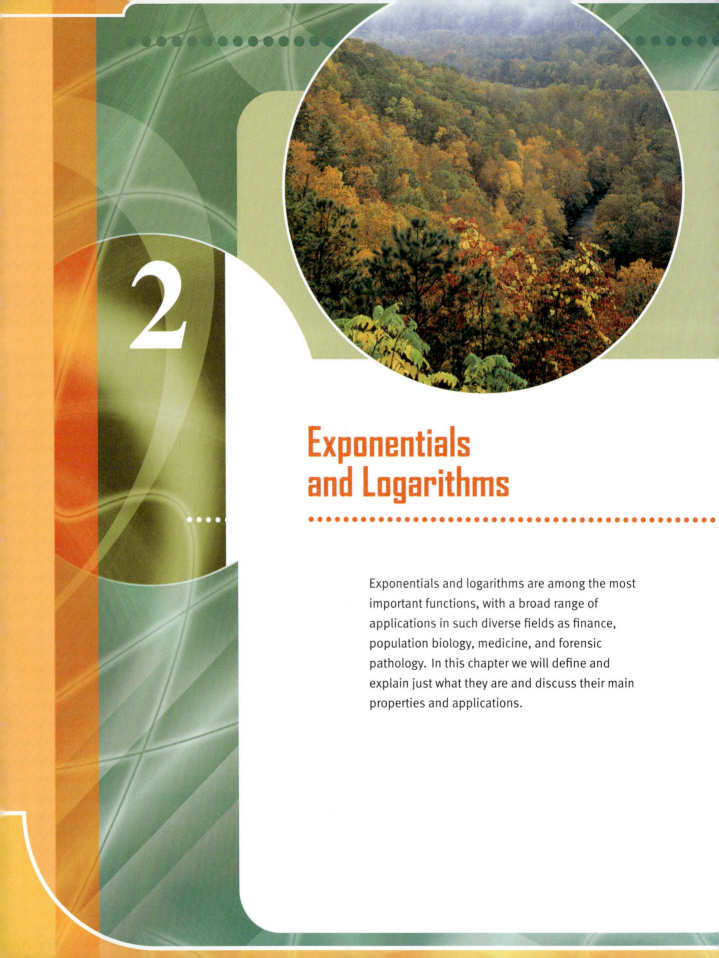

2

Exponentials and Logarithms

Exponentials and logarithms are among the most important functions, with a broad range of applications in such diverse fields as finance, population biology, medicine, and forensic pathology. In this chapter we will define and explain just what they are and discuss their main properties and applications.

An **exponential function** is a function of the form

$$f(x) = b^x, \tag{1}$$

where x is an independent variable and b is a positive constant (other than 1), called the **base**.

In Section 0.6 we defined the meaning of b^x when x is a rational number, say, $x = p/q$, where p and q are integers. To review:

- if $p/q > 0$, then $b^{p/q} = \sqrt[q]{b^p} = (\sqrt[q]{b})^p$, and
- $b^{-p/q} = 1/b^{p/q}$.

If x is irrational, we define b^x as a limiting value of exponentials with rational exponents. That is, $b^x = \lim_{r \to x} b^r$, where r is a sequence of rational numbers approaching x, such as the successive decimal approximations arising from the infinite decimal expansion of x.

For example, by using the decimal expansion $\sqrt{2} = 1.414213562\ldots$, we can define $3^{\sqrt{2}}$ as the limit of the sequence of numbers

$$3^1, \quad 3^{1.4}, \quad 3^{1.41}, \quad 3^{1.414}, \quad 3^{1.4142}, \quad 3^{1.41421}, \quad 3^{1.414213}, \quad 3^{1.4142135}, \quad \ldots$$

By computing each of the numbers in this sequence, we get the following table, whose bottom row (rounded to six decimal places) gives better and better approximations to $3^{\sqrt{2}}$:

r	1	1.4	1.41	1.414	1.4142	1.41421	1.414213	1.4142135
3^r	3	4.655537	4.706965	4.727695	4.728734	4.728786	4.728801	4.728804

From the table, we infer that $3^{\sqrt{2}} \approx 4.7288$, and, indeed a scientific calulator gives $3^{\sqrt{2}} = 4.728804388$, correct to nine decimal places.

The graph of b^x for $b > 1$

To draw the graph of $y = b^x$ for $b > 1$, we construct a table of values with the help of a calculator. For instance, for $b = 3$ and x between -2.5 and 2.5 in steps of 0.5 we get

x	-2.5	-2	-1.5	-1	-0.5	0	0.5	1	1.5	2	2.5
3^x	0.06	0.11	0.19	0.33	0.58	1.00	1.73	3.00	5.20	9.00	15.59

where the values in the bottom row have been rounded off to two decimal places. (The values corresponding to $x = 0, 1,$ and 2 are integers and do not need to be rounded off.) Figure 2.1.1 shows these points plotted in the x, y-plane.

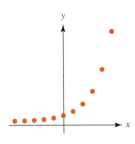

Figure 2.1.1

Adding more values to the table yields more points. As we keep adding more points, they fit into a continuous curve that is the graph of the function $f(x) = 3^x$. It is shown in Figure 2.1.2. In fact, it can be proved that for any positive base b, the graph of $y = b^x$ is *continuous* everywhere.

Using $b = 3$ gives us one exponential function out of an infinite family of such functions. There is a different exponential function for every positive choice of base b. The functions are similar in many ways, as you can see in Figure 2.1.3, which contains the graphs of three exponential functions drawn on the same set of axes.

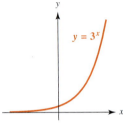

Figure 2.1.2

> **Properties of the exponential function with base $b > 1$**
>
> - Its graph goes through the point $(0, 1)$.
> - Its domain is $(-\infty, \infty)$ and its range is $(0, \infty)$.
> - It is continuous and increasing.
> - $\lim_{x \to \infty} b^x = \infty$ and $\lim_{x \to -\infty} b^x = 0$.

This first property reflects an important fact: *No matter what base b we use, we always have $b^0 = 1$.* That is one of the well-known laws of exponents, and at this point it might be a good idea to review the others.

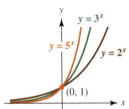

Figure 2.1.3

The laws of exponents

For any positive base b and any numbers u and v, the following rules apply:

> 1. $b^{u+v} = b^u b^v$.   4. $b^0 = 1$.
>
> 2. $b^{u-v} = \dfrac{b^u}{b^v}$.   5. $b^{-v} = \dfrac{1}{b^v}$.
>
> 3. $b^{ru} = \left(b^u\right)^r$ for any number r.

These rules are valid for *any* exponents u and v, rational or irrational, positive or negative—but they are most easily grasped and remembered in the cases when u and v are positive integers. For then, as you know, b^u simply means b multiplied by itself u times, that is,

$$b^u = \underbrace{b \cdot b \cdots b}_{u},$$

and, similarly,

$$b^v = \underbrace{b \cdot b \cdots b}_{v}.$$

Therefore,

$$b^u b^v = \underbrace{b \cdot b \cdots b}_{u} \cdot \underbrace{b \cdot b \cdots b}_{v} = \underbrace{b \cdot b \cdots b}_{u+v} = b^{u+v},$$

which is just rule 1 with the left- and right-hand sides switched.

Rules 2 and 3 have similar explanations. For instance, if $u = 5$ and $v = 2$, cancellation gives

$$\frac{b^5}{b^2} = \frac{b \cdot b \cdot b \cdot b \cdot b}{b \cdot b} = b \cdot b \cdot b = b^3,$$

which illustrates rule 2; and

$$\left(b^5\right)^2 = \underbrace{(b \cdot b \cdot b \cdot b \cdot b)}_{5} \cdot \underbrace{(b \cdot b \cdot b \cdot b \cdot b)}_{5} = \underbrace{b \cdot b \cdot b \cdot b \cdot b \cdot b \cdot b \cdot b \cdot b \cdot b}_{10} = b^{10},$$

which illustrates rule 3.

Rules 4 and 5 follow from the first three. For instance, taking u and v to be the same number in rule 2 gives rule 4, that is,

$$b^0 = b^{u-u} = \frac{b^u}{b^u} = 1,$$

and taking $u = 0$ in rule 2 yields rule 5.

The graph of b^x for $0 < b < 1$

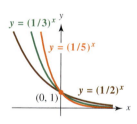

Figure 2.1.4

If the base b is *less* than 1—if $0 < b < 1$, that is—then the graph of the function b^x is different from those we observed above. The graphs for the cases $b = 1/2$, $b = 1/5$, and $b = 1/3$ are shown on a common set of axes in Figure 2.1.4. If you compare these graphs with those of Figure 2.1.3, you may observe that they are "mirror images" of the previous ones. For instance, the graph of $y = (1/3)^x$ looks like the graph of $y = 3^x$ reflected in the vertical axis—and, in fact, that's exactly what it is. That is because $(1/3)^x = 1/3^x = 3^{-x}$, which means that the point on the graph of $y = (1/3)^x$ over $x = a$ (for any number a) has the same height as the point on the graph of $y = 3^x$ over $x = -a$. For instance, $(1/3)^2$ is the same as 3^{-2}, and $(1/3)^{-4}$ is the same as 3^4.

The same is true for the graphs of $y = 2^x$ and $y = (1/2)^x$ and the graphs of $y = 5^x$ and $y = (1/5)^x$. In general, for any base b, the graph of $y = b^x$ is the reflection in the vertical axis of the graph of $y = (1/b)^x$.

Properties of the exponential function with base $0 < b < 1$

- Its graph goes through the point $(0, 1)$.

- Its domain is $(-\infty, \infty)$ and its range is $(0, \infty)$.

- It is continuous and decreasing.

- $\lim_{x \to \infty} b^x = 0$ and $\lim_{x \to -\infty} b^x = \infty$.

You should compare them with the corresponding properties, listed earlier, for the case when $b > 1$.

Some applications: interest, growth, and decay

Exponential functions play an important role in the mathematics of finance, population growth, and radioactive decay. We start with simple but important models that we will build on as we go along.

Compound interest   Suppose that you invest an amount of money P, called the **principal**, into an account paying an annual interest rate r. What will be the amount in your account after t years? To be more concrete, let us assume that $r = 0.06$. Then at the end of a year an extra 6% is added to your account, so that it contains the original P dollars plus $0.06P$ interest—for a total of $P + 0.06P$. If we factor out P, we get the amount $A(1)$ at the end of 1 year in the form

$$A(1) = P(1.06).$$

Now suppose you leave the money in the account for a second year. The amount at the beginning of the second year is $P(1.06)$ (the same as the amount at the end of the first year). By the end of the second year, that increases by 6%. In other words, at the end of the second year, you have $P(1.06) + (0.06)P(1.06)$ (the amount at the beginning of the year plus an additional 6%). Factoring out $P(1.06)$ gives the amount $A(2)$ at the end of 2 years:

$$A(2) = P(1.06)^2.$$

If we continue in this way, we get the formula for the amount $A(t)$ at the end of t years:

$$\boxed{A(t) = P(1.06)^t.} \qquad (2)$$

This method of computing interest is called **compounding**. What that means is that the amount at the end of any period is taken as the starting amount for the next period, and the interest is then computed on that amount. In this example the period is a year, and we say that the interest is **compounded once a year** or **compounded annually**. (In the next section we will compute compound interest using other periods, such as a month or a week.)

As you can see, formula (2) defines an exponential function—to be exact, an exponential function multiplied by the coefficient P. The base is 1.06. Figure 2.1.5 shows a table of values of $A(t)$ for $P = 100$ and t from 1 to 50 and a graph created by an Excel spreadsheet.

If we use a different principal P and interest rate r, we have to modify formula (2) accordingly. For any P and r, the interest at the end of 1 year equals rP, so the total in the account is $P + rP$. Factoring out P gives the amount at the end of 1 year:

$$A(1) = P(1 + r),$$

which is also the amount at the beginning of the second year. To compute the interest at the end of the second year, we multiply that amount by r, getting $rP(1 + r)$. Adding it on and factoring give the amount $A(2)$ at the end of 2 years:

$$A(2) = P(1 + r) + rP(1 + r) = P(1 + r)^2.$$

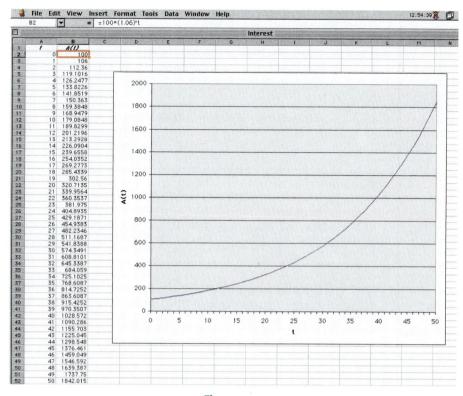

Figure 2.1.5

We can continue in this way from one year to another. The amount at the end of any year is the same as the amount at the beginning of the next year. That amount then grows to $(1 + r)$ times its size by the end of the next year. The following table gives the amounts at the beginning and end of each of the first 6 years:

Year	Amount at beginning	Amount at end
1	P	$P(1+r)$
2	$P(1+r)$	$P(1+r)^2$
3	$P(1+r)^2$	$P(1+r)^3$
4	$P(1+r)^3$	$P(1+r)^4$
5	$P(1+r)^4$	$P(1+r)^5$
6	$P(1+r)^5$	$P(1+r)^6$

Continuing in this way, we obtain the following formula:

The amount $A(t)$ at the end of t years on an initial investment of P dollars (principal) at annual interest rate r, compounded annually:

$$A(t) = P(1+r)^t. \qquad (3)$$

> ### EXAMPLE 2.1.1

Suppose you invest $10,000 in an account that pays 4.5% interest, compounded annually. How much will there be in the account at the end of 6 years? (Assume that you do not make any further deposits or withdrawals.)

SOLUTION   We apply formula (3) with $P = 10,000$ and $r = 0.045$ to get

$$A(6) = 10,000(1.045)^6 = 13,022.60.$$

Among other things, formula (3) says the amount at the end of t years is directly proportional to $(1 + r)^t$. Any quantity that is directly proportional to an exponential function b^t, with $b > 1$ and t representing time, is said to **grow exponentially** or have **exponential growth**. Next, we consider another example of a quantity that grows exponentially.

Population growth   The population of a colony of simple microorganisms, such as bacteria, grows exponentially, at least over a relatively short period of time. In other words, the population is given by a formula of the type

$$P(t) = P_0 b^t \quad \text{for } t \geq 0, \tag{4}$$

where t denotes time, $P(t)$ is the size of the population after t years, and P_0 and b are positive constants. P_0 has a simple interpretation—it is the **initial population** $P(0)$, as we see by setting $t = 0$ in (4).

> ### EXAMPLE 2.1.2

Suppose a certain colony of bacteria is growing exponentially. At the beginning of the first day there are 1,200 bacteria in the colony, and at the end of 2 days there are 1,800. Find a formula describing the population as a function of time.

SOLUTION   We will use formula (4), but we need to find explicit values for P_0 and b. Since $P = 1,200$ when $t = 0$, we have $P_0 = 1,200$ (the initial population), and we can rewrite (4) in the form

$$P(t) = 1,200b^t, \tag{5}$$

where t is the time in days.

We also know that $P = 1,800$ when $t = 2$. Substituting these values into (5) gives

$$1,800 = 1,200\, b^2,$$

which can be solved for b to give $b = \sqrt{3/2}$. Thus, we get the formula we are looking for:

$$P(t) = 1,200 \left(\sqrt{\frac{3}{2}} \right)^t .$$

Using the laws of exponents, we can also write this equation in the form

$$P(t) = 1{,}200 \left(\frac{3}{2}\right)^{t/2}.$$

Population growth and compound interest are examples of exponential growth, which means that they are directly proportional to a function of the form b^t, where $b > 1$. There are other applications involving quantities proportional to b^t with $0 < b < 1$. In those cases the quantity *decreases* as time goes on, and we say that it **decays exponentially** or obeys an **exponential decay** law. A basic example of exponential decay is the decomposition of a radioactive substance.

Decay of radioactive substances   Certain substances, such as radium and uranium, decompose into other substances by emitting subatomic particles. Such substances are called **radioactive**. They can be very useful, but they can also be dangerous. For instance, radon gas is formed by the decomposition of radium in the earth's crust. It is radioactive and can be a health hazard if it accumulates in the basement of a home.

If we let y stand for the amount of a particular radioactive substance, then y is a function of time, given by an equation of the form

$$y = y_0 b^t \quad \text{for } t \geq 0. \tag{6}$$

In this equation t stands for time, measured in appropriate units (years, days, minutes, etc., depending on the particular substance). The base satisfies $0 < b < 1$. Its exact value depends on the substance involved. The constant y_0 is the initial amount.

In the case of radon gas, for example, t is usually measured in days, and the base is approximately 0.835, so the decay equation for radon gas is

$$y = y_0 (0.835)^t \quad \text{for } t \geq 0, \tag{7}$$

where y_0 is the initial amount.

EXAMPLE 2.1.3

Starting with an initial amount of 1,000 cubic centimeters of radon gas, how much will there be at the end of 1 day? after 2 days?

SOLUTION   Using formula (7) with $y_0 = 1{,}000$ and $t = 1$ gives the amount of radon gas remaining at the end of 1 day:

$$1{,}000(0.835)^1 = 835 \text{ cubic centimeters.}$$

After 2 days, the amount remaining is

$$y(2) = 1{,}000(0.835)^2 \approx 697.2 \text{ cubic centimeters.}$$

As you can see, Eq. (6) is the same as formula (4), except that its base is smaller than 1 rather than larger. Indeed, exponential decay behaves very much like simple population growth (and like compound interest), but in the opposite direction. Think of it this way: An initial amount of radon gas *loses* 16.5% every day, so that 83.5% is left. Starting with an initial amount of y_0 units, we have $y_0(0.835)$ at the end of the first day. That amount becomes the starting amount for the second day, so by the end of the second day there is $y_0(0.835)^2$. At the end of the third day there is 83.5% of that amount left—in other words, $y_0(0.835)^3$. Continuing this process, we see that the amount left at the end of t days is $y_0(0.835)^t$.

How many days will it take the original amount to reduce by half—that is, to go from y_0 to $y_0/2$? Surprisingly, the answer to that question does not depend on the initial amount—in other words, it is the same for every y_0. It is called the **half-life** of the substance. In order to find it, we need to know something about **logarithms**, which we will study in Section 2.3.

Practice Exercises 2.1

1. Predict the population of the United States for the year 2020, assuming that it is modeled by the exponential function $p(t) = 281(1.02)^t$ in millions, where t is the number of years after 2000.

2. Suppose you put \$8,000 in an account paying 9% interest, compounded annually. Find the balance in your account at the end of t years. Then, compute your balance for $t = 10$.

Exercises 2.1

In Exercises 1–4, use a calculator or spreadsheet to make a table of values of the function for x from −3 to 3 in steps of 0.5. Then plot the points and connect them to get a rough sketch of the graph.

1. $f(x) = \left(\frac{3}{2}\right)^x$
2. $f(x) = \left(\frac{3}{5}\right)^x$
3. $f(x) = (0.4)^x$
4. $f(x) = 4^x$

In Exercises 5–12, use the tables from the preceding problems but not a calculator or spreadsheet to make a table of values of each of the following functions for x from −3 to 3 in steps of 0.5. (Round the values to two decimal places.) Then plot the points and connect them to get a rough sketch of the graph.

5. $f(x) = \left(\frac{2}{3}\right)^x$
6. $f(x) = \left(\frac{5}{3}\right)^x$
7. $f(x) = \left(\frac{5}{2}\right)^x$
8. $f(x) = (0.25)^x$
9. $f(x) = \left(\frac{2}{3}\right)^{-x}$
10. $f(x) = \left(\frac{3}{5}\right)^{-x}$
11. $f(x) = \left(\frac{5}{3}\right)^{-x}$
12. $f(x) = 4^{-x}$

13. The number $\sqrt{3}$ is irrational, and it has the infinite decimal expansion $\sqrt{3} = 1.7320508\ldots$. Compute the sequence of numbers 2^1, $2^{1.7}$, $2^{1.73}$, $2^{1.732}$, and so forth and use it to estimate $2^{\sqrt{3}}$. Then use a calculator to find $2^{\sqrt{3}}$ and compare it with your estimate. Is the answer your calculator gives the precise value of $2^{\sqrt{3}}$?

14. The number π is famous in elementary geometry as the ratio of the circumference of a circle to its diameter. It is an irrational number with infinite decimal expansion $\pi = 3.141592654\ldots$. Use the decimal expansion to construct a table of values for 2^r, where r is a sequence of rational numbers tending toward π. What do you infer from your table about the value of 2^π? Compare your estimate to the answer you get from a calculator.

In Exercises 15–22, assume that b is a positive number and that u and v are fixed numbers so that $b^u = 3$ and $b^v = 4$. Find a specific numerical value for each of the given expressions.

15. b^{2u}
16. b^{-v}
17. b^{v-u}
18. b^{u+2v}
19. b^0
20. $b^{v/2}$
21. $\dfrac{1}{b^{-3u}}$
22. $b^{-(u+v)}$

23. Suppose that \$500 is put in an account paying 7% interest, compounded annually.
    (a) Write a formula giving the amount in the account at the end of t years.
    (b) Determine how much there will be in the account at the end of 10 years.

24. Suppose that \$2,000 is put in an account paying 4.5% interest, compounded annually.
    (a) Write a formula giving the amount in the account at the end of t years.
    (b) Determine how much there will be in the account at the end of 3 years.

25. Suppose you invest a certain amount of money at 6% interest, compounded annually. By what factor will your investment increase after 7 years? after 12 years?

26. How much money should be put in an account paying 5% interest, compounded annually, in order to have $5,000 eight years from now? (*Hint:* Write the formula for the amount after eight years on an initial investment of P dollars, set it equal to 5,000, and solve for P.)

27. How much money should be put in an account paying 4% interest, compounded annually, in order to have $50,000 twenty years from now?

28. Which of the following will yield the greater amount: (a) putting $1,000 in an account paying 3% interest, compounded annually, and leaving it for 10 years, or (b) putting $1,000 in an account paying 6% interest, compounded annually, and leaving it for 5 years?

29. A certain bacteria culture grows exponentially. In 1 hour the population grows from 500,000 to 800,000. Write a formula expressing the population P as a function of the time t in hours.

30. A colony of fruit flies is growing exponentially. At the beginning there were 3,000 flies, and the end of 3 days there were 7,000 flies.
(a) Write a formula expressing the population P as a function of the time t in days.
(b) Determine how many flies there will be at the end of 5 days.

31. A colony of ants is increasing at a rate of 11% a day. At the end of the fifth day there are 580 ants. How many were there at the beginning of the first day? at the beginning of the second day?

32. Suppose a certain quantity of radon gas decayed to 400 cubic centimeters over a period of 8 days. How much was there to begin with? (*Hint:* Refer to Eq. (7).)

33. A certain radioactive substance decays in accordance with the formula $y = y_0(0.88)^t$, where y_0 is the initial amount and y is the amount after t years. What percentage of the original amount will be left after 5 years?

34. A certain radioactive substance decays in accordance with the formula $y = y_0(0.93)^t$, where y_0 is the initial amount and t is the time in days. If there are 200 grams of the substance at the end of the third day, how many will there be at the end of the fifth day?

35. Plutonium-230 is a radioactive substance, whose decay equation has the form $y = y_0 b^t$, where y is the amount at the end of t years. An initial amount of 50 milligrams of plutonium-230 decays to approximately 30 milligrams in 10 years.
(a) Use the given data to find the constants y_0 and b. Round your answer to two decimal places. (You can use a calculator for b, but you shouldn't need one to find y_0.)
(b) How much of the initial 50 milligrams will be left at the end of 20 years?
(c) Suppose the initial amount is 80 milligrams. How much will be left at the end of 15 years?

36. Suppose $1,000 is invested at an annual interest rate of 4.5%. Create a table of values and a chart of $A(t)$, the amount after t years, for t from 1 to 25.

37. Suppose $5,000 is invested at an annual interest rate of 7%. Create a table of values and a chart of the amount after t years for t from 1 to 30.

38. Suppose $1,000 is invested at an annual interest rate r. Create a table of values and a chart showing the amount at the end of 10 years for values of r from 0.03 to 0.05 in steps of 0.0025.

39. The world population in the year 2000 was approximately 6.16 billion. Assume it is growing exponentially at a rate of 1.0145% a year. Make a table of the projected population for each of the years from 2001 to 2020.

Solutions to practice exercises 2.1

1. To predict the U.S. population for 2020, we simply compute $p(20)$. We have $p(t) = 281(1.02)^{20} = 417.551$ million.

2. If $y(t)$ denotes the balance t years later, then using formula (3), we have $y(t) = 8,000(1 + 0.09)^t$, or $y(t) = 8,000(1.09)^t$, and 10 years later it will be equal to $y(10) = 8,000(1.09)^{10} = 18,938.91$.

2.2 Compound Interest and the Number e

Compounding interest repeatedly

Interest compounded n times over 1 year   Let us review what we know about interest. Suppose you invest a certain amount of money—say, $10,000—at a certain annual interest rate—say, 6%. If the interest is compounded annually, the amount at the end of 1 year will be the original 10,000 plus an additional 6% for a total of $10,600. In general, if you invest P dollars at annual interest rate r, compounded annually, the amount at the end of the year will be $P(1 + r)$, the principal P plus the interest rP.

But in most financial transactions the interest is compounded more frequently than once a year. Here is how it works. If the interest is compounded n times a year, the bank computes the interest at the end of each of the n periods by multiplying the balance in your account by r/n (one nth of the annual interest) and adding that interest to the total. In other words, if your account has a balance of X dollars at the beginning of any period, then the bank adds $(r/n)X$ interest at the end of that period for a total of $X + (r/n)X$. By factoring an X from both terms, we get $X(1 + r/n)$. We now have a simple rule for computing the total balance at the end of any period.

Compound Interest Rule

Multiply the amount at the beginning of any period by $(1 + r/n)$ to get the amount at the end of that period.

Let's use this rule to compute the total at the end of 1 year. Suppose you start with an initial amount of P dollars.

- At the end of the first period you will have $P(1 + r/n)$. That becomes your initial amount for the second period.
- Multiplying that by $(1 + r/n)$ gives $P(1 + r/n)^2$ at the end of the second period. That becomes your initial amount for the third period.
- Multiplying that by $(1 + r/n)$ gives $P(1 + r/n)^3$ at the end of the third period. That becomes your initial amount for the fourth period.
- Continuing in this way up to the end of n periods (1 full year) gives you a total of $P(1 + r/n)^n$.

We have now arrived at a formula for *the amount $A(1)$ in the account at the end of 1 year on an investment of P dollars at an annual interest rate r compounded n times a year.* It is

$$A(1) = P\left(1 + \frac{r}{n}\right)^n. \tag{8}$$

EXAMPLE 2.2.1

Suppose you invest \$10,000 at an annual rate of 6%. Find the balance at the end of 1 year if the interest is compounded weekly.

SOLUTION Applying formula (8) with $P = 10,000$, $r = 0.06$, and $n = 52$ gives

$$A(1) = 10,000 \left(1 + \frac{0.06}{52}\right)^{52} \approx 10,618.00.$$

Interest compounded n times a year over t years   By using formula (8) repeatedly, we can compute the amount in the account at the end of t years starting with an initial principal P.

- Multiplying P by $(1 + r/n)^n$ gives $A(1) = P(1 + r/n)^n$ at the end of the first year. That becomes your initial amount for the second year.
- Multiplying $A(1)$ by $(1 + r/n)^n$ gives

$$A(2) = \left(1 + \frac{r}{n}\right)^n P \left(1 + \frac{r}{n}\right)^n = P \left(1 + \frac{r}{n}\right)^{2n}$$

  at the end of the second year. That becomes your initial amount for the third year.
- Multiplying $A(2)$ by $(1 + r/n)^n$ gives

$$A(3) = \left(1 + \frac{r}{n}\right)^n P \left(1 + \frac{r}{n}\right)^{2n} = P \left(1 + \frac{r}{n}\right)^{3n}$$

  at the end of the third year. That becomes your initial amount for the fourth year.

Continuing in this way up to the end of t years, we obtain the following.

> The amount $A(t)$ in the account at the end of t years on an investment of P dollars at an annual interest rate r compounded n times a year is given by
>
> $$A(t) = P \left(1 + \frac{r}{n}\right)^{tn}. \tag{9}$$

EXAMPLE 2.2.2

Suppose you invest \$4,000 at an annual rate of 7%. Find the balance at the end of 5 years if the interest is compounded daily. Ignore leap years and assume every year has exactly 365 days.

SOLUTION Applying formula (9) with $P = 4{,}000$, $r = 0.07$, $t = 5$, and $n = 365$ gives

$$A(5) = 4{,}000 \left(1 + \frac{0.07}{365} \right)^{5 \cdot 365} \approx 5{,}676.08.$$

The number *e*

Now let us assume that we start with a principal P of just \$1 but the annual interest rate r is 1 (that is 100%) compounded n times a year. According to formula (8), the total amount accrued after 1 year is

$$A(1) = \left(1 + \frac{1}{n} \right)^n.$$

The following table gives the value of $A(1)$ in dollars (up to five decimal places) for a number of different choices of n.

n (number of periods)	$(1 + 1/n)^n$ (total accrued in 1 year)
1	2
2	2.25
4	2.44141
12	2.61304
24	2.66373
52	2.69260
365	2.71457
1,000	2.71692
10,000	2.71815
100,000	2.71827
1,000,000	2.71828

As you can see in the table, the total amount accrued in 1 year gets larger as the number of compounding periods increases. But the *rate* of increase seems to diminish—the difference between compounding 4 times and 12 times is about 0.17163, while the difference between compounding 10,000 times and 100,000 times is only 0.00012! In addition, the table indicates that there is a ceiling, or upper bound, on how large the numbers in the right-hand column can get. Indeed, it can be shown that $(1 + 1/n)^n < 3$ for all n. Combining these facts, we conclude that the quantity $(1 + 1/n)^n$ has a limit as n goes to infinity. This limit is an important irrational number, which is denoted by the letter e in honor of the great mathematician Leonhard Euler (1707–1783). To summarize, we have the following:

Definition 2.2.1

Definition of *e*:

$$e = \lim_{n \to \infty} \left(1 + \frac{1}{n} \right)^n. \tag{10}$$

By taking n large enough, we can approximate e to any desired degree of accuracy. The approximation to 40 decimal places is

$$e \approx 2.7182818284590452353602874713526624977757.$$

The larger the number n of compounding periods, the shorter the length of each period, and vice versa. For instance, compounding interest every second means compounding 31,536,000 times per year (in a nonleap year). Taking the limit as $n \to \infty$, we say the interest is **compounded continuously**. This leads to the following **business definition of e**:

> e is the balance at the end of 1 year of an investment of \$1
> at an annual interest rate of 100%, compounded continuously.

Continuously compounded interest

Let's recall what we know about compound interest. If P dollars are invested at annual interest rate r compounded n times a year, then the amount in the account at the end of t years is equal to

$$P\left(1 + \frac{r}{n}\right)^{nt},$$

as we saw in formula (9). Taking the limit as $n \to \infty$ gives us the amount after t years if the interest is compounded continuously:

$$A(t) = \lim_{n \to \infty} P\left(1 + \frac{r}{n}\right)^{nt}.$$

We can simplify this formula as follows. First, we let $m = n/r$ or, equivalently, $n = mr$.

Since $m \to \infty$ if and only if $n \to \infty$, we can rewrite the equation as follows:

$$A(t) = P \lim_{m \to \infty} \left(1 + \frac{1}{m}\right)^{mrt}$$

$$= P \left[\lim_{m \to \infty} \left(1 + \frac{1}{m}\right)^{m}\right]^{rt}. \qquad \text{(Why?)}$$

Since $(1 + 1/m)^m \to e$ as $m \to \infty$, we conclude that $A(t) = Pe^{rt}$. Thus, we arrive at the following important result.

Continuously compounded interest is calculated by the formula

$$A(t) = Pe^{rt}, \tag{11}$$

where P = principal

r = annual interest rate

t = time in years

$A(t)$ = accumulated amount after t years.

Setting $t = 1$ in (11) gives us the following very important formula:

$$\lim_{n \to \infty} \left(1 + \frac{r}{n}\right)^n = e^r \tag{12}$$

EXAMPLE 2.2.3

Suppose you put $5,000 in an account paying 4% annual interest, and you leave it there without adding or withdrawing anything. How much will you have at the end of 3 years if the interest is compounded:

(i) 6 times a year,   (ii) 24 times a year,   (iii) continuously?

SOLUTION   To answer question (i), we use formula (9) with $P = 5,000$, $r = 0.04$, $t = 3$, and $n = 6$. Rounding off to two decimal places gives the amount

$$5,000 \left(1 + \frac{0.04}{6}\right)^{18} \approx 5,635.24.$$

For question (ii) we use the same formula with the same values of P, r, and t, but this time with $n = 24$, to get the amount

$$5,000 \left(1 + \frac{0.04}{24}\right)^{72} \approx 5,636.92.$$

For question (iii) we use formula (11) with $P = 5,000$, $r = 0.04$, and $t = 3$ to obtain the amount

$$5,000e^{0.12} \approx 5,637.48.$$

Instead of specifying the initial amount as we just did, we might turn the problem around by specifying the *final* amount and asking for the initial amount that will produce it.

EXAMPLE 2.2.4

How much money must be put into an account paying 6% annual interest, compounded continuously, in order to have $10,000 at the end of 5 years? Assume that after the initial deposit no money is deposited or withdrawn.

SOLUTION Once again, we use formula (11), this time with P as the unknown:

$$10,000 = Pe^{5(0.06)} = Pe^{0.3}.$$

To solve for P, we divide both sides by $e^{0.3}$, or, what amounts to the same thing, multiply both sides by $e^{-0.3}$. The initial amount, rounded off to the nearest cent, is

$$P = 10,000e^{-0.3} \approx 7,408.18.$$

Formula (11) is also valid when t is not an integer.

EXAMPLE 2.2.5

If $1,000 is put in an account paying 4.5% annual interest, compounded continuously, how much will there be at the end of 18 months?

SOLUTION Once again, we use formula (11), this time with $P = 1,000$, $r = 0.045$, and $t = 1.5$, to get

$$A(1.5) = 1,000e^{0.0675} \approx 1,069.83.$$

The natural exponential function

As we saw in the last section, any positive number different from 1 can be used as the base of an exponential function. In practice, however, certain numbers occur more frequently in applications. Among integers, the numbers 2 and 10 appear most often. But, in general, the most important and widely used base is the number e, which, as we saw earlier, arises naturally in the computation of continuously compounded interest.

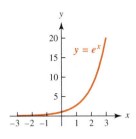

Figure 2.2.1

> ## Definition 2.2.2
>
> The **natural exponential function** is the exponential function with base e. That is,
>
> $$f(x) = e^x.$$
>
> It is often simply called the **exponential function**.

The graph of $y = e^x$ is shown in Figure 2.2.1. Of course, its general appearance is very similar to other exponential graphs with a base greater than 1, and the function has the same properties as the other exponential functions with base greater than 1.

Main properties of e^x

- It is continuous and increasing.
- $e^0 = 1$.
- $\lim_{x \to \infty} e^x = \infty$ and $\lim_{x \to -\infty} e^x = 0$.

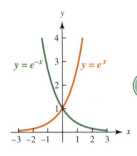

Figure 2.2.2

EXAMPLE 2.2.6

Sketch the graphs of each of these functions:

     (i) $f(x) = e^{-x}$,   (ii) $f(x) = 2e^{0.5x}$,   (iii) $f(x) = 2e^{-0.5x}$.

SOLUTION   The graph of $y = e^{-x}$ is obtained by reflecting the graph of $y = e^x$ in the y-axis, as shown in Figure 2.2.2.

     Similarly, the graph of $y = 2e^{-0.5x}$ is obtained by reflecting the graph of $y = 2e^{0.5x}$ in the y-axis. Both are shown in Figure 2.2.3. The shape of the graph of $y = 2e^{0.5x}$ is similar to that of $y = e^x$. Using the following table of values as a guide helps us to sketch its graph:

Figure 2.2.3

x	-6	-4	-2	0	2	4	6
$2e^{0.5x}$	0.0995741	0.270671	0.735759	2	5.43656	14.7781	40.1711

Practice Exercise 2.2

Suppose you invest \$5,000 at an annual rate of 8%. Find the balance at the end of 4 years if the interest is compounded

   (a) daily
   (b) continuously.

Exercises 2.2

In Exercises 1–12, assume that no money is deposited or withdrawn after the initial investment.

1. If \$500 is invested in an account paying 8% interest, how much will it grow to in 1 year if the interest is compounded **(a)** annually? **(b)** monthly? **(c)** continuously?

2. If $6,000 is invested in an account paying 6.5% interest, how much will it grow to in 7 years if the interest is compounded
(a) quarterly? (b) 24 times per year?
(c) continuously?

3. If $100 is invested in an account paying 7% interest, how much will it grow to in 20 years if the interest is compounded
(a) 365 times a year? (b) continuously?

4. Suppose that on the day when his granddaughter is born, a man invests $5,000 in her name in an account paying 4.5% interest. How much will there be on her 21st birthday if the interest is compounded
(a) annually? (b) quarterly?
(c) continuously?

5. Which of the following will provide a greater return after one year on an initial investment: a 6% interest rate compounded four times a year, or a 4% interest rate compounded six times a year?

6. Which of the following will provide a greater return on an initial investment at an annual interest rate of 5%: compounded 12 times a year for 1 year, or compounded six times a year for 2 years?

7. Suppose you put $1,000 in an account paying 7.5% interest, compounded continuously. How much will there be at the end of
(a) 6 months? (b) 8 months? (c) 27 months?

8. Suppose you make an initial deposit in an account paying 6.25% interest, compounded continuously. By what factor does your investment increase
(a) after 6 months (b) after 1 year
(c) after of 18 months?

9. Suppose you put $10,000 in an account paying 4.5% interest, compounded continuously. How much will there be at the end of
(a) 6 months? (b) 18 months? (c) 36 months?

10. Suppose that you open an account paying 5% interest, compounded continuously. How much should you deposit to ensure that there will be $10,000 in 7 years?

11. Suppose that you open an account paying 12% interest, compounded continuously. How much should you deposit to ensure that there will be $2,000 in 6 months?

12. Suppose that you open an account paying 7% interest, compounded annually. How much should you deposit to ensure that there will be $5,000 in 10 years?

In Exercises 13–21, find the given limit. First express your answer as an exponential expression without using a calculator, then use a calculator to evaluate it to four decimal places.

13. $\lim_{n\to\infty}\left(1+\dfrac{3}{n}\right)^n$

14. $\lim_{n\to\infty}\left(1+\dfrac{1}{4n}\right)^n$

15. $\lim_{n\to\infty}\left(1-\dfrac{2}{n}\right)^n$

16. $\lim_{n\to\infty}100\left(1-\dfrac{1}{2n}\right)^n$

17. $\lim_{n\to\infty}\left(1+\dfrac{3}{4n}\right)^n$

18. $\lim_{n\to\infty}\left(1+\dfrac{1}{n}\right)^{3n}$

19. $\lim_{n\to\infty}10\left(1-\dfrac{1}{n}\right)^{2n}$

20. $\lim_{n\to\infty}\left(1+\dfrac{1}{2n}\right)^{2n}$

21. $\lim_{n\to\infty}1000\left(1+\dfrac{1}{n}\right)^{n/2}$

22. Make a table of values of $(1+1/(2n))^n$ for the following choices of n: 10, 50, 100, 500, 1,000, 5,000. Next, use your calculator to find $\sqrt{e}$. What do you observe, and how do you explain it?

23. Make a table of values of $(1-1/n)^n$ for the following choices of n: 10, 50, 100, 500, 1,000, 5,000. Next, use your calculator to find $1/e$. What do you observe, and how do you explain it?

24. Make two tables: the first for $(1+1/(5n))^n$ and the second for $(1+1/n)^{n/5}$, both for n from 1,000 to 10,000 in steps of 1,000. What do you observe when you compare the tables? How do you explain it?

25. Make a table of values of $(1+3/n)^{-n/2}$ for n from 1,000 to 10,000 in steps of 1,000. Toward what limit are these numbers tending?

In Exercises 26–28, treat each month as if it were exactly $1/12$ of a year.

26. Suppose that on the first of next month you deposit $3,000 in an account paying 4.5% annual interest, compounded monthly. Make a table of the balance in the account at the end of each of the next 48 months.

27. Suppose that on the first of every month you deposit $100 in an account paying 3.6% annual interest, compounded monthly. Assuming that you never make any withdrawals, make a table of the balance in the account immediately after your monthly deposit for each of the first 36 months.

28. A retired person has $120,000 in an account that pays 5% annual interest, compounded continuously. She never adds to the balance and on the first day of each month she withdraws $300. Make a table of the balance in the account immediately after her withdrawal for each of the next 36 months.

29. Solve the equation $e^{-t}=t$.

30. Solve the equation $e^t=2-t$.

Solutions to practice exercise 2.2

(a) Applying formula (9) with $P = 5{,}000, r = 0.08, n = 365$, and $t = 4$ gives the balance

$$5{,}000 \left(1 + \frac{0.08}{365}\right)^{365 \cdot 4} \approx 6{,}885.40.$$

(b) Now we apply formula (11) with $P = 5{,}000$, $r = 0.08$, and $t = 4$ to obtain

$$5{,}000 e^{0.08 \cdot 4} \approx 6{,}885.64.$$

2.3   Logarithmic Functions

Suppose you invest \$1,000 in an account paying 6% interest, compounded continuously. How long will it take for your money to double? From our work in the last section—and from formula (11), in particular—we know that the amount at the end of t years is $1{,}000 e^{0.06t}$, so we have to solve the equation

$$2{,}000 = 1{,}000 e^{0.06t},$$

which simplifies, after dividing both sides by 1,000, to

$$2 = e^{0.06t}.$$

But here we run into a difficulty we have not met before—the unknown t is in the exponent! How do we solve such an equation? We will answer that question in Section 2.4 after we have learned about logarithms.

What is a logarithm?

A **logarithmic function** is the inverse of an exponential function. We recall from formula (4) of Section 0.2 that two functions, f and g, are *inverses* if they satisfy the relation

$$y = f(x) \iff x = g(y),$$

where $\iff$ means that any pair (x, y) satisfies the first equation if and only if it satisfies the second. In order for f to have an inverse, it must be one-to-one, which means that its graph must pass the horizontal line test: *No horizontal line meets the graph more than once* (see Section 0.2). Every exponential function $f(x) = b^x$ passes that test, as we saw in Section 2.1, since it is increasing if $b > 1$ and decreasing if $0 < b < 1$. Therefore, we can invert the exponential function to obtain a new function, denoted by $\log_b$ and called the **logarithmic function with base b**. This discussion leads to the following definition.

Definition 2.3.1

The *logarithmic function with base b*, where b is any positive number other than 1, is defined by

$$\log_b x = y \iff b^y = x, \quad x > 0. \tag{13}$$

The output value $\log_b x$ is called the **logarithm of x to the base b**. Informally, we can think of $\log_b x$ as the **exponent** to which the base b must be raised to give x.

For example, to compute $y = \log_2 8$, we must solve the equation $2^y = 8$, which gives $y = 3$. Similarly, to compute $y = \log_2 \frac{1}{8}$, we must solve the equation $2^y = \frac{1}{8}$, which gives $y = -3$. Observe that the input of a logarithmic function must be a positive number because the output of an exponential function is positive.

The following table emphasizes the idea of the logarithm as an exponent:

x	10^4	10^3	10^2	10	1	10^{-1}	10^{-2}	10^{-3}	10^{-4}
$\log_{10} x$	4	3	2	1	0	-1	-2	-3	-4

EXAMPLE 2.3.1

Compute $\log_5 25$, $\log_3 \frac{1}{3}$, and $\log_2 \sqrt{2}$.

SOLUTION   We have

$$y = \log_5 25 \iff 25 = 5^y. \quad \text{Therefore, } \log_5 25 = 2.$$

$$y = \log_3 \tfrac{1}{3} \iff \tfrac{1}{3} = 3^y. \quad \text{Therefore, } \log_3 \tfrac{1}{3} = -1.$$

$$y = \log_2 \sqrt{2} \iff \sqrt{2} = 2^y. \quad \text{Therefore, } \log_2 \sqrt{2} = \tfrac{1}{2}.$$

Most logarithms are not as simple to compute, and we have to use a calculator. To find $\log_{10} 13$, for instance, we use a calculator, which gives $\log_{10} 13 \approx 1.1139434$. (Here we have used the symbol $\approx$ because the answer is not exact. The exact answer is an infinite decimal—an irrational number, in fact. We have rounded it off to seven decimal places.)

The graph of $\log_b x$ for $b > 1$

Since $y = \log_b x$ is the inverse of the exponential function $y = b^x$, its graph is obtained by reflecting the graph of $y = b^x$ across the diagonal line $y = x$. The result is illustrated by Figure 2.3.1, where the graphs of $y = b^x$ and $y = \log_b x$ are shown for $b = 2$ and $b = 10$.

Because of the symmetric relationship between the exponential and logarithmic functions, there is a duality between their properties. It is instructive to compare the following properties of the logarithmic function—as displayed in the graphs of Figure 2.3.1—with the corresponding properties of the exponential function with base $b > 1$.

Properties of the logarithmic function with base $b > 1$

- Its graph goes through the point $(1, 0)$, since $\log_b 1 = 0$.
- Its domain is $(0, \infty)$ and its range is $(-\infty, \infty)$.
- It is continuous and increasing.
- $\lim_{x \to \infty} \log_b x = \infty$ and $\lim_{x \to 0^+} \log_b x = -\infty$.

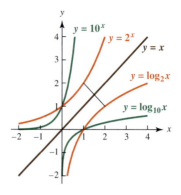

Figure 2.3.1

Remark The graph and properties of the logarithmic function $\log_b x$ for $0 < b < 1$ are discussed in the exercises that follow. However, the most frequently used bases are 2, 10, and e, which are all greater than 1.

The laws of logarithms

The basic properties of logarithms are listed below. Because the input in a logarithm has to be positive, the numbers s and t in the following formulas have to be positive:

1. $\log_b(st) = \log_b s + \log_b t.$
2. $\log_b\left(\dfrac{s}{t}\right) = \log_b s - \log_b t.$
3. $\log_b(t^r) = r \log_b t$ for any number r.
4. $\log_b 1 = 0.$
5. $\log_b\left(\dfrac{1}{t}\right) = -\log_b t.$

Each of these rules is the counterpart of one of the laws of exponents, to which it is related by an input-output switch. For example, the first law of exponents, $b^{u+v} = b^u b^v$, says that the output is a product when the input is a sum. By contrast, the first law of logarithms says that the output is a sum when the input is a product.

Similarly, we can interpret the third law of exponents, $b^{ru} = (b^u)^r$, as saying that the output is raised to a power when the input is multiplied by a coefficient, whereas the corresponding law of logarithms—rule 3 above—says that the output is multiplied by a coefficient when the input is raised to a power.

A more precise way to derive each of the logarithm laws from the corresponding rule for exponents is by applying the basic logarithm-exponential relation expressed in formula (13). For instance,

$$b^0 = 1 \iff \log_b 1 = 0,$$

from which we see that rule 4 for logarithms is equivalent to the fourth law of exponents given in Section 2.1.

Deriving the other rules is a little more complicated, but not much. For example, here is how to get rule 1 for logarithms from the corresponding rule for exponents. If s and t are positive numbers, we let

$$u = \log_b s \quad \text{and} \quad v = \log_b t. \tag{14}$$

According to formula (13), these are the same as the equations

$$s = b^u \quad \text{and} \quad t = b^v.$$

Combining this with the first law of exponents gives

$$st = b^u b^v = b^{u+v}.$$

Using formula (13) once again, we see that this last equation is equivalent to

$$\log_b st = u + v.$$

Finally, we can replace u and v by using (14). The result is

$$\log_b st = \log_b s + \log_b t,$$

which is exactly rule (1) for logarithms.

The other laws of logarithms can be derived in a similar way from the laws of exponents by using the basic logarithm-exponential relation.

Logarithms with base 10

As with exponentials, any positive number other than 1 can serve as the base of a logarithmic function, but certain bases are more commonly found in applications—the most widely used being the numbers 10 and e. Logarithms with base 10 are sometimes called **common logarithms**, and many calculators have a special key for them—usually marked simply as **log**, with no base indicated. In the days before hand calculators and high-speed computers, common logarithms were an indispensable aid to calculating. For example, the monumental calculations made by Johannes Kepler in his studies of planetary motion would have been impossible to achieve within one lifetime without using logarithms. The reason that logarithms make calculations so much easier is that they convert products into sums (rule 1), and addition is a lot faster than multiplication. (Think how long it would take you to multiply two large numbers, such as 1,233,659 and 835,414, compared to the time it would take you to add them.) They also convert exponents into coefficients (rule 3), which greatly reduces the complexity of calculations.

Common logarithms are used in a number of well-known applications. Here are a few examples.

The Richter scale   As you probably know, this is the standard index used in assessing the severity of an earthquake. Here is how it works—actually, a somewhat simplified version of how it works, but it shows the main idea.

We take an earthquake of minimal intensity as reference point—think of it as being at zero level—and let A be the amplitude of its seismic wave. (You do not have to know precisely what that means. Just think of it as a measure of the intensity of the quake.) The Richter value of any other earthquake is then defined by the formula

$$\text{Richter value} = \log_{10}\left(\frac{x}{A}\right),$$

where x is the amplitude of the seismic wave of the earthquake in question.

In other words, we compare the given earthquake to the reference one by taking the ratio of its amplitude, x, to the amplitude of the reference quake, A. Then we take the common logarithm.

The reference earthquake—that is, the one with minimal amplitude—has value zero on the Richter scale. That is because $x = A$ in that case, so the Richter value is $\log_{10} 1$, which equals zero.

An increase of 1 on the Richter scale indicates a tenfold increase in the amplitude of the seismic wave. To see why, suppose one earthquake (call it earthquake E_1) has amplitude x and a second earthquake (call it earthquake E_2) has amplitude $10x$; then,

$$\text{Richter value of earthquake } E_2 = \log_{10}\left(\frac{10x}{A}\right)$$

$$= \log_{10} 10 + \log_{10}\left(\frac{x}{A}\right)$$

$$= 1 + \text{Richter value of earthquake } E_1.$$

EXAMPLE 2.3.2

One of the worst earthquakes in history occurred in Tokyo in 1923 and registered 8.3 on the Richter scale. A more recent earthquake in California in 1989 registered 7.2. How do these earthquakes compare, as measured by the amplitudes of their seismic waves?

SOLUTION　If we let x_T be the amplitude of the seismic wave of the Tokyo quake, then $8.3 = \log_{10}(x_T/A)$, which is the same as

$$\frac{x_T}{A} = 10^{8.3}. \tag{15}$$

Similarly, using x_C for the amplitude of the seismic wave of the California quake, we get

$$\frac{x_C}{A} = 10^{7.2}. \tag{16}$$

If we divide Eq. (15) by Eq. (16), the A's cancel, and we get

$$\frac{x_T}{x_C} = \frac{10^{8.3}}{10^{7.2}} = 10^{1.1} \approx 12.59.$$

In other words, the 1933 Tokyo quake was about 12.59 times more intense than the one in California in 1989.

HISTORICAL PROFILE

John Napier (1550–1617) is credited with inventing the concept of the logarithm. He was the son of one of the most important families in Scotland and held extensive estates, which he managed with great skill. He was also known for being a great genius at inventing as well as for his religious fervor, which involved him in a number of heated controversies. Mathematics to him was a hobby, to be pursued in the time off from his managerial and religious pursuits. Among his other mathematical contributions is a device called **Napier's bones**, for mechanically multiplying, dividing, and taking square roots and cube roots. In the preface to his treatise on logarithms, he wrote, "Seeing that there is nothing so troublesome to mathematical practice nor that doth more molest and hinder calculators, than the multiplications, divisions, square and cubical extractions of great numbers, I began therefore to consider in my mind by what art I could remove these hindrances."

A similar use of common logarithms occurs in the decibel scale that is used as a measure of noise level.

The decibel scale   This works in much the same way as the Richter scale. We take a minimal audible sound wave as reference point and let I be its amplitude. For any sound of different amplitude, say, x, we measure the noise level by the formula

$$\text{Noise level in decibels } = 10\log_{10}\left(\frac{x}{I}\right).$$

This differs from the Richter scale formula only in the coefficient 10 on the right-hand side, which is simply a convenience factor to keep the numbers on the scale from being too small.

In this case, a tenfold increase in sound amplitude, say, from x to $10x$, adds 10 to the decibel rating. That is because

$$10\log_{10}\left(\frac{10x}{I}\right) = 10\left[\log_{10}10 + \log_{10}\left(\frac{x}{I}\right)\right]$$
$$= 10\left[1 + \log_{10}\left(\frac{x}{I}\right)\right]$$
$$= 10 + 10\log_{10}\left(\frac{x}{I}\right).$$

EXAMPLE 2.3.3

An advertisement for an electric generator claims that the noise level is one-fourth that of a competing model. What is the difference in decibels?

SOLUTION   Let x be the amplitude of the sound wave given off by the competing model. According to the manufacturer's claim, its generator produces a soundwave with amplitude $0.25x$. The decibel levels compare as follows:

$$10 \log_{10}\left(\frac{0.25x}{I}\right) = 10 \log_{10}(0.25) + 10 \log_{10}\left(\frac{x}{I}\right) = -6.02 + 10 \log_{10}\left(\frac{x}{I}\right).$$

In other words, the decibel level of the advertised model is 6.02 less than the competing brand.

The next application of common logarithms is important in biology.

The pH scale   This scale measures the acidity of a chemical solution, which depends on the concentration of hydrogen ions in the solution. That concentration is symbolized by $[H^+]$, and it is measured in moles of hydrogen ions per liter. Once again, you do not need to know the precise meaning of those words. What counts is that the larger $[H^+]$ is, the greater the acidity of the solution. The pH formula is

$$\text{pH value} = -\log_{10}[H^+]. \tag{17}$$

The reason for the minus sign is that $[H^+]$ is usually very small, so that its logarithm is negative, and the minus sign makes the pH value positive. It has the effect of giving the more acidic solutions—those for which $[H^+]$ is larger—a lower pH value.

Pure water has $[H^+] = 10^{-7}$, which means that its pH value is 7. Substances with a pH value under 7, such as vinegar and lemon juice, are called acids. They taste sour and react strongly with metals. Those with a pH value above 7, such as lye and ammonia, are called bases. They taste bitter and feel slippery. Both acids and bases at extreme levels can be dangerous, causing skin burns among other things.

EXAMPLE 2.3.4

How much greater is the concentration of hydrogen ions in lemon juice, which has a pH value of 2.2, than in pure water?

SOLUTION   As we noted above, pure water has $[H^+] = 10^{-7}$. To find the corresponding number for lemon juice, we use Eq. (17):

$$2.2 = -\log_{10}[H^+],$$

which means that $[H^+] = 10^{-2.2}$ for lemon juice. Taking the ratio of these numbers gives

$$\frac{10^{-2.2}}{10^{-7}} = 10^{4.8} \approx 63{,}096.$$

In other words, the concentration of hydrogen ions in lemon juice is more than 63,000 times greater than in pure water.

Practice Exercises 2.3

1. Given the approximations $\log_3 2 \approx 0.631$ and $\log_3 5 \approx 1.465$, approximate the following without using your calculator:

(a) $\log_3(12)$ (b) $\log_3(0.4)$

2. Use a calculator to find x correct to at least seven decimal places *without using the equation solver*.

(a) $\log_{10}(10x) = 3.17$ (b) $(0.1)^x = 42.03$

Exercises 2.3

In Exercises 1–16, express the given logarithm as an integer or fraction without using a calculator.

1. $\log_2 8$

2. $\log_2 \frac{1}{4}$

3. $\log_3 1$

4. $\log_9 3$

5. $\log_9 27$

6. $\log_8 4$

7. $\log_{1/2} 2$

8. $\log_{(0.1)} 100$

9. $\log_{10}(0.001)$

10. $\log_{(0.01)} 1$

11. $\log_7 7$

12. $\log_5 \sqrt{5}$

13. $\log_2 \sqrt{8}$

14. $\log_{1/3}(3\sqrt{3})$

15. $\log_{10}\left(\dfrac{1}{\sqrt{10}}\right)$

16. $\log_{(0.1)}(10^{-1/2})$

17. Use the approximation $\log_{10} 4 \approx 0.602$ to estimate the following without using a calculator:

(a) $\log_{10} 40$ (b) $\log_{10} 0.4$
(c) $\log_{10} 0.25$ (d) $\log_{10} 2$

18. Use the approximation $\log_{10} 0.5 \approx -0.301$ to estimate the following without using a calculator:

(a) $\log_{10} 50$ (b) $\log_{10} 25$ (c) $\log_{10} 20$ (d) $\log_{10} 8$

In Exercises 19–24, use a calculator to find x correct to at least seven decimal places without using the equation solver or trace function. Then use the equation solver to check your answer.

19. $\log_{10} x = 1.463$

20. $10^x = 214.11$

21. $\log_{10}\left(\dfrac{x}{100}\right) = 0.418$

22. $\log_{10}\left(\dfrac{10}{x}\right) = 0.103$

23. $100^x = 204.5$

24. $\log_{10}\sqrt{x} = 1.233$

In Exercises 25–36, use the approximations $\log_2 3 \approx 1.585$ and $\log_2 5 \approx 2.322$ to approximate the given logarithm. Do not use a calculator.

25. $\log_2 9$

26. $\log_2 \frac{1}{5}$

27. $\log_2 \frac{5}{9}$

28. $\log_2 \sqrt{5}$

29. $\log_2 6$

30. $\log_2 0.75$

31. $\log_2 \sqrt{10}$

32. $\log_2 24$

33. $\log_2 30$

34. $\log_2 75$

35. $\log_2 \sqrt{15}$

36. $\log_2 \frac{9}{10}$

In Exercises 37–45, suppose that A and b are positive numbers with $\log_3 A = b$. Write the given logarithm as a function of b.

37. $\log_3(9A)$

38. $\log_3\left(\dfrac{A^2}{3}\right)$

39. $\log_3(\sqrt{3A})$

40. $\log_3\left(\dfrac{3}{\sqrt{A}}\right)$

41. $\log_9 A$ (*Hint:* Use the relation $\log_9 A = v \Longleftrightarrow 9^v = A$.)

42. $\log_{1/3} A$

43. $\log_{\sqrt{3}} A$

44. $\log_A 3$

45. $\log_{\sqrt{A}} 3$

46. In August 1999, a severe earthquake in Turkey registered 7.4 on the Richter scale. In September of that same year, an even more severe one in Taiwan registered 7.6. How much more severe was the earthquake in Taiwan in terms of the amplitude of its seismic wave?

47. If an earthquake registers 6.2 on the Richter scale, what will an earthquake whose seismic amplitude is four times larger register?

48. If an earthquake has a seismic amplitude 5,000 times as large as the minimal amplitude of the reference earthquake, what will it register on the Richter scale?

49. A stereo headset whose volume is set at 6 has a sound level of 115 decibels. How much bigger is the amplitude of its sound wave than that of normal conversation, which registers 60 decibels?

50. A company that manufactures back-up alarm signals for trucks has models whose sound levels range from 87 to 107 decibels. How much more intense is the sound of the loudest model than that of the softest?

51. The sound of a leaf blower is 300 times stronger at the operator's ear than at a distance of 50 feet. If it registers 75 decibels at 50 feet, what does it register at the operator's ear?

52. The pH value of beer is approximately 4, and that of vinegar is approximately 3. Which is more acidic and how much greater is the concentration of hydrogen ions?

53. In testing for acid rain at the Great Smoky Mountains National Park, field workers got pH readings as low as 3.6. How many times more acidic is that, in terms of the hydrogen ion concentration, than normal rain, which has a pH value of 5.6?

54. **(a)** Show that $\log_{1/b} x = -\log_b x$ for any positive base b and any positive number x. (*Hint:* Use formula (13).)
(b) If $0 < a < 1$, let $b = 1/a$. Then $b > 1$, and from step (i) we have $\log_b x = -\log_a x$ for all $x > 0$. Use this formula to derive the properties of $\log_a x$ for $0 < a < 1$ from those of $\log_b x$ with $b > 1$.
(c) Refer to the graphs of $\log_2 x$ and $\log_{10} x$ shown in Figure 2.3.1 and use them as a guide to sketch the graphs of $\log_{1/2} x$ and $\log_{0.1} x$.

Solutions to practice exercises 2.3

1. (a) Using the laws of logarithms gives

$$\log_3(12) = \log_3(3 \cdot 2^2) = \log_3(3) + \log_3(2^2)$$
$$= 1 + 2\log_3(2) = 1 + 2 \cdot (0.631) \approx 2.262.$$

(b) $\log_3(0.4) = \log_3 \frac{2}{5} = \log_3 2 - \log_3 5$
$$= 0.631 - 1.465 \approx -0.834.$$

2. (a)

$$\log_{10}(10x) = 3.17 \iff 10x = 10^{3.17} \iff x = 10^{2.17}.$$

A calculator gives $10^{2.17} = 147.9108388$ correct to seven decimal places.

(b)

$$(0.1)^x = 42.03 \iff 10^{-x}$$
$$= 42.03 \iff x = -\log_{10} 42.03.$$

A calculator gives $\log_{10} 42.03 = 1.6235594$, correct to seven decimal places, which means that $x = -1.6235594$.

2.4 Natural Logarithm and Applications

The logarithmic function with base e is called the **natural logarithm**. It plays a prominent role in calculus, and is by far the most important logarithm in applications. Every scientific calculator has a natural logarithm key, usually labeled **ln**.

Definition 2.4.1

The logarithmic function with base e is called *the natural logarithm* and is denoted by ln. That is,

$$\ln x = \log_e x, \quad x > 0.$$

It is the inverse of the natural exponential function. That is,

$$y = \ln x \iff e^y = x, \quad x > 0. \tag{18}$$

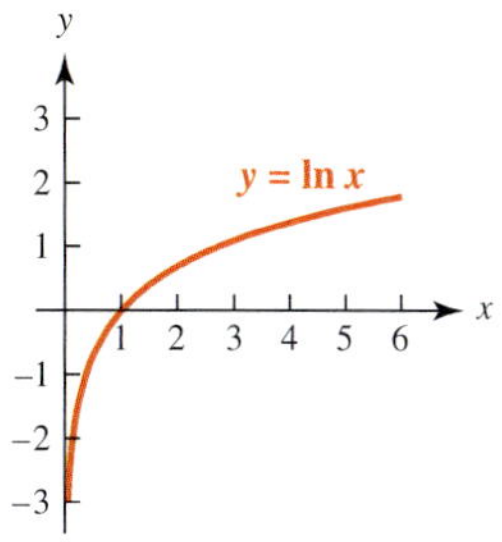

The fact that the exponential and logarithm functions are inverses of each other can also be expressed by the following pair of useful formulas:

$$y = e^{\ln y}, \quad y > 0 \qquad \text{and} \qquad \ln(e^x) = x, \quad \text{any } x. \tag{19}$$

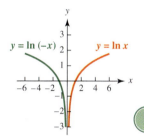

y = ln x

The graph of $y = \ln x$ is shown in Figure 2.4.1. Its properties and general appearance are very similar to the graphs of the other logarithmic functions with base greater than 1.

Figure 2.4.1

> **Properties of ln x**
>
> - Its domain is $(0, \infty)$ and its range is $(-\infty, \infty)$.
> - It is continuous and increasing.
> - $\lim_{x \to \infty} \ln x = \infty$ and $\lim_{x \to 0^+} \ln x = -\infty$.
> - $\ln 1 = 0$, $\ln e = 1$, and $\ln(1/e) = -1$.

These last three values of ln are important enough that you should commit them to memory. Other values are more difficult to compute, and we usually rely on a calculator. Of course, the answers in that case are only approximations, accurate to the number of decimal places given by the particular calculator. For instance, a calculator with accuracy to seven places gives $\ln 2 \approx 0.6931472$, while one with nine-place accuracy yields $\ln 2 \approx 0.693147181$. You may be curious to know what procedure is used to get these approximations. In fact, there is an active area of mathematical research, known as **numerical analysis**, devoted to the study of approximation methods. Later on in this text you will learn a few of these techniques and, in particular, a method for approximating $\ln 2$.

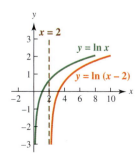

$y = \ln(-x)$ $y = \ln x$

The two branches of $y = \ln|x|$

Figure 2.4.2

EXAMPLE 2.4.1

Sketch the graphs of each of these functions:

> (i) $f(x) = \ln|x|, \quad x \neq 0$ and   (ii) $g(x) = \ln(x - 2), \quad x > 2.$

SOLUTION (i) The graph of $y = \ln|x|$, shown in Figure 2.4.2, has two branches. One is over the interval $(0, \infty)$ and is the graph of the function $y = \ln x$. The other is over the interval $(-\infty, 0)$ and is the graph of $y = \ln(-x), \ x < 0$. It is obtained by reflecting the graph of $y = \ln x$ in the y-axis.

(ii) The graph of $y = \ln(x - 2)$ is the horizontal translation of the graph of $y = \ln x$ by two units to the right. It is shown in Figure 2.4.3.

Converting from base b to base e

Figure 2.4.3

The first important application of the natural logarithm is to convert an exponential expression with base $b > 0$ into an equivalent expression with base e. Using (19), we have $b = e^{\ln b}$. Now the third law of exponents gives the **conversion formula**

$$b^x = e^{x \ln b}. \tag{20}$$

As you will see later, this conversion is very convenient because it simplifies a number of calculations.

EXAMPLE 2.4.2

Convert $3^{1/2}$ to an exponential with base e.

SOLUTION   Applying formula (20) with $b = 3$ and $x = \frac{1}{2}$ gives

$$3^{1/2} = e^{(\ln 3)/2}.$$

Exponential growth and decay

We can use Eq. (20) to rewrite the exponential growth/decay formula of Section 2.1 —that is,

$$y = y_0 b^t \quad \text{for } t \geq 0,$$

in the form

$$y = y_0 e^{(\ln b)t} \quad \text{for } t \geq 0.$$

If $b > 1$, this equation describes exponential growth, which we encountered in continuously compounded interest and simple population growth. The constant $\ln b$ is called the **growth constant**. On the other hand, if $b < 1$, the equation describes exponential decay, as in the case of radioactive material. In that case, the constant $|\ln b|$ is called the **decay constant**.

EXAMPLE 2.4.3

In the last section we considered the decay equation of radon gas in the form

$$y = y_0 (0.835)^t \quad \text{for } t \geq 0,$$

where y_0 is the initial amount. Convert this formula into an exponential with base e and find the decay constant of radon gas.

SOLUTION   Using formula (20), we can rewrite the decay equation in the form

$$y = y_0 e^{t \ln(0.835)}.$$

Using a calculator, we get $\ln(0.835) \approx -0.1803$ (to four decimal places), which gives the formula

$$y \approx y_0 e^{-0.1803t}. \tag{21}$$

The constant $|\ln(0.835)|$, which is approximately 0.1803, is the decay constant for radon gas.

The decay constant of a radioactive substance is usually determined experimentally. It depends on the substance and also on the unit of time used. If we know the decay constant k for a particular substance, we can write the decay equation in the form

$$y = y_0 e^{-kt} \quad \text{for } t \geq 0. \tag{22}$$

EXAMPLE 2.4.4

The radioactive substance plutonium-241 has decay constant $k = 0.0525$, with time measured in years. Starting with an initial amount of 50 milligrams, how much will remain after 10 years?

SOLUTION The decay equation is

$$y = y_0 e^{-0.0525t}, \tag{23}$$

where y_0 is the initial amount and y is the amount after t years.

Substituting $y_0 = 50$ and $t = 10$ in Eq. (23) gives the amount at the end of 10 years:

$$y = 50e^{-0.525} \approx 29.58 \text{ milligrams}.$$

Converting logarithms to base e

Just as there is a formula for converting exponentials from base b to base e, there is also a formula for converting logarithms with base b to a quotient of natural logarithms. It reads as follows:

$$\boxed{\log_b x = \frac{\ln x}{\ln b}.} \tag{24}$$

To derive this formula, we let $y = \log_b x$, which is equivalent to $x = b^y$. Taking the natural logarithm on both sides of this last equation gives

$$\ln x = \ln(b^y) = y \ln b,$$

and solving for y yields

$$y = \frac{\ln x}{\ln b}.$$

Finally, replacing y by $\log_b x$ gives formula (24).

Formula (24) gives you a way to compute logarithms with any base even though your calculator may only handle bases 10 and e.

EXAMPLE 2.4.5

Use a calculator to find $\log_3 5$.

SOLUTION   According to formula (24), $\log_3 5 = \ln 5 / \ln 3$. Both the numerator and denominator can be found on a calculator, and taking their quotient gives

$$\log_3 5 \approx \frac{1.6094379}{1.0986123} \approx 1.4649735.$$

Solving equations—doubling time and half-life

An important use of logarithms is in solving equations that involve exponentials. To solve an equation of the form

$$b^t = c$$

where b and c are fixed positive numbers and t is an unknown, we simply take the natural logarithm of both sides of the equation and use the third law of logarithms to get

$$t \ln b = \ln c.$$

Dividing by $\ln b$ gives the solution

$$t = \frac{\ln c}{\ln b}.$$

EXAMPLE 2.4.6

Solve the following equations for t:

$$\text{(i)}\ \ 2 = 5^t \quad \text{and} \quad \text{(ii)}\ \ 4 = \ln(1 + 3t).$$

SOLUTION   (i) Taking the natural logarithm on both sides gives

$$\ln 2 = \ln(5^t) = t \ln 5.$$

Therefore, $t = \ln 2 / \ln 5 \approx 0.4307$.

(ii) Exponentiating both sides, we get

$$e^4 = e^{\ln(1+3t)} = 1 + 3t.$$

Solving for t gives $t = (e^4 - 1)/3 \approx 17.866$.

PROBLEM-SOLVING TACTIC

- To solve an equation with the unknown in an exponent, take the logarithm of both sides.

- To solve an equation with the unknown inside a logarithm, exponentiate both sides.

Equations involving exponentials, in which the unknown appears in the exponent, arise in problems involving compound interest, radioactive decay, and other exponential growth/decay problems. One such problem was posed at the beginning of Section 2.3, and we are now in a position to solve it.

EXAMPLE 2.4.7

Suppose you deposit \$1,000 in an account paying an annual interest rate of 6%, compounded continuously. How long will it take for the account to reach \$2,000?

SOLUTION The total amount at the end of t years is given by the formula

$$y = 1{,}000e^{0.06t}. \tag{25}$$

We want to know when $y = 2{,}000$, which means we have to solve the equation

$$1{,}000e^{0.06t} = 2{,}000.$$

Dividing both sides by 1,000 gives

$$e^{0.06t} = 2,$$

and taking the natural logarithm gives

$$(0.06)t = \ln 2.$$

Dividing by 0.06 and using a calculator, we get the answer

$$t = \frac{\ln 2}{0.06} \approx 11.55.$$

Thus, it takes a little over $11\frac{1}{2}$ years to double the initial investment.

There are a couple of things worth noting about the last example. First, the time it takes to double your investment does not depend on the amount you start with! That may surprise you, but you can easily check it out by redoing all the calculations we just did, replacing 1,000 by y_0 and 2,000 by $2y_0$. No matter what y_0 you start with, you get the equation

$$e^{0.06t} = 2,$$

whose solution is the one we just found. The point is that y_0 gets cancelled out on both sides of the equation, so the solution does not depend on it.

Second, instead of just finding out how long it takes your investment to double, you can use the same method to see how long it takes to triple or quadruple it, and so forth. As long as the question is posed in terms of a *multiple* of the original amount, the answer does not depend on the size of the initial investment.

EXAMPLE 2.4.8

Suppose you invest a certain sum at 5.5% annual interest, compounded continuously. How long does it take to double your initial investment? to triple it?

SOLUTION   The amount at the end of t years is given by the formula

$$y = y_0 e^{0.055t} \quad \text{for } t \geq 0.$$

To find the time required to double the initial amount, we must solve the equation

$$2y_0 = y_0 e^{0.055t}.$$

By canceling the y_0's, we get $2 = e^{0.055t}$, which is equivalent to $\ln 2 = 0.055t$. Solving for t gives

$$t = \frac{\ln 2}{0.055} \approx 12.6.$$

Similarly, to find the time required for the initial investment to triple, we have to solve the equation $3y_0 = y_0 e^{0.055t}$. Once again, we can cancel the y_0's, take the natural logarithm on both sides, and solve for t. We get

$$t = \frac{\ln 3}{0.055} \approx 19.97.$$

The same method can be used to answer questions about radioactive decay, only in that case the initial amount *decreases* as time goes on. So, instead of asking how long it takes to double, one can ask how long it takes for an initial amount of radioactive material to be reduced by half. That number is called the **half-life** of the substance. It depends on the decay constant but not on the initial amount.

EXAMPLE 2.4.9

Radium-226 is a radioactive substance whose decay constant is 0.000436, with time measured in years. How long does it take a given quantity of radium to decay to half the initial amount? In other words, what is the half-life of radium-226?

SOLUTION   The decay equation is

$$y = y_0 e^{-0.000436t}.$$

To find the half-life, we have to solve the equation

$$\frac{y_0}{2} = y_0 e^{-0.000436t}.$$

Dividing both sides by y_0 and taking the natural logarithm give

$$\ln \tfrac{1}{2} = -0.000436t,$$

so that

$$t = -\frac{\ln \tfrac{1}{2}}{0.000436} \approx 1{,}590 \text{ years}.$$

Notice that the answer is positive despite the minus sign. That is because $\ln(\tfrac{1}{2})$ is negative. In fact, $\ln(\tfrac{1}{2}) = -\ln 2$, so we could also write the solution as

$$t = \frac{\ln 2}{0.000436} \approx 1{,}590.$$

In Example 2.4.9 we found the half-life to be equal to $\ln 2$ divided by the decay constant. That formula is true for any radioactive substance:

$$\text{Half-life} = \frac{\ln 2}{\text{decay constant}}. \tag{26}$$

HISTORICAL PROFILE

Marie Sklodowska Curie (1867–1934) was one of the great scientists of the twentieth century. Among her many discoveries were the radioactive substances radium and polonium (which she named in honor of Poland, her native country). She was born in Warsaw, where her mother was a musician and teacher, and her father was a professor of mathematics and physics. After working as a teacher and governess, she went to Paris in 1891 to further her education in science. Her marriage to Pierre Curie 3 years later was the start of a scientific partnership whose results profoundly changed our understanding of the physical world. Together, they did groundbreaking work in the study of radioactivity, for which they won the Nobel Prize in physics in 1903. After Pierre's untimely death in 1906, she continued her research on radioactive elements and was awarded the Nobel Prize for chemistry in 1911.

She was the first woman to teach at the Sorbonne in Paris, the first woman to receive a Nobel prize, and the first person to receive two of them. She also became the only Nobel prize winner to be the mother of another Nobel Laureate, when her daughter Irene Joliot-Curie won the prize for chemistry in 1935. Her death from leukemia is believed to have been caused by her repeated exposure to radioactive substances. The spirit that guided her life is suggested by the following words attributed to her: *Nothing in life is to be feared. It is only to be understood.*

To see why, we simply copy what we did in Example 2.4.9. For any radioactive substance, the decay equation is

$$y = y_0 e^{-kt},$$

where k is the decay constant. To find the half-life, we solve the equation

$$\frac{y_0}{2} = y_0 e^{-kt}.$$

Just as before, we divide by y_0, take ln, and then divide by $-k$, to get

$$t = -\frac{\ln \frac{1}{2}}{k} = \frac{\ln 2}{k}.$$

We can also use Eq. (26) to find the decay constant if we know the half-life.

EXAMPLE 2.4.10

The element carbon has a radioactive form know as carbon-14, whose half-life is approximately 5,730 years. Find its decay constant and decay equation.

SOLUTION Let k be the decay constant of carbon-14. Equation (26) says that

$$5{,}730 = \frac{\ln 2}{k}.$$

Solving for k, we get

$$k = \frac{\ln 2}{5{,}730} \approx 0.000121.$$

The decay equation is

$$y = y_0 e^{-(t \ln 2)/5{,}730} \approx y_0 e^{-0.000121t}.$$

Once we know the decay equation, we can answer many questions about the decay process.

EXAMPLE 2.4.11

How long does it take an initial amount of carbon-14 to diminish by 15%?

SOLUTION Another way to phrase the question is: How long does it take for a given amount y_0 to decay to $(0.85)y_0$. To answer it, we have to solve the equation

$$(0.85)y_0 = y_0 e^{-(t \ln 2)/5{,}730}.$$

Dividing both sides by y_0 and taking logarithms give

$$-\frac{t \ln 2}{5,730} = \ln(0.85).$$

Solving for t gives

$$t = -\frac{5,730 \ln(0.85)}{\ln 2} \approx 1,343.5 \text{ years.}$$

The decay of carbon-14 is the basis of the method of **carbon dating**, invented in the 1940s by Willard Libby, an American chemist and Nobel Laureate. We will come back to that method in Chapter 4.

Practice Exercises 2.4

1. In 1980 the U.S. population was about 225 million, and in 1990 it was about 250 million. Assuming that the population is modeled by an exponential function $y = y_0 e^{kt}$, where y is the population in millions and t is the number of years after 1980, determine y_0 and k. Also, predict the population in 2020.

2. Plutonium-241 decays according to the formula $y = y_0 e^{-0.0525t}$, where t is measured in years. Find its half-life.

3. If you put money in an account that pays 10% interest, compounded continuously, how long will it take for your money to quadruple?

Exercises 2.4

In Exercises 1–8, rewrite the given expression as an exponential with base e.

1. 3^7 **2.** 5^x **3.** x^5 **4.** $\sqrt{x}$

5. a^{-3} **6.** 3^{-a} **7.** $a^{1/3}$ **8.** $3^{1/a}$

In Exercises 9–16, evaluate the given logarithm as a number in decimal form. Do not use a calculator.

9. $\ln\left(\dfrac{1}{\sqrt{e}}\right)$ **10.** $\ln(e^{-2})$

11. $e^{\ln 5}$ **12.** $\ln(e^{0.23})$

13. $\ln\left(\dfrac{1}{e^{0.4}}\right)$ **14.** $e^{-\ln 2}$

15. $e^{2 \ln 3}$ **16.** $e^{3 \ln 2}$

In Exercises 17–22, use the approximation $\ln 2 \approx 0.693$ to approximate the given quantity. Do not use a calculator.

17. $\ln(2e)$ **18.** $\ln(\sqrt{2e})$ **19.** $\ln\frac{1}{4}$

20. $\ln(e^{2-\ln 2})$ **21.** $\ln(e^{1+\ln 4})$ **22.** $\ln 8$

In Exercises 23–34, simplify the given expression as much as possible.

23. $e^{(\ln 4)/2}$ **24.** $e^{-4 \ln 2}$ **25.** $e^{\ln 3 + 3 \ln t}$

26. $e^{t - \ln t}$ **27.** $\ln\left(\dfrac{1}{e^{3x}}\right)$ **28.** $e^{-\ln x}$

29. $\ln(e^{1/x})$ **30.** $e^{\ln(x^2+1)}$ **31.** $e^{2 \ln(x-1)}$

32. $\ln(e^{2x+1}) - x$ **33.** $e^{\ln(2x)+\ln(2/x)}$

34. $\ln(e^{\sqrt{x}}) - e^{(\ln x)/2}$

In Exercises 35–46, solve the given equation for x. Do not use a calculator. Your answer may involve natural logarithms or the number e.

35. $e^{2x+1} = 7$ **36.** $e^{x/2} = 7$

37. $e^{x-2} = 1$ **38.** $e^{x^2} = 10$

39. $\ln(x+1) = 1$ **40.** $\ln(2x) = 3$

41. $\ln(x^3) = 5$ **42.** $\ln(2x-3) = -1$

43. $e^{3 \ln x} = 8$ **44.** $e^{x \ln 3} = 8$

45. $\ln(2x-2) = 0$ **46.** $\ln(\ln x) = 0$

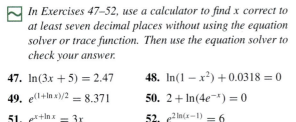

In Exercises 47–52, use a calculator to find x correct to at least seven decimal places without using the equation solver or trace function. Then use the equation solver to check your answer.

47. $\ln(3x + 5) = 2.47$ **48.** $\ln(1 - x^2) + 0.0318 = 0$

49. $e^{(1+\ln x)/2} = 8.371$ **50.** $2 + \ln(4e^{-x}) = 0$

51. $e^{x+\ln x} = 3x$ **52.** $e^{2\ln(x-1)} = 6$

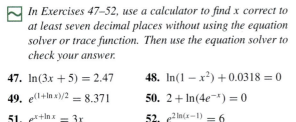

In Exercises 53–56, write the given expression as a quotient of natural logarithms. Then use a calculator to find a decimal approximation.

53. $\log_5 12$ **54.** $\log_{12} 5$

55. $\log_2 10$ **56.** $\log_2 e$

57. Use formula (24) to show that if a and b are positive numbers,

$$\log_a b = \frac{1}{\log_b a}.$$

(*Hint:* Write both sides as quotients of natural logarithms.)

In Exercises 58–61, sketch the graph of the given function by referring to the graph of $y = \ln x$ (shown in Figure 2.4.1).

58. $y = \ln(1 + x)$ **59.** $y = 1 + \ln x$

60. $y = \ln\left(\dfrac{1}{x}\right)$ **61.** $y = \ln\left(\dfrac{x}{e}\right)$

In Exercises 62–65, match the equation to one of the graphs in Figure 2.4.4.

62. $y = \ln(\sqrt{x})$ **63.** $y = \ln(3e^{2x})$

64. $y = \ln(3x)$ **65.** $y = \ln\left(\dfrac{1}{x^2}\right)$

66. If \$4,000 is deposited in an account paying 6% interest per year, compounded continuously, how long will it take for the balance to reach \$6,000?

67. If you put money in an account that pays 7.5% interest, compounded continuously, how long will it take for your money to triple?

68. Suppose you invest \$5,000 in an account that pays interest, compounded continuously.
(a) How much money is in the account after 8 years if the annual interest rate is 4%?
(b) If you want the account to contain \$8,000 after 8 years, what annual interest rate is needed?

69. In an experiment on learning patterns, the psychologists Miller and Dollard recorded the time it took a 6-year-old girl to find a hidden piece of candy in a series of tries. It took her 210 seconds to find the candy on the first try, and it took 86 seconds on the second try. Assume that the

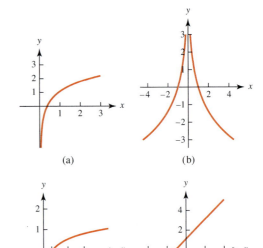

(a)   **(b)**

(c)   **(d)**

Figure 2.4.4

amount of time it took is modeled by $T = Ae^{-kn}$, where n is the number of tries and k is a constant.
(a) Find the constants A and k.
(b) If the model is correct, how much time should it take the girl on the ninth try? (The actual experimental result was 2 seconds.)

Source: N. Miller and J. Dollard, *Social Learning and Imitation,* Yale University Press, 1941.

70. A crime scene investigator knows that h hours after death, the human body has a temperature of

$$T = T_a + (98.6 - T_a)(0.6)^h,$$

where T_a is the temperature (in degrees Fahrenheit) of the air surrounding the body. Find the time of death of a person found in a room of 72°F constant air temperature, when its body temperature is 78°F.

71. A colony of insects had a population of 4,000 when first observed, and 5 days later it had grown to 6,000. Assume the population is growing exponentially.
(a) Find a formula of the form $y = Ae^{kt}$, where y is the population and t is the number of days after the initial observation. Use a calculator to determine k to four decimal places.
(b) According to this formula, what will the population be 8 days after the initial observation?
(c) According to this formula, how many days after the initial observation will the population reach 12,000?

72. In 1940 the U.S. population was 131.67 million, and in 2000 it was 281 million.

(a) Assuming that the population growth is exponential, find a formula of the form $y = Ae^{kt}$, where y is the population in millions and t is the number of years after 1900. Use a calculator to determine k to five decimal places and A to two decimal places.

(b) According to this equation, what will the U.S. population be in the year 2020?

(c) According to this equation, what was the U.S. population in the year 1900? (According to census figures, the actual population was 75.99 million.)

73. Assume that for the past 10 years you have been studying a certain population and your data indicate that it could be modeled by a function of the form

$$p(t) = a - be^{-rt},$$

with p in millions. Choose the parameters a, b, and r so that the model gives an initial population of 5 million, a population of 6 million at the end of 10 years, and a limiting value of 8 million for the population as t goes to ∞.

74. In analyzing human population growth, a type of equation known as a **logistic equation** gives a more realistic model than exponential growth. For example, one estimate of the world's population is given by the following logistic equation:

$$P = \frac{73.2}{6.1 + 5.9e^{-0.016t}},$$

where t is the number of years after 2000 and P is the size of the world's population in billions. For $t = 0$, this gives $P = 6.1$, which was the approximate size of the world's population in 2000.

(a) According to this logistic equation, what will the earth's population be in the year 2050?

(b) When will the earth's population reach 10 billion?

(c) What is the long-range tendency of the world's population? In other words, what is $\lim_{t \to \infty} P$?

75. Using the logistic equation of the previous exercise to model the world's population, make a table showing the population (in billions) in steps of 25 years from the year 2000 to 2500. Toward what limit does the population seem to be tending? Compare that to the answer in part (c) of the last exercise. Also, make a chart or graph of the population.

76. Strontium-90 is one of the radioactive substances produced by nuclear explosions. Its decay constant is approximately 0.0244, with time measured in years.

(a) Starting with an initial amount of 100 milligrams of strontium-90, how much will be left after 12 years?

(b) How long will it take an initial amount of 100 milligrams to decay to 30 milligrams?

(c) Starting with *any* quantity of strontium-90, how long will it take for it to decay to 30% of the initial amount?

(d) Find the half-life of strontium-90.

77. Referring to the previous exercise, make a table showing the amount of strontium-90 over the next 30 years in 2 year steps, starting with an initial amount of 100 grams. Use the table to estimate the half-life and compare it to the answer in part (c) of the previous exercise. Also, make a graph of the amount of strontium-90.

78. Thorium-234 is a radioactive substance with a half-life of 24.5 days.

(a) Given an initial amount A of thorium-234, write a formula that gives the amount at the end of t days.

(b) How long will it take for an initial amount to decrease to one-third of its mass?

79. One hundred milligrams of a certain radioactive substance will decay to 67 milligrams in 5 days. How many milligrams will there be after 7 days?

80. Iodine-131 is a radioactive substance that is used in treating thyroid disorders. An initial amount will diminish by 20% after approximately 2.6 days. Find the half-life of iodine-131.

Solutions to practice exercises 2.4

1. When $t = 0$ (i.e., in 1980), the population was $y(0) = y_0 = 225$ million. When $t = 10$ (in 1990), it was $250 = 225e^{10k}$. Dividing by 225 and taking the logarithm, we have $\ln(250/225) = \ln(e^{10k}) = 10k$. Therefore, $k = 0.1 \ln(250/225) \approx 0.01054$, and $y = 225e^{0.01054t}$. Finally, for the year 2020, $t = 40$, and we get the population $y = 225 \, e^{(0.01054) \cdot 40} \approx 342.99$ million.

2. To find the half-life, we must solve the equation $0.5y_0 = y_0 \, e^{(-0.0525)t}$. After cancelling y_0, we obtain

$0.5 = e^{(-0.0525)t}$. Taking the logarithm, we get $\ln(0.5) = (-0.0525)t$, or $-0.693147 = (-0.0525)t$, which gives $t = 13.20$ years.

3. We use the formula $y = y_0 e^{rt}$. Letting $y = 4y_0$ and $r = 0.1$, we obtain $4y_0 = y_0 e^{0.1t}$. Dividing by y_0 and taking the logarithm gives $\ln 4 = 0.1 \, t$. Therefore, $t = \ln 4 / 0.1 \approx 13.86$ years.

Chapter 2 Summary

- An **exponential function** is of the form $f(x) = b^x$, where b is a positive constant other than 1, called the **base**. It satisfies the basic laws of exponents.

- The most **natural base** is the number

$$e = \lim_{n \to \infty} \left(1 + \frac{1}{n}\right)^n = 2.7182818284590452353602874713526624977757\ldots.$$

  which is the balance at the end of 1 year on an initial investment of \$1 at an annual interest rate of 100%, compounded continuously.

- The balance at the end of t years on an initial investment A at an annual interest rate r compounded continuously is Ae^{rt}.

- A **logarithmic function** is the inverse of an exponential function. That is, for $x > 0$ we have $y = \log_b x \Longleftrightarrow x = b^y$. For each one of the five laws of exponents, there is a corresponding law of logarithms.

- The **natural logarithm** has base e and is denoted by $\ln x$. That is, $y = \ln x \Longleftrightarrow x = e^y$. Therefore, we have $e^{\ln x} = x$ and $\ln e^x = x$. Also, we have $\ln e = 1$ and $\ln 1 = 0$.

- The formula $b^x = e^{(\ln b)x}$ converts an exponential function from base b to base e (it naturalizes).

Chapter 2 Review Questions

- How do you solve the equation $e^{0.05t} = 2$?

- What are the domain and range of e^x and $\ln x$?

- Explain the five laws of logarithms using the laws of exponents.

- How does the graph of $y = \ln x$ relate to the graph of $y = e^x$?

- How much stronger is an earthquake that registers 7 on the Richter scale than one that registers 5?

Chapter 2 Review Exercises

1. Compute each of the following without a calculator. Your answer should be an integer or fraction.
   (a) $\left(\frac{1}{9}\right)^{1/2}$
   (b) $\left(\frac{1}{9}\right)^{-2}$
   (c) $(0.1)^3$
   (d) $(0.1)^{-3}$

2. Suppose you put \$100 into an account that pays 4% interest, compounded annually.

(a) If you leave the money in the account without making any further deposits or withdrawals, how much will you have at the end of t years? (Your answer should be a function of t.)

(b) Use a calculator to find the amount you will have at the end of 4 years and at the end of 10 years.

3. A colony of insects is growing exponentially. Its population grows from 600 to 1,000 in 3 days.

(a) Write a formula giving the population at the end of t days.

(b) How large will the colony be at the end of 6 days? (Round your answer to the nearest integer.)

4. An initial amount of 100 milligrams of a certain radioactive substance will decay to 91 milligrams in 4 years. Write the decay equation in the form $y = Ab^t$, where y is the amount at the end of t years. You can use a calculator to find b, but you should not need one for A.

5. Suppose you put $5,000 into an account that pays 6% annual interest and make no further deposits or withdrawals.

(a) If the interest is compounded 12 times a year, how much will be in the account at the end of 1 year?

(b) If the interest is compounded 4 times a year, how much will be in the account at the end of 3 years?

6. Suppose you put $600 into an account that pays 4.5% annual interest, compounded continuously, and make no further deposits or withdrawals.

(a) How much will be in the account at the end of 1 year?

(b) How much will be in the account at the end of 3 years?

7. Suppose you put $100 into an account that pays 6% annual interest, compounded n times a year. How much do you have at the end of 1 year?

8. Suppose you put $200 into an account that pays 5% annual interest, compounded continuously. How much do you have at the end of 3 years?

9. Suppose an account pays 5% annual interest. How much should you put into the account now in order to have $1,000 at the end of 2 years, if

(a) the interest is compounded once a year?

(b) the interest is compounded 6 times a year?

(c) the interest is compounded continuously?

In Exercises 10–13, compute the given limit.

10. $\lim_{n \to \infty} \left(1 + \dfrac{2}{3n}\right)^n$ **11.** $\lim_{n \to \infty} \left(1 - \dfrac{1}{n}\right)^{2n}$

12. $100 \lim_{n \to \infty} \left(1 + \dfrac{4}{100n}\right)^{3n}$ **13.** $\lim_{h \to 0} (1 - h)^{2/h}$

14. Write each of the following as an integer or fraction. Do not use a calculator.

(a) $\log_3 (1/9)$ **(b)** $\log_4 2$ **(c)** $\log_{1/3} 3$
(d) $\log_{1/2} 1$ **(e)** $\log_7 7$

15. Use the approximations $\log_3 2 \approx 0.631$ and $\log_3 7 \approx 1.771$ to approximate the following. Do not use a calculator.

(a) $\log_3 (1/4)$ **(b)** $\log_3 \sqrt{14}$ **(c)** $\log_3 18$
(d) $\log_9 2$ **(e)** $\log_{1/3} 7$

Exercises 16–18 refer to the following graphs. Use the letter below the graph.

16. Which graph most closely resembles the graph of $y = 2^x$?

17. Which graph most closely resembles the graph of $y = (1/2)^x$?

18. Which graph most closely resembles the graph of $y = -2^x$?

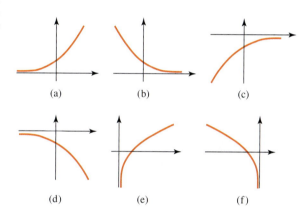

(a)   (b)   (c)

(d)   (e)   (f)

19. Which graph most closely resembles the graph of $y = 2^{-x}$?

In Exercises 20–25, solve for x. Do not use a calculator, but simplify as much as you can. Answers involving e or $\ln$ can be left that way.

20. $e^{2x} = 3$ **21.** $\ln(1 + \ln x) = 0$

22. $\ln(x^2) = 8$ **23.** $e^{\ln(2x+3)} = 7$

24. $e^{x^2 - 2x - 3} = 1$ **25.** $\ln(e^x) = 2x - 3$

In Exercises 26–29, write the number as an integer or fraction.

26. $e^{-2\ln 3}$ **27.** $\ln \sqrt{e}$

28. $\ln(\ln e)$ **29.** $\sqrt{e^{\ln 4}}$

30. Simplify the expression $e^{-t \ln 2} + \ln e^{2t}$.

31. The growth of a certain bacteria culture satisfies an exponential growth equation $y = Ae^{kt}$, where t is the time and y is the size of the population. Suppose that the culture contains 10,000 bacteria at noon and 40,000 bacteria at 2:00 P.M. How many bacteria will it contain at 3:00 P.M.?

32. What is $e^{(\ln 3)/2}$?

33. Suppose you put $250 into a bank account that receives 8% annual interest, compounded continuously. How many years will it take before your bank account has $1,000?

34. A certain radioactive substance satisfies the exponential decay law, $y = Ae^{-kt}$, where t is the time (in days) and y is the amount (in grams) at time t. Suppose an inital amount of 100 grams decays to 20 grams at the end of 4 days. What is the value of k?

35. Find the following limits:

    (a) $\lim\limits_{x \to \infty} e^{0.1x}$ **(b)** $\lim\limits_{x \to \infty} e^{-0.1x}$ **(c)** $\lim\limits_{x \to -\infty} e^{0.1x}$

    (d) $\lim\limits_{x \to -\infty} e^{-0.1x}$ **(e)** $\lim\limits_{x \to \infty} \ln(2x)$ **(f)** $\lim\limits_{x \to 0^+} \ln(2x)$

36. Solve the equation $5x^2 e^{-x} - 9e^{-x} = 0$.

37. Find the solution to the equation $2^{(x+3)} = 31$, correct to five decimal places, without using the equation solver or trace function of your calculator.

38. Solve the equation $0.2y\, e^{1-y/5} - 0.1y = 0$.

39. Solve the equation $0.1y\, \ln(8/y) - 0.01y = 0$.

40. Sketch the graph of the functions

    (a) $y = 2 + e^{x-3}$ **(b)** $y = 2 - \ln(x - 5)$

41. Suppose $10,000 is deposited into an account paying an annual interest rate of 4%, compounded continuously. How long will it take for the account to grow to $25,000?

42. The radioactive substance einsteinium-253 decays exponentially—in other words, the amount y remaining after t days is given by a formula of the form $y = Ae^{-kt}$, where A is the initial amount and k is the decay constant. Suppose an initial amount of 30 milligrams of the substance decays to 10 milligrams in 11.7 days.

    (a) Find the decay constant of einsteinium-253.

    (b) Find its half-life.

    (c) If you start with 100 milligrams of einsteinium-253, how much will there be after 20 days?

43. A certain radioactive substance satisfies the exponential decay law, $y = Ae^{-kt}$, where t is the time (in days) and y is the amount (in grams) at time t. Suppose an initial amount of 100 grams decays to 25 grams at the end of 10 days. How much substance is left after 15 days?

44. An account paying 4% annual interest, compounded continuously, starts out with an initial deposit, after which no further deposits or withdrawals are made. How many years will it take for the initial amount to double?

45. Suppose you invest a certain amount in an account that offers continuously compounded interest and your money triples in exactly 14 years. (After the initial investment you do not add or withdraw from the account until that time.) What is the annual interest rate?

Chapter 2 Practice Exam

1. A student buys a $1,000 savings bond at 4% compounded monthly. What will the value be after 3 years?

2. How much should you deposit in a savings account with an interest rate of 5%, compounded annually, in order to have $5,000 at the end of 6 years?

3. On her 20th birthday a woman deposits $1,000 in a bank account with interest compounded continuously at a rate of 6.5%. How much money will she have on her 40th birthday? (Round to the nearest cent.)

4. If you deposit $1 in the bank with an interest rate of 100%, compounded continuously, what will the balance be at the end of the year?

5. Find

    (a) $\lim\limits_{n \to \infty} \left(1 + \dfrac{1}{4n}\right)^n$ **(b)** $\lim\limits_{n \to \infty} \left(\dfrac{n-3}{n}\right)^{2n}$

6. Find

    (a) $\log_9 3$ **(b)** $\log_3\left(\frac{1}{9}\right)$

7. Use the approximation $\log_3 5 \approx 1.465$ to approximate the following without a calculator:

    (a) $\log_3\left(\frac{1}{25}\right)$ **(b)** $\log_3(45)$

8. Let $f(x) = e^{x/2}$. Find $f[\ln(9)]$.

9. Simplify the following expressions:

    (a) $e^{3\ln(2x+1)}$

    (b) $\ln\left(\frac{1}{2}e^{3x^2+4}\right)$

10. Starting from the graph of the function $f(x) = \ln x$, sketch by hand, on the same axes, the graphs of the functions

    (a) $y = \ln(x + 2)$

    (b) $y = \ln(x - 2)$

    (c) $y = 3 - 4\ln(x - 2)$

11. Starting from the graph of the function $f(x) = e^x$, sketch by hand, on the same axes, the graphs of the functions
   (a) $y = e^{x+2}$
   (b) $y = e^{x-2}$
   (c) $y = -1 - e^{x-2}$

In Exercises 12–14, do not use the trace function or equation solver on your calculator.

12. Find the solution to the equation $5^{(x-1)} = 23$, correct to five decimal places.

13. Solve the equation $e^{2x} - 2e^x - 8 = 0$.

14. Solve the equation $\ln(x) + \ln(x + 2) = 0$.

15. Find the vertical and horizontal asymptotes, if any, of the following functions:
   (a) $f(x) = \dfrac{2}{1 + 2e^{-x}}$ **(b)** $g(x) = \dfrac{2}{2e^{-x} - 1}$

16. A chemist knows that a certain radioactive substance has a half-life of 8 days. If he has 500 grams of this substance today, how much substance will he have after 16 days?

17. If you put money in an account that pays 7% interest, compounded continuously, how long will it take for your money to quadruple?

18. Someone deposits $300 with interest compounded continuously. After 10 years, he has $500. How many dollars will he have after 30 years?

19. Suppose the algae population under the diving board grows exponentially. If there are 4,000 algae initially and 9,000 algae 2 days later, how many days will it take for the population to reach 64,000?

20. The earthquake that occurred in Washington State on February 28, 2001, registered 6.8 on the Richter scale, while the earthquake that occurred in California on October 19, 1989, registered 7.2 on the Richter scale. How much more severe was the California earthquake?

········ **Chapter 2 Projects** ·····························

1. **Computing mortgage payments and balances** Suppose you borrow L dollars, which you are to repay over a period of 20 years at 6% interest, compounded monthly. How much will your monthly payments be? To answer that, let p be the monthly payment and A_k be the balance remaining on the loan at the end of k months.

   (a) The key formula is

   $$A_k = 1.005 A_{k-1} - p. \tag{27}$$

   Explain in words what this says and where it comes from.

   (b) For $k = 1$, we have to take $A_{k-1} = L$ in (27). Applying formula (27) recursively for $k = 1, 2, 3$ gives

   $$A_1 = 1.005L - p$$
   $$A_2 = 1.005 A_1 - p = (1.005)^2 L - 1.005 p - p$$
   $$A_3 = 1.005 A_2 - p = (1.005)^3 L - (1.005)^2 p - 1.005 p - p.$$

   Continuing in this way, write a formula for A_k in terms of L and p.

   (c) To simplify your formula, use the following formula, known as the **geometric series formula**: For any number $t \neq 1$ and any positive integer m,

   $$1 + t + t^2 + \cdots + t^m = \frac{t^{m+1} - 1}{t - 1}.$$

   If you are unfamiliar with this formula, get hold of an algebra book and see how to derive it. It's not hard.

(d) If the loan is to be paid off in 20 years, we need $A_{240} = 0$. This leads to an equation in L and p, and solving for p gives you the size of your monthly payment.

(e) If you borrow \$120,000 to buy a house, to be paid off in 20 years at 6% interest, compounded monthly, what will your monthly payment be?

(f) Rework the basic formula to cover the case of interest rate r, compounded n times a year, with the loan to be paid off in Y years.

(g) Turn the formula around by solving for L as a function of p to find out how much you may borrow if you can afford to pay p dollars per month.

(h) Suppose you want to buy a house, with the loan to be paid off in 30 years at 5% interest, compounded monthly. If you can afford to pay \$800 per month, what size loan can you take?

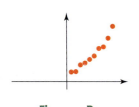

Figure 2.P.1

2. **Exponential regression**   Sometimes a set of data appears to come from an exponential curve, and we seek the function of the form

$$y = Ae^{kx} \tag{28}$$

that best fits the given data. For instance, the two scatter diagrams in the margin seem to come from data that are approximately related by an equation such as (28), the one in Figure 2.P.1 for $k > 0$, and the one in Figure 2.P.2 for $k < 0$.
Suppose we have a set of data points, say,

$$(x_1, y_1), \quad (x_2, y_2), \quad \ldots, \quad (x_n, y_n). \tag{29}$$

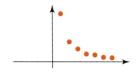

Figure 2.P.2

To find the exponential curve of best fit, we linearize the problem by taking the natural logarithm on both sides of (28), to get an equation of the form

$$u = kx + b, \tag{30}$$

where $u = \ln y$ and $b = \ln A$. We now use *linear* regression to find the coefficients k and b so that Eq. (30) gives the best linear fit to the data points $(x_1, \ln y_1)$, $(x_2, \ln y_2)$, $\ldots, (x_n, \ln y_n)$.

(a) Apply formulas (17) and (18) of Chapter 0 with y_i replaced by $\ln y_i$ to find the coefficients k and b. Then set $A = e^b$ to find the constants in (28) that give the exponential curve that best fits the data points of (29).
This process is called **exponential regression**.

(b) In testing how a body metabolizes calcium, a researcher injects a small amount of calcium that is chemically "labeled" into a patient's bloodstream to measure how fast it is removed from the blood. Suppose the concentration c of labeled calcium t hours after the injection is given by the following table:

t	0	1	2	3	4
c	0.026	0.015	0.0052	0.0026	0.001

Draw a scatter diagram of these data points. Then apply exponential regression to find the equation of the exponential curve that best fits the data.

(c) Graphing calculators, such as the TI-83 Plus, are programmed to do exponential regression. Use your calculator to find the exponential function that best fits the following data:

x	0	0.2	0.4	0.6	0.8	1	1.2	1.4	1.6	1.8	2
y	1.9	2.3	2.7	3.1	3.8	4.4	5.1	6.3	7.1	8.7	10.4

Consult your calculator manual for specific instructions concerning exponential regression.

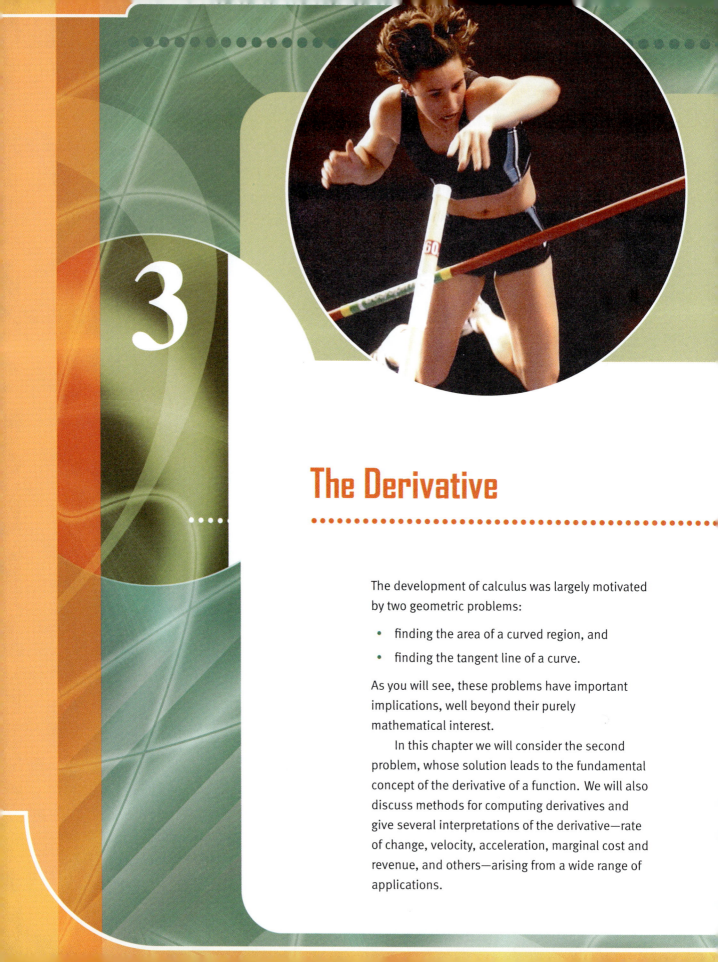

3

The Derivative

The development of calculus was largely motivated by two geometric problems:

- finding the area of a curved region, and
- finding the tangent line of a curve.

As you will see, these problems have important implications, well beyond their purely mathematical interest.

In this chapter we will consider the second problem, whose solution leads to the fundamental concept of the derivative of a function. We will also discuss methods for computing derivatives and give several interpretations of the derivative—rate of change, velocity, acceleration, marginal cost and revenue, and others—arising from a wide range of applications.

Let's go back for a moment to the **slope** of a linear function introduced in Section 0.4. If x and y are related by a linear equation

$$y = mx + b,$$

then the slope m measures the **rate** at which the dependent variable y changes with respect to the independent variable x—in other words, the change in output per unit change of input.

As an example, let's recall the formula for converting Celsius to Fahrenheit:

$$F = \frac{9}{5}C + 32.$$

The slope is $\frac{9}{5}$, and the dependent variable F (the Fahrenheit reading) changes by $\frac{9}{5}$th of a degree for every 1-degree change in the independent variable C (the Celsius reading). In this case, we say the *rate of change* of F *with respect to* C equals $\frac{9}{5}$.

What about a relation of the form

$$y = f(x),$$

where $f(x)$ is *not* linear? For example, the height of an object moving under the influence of gravity is given by an equation of the form

$$h = -16t^2 + v_0 t + h_0$$

(as we saw in Section 0.5), where t is the time in seconds, h is the height in feet, and h_0 and v_0 are constants. The rate at which h changes with respect to t—in other words, the change of height in feet per second—is called the **velocity**.

There is an important difference, however, between this rate and the Celsius-to-Fahrenheit change. In this case, *the rate of change is not constant.* It is a fact of nature that a falling object increases its speed as it falls. This difference is exactly what distinguishes nonlinear from linear functions. In the linear case the rate is a constant (the slope of the line). In the nonlinear case it is not.

In both cases, linear and nonlinear, the graph of $y = f(x)$ gives a geometric picture of how y changes with respect to x, and the rate of change is related to the *steepness* of the graph and whether it rises or falls to the right. We already saw that in the linear case, in which the slope measures both the rate of change of y with respect to x and the steepness of the line. With that in mind, we will try to define the slope of a nonlinear function by considering how to measure the steepness of its graph.

To start, let's consider the simplest nonlinear case—the graph of $y = x^2$, shown in Figure 3.1.1. How steep is this graph?

As you can see, the steepness varies from point to point. The right side of the graph goes uphill and gets steeper as it goes further to the right. The left side goes downhill—and the further to the left, the more negative the slope. To begin with, let's concentrate on a single point—the point $(1, 1)$, say—and try to determine how steep the graph is at that point.

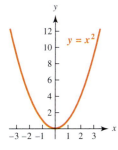

Figure 3.1.1

The slope of the graph of $y = x^2$ at the point (1, 1)

Start with the given point (1, 1). Think of it as a fixed point. To measure the steepness of the graph at that point, we follow a three-step procedure.

Step 1   Take a second point on the graph, close to the fixed point. We will write the horizontal coordinate of this second point in the form $1 + h$, to emphasize that it is close to the fixed point. (Think of h as some small amount we have added to 1, a small *horizontal* change.) Since the second point is also on the graph of $y = x^2$, its vertical coordinate is $(1 + h)^2$.

Step 2   Compute the slope of the line connecting the fixed point and the nearby point. This line, shown in Figure 3.1.2, is called a **secant line**. Its slope is computed by our usual formula for the slope of a line—the difference in the vertical coordinates divided by the difference in the horizontal coordinates.
   Using the coordinates of our two points (shown in Figure 3.1.2), we get

$$\text{Slope of the secant line} \quad = \frac{(1 + h)^2 - 1}{h}. \tag{1}$$

We can use the slope of the secant line to estimate the slope of the graph. Of course, it is not an exact measure. In this case, as you can see in the figure, the secant line is steeper than the graph at the point (1, 1). But the slope of the secant line does give us an *approximation* to the steepness of the graph. And the smaller we choose h, the closer we come to measuring the steepness of the graph. That is a key point—*if h is small, the slope of the secant line can be used to approximate the steepness of the graph, and we can make the approximation as close as we want by taking h small enough.* That leads to the third step of our procedure.

Step 3   Compute $\displaystyle\lim_{h \to 0} \frac{(1 + h)^2 - 1}{h}$.

In other words, we consider the slope of the secant line as a function of h and take its limit as $h \to 0$. By way of illustration, consider Figures 3.1.3 and 3.1.4. The first shows the secant line we get by choosing $h = 0.5$. In that case, the second point has coordinates (1.5, 2.25), and the slope of the secant line is 2.5, as we see by using formula (1) with $h = 0.5$.
   Figure 3.1.4 shows the secant line corresponding to the choice of $h = 0.2$. In that case, the second point has coordinates (1.2, 1.44), and the secant line has slope 2.2. As

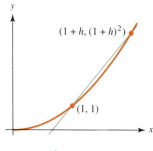

Figure 3.1.2

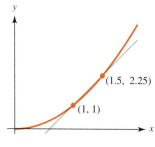

Figure 3.1.3

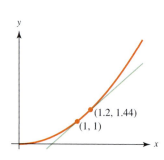

Figure 3.1.4

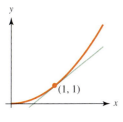

Figure 3.1.5

you can see, the secant line in Figure 3.1.4 fits the graph better near the point $(1, 1)$ than that of Figure 3.1.3, and its slope is a better measure of the graph's steepness.

If we continue this process, choosing h smaller and smaller, the secant line moves closer and closer to a tangent position. The line tangent to the graph at the point $(1, 1)$ is shown in Figure 3.1.5.

We say that the tangent line is the **limiting** position of the secant line as h goes to zero, and the slope of the tangent line is the **limit** of the slope of the secant line. We conclude that *the slope of the graph equals the slope of its tangent line and is given by the limit as $h \to 0$ of the slope of the secant line.* That is,

$$\text{Slope of graph} = \text{tangent line slope} = \lim_{h \to 0} \frac{(1+h)^2 - 1}{h}. \qquad (2)$$

This quotient is indeterminate—both the numerator and denominator go to zero. However, the following table of values suggests that the limit exists and is equal to 2:

h	-0.003	-0.002	-0.001	0	0.001	0.002	0.003
secant line slope	1.997	1.998	1.999	?	2.001	2.002	2.003

To get a precise evaluation of the limit, we need to use some algebra to simplify the quotient. From the identity $(1 + h)^2 = 1 + 2h + h^2$, we get

$$\frac{(1+h)^2 - 1}{h} = \frac{2h + h^2}{h} = 2 + h,$$

for any $h \neq 0$. That gets rid of the indeterminacy, and we can now compute the limit:

$$\lim_{h \to 0} \frac{(1+h)^2 - 1}{h} = \lim_{h \to 0} (2 + h) = 2.$$

By combining this result with formula (2), we conclude that

- the slope of the graph of $y = x^2$ at $(1, 1)$ is equal to the slope of its tangent line at that point, and
- both are equal to 2.

The slope of the graph of $y = x^2$ at the point (x, x^2)

What about the slope of the parabola at other points? Since the equation of the graph is $y = x^2$, a point on the graph with horizontal coordinate x has vertical coordinate x^2. Thinking of (x, x^2) as the coordinates of a fixed point, we modify the procedure we just used as follows.

Step 1 Take a second point on the graph, close to the fixed point. Write its horizontal coordinate in the form $x + h$ to emphasize that it is close to the fixed point—thinking of h as some small amount we have added to x. Since the second point is also on the graph, its vertical coordinate is $(x + h)^2$.

Step 2   Compute the slope of the secant line connecting these two points. The usual slope formula gives

$$\text{Slope of the secant line } = \frac{(x+h)^2 - x^2}{h}. \tag{3}$$

This approximates the steepness of the graph at the point (x, x^2).

Step 3   Compute

$$\lim_{h \to 0} \frac{(x+h)^2 - x^2}{h}$$

to find the slope of the tangent at the point (x, x^2), or—what amounts to the same thing—the slope of the graph.

This limit is indeterminate, and we have to simplify it by expanding and canceling terms. If $h \neq 0$,

$$\frac{(x+h)^2 - x^2}{h} = \frac{(x^2 + 2xh + h^2) - x^2}{h}$$

$$= \frac{2xh + h^2}{h}$$

$$= 2x + h.$$

Therefore,

$$\lim_{h \to 0} \frac{(x+h)^2 - x^2}{h} = \lim_{h \to 0} (2x + h) = 2x.$$

Conclusion   _The line tangent to the graph of $y = x^2$ at the point (x, x^2) has slope $2x$; or, what amounts to the same thing, the graph itself has slope $2x$ at the point (x, x^2)._

Figures 3.1.6 and 3.1.7 illustrate the procedure—the first showing a typical secant line and the second the tangent line.

Once we know the slope, we can write the equation of the tangent line at any point on the parabola.

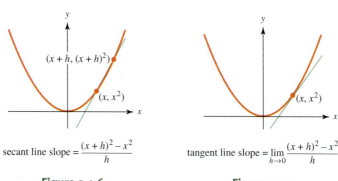

Figure 3.1.6   **Figure 3.1.7**

EXAMPLE 3.1.1

What is the equation of the line tangent to the graph of $y = x^2$ at the point $(-3, 9)$?

SOLUTION   As we have seen, the line tangent to the parabola at the point (x, x^2) has slope $2x$. When $x = -3$, the slope is -6, so we need to find the equation of the line of slope -6 through the point $(-3, 9)$. That is easily seen to be $y = -6x - 9$.

The slope of the graph of $y = f(x)$

The procedure we just used to determine the slope of the graph of $y = x^2$ can be applied to a wide range of functions. Consider an equation of the form $y = f(x)$, where f is some given function. How steep is the graph of this equation? If f is nonlinear, the answer will vary from point to point along the graph. Let's fix one such point—say, with horizontal coordinate x. Since the point is on the graph, its vertical coordinate is $f(x)$. Thinking of $(x, f(x))$ as a fixed point, we follow the same three steps we used to find the slope of the parabola, with suitable modifications.

Step 1   Take a second point on the graph, close to the fixed point. To emphasize that it is close to the fixed point, we will write the horizontal coordinate of this second point in the form $x + h$, where h is some small amount we have added to x (a small horizontal change). The vertical coordinate is $f(x + h)$.

Step 2   Compute the slope of the secant line connecting the two points. The usual slope formula gives

$$\text{Slope of the secant line} = \frac{f(x + h) - f(x)}{h}, \quad \left(= \frac{\text{rise}}{\text{run}} \right) \tag{4}$$

which approximates the slope of the graph at the point $(x, f(x))$.

Step 3   Compute

$$\lim_{h \to 0} \frac{f(x + h) - f(x)}{h}$$

to find the slope of the tangent line, which is the same as the slope of the graph.

Summary for slope of a graph

- The slope of the graph of $y = f(x)$ at a given point $(x, f(x))$ is the same as the slope of its tangent line at that point.
- We can approximate the slope of the tangent by the slope of the secant line connecting the given point $(x, f(x))$ to a nearby point $(x + h, f(x + h))$.
- We can make the approximation as good as we want by taking h small enough.

• This leads to the formula:

$$\text{Slope of the graph} = \lim_{h \to 0} \frac{f(x+h) - f(x)}{h}, \tag{5}$$

provided that this limit exists.

Figures 3.1.8 and 3.1.9 illustrate the procedure—the first showing a typical secant line and the second the tangent line.

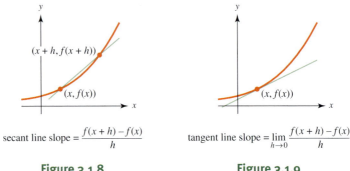

secant line slope $= \dfrac{f(x+h) - f(x)}{h}$ tangent line slope $= \lim\limits_{h \to 0} \dfrac{f(x+h) - f(x)}{h}$

Figure 3.1.8   **Figure 3.1.9**

As you might expect, the hard part of the procedure is actually finding the limit as h goes to zero. Later in this chapter we will study some rules and shortcuts for avoiding that difficulty. But even with shortcuts we sometimes have to go back to basics and work directly with formula (5), and it is important that you gain some experience in using it. Here are a few examples you can use as models.

EXAMPLE 3.1.2

Find the slope of the graph of $y = x^3$ at the point (x, x^3).

SOLUTION   Taking $f(x) = x^3$ in formula (5), we get the following expression for the slope of the graph:

$$\text{Slope of the graph} = \lim_{h \to 0} \frac{(x+h)^3 - x^3}{h}. \tag{6}$$

Once again, the limit is indeterminate, and we have to simplify the quotient in order to see what happens. The key fact from algebra we use here is the formula

$$(x+h)^3 = x^3 + 3x^2h + 3xh^2 + h^3. \tag{7}$$

If we substitute this into the quotient in (6), we get

$$\frac{(x+h)^3 - x^3}{h} = \frac{3x^2h + 3xh^2 + h^3}{h}$$
$$= 3x^2 + 3xh + h^2,$$

for all $h \neq 0$. Taking the limit as $h \to 0$, we obtain

$$\lim_{h \to 0} \frac{(x+h)^3 - x^3}{h} = \lim_{h \to 0}(3x^2 + 3xh + h^2) = 3x^2.$$

Therefore, the slope of the graph equals $3x^2$.

Figures 3.1.10 and 3.1.11 show the graph of $y = x^3$, the first displaying the secant line and the second the tangent line.

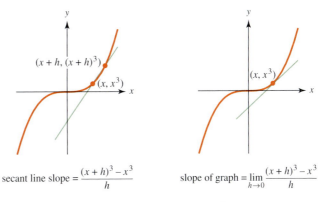

secant line slope $= \dfrac{(x+h)^3 - x^3}{h}$ slope of graph $= \lim\limits_{h \to 0} \dfrac{(x+h)^3 - x^3}{h}$

Figure 3.1.10   **Figure 3.1.11**

EXAMPLE 3.1.3

Find the slope of the graph of $y = \sqrt{x}$ at the point $(x, \sqrt{x})$.

SOLUTION Taking $f(x) = \sqrt{x}$ in formula (5) leads to the formula

$$\text{Slope of the graph} = \lim_{h \to 0} \frac{\sqrt{x+h} - \sqrt{x}}{h}. \tag{8}$$

A typical secant line is drawn in Figure 3.1.12 and the tangent line in Figure 3.1.13. (Notice that in drawing the secant line, the second point was chosen to the *left* of x, which means that h was chosen to be negative. But that doesn't matter. Our method works for both positive *and* negative choices of h. In other words, if the limit exists, we get the same result as $h \to 0$ from both the positive and negative directions.)

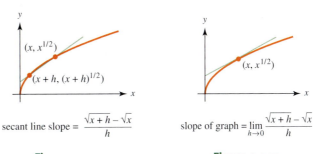

secant line slope $= \dfrac{\sqrt{x+h} - \sqrt{x}}{h}$ slope of graph $= \lim\limits_{h \to 0} \dfrac{\sqrt{x+h} - \sqrt{x}}{h}$

Figure 3.1.12   **Figure 3.1.13**

We have already encountered a limit similar to (8) in Example 1.1.6 of the last chapter. Let's review how to handle it. We start with the equation

$$(u - v)(u + v) = u^2 - v^2. \tag{9}$$

Replacing u by $\sqrt{x + h}$ and v by $\sqrt{x}$, we get

$$\left(\sqrt{x + h} - \sqrt{x}\right)\left(\sqrt{x + h} + \sqrt{x}\right) = (x + h) - x = h.$$

Multiplying both the numerator *and* denominator on the right-hand side of Eq. (8) by $(\sqrt{x + h} + \sqrt{x})$ does not change the fraction. And it removes the indeterminacy, as follows:

$$\frac{\sqrt{x + h} - \sqrt{x}}{h} = \frac{\sqrt{x + h} - \sqrt{x}}{h} \cdot \frac{\left(\sqrt{x + h} + \sqrt{x}\right)}{\left(\sqrt{x + h} + \sqrt{x}\right)}$$

$$= \frac{h}{h\left(\sqrt{x + h} + \sqrt{x}\right)}$$

$$= \frac{1}{\sqrt{x + h} + \sqrt{x}}.$$

Taking the limit gives

$$\text{Slope of the graph} = \lim_{h \to 0} \frac{\sqrt{x + h} - \sqrt{x}}{h}$$

$$= \lim_{h \to 0} \frac{1}{\sqrt{x + h} + \sqrt{x}} = \frac{1}{2\sqrt{x}}.$$

Slope as a rate of change

For nonlinear functions as well as linear functions, the slope measures the *rate* at which the dependent variable changes with respect to the independent variable. More precisely, if a quantity y is a function of another quantity x, say, $y = f(x)$, then formula (4) for the slope of the secant line also gives the **average rate of change of y with respect to x** over the interval $[x, x + h]$, and formula (5) for the slope of the tangent line gives the **(instantaneous) rate of change of y with respect to x**. To summarize:

$$\text{Average rate of change over } [x, x + h] = \frac{f(x + h) - f(x)}{h} \tag{10}$$

and

$$\text{Instantaneous rate of change at } x = \lim_{h \to 0} \frac{f(x + h) - f(x)}{h}. \tag{11}$$

EXAMPLE 3.1.4

The consumption of electricity E, in millions of kilowatt-hours, in a city during 24 hours of a typical day is given by the graph $E = E(t)$ displayed in Figure 3.1.14.

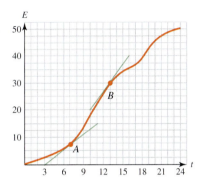

Figure 3.1.14

 (i) Estimate the average rate of electricity consumption between 3 A.M. and 8 A.M. Also estimate the average rate over a 24-hour period from midnight to midnight.
 (ii) Estimate the instantaneous rate of electricity consumption at $t = 7$ and $t = 13$.

SOLUTION (i) Here, the consumption of electricity E is a function of the time t, where t is measured in hours, and the rate of consumption is the rate of change of $E(t)$ with respect to t. Figure 3.1.14 shows that $E(3) = 2.5$ and $E(8) = 10$. Therefore, using formula (10), we find that the average rate of change of consumption over the time interval [3, 8] is

$$\frac{E(8) - E(3)}{8 - 3} = \frac{10 - 2.5}{5} = 1.5.$$

Similarly, we find that the average rate of change of consumption over the time interval [0, 24] is $[E(24) - E(0)]/(24 - 0) = (50 - 0)/24 \approx 2.08$.

 (ii) The instantaneous rate of electricity consumption is the same as the instantaneous rate of change of $E(t)$, which is the slope of the tangent line to the graph. Figure 3.1.14 shows that the tangent line at $t = 7$ (through the point A) also passes through the points $(3, 0)$ and $(11, 15)$. Therefore, its slope is $(15 - 0)/(11 - 3) = 15/8 = 1.875$, which means that the instantaneous rate of change of E at $t = 7$ is 1.875 millions of kilowatt-hours per hour.

Similarly, using the tangent line to the graph at $t = 13$ (through the point B), we find that the instantaneous rate of change of E at $t = 13$ is $(40 - 20)/(16 - 10) = 10/3$ millions of kilowatt-hours per hour.

Practice Exercise 3.1

1. (a) Find a formula for the slope of the graph of $y = x^2 - 3x + 8$ at any x.

(b) Write the equation of the tangent line at the point where $x = 4$.

Exercises 3.1

Figure 3.1.15 shows the graph of a function and four of its points, A, B, C, and D. Exercises 1–4 refer to these points.

1. Find the slope of the graph at the point C.

2. At which points A, B, C, and D is the slope of the graph negative?

3. At which points A, B, C, and D is the slope of the graph positive?

4. Put the slopes of the graph at the given points in increasing order.

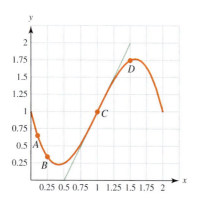

Figure 3.1.15

Figure 3.1.16 shows the graph of a function and four of its points, A, B, C, and D. Exercises 5–8 refer to these points.

5. Find the slope of the graph at the points B, C, and D.

6. At which points A, B, C, and D is the slope of the graph zero?

7. Put the slopes of the graph at the points A, B, and C in increasing order.

8. Estimate the slope of the graph at the point A.

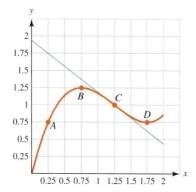

Figure 3.1.16

9. Find the equation of the line tangent to the graph of $y = x^2$ at each of the following points:
   (a) $(\frac{1}{2}, \frac{1}{4})$ (b) $(0, 0)$
   (c) the point with $x = -3$

10. Find the following points on the graph of $y = x^2$:
    (a) The point where the graph has slope 3
    (b) The point where the tangent line is parallel to the line $2x + y = 3$

In Exercises 11–18, use the result of Example 3.1.2.

11. Find the equation of the line tangent to the graph of $y = x^3$ at the point $(-2, -8)$.

12. Check your answer to the previous problem visually by graphing both the function and the tangent line in a $[-3, 0] \times [-20, 0]$ window.

13. Find the equation of the line tangent to the graph of $y = x^3$ at the point with $x = \frac{1}{2}$.

14. Check your answer to the previous problem visually by graphing both the function and the tangent line in a $[-1, 1] \times [-0.5, 0.5]$ window.

15. Find all points, if any, on the graph of $y = x^3$ where the slope equals 3.

16. Find all points, if any, on the graph of $y = x^3$ where the slope equals -1.

17. Find all points, if any, on the graph of $y = x^3$ where the tangent line is horizontal.

18. Find all points, if any, on the graph of $y = x^3$ where the tangent line is parallel to the line $3x - 4y = 7$.

In Exercises 19–23, use the result of Example 3.1.3.

19. Find the equation of the line tangent to the graph of $y = \sqrt{x}$ at the point $(9, 3)$.

20. Check your answer to the previous problem visually by graphing both the function and the tangent line in a $[0, 16] \times [0, 4]$ window.

21. Find all points, if any, on the graph of $y = \sqrt{x}$ where the slope equals 2.

22. Find all points, if any, on the graph of $y = \sqrt{x}$ where the tangent line is parallel to the line $x - y = 3$.

23. Find all points, if any, on the graph of $y = \sqrt{x}$ where the tangent line is horizontal.

24. Suppose $f(x) = mx + b$, where m and b are constants.
    (a) Without doing any computations, determine $\lim_{h \to 0} (f(x + h) - f(x))/h$.
    (b) Check your answer by computing the quotient and taking the limit.

25. (a) Find a formula (in terms of h) for the slope of the secant line of the graph of $y = x^2 + x$ between the point where $x = 1$ and the point where $x = 1 + h$. Simplify the formula as much as possible.
(b) Find the slope of the graph at the point $(1, 2)$ by determining what happens as $h \to 0$.
(c) Use a similar procedure to find the slope of the graph of $y = x^2 + x$ for every x.
(d) Write the equation of the line tangent to the graph of $y = x^2 + x$ at each of the following points:
(i) $(1, 2)$ (ii) $(-2, 2)$ (iii) the point where $x = -1$

26. (a) Find a formula (in terms of x and h) for the slope of the secant line of the graph of $y = 3x^2$ between a point with horizontal coordinate x and one with horizontal coordinate $x + h$. Simplify the formula as much as possible.
(b) Find the slope at any point of the graph by taking the limit as $h \to 0$.
(c) Write the equation of the line tangent to the graph of $y = 3x^2$ at each of the following points:
(i) $(1, 3)$ (ii) $(-1, 3)$ (iii) the point where $x = \frac{1}{2}$

27. (a) Write a formula, in terms of x, for the slope of the graph of $y = x^2 + 3x - 1$ at any point.
(b) Find the slope of the tangent line at the point $(1, 3)$.
(c) Find the equation of the tangent line at the point with $x = -2$.

28. (a) Using Example 3.1.2 as a model, find a formula, in terms of x, for the slope of the graph of $y = 2x^3 + 1$ at any point.
(b) Write the equation of the line tangent to the graph of $y = 2x^3 + 1$ at each of the following points:
(i) $(1, 3)$ (ii) $(\frac{1}{2}, \frac{5}{4})$ (iii) the point with $x = -1$

29. (a) Using Example 3.1.3 as a model, find a formula for the slope of the graph of $y = \sqrt{x + 4}$ at any $x > -4$.
(b) Write the equation of the line tangent to the graph of $y = \sqrt{x + 4}$ at each of the following points:
(i) $(0, 2)$ (ii) $(-3, 1)$ (iii) the point with $x = 5$

30. (a) Find a formula for the slope of the graph of $y = \sqrt{2x + 1}$ at any $x > -\frac{1}{2}$.
(b) Write the equation of the tangent line at the point $(4, 3)$.

31. (a) Find a formula for the slope of the graph of $y = x^3 - 2x + 1$ at any x.
(b) Write the equation of the tangent line at the point where $x = 0$.

32. (a) If $f(x) = 2/x$, simplify $f(x + h) - f(x)$ by combining the two fractions over a common denominator.
(b) Find $\lim_{h \to 0}(f(x + h) - f(x))/h$.
(c) Find the slope of the graph of $y = 2/x$ at the point $(\frac{1}{2}, 4)$.
(d) Write the equation of the tangent line at the point where $x = 1$.

33. Astronauts experiencing microgravity begin to lose bone matter. Large amounts of loss can pose a serious medical problem. Assume that the percent of total bone matter loss in an astronaut is given by $L(t) = 0.01t^2$, where t is the time (in months) spent in space.
(a) Use formula (5) to find the slope of the graph of $L(t)$ as a function of t.
(b) At what time is an astronaut losing bone matter at the rate of 0.08% of total bone matter per month?

34. Assume that sales per week of a music CD for several weeks after media advertisement has ceased are approximated by the function $s(t) = -t^2 + 100$, where s is the number of sales in thousands per week and t is the time in weeks.
(a) Sketch the graph of the function and use formula (5) to find its slope as a function of t.
(b) Find the slope at $t = 4$. What does it tell you about sales?
(c) If the sales per week from the 4th week on stay at the same constant rate, when will they end? (*Hint:* Where does the tangent line at $t = 4$ intersect the t-axis?)

35. Suppose that the amount of grapes (in thousands of pounds) grown in California t years after they were first introduced is given by $g = 12t^3$.
(a) Use formula (5) to find the slope of the graph this function.
(b) At what rate (in thousands of pounds per year) was the grape crop increasing 5 years after it was started?
(c) Suppose the amount of wine (in thousands of liters) produced during the same period is given by $w = t^3$. What is the relation between the rates of grape production and wine production?

36. A cup of tea initially at $200°$ is placed in a room where the temperature is kept at $75°$ (both temperatures in Fahrenheit). The graph in Figure 3.1.17 shows how the temperature of the tea T, measured in degrees Fahrenheit, decreases toward room temperature as the time t, measured in minutes, increases.
(a) Use this graph to estimate the instantaneous rate of change of T with respect to t when $t = 20$ minutes.
(b) Compare the rate of change at $t = 20$ with the instantaneous rate of change when $0 < t < 20$ and when $t > 20$.

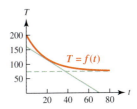

Figure 3.1.17

37. The number of cell phones in millions of units sold per year in a region over the last 10 years is approximated by the graph in Figure 3.1.18.

(a) Find the slope of the secant lines AB and AC and the average rate of sales per year over the time intervals $[2, 3]$ and $[2, 5]$. Then estimate the slope of the graph at A and the instantaneous rate of change at $t = 2$.

(b) Find the slope of the secant lines ED and EF, and the average rate of sales per year over the time intervals $[6, 7]$ and $[7, 8]$. Then estimate the slope of the graph at E and the instantaneous rate of change at $t = 7$.

(c) Compare the rates of change for $t = 2, 3, 5, 7$, and 8. When are sales per year increasing fastest and when are they decreasing fastest? What happens at $t = 5$?

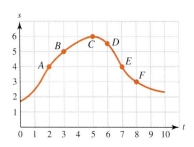

Figure 3.1.18

38. Research has shown that species diversity in a creek declines when water from a treatment plant is released into the creek. Figure 3.1.19 shows the number of species N in hundreds as a function of the distance d in miles from a certain point in the creek.

(a) At what distance is the treated water released into the creek?

(b) At which one of the points A, B, C, D, and E is the slope of the graph most negative?

(c) What is the slope at D, and what happens to the rate of change there?

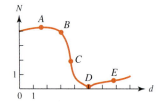

Figure 3.1.19

39. According to the Economic Growth and Tax Relief Reconciliation Act of 2001 passed by the U.S. Congress and signed by President George W. Bush, the amount of tax relief for a single person is the lesser of $300 or 5% of that person's year 2000 taxable income.

(a) Write the amount of tax relief R as a function of the taxable income i.

(b) Sketch its graph and find the slope for $i = 4,000$ and $i = 8,000$.

(c) What are all possible values of the slope? Explain their meaning.

40. During the 24 hours of October 5, 2001, the U.S. National Weather Service recorded the temperature readings in Flagstaff, Arizona shown below in Table 3.1.Ex.40, where t is the time in hours and C is the temperature in degrees Celsius:

(a) Using the given values, draw the graph of the function $C = C(t)$ as a continuous curve that has a tangent line at every point.

(b) Estimate the average rate of change of temperature from 8 A.M. to noon and also over the entire 24-hour period.

(c) Estimate the instantaneous rate of change of temperature at 8 A.M. and at 8 P.M.

41. Suppose the population (in thousands) of a colony of bacteria is given by the formula

$$p = \frac{4.8}{0.62 + 5.1e^{-0.06t}},$$

where t is the time in hours.

(a) Graph the population curve over the first 100 hours in a $[0, 100] \times [0, 10]$ window.

(b) Using the zoom function on your calculator, estimate the time at which the population is increasing fastest.

(c) What is happening to the population as more and more time passes? Express your answer as a limit as $t \to \infty$ and interpret what it means in terms of the population growth. What happens to the slope?

42. Write a formula (in terms of h) for the slope of the secant line between the points $(0, 1)$ and (h, e^h) on the graph of $y = e^x$. Then make a table of the slopes corresponding to $h = 10^{-k}$ for $k = 1, 2, \ldots, 6$, and try to guess the slope of the graph at $x = 0$.

t	0	1	2	3	4	5	6	7	8	9	10	11	12
C	10.6	8.3	7.8	6.7	5.0	6.7	5.0	7.8	9.4	10.0	13.3	18.9	21.1
t	13	14	15	16	17	18	19	20	21	22	23	24	
C	22.2	22.8	22.2	21.7	22.8	22.2	21.7	20.0	17.8	16.1	15.0	14.4	

Table 3.1.Ex.40

Solutions to practice exercise 3.1

1. **(a)** Using the formula (5) with $f(x) = x^2 - 3x + 8$, we get the following expression for the slope of the graph:

slope of the graph

$$= \lim_{h \to 0} \frac{(x+h)^2 - 3(x+h) + 8 - (x^2 - 3x + 8)}{h}$$

$$= \lim_{h \to 0} \frac{x^2 + 2xh + h^2 - 3x - 3h + 8 - x^2 + 3x - 8}{h}$$

$$= \lim_{h \to 0} \frac{2xh + h^2 - 3h}{h} = \lim_{h \to 0} (2x + h - 3) = 2x - 3.$$

(b) By (a) the tangent to our graph at $(x, x^2 - 3x + 8)$ has slope $2x - 3$. When $x = 4$, the slope is 5, so we need to find the equation of the line of slope 5 through the point $(4, 4^2 - 3 \cdot 4 + 8) = (4, 12)$. Using the point-slope formula, we have $y - 12 = 5(x - 4)$, or $y = 5x - 8$.

3.2 The Derivative of a Function

The procedure we used in the last section to determine the slope of a graph is at the heart of calculus. The slope of the graph is called the **derivative** of the function. In other words, the derivative of the function $f(x)$ is a new function that gives the slope of the graph. We represent this new function by the symbol $f'(x)$, which is read as *f prime of x*. The material of the last section—formula (5), in particular—leads to the following definition, which is one of the most fundamental formulas of calculus:

Definition 3.2.1

Definition of the derivative

$$f'(x) = \lim_{h \to 0} \frac{f(x+h) - f(x)}{h}, \text{ provided the limit exists.}$$

A function that has a derivative at a given point is said to be **differentiable** at that point.

The following table summarizes some of the results of the last section in terms of the derivative:

$f(x)$	x^2	x^3	$x^{1/2}$
$f'(x)$	$2x$	$3x^2$	$\frac{1}{2}x^{-1/2}$

This table suggests the following formula for computing the derivative of a power function:

The Power Rule

$$(x^m)' = mx^{m-1}.$$

In other words, if $f(x) = x^m$, then $f'(x) = mx^{m-1}$.

This rule is valid for any fixed exponent m, whether positive or negative, rational or irrational. We will not prove the power rule, relying instead on one more illustrative example and several exercises to make it convincing.

EXAMPLE 3.2.1

Using the definition, show that if $f(x) = x^{-2}$ for $x \neq 0$, then $f'(x) = -2x^{-3}$.

SOLUTION In this case, $f(x) = 1/x^2$ and $f(x+h) = 1/(x+h)^2$. We have

$$f(x+h) - f(x) = \frac{1}{(x+h)^2} - \frac{1}{x^2}$$

$$= \frac{x^2 - (x+h)^2}{x^2(x+h)^2} \quad \text{(by combining fractions)}$$

$$= \frac{x^2 - (x^2 + 2xh + h^2)}{x^2(x+h)^2} \quad \text{(by expanding)}$$

$$= \frac{-2xh - h^2}{x^2(x+h)^2}$$

$$= \frac{-h(2x+h)}{x^2(x+h)^2} \quad \text{(by factoring out } -h\text{)}.$$

Dividing by h gives

$$\frac{f(x+h) - f(x)}{h} = \frac{-(2x+h)}{x^2(x+h)^2},$$

for all $h \neq 0$. The numerator and denominator both have nonzero limits, since

$$\lim_{h \to 0} -(2x+h) = -2x \quad \text{and} \quad \lim_{h \to 0} x^2(x+h)^2 = x^2 x^2 = x^4.$$

Therefore,

$$\lim_{h \to 0} \frac{f(x+h) - f(x)}{h} = \frac{-2x}{x^4} = \frac{-2}{x^3},$$

and we have verified that $f'(x) = -2x^{-3}$.

> **PROBLEM-SOLVING TACTIC**
>
> To find the derivative $f'(x)$ of a function $f(x)$,
>
> - Write $f(x)$ and $f(x+h)$ and then form the quotient $(f(x+h) - f(x))/h$.
> - Next, expand and simplify.
> - Finally, compute the limit as $h \to 0$.

Finally, a small but useful observation: *The power rule is also valid if the exponent m is zero.* Since $x^0 = 1$, the function is constant, $f(x) = 1$, and its graph is a horizontal line. Therefore, the slope is zero everywhere, so that $f'(x) = 0$. And, indeed, the same holds true for *any* constant function. Its graph is a horizontal line, and the slope is zero. To summarize:

> The derivative of a constant function is zero: $(c)' = 0$.

Notation for the derivative

In addition to writing $f'(x)$ for the derivative of the function $f(x)$, we sometimes write it as either

$$\frac{d}{dx} f(x) \quad \text{or} \quad \frac{df}{dx}.$$

The symbol d/dx stands for "the derivative with respect to x of . . . ," so that df/dx

means "the derivative of f with respect to x," to specify that x is the independent variable. If we write $y = f(x)$, we may also denote the derivative by y' or dy/dx. For instance, if $y = x^2$, then $dy/dx = 2x$. We can express this even more concisely by writing

$$\frac{d}{dx} x^2 = 2x.$$

This notation was invented in the seventeenth century by one of the founders of calculus—the German mathematician and philosopher, Gottfried Wilhelm Leibniz—and it proved to be a very natural and useful device for calculations involving derivatives. It arose from a slight variation on the formula defining the derivative, one that was used in Leibniz's time and is still found in many calculus books. It uses the symbol Δx instead of h, so that the definition of the derivative takes the form

$$f'(x) = \lim_{\Delta x \to 0} \frac{f(x + \Delta x) - f(x)}{\Delta x}. \tag{12}$$

Of course, there is no real difference between using Δx and h. Both are used to represent a small quantity that is added to x (or subtracted from it). In fact, the symbol Δ is the capital delta—the fourth letter of the Greek alphabet. It corresponds to our letter D, which in this case stands for the word "difference" (written as $\Delta\iota\alpha\varphi o\rho\acute{\alpha}$ in Greek), so that Δx represents a small difference, or change, in x.

If we let $y = f(x)$, we can also think of Δy as representing the change in y that results from adding Δx to x. To be specific,

$$\Delta y = f(x + \Delta x) - f(x).$$

Using this notation, we can rewrite formula (12) as follows:

$$f'(x) = \lim_{\Delta x \to 0} \frac{\Delta y}{\Delta x},$$

and Leibniz found it natural to let dy/dx represent the limit on the right.

HISTORICAL PROFILE

Sir Isaac Newton (1643–1727) is one of the two people credited with the invention of calculus. His father was a farmer who died 3 months before his son was born. As a youngster, he was reputed to have great mechanical ability, and stories abound of his skill in making models of clocks and windmills while at school. He entered Cambridge University in 1661 with the intention of earning a law degree, but when the university closed because of the plague in 1665, he returned to his home. In less than 2 years, before turning 25, he invented calculus, discovered the law of gravitation, and made other revolutionary contributions to physics, optics, and astronomy. He returned to Cambridge when it reopened in 1667, and was appointed professor of mathematics in 1669. In 1687 he summarized his full treatment of physics and astronomy in the *Principia naturalis*, arguably the greatest scientific book ever written. After suffering two nervous breakdowns, he decided to leave Cambridge for a government position, becoming Warden of the Royal Mint in 1696 and Master in 1699, a position to which he made substantial contributions and that made him a very rich man. He was elected president of the Royal Society in 1703 and every year thereafter until his death.

Finally, another notation that is frequently used for defining the derivative at a number a is the following:

$$f'(a) = \lim_{x \to a} \frac{f(x) - f(a)}{x - a}. \tag{13}$$

In this case, a is fixed. If we let $x = a + h$, then $h \to 0$ is equivalent to $x \to a$. Therefore,

$$f'(a) = \lim_{h \to 0} \frac{f(a + h) - f(a)}{h} = \lim_{x \to a} \frac{f(x) - f(a)}{x - a},$$

which gives us formula (13).

The tangent line

Saying that a function $f(x)$ is differentiable at a is equivalent to saying that its graph has a nonvertical tangent line at $(a, f(a))$, and the derivative $f'(a)$ is the slope of that tangent line. Therefore, using the point-slope formula, we determine the following:

The Equation of the Tangent Line at $(a, f(a))$ is

$$y = f(a) + f'(a)(x - a). \tag{14}$$

HISTORICAL PROFILE

Gottfried Wilhelm von Leibniz (1646–1716) is the other person credited with the invention of calculus. He entered the University of Leipzig at the age of 14 and studied both philosophy and law. For a time he served as secretary of the Nuremberg Alchemical Society and later practiced law in the city of Mainz. His interests were exceptionally broad, encompassing law, religion, science, and literature, and one of his lifelong goals was to collate all human knowledge. It was that interest that led to his study of mathematics and physics and to his invention of calculus. Leibniz's work was several years behind Newton's, completely independent of it and without any knowledge of Newton's previous discoveries. Unfortunately, however, both men's competing claims considerable friction and bitterness between them. It is Leibniz who is responsible for the common notation used to this day, such as dy/dx for the derivative and $\int$ for the integral, which was much more natural and provided greater flexibility than Newton's notation. He made a number of other important mathematical contributions, including work on determinants and the development of the binary system of arithmetic. He also continued his philosophical studies, attempting to develop an algebra that would encompass all reasoning.

EXAMPLE 3.2.2

Find the equation of the line tangent to the graph of $y = 1/\sqrt{x}$ at the point $(9, \frac{1}{3})$.

SOLUTION   Since $1/\sqrt{x} = x^{-1/2}$, we can use the power rule: $f'(x) = -\frac{1}{2}x^{-3/2}$, and the slope of the graph at the point $(9, \frac{1}{3})$ is given by

$$f'(9) = -\left(\frac{1}{2}\right)9^{-3/2} = -\frac{1}{54}.$$

To find the equation of the tangent line, we simply write the equation of the line with slope $-\frac{1}{54}$ through the point $(9, \frac{1}{3})$. That equation is

$$y = \frac{1}{3} - \frac{1}{54}(x - 9) \quad \text{or} \quad y = -\frac{1}{54}x + \frac{1}{2},$$

or, equivalently, $x + 54y = 27$. Figure 3.2.1 shows the graph and the tangent line.

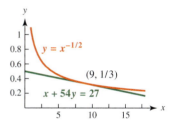

Figure 3.2.1

Some rules for computing derivatives

We can use the power rule to find the derivative of any polynomial. Recall that a polynomial is a sum of powers of x (with positive integers as exponents), each multiplied by a constant coefficient—for example, an expression such as

$$f(x) = 2x^5 - x^4 + 7x^3 + 5x^2 + 8x - 1.$$

To find the derivative of such a function, we have to know how to handle the sums and the coefficients. The following rule tells us how to deal with the coefficients:

The Constant Multiple Rule

If c is a constant, then

$$\frac{d}{dx}[cf(x)] = c \cdot \frac{d}{dx}f(x).$$

In other words, if we change a function simply by multiplying it by a constant, the derivative changes by the same constant multiple.

EXAMPLE 3.2.3

Find the derivative of $5x^{1/3}$.

SOLUTION   The power rule tells us that

$$\frac{d}{dx} x^{1/3} = \frac{1}{3} x^{-2/3}.$$

By combining that with the constant multiple rule, we get

$$\frac{d}{dx} 5x^{1/3} = \frac{5}{3} x^{-2/3}.$$

The next rule tells us how to deal with sums.

The Sum Rule

$$\frac{d}{dx} [f(x) + g(x)] = \frac{d}{dx} f(x) + \frac{d}{dx} g(x).$$

In other words, if we add two functions together to form a new function, the derivative of the sum is the sum of the derivatives.

EXAMPLE 3.2.4

If $f(x) = 3x^2 + \dfrac{2}{x}$, what is $f'(x)$?

SOLUTION   By combining the sum, constant multiple, and power rules, we get

$$\frac{d}{dx} \left(3x^2 + \frac{2}{x} \right) = \frac{d}{dx} (3x^2) + \frac{d}{dx} (2x^{-1})$$

$$= 3 \frac{d}{dx} x^2 + 2 \frac{d}{dx} \left(x^{-1} \right)$$

$$= 6x - 2x^{-2}$$

$$= 6x - \frac{2}{x^2}.$$

The sum rule is also valid for sums of three or more functions.

EXAMPLE 3.2.5

Find $\dfrac{d}{dx}(x^2 - 3x + 5)$.

SOLUTION   By combining the sum, constant multiple, and power rules, we get

$$\frac{d}{dx}(x^2 - 3x + 5) = \frac{d}{dx}x^2 + \frac{d}{dx}(-3x) + \frac{d}{dx}(5)$$
$$= 2x - 3.$$

The same combination of rules enables us to find the derivative of any polynomial, by calculating each term separately, carrying along the coefficient, and using the power rule.

EXAMPLE 3.2.6

Find the equation of the line tangent to the graph of

$$y = x^3 - 2x^2 + 5x - 7$$

at the point $(2, 3)$.

SOLUTION   To begin with, you should check that the point $(2, 3)$ is indeed on the graph. That amounts to checking that substituting $x = 2$ gives $y = 3$, which you can easily verify. The slope at any point of the graph is given by the derivative:

$$\frac{d}{dx}(x^3 - 2x^2 + 5x - 7) = 3x^2 - 4x + 5.$$

For $x = 2$ this equals 9, so we are looking for the equation of the line through the point $(2, 3)$ with slope 9. That is easily found to be

$$y = 9x - 15.$$

Instantaneous rate of change

By comparing the definition of the derivative given by formula (12) with the instantaneous rate of change given by formula (11) of the last section, we arrive at another very important interpretation of the derivative. That is,

$$f'(x) = \text{instantaneous rate of change at } x = \lim_{\Delta x \to 0} \frac{\Delta y}{\Delta x}. \tag{15}$$

Rewriting formula (15) in the form $\Delta y \approx f'(x) \cdot \Delta x$, we can think of it as saying that if Δx is small, a change of Δx in the independent variable causes a change of approximately $f'(x) \cdot \Delta x$ in the dependent variable.

EXAMPLE 3.2.7

Assume that the annual demand function for a popular model of scientific calculator is given by

$$p = D(q) = 0.01q^2 - 2.4q + 144,$$

where the price p is measured in dollars and the quantity q is measured in units of hundred thousands. Find the instantaneous rate of change of p with respect to q when the quantity demanded is 2 million units. Explain its meaning.

SOLUTION   As formula (15) states, the instantaneous rate of change of p with respect to q is given by the derivative

$$\frac{dp}{dq} = 0.02q - 2.4.$$

In particular, when $q = 20$, which is our case, the instantaneous rate of change is equal to $0.02 \cdot 20 - 2.4 = -2$. This means that, at the demand level of 2 million calculators, a one-unit increase in the supply of calculators (i.e., an increase of 100,000 calculators) will cause the price to drop by approximately \$2.

     The graph of $D(q)$ is shown in Figure 3.2.2.

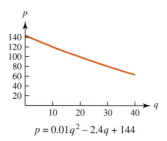

$$p = 0.01q^2 - 2.4q + 144$$

Figure 3.2.2

How to derive the sum and constant multiple rules

The sum and constant multiple rules can be derived from the basic definition of the derivative and the corresponding rules for limits. First, suppose we add two functions, say, $f(x)$ and $g(x)$. To find the derivative of the sum, $f(x) + g(x)$, we think of it as a single function and apply the definition of the derivative to it:

$$\frac{d}{dx}\left[f(x) + g(x)\right] = \lim_{h \to 0} \frac{[f(x+h) + g(x+h)] - [f(x) + g(x)]}{h} \tag{16}$$

The right-hand side looks rather imposing, but we can rearrange the terms and break up the quotient, as follows:

$$\frac{[f(x+h)+g(x+h)]-[f(x)+g(x)]}{h} = \frac{[f(x+h)-f(x)]+[g(x+h)-g(x)]}{h}$$

$$= \frac{f(x+h)-f(x)}{h} + \frac{g(x+h)-g(x)}{h}.$$

Therefore, by making this substitution in Eq. (16) and applying the addition formula for limits (see Theorem 1.1.1 in Chapter 1), we get

$$\frac{d}{dx}[f(x)+g(x)] = \lim_{h\to 0}\frac{f(x+h)-f(x)}{h} + \lim_{h\to 0}\frac{g(x+h)-g(x)}{h}. \qquad (17)$$

You may recognize these last two limits. They are exactly the derivatives of $f(x)$ and $g(x)$. That is,

$$f'(x) = \lim_{h\to 0}\frac{f(x+h)-f(x)}{h} \quad \text{and} \quad g'(x) = \lim_{h\to 0}\frac{g(x+h)-g(x)}{h},$$

and substituting these on the right-hand side of (17) gives

$$\frac{d}{dx}[f(x)+g(x)] = f'(x)+g'(x),$$

which is the sum rule.

The constant multiple rule can be proved in a similar way. If $f(x)$ is a function and c is a constant, we can think of the product $cf(x)$ as a single function and apply the definition of the derivative to it:

$$\frac{d}{dx}[cf(x)] = \lim_{h\to 0}\frac{cf(x+h)-cf(x)}{h}.$$

By factoring c out of the numerator, we can rewrite the right-hand quotient as

$$c\,\frac{f(x+h)-f(x)}{h},$$

and applying the constant multiple formula for limits (see Theorem 1.1.1 in Chapter 1) gives

$$\frac{d}{dx}[cf(x)] = \lim_{h\to 0}\frac{cf(x+h)-cf(x)}{h}$$

$$= \lim_{h\to 0} c\,\frac{f(x+h)-f(x)}{h}$$

$$= c\lim_{h\to 0}\frac{f(x+h)-f(x)}{h} = cf'(x),$$

which is the constant multiple rule.

Estimating the derivative

In many applications, particularly when dealing with experimental data, a function may be given as a table of values rather than a formula. In such cases—assuming that the data represent the values of a differentiable function—we can use the information in the table to estimate the derivative.

EXAMPLE 3.2.8

Given the following values of a function $y = f(x)$:

x	3.97	3.98	3.99	4	4.01	4.02	4.03
$f(x)$	8.79	8.87	8.95	9	9.06	9.09	9.11

estimate $f'(4)$.

SOLUTION   From the definition of the derivative, we have

$$f'(4) = \lim_{h \to 0} \frac{f(4 + \Delta x) - f(4)}{\Delta x},$$

which means that

$$f'(4) \approx \frac{f(4 + \Delta x) - f(4)}{\Delta x}$$

if $|\Delta x|$ is small. In the given table, $|\Delta x| = 0.01$. Using $\Delta x = 0.01$ gives the estimate

$$f'(4) \approx \frac{f(4.01) - f(4)}{0.01} = \frac{9.06 - 9}{0.01} = \frac{0.06}{0.01} = 6.$$

We could also use $\Delta x = -0.01$, which gives

$$f'(4) \approx \frac{f(3.99) - f(4)}{-0.01} = \frac{8.95 - 9}{-0.01} = \frac{0.05}{0.01} = 5.$$

Neither estimate is more correct than the other. A common practice is to average them, which in this case gives $f'(4) \approx 5.5$.

In general, given a table of values for f around $x = a$ in steps of size Δx, where Δx is some small positive number—that is, given the data $f(a)$ and $f(a \pm \Delta x)$, we have two possible estimates for $f'(a)$, namely,

$$f'(a) \approx \frac{f(a + \Delta x) - f(a)}{\Delta x}, \tag{18}$$

which is called a **forward difference formula**, and

$$f'(a) \approx \frac{f(a) - f(a - \Delta x)}{\Delta x}, \tag{19}$$

which is called a **backward difference formula**.

In Example 3.2.8, with $\Delta x = 0.01$, these formulas gave the estimates 6 and 5. We also got a third estimate by averaging these two approximations. In general, if we add the right-hand sides of (18) and (19) and divide by 2, we obtain the average:

$$f'(a) \approx \frac{f(a + \Delta x) - f(a - \Delta x)}{2\Delta x}, \tag{20}$$

which is called a **central difference formula**. In Example 3.2.8, the central difference formula gave the approximation $f'(4) \approx 5.5$.

EXAMPLE 3.2.9

The California condor (*Gymnogyps californianus*) is the largest North American bird and an endangered species. From 1965 to 1982 its population dropped from 60 to 24. After that, thanks to conservation efforts, the population increased. The table below shows the condor's population levels in subsequent years.*

1982	1983	1984	1985	1986	1987	1988	1989	1990	1991
24	25	27	27	27	27	28	32	40	52

*Source: http://biology.usgs.gov and the Los Angeles Zoo.

Use the central difference formula to estimate the rate at which the condor population changed in the years 1983 and 1990.

SOLUTION   If $f(t)$ is the condor population, where t is the time in years, then $f'(t)$ is the rate of change of the population with respect to time. With $a = 1983$ and $\Delta t = 1$, formula (20) gives

$$f'(1983) \approx \frac{f(1984) - f(1982)}{2} = \frac{27 - 24}{2} = 1.5 \text{ condors per year.}$$

With $a = 1990$, we get

$$f'(1990) \approx \frac{f(1991) - f(1989)}{2} = \frac{52 - 32}{2} = 10 \text{ condors per year.}$$

Warning   These estimates are based on the assumption that the given table represents the values of a differentiable function, and they may give misleading results if the derivative does not exist at a.

APPLYING TECHNOLOGY

Using a calculator or spreadsheet can save a great deal of tedious calculation in estimating a derivative from a table of values. Figure 3.2.3 shows an Excel spreadsheet in which all three difference formulas—forward, backward, and central—have been computed for a

Figure 3.2.4

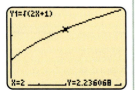

Figure 3.2.5

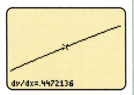

Figure 3.2.6

File Edit View Insert Format Tools Data Window Help													11:16:05
C6 ▼ = =(D2-B2)/(D1-B1)													

	A	B	C	D	E	F	G	H	I	J	K	L	M	N
1	x	0.4	0.5	0.6	0.7	0.8	0.9	1	1.1	1.2	1.3	1.4	1.5	1.6
2	f(x)	0.30902	0.38268	0.45399	0.5225	0.58779	0.64945	0.70711	0.76041	0.80902	0.85264	0.89101	0.92388	0.95106
3														
4	Forward	0.73666	0.71307	0.68508	0.65287	0.61663	0.57659	0.53299	0.48611	0.43623	0.38366	0.32873	0.27177	
5	Backward		0.73666	0.71307	0.68508	0.65287	0.61663	0.57659	0.53299	0.48611	0.43623	0.38366	0.32873	0.27177
6	Central		0.72487	0.69908	0.66897	0.63475	0.59661	0.55479	0.50955	0.46117	0.40995	0.3562	0.30025	

Figure 3.2.3

given set of data. The central difference formula, as applied to the selected cell (C6), is shown at the top of the spreadsheet, preceded by an "=" sign.

A standard graphing calculator has a function (usually called **nDeriv** or **nDer**, standing for "numerical differentiation") that estimates the derivative of a function whose formula is known, by first creating a table of values and then applying the central difference formula. Figure 3.2.4 shows the **nDeriv** function applied to the function $f(x) = 1/\sqrt{x}$ to estimate $f'(4)$ (using a TI-83 Plus). The information entered consists of the function, the variable, the point at which the derivative is to be estimated, and the step size. (The last may be omitted, since the program contains a default step size.)

In this case, we can check the accuracy of the approximation by computing the derivative. Since $f(x) = x^{-1/2}$, the power rule gives $f'(x) = -\frac{1}{2}x^{-3/2}$. Therefore, $f'(4) = -\frac{1}{16} = -0.0625$.

Another method of estimating the derivative with a graphing calculator is by zooming in on the point in question and using the numerical derivative function within the CALCULATE mode. Figure 3.2.5 shows the graph of $f(x) = \sqrt{2x+1}$ with the trace cursor at $x = 2$. Figure 3.2.6 shows the result of zooming in and then using the numerical derivative (on a TI-83 Plus, the **dy/dx** function in the CALC mode). The result shows that $f'(2) \approx 0.4472136$. In fact, as we will see later, $f'(2) = 1/\sqrt{5} \approx 0.4472135955$.

Practice Exercises 3.2

1. Using the definition, show that if $f(x) = x^{-1}$ for $x \neq 0$, then $f'(x) = -x^{-2}$.

2. Find the equation of the line tangent to the graph of $y = 3x^4 + 1/x^4$ at the point $(1, 4)$.

Exercises 3.2

In Exercises 1–3, use the definition of the derivative and no other derivative formula to find $f'(x)$.

1. $f(x) = \dfrac{1}{3x}$

2. $f(x) = \dfrac{1}{2x+1}$

3. $f(x) = \dfrac{1}{2-x}$

4. **(a)** Use the definition of derivative and no other derivative formula to find $f'(x)$ if $f(x) = 1/x^3$. (*Hint:* Follow the model of Example 3.2.1, using formula (7).)
   (b) Use the power rule to find $f'(x)$ and check that you get the same answer.

In Exercises 5–13, use the power rule to find the derivative.

5. $f(x) = x^{5/3}$

6. $f(x) = \dfrac{1}{x^5}$

7. $f(x) = x^{1.4}$

8. $f(x) = x^{-1.4}$

9. $f(x) = \dfrac{1}{\sqrt{x}}$

10. $f(x) = x^{3/2}$

11. $f(x) = x^{-4}$

12. $f(x) = x^{1/4}$

13. $f(x) = x^{-1/4}$

14. Find $f'(4)$ if
   (a) $f(x) = 1/x$
   (b) $f(x) = 1/x^2$
   (c) $f(x) = x^{3/2}$
   (d) $f(x) = x^{-3/2}$

15. Figure 3.2.7 shows the tangent line to the graph of a function $f(x)$ and its equation at a point A. Find the derivative of the function at $x = a$.

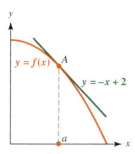

Figure 3.2.7

In Exercises 16–17, estimate the derivative of the function $f(x)$, whose graph is shown in Figure 3.2.8, at the indicated points.

16. $f'(0.6)$, $f'(0.8)$, and $f'(1)$

17. $f'(0.7)$, $f'(0.9)$, and $f'(1.1)$

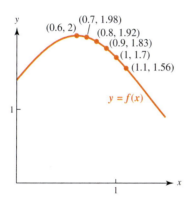

Figure 3.2.8

18. Find the slope of the graph of $y = x^5$ at the point $(-1, -1)$.

19. Find the slope of the graph of $y = x^{-3}$ at the point with $x = \frac{1}{2}$.

20. Find the equation of the line tangent to the graph of $y = x^{1/3}$ at the point $(8, 2)$.

21. Find the equation of the line tangent to the graph of $y = x^3$ at the point $(2, 8)$.

22. Find all points where the graph of $y = x^{3/2}$ has slope 3.

23. If $f(x) = x^{2/3}$, find all points where $f'(x) = \frac{1}{6}$.

24. Find all points on the graph of $y = 1/x^2$ where the tangent is parallel to the line $x - 4y = 3$.

25. Suppose the number of new math concepts a business calculus student can grasp in a 1-hour class is represented by $m(t) = 0.92\sqrt{t}$, where t is measured in minutes.
   (a) Find $m'(t)$.
   (b) Compare $m'(5)$ to $m'(50)$ and explain what is happening to the student's ability to learn.

26. A hospital patient who suffered brain damage from an accidental fall has received therapy to help him pronounce words more clearly. His progress is monitored weekly, and when his rate of increase in improvement falls below 0.9, he is released to continue therapy as an outpatient. Assuming that his progress, as measured in words he can pronounce after w weeks, is modeled by $P(w) = 5.7w^{1/3}$, decide whether the patient is ready to be released at week three.

27. Suppose that $f(x)$ is a function whose graph goes through the point $(2, -3)$ and whose tangent line at that point has the equation $x + y = -1$. Without computing, find each of the following limits:
   (a) $\lim\limits_{h \to 0} \dfrac{f(2 + h) + 3}{h}$ **(b)** $\lim\limits_{\Delta x \to 0} \dfrac{f(2 + \Delta x) + 3}{\Delta x}$
   (c) $\lim\limits_{x \to 2} \dfrac{f(x) + 3}{x - 2}$

28. Suppose $f(x)$ is a function that has a horizontal tangent at the point $(1, 0)$. Without computing, find each of the following limits:
   (a) $\lim\limits_{h \to 0} \dfrac{f(1 + h)}{h}$ **(b)** $\lim\limits_{\Delta x \to 0} \dfrac{f(1 + \Delta x)}{\Delta x}$
   (c) $\lim\limits_{x \to 1} \dfrac{f(x)}{x - 1}$

29. Suppose $f(-1) = 3$ and $f'(-1) = -2$. Find each of the following limits:
   (a) $\lim\limits_{x \to -1} \dfrac{f(x) - 3}{x + 1}$ **(b)** $\lim\limits_{h \to 0} \dfrac{f(-1 + h) - 3}{h}$
   (c) $\lim\limits_{\Delta x \to 0} \dfrac{f(-1 + \Delta x) - 3}{\Delta x}$

In Exercises 30–35, find the derivative of the given function.

30. $f(x) = x^3 - 5x^2 + 4x - 1$

31. $g(t) = 3t^5 + 2t^3 - 5t + 7$

32. $f(t) = t^{-2} - 3t^{-1} + 4$

33. $g(x) = \frac{1}{2}(x^3 + x^{-3})$

34. $f(s) = s - \dfrac{1}{s}$ **35.** $f(s) = s^2 - 3s^{-2} + 5s^{1/2}$

In Exercises 36–39, find the derivative.

36. $\dfrac{d}{dx}\left(x^4 + \dfrac{3}{x^2}\right)$ **37.** $\dfrac{d}{du}\left(\dfrac{\sqrt{u}}{2} + \dfrac{2}{\sqrt{u}}\right)$

38. $\dfrac{d}{dt}\left(\dfrac{t^2+3t+1}{3}\right)$

39. $\dfrac{d}{dz}\left(\dfrac{z^3}{3}-\dfrac{z^2}{4}+\dfrac{z}{2}-\dfrac{3}{2}\right)$

40. Find the equation of the line tangent to the graph of $y = x^3 - x^2 - 7x - 3$ at the point $(-1, 2)$.

41. Find the equation of the line tangent to the graph of $y = x^4 + 2x^3 + 5x^2 + 4x - 2$ at the point with $x = -1$.

42. Find the equation of the line tangent to the graph of $y = 2x + (1/x)$ at the point $(1, 3)$.

43. Find the equation of the line tangent to the graph of $y = 3x^2 - 5x + 1$ at the point $(2, 3)$.

44. Find all points, if any, where the tangent to the graph of $y = 2x^3 - 3x^2 + 1$ is horizontal.

45. Find all points where the graph of $y = 3x^2 - 4x - 5$ has slope 8.

46. If $f(x) = x + (1/x)$ for $x \neq 0$, find all solutions of the equation $f'(x) = 0$.

47. Find all points where the tangent to the graph of $y = x^3 + x$ is parallel to the line $7x - 4y = 2$.

48. A farmer's grain silo is cylindrical in shape, with a base diameter of 30 feet. The farmer wants to know how much the volume of grain changes for every 1-foot layer of grain he adds. What do you tell the farmer? (Recall that the volume of a cylinder is $V = \pi r^2 h$, where r is the radius of the base and h is the height.)

49. A river carries soil particles down to the ocean where they are deposited into layers of sediment. A metal plate is placed on the ocean floor; then the depth of accumulated sediments is measured each month. The sediment depth in millimeters at any given month m is given by

$$D(m) = 12.5m - 5.08m^2 + 0.712m^3.$$

Examine the rate of change of sediment depth at $m = 1, 3$, and 7. In which of those months is the depth increasing fastest? slowest?

50. Water treatment plants take dirty sewer effluents, remove the solids, evaporate off volatile chemicals, and kill bacteria to produce reasonably clean water. Suppose the total cost in hundreds of dollars to produce x thousand gallons of clean water is $C(x) = 4x^2 + 100x + 500$.
   (a) Is there any value of x for which $C(x)$ is not differentiable?
   (b) Find $C'(5)$ and estimate the additional cost for producing 6,000 gallons. (*Hint:* See the remarks following formula (15).)
   (c) Find $C'(30)$ and estimate the additional cost for producing 31,000 gallons.
   (d) Compare $C'(5)$ to $C'(30)$.

51. In economics, capital stock is defined as the amount of capital invested per worker per year (in machinery, computers, etc.). A company estimates that its production function is given by $q = 18k^{0.3}$, where k is the capital stock and q is the output per worker, both measured in thousands of dollars. Compute $q'(k)$ and explain its meaning, in particular when $k = 8$.

52. Given the following table of values for a differentiable function $f(x)$:

x	-0.4	-0.3	-0.2	-0.1	0
$f(x)$	2.36	1.98	1.64	1.31	1.0
x	0.1	0.2	0.3	0.4	
$f(x)$	0.71	0.44	0.19	-0.04	

Estimate each of the following derivatives, using the forward, backward, and central difference formulas.
   (a) $f'(0)$ **(b)** $f'(0.2)$ **(c)** $f'(-0.1)$

53. The following table gives the coordinates of a set of points on the graph of a differentiable function:

x	0.97	0.98	0.99	1	1.01	1.02	1.03
y	6.96	7.12	7.26	7.39	7.54	7.66	7.87

Estimate the slope of the tangent at $x = 1$.

54. During the 24 hours of October 14, 2001, the U.S. National Weather Service recorded the temperature readings, in degrees Celsius, in Durango, Colorado:

t	0	1	2	3	4	5	6
T	8.6	8.3	8.9	5	1.7	1.1	-0.6
t	7	8	9	10	11	12	
T	-0.6	-3.3	-0.6	2.8	7.8	10	

t	13	14	15	16	17	18
T	12.8	16.1	17.8	20	20	19.4
t	19	20	21	22	23	24
T	17.2	15	13.9	12.8	11.1	6.1

   (a) Using the given values, draw the graph of the function $T = T(t)$ as a continuous curve that has a derivative at every point.

(b) Use the central difference formula to make a table of values of the derivative for the given values of t.
(c) Estimate the maximum and minimum instantaneous rate of change of the temperature during the given 24-hour period.

55. The table for predicting a girl's future height given her present age and height is as follows:

age (in years)	1	2	3	4	5
multiplier	2.30	2.01	1.76	1.62	1.51
age (in years)	6	7	8	9	
multiplier	1.43	1.35	1.29	1.23	

age (in years)	10	11	12	13	14
multiplier	1.17	1.12	1.07	1.03	1.01
age (in years)	15	16	17	18	
multiplier	1.002	1.001	1.001	1.00	

In other words, a girl at age 11 would expect to be 1.12 times taller at full adult height than she is now.
Use the central difference formula to estimate the rate of change of the multiplier $M(t)$ at $t = 2, 4, \ldots, 16$. Make a table of these rates. What do you observe?

Source: The Encyclopedia of Education, Vol. 7, Lee C. Deighton, ed., The Macmillian Company and the Free Press, 1971, page 126.

56. The following table gives the U.S. population in millions of people from 1960 to 2000. Use the central difference formula to estimate how fast the population was increasing (in millions of people per year) in 1970 and 1990.

	A	B	C	D	E	F
Year		1960	1970	1980	1990	2000
Population		179	203	227	249	281

57. Let $f(x) = \sqrt[3]{x}$. Use the numerical differentiation function to estimate $f'(8)$, first with step size 0.1, then with step size 0.001. Then compute $f'(8)$ directly and compare it to the estimates.

58. The function $f(x) = \ln(2 - 3x)$ is differentiable throughout its domain. Use the numerical differentiation function with step size 0.001 to estimate $f'(0.1)$ and $f'(-0.1)$.

59. The function $f(x) = 3^x$ is differentiable throughout its domain.
(a) Use the numerical differentiation function with step size 0.01 to estimate $f'(1)$.
(b) Use the result of part (a), rounded to three decimal places, to write an approximate equation of the tangent line at $x = 1$.
(c) Verify your result by graphing both the function and its tangent line in a $[0, 2] \times [1, 5]$ window.

60. Let $f(x) = \sqrt{2x^2 + 1}$.
(a) Calculate $\Delta y / \Delta x$ for $x = 2$ and $\Delta x = 0.1, 0.01, 0.001, 0.0001$.
(b) Use the result of part (a) to guess the value of $f'(2)$ to five decimal places.
(c) Compare your guess with the estimate obtained by using your calculator's numerical differentiation function with step size 0.001.

61. A drug is being administered intravenously, with the amount A (in milligrams) in the patient's bloodstream at the end of t hours being given by the formula $A = 800 - 760e^{-0.12t}$. Graph this function in a $[0, 20] \times [0, 800]$ window and answer the following questions.
(a) At what time in the first 20 hours is the amount increasing fastest?
(b) What is happening to the rate of increase as t goes from 0 to 20?
(c) Use the numerical differentiation function on your calculator to estimate the rate at which the amount is increasing at the end of 20 hours.

62. For the data shown in Figure 3.2.9, assume that f is differentiable for all x. Estimate the derivative $f'(x)$
(a) for x from 0.0 to 1.1 by the forward difference formula,
(b) for x from 0.1 to 1.2 by the backward difference formula, and
(c) for x from 0.1 to 1.1 by the central difference formula.

63. For the data shown in Figure 3.2.10, assume that f is differentiable for all x. Make a table of values of $f'(x)$ for all possible x by the central difference formula.

	A	B	C	D	E	F	G	H	I	J	K	L	M	N
x	0.0	0.1	0.2	0.3	0.4	0.5	0.6	0.7	0.8	0.9	1.0	1.1	1.2	
f(x)	-1.0000	-0.6990	-0.5229	-0.3979	-0.3010	-0.2218	-0.1549	-0.0969	-0.0458	0.0000	0.0414	0.0792	0.1139	

Figure 3.2.9

	A	B	C	D	E	F	G	H	I	J	K	L	M	N	O	P	Q	R
1	*x*	1.92	1.93	1.94	1.95	1.96	1.97	1.98	1.99	2	2.01	2.02	2.03	2.04	2.05	2.06	2.07	2.08
2	*f(x)*	3.3372	3.3722	3.4075	3.4432	3.4792	3.5156	3.5523	3.5894	3.6269	3.6647	3.7028	3.7414	3.7803	3.8196	3.85926	3.89932	3.93977

Workbook1

Figure 3.2.10

Solutions to practice exercises 3.2

1. In this case, $f(x) = 1/x$ and $f(x+h) = 1/(x+h)$. The definition of the derivative tells us that

$$f'(x) = \lim_{h \to 0} \frac{\dfrac{1}{x+h} - \dfrac{1}{x}}{h}. \qquad (21)$$

We can simplify the numerator by combining the two fractions over a common denominator, as follows:

$$\frac{1}{x+h} - \frac{1}{x} = \frac{x - (x+h)}{(x+h) \cdot x} = \frac{-h}{(x+h) \cdot x}.$$

Substituting this into the numerator of formula (21) and canceling h's give

$$f'(x) = \lim_{h \to 0} \frac{-1}{(x+h) \cdot x} = -\frac{1}{x^2} = -x^{-2}.$$

2. Substituting $x = 1$ gives $y = 4$, which verifies that $(1, 4)$ is on the graph. The slope at any point of the graph is given by the derivative

$$\frac{d}{dx}\left(3x^4 + \frac{1}{x^4}\right) = 12x^3 - 4x^{-5}.$$

For $x = 1$, this equals 8, so we are looking for the equation of the line through the point $(1, 4)$ with slope 8. Using the point-slope formula, we obtain $y - 4 = 8(x - 1)$, or $y = 8x - 4$.

3.3 The Derivative as a Rate

As we saw in the last section, the derivative can be interpreted as a *rate*—the rate at which the dependent variable changes with respect to the independent variable—and it is that interpretation that makes it important in applications. In this section we will take up a few of the most important examples.

Average and instantaneous velocity

Imagine an object moving in a straight line—such as a baseball falling straight down or a car moving on a straight road. Its position on the line changes with time. To measure its position, we need to fix a reference point and decide which direction to call positive. For a car moving on a road, we might choose a particular mileage marker as the reference point and label it as the zero point. We can choose either direction as positive. That does not mean the object has to be moving in that direction. It is simply an orientation of the line. In Figure 3.3.1 east was chosen as the positive direction. By convention, a horizontal line is oriented with the positive direction to the right—just as we did with the x-axis—and we will follow that rule.

The **position** of the object at time t, denoted by s or $s(t)$, is the oriented distance from the reference point—meaning that we take the distance with a plus or minus sign, depending on whether the object is on the positive or negative side of the reference point. For instance, if at $t = 5$ minutes the car is 4 miles east of the reference point, then $s(5) = 4$; if it is 4 miles west of the reference point, then $s(5) = -4$.

Figure 3.3.1

The **average velocity** of an object over a time interval $a \le t \le b$ is the change in its position divided by the change in time. In short,

$$\text{Average velocity} = \frac{\text{change in position}}{\text{change in time}} = \frac{s(b) - s(a)}{b - a}. \tag{22}$$

Keep in mind that the average velocity can be positive or negative, according to whether the position change was in the positive or negative direction. For instance, if the car is 5 miles east of the zero point at 8:00 A.M. and 115 miles east of the zero point at 10:00 A.M., then its average velocity over that time period is given (in miles per hour) by

$$\frac{s(10) - s(8)}{10 - 8} = \frac{115 - 5}{2} = 55.$$

On the other hand, if it is 5 miles west of the zero point at 8:00 A.M. and 115 miles west of the zero point at 10:00 A.M., then its average velocity is given by

$$\frac{s(10) - s(8)}{10 - 8} = \frac{-115 - (-5)}{2} = -55.$$

A related concept, one which we use in everyday life, is the average speed.

The **average speed** is the total distance traveled divided by the time elapsed. In short,

$$\text{Average speed} = \frac{\text{distance traveled}}{\text{time elapsed}}, \tag{23}$$

where the distance is taken to be positive regardless of the direction.

For instance, in both of the situations described above—moving east or west—the average speed of the car is 55 miles per hour.

EXAMPLE 3.3.1

A car is traveling west on a straight toll road. At 9:00 A.M. it is 10 miles east of a particular exit, and at 9:30 A.M. it is 15 miles west of it. Find its average velocity and its average speed.

SOLUTION Taking the exit as the zero point and using $s(t)$ as the position function, where t is the number of hours after 9:00 A.M., we can summarize the given information as follows: $s(0) = 10$ and $s(0.5) = -15$. Therefore,

$$\text{Average velocity} = \frac{-15 - 10}{0.5} = -50 \text{ miles per hour}$$

and

$$\text{Average speed} = \frac{25}{0.5} = 50 \text{ miles per hour.}$$

While defining the average velocity over a time interval is simple, defining the velocity of a moving object at a precise instant is a more challenging problem, whose solution leads us once again to the notion of the derivative. Indeed, the problem of "instantaneous" velocity was, together with that of finding the tangent line of a graph, one of the motivating factors in the development of calculus.

Here is the idea. Suppose an object is moving in a straight line, and let $s(t)$ be its position at time t. We want to define its velocity $v(t)$ at time t. To do that,

- we consider a small time change Δt,
- take the average velocity between times t and $t + \Delta t$, and
- take the limit as $\Delta t \to 0$.

Since $s(t + \Delta t)$ is the position of the object at time $t + \Delta t$ and $s(t)$ is its position at time t, its average velocity over the time period is

$$\frac{\text{Change in position}}{\text{Change in time}} = \frac{s(t + \Delta t) - s(t)}{\Delta t}.$$

Letting $\Delta t \to 0$, we obtain both

- the velocity $v(t)$ at time t, and
- the derivative ds/dt,

which leads to the following definition.

Definition 3.3.1

The **velocity** $v(t)$ of an object moving in a straight line (also called its **instantaneous velocity**) is defined as

$$v(t) = \frac{ds}{dt}.$$

It is the rate at which its position changes with respect to time.

The **speed** (also called the **instantaneous speed**) is the magnitude (absolute value) of the velocity.

Remark The words *velocity* and *speed* with no qualifying adjective always refer to the *instantaneous* velocity and speed.

EXAMPLE 3.3.2

If a ball is dropped from a height of 100 feet, its height is given by $s = -16t^2 + 100$ for $0 \le t \le 2.5$, where s is in feet and t in seconds.
    (i) Find its velocity as a function of t.
    (ii) Find its velocity and speed at the end of 0.5 seconds.
    (iii) With what speed does it hit the ground?

SOLUTION (i) In this case, the ball's position is its height. Therefore, its velocity is given by

$$v(t) = \frac{ds}{dt} = -32t.$$

(Notice that the velocity is negative, reflecting the fact that the positive direction is upward, and the ball is falling downward.)

(ii) When $t = 0.5$, the velocity is given by $v(0.5) = -16$ feet per second, and its speed is 16 feet per second.

(iii) To find the time when the ball hits the ground, we set $s = 0$ and solve for t, which gives $t = 2.5$. Its velocity at that instant is given by $v(2.5) = -80$ feet per second. Therefore, its speed is 80 feet per second.

EXAMPLE 3.3.3

Suppose a small particle is moving along the x-axis in such a way that its position at time t is given by the formula

$$x = t^{5/2} + 1.$$

What is its velocity when $t = 1$? when $t = 4$? What is its average velocity between times 1 and 4?

SOLUTION The velocity is given by

$$v(t) = \frac{dx}{dt} = \frac{5}{2}t^{3/2}.$$

When $t = 1$, we have $v(1) = \frac{5}{2}$, and when $t = 4$, we have $v(4) = 20$. For the average velocity, we use formula (22). Since $x = 2$ when $t = 1$ and $x = 33$ when $t = 4$, the average velocity between these times is $\frac{33-2}{4-1}$, which equals $\frac{31}{3}$.

This example illustrates a point worth noting—in general, *computing the average velocity between times a and b is not the same as averaging the instantaneous velocities at those two times.* As we have just seen, the average velocity between $t = 1$ and $t = 4$ is $\frac{31}{3}$ (or 10.33 to two decimal places). On the other hand, the instantaneous velocity when $t = 1$ is $\frac{5}{2}$, and it is 20 when $t = 4$. Averaging these two numbers gives $\frac{45}{4}$, or 11.25.

Acceleration

Once again, let's consider an object moving in a straight line, with its position given as a function of time by $s(t)$ and its velocity given by

$$v(t) = \frac{ds}{dt}.$$

The velocity is also a function of time, and its derivative measures the *rate* at which the velocity is changing.

Definition 3.3.2

The rate at which the velocity changes with respect to time is called the **acceleration**, and is often denoted by $a(t)$. That is,

$$\text{acceleration at time } t = a(t) = \frac{dv}{dt}.$$

The velocity of a moving object is increasing if $a(t) > 0$ and decreasing if $a(t) < 0$.

EXAMPLE 3.3.4

Suppose a particle is moving along the x-axis so that its position at time t is given by the formula

$$x = 3t^2 + 8t - 2t^{5/2} \quad \text{for } t \geq 0.$$

Compute its velocity and its acceleration as functions of t. Next, decide in which direction (left or right) the particle is moving when $t = 1$ and whether its velocity and speed are increasing or decreasing. Finally, answer the same questions for $t = 4$.

SOLUTION The velocity is

$$v(t) = \frac{dx}{dt} = 6t + 8 - 5t^{3/2},$$

and the acceleration is

$$a(t) = \frac{dv}{dt} = 6 - \frac{15}{2}t^{1/2}.$$

Since $v(1) = 9$ and $a(1) = -\frac{3}{2}$, we conclude that the particle is moving in the positive direction when $t = 1$, and its velocity is decreasing. Since the velocity is both positive and decreasing when $t = 1$, its absolute value, the speed, is also decreasing.

When $t = 4$, we have $v(4) = -8$ and $a(4) = -9$. It follows that the particle is moving to the left and its velocity is decreasing. In this case, however, the speed is increasing, because the velocity is negative and decreasing. If a negative quantity decreases, its absolute value increases.

In the special case of an object moving under the influence of gravity, its position is given by the formula

$$s = -16t^2 + v_0 t + h_0,$$

which we previously encountered as formula (28) in Section 0.5. Here, t is the time (in seconds), and v_0 and h_0 are the initial velocity and height (in feet per second and feet, respectively). The velocity is given by

$$v(t) = \frac{ds}{dt} = -32t + v_0. \tag{24}$$

Notice that $v(0) = v_0$, confirming the fact that v_0 is the initial velocity. To find the acceleration, we take the derivative of v and get

$$a = \frac{dv}{dt} = -32. \tag{25}$$

This formula states a physical law first discovered by Galileo: *If an object is moving under the influence of the earth's gravity, with no other forces acting on it, then its acceleration is constant.* The particular constant depends on the units used to measure height and time. In this case, with height measured in feet and time in seconds, the constant is -32. If height is measured in meters, the height is given by

$$s = -4.9t^2 + v_0 t + h_0,$$

and $a = -9.8$. As before, h_0 and v_0 are the initial height and velocity, but this time h_0 is in meters and v_0 in meters per second.

With time in seconds and height in feet, velocity is measured in feet per second. Acceleration is measured in velocity units per second—in other words, in feet per second per second. The standard notation for acceleration units in that case is ft/sec^2.

The second derivative

In studying acceleration and velocity, we took the derivative of s (the distance) to get v and then took the derivative of v to get a. For that reason we say a is the **second derivative** of s. In other words, it is the derivative of the derivative.

The idea of taking the derivative twice has many applications other than acceleration. If $f(x)$ is a function, we denote its second derivative by $f''(x)$. That is,

$$f''(x) = \frac{d}{dx} f'(x).$$

EXAMPLE 3.3.5

If $f(x) = x^4 - 5x^2 + 3x^{-1}$, find $f''(x)$.

SOLUTION   Taking the first derivative gives $f'(x) = 4x^3 - 10x - 3x^{-2}$. Therefore,

$$f''(x) = \frac{d}{dx} f'(x) = 12x^2 - 10 + 6x^{-3}.$$

If we write $y = f(t)$, then we sometimes denote the second derivative by y'' and also by $d^2 y/dt^2$. For instance, we can write the acceleration as

$$a = \frac{d^2 s}{dt^2}.$$

The idea behind this notation is to think of the symbol

$$\frac{d}{dt}$$

as representing the operation of taking the derivative. In that case, taking the derivative of s twice would be represented by

$$\frac{d}{dt} \cdot \frac{d}{dt} \cdot s = \frac{d^2}{dt^2} \cdot s = \frac{d^2 s}{dt^2},$$

which is why the second derivative is written that way.

As you might guess, we do not have to stop with the second derivative. We can take the derivative of $f''(x)$ to get a **third derivative**, denoted by $f'''(x)$ or $d^3 f/dx^3$. And we can take the derivative of $f'''(x)$ to get the fourth derivative, and so forth. If we continue Example 3.3.5 one step further, we get

$$f'''(x) = 24x - 18x^{-4}.$$

Higher-order derivatives have many uses, but for now we will work only with the first and second derivatives.

Marginal analysis

The interpretation of the derivative as a rate also occurs in applications to economics, in which the terms **marginal cost**, **marginal revenue**, and **marginal profit** are used to denote the rates at which the cost, revenue, and profit are changing.

These terms actually have two different but closely related meanings. If $C(x)$ is the cost of producing x units of a certain item, the term marginal cost is used to denote the increase in cost for producing an additional unit. That is, the marginal cost equals $C(x + 1) - C(x)$, or

$$\frac{C(x + 1) - C(x)}{1}. \tag{26}$$

However, in real situations x is a much bigger number than 1, and economists treat a change of one unit as a small change Δx and replace (26) by

$$\frac{C(x + \Delta x) - C(x)}{\Delta x},$$

which is approximately equal to $C'(x)$, the instantaneous rate of change of the cost function. Similar considerations hold for the revenue and profit functions, $R(x)$ and $P(x)$. Therefore, we arrive at the following definitions.

- The **marginal cost** is the derivative of the cost function:
$$MC(x) = C'(x).$$

- The **marginal revenue** is the derivative of the revenue function:
$$MR(x) = R'(x).$$

- The **marginal profit** is the derivative of the profit function:
$$MP(x) = P'(x).$$

Since $P(x) = R(x) - C(x)$, we have the relation $P'(x) = R'(x) - C'(x)$. In other words,

$$MP(x) = MR(x) - MC(x).$$

EXAMPLE 3.3.6

A furniture company producing coffee tables has fixed costs of $5,000 per month, and the cost per unit item is $80. The company estimates that the price of a table starts at $240 and goes down by 50 cents for each one available. Find the cost, revenue, and profit functions and their marginals.

SOLUTION If x units per month are produced and sold, the cost function is

$$C(x) = 5,000 + 80x.$$

The revenue function is

$$R(x) = x\left(240 - \frac{x}{2}\right) = 240x - \frac{x^2}{2},$$

and the profit function is

$$P(x) = R(x) - C(x) = -\frac{x^2}{2} + 160x - 5,000.$$

Taking the derivatives gives the marginals:

$$MC(x) = 80, \quad MR(x) = 240 - x, \quad MP(x) = -x + 160.$$

Many economic decisions are made by analyzing the marginal cost and revenue. One basic rule is this:

- If the marginal profit is positive, increasing production is beneficial.
- If the marginal profit is negative, decreasing production is beneficial.

To see the reason for these statements, suppose that the production level is currently at $x = a$. If we assume that $P(x)$ is differentiable at a, the definition of the derivative says that

$$P'(a) = \lim_{x \to a} \frac{P(x) - P(a)}{x - a}.$$

Therefore, for all x close to a, the derivative and quotient have the same sign.

Now suppose we want to choose x so that the profit increases—that is, $P(x) > P(a)$. Then the sign of $P'(a)$ is the same as that of $x - a$. Therefore, we conclude that for x close to a,

- $P'(a) > 0$ means that $P(x) > P(a)$ when $x > a$, and increasing production increases profit, while

- $P'(a) < 0$ means that $P(x) > P(a)$ when $x < a$, and decreasing production increases profit.

EXAMPLE 3.3.7

A furniture company producing coffee tables is operating at a production level of 150 units per month. It has estimated that its profit function is $P(x) = -0.5x^2 + 160x - 5,000$. Should the company increase its production level?

SOLUTION   Its marginal profit is $P'(x) = -x + 160$. Since

$$P'(150) = -150 + 160 = 10,$$

the profit will increase if the company raises its production level. In other words, increasing production is economically beneficial.

Remark   A natural question that arises in connection with the last example is: What production level will *maximize* the profit? In the next chapter we will answer that question in considerable generality. In the present case, the production function is quadratic, and we have already seen (in Section 0.5) how to maximize a quadratic. The answer, which you should verify, is that the profit is maximum at a production level of 160 units per month.

Modeling population growth

An important application of calculus is the study of population growth. To see the role played by the derivative, suppose that a certain population varies over time, which we can express by writing $P = f(t)$, where P stands for the size of the population, t stands for time (measured in some units such as days or years), and f is a function linking the two.

How fast does the population grow (or decline)? The *rate* at which the population changes is its *derivative* with respect to time. Therefore, the rate of growth is dP/dt. In general, this rate will vary as time changes, so that dP/dt is a nonconstant function of t. However, for very simple populations—such as bacteria or fruit flies—over short periods of time, the *relative* growth varies at a constant rate. To see what that means, let

$$\Delta P = f(t + \Delta t) - f(t),$$

which represents the change in population over a short time period Δt starting at a given time t. The relative change in population is defined as the ratio $(\Delta P)/P$, which measures the fraction by which the population has changed. To say that it varies at a constant rate means that it is approximately proportional to the length of the time interval—that is,

$$\frac{\Delta P}{P} \approx k \Delta t,$$

where k is a constant. The smaller Δt is—in other words, the smaller the time period— the better the approximation. If we multiply both sides by P and divide by Δt, we get

$$\frac{\Delta P}{\Delta t} \approx k P,$$

and by taking shorter and shorter time periods, we obtain

$$\lim_{\Delta t \to 0} \frac{\Delta P}{\Delta t} = k P.$$

But this limit is exactly the derivative, dP/dt. That is,

$$\lim_{\Delta t \to 0} \frac{\Delta P}{\Delta t} = \lim_{\Delta t \to 0} \frac{f(t + \Delta t) - f(t)}{\Delta t} = f'(t) = \frac{dP}{dt}.$$

This leads us to an important equation, called a **differential equation**, describing simple population growth:

$$\frac{dP}{dt} = kP. \tag{27}$$

In words, it says that the *growth rate of the population is proportional to its size* at any given time. This same equation governs a number of other phenomena, including the decay of radioactive material and the growth of an investment under continuously compounded interest. We will come back to it in a later section, after we have studied the derivatives of logarithms and exponentials.

Other examples of rates

The interpretation of the derivative as a rate plays a role in many other applications. In fact, many of the laws of physical science and economics are statements about rates, which can be translated into mathematical equations involving derivatives.

EXAMPLE 3.3.8

A test for diabetes involves the concentration of glucose in a person's bloodstream over a period of time. Suppose that t hours after an initial glucose injection, the concentration of glucose in the blood is given by the formula

$$q = 1.8 + \frac{3.6}{\sqrt{t}},$$

where q is the number of milligrams of glucose per cubic centimeter of blood. How fast is the glucose concentration changing at the end of 4 hours?

SOLUTION   Once again, we are being asked for a rate—the rate at which the glucose concentration is changing. To answer the question, we first take the derivative

$$q' = -1.8t^{-3/2}$$

and then substitute $t = 4$ to get the rate at which the concentration is changing at that moment:

$$q'(4) = -\frac{1.8}{8} = -0.225 \text{ milligrams per cubic centimeter per hour.}$$

Notice that in this case the derivative is negative, reflecting the fact that the concentration is decreasing.

Practice Exercises 3.3

1. The position (in meters) of a particle moving along a straight line is given by $s(t) = t^3 - 4t^{5/2} - 2$. What are its velocity and acceleration at the end of 4 seconds?

2. A company operating at the production level of 1,000 units per week estimates that its profit function is $P(x) = -0.25x^2 + 480x - 12{,}000$. Should the company increase or decrease its production level?

Exercises 3.3

1. A miniature rocket is launched straight upward from the surface of the earth. Its height after t seconds is given by the formula $y = -16t^2 + 128t$ for $0 \leq t \leq 8$, where y is the height in feet.
   (a) Find the height, velocity, and speed of the rocket at the end of 5 seconds.
   (b) When is its velocity equal to 112 feet per second?
   (c) When is its velocity equal to -128 feet per second?

2. A ball is tossed into the air from a window, and its height y (in feet) above the ground t seconds after it is thrown is given by $y = f(t) = -16t^2 + 50t + 36$.
   (a) At what height did the ball start?
   (b) With what velocity did it start?
   (c) What is its velocity at $t = 1$ second?

   (d) What is its velocity at $t = 2$ seconds?
   (e) The answers to (c) and (d) have different signs. Explain the significance of the signs.
   (f) When does the ball have zero velocity? What does that mean in terms of where it is?

3. If an object is dropped from a platform 64 feet high, its height at the end of t seconds is given by $y = -16t^2 + 64$, for $0 \leq t \leq 2$, where y is the height in feet.
   (a) Use a calculator to find its average velocities between the following time periods: between 1 and 1.1 seconds; between 1 and 1.01 seconds; and between 1 and 1.001 seconds.
   (b) Find its instantaneous velocity at $t = 1$.

4. Suppose a particle is moving along the x-axis so that its coordinate at time t is given by the formula $x = t^3 - t^2 - 5t + 10$ for $t \geq 0$.
   (a) What is its instantaneous velocity when $t = 2$? when $t = 4$?
   (b) What is its average velocity between $t = 2$ and $t = 4$?
   (c) In which direction (left or right) is the particle moving when $t = 1$? when $t = 3$?
   (d) At what time does the particle change direction?

5. If a ball is dropped from the top of a 100-meter tower on the surface of the moon, its height s (in meters) at the end of t seconds will satisfy the equation $s = 100 - 0.83t^2$ until the moment it hits the ground.
   (a) What is its velocity at the end of 2 seconds?
   (b) How many seconds after it is dropped will it hit the ground?
   (c) What are its velocity and speed at that moment?

6. (Continuing Exercise 5) If, instead of being dropped, the ball is thrown upward with an initial velocity of 8 meters per second, its height is given by the formula $s = 100 + 8t - 0.83t^2$ until it hits the ground.
   (a) Write a formula giving the ball's velocity until it hits the ground.
   (b) In what direction (up or down) is it moving at the end of 4 seconds? at the end of 5 seconds?
   (c) What is the ball's acceleration at the end of 1 second? at the end of five seconds?

7. The peregrine falcon (*Falco peregrinus*) is a fast, accurate bird-of-prey that hunts other birds, such as ducks. A peregrine might fly high above a lake until it spots a likely duck sitting in the water. Then the falcon folds in its wings and drops into a dive or stoop, plummeting toward its prey until spreading its feathers at the last moment to brake and extend its sharp talons. You estimate that the falcon's height above the water at time t is given by

$$H(t) = -t^4 - 1.36t^3 - 1.2t + 138,$$

   where t is in seconds and H is in feet.
   (a) What is the peregrine's instantaneous velocity at $t = 1$ second? at about $t = 3$ seconds?
   (b) Peregrines are known to achieve speeds of up to 200 miles per hour in a stoop. Is your model accurate within these guidelines in estimating the falcon's flight velocity? (*Hint:* Convert the units from feet per second to miles per hour. There are 5,280 feet in a mile.)

8. You are studying the flow of water in the middle of a large, straight irrigation canal. To do so, you drop in a floating ball at point A and record the time it takes for the ball to pass several other points. The position of the ball from point A after t minutes is then approximated by

$$P(t) = 0.018t^{5/2} + 0.388t^2,$$

   where P is in hundreds of feet.

   (a) Calculate the velocity function and evaluate it at $t = 1$.
   (b) Find the acceleration function and evaluate it at $t = 1$.

9. Two bicycles are moving in the same (positive) direction along a straight track. Let $A(t)$ and $B(t)$ be the distances (in feet) of bikes A and B from the starting point at the end of t seconds. Translate each of the following statements into a formula involving the functions $A(t)$ and $B(t)$ and/or their derivatives.
   (a) At the end of 30 seconds, bike A is 10 feet ahead of bike B.
   (b) At the end of 10 seconds, bike B is traveling 5 feet per second faster than bike A.
   (c) At the end of 20 seconds, bike A is traveling twice as fast as bike B.
   (d) At the end of 1 second, the bikes have equal acceleration.

10. For the rocket of Exercise 1, find its acceleration at the end of
    (a) 10 seconds   (b) 15 seconds   (c) 25 seconds.

11. For the ball of Exercise 2, find its acceleration
    (a) at the moment when the velocity is zero and
    (b) at the moment when the ball hits the ground.

12. Suppose a particle is moving along the x-axis so that its coordinate at time t is given by the formula $x = t^2 + 4t - 2t^{5/2}$ for $t \geq 0$.
    (a) What is its instantaneous velocity when $t = 0$? when $t = 1$?
    (b) What is its average velocity between $t = 0$ and $t = 1$?
    (c) What is its acceleration at $t = 1$?
    (d) In which direction (left or right) is the particle moving when $t = 1$? Is the velocity increasing or decreasing? What about the speed?

13. An object is moving under the influence of gravity, with its height (in meters) being given by the formula $y = -4.9t^2 + 16.4t$. Determine whether its velocity and speed are increasing or decreasing at each of the following times:
    (a) $t = 1$ (b) $t = 2$.

14. The velocity v and acceleration a of a particle moving along the x-axis are given at various times by the following table:

t	2	4	6	8	10	12
v	3.1	1.7	−0.8	−1.2	−0.1	1.4
a	0.06	−0.14	−0.22	0.01	0.08	−0.02

For each t, state whether the particle is moving to the right or left at that instant, whether its velocity is increasing or decreasing, and whether its speed is increasing or decreasing.

15. A particle is moving along the x-axis with $x(t)$ being its coordinate at time t. Represent each of the following statements by a pair of inequalities involving dx/dt and d^2x/dt^2:
(a) The particle is moving toward the right with increasing speed.
(b) The particle is moving to the right with decreasing speed.
(c) The particle is moving to the left with increasing speed.
(d) The particle is moving to the left with decreasing speed.

Exercises 16–21 are based on the following situation. At $t = 0$ (the initial time of observation), a car is 80 miles east of Chicago and traveling east toward Detroit at a speed of 60 miles per hour. In 16–19, match each statement about the motion of the car with the corresponding graph in Figure 3.3.2 of the distance function $s(t)$ over the next 5 minutes. Explain your choice in each case. Then, in 20–21, draw the graph as instructed.

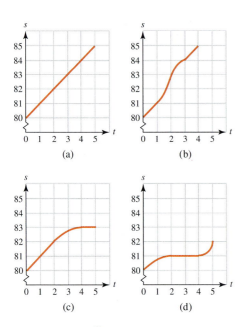

Figure 3.3.2

16. The car travels at the same speed for 2 more minutes, then decelerates for the next 2 minutes, and reaches a full stop at the 4th minute.

17. The car keeps traveling at the same speed for the next 5 minutes.

18. The car decelerates for 2 minutes, at which time it reaches a full stop and stays still for 2 more minutes. Then it starts moving again, accelerating continuously during the 5th minute.

19. The car continues moving at the same speed for 1 minute, accelerates during the 2nd minute, and decelerates during the 3rd minute. Then it moves at a steady speed of 60 miles per hour for 1 more minute.

20. Draw a graph illustrating the following situation. The car decelerates for 1 minute, reaches a full stop, and stays still for 2 minutes. Then it backs up for 2 minutes at a speed of 15 miles per hour.

21. Draw a graph illustrating the following situation. The car decelerates for 1 minute, reaches a full stop, and stays still for 1 minute. Then the car backs up for 1 minute at a speed of 15 miles per hour and stops for an additional minute. It finally accelerates, moving forward for the 5th minute.

22. Example 3.3.3 showed that, in general, the average velocity between two times, t_1 and t_2, is not the same as the average of the instantaneous velocities at those times. There is a case, however, when they are the same. Show that if the position $s(t)$ function is quadratic, say, $s(t) = at^2 + bt + c$, the average velocity between times t_1 and t_2 is equal to $\frac{1}{2}[s'(t_1) + s'(t_2)]$.

In Exercises 23–28, find $f''(t)$.

23. $f(t) = t^2 - 17t + 51$

24. $f(t) = t^4 - \frac{1}{4}t^3 - \frac{1}{3}t^2 + \frac{1}{12}t - \frac{1}{8}$

25. $f(t) = \dfrac{1}{t}$ **26.** $f(t) = \sqrt{t}$

27. $f(t) = 3t - 8$ **28.** $f(t) = t^3 - 8$

29. If $f(x) = x^5 + 4x^3 - 7x + 1$, what is $f''(x)$? What is $f''(1)$?

30. If $f(u) = u^{3/2}$, what is $f''(4)$?

31. Find $\dfrac{d^2}{dx^2}x^{1/3}$.   **32.** Find $\dfrac{d^2}{dt^2}\left(t^2 + \dfrac{1}{t^2}\right)$.

In Exercises 33–36, find the third derivative.

33. $f(x) = x^5 - 3x^4 + 2x^3 - 7x^2 + x - 10$

34. $g(t) = t^3 + t^{-3}$

35. $f(u) = \dfrac{1}{\sqrt{u}}$ **36.** $h(x) = x^4 - x^{1/4}$

37. A small company that produces stuffed toys has fixed costs of $1,000 per week. In addition, each toy costs $4.50 to make. The weekly revenue is given by the formula $R(x) = 10x - 0.01x^2$ for $0 \le x \le 800$, where x is the number produced and sold each week. Find the marginal cost, revenue, and profit.

38. The marginal cost and revenue arising in the production and sale of a certain item are given by $MC(x) = 350$ and $MR(x) = 800 - 2x$, where x is the number produced and sold per week.
(a) What is the marginal profit?

(b) If the company is currently producing 160 items per week, should it increase or decrease production in order to raise its profit? Explain your answer.

(c) What if the company is currently producing 200 items per week? 240 items per week?

39. A cruise line estimates that it can sell 3,400 tours to Alaska at a price of $900 each, but it will lose 100 sales for each $50 increase in its price. Let p be the price and q the number of tours sold.

(a) Write q as a linear function of p.

(b) Write the revenue R as a function of p and find the marginal revenue with respect to p (i.e., the rate of change of R with respect to p).

(c) Suppose it costs $400 per passenger to operate the tour. Write the cost as a function of the price p and find the marginal cost with respect to p.

(d) Find the marginal profit with respect to p.

(e) The company priced its tour at $1,050 last year. If it increases the price this year, will it increase or decrease its profit? Explain your answer.

(f) Answer the same question and explain your answer if last year's price was $1,550.

40. A 300-gallon tank is leaking water at a certain rate. The volume (in gallons) of water in the tank at the end of t minutes is given by a formula of the form $V = f(t)$.

(a) What is the sign of $f'(t)$?

(b) At the end of 6 minutes, the water is leaking at the rate of 9 gallons per minute. Write an equation expressing that fact.

(c) Suppose $f(t) = (t^2/12) - 10t + 300$ for $0 \le t \le 60$. How much water is in the tank at the end of 12 minutes and how fast is it leaking?

41. To test how a body metabolizes calcium, a researcher may inject a sample of calcium that is chemically "labeled" into the bloodstream in order to measure how fast it is removed from the blood. Suppose that

$$A(t) = 2 - 0.06t + 0.03t^2 - 0.01t^3$$

gives the amount (in milligrams) of labeled calcium in the bloodstream at the end of t hours. At what rate is the calcium being eliminated from the bloodstream at the end of 2 hours?

42. Suppose the weight of a tumor in a laboratory animal is given by the formula $w = 0.2 + 0.3t + 0.015t^2$, where t is the time in days. How fast is the weight increasing when $t = 3$? when $t = 5$?

43. Suppose $V_A(t)$ and $V_B(t)$ are the volumes (in cubic inches) of two spherical balloons, both of which are functions of time (in minutes). Translate each of the following statements into a formula involving the functions and/or their derivatives:

(a) At the end of 1 minute, the volume of balloon A is decreasing at the rate of 2.4 cubic inches per minute.

(b) At the end of 2 minutes, the volume of balloon A is increasing at the rate of 3 cubic inches per minute.

(c) At the end of 3 minutes, the volume of balloon B is increasing 1.1 cubic inches faster than that of balloon A.

(d) At the end of 4 minutes, the volume of balloon B is increasing twice as fast as that of balloon A.

44. Suppose $P_A(t)$ and $P_B(t)$ are the population sizes of two colonies of insects, both of which are functions of time. Translate each of the following statements into a formula involving the functions and/or their derivatives:

(a) Colony A is twice as large as colony B.

(b) Colony B is growing at twice the rate of colony A.

(c) The rate at which colony A grows is proportional to its size.

45. Suppose that C is the concentration of a certain drug in a patient's bloodstream, measured in milligrams per cubic centimeter. Translate the following statements into formulas involving C and/or dC/dt. Assume that t is measured in hours.

(a) At the end of 2 hours, there are 50 milligrams per cubic centimeter in the patient's bloodstream.

(b) At that same time, the concentration is decreasing at the rate of 3.5 milligrams per cubic centimeter per hour.

(c) The drug is being eliminated from the patients bloodstream at a rate that is proportional to the concentration present at any given time.

46. If a cup of hot tea is left standing in a room, it will cool down as time goes on. Let $H(t)$ be the temperature of the tea (in Celsius) at the end of t minutes.

(a) What is the sign of $H'(t)$?

(b) Write an equation expressing the fact that the initial temperature of the tea is 90 degrees.

(c) Write an equation expressing the fact that at the end of 10 seconds the tea is cooling at a rate of 2.5 degrees per minute.

(d) A famous law of physics, known as **Newton's law of cooling**, says that the rate at which the tea cools down is proportional to the difference between its temperature and the temperature of the room. Suppose the temperature of the room is maintained at a steady 20°C. Write an equation involving $H(t)$, $H'(t)$, and a constant of proportionality, say, k, to express the law in this case.

47. Figure 3.3.3 shows a car's velocity (in feet per second) at every quarter second for a period of 5 seconds. Assuming that v is a differentiable function of t, estimate its acceleration at every quarter second from 0.25 to 4.75. Use the central difference formula.

48. Suppose that the world population is modeled by the logistic formula

$$P = \frac{73.2}{6.1 + 5.9e^{-0.016t}},$$

with P being the population (in billions) and t being the

	A	B
1	*t*	*y*
2	0.00	0.00
3	0.25	0.23
4	0.50	0.66
5	0.75	1.21
6	1.00	1.86
7	1.25	2.60
8	1.50	3.42
9	1.75	4.31
10	2.00	5.26
11	2.25	6.28
12	2.50	7.35
13	2.75	8.48
14	3.00	9.66
15	3.25	10.90
16	3.50	12.18
17	3.75	13.51
18	4.00	14.88
19	4.25	16.30
20	4.50	17.76
21	4.75	19.26
22	5.00	20.80

Figure 3.3.3

number of years after 2000. Use the numerical differentiation function to estimate how fast (in billions per year) the population will be changing in 2010 and 2020.

49. Figure 3.3.4 shows the (x, y) coordinates of points on the famous bell-shaped curve of great importance in statistics. This curve is the graph of a differentiable function. Use the central difference formula to estimate its slope for each x from -1.4 to 1.4 in steps of 0.2.

50. The bell-shaped curve of statistics is the graph of the function $f(x) = (1/\sqrt{2\pi})e^{-x^2/2}$. Use the numerical differentiation function to estimate the derivative for each x from -1 to 1 in steps of point 0.2. (*Note:* You can answer this question fully by only computing the slopes for x from 0 to 1. Why?)

51. A cell-phone manufacturer plans to introduce a new model, and on the basis of marketing surveys it estimates that the sales over the next 36 months will be modeled by the equation

$$S = \frac{1}{1 + [\ln(t + 0.5) - 2]^2} - 0.12,$$

where t is the time in months and S is the number of phones sold (in millions) per month. Graph this function in a $[0, 34] \times [0, 1.25]$ window and use the numerical derivative operation to answer the following questions.
(a) How fast is the rate of sales changing at the end of 3 months? At the end of 17 months?
(b) Explain the significance of the sign of each of the previous answers.
(c) Find the point where the rate of sales changes from positive to negative.

52. Suppose the spread of a disease through a community is modeled by the equation

$$y = \frac{1}{1 + 50e^{-0.8t}},$$

where t is the time in days and y is the number (in thousands) of people infected. Graph this function in a $[0, 12] \times [0, 1]$ window and answer the following questions:
(a) How many people are infected at the end of 4 days and how fast is the disease spreading (in number of infected persons per day)?
(b) How many people are infected at the end of 8 days and how fast is the disease spreading (in number of infected persons per day)?
(c) When is the disease spreading fastest?

53. For an object falling from rest, show that the distance traveled during each successive second increases in the ratio of successive odd integers, as was first shown by Galileo.

BellCurve

	A	B	C	D	E	F	G	H	I	J	K	L	M	N	O	P	Q	R
1	*x*	-1.6	-1.4	-1.2	-1.0	-0.8	-0.6	-0.4	-0.2	0.0	0.2	0.4	0.6	0.8	1.0	1.2	1.4	1.6
2	*y*	0.1109	0.1497	0.1942	0.2420	0.2897	0.3332	0.3683	0.3910	0.3989	0.3910	0.3683	0.3332	0.2897	0.2420	0.1942	0.1497	0.1109
3																		

Figure 3.3.4

Solutions to practice exercises 3.3

1. Its velocity is equal to $s'(4)$, and its acceleration is equal to $s''(4)$. Since $s'(t) = 3t^2 - 10t^{3/2}$, we have $s'(4) = 48 - 80 = -32$ meters per second, and since $s''(t) = 6t - 15t^{1/2}$, we have $s''(4) = 24 - 30 = -6$ m/sec².

2. Its marginal profit is given by $P'(x) = -0.5x + 480$. At a production level of 1,000, we have $P'(1,000) = -500 + 480 = -20$, and the profit will increase if the company decreases its production level.

We have seen that if a function $f(x)$ is differentiable at $x = a$, its graph has a nonvertical tangent line at the point $(a, f(a))$. Zooming in toward the tangency point, we see a magnified picture of both the graph of the function and the tangent line, as shown in Figure 3.4.1.

The more we zoom in, the more the graph of the function coincides with the tangent line, and eventually near the tangency point the two graphs become almost identical. This suggests that for x very close to a, we can replace the graph with its tangent line, and the function $f(x)$ with the linear function corresponding to that line.

To be more precise, suppose $f(x)$ is differentiable at $x = a$. Then its tangent line at $(a, f(a))$ has slope $f'(a)$ and equation

$$y = f(a) + f'(a)(x - a), \tag{28}$$

as we saw in Section 3.2. We wish to use the linear function defined by (28) to approximate $f(x)$ for x close to a, that is,

$$f(x) \approx f(a) + f'(a)(x - a).$$

The **approximation error** $E(x)$ is the difference between the two sides of this formula—that is,

$$E(x) = f(x) - f(a) - f'(a)(x - a).$$

It is not hard to see that $E(x) \to 0$ as $x \to a$, but we can say something even stronger by observing that

$$\frac{E(x)}{x - a} = \frac{f(x) - f(a) - f'(a)(x - a)}{x - a} = \frac{f(x) - f(a)}{x - a} - f'(a).$$

Since $f'(a) = \lim_{x \to a} (f(x) - f(a))/(x - a)$, we have

$$\lim_{x \to a} \frac{E(x)}{x - a} = \lim_{x \to a} \left[\frac{f(x) - f(a)}{x - a} - f'(a) \right] = 0.$$

The point here is that not only does $E(x) \to 0$ as $x \to a$, but it goes to zero much faster

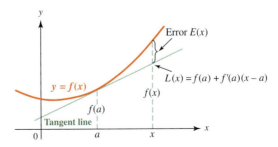

Figure 3.4.1

than $x - a$, which means that the approximation error will be very small in comparison with $x - a$.

The following theorem summarizes this useful information.

Theorem 3.4.1

Linear approximation

If $f(x)$ is a differentiable function at $x = a$, then

$$f(x) \approx f(a) + f'(a)(x - a). \tag{29}$$

Moreover, the error $E(x) = f(x) - f(a) - f'(a)(x - a)$ satisfies

$$\lim_{x \to a} \frac{E(x)}{x - a} = 0.$$

The linear function $f(a) + f'(a)(x - a)$ is called the **linear approximation** or the **tangent line approximation** of the function $f(x)$ near the point a.

A few examples will illustrate how to apply this result.

EXAMPLE 3.4.1

Using the fact that $\sqrt{49} = 7$, approximate $\sqrt{50}$.

SOLUTION If we take $f(x) = \sqrt{x}$, then, $f'(x) = 1/(2\sqrt{x})$, and formula (29) becomes

$$\sqrt{x} \approx \sqrt{a} + \frac{1}{2\sqrt{a}} \cdot (x - a).$$

Taking $a = 49$ and $x = 50$, we get

$$\sqrt{50} \approx \sqrt{49} + \frac{1}{2\sqrt{49}} \cdot (50 - 49)$$

$$= 7 + \frac{1}{14} \approx 7.0714286.$$

By way of comparison, the value of $\sqrt{50}$ to seven decimal places, as given by a calculator, is 7.0710678.

EXAMPLE 3.4.2

Use linear approximation to estimate $\dfrac{1}{(9.875)^2}$ and $\dfrac{1}{(5.0112)^2}$.

SOLUTION In both cases we take $f(x) = 1/x^2$, so that $f'(x) = -2/x^3$. This time the linear approximation formula, Eq. (29), takes the form

$$\frac{1}{x^2} \approx \frac{1}{a^2} - \frac{2}{a^3} \cdot (x - a).$$

With $x = 9.875$, we choose $a = 10$ because it is the closest point to 9.875 at which $f(a)$ and $f'(a)$ are easy to compute. Then,

$$\frac{1}{(9.875)^2} \approx \frac{1}{10^2} - \frac{2}{10^3} \cdot (9.875 - 10)$$

$$\approx \frac{1}{100} + \frac{0.25}{1,000}$$

$$= \frac{10.25}{1,000} \approx 0.01025.$$

By way of comparison, a calculator gives $1/(9.875)^2 \approx 0.0102548$, to seven decimal places.

Next, using $x = 5.0112$, we choose $a = 5$ as the closest point at which $f(a)$ and $f'(a)$ are easy to compute. Then,

$$\frac{1}{(5.0112)^2} \approx \frac{1}{5^2} - \frac{2}{5^3} \cdot (5.0112 - 5)$$

$$\approx \frac{1}{25} - \frac{0.0224}{125}$$

$$= \frac{4.9776}{125} \approx 0.0398208.$$

By way of comparison, a calculator gives $1/(5.0112)^2 \approx 0.0398214$, to seven decimal places.

In most cases, it is impossible to precisely determine the size of the error. Here is one example, however, in which we can do that.

EXAMPLE 3.4.3

Compute the approximation error for $f(x) = x^2$.

SOLUTION Using $E(x) = f(x) - f(a) - f'(a)(x - a)$ with $f'(a) = 2a$ gives

$$E(x) = x^2 - a^2 - 2a(x - a) = x^2 - 2ax + a^2 = (x - a)^2.$$

In this case, we can see directly that $\lim_{x \to a} E(x)/(x - a) = \lim_{x \to a} (x - a) = 0.$

PROBLEM-SOLVING TACTIC

To approximate $f(x)$, we choose a so that

- $f(a)$ and $f'(a)$ are easy to compute, and
- $x - a$ is relatively small.

Then, we evaluate $f(a) + f'(a)(x - a)$.

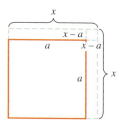

Figure 3.4.2

In the case of $f(x) = x^2$, linear approximation has a simple geometric interpretation. In Figure 3.4.2, the big square has sides of length x, and its area is x^2. If we use linear approximation to estimate that area, we get $a^2 + 2a(x - a)$. The first part a^2 is the area of the smaller square of side a that occupies the lower left-hand corner. Above and to the right of that square are two rectangular strips, each with dimensions a by $x - a$. That means that each strip has an area of $a(x - a)$, so that $2a(x - a)$ is the combined area of the two strips. Thus, $a^2 + 2a(x - a)$ is the combined areas of the lower left square and the two rectangular strips. The error $E(x)$ is the area of the small $(x - a)$ by $(x - a)$ square in the upper right-hand corner, which equals $(x - a)^2$.

In these days of hand-held calculators, linear approximation is not so useful as it once was for simplifying computations. But it is an important theoretical tool in studying derivatives. It is also useful in estimating changes in the values of a function when we have limited data, as the following example shows.

EXAMPLE 3.4.4

Suppose the profit from producing and selling 1,000 CDs is $2,500 and the marginal profit at that production level is 75 cents. Estimate the profit from producing and selling 1,020 CDs.

SOLUTION   Applying linear approximation to the profit function, we have

$$P(x) \approx P(a) + P'(a)(x - a).$$

Since 1,020 is fairly close to 1,000, it seems reasonable to use this formula with $a = 1,000$ and $x = 1,020$. Recalling that $P'(x)$ is the same as the marginal profit, we get

$$P(1,020) \approx 2,500 + (0.75) \cdot (20) = 2,515.$$

Differentiability and continuity

Recall that a function $f(x)$ is **continuous** at $x = a$ if $\lim_{x \to a} f(x) = f(a)$. An intuitive description of a continuous function is one whose graph can be drawn without lifting your pencil from the paper.

There is also an intuitive criterion for a function to be differentiable—its graph must be *smooth* and have no sharp *corners*. In terms of the tangent line, that means there is a *unique* tangent line at every point, which provides a good linear approximation to the function. In addition, we require that the tangent line is *not vertical*, for in that case the slope would be infinite.

The following example shows the simplest case of a function whose graph has a corner and is therefore not differentiable.

EXAMPLE 3.4.5

Show that the function $f(x) = |x|$ is not differentiable at $x = 0$.

SOLUTION The graph of $f(x) = |x|$ is shown in Figure 3.4.3. As you can see, it consists of two lines meeting in a corner at $(0, 0)$. The right-hand branch has slope 1, and the left-hand branch has slope -1, which means there is neither a unique slope nor unique tangent at $x = 0$.

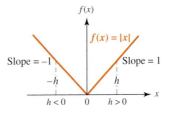

Figure 3.4.3

To be more precise, we can apply the definition of derivative to $f(x) = |x|$ to show that $f'(0)$ does not exist. First, we observe that

$$\frac{f(0+h) - f(0)}{h} = \frac{|h|}{h}.$$

Since $|h| = h$ for $h > 0$, we get the one-sided limit

$$\lim_{h \to 0^+} \frac{|h|}{h} = \lim_{h \to 0^+} \frac{h}{h} = 1.$$

And since $|h| = -h$ for $h < 0$, we also get

$$\lim_{h \to 0^-} \frac{|h|}{h} = \lim_{h \to 0^-} \frac{-h}{h} = -1.$$

Since the two one-sided limits do not agree, we conclude that the limit as $h \to 0$ does not exist, which means the function does not have a derivative at $x = 0$.

EXAMPLE 3.4.6

According to the Economic Growth and Tax Relief Reconciliation Act of 2001 passed by the U.S. Congress and signed by President George W. Bush, the amount of tax relief for a married couple filing jointly is the lesser of $600 or 5% of that couple's taxable income for the year 2000.

(i) Write the amount of tax relief $R(i)$ as a function of the taxable income i and draw its graph.

(ii) Find all points where $R(i)$ is not differentiable and then sketch the graph of the derivative of R where it is differentiable.

SOLUTION (i) We first ask, for what i is $0.05i \leq 600$? Multiplying both sides of the inequality by 20 gives $i \leq 12{,}000$. For those values of i, the refund amounts to 5% of the taxable income, whereas for $i > 12{,}000$ the refund is \$600. In other words,

$$R(i) = \begin{cases} 0.05i & \text{if } 0 \leq i \leq 12{,}000 \\ 600 & \text{if } i > 12{,}000. \end{cases}$$

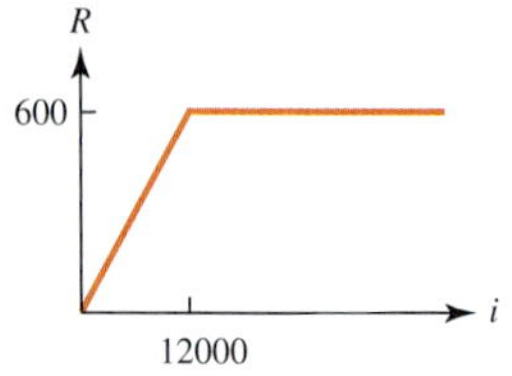

The graph is shown in Figure 3.4.4.

(ii) The graph of $R(i)$ on the interval $(0, 12{,}000)$ is a straight line of slope 0.05. On the interval $(12{,}000, \infty)$, it is a horizontal line (line of slope zero). Therefore, $R(i)$ is differentiable for all $i \neq 12{,}000$. It is not differentiable at $i = 12{,}000$ because

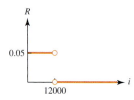

Figure 3.4.4

$$\lim_{h \to 0^-} \frac{R(12{,}000 + h) - R(12{,}000)}{h} = \lim_{h \to 0^-} \frac{0.05(i + h) - 0.05i}{h} = 0.05,$$

whereas

$$\lim_{h \to 0^+} \frac{R(12{,}000 + h) - R(12{,}000)}{h} = \lim_{h \to 0^+} \frac{600 - 600}{h} = 0,$$

which means that $R'(12{,}000)$ is not defined. (In terms of the graph, there is a "corner" at $i = 12{,}000$, where the two lines of different slopes meet.) The graph of the derivative $R'(i)$ for $i \neq 12{,}000$ is shown in Figure 3.4.5.

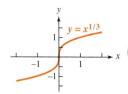

Figure 3.4.5

A function may also fail to be differentiable because its tangent line is vertical. Here is an example of such a function, taken from mathematical economics.

Vertical tangent at $x = 0$

Figure 3.4.6

EXAMPLE 3.4.7

Show that the Cobb-Douglas production function

$$f(x) = x^{1/3}$$

is not differentiable at $x = 0$.

PROBLEM-SOLVING TACTIC

A continuous function is *not* differentiable at any point where

• the graph has a corner or
• the tangent is a vertical line.

SOLUTION The graph of $f(x) = x^{1/3}$, displayed in Figure 3.4.6, has a vertical tangent line at $(0, 0)$. Therefore, its slope at $x = 0$ is infinite, which means that the function is not differentiable at zero.

To be more precise, we can apply the definition of the derivative, as follows:

$$\lim_{h \to 0} \frac{f(0 + h) - f(0)}{h} = \lim_{h \to 0} \frac{h^{1/3} - 0}{h} = \lim_{h \to 0} \frac{1}{h^{2/3}} = \infty.$$

The previous two examples showed that there are functions that are continuous and not differentiable. Are there functions that are differentiable but not continuous? The following theorem shows that the answer is *no*.

Theorem 3.4.2

If a function f is differentiable at a, then it is also continuous at a.

In fact, if f is differentiable at a, then

$$\lim_{x \to a}[f(x) - f(a)] = \lim_{x \to a}\frac{f(x) - f(a)}{x - a} \cdot (x - a) = f'(x) \cdot 0 = 0,$$

which says that $\lim_{x \to a} f(x) = f(a)$. Therefore, f is continuous at a.

Practice Exercises 3.4

1. Use linear approximation to estimate $\sqrt{97}$.

2. At a particular moment, the temperature of a cup of coffee is 94.7°C and is decreasing at a rate of 0.5 degrees per minute. Estimate its temperature 10 minutes later.

Exercises 3.4

1. Use linear approximation with $f(x) = \sqrt{x}$ and $a = 36$ to approximate
   (a) $\sqrt{38}$　　　　　(b) $\sqrt{35}$

2. Approximate $\sqrt[3]{25}$ by linear approximation, using $f(x) = x^{1/3}$ and $a = 27$.

In Exercises 3–6, apply linear approximation to the given function to get the desired approximations. Express your answer in decimal or fraction form.

3. Use $f(x) = \sqrt{x}$ to approximate
   (a) $\sqrt{405}$　　(b) $\sqrt{3.95}$　　(c) $\sqrt{25.03}$

4. Use $f(x) = x^{-1/2}$ to approximate
   (a) $\dfrac{1}{\sqrt{98}}$　　(b) $\dfrac{1}{\sqrt{50}}$　　(c) $\dfrac{1}{\sqrt{3.9}}$

5. Use $f(x) = x^2$ to approximate
   (a) $(5.0031468)^2$　(b) $\left(\dfrac{17}{32}\right)^2$　(c) $(0.999973)^2$

6. Use $f(x) = 1/x$ to approximate
   (a) $\dfrac{1}{9.95}$　　(b) $\dfrac{1}{10.03}$　　(c) $\dfrac{1}{102}$

In Exercises 7–15, use linear approximation to estimate the given quantity.

7. $\sqrt[3]{997}$　　　8. $(1.0427)^3$　　9. $\dfrac{1}{\sqrt[3]{7.24}}$

10. $\dfrac{1}{(1.01)^4}$　　11. $\dfrac{1}{102}$　　12. $\dfrac{2}{99}$

13. $\sqrt[4]{80}$　　14. $(33)^{-0.2}$　　15. $(0.99)^{0.3}$

Exercises 16–17 refer to Figure 3.4.7, which shows the graph of $y = f(x)$ of a differentiable function f and two of its tangent lines.

16. Estimate $f'(1)$ and use it to estimate $f(1.01)$ and $f(1.1)$. Which of these estimates would you expect to be more accurate? Would you guess they are greater or smaller than the true values?

17. Estimate $f'(3)$ and use it to estimate $f(2.9)$ and $f(2.99)$. Which of these estimates would you expect to be more accurate? Would you guess they are greater or smaller than the true values?

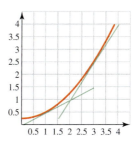

Figure 3.4.7

Exercises 18–19 refer to Figure 3.4.8, which shows the graph of $y = f(x)$ for a differentiable function f and two of its tangent lines.

18. Estimate $f'(1)$ and use it to estimate $f(1.1)$ and $f(1.02)$. Which of these estimates would you expect to be more accurate? Would you guess they are greater or smaller than the true values?

19. Estimate $f'(5)$ and use it to estimate $f(4.9)$ and $f(4.95)$. Which of these estimates would you expect to be more accurate? Would you guess they are greater or smaller than the true values?

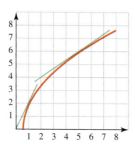

Figure 3.4.8

20. The cost of producing 100 units of a certain item is \$2,400, and the marginal cost of producing 100 units is \$62; in other words, $C(100) = 2,400$ and $MC(100) = 62$. Use linear approximation to estimate the cost of producing 102 units.

21. The revenue from the sale of 10,000 units is \$8,750, and the marginal revenue at that level is 64 cents. Use linear approximation to estimate the revenue if the sales level drops to 9,800 units.

22. A company produces 1,000 tennis balls per day. At that production level, its profit (in dollars) is 1,870, its marginal revenue is 3.20, and its marginal cost is 1.80. Estimate how much the company's profit will increase if it increases production by 10 balls a day.

23. A manufacturing company estimates the cost of hiring and training x new employees is

$$C(x) = 500(x - \sqrt{x}).$$

(a) Find $C'(16)$.
(b) The company is trying to decide whether to hire 16 or 17 new employees. Approximate the difference in cost using linear approximation.
(c) Use a calculator to compute the actual difference and compare that to your approximation.

24. A company's revenues during a given quarter is \$1.8 billion, and its marginal revenue is \$0.2 billion per quarter. Estimate its revenue over the subsequent three quarters. For which quarter is the estimate most accurate?

25. An object is moving in a straight line, and after 5 seconds its velocity is 40 feet per sec and its acceleration is -3 ft/sec². Use linear approximation to estimate its velocity after 5.2 seconds.

26. A particle is moving along the x-axis so that its velocity at t seconds is given by $v = 2t - \sqrt{t}$. Its coordinate at 4 seconds is $x = -3$. Use linear approximation to estimate its coordinate at 4.02 seconds.

27. If the volume of a cube decreases from 8 to 7.88 cubic inches, estimate how much the length of each edge decreases.

28. At noon, there were 9.42 milligrams of medicine in a patient's bloodstream, and the amount was decreasing at a rate of 0.28 milligrams per hour. Estimate the amount in the patient's bloodstream at 12:15 P.M.

29. In 1998 the world population was about 5.9 billion people and was increasing at the rate of approximately 2% per year. Use linear approximation to estimate its size in the year 2008.

30. Our sun is composed of gases maintained at very high temperatures by nuclear reactions. Scientists have calculated the temperature along the sun's radius. These measurements, in millions of degrees Kelvin, are given by the following table, with zero being the sun's center and its radius being the unit:

r	0.0	0.1	0.2	0.3	0.4	0.5
$T(r)$	13.6	11.6	8.5	6.0	4.2	2.8
r	0.6	0.7	0.8	0.9	1	
$T(r)$	1.9	1.2	0.68	0.31	0.006	

Assuming that $T'(0.7) = -6.1$, use linear approximation to estimate the temperature at $r = 0.72$.

Source: Encyclopedia of Astronomy, Gilbert E. Satterthwaite, ed., The Hamlyn Publishing Group, 1973, page 449.

31. A carbon-molybdenum steel pipe is being installed in an area where it is routinely exposed to an operating temperature of 710°F. Since it is installed at a cooler temperature, the pipe's length will expand at the high heat. The expansion multipliers for an installation temperature of 60°F are given by

operating temperature t	650	700	750
expansion multiplier, $M(t)$	0.0525	0.0575	0.0624

(a) Calculate the slope to the left of $t = 700$ by using $\Delta y = M(700) - M(650)$ and $\Delta x = 700 - 650$.
(b) Calculate the slope to the right.

(c) Next, average the two slopes and label the estimate $M'(700)$. (This is the **central difference formula** method.)

(d) Estimate the multiplier for 710°F.

Source: Standard Handbook of Engineering Calculations, Tyler G. Hicks, ed., McGraw-Hill, 1972, pages 3–395.

32. You are designing a cold storage room whose temperature will be maintained at a constant 38°F. You need to know what the moisture content of the air will be, assuming that the air is saturated, that is, holds as much water as it can. An engineering handbook gives you the following table, where t is the temperature (°F) and $M(t)$ is the grains of water per cubic feet of air:

t	25	30	35	40	45
$M(t)$	1.558	1.946	2.376	2.863	3.436

(a) Using the estimate $M'(35) = 0.092$, approximate the moisture content at 38°F.

(b) Would you expect your estimate to be more or less accurate if you were given $M'(40)$? Why?

Source: Standard Handbook of Engineering Calculations, Tyler G. Hicks, ed., McGraw-Hill, 1972, pages 2–107.

33. If you use linear approximation to estimate $(1.04)^2$, how large is the error? Give a precise answer without actually computing $(1.04)^2$. (*Hint:* look at Example 3.4.3.)

34. Figure 3.4.9 is a table of values of $f'(x)$. Taking $f(0) = 0$, use repeated linear approximation to estimate $f(x)$ for x from 0.1 to 1 in steps of 0.1.

35. Figure 3.4.10 is a table of the velocity (in feet per second) of a car at 0.25-second intervals from 0 to 5 seconds. We can use it to estimate the car's position $s(t)$ at each of those times by taking $s(0) = 0$ and proceeding as follows. Linear approximation says that $s(0.25) \approx s(0) + v(0) \cdot (0.25) = 0$, and we enter that in the cell corresponding to $s(0.25)$. We then use that value in place of $s(0.25)$ in the next linear approximation

$$s(0.5) \approx s(0.25) + v(0.25) \cdot (0.25) \approx 0.0575.$$

We enter that value in the cell corresponding to $s(0.5)$, as shown, and use it in the next linear approximation to estimate $s(0.75)$. Continuing in this way, we can approximate $s(t)$ for each of the time values in the table. Do it.

36. (a) Show that the tangent line to the graph of $y = \sqrt{x}$ at $x = 4$ has equation $y = 0.25x + 1$. What is the linear approximation to $\sqrt{x}$ for x near 4?

	A	B	C
1	t	v	s
2	0.00	0.00	0
3	0.25	0.23	0
4	0.50	0.66	0.0575
5	0.75	1.21	
6	1.00	1.86	
7	1.25	2.60	
8	1.50	3.42	
9	1.75	4.31	
10	2.00	5.26	
11	2.25	6.28	
12	2.50	7.35	
13	2.75	8.48	
14	3.00	9.66	
15	3.25	10.90	
16	3.50	12.18	
17	3.75	13.51	
18	4.00	14.88	
19	4.25	16.30	
20	4.50	17.76	
21	4.75	19.26	
22	5.00	20.80	
23			

Figure 3.4.10

(b) Find an interval around $x = 4$ in which the linear approximation $\sqrt{x} \approx (0.25x + 1)$ has an error less than or equal to 0.1 in the following steps:

(i) Observe that the error is less than or equal to 0.1 if and only if

$$|\sqrt{x} - (0.25x + 1)| \le 0.1,$$

which is the same as

$$\sqrt{x} - 0.1 \le 0.25x + 1 \le \sqrt{x} + 0.1 \qquad (30)$$

(ii) Graph $y = \sqrt{x} - 0.1$, $y = \sqrt{x} + 0.1$, and $y = 0.25x + 1$ in the same window and determine the largest interval in which the line stays between the other graphs. (Figure 3.4.11 shows the three graphs in a $[1, 8] \times [1, 3]$ window, with the cursor approximately at the leftmost point of that interval.) Use the TRACE cursor to estimate an interval in which (30) is valid.

(iii) Use the ZOOM function to improve the estimation.

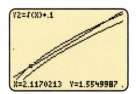

Figure 3.4.11

	A	B	C	D	E	F	G	H	I	J	K	L
							Derivative					
1	x	0	0.1	0.2	0.3	0.4	0.5	0.6	0.7	0.8	0.9	1
2	df/dx	1.0033	2.0271	3.0934	4.2279	5.4630	6.8414	8.4229	10.2964	12.6016	15.5741	19.6476
3												

Figure 3.4.9

37. Verify that the tangent line to the graph of $y = x^{1/3}$ at $x = 8$ has the equation $y = (x/12) + \frac{4}{3}$. Then use the method of the previous exercise to find an interval around $x = 8$ in which the linear approximation has an error less than or equal to 0.1.

38. Use the definition of the derivative to show that the function $f(x) = x^{2/3}$ is not differentiable at $x = 0$. Is it continuous there?

39. Let $f(x) = \begin{cases} x^2 + x & \text{if } x \geq 0 \\ x^2 - x & \text{if } x < 0. \end{cases}$

Use the definition of the derivative to show that $f(x)$ is not differentiable at $x = 0$. Is it continuous there?

40. Let $f(x) = \begin{cases} x^2 & \text{if } x \geq 0 \\ -x^2 & \text{if } x < 0. \end{cases}$

Use the definition of the derivative to show that $f(x)$ is differentiable at $x = 0$. Is it continuous there?

41. Let $f(x) = \begin{cases} x^2 & \text{if } x \geq 0 \\ 1 - x^2 & \text{if } x < 0. \end{cases}$

Without using the definition of the derivative, show that $f(x)$ is not differentiable at $x = 0$. (*Hint:* Look at Theorem 3.4.2.)

42. Let $f(x) = \begin{cases} x/|x| & \text{if } x \neq 0 \\ 0 & \text{if } x = 0. \end{cases}$

Show that $f'(0)$ does not exist in two ways:
(a) by using the definition of the derivative, and
(b) by appealing to Theorem 3.4.2.

43. Suppose the Interior Department of the State of Montana monitored a small, isolated herd of wild horses over a 10-year period, reporting on the population at the beginning of each July. The yearly population levels are as follows:

year	1	2	3	4	5
no. of horses	40	50	58	60	50
year	6	7	8	9	10
no. of horses	50	50	54	64	70

(a) Considering the population in any particular year to be constant throughout that year, graph the population as

a function of time (in years). Is the graph continuous? If not, where is it discontinuous?
(b) What is the slope at those points where the function is differentiable?

44. A hardware store rents a machine for sanding hardwood floors at a rate of $30 for the first 24 hours and $20 for each additional 12-hour period. Let x be the number of hours a renter keeps the machine, and let $f(x)$ be the rental fee.
(a) Write a (multiline) formula for $f(x)$ over the interval $(0, 60)$ and sketch its graph. At what points, if any, is $f(x)$ discontinuous?
(b) What is the slope at those points where the function is differentiable?

45. A real estate broker charges a commission of 7% on any house sale of $100,000 or less. On sales that exceed $100,000, the broker charges a flat fee of $3,000 plus 4% of the sale price. Let x be the amount of a given sale and $f(x)$ the brokerage fee.
(a) Write a (multiline) formula for $f(x)$ and sketch the graph.
(b) At what points, if any, is $f(x)$ discontinuous?
(c) At what points, if any, does $f(x)$ fail to be differentiable?

46. The 1998 federal tax rate schedule for a single person is shown in Table 3.4.Ex.46:
(a) Letting x be the person's income and $f(x)$ be the tax owed, write a multiline formula for $f(x)$, with each line in the form $mx + b$.
(b) At what points, if any, is $f(x)$ discontinuous?
(c) At what points, if any, does $f(x)$ fail to be differentiable?

47. Figure 3.4.12 shows the graph of a function $f(x)$. What can you say about the continuity and differentiability of the function at each of the points $x = -2, -1, 0, 1, 2, 3, 4, 5, 6, 7, 8$?

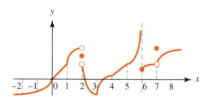

Figure 3.4.12

For income over	but not over	tax owed is	of the amount over
$0	$25,350	15%	$0
25,350	61,400	$3, 802.50 + 28\%$	25,350
61,400	128,100	$13, 896.50 + 31\%$	61,400
128,100	278,450	$34, 573.50 + 36\%$	128,100
278,450	$\cdots$	$88,699.50 + 39.6\%$	278,450

Table 3.4.Ex.46

Solutions to practice exercises 3.4

1. We shall use the linear approximation of $f(x) = \sqrt{x}$ with $a = 100$ and $x = 97$. Since $f'(x) = 1/(2\sqrt{x})$, we have

$$\sqrt{97} \approx \sqrt{100} + \frac{1}{2\sqrt{100}} \cdot (97 - 100) = 10 - \frac{3}{20} = 9.85.$$

By way of comparison, a calculator gives $\sqrt{97} \approx 9.8488578$, to seven decimal places.

2. We denote by $H(t)$ the temperature of the coffee at any time t after the time $t = 0$ when we know that $H(0) = 94.7$ and $H'(0) = -0.5$. Then applying the linear approximation formula $H(t) \approx H(a) + H'(a)(t - a)$ with $t = 10$ and $a = 0$, we obtain

$$H(10) \approx H(0) + H'(0)(10 - 0) = 94.7 - 0.5 \cdot 10 = 89.7.$$

3.5   Derivative of Logarithms and Exponentials

The main objective of this section is to introduce the basic derivative rules for the logarithm and exponential functions, together with a few additional formulas and some applications. The first new rule we take up is a rather surprising formula, whose explanation appears at the end of this section.

The Derivative of the Logarithm Function

$$\frac{d}{dx} \ln x = \frac{1}{x}. \tag{31}$$

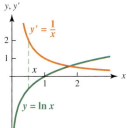

Figure 3.5.1

What is surprising about this formula is that the right-hand side is so simple—a power of x, in fact—whereas the function $\ln x$ is not. Figure 3.5.1 shows the graph of $\ln x$ together with the graph of its derivative. Looking at the graph of $\ln x$, you can see that the behavior of its slope is indeed compatible with formula (31). To begin with, the slope is always positive, as is $1/x$. (Recall, by the way, that $\ln x$ is only defined for $x > 0$.) As x gets larger, the graph flattens out and the slope becomes smaller and smaller, which is compatible with the fact that $1/x \to 0$. As x gets close to zero, the graph gets steeper and steeper, which means the slope becomes greater and greater, which is compatible with the fact that $1/x \to \infty$ as $x \to 0^+$.

Of course, these observations are far from a precise derivation or proof of formula (31). That will be done at the end of this section. For now, we will discuss some of its consequences. First, we note that it can be combined with our previous derivative formulas, as in the following example.

EXAMPLE 3.5.1

Find $f'(x)$ if $f(x) = x^2 - 3 \ln x$.

SOLUTION   By combining the power rule, sum rule, and constant multiple rule, we get

$$f'(x) = \frac{d}{dx}(x^2 - 3 \ln x)$$

$$= \frac{d}{dx}(x^2) - 3\frac{d}{dx}(\ln x) = 2x - \frac{3}{x}.$$

We can also combine the derivative formula with the basic properties of the logarithm.

EXAMPLE 3.5.2

Find $f'(x)$ if $f(x) = \ln(5x)$.

SOLUTION   By the first law of logarithms, $\ln(5x) = \ln 5 + \ln x$. However, $\ln 5$ is a constant, so its derivative is zero, and we get

$$f'(x) = \frac{d}{dx}(\ln 5 + \ln x) = \frac{1}{x}.$$

You may find it a little surprising that $\ln(5x)$ has the same derivative as $\ln x$, but that's because the two functions only differ by a constant, which does not change the derivative.

Next, we take up the exponential function. In this case, we have the following fundamental result, whose explanation appears at the end of this section.

The Derivative of the Exponential Function

$$\frac{d}{dx}e^x = e^x. \tag{32}$$

You may find this formula even more surprising than the one for $\ln x$. It says that the exponential function e^x is its own derivative! In terms of the graph, it says that at any point on the graph, the slope is equal to the height. (*Remember:* The value of the function is the height of the graph, and the derivative is its slope.) It is certainly not obvious why this should be so, and it is a very important property of the exponential function. We will derive formula (32) at the end of this section, but for now let's just see how to use it.

As before, we can combine our new derivative formula with all of our previous ones.

EXAMPLE 3.5.3

Find $f'(x)$ if $f(x) = 3e^x - x^3 + 2\ln x$.

SOLUTION   By combining the sum, constant multiple, and power rules, we get

$$f'(x) = \frac{d}{dx}(3e^x - x^3 + 2\ln x)$$

$$= 3\frac{d}{dx}(e^x) - \frac{d}{dx}(x^3) + 2\frac{d}{dx}(\ln x)$$

$$= 3e^x - 3x^2 + \frac{2}{x}.$$

Formula (32) is a special case of a more general derivative formula, one that is very useful. It reads as follows:

$$\frac{d}{dx} e^{cx} = c\, e^{cx}, \quad \text{where } c \text{ is any constant.} \tag{33}$$

EXAMPLE 3.5.4

Find $\dfrac{d}{dx} e^{-x}$ and $\dfrac{d}{dx} e^{x/2}$.

SOLUTION   Using formula (33) with $c = -1$ gives

$$\frac{d}{dx} e^{-x} = -e^{-x}.$$

Using the same formula with $c = \frac{1}{2}$ yields

$$\frac{d}{dx} e^{x/2} = \left(\frac{1}{2}\right) e^{x/2}.$$

We will defer any explanation of how to derive formula (33) until the end of this section. In the meantime, we will take up a number of its consequences.

EXAMPLE 3.5.5

Find $f'(x)$ if $f(x) = 2^x$.

SOLUTION   Recall that $2^x = e^{x \ln 2}$. Since $\ln 2$ is a constant, we can apply formula (33) with $c = \ln 2$ to get

$$\frac{d}{dx} 2^x = \frac{d}{dx} e^{x \ln 2} = (\ln 2) \cdot e^{x \ln 2}.$$

Next, replacing $e^{x \ln 2}$ by 2^x on the right-hand side of this last formula gives

$$\frac{d}{dx} (2^x) = (\ln 2) \cdot 2^x.$$

The same procedure works with any positive base b, and we get the general formula

$$\frac{d}{dx} b^x = (\ln b) \cdot b^x. \tag{34}$$

Differential equations as models for exponential growth and decay

In Section 2.4 of Chapter 2, we studied several applications of the exponential growth and decay equation

$$y(t) = Ae^{rt}, \qquad (35)$$

where A and r are constants, t represents time (in appropriate units), and y is a quantity that is changing over time.

Taking the derivative on both sides of (35) and using formula (33), we get

$$\frac{dy}{dt} = rAe^{rt}.$$

Comparing this equation with (35), we see that

$$\frac{dy}{dt} = ry. \qquad (36)$$

This is another form of the exponential growth-decay equation. It provides a very simple and intuitive way of thinking of exponential growth or decay, for it says that *a quantity is growing or decaying exponentially if and only if its rate of change is proportional to the amount present at any time.* The quantity is growing if $r > 0$ and decaying if $r < 0$.

An equation like (36), relating a function and its derivative, is called a **differential equation**. The function given by (35) is called the **solution** of the differential equation with **initial value** $y(0) = A$.

If a differential equation describes a certain process in science or economics, we say that the equation **models** the process. Constructing and solving differential equation models is one of the most important applications of calculus. We have already seen a number of phenomena that are modeled by Eq. (36). Though seemingly unrelated, they all follow the same law: *The rate of growth (or decay) is proportional to the amount present at any time.*

Modeling population growth   Earlier in this chapter, in Section 3.3, we obtained the equation

$$\frac{dP}{dt} = rP, \qquad (37)$$

where P is the size of a population at time t and r is the relative growth rate.

Referring to Eqs. (35) and (36), we see that the solution of (37) is given by

$$P(t) = Ae^{rt}, \qquad (38)$$

where A is the population when $t = 0$.

EXAMPLE 3.5.6

Assume that in 1990 the U.S. population was 250 million, and that after 1990 it is modeled by the differential equation

$$\frac{dP}{dt} = 0.02P,$$

where $P(t)$ denotes the population t years after 1990. Find the formula for $P(t)$ and the population in the year 2010.

SOLUTION $P(t)$ is the solution of the given differential equation with initial value $P(0) = 250$. Thus, according to formula (38),

$$P(t) = 250e^{0.02t}.$$

In the year 2010 the population will equal $P(20) = 250e^{0.4} \approx 372.956$ million.

Modeling compound interest   The same differential equation applies to continuously compounded interest. If an initial deposit of A dollars earns interest at rate r, continuously compounded, the rate at which the investment grows is proportional to the size of the balance at any time. In this case the rate is the annual interest rate, and we are once again led to the differential equation

$$\frac{dy}{dt} = ry,$$

whose solution is

$$y = Ae^{rt}.$$

EXAMPLE 3.5.7

A bank account earns interest at a rate of 6%, compounded countinuously. Assuming that the initial deposit is $2,000, write a differential equation modeling the balance $y(t)$ at any time after the account is opened. Then find its solution and compute the balance when $t = 5$ years.

SOLUTION   Since $r = 0.06$, the differential equation is

$$\frac{dy}{dt} = 0.06y,$$

and its solution is $y = 2,000e^{0.06t}$. The balance at $t = 5$ is

$$y(5) = 2,000e^{0.3} \approx \$2,699.72.$$

Modeling radioactive decay   The decay of a radioactive substance is modeled by the same differential equation, this time with $r = -k$, where k is the decay constant

$$\boxed{\frac{dy}{dt} = -ky.}\qquad (39)$$

The solution is the familiar decay equation

$$\boxed{y = Ae^{-kt}.}\qquad (40)$$

Newton's law of cooling   If a warm object is put in cooler surroundings, its temperature will steadily decrease. A law of physics known as **Newton's law of cooling** says that *the rate at which the object cools is proportional to the difference between its temperature and the surrounding temperature.* This law is modeled by the differential equation

$$\boxed{\frac{dH}{dt} = k(H - M),}\qquad (41)$$

where t is the time, H is the temperature of the object, k is a proportionality constant, and M is the temperature of the surrounding medium, which we assume to be constant.

A solution of Eq. (41) is given by the following formula describing the temperature H as a function of t:

$$\boxed{H = M + Ae^{kt},}\qquad (42)$$

where A is a constant. To check that (42) is the solution, take the derivative

$$\frac{dH}{dt} = kAe^{kt}.\qquad (43)$$

Comparing (42) and (43) shows that H and dH/dt do indeed satisfy (41).

Notice that there is a new constant A in the solution. To find its value, we need to know the temperature of the object at one particular time. To find the value of k, we need a second piece of data.

EXAMPLE 3.5.8

Suppose a beaker of water whose temperature is 100°C is put into a cooler whose temperature is kept fixed at 10 degrees. At the end of 30 minutes, it has cooled to 85 degrees. Find its temperature at the end of 1 hour.

SOLUTION   The temperature of the water satisfies Eq. (42) with $M = 10$ (the temperature of the surrounding medium). That is,

$$H = 10 + Ae^{kt}.$$

The given information says that $H = 100$ when $t = 0$. Putting those values into the last equation gives $100 = 10 + A$, so that $A = 90$. We can now rewrite the temperature equation in the form

$$H(t) = 10 + 90e^{kt}.$$

To find k, we use the additional information that $H = 85$ when $t = 30$. Putting those values into the last equation gives

$$85 = 10 + 90e^{30k}.$$

To solve this for k, we collect terms to get $e^{30k} = \frac{75}{90}$ and take the natural logarithm of both sides:

$$30k = \ln\left(\frac{75}{90}\right).$$

Dividing both sides by 30 gives $k = \ln(\frac{75}{90})/30$. The equation for the temperature now has its final form:

$$H(t) = 10 + 90e^{t\ln(\frac{75}{90})/30}.$$

We can use this equation to compute the water temperature at any time by setting t equal to that time (in minutes). In particular, the temperature at the end of 60 minutes is given by

$$H(60) = 10 + 90e^{2\ln(\frac{75}{90})} = 72.5.$$

Carbon dating

An interesting application of the radioactive decay formulas (40) and (39) is the method of **carbon dating**, used to determine the age of animal and plant samples—such as pieces of wood, parchment, or bone—found at archaeological sites. It was invented by the American chemist and Nobel Laureate Willard Libby and is based on the decay of radioactive carbon.

As you probably know, carbon is the key element in all organic matter. It is found in nature in two forms—a radioactive form, known as C-14, and a more common, nonradioactive form, known as C-12. (There is also a third form, which is known as C-13 and is also nonradioactive, but it can be lumped together with C-12 for the purposes of this application.)

Radioactive carbon is produced by the action of cosmic rays, then oxidized to form radioactive carbon dioxide, which mixes in the atmosphere with the more common, nonradioactive kind containing C-12. Plants take in the mixture by photosynthesis, and animals ingest it by eating plants. Each living organism has a mixture of both types of carbon, and the relative proportions stay the same during the life of the organism. When the organism dies, however, the C-14 decays without being replenished. By comparing the concentration of C-14 in an ancient sample—say, a piece of charcoal, tree bark, or bone—with the concentration in a fresh sample, we can get an idea of the age of the ancient sample. That is the idea behind carbon dating.

In practice, however, we cannot actually measure the concentration of C-14 in each of the samples. What can be measured instead, by a device called a **Geiger counter**, is the *rate* at which C-14 is decaying in each of the samples. So instead of comparing the concentrations of radioactive carbon in the two samples, the carbon dating method compares the *rates* at which they decay.

If the ancient sample has been dead for exactly t years, the amount of C-14 it contains is given by the radioactive decay formula

$$y(t) = Ae^{-kt},$$

where k is the decay constant for radioactive carbon and A is the amount in the sample at the time it died. To find the rate at which the C-14 is decaying, we take the derivative:

$$y'(t) = -kAe^{-kt}.$$

What about the rate of C-14 decay in a fresh sample? To find it, we simply substitute $t = 0$ into the last formula, getting

$$y'(0) = -kA.$$

In other words, the decay rate of a fresh sample is the same as it was for the ancient sample at time zero—the start of the decay period.

Writing R for the ratio of these two rates, we have

$$R = \frac{y'(t)}{y'(0)} = \frac{-kAe^{-kt}}{-kA} = e^{-kt}. \tag{44}$$

If we know k and can determine R from measurements, then we simply have to solve the equation

$$R = e^{-kt}$$

for t in order to find the age of the sample:

$$t = -\frac{\ln R}{k}. \tag{45}$$

The value of k used by Libby is based on an estimate of 5,568 years for the half-life of radioactive carbon. This means that $k = (\ln 2)/5{,}568$, and formula (45) then takes the form

$$t = -\frac{5{,}568\ln R}{\ln 2} \approx -8{,}033\ln R, \tag{46}$$

which is known as the **conventional radiocarbon age**.

EXAMPLE 3.5.9

The radioactive carbon in a piece of charcoal taken from a cave in Lascaux, France (the site of a remarkable collection of ancient cave paintings) was discovered to be decaying at about 15% of the rate of a fresh sample. Find the age of the sample.

SOLUTION In our notation, this means that $R = 0.15$, so Eq. (46) gives the age (in years) of the sample as

$$t = -\frac{5{,}568 \ln(0.15)}{\ln 2} \approx 15{,}240.$$

One point about carbon dating should be mentioned. In recent years it has been found not to be entirely accurate in all cases. One reason is that it is based on the assumption that the proportion of C-14 in the atmosphere is stable, so that the concentration in a living organism is constant. This is not, in fact, the case. The concentration changes slowly under the influence of such factors as the sun's activity, the earth's magnetic field, and environmental changes caused by human activity. However, the errors are small and can be adjusted for samples less than 9 to 10 thousand years old, so that for samples within that age range the method is accurate. For samples between 10 and 30 thousand years old, on the other hand, the error may be as high as 10%. Methods other than conventional radiocarbon dating have been developed to give more accurate results in those cases.

Substances other than radioactive carbon have also been used in determining age. One of the projects at the end of this chapter discusses a case in which a radioactive form of lead was used to date certain paintings, attributed to the Dutch master Jan Vermeer, and to show that they were actually forgeries.

Appendix: deriving the basic derivative formulas

The derivative of the natural logarithm   From the definition of the derivative, we get

$$\frac{d}{dx} \ln x = \lim_{h \to 0} \frac{\ln(x + h) - \ln x}{h}. \tag{47}$$

For simplicity, we will restrict our attention to positive h and take the limit as $h \to 0^+$. The limit as $h \to 0^-$ only requires a minor modification.

Using the laws of logarithms and some elementary algebra, we can simplify the right-hand side of (47) as follows:

$$\frac{\ln(x + h) - \ln x}{h} = \frac{1}{h} [\ln(x + h) - \ln x]$$

$$= \frac{1}{h} \ln \left(\frac{x + h}{x} \right) \quad \text{(second law of logarithms)}$$

$$= \frac{1}{h} \ln \left(1 + \frac{h}{x} \right)$$

$$= \ln \left(1 + \frac{h}{x} \right)^{1/h} \quad \text{(third law of logarithms)}.$$

Substituting this into formula (47), we get

$$\frac{d}{dx} \ln x = \lim_{h \to 0} \ln \left(1 + \frac{h}{x} \right)^{1/h}. \tag{48}$$

To find the limit on the right-hand side of this equation, we need to use the formula

$$\lim_{n \to \infty} \left(1 + \frac{r}{n}\right)^n = e^r, \tag{49}$$

which we first encountered in Chapter 2 (see Section 2.2). If we substitute $r = 1/x$ and $n = 1/h$ in this formula, the left-hand side becomes

$$\left(1 + \frac{1/x}{1/h}\right)^{1/h} = \left(1 + \frac{h}{x}\right)^{1/h}.$$

Moreover, $n \to \infty$ is the same as $h \to 0$, because $n = 1/h$. Therefore, we can rewrite formula (49) as

$$\lim_{h \to 0} \left(1 + \frac{h}{x}\right)^{1/h} = e^{1/x}. \tag{50}$$

Taking the natural logarithm on each side of this equation and using the continuity of the logarithm function lead to the equation

$$\lim_{h \to 0} \ln\left(1 + \frac{h}{x}\right)^{1/h} = \ln(e^{1/x}) = \frac{1}{x}.$$

Putting it all together, we get

$$\begin{aligned}
\frac{d}{dx} \ln x &= \lim_{h \to 0} \frac{\ln(x + h) - \ln x}{h} \\
&= \lim_{h \to 0} \ln\left(1 + \frac{h}{x}\right)^{1/h} \\
&= \ln(e^{1/x}) = \frac{1}{x},
\end{aligned} \tag{51}$$

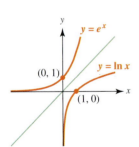

Figure 3.5.2

which is precisely the formula for the derivative of $\ln x$.

The derivative of the exponential　The first step is to determine the slope of the graph of $y = e^x$ at the point $(0, 1)$. In fact, the slope equals 1 at that point. We might guess that by using the numerical differentiation operator of a graphing calculator (see Exercise 37), but for a more precise justification, we need to recall how we determined the graph of the logarithm in the last chapter. Because of the inverse relationship

$$y = \ln x \iff x = e^y,$$

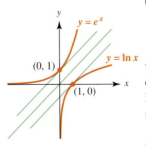

Figure 3.5.3

we know that the graphs of $y = e^x$ and $y = \ln x$ are mirror images of one another—one comes from the other by reflection in the 45-degree line $y = x$. You can see that in Figure 3.5.2, in which the dashed line is the 45 degree line, with the two graphs on either side of it.

In particular, the point $(0, 1)$ on the exponential graph is the mirror image of the point $(1, 0)$ on the logarithm graph. Figure 3.5.3 shows the tangent lines to the graphs at those points. Since the graphs are mirror images, so are the tangent lines. That means we should be able to figure out the slope of one of the lines if we know the slope of the other.

We know the slope of the logarithm graph: It is the derivative of $\ln x$, which equals $1/x$. At the point $(1, 0)$ that equals 1, which means that the tangent line is parallel to the 45-degree line. But then its reflection must also be parallel to the 45-degree line. That means the tangent to the graph of $y = e^x$ at $(0, 1)$ also has slope 1.

To repeat: *The graph of $y = e^x$ has slope 1 at the point $(0, 1)$*. If we write $f(x) = e^x$, this means that $f'(0) = 1$. By definition,

$$f'(x) = \lim_{h \to 0} \frac{f(x + h) - f(x)}{h}.$$

If we apply that to the present case, using $f(x) = e^x$ with $x = 0$, it becomes

$$f'(0) = \lim_{h \to 0} \frac{e^h - 1}{h}.$$

Since, as we have just seen, $f'(0) = 1$, we get the following useful limit formula:

$$\lim_{h \to 0} \frac{e^h - 1}{h} = 1. \tag{52}$$

Once we know formula (52), it is not hard to find the derivative of e^x for any x. If we apply the basic derivative formula to the function $f(x) = e^x$, we get

$$f'(x) = \lim_{h \to 0} \frac{f(x + h) - f(x)}{h}$$

$$= \lim_{h \to 0} \frac{e^{x+h} - e^x}{h}$$

$$= \lim_{h \to 0} \frac{e^x e^h - e^x}{h} \quad \text{(first law of exponents)}$$

$$= \lim_{h \to 0} \frac{e^x (e^h - 1)}{h} \quad \text{(by factoring)}$$

$$= e^x \lim_{h \to 0} \frac{e^h - 1}{h} \quad \text{(because e^x is fixed as $h \to 0$)}$$

$$= e^x \quad \text{[using formula (52)].}$$

The more general formula $d/dx\,(e^{cx}) = ce^{cx}$ can be derived in a similar way and is left as an exercise. However, in the next section we will see an easier way to get this formula by using a rule for derivatives called the **chain rule**.

Practice Exercises 3.5

1. Find the slope of the graph of $f(x) = 2x^2 - \ln 2x$ at the point $(\frac{1}{2}, \frac{1}{2})$.

2. At the start of a certain experiment there are 100 milligrams of a radioactive substance that is decaying at the rate of 2.83 milligrams per day. Find an expression for the amount present at any time. How much will be left after 30 days?

Exercises 3.5

In Exercises 1–6, find the derivative of the given function.

1. $x^2 + 4x + \ln x$

2. $2t^3 - 3\ln t$

3. $10 - \ln x$

4. $2\ln x - \dfrac{1}{x}$

5. $\dfrac{x\ln x + 1}{x}$

6. $\dfrac{\ln x - 1}{2}$

7. Find the slope of the graph of $y = \ln x$ at $x = \frac{1}{2}$.

8. Find the equation of the line tangent to the graph of $y = \ln x$ at $x = 1$.

9. Find the equation of the line tangent to the graph of $y = x^2 + \ln x$ at the point $(1, 1)$.

10. Find the equation of the line tangent to the graph of $y = x - 2\ln x + 1$ at $x = 2$.

In Exercises 11–13, find both coordinates of the point in question.

11. The point where the graph of $y = \ln x$ has slope 2

12. The point where the tangent to the graph of $y = \ln x$ is parallel to the line $x - 3y = 7$

13. The point where the tangent to the graph of $y = x^2 - \ln x$ is horizontal

In Exercises 14–19, use the derivative formula for $\ln x$ and the laws of logarithms to find the derivative of the given function.

14. $\ln\left(\dfrac{x}{5}\right)$

15. $\ln\left(\dfrac{5}{x}\right)$

16. $\frac{1}{5}\ln 2x$

17. $\ln\sqrt{x}$

18. $\ln(2t^3)$

19. $\ln\left(\dfrac{2}{t^3}\right)$

20. If m and c are constants, with $c > 0$, show that $d/dx\,\ln(cx^m) = m/x$.

21. **(a)** Find the equation of the line tangent to the graph of $y = \ln x$ at $x = 1$.
(b) Use linear approximation to find approximate values for $\ln(1.1)$, $\ln(1.2)$, $\ln(0.9)$, and $\ln(0.8)$, and compare them with the values you get from your calculator.
(c) If you did the approximations in part (b) correctly, they were all greater than the values given by your calculator. Can you explain why? (*Hint:* Use your calculator to graph both $\ln x$ and its tangent line at $x = 1$ in a $[0.25, 3] \times [-1.5, 1.5]$ window. What do you observe about the graphs?)

22. Let $f(x) = \ln(\sqrt{x})$. Find the equation of the tangent line at $x = 1$ and verify it visually by graphing both the function and its tangent line in a $[0.5, 2] \times [-0.5, 0.5]$ window.

23. Use your calculator to approximate $\ln 2$ to five decimal places, and then use that approximation to answer the following:
(a) Estimate $\ln(2.01)$ and $\ln(1.9)$ by linear approximation.
(b) Write the equation of the tangent line to the graph of $\ln x$ at $x = 2$ and verify it visually by graphing both $\ln x$ and the tangent line in a $[1, 3] \times [0, 1.5]$ window.
(c) Based on the graphs of part (b), do you think the linear approximations of part (a) are greater or smaller than the actual values? Explain.

In Exercises 24–29, find the derivative of the given function.

24. $3e^t - t^2 + 1$

25. $2e^x - \dfrac{1}{\sqrt{x}}$

26. $e^x - \ln x + 1$

27. $\dfrac{e^t + 1}{2}$

28. $\dfrac{t^2 e^t + t + 1}{t^2}$

29. $e^{\ln x} + e^x + \ln(e^x)$

30. Find the slope of the graph of $y = e^x - x$ at $x = -1$.

31. Find the equation of the line tangent to the graph of $y = e^x - e$ at $x = 1$.

32. Find the equation of the line tangent to the graph of $y = e^x + x^2 - 2x$ at the point $(0, 1)$.

In Exercises 33–35, find both coordinates of the point in question.

33. The point where the graph of $y = e^x$ has slope 3

34. The point where the line tangent to the graph of $y = e^x$ at the point $(0, 1)$ meets the x-axis

35. The point where the graph of $y = 2e^x - x$ has a horizontal tangent

36. Show that if the graph of $y = e^x$ has slope m at $x = a$, it has slope $1/m$ at $x = -a$.

37. Verify that the graph of $y = e^x$ has slope 1 at the point $(0, 1)$ by using the numerical differentiation operation of your calculator.

38. Find the equation of the line tangent to the graph of $y = e^x + x$ at $x = 0$. Then verify it visually by graphing the function and the tangent line in a $[-1, 1] \times [-1, 3]$ window.

In Exercises 39–44, use the derivative formula for e^{cx} and the laws of exponents to find the derivative of the given function.

39. $e^{x/3}$

40. e^{-2x}

41. e^{3x+1}

42. $(e^x)^4$

43. $\dfrac{1}{e^t}$

44. $\sqrt{e^t}$

45. Show that the slope of the graph of $y = e^x$ at $x = a$ is the negative of the slope of the graph of $y = e^{-x}$ at $x = -a$.

46. Write the equation of the line tangent to the graph of $y = e^x - e^{-x}$ at $x = 0$.

47. For what x does the graph of $y = e^{2x} + e^{-x}$ have slope zero?

In Exercises 48–53, find the derivative of the given function.

48. $3^t + t^3$

49. $\dfrac{2^x}{3} + \dfrac{2}{3x}$

50. $5^x + 5 \cdot 4^x$

51. $\dfrac{1}{2^t}$

52. 3^{2t}

53. 2^{1+t}

54. If the inflation rate is 4% per year over a period of years, an object costing $100 today will cost $100(1.04)^t$ in t years. How much will the object cost (in dollars) and how fast will its price be rising (in dollars per year) 2 years from today?

55. A colony of insects increases its population by 2.3% per day. If the initial population consists of 1,000 insects, its size at the end of t days is given by the formula $P = 1,000(1.023)^t$. How large will the population be at the end of 10 days and at what rate (in insects per day) will it be increasing?

56. Atmospheric pressure varies according to the height above sea level. Under ideal conditions the rate at which the pressure changes with respect to the height is proportional to the pressure.
(a) Letting $P(s)$ denote the atmospheric pressure at height s, write a differential equation that models the atmospheric pressure.
(b) If the pressure is 100 at sea level and 87 at a height of 3,000 meters, what will it be at 3,500 meters?

57. Farm-raised trout are released into a lake the day before a month-long fishing contest begins. The event coordinators monitor the overall daily catch and notice that the percentage of trout remaining in the lake for the first several days of the contest is best approximated by $P(t) = 100e^{-0.2t}$, where t is the time in days.
(a) At what rate are the trout being removed?
(b) If the removal rate follows this formula for the first 5 days and then remains constant after that, how many days will it take to remove all the fish? (*Hint:* Look at the linear approximation at $t = 5$ and see where it intersects the t-axis.)

58. Winning heights in the Olympic sport of pole vaulting are dependent not only on the ability of the athlete but on the technology of the pole. The introduction of fiberglass poles in the 1960s caused dramatic increases in the winning heights compared to those previously made with bamboo poles. The Olympic gold medal heights in meters

from 1956 to 1992 are modeled by:

$$H(t) = 6.5 - 2e^{-.035t},$$

where t is time in years after 1956.
(a) What was the winning height in 1956?
(b) Compare the rate of increase in winning heights at 1956 with that of 1984.
(c) Physicists calculate that the maximum height an athlete can attain with a fiberglass pole is 6.5 meters. Compare this to the limit of $H(t)$ as t approaches infinity.

Source: Steve Haake, Physics, technology, and the Olympics, *Physics World*, IOP Publishing, Sept. 2000.

59. The American psychologist C. L. Hull (1884–1952) believed that the strength of a habit is a function of the length of time the subject has had the habit. One possible model for the strength of a particular habit might be

$$H = 1 - e^{-0.1t},$$

where t is the number of years a subject has lived with the habit and H is an index of the difficulty of quitting it. When $H = 1$, it is virtually impossible to quit the habit, while the closer H is to zero, the easier it is to quit.
(a) Without doing any calculation, can you predict what the sign of the derivative is? Why?
(b) Now calculate H'. Is the sign what you expected?
(c) How would you modify the equation if your data showed that the graph has the same general shape but the difficulty index increases more slowly? More quickly?

60. The graph of a certain function lies entirely above the axis, goes through the point $(0, 3)$, and has the property that its slope at any point is twice its height. What is the function?

61. A carafe of water whose temperature is 65°F is put into a refrigerator that is kept at a constant temperature of 40 degrees. After 20 minutes the water cools down to 55 degrees.
(a) Write a formula giving the temperature of the water as a function of time.

(b) What will the temperature of the water be after 30 minutes? At what rate will the temperature be changing?
(c) How many minutes will it take for the water to cool to 45 degrees?

62. A cup of coffee is brought into a room whose temperature is kept at a steady 65°F. Five minutes later the temperature of the coffee is 90° and decreasing at a rate of 3° per minute. Find its temperature as a function of time.

63. Newton's law of cooling is also valid if the object is *cooler* than the surrounding medium. In that case, the rate at which the temperature of the object *increases* is proportional to the difference between its temperature and that of the surrounding medium.

Suppose you put a baking potato in a hot oven maintained at a constant temperature of 400°F, and at the end of 20 minutes the potato's temperature is 320° and is increasing at the rate of 6° per minute. Newton's law of cooling (or, in this case, heating) says that the temperature H of the potato is increasing at a rate proportional to the difference between its temperature and that of the oven.
(a) Write a formula giving the temperature of the potato as a function of time.
(b) What was the temperature of the potato and how fast was it increasing when $t = 0$?
(c) The temperature of the potato increases over time. What about the *rate* at which the temperature increases— does it increase, decrease, fluctuate, or remain steady?

64. City officials became concerned that too many new homes were being built locally each year. They felt that the infrastructure of roads would not be able to handle the increase in population and traffic flow. To control this, they allowed only 175 building permits to be sold each year. After the restriction took effect, a study showed that 35% of all new homes are sold annually and that there were 1,500 new homes available for sale. The differential equation used to study the effect of this change is

$$\frac{dH}{dt} = 175 - 0.35H,$$

where H is the number of new houses available and t is the number of years after the the study took place.
(a) What is dH/dt at $t = 0$?
(b) Verify that $H(t) = 500 + 1{,}000e^{-0.35t}$ is a solution.
(c) What is the limiting value of H when t goes to ∞?
(d) Show that the differential equation can be put into the same form as Newton's law of cooling.

65. A crime scene investigator knows that h hours after death a human body has a temperature of

$$T = T_a + (98.6 - T_a)(0.6)^h$$

where T_a is the temperature (in degrees Fahrenheit) of the air surrounding the body.

(a) Use T' to determine how fast the body's temperature is decreasing after $4\frac{1}{2}$ hours, assuming the air temperature is a constant 72°F.
(b) What is it if the air temperature is a constant 98°F?
(c) Show that T satisfies the differential equation of the form $T' = k(T - T_a)$, where k is a constant. Compare this to Newton's law of cooling.

66. In 1950, carbon dating was used to determine the age of wood samples excavated from a city in ancient Babylon. The rate of radioactive carbon decay of these samples was measured at 4.09 disintegrations per minute (dpm). By comparison the decay rate from fresh wood samples was measured at 6.68 dpm. Using 5,568 years as the half-life of radioactive carbon, estimate the age of the samples.

67. In one of the earliest tests of the carbon dating method, it was applied to samples of acacia wood found in the tomb of Djoser, an Egyptian Pharaoh of the Third Dynasty. The samples were known to be about 4,600 years old. The rate of radioactive carbon decay in these samples was measured and compared with the decay rate measured in fresh samples. According to the carbon dating method, what should the ratio of these two rates be for a 4,600-year-old sample?

68. The school coorporation of a growing region estimates that at any given time the transfer students and new first-year students comprise 30% of the student body of its high schools, and that graduating seniors and dropouts make up 20%.
(a) The rate of change of the student body per year ds/dt is the difference between the increase and decrease rates. Write a differential equation expressing that fact.
(b) If in a certain year the total high-school population was 1,600, find $s(t)$ after that year.
(c) If the maximum capacity of the district high schools is 2,400 students, when will they become overcrowded?

69. It is estimated that in 1999 there were about 2,000 million hectares of tropical forests in the world and about 16 million hectares* were converted to other uses.
(a) Use the given data to find the percentage of hectares that were converted in 1999.
(b) Consider that percentage to be the rate of annual tropical forest decay. Write a differential equation to represent this. What is the initial condition?
(c) Write the solution.

*A unit of land measure equal to 10,000 square meters, equivalent to 2.47 acres.

70. Show that $(d/dx)\, e^{cx} = ce^{cx}$ in the following sequence of steps:
(a) Starting with the definition of the derivative and using the laws of exponents, show that $(d/dx)\, e^{cx} = e^{cx} \lim_{h \to 0} (e^{ch} - 1)/h$.

(b) Using the substitution $h = t/c$, show that $\lim_{h \to 0} (e^{ch} - 1)/h = c \lim_{t \to 0} (e^t - 1)/t$.

(c) Applying formula (52) with t in place of h, conclude that $\lim_{t \to 0} (e^t - 1)/t = 1$ and, therefore, $\lim_{h \to 0} (e^{ch} - 1)/h = c$.

Solutions to practice exercises 3.5

1. Writing $\ln 2x = \ln 2 + \ln x$ and differentiating give $f'(x) = 4x - (1/x)$. Setting $x = \frac{1}{2}$, we obtain $f'(\frac{1}{2}) = 2 - 2 = 0$. Therefore, the slope of the graph at the point $(\frac{1}{2}, \frac{1}{2})$ is equal to zero.

2. According to the radioactive decay equation, the amount $Q(t)$ of the material after t days is $Q(t) = Ae^{-kt}$. Since

the initial amount is $Q(0) = 100$, we must have $A = 100$, and so $Q(t) = 100e^{-kt}$. Taking the derivative, we find $Q'(t) = -100ke^{-kt}$. Since we know that $Q'(0) = -2.83$, we must have $-100k = -2.83$ or $k = 0.0283$. Thus, $Q(t) = 100e^{-0.0283t}$ milligrams, where t is measured in days. The amount of the material after 30 days is given by $Q(30) = 100e^{-0.0283 \cdot 30} \approx 42.78$.

3.6 The Product and Quotient Rules

In this section we will take up the product and quotient rules for differentiation that will expand considerably our ability to compute derivatives.

The product rule

Sometimes a function is given as the product of two other functions. For example, if $f(x) = x^2$ and $g(x) = e^{3x}$, their product is

$$f(x) \cdot g(x) = x^2 e^{3x}.$$

If we know the derivatives $f'(x)$ and $g'(x)$, it is easy to find the derivative of the product by using the following rule:

> **The Product Rule**
>
> $$\frac{d}{dx}[f(x) \cdot g(x)] = f(x) \cdot g'(x) + g(x) \cdot f'(x).$$

Notice that the product rule is symmetric with respect to $f(x)$ and $g(x)$. That is,

$$\frac{d}{dx}[f(x) \cdot g(x)] = \frac{d}{dx}[g(x) \cdot f(x)].$$

That is because the product itself is the same no matter which function is written first.

EXAMPLE 3.6.1

Find $\dfrac{d}{dx}(x^2 e^{3x})$.

SOLUTION With $f(x) = x^2$ and $g(x) = e^{3x}$, we have $f'(x) = 2x$ and $g'(x) = 3e^{3x}$. The product rule says that

$$\frac{d}{dx}(x^2 e^{3x}) = 3x^2 e^{3x} + 2x e^{3x}.$$

EXAMPLE 3.6.2

Find the slope of the graph of $y = x \ln x$ at the point $(1, 0)$.

SOLUTION Applying the product rule with $f(x) = x$ and $g(x) = \ln x$, we get

$$\frac{d}{dx} x \ln x = x \cdot \frac{d}{dx} \ln x + (\ln x) \cdot \frac{d}{dx} x$$

$$= x \cdot \left(\frac{1}{x}\right) + (\ln x) \cdot 1$$

$$= 1 + \ln x.$$

If $x = 1$, this derivative takes the value $1 + \ln 1$, which equals 1.

The quotient rule

Another way to construct a new function out of two given ones is by taking the quotient. For example, if $f(x) = \ln x$ and $g(x) = x^2$ for $x > 0$, then

$$\frac{f(x)}{g(x)} = \frac{\ln x}{x^2}, \quad x > 0.$$

To differentiate a quotient, we use the following rule:

The Quotient Rule

$$\frac{d}{dx}\left[\frac{f(x)}{g(x)}\right] = \frac{g(x)f'(x) - f(x)g'(x)}{[g(x)]^2}.$$

EXAMPLE 3.6.3

Find $\dfrac{d}{dx}\left(\dfrac{\ln x}{x^2}\right)$.

SOLUTION   If we apply the quotient with $f(x) = \ln x$ and $g(x) = x^2$, we get

$$\frac{d}{dx}\left(\frac{\ln x}{x^2}\right) = \frac{x^2 \dfrac{d}{dx}\ln x - \ln x \dfrac{d}{dx}x^2}{(x^2)^2}$$

$$= \frac{x - 2x \ln x}{x^4}.$$

Unlike the product rule, the quotient rule is *not* symmetric with respect to $f(x)$ and $g(x)$. The derivative of $f(x)/g(x)$ is quite different from that of $g(x)/f(x)$. This is not surprising since the two quotients are different functions.

EXAMPLE 3.6.4

Find $\dfrac{d}{dx}\left(\dfrac{x^2}{\ln x}\right)$.

SOLUTION   If we apply the quotient with $f(x) = x^2$ and $g(x) = \ln x$, we get

$$\frac{d}{dx}\left(\frac{x^2}{\ln x}\right) = \frac{2x \ln x - x}{(\ln x)^2}.$$

With the quotient rule, we can differentiate any **rational** function—that is, any quotient of polynomials.

EXAMPLE 3.6.5

Find $\dfrac{d}{dx}\left(\dfrac{x-1}{x^2+1}\right)$.

SOLUTION   We apply the quotient rule, with $f(x) = x - 1$ and $g(x) = x^2 + 1$, to get

$$\frac{d}{dx}\left(\frac{x-1}{x^2+1}\right) = \frac{(x^2+1)\dfrac{d}{dx}(x-1) - (x-1)\dfrac{d}{dx}(x^2+1)}{(x^2+1)^2}$$

$$= \frac{(x^2+1)\cdot 1 - (x-1)\cdot 2x}{(x^2+1)^2}$$

$$= \frac{1 + 2x - x^2}{(x^2+1)^2}.$$

EXAMPLE 3.6.6

The American psychologist L. L. Thurstone (1887–1955) studied the process of learning certain repetitive tasks such as typing, and he proposed a model based on an equation of the form

$$y = \frac{kt}{t+b} \quad \text{for } t > 0,$$

where k and b are positive constants, t denotes the amount of practice time (say, in hours), and y is a measure of the skill level. In the case of typing, for example, y could be taken as the number of words per minute the student can type after t hours of practice.

(i) Compute dy/dt.

(ii) Thurstone said the learning curve followed a "law of diminishing returns." Explain that phrase in terms of the behavior of dy/dt.

SOLUTION   (i) Using the quotient rule, we obtain

$$\frac{dy}{dt} = \frac{(t+b) \cdot k - kt}{(t+b)^2} = \frac{kb}{(t+b)^2}.$$

(ii) As t gets larger, dy/dt decreases, and, in fact $\lim_{t \to \infty} dy/dt = 0$. The phrase diminishing returns refers to the fact that the rate of improvement in skill (i.e., dy/dt) decreases over time.

Deriving the product rule

By applying the definition of the derivative to the product $f(x)g(x)$, we get

$$\frac{d}{dx}[f(x)g(x)] = \lim_{h \to 0} \frac{f(x+h)g(x+h) - f(x)g(x)}{h}.$$

In order to convert the right-hand side to a more useful form, we add and subtract the same term $f(x)g(x+h)$ in the numerator. That does not change the value of the quotient, but it does enable us to factor it, as follows:

$$\frac{d}{dx}[f(x)g(x)] = \lim_{h \to 0} \frac{f(x+h)g(x+h) - f(x)g(x+h) + f(x)g(x+h) - f(x)g(x)}{h}$$

$$= \lim_{h \to 0}\left[g(x+h)\frac{f(x+h) - f(x)}{h} + f(x)\frac{g(x+h) - g(x)}{h}\right]$$

$$= \lim_{h \to 0} g(x+h) \cdot \lim_{h \to 0} \frac{f(x+h) - f(x)}{h} + f(x) \cdot \lim_{h \to 0} \frac{g(x+h) - g(x)}{h},$$

where the third line comes from applying the sum and product rules for limits. Now, since g is continuous, $\lim_{h \to 0} g(x+h) = g(x)$. And, since f and g are differentiable,

$$\lim_{h \to 0} \frac{f(x+h) - f(x)}{h} = f'(x) \quad \text{and} \quad \lim_{h \to 0} \frac{g(x+h) - g(x)}{h} = g'(x).$$

By making these substitutions, we get the product formula:

$$\frac{d}{dx}[f(x)g(x)] = g(x)f'(x) + f(x)g'(x).$$

Remark The **quotient rule** is derived in a similar way. First, one shows that the derivative of $1/g(x)$ is $-g'(x)/g^2(x)$ and then applies the product rule to the functions $f(x)$ and $1/g(x)$.

Practice Exercises 3.6

1. Find $\dfrac{d}{dx}(x^4 \ln x)$.

2. Find the equation of the line tangent to the graph of $y = (4x + 5)/(x^2 + 2)$ at $x = 1$.

Exercises 3.6

In Exercises 1–6, use the product rule to find the derivative.

1. xe^{-x}

2. $(x^2 + 5x + 1)(x^3 - x^2 + 4)$

3. $\dfrac{1}{x} \ln x$

4. $x^4 e^{2x}$

5. $e^{-3x} \ln x$

6. $\sqrt{x}\left(x - \dfrac{1}{x}\right)$

In Exercises 7–12, use the quotient rule to find the derivative.

7. $\dfrac{x^2 - 2}{x^2 + 4}$

8. $\dfrac{\ln x}{x + 1}$

9. $\dfrac{x}{e^x + 1}$

10. $\dfrac{3x + 1}{2x - 3}$

11. $\dfrac{e^x - 1}{e^x + 1}$

12. $\dfrac{x}{\ln x}$

In Exercises 13–15, find the derivative.

13. $x^2 \ln x$

14. $\dfrac{x}{x^2 + 1}$

15. $\dfrac{x \ln x}{x + 1}$

In Exercises 16–18, suppose $f(x)$ and $g(x)$ are functions satisfying

$$f(2) = 3, \quad g(2) = \tfrac{1}{2}, \quad f'(2) = -1, \quad g'(2) = \tfrac{2}{3}.$$

16. Find the slope of the graph of $f(x)/g(x)$ at $x = 2$.

17. Find the slope of the graph of $g(x)/f(x)$ at $x = 2$.

18. Find the equation of the line tangent to the graph of $f(x)g(x)$ at $x = 2$.

19. In the next section we will see that $(d/dx)(2x + 1)^{1/2} = (2x + 1)^{-1/2}$. Use that fact to find the following:
(a) the slope of the graph of $y = x\sqrt{2x + 1}$ at $x = 12$
(b) the slope of the graph of $y = x/\sqrt{2x + 1}$ at $x = 4$

In Exercises 20–23, find the equation of the tangent line at the given point.

20. $y = \dfrac{3x + 2}{2x + 3}$ at $x = 1$

21. $y = (x^2 + 2x + 3)e^x$ at $x = 0$

22. $y = \dfrac{1 - \ln x}{1 + \ln x}$ at $x = 1$ **23.** $y = \dfrac{x \ln x}{1 + \ln x}$ at $x = e$

24. Find the equation of the line tangent to the graph of $y = x^2/(x^2 + 1)$ at $x = 1$. Verify your answer visually by graphing the function and the tangent line in the same $[0, 2] \times [0, 1]$ window.

25. Find the equation of the line tangent to the graph of $y = x^2 e^x$ at $x = 1$. Verify your answer visually by graphing the function and the tangent line in the same $[0.5, 1.5] \times [0, 10]$ window.

26. For what x does the graph of $y = (\ln x)/x$ have slope zero?

27. For what x does the graph of $y = xe^x$ have slope zero?

28. Suppose the demand for a certain product is given by $q = f(p)$, where p is the price per unit and q is the number sold. The revenue is given by $R = pq$.
(a) If $f(100) = 12,000$ and $f'(100) = -40$, find dR/dp when $p = 100$.
(b) If the product is currently selling for \$100 per unit, should the company increase or decrease the price in order to raise the revenue?

29. Suppose the demand function for certain product is given by

$$q = 5,000e^{-0.01p}.$$

(a) Find the revenue as a function of the price.
(b) Find the marginal revenue function.

(c) Find the revenue and marginal revenue when $p = 200$.

(d) If the item is currently selling for $200, should the company increase or decrease the price in order to raise the revenue?

30. A predatory animal gains energy (in calories) from consuming prey. Suppose that for a certain predator-prey pair the caloric gain is modeled by

$$G(w) = 0.12we^{0.001w}, \qquad (53)$$

where w is the prey's weight in grams. Find $G'(w)$ (the marginal caloric gain).

31. A predator uses up energy (in calories) in catching its prey. Suppose that for a certain predator-prey pair the caloric expense is modeled by

$$E(w) = \frac{0.5w^2}{9 + 0.3w}, \qquad (54)$$

where w is the prey's weight in grams. Find $E'(w)$ (the marginal caloric expense).

32. Caloric gain and expense are analogous to revenue and cost in business, and **caloric profit** can be defined as the difference between caloric gain and expense. As in business applications, the **marginal profit** is the derivative of the profit function.

(a) If the caloric gain and expense are modeled by Eqs. (53) and (54), find the marginal caloric profit.

(b) If the marginal profit is positive at a certain weight, it is worthwhile for the predator to go after heavier prey. If it is negative, on the other hand, it is worthwhile for

the predator to go after lighter prey. Find the marginal caloric profit at $w = 200$ and determine if it is worthwhile capturing heavier prey. Do the same for $w = 300$.

33. Molten lava can fill a chamber in the earth's crust before it builds up enough pressure to erupt. Let the pressure be modeled by

$$P(t) = 0.47t^2 e^{0.0035t},$$

where t is the time in months.

(a) At what rate is the pressure changing with respect to time?

(b) The model says that an eruption is highly likely to occur if the rate of change of pressure is greater than 20. Is an eruption likely at the end of 30 months?

34. The rate of photosynthesis in plants is a function of the intensity of the light. If we let s denote the light intensity (in suitable units) and $p(s)$ the rate of photosynthesis, the function has the form

$$p(s) = \frac{s}{as + b},$$

where a and b are positive constants (depending on the type of plant and the environmental conditions).

(a) Find $p'(s)$.

(b) Use a linear approximation to estimate the relative increase in $p(s)$ caused by a 10% increase in light intensity. In other words, estimate

$$\frac{p(s + 0.1s) - p(s)}{p(s)}$$

in terms of s, a, and b.

Solutions to practice exercises 3.6

1. Applying the product rule, we obtain

$$\frac{d}{dx}(x^4 \ln x) = \ln x \cdot \frac{d}{dx}(x^4) + x^4 \cdot \frac{d}{dx}(\ln x)$$

$$= \ln x \cdot 4x^3 + x^4 \cdot \frac{1}{x}$$

$$= 4x^3 \ln x + x^3.$$

2. Let $f(x) = (4x + 5)/(x^2 + 2)$. Applying the quotient rule, we obtain

$$f'(x) = \frac{(x^2 + 2) \cdot \dfrac{d}{dx}(4x + 5) - (4x + 5) \cdot \dfrac{d}{dx}(x^2 + 2)}{(x^2 + 2)^2}$$

$$= \frac{(x^2 + 2) \cdot 4 - (4x + 5) \cdot 2x}{(x^2 + 2)^2} = \frac{(8 - 10x - 4x^2)}{(x^2 + 2)^2}.$$

Therefore, $f'(1) = -\frac{6}{9} = -\frac{2}{3}$, and, since $f(1) = 3$, the equation of the tangent line is $y - 3 = -\frac{2}{3}(x - 1)$, or

$$y = -\frac{2}{3}x + \frac{11}{3}.$$

3.7 The Chain Rule

Another method of combining functions is called **composition**. If $f(x)$ and $g(x)$ are two functions, we write

$$f[g(x)]$$

for the function we get by applying f to the output of g. In other words, we use the output of g as the input of f. For instance, if $f(x) = \ln x$ and $g(x) = x^2 + 1$, then

$$f[g(x)] = \ln(x^2 + 1).$$

By using composition, we greatly increase the number of functions we can construct out of the elementary ones we know. The following important rule tells how to take the derivative of a composition:

The Chain Rule

$$\frac{d}{dx} f[g(x)] = f'[g(x)] \cdot g'(x).$$

EXAMPLE 3.7.1

Find $\dfrac{d}{dx} \ln(x^2 + 1)$.

SOLUTION   In this case, f is the natural logarithmic function ln, and $g(x) = x^2 + 1$. Writing $f(u) = \ln u$, we get the derivative $f'(u) = 1/u$, and substituting $g(x)$ for u gives

$$f'[g(x)] = \frac{1}{g(x)} = \frac{1}{x^2 + 1}.$$

According to the chain rule, we have to multiply that by $g'(x)$, which, in this case, equals $2x$. Putting these together gives

$$\frac{d}{dx} \ln(x^2 + 1) = \frac{1}{x^2 + 1} \cdot (2x) = \frac{2x}{x^2 + 1}.$$

In the next example, we will verify a formula that was introduced in Section 3.5.

EXAMPLE 3.7.2

If c is a constant, then

$$\frac{d}{dx}\left(e^{cx}\right) = ce^{cx}.$$

SOLUTION   In this case, f is the exponential function—that is, $f(u) = e^u$ and $g(x) = cx$. The derivatives are as follows:

$$f'(u) = e^u \quad \text{and} \quad g'(x) = c,$$

and the chain rule gives

$$\frac{d}{dx}(e^{cx}) = f'[g(x)] \cdot g'(x) = e^{cx} \cdot c = ce^{cx}.$$

Another way of writing the chain rule is to use an auxiliary variable in place of the function g. That is essentially what we did in the previous two examples by using the variable u. By letting $u = g(x)$ and $y = f(u)$, we can write the chain rule in the following form:

Chain Rule in Leibniz Notation

$$\frac{dy}{dx} = \frac{dy}{du} \cdot \frac{du}{dx}. \tag{55}$$

In this formula, x is the primary input, y is the final output, and u is an auxiliary variable that serves both as output for the function g and input for the function f. For a simple example that provides insight into formula (55), consider three cars y, u, and x. If u travels twice the speed of x and y travels three times the speed of u, then y travels $3 \cdot 2$ times the speed of x.

EXAMPLE 3.7.3

Find $\dfrac{d}{dx}\sqrt{x^2 + 3x + 1}$.

SOLUTION   We set $y = \sqrt{u}$ and $u = x^2 + 3x + 1$. Then $y = \sqrt{x^2 + 3x + 1}$, and we want to find dy/dx. Using formula 55, we get

$$\frac{dy}{dx} = \frac{dy}{du} \cdot \frac{du}{dx}$$

$$= \frac{1}{2\sqrt{u}} \cdot (2x + 3)$$

$$= \frac{2x + 3}{2\sqrt{x^2 + 3x + 1}}.$$

Notice that in the last step above we replaced u by its expression in terms of x, so that the auxiliary variable is eliminated and does not appear in the final answer.

One important point: Unlike products, compositions are not symmetric in f and g. That is, in general,

$$f[g(x)] \neq g[f(x)],$$

and the derivatives may be very different.

EXAMPLE 3.7.4

Let $f(x) = x^5$ and $g(x) = x^2 + 1$. Compute $\frac{d}{dx} f[g(x)]$ and $\frac{d}{dx} g[f(x)]$.

SOLUTION We have

$$f[g(x)] = (x^2 + 1)^5,$$

whereas

$$g[f(x)] = (x^5)^2 + 1 = x^{10} + 1.$$

These are very different functions. For instance, with $x = 1$ we get

$$g[f(1)] = 2 \quad \text{and} \quad f[g(1)] = 32.$$

The derivatives are also very different. Using the chain rule, we have

$$\frac{d}{dx} f[g(x)] = 5(x^2 + 1)^4 \cdot (2x) = 10x(x^2 + 1)^4.$$

For the other derivative we have a choice of computing directly:

$$\frac{d}{dx} g[f(x)] = \frac{d}{dx} (x^{10} + 1) = 10x^9,$$

or using the chain rule:

$$\frac{d}{dx} g[f(x)] = g'[f(x)] \cdot f'(x) = 2x^5 \cdot 5x^4 = 10x^9.$$

Of course, both methods give the same answer.

The chain rule may also be applied to several successive compositions, as in the following example.

EXAMPLE 3.7.5

An early use of mathematics in psychology was made by Hermann Ebbinghaus, who proposed the following equation[1] as a model for the process of forgetting information:

$$y = \frac{100k}{(\log_{10} t)^c + k}, \quad t \geq 1.$$

In this model, t is the time in minutes after memorizing a given list. The dependent variable y is the percentage of items from the list retained in memory, and c and k are constants. On the basis of data, Ebbinghaus assigned the values of $c = 1.25$ and $k = 1.84$ for an experimental situation in which a subject memorized a list of nonsense syllables.

Using these constants, find the rate at which the percentage of information retained varies over time. In particular, how fast is the percentage changing at the end of 10 minutes?

SOLUTION   Substituting the values assigned to the constants c and k, we get

$$y = \frac{184}{(\log_{10} t)^{1.25} + 1.84}.$$

We can think of this function as a composition, with

$$u = \log_{10} t, \quad w = u^{1.25} + 1.84, \quad \text{and} \quad y = \frac{184}{w}.$$

Then the chain rule gives

$$\frac{dy}{dt} = \frac{dy}{dw} \frac{dw}{du} \frac{du}{dt},$$

and (recalling that $\log_{10} t = \ln t / \ln 10$) we obtain

$$\frac{dy}{dw} = -\frac{184}{w^2}, \quad \frac{dw}{du} = (1.25)u^{0.25}, \quad \text{and} \quad \frac{du}{dt} = \frac{1}{t \ln 10}.$$

Therefore,

$$\frac{dy}{dt} = -\frac{184(1.25)u^{0.25}}{w^2(t \ln 10)}$$

$$= -\frac{184(1.25)u^{0.25}}{(u^{1.25} + 1.84)^2 \, (t \ln 10)} \quad \text{(substituting } w = u^{1.25} + 1.84\text{)}$$

$$= -\frac{184(1.25)(\log_{10} t)^{0.25}}{[(\log_{10} t)^{1.25} + 1.84]^2 \, (t \ln 10)} \quad \text{(substituting } u = \log_{10} t\text{)}$$

[1]Ebbinghaus, H., *Memory* (trans. by H. A. Ruger), Teachers College, Columbia University, New York, 1913.

At the end of 10 minutes, we have

$$\frac{dy}{dt}\bigg|_{t=10} = -\frac{184(1.25)}{(2.84)^2 10 \ln 10} \approx -1.24.$$

In other words, at the end of 10 minutes the subject is forgetting the syllables he learned at the rate of 1.24% per minute.

Deriving the chain rule

Again we start from the definition of the derivative, this time applied to $f[g(x)]$. We have

$$\frac{d}{dx} f[(g(x)] = \lim_{h \to 0} \frac{f[g(x+h)] - f[g(x)]}{h}.$$

Now the trick is to divide and multiply the quotient on the right by $g(x+h) - g(x)$ [assuming $g(x+h) \neq g(x)$, to avoid a technical complication]. Since dividing and multiplying by the same quantity does not change the quotient, we get

$$\frac{d}{dx} f(g(x)) = \lim_{h \to 0} \left\{ \frac{f[g(x+h)] - f[g(x)]}{g(x+h) - g(x)} \cdot \frac{g(x+h) - g(x)}{h} \right\}$$

$$= \lim_{h \to 0} \frac{f[g(x+h)] - f[g(x)]}{g(x+h) - g(x)} \cdot \lim_{h \to 0} \frac{g(x+h) - g(x)}{h},$$

where the second line follows from the product rule for limits.

The second limit is clearly equal to $g'(x)$, by the definition of derivative. To evaluate the first limit, we first note that $g(x+h) \to g(x)$ as $h \to 0$, since g is continuous. Then, writing $u = g(x+h)$ and $a = g(x)$, we see that

$$\lim_{h \to 0} \frac{f[g(x+h)] - f[g(x)]}{g(x+h) - g(x)} = \lim_{u \to a} \frac{f(u) - f(a)}{u - a}$$

$$= f'(a) = f'[g(x)]$$

Putting the steps together, we get the chain rule:

$$\frac{d}{dx} f[g(x)] = f'[g(x)] \cdot g'(x).$$

Practice Exercises 3.7

1. Find the derivative of each of the following:

   (a) $\ln(e^x + e^{-x})$ (b) $\left(\dfrac{x-1}{x+1}\right)^{3/2}$

2. For what x, if any, does the graph of $y = x^2 e^{-x^2}$ have a horizontal tangent?

Exercises 3.7

• •

In Exercises 1–6, use the chain rule to find the derivative.

1. $e^{1/x}$ **2.** $(\ln x)^2$ **3.** $\left(e^x + e^{-x}\right)^2$

4. $\sqrt{x^2 + 2}$ **5.** $\ln(1 + e^x)$ **6.** $(3x + 2)^{-4}$

In Exercises 7–10, write y as a function of x and find dy/dx by the chain rule.

7. $u = x^2 + x + 1;\ y = u^{10}$ **8.** $u = 3x + 4;\ y = \sqrt{u}$

9. $u = x^4 + 1;\ y = \ln u$ **10.** $u = -x^2;\ y = e^u$

In Exercises 11–25, find the derivative.

11. $x\sqrt{1 + x^4}$ **12.** $\dfrac{1}{\sqrt{3x + 1}}$ **13.** $e^{-\sqrt{x}}$

14. xe^{-x^2} **15.** $x\ln(2x + 1)$ **16.** $(e^x + 1)^{-2}$

17. $\dfrac{1}{e^{2x} + e^x + 1}$ **18.** $(x^3 - 3x^2 + 2x - 1)^{12}$

19. $\dfrac{e^{2x} - 1}{e^{2x} + 1}$ **20.** $\left(\dfrac{x - 1}{x + 1}\right)^3$

21. $\ln(x^2 + 3x + 5)$ **22.** $e^{x/(x+1)}$

23. $\ln(1 + \sqrt{x})$ **24.** $[\ln(3x + 1)]^2$ **25.** $x^2\sqrt{2x + 1}$

In Exercises 26–29, find the equation of the tangent line at the given point.

26. $y = \ln(3x - 2)$ at $x = 1$

27. $y = \sqrt{5 + x^2}$ at $x = 2$

28. $y = e^{x^2 - 3x + 2}$ at $x = 1$ **29.** $y = xe^{2x-1}$ at $x = \frac{1}{2}$

30. Find the equation of the line tangent to the graph of $y = e^{x^2 - 1}$ at $x = 1$. Verify your answer visually by graphing the function and the tangent line in the same $[0, 2] \times [0, 2]$ window.

31. Find the equation of the line tangent to the graph of $y = \sqrt{4x + 3}$ at $x = 3$. Verify your answer visually by graphing the function and the tangent line in the same $[0, 6] \times [0, 6]$ window.

32. For what x does the graph of $y = e^{x^2 - 2x}$ have slope zero?

33. For what x does the graph of $xe^{-x^2/2}$ have slope zero?

In Exercises 34–35, suppose $f(x)$ and $g(x)$ are functions satisfying

$$f(2) = 3, \quad g(2) = \tfrac{1}{2}, \quad f'(2) = -1, \quad g'(2) = \tfrac{2}{3}.$$

34. Find the equation of the line tangent to the graph of $\ln[g(x)]$ at $x = 2$.

35. Find the slope of the graph of $f(1 + x^3)$ at $x = 1$.

In Exercises 36–39, suppose $f(x)$ and $g(x)$ are functions satisfying

$$f(0) = 2, \quad g(0) = 1, \quad f'(0) = -3, \quad g'(0) = 5.$$

36. Find the slope of the graph of $e^{g(x)-1}$ at $x = 0$.

37. Find the equation of the line tangent to the graph of $\sqrt{2 + f(x)}$ at $x = 0$.

38. Find the slope of the graph of $\ln[f(x) + g(x)]$ at $x = 0$.

39. Find the slope of the graph of $f(e^{2x} - 1)$ at $x = 0$.

In Exercises 40–42, use the following table of values to find the derivative of the given function at the point indicated:

x	$f(x)$	$g(x)$	$f'(x)$	$g'(x)$
0	2	1	$-1/3$	3
1	$-2/5$	2	$1/2$	-3
2	$2/3$	$-1/4$	-1	$3/5$

40. $f(g(x))$ at $x = 1$ **41.** $f(g(x))$ at $x = 0$

42. $g(f(x))$ at $x = 0$

In Exercises 43–47, use the following table of values to find the derivative of the given function at the point indicated:

x	$f(x)$	$g(x)$	$f'(x)$	$g'(x)$
1	2	3	$-1/2$	-2
2	5	$-1/3$	1	$3/5$
3	-1	2	$1/4$	$3/2$

43. $[f(x) \cdot g(x)]^2$ at $x = 3$ **44.** $f[g(x)]$ at $x = 1$

45. $g[f(x)]$ at $x = 1$

46. $\ln[f(x) + g(x)]$ at $x = 1$

47. $\dfrac{1}{x + f(x)}$ at $x = 2$

48. A manufacturer's monthly revenue from the sale of x items, for $x \le 20$, is $R(x) = 20\sqrt{100x - x^2}$. Find the marginal revenue.

49. Wine is one of the important products of California. Suppose that the annual grape crop since 1900 is modeled by the equation $w = 200t^{3/2}$, where t is the time (in years after 1900) and w is the weight (in thousands of pounds). Furthermore, suppose that the quantity of wine produced is a function of the total grape crop (in thousands of pounds), given by the equation

$$g = \frac{3w^2}{32w + 10},$$

where g is the amount of wine (in thousands of gallons). At what rate was wine production (in thousands of gallons per year) increasing in the year 2000?

50. Diatoms are microscopic algae surrounded by a silica shell that are found both in salt and fresh water, and they are a major source of atmospheric oxygen. The size of a diatom colony depends on many factors, including temperature. Suppose that samples taken in a midwestern lake showed that the concentration of diatoms was modeled as a function of temperature by the equation

$$C = 1.4 - e^{-0.001h^2} \quad \text{for } 0 < h < 40,$$

where C is the concentration of diatoms (in millions per cubic centimeter) and h is the temperature of the water (in degrees Celsius).
(a) Find dC/dh.

(b) Suppose the temperature of the lake is $10°$ and falling at the rate of 2 degrees per hour. At what rate is the concentration of diatoms changing with respect to time?

51. A toy rocket is fired straight upward, and its height (in meters) is given by

$$h = t + 10 - \sqrt{2t^2 + 100}, \quad 0 \le t \le 20,$$

where t is the time in seconds. Find a formula for the velocity in terms of t.

52. A chain of gourmet food stores sells a delicacy prepared from a rare fish species. Suppose that the amount of the delicacy available at any time during the 16-week season is given by

$$w = 1{,}000te^{-0.02t^2}, \quad 0 \le t \le 16,$$

where w is the number of pounds and t is the time in weeks.
(a) Find dw/dt.
(b) Suppose the price per pound is $p = 500 - 0.08w$. How fast (in dollars per week) is the revenue from this delicacy changing at the end of 8 weeks?

Solutions to practice exercises 3.7

1. (a) Using the chain rule, we get

$$\frac{d}{dx} \ln(e^x + e^{-x}) = \frac{1}{(e^x + e^{-x})} \cdot \frac{d}{dx}(e^x + e^{-x}) = \frac{e^x - e^{-x}}{e^x + e^{-x}}.$$

(b) Let $u = (x - 1)/(x + 1)$ and $y = u^{3/2}$. Then $dy/du = \frac{3}{2}u^{1/2}$, and the quotient rule gives

$$\frac{du}{dx} = \frac{(x + 1) \cdot 1 - (x - 1) \cdot 1}{(x + 1)^2} = \frac{2}{(x + 1)^2}.$$

Using the chain rule, we get

$$\frac{d}{dx}\left(\frac{x - 1}{x + 1}\right)^{3/2} = \frac{3}{2}u^{1/2} \cdot \frac{2}{(x + 1)^2}$$

$$= \frac{3}{2}\left(\frac{x - 1}{x + 1}\right)^{1/2} \cdot \frac{2}{(x + 1)^2}$$

$$= \frac{3(x - 1)^{1/2}}{(x + 1)^{5/2}}.$$

2. We must solve the equation $dy/dx = 0$. The product and chain rules give

$$\frac{dy}{dx} = x^2 \frac{d}{dx}e^{-x^2} + 2xe^{-x^2} = e^{-x^2}(-2x^3 + 2x).$$

Therefore, $dy/dx = 0$ for $x = 0$, 1, and -1.

3.8 Implicit Differentiation and Related Rates

In this section we will learn how to use the rules of differentiation—in particular, the chain rule—to compute the derivative of a function that is defined implicitly by an equation. Then we will consider problems where two quantities, both functions of the same independent variable, are related by an equation. By using implicit differentiation, we can determine the rate of change of one quantity in terms of the rate of change of the other.

Implicit differentiation

Up until now we have been learning various techniques for differentiating a function given by an explicit formula. For instance, the formula

$$y = \sqrt{25 - x^2}, \quad -5 < x < 5,$$

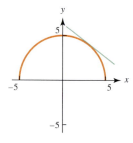

Figure 3.8.1

defines y as a function of x **explicitly**. Its graph is the semicircle of radius 5 shown in Figure 3.8.1. To compute dy/dx, we apply the chain rule and power rule to get

$$\frac{dy}{dx} = -\frac{x}{\sqrt{25 - x^2}}, \quad -5 < x < 5. \tag{56}$$

However, the same function is also defined **implicitly** by adding the restrictions $-5 < x < 5$ and $y \geq 0$ to

$$y^2 + x^2 = 25, \tag{57}$$

which is the equation of the complete circle shown in Figure 3.8.2. We can find dy/dx implicitly by taking the derivative (with respect to x) on both sides of Eq. (57),

$$\frac{d}{dx}(y^2 + x^2) = \frac{d}{dx}(25),$$

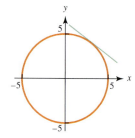

Figure 3.8.2

and using the sum and chain rules—treating y as a function of x. That gives

$$2y\frac{dy}{dx} + 2x = 0,$$

and solving for dy/dx gives the following implicit formula for the derivative:

$$\frac{dy}{dx} = -\frac{x}{y}. \tag{58}$$

Since $y = \sqrt{25 - x^2}$, we see that formulas (56) and (58) are equivalent. For example, using (56) gives

$$\frac{dy}{dx}\bigg|_{x=3} = -\frac{3}{\sqrt{25 - 3^2}} = -\frac{3}{4},$$

and using (58) with $x = 3$ and the corresponding $y = 4$ yields

$$\frac{dy}{dx}\bigg|_{x=3} = -\frac{3}{4}.$$

In this case, we were fortunate to have a function that was easily expressed both in explicit and implicit forms. In many situations, however, we encounter a function that is implicitly defined by an equation that involves both the dependent and independent variables and is difficult, if not impossible, to solve for the dependent variable. In such situations we rely on the following general procedure, in which we let x denote the independent variable and y the dependent variable.

> **Method for Implicit Differentiation**
>
> - Differentiate both sides of the equation with respect to x, using the differentiation rules learned thus far.
> - Apply the chain rule whenever you encounter an expression involving y.
> - Move all terms involving dy/dx to the left side of the equation and move the rest of the terms to the right side of the equation.
> - Factor out dy/dx on the left side of the equation.
> - Solve for dy/dx by dividing both sides of the equation by the other factor on the left.

EXAMPLE 3.8.1

Let $y = y(x)$ be the function defined implicitly by the equation $x^3 + y^3 = 9xy$.
  (i) Find dy/dx.
  (ii) Find the line tangent to the graph of $y = y(x)$ at the point $(2, 4)$.

SOLUTION (i) Differentiating both sides of the equation $x^3 + y^3 = 9xy$ with respect to the independent variable x, using the chain rule for the term y^3 and the product rule for the term $9xy$, we obtain

$$3x^2 + 3y^2 \frac{dy}{dx} = 9y + 9x \frac{dy}{dx}.$$

Moving $9x(dy/dx)$ to the left side and $3x^2$ to the right, we obtain

$$3y^2 \frac{dy}{dx} - 9x \frac{dy}{dx} = 9y - 3x^2.$$

Factoring out dy/dx on the left side gives

$$(3y^2 - 9x) \frac{dy}{dx} = 9y - 3x^2.$$

Finally, dividing both sides of the last equation by $3y^2 - 9x$, we obtain the formula

$$\frac{dy}{dx} = \frac{9y - 3x^2}{3y^2 - 9x} = \frac{3y - x^2}{y^2 - 3x}. \tag{59}$$

  (ii) Since $2^3 + 4^3 = 9 \cdot 2 \cdot 4$, we see that the point $(2, 4)$ is indeed on the graph of the function $y(x)$. Next, substituting $x = 2$ and $y = 4$ into formula (59), we get

$$\frac{dy}{dx}\bigg|_{x=2} = \frac{3 \cdot 4 - 2^2}{4^2 - 3 \cdot 2} = \frac{4}{5}.$$

Finally, we write the equation of the line tangent to the graph at the point $(2, 4)$ in point-slope form:

$$y - 4 = \frac{4}{5}(x - 2) \quad \text{or} \quad y = \frac{4}{5}x + \frac{12}{5}.$$

The graph and tangent line are shown in Figure 3.8.3.

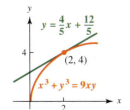

Figure 3.8.3

Related Rates

When two variables are related by an equation and both are functions of a third variable, such as time, we can find a relation between their rates of change. In such cases, we say the rates are **related**, and we can compute one if we know the other. The following example is typical.

EXAMPLE 3.8.2

Suppose that air is being pumped into a spherical balloon so that its volume is increasing at a rate of 400 cubic centimeters per second. How fast is the radius of the balloon increasing when it is 100 centimeters?

SOLUTION   The formula linking the volume V and radius r of a sphere is

$$V = \frac{4}{3}\pi r^3.$$

In this example, both V and r are functions of t. If we differentiate with respect to t on both sides, using the chain rule to differentiate r^3, we get

$$\frac{dV}{dt} = 4\pi r^2 \frac{dr}{dt}.$$

Thus, we have an equation relating the rates dV/dt and dr/dt. Substituting $r = 100$ and $dV/dt = 400$ gives

$$400 = 4\pi (100)^2 \frac{dr}{dt}.$$

Finally, by solving for dr/dt, we find that the radius of the balloon is increasing at the rate of $1/(100\pi)$ centimeters per second.

The procedure used in this example indicates the following guidelines for solving other problems involving related rates.

Method for Solving Related Rates Problems

- Identify the independent variable (usually time) on which the other quantities depend and assign it a symbol, such as t. Also assign symbols to the variable quantities that depend on t.
- Find an equation that relates the dependent variables.
- Differentiate both sides of the equation with respect to t, using the chain rule if necessary, to get the related rates equation.
- Substitute the given information into the related rates equation and solve for the unknown rate.

EXAMPLE 3.8.3

Suppose that the demand function for a certain product is given by

$$q = 4{,}000e^{-0.01p}.$$

If the item is currently selling for \$100 per unit, and the quantity supplied is decreasing at a rate of 80 units per week, find the rate at which the price of the product is changing.

SOLUTION   In this case, both variables, the price p and the quantity q, are functions of the time t (measured in weeks), and the equation relating them is given. Taking the derivative with respect to t on both sides of the demand equation and using the chain rule to differentiate $e^{-0.01p}$ give

$$\frac{dq}{dt} = 4{,}000 \cdot (-0.01)e^{-0.01p} \frac{dp}{dt}.$$

This equation relates the rates dq/dt and dp/dt. Putting the given information $p = 100$ and $dq/dt = -80$ into our equation, we obtain

$$-80 = -40e^{-1}\frac{dp}{dt} \quad \text{or} \quad \frac{dp}{dt} = 2e \approx 5.44.$$

Thus, we see that the unit price of the product is increasing at the rate of about \$5.44 per week.

Practice Exercises 3.8

1. The equation $x^2y + y^4 - x = 3$ defines y implicitly as a function of x. Find dy/dx in terms of x and y, and find the slope of the graph at $(2, 1)$.

2. Suppose water is pouring into a rectangular tank at the rate of $\frac{1}{2}$ cubic feet per minute. The tank is 4 feet long and 2 feet wide. How fast is the water level rising?

Exercises 3.8

In Exercises 1–6, find dy/dx in two ways: first, by solving explicitly for y in terms of x and taking the derivative, and, second, by implicit differentiation. Check that the answers are equivalent.

1. $x - y^3 = 0$

2. $x^2y = 1$

3. $x^2 - y^2 = 1, y > 0$

4. $\dfrac{y+1}{y-1} = x$

5. $x^2 + 1 = e^y$

6. $\sqrt{y} = e^x$

In Exercises 7–12, find dy/dx in terms of x and y by implicit differentiation.

7. $x^2 + xy + y^2 = 4$

8. $x^3y - xy^3 = 1$

9. $e^y - e^{-y} = 2x$

10. $y^2 + ye^x + e^{2x} = 3$

11. $y + \ln y = x^2$

12. $e^{y^2} - x^2 - y^2 = 0$

13. The curve in Figure 3.8.4 (called an ellipse) has the equation $x^2 + 4y^2 = 4$. Using implicit differentiation, find
(a) the slope of the tangent line at the point $(1, \sqrt{3}/2)$
(b) all points on the curve where the slope of the tangent line equals $\frac{1}{2}$

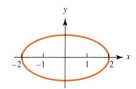

Figure 3.8.4

14. The curve in Figure 3.8.5 (called a hyperbola) has the equation $y^2 - x^2 = 1$. Using implicit differentiation, find
(a) the slope of the tangent line at the point $(2, \sqrt{5})$
(b) all points on the curve where the slope of the tangent line equals $\frac{1}{3}$.

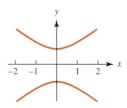

Figure 3.8.5

15. (a) Using implicit differentiation, show that at any point of the circle of radius r centered at the origin the tangent line has slope $-x/y$.
(b) Verify the following theorem of elementary geometry: The radial and tangent lines at any point of a circle are perpendicular (see Figure 3.8.6). (*Hint:* Recall that two nonvertical lines are perpendicular if and only if the product of their slopes equals -1.)

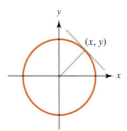

Figure 3.8.6

Exercises 16–18 refer to the curve shown in Figure 3.8.7. It is called a lemniscate and its equation is

$$(x^2 + y^2)^2 = x^2 - y^2.$$

16. Find a formula for the slope of the tangent line as a function of x and y.

17. Find the slope of the tangent line at the point $(\sqrt{3/8}, \sqrt{1/8})$.

18. Find all points where the tangent line is horizontal.

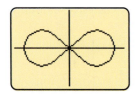

Figure 3.8.7

19. (a) Show that the curves defined by the equations $y^2 - x^2 = 3$ and $xy = 2$ intersect at the points $(1, 2)$ and

$(-1, -2)$, and their tangent lines are perpendicular at each of those points.
(b) Given any pair of curves $y^2 - x^2 = a$ and $xy = b$, where a and b are positive constants, show that their tangents are perpendicular at any point where they intersect.

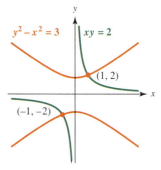

20. Based on studying fossil skeletons, biologists concluded that for a certain marine vertebrate the skull length S and backbone length B satisfy an equation of the form $S = AB^k$, where A and k are positive constants.
(a) Both S and B are functions of the individual's age t in years. Write an eqation linking the growth rates of the two quantities.
(b) Show that the *relative* growth rates (defined as the rate of growth of a function divided by its value) satisfy the equation $\dfrac{dS/dt}{S} = k\dfrac{dB/dt}{B}$.
(c) In many vertebrates, newborns have oversized heads relative to their bodies, and the skulls grow more slowly. In that case, what can you conclude about the constant k?

21. The demand function for a certain petroleum byproduct is given by

$$q = 5000e^{-0.08p},$$

where p is the price per barrel (in dollars) and q is the demand (in barrels). Suppose that the unit price is \$10 and is decreasing at the rate of 50 cents per day. At what rate is the demand changing?

In Problems 22–24, use the formula for the volume of a cylinder, $V = \pi r^2 h$, where h is the height, r is the base radius, and $\pi \approx 3.14$.

22. Suppose a cylindrical tank, whose base is a circle of radius 1.5 feet, is filling with water at the rate of 0.25 cubic feet per minute. How fast is the water level rising?

23. Water is leaking out of a cylindrical tank, whose base is a circle of radius 0.5 feet, at the rate of 0.1 cubic feet per minute. Find dh/dt.

24. The water level in a cylindrical tank with a base radius of 5 feet is rising at a rate of 3 inches per hour. At what rate (in cubic feet per hour) is water entering the tank?

25. Suppose that a circular metal disc is expanding uniformly upon being heated. If the radius increases at the rate of

0.02 centimeters per second, how fast is the circumference increasing? (Recall that the circumference is related to the radius by the formula $c = 2\pi r$.)

26. If you place a solution of grape juice and yeast in a bottle kept at a fixed temperature, say, 80°F, after about half an hour you will see changes taking place in the bottle. Carbon dioxide gas is formed, which you can collect by attaching an empty balloon to the bottle (see Figure 3.8.8). This experiment was done and the following data were collected:

time t	0.5	1.0	1.5	2.0	2.5
circumference C	16.8	27.2	32.7	36.6	38.6
time t	3.0	3.5	4.0	4.5	
circumference C	39.5	40.8	41.3	42	

Source: Eirene G. Alexandrou, science project 2001.

where t is in hours and C (the circumference of the balloon) is in centimeters.
(a) Estimate the rate of change of C at $t = 3$.
(b) Estimate the rate of change of V at $t = 3$.

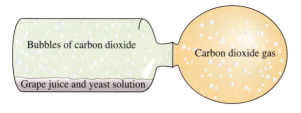

Figure 3.8.8

27. A boat is being pulled toward a pier by means of a rope attached to a winch (see Figure 3.8.9). Suppose the winch is 4 feet higher than the boat deck, and the rope is being reeled in at a steady rate of 2 feet per minute. How fast is the boat approaching the pier when it is 12 feet away?

Figure 3.8.9

28. If a gas is kept at a constant temperature in an engine cylinder, its pressure and volume are related by an equation of the form $P = k/V$, where k is a positive constant.

(a) Use the chain rule to write a formula relating dP/dt (the rate at which the pressure is changing) and dV/dt (the rate at which the volume is changing).
(b) If the volume is increasing, is the pressure increasing or decreasing? Explain your answer.

29. If a pebble is dropped into a pail of water, it creates an expanding circular ripple. The radius and area of the circle are both functions of time, and they are related to each other by the formula

$$A = \pi r^2.$$

(a) Use the chain rule to write a formula relating dA/dt (the rate at which the area is increasing) to dr/dt (the rate at which the radius is increasing).
(b) Suppose that at a certain moment the radius is 3 centimeters and is increasing at a rate of 0.2 centimeters per second. What is the area at that moment and how fast is it increasing?

30. Suppose that air is being pumped into a spherical balloon, so that it is expanding. The surface area of the balloon and the volume of air inside it are both functions of time. They are related to each other by the formula

$$S = kV^{2/3},$$

where V is the volume, S is the surface area, and k is a constant that is approximately 4.836. (Strictly speaking, $k = \sqrt[3]{36\pi}$, but for the purposes of this problem you can ignore that.)
(a) Use the chain rule to write a formula relating dS/dt (the rate at which the surface area is increasing) and dV/dt (the rate at which the volume is increasing).
(b) Suppose that at a certain moment the volume of air inside the balloon is 270 cubic inches and is increasing at a rate of 18 cubic inches per second. What is the surface area at that moment and how fast is it increasing?

31. Two cars start from the same intersection at the same time. Car A heads east at a constant speed of 40 miles per hour, and car B heads north at a constant speed of 30 miles per hour. (See Figure 3.8.10.)
(a) How far apart are they at the end of 1 hour?
(b) How fast is the distance between them changing at that time?

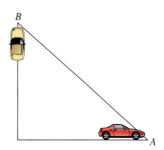

Figure 3.8.10

Solutions to practice exercises 3.8

1. Taking the derivative with respect to x on both sides and using the product rule for $x^2 y$ and the chain rule for y^4 give

$$x^2 \frac{dy}{dx} + 2xy + 4y^3 \frac{dy}{dx} - 1 = 0.$$

By collecting terms, factoring, and solving for dy/dx, we get

$$\frac{dy}{dx} = \frac{1 - 2xy}{x^2 + 4y^3}.$$

Substituting $x = 2$ and $y = 1$ gives

$$\frac{dy}{dx}\bigg|_{x=2} = \frac{1-4}{4+4} = -\frac{3}{8}.$$

2. When the water is h feet high, the volume of water is $8h$ cubic feet—that is, 4 feet long by 2 feet wide by h feet high. In other words,

$$V = 8h,$$

where V is the volume of water. Both V and h are functions of t, and taking derivatives on both sides gives

$$\frac{dV}{dt} = 8 \frac{dh}{dt}. \tag{60}$$

We can interpret these derivatives as follows: dh/dt is the rate at which the height is changing—that is, the rate at which the water is rising, which is what we are trying to find. Second, dV/dt is the rate at which the volume is changing, which is the same as the rate at which the water is entering the tank. According to the given data, $dV/dt = 1/2$. (Notice that this derivative is positive because the volume of water is increasing.) Substituting this into Eq. (60) and solving for h give

$$\frac{dh}{dt} = \frac{1}{16} \text{ feet per minute,}$$

which answers the question.

Chapter 3 Summary

- The **derivative** of a function $f(x)$ is a new function that gives the **slope** of the tangent line to the graph at every point. It is the following limit of the **difference quotient**:

$$f'(x) = \lim_{h \to 0} \frac{f(x+h) - f(x)}{h},$$

provided this limit exists. If it does, the function is said to be **differentiable** at x.

- The equation of the **tangent line** at $(a, f(a))$ is $y = f(a) + f'(a)(x - a)$.

- And the **linear approximation** is $f(x) \approx f(a) + f'(a)(x - a)$.

- The basic differentiation formulas are:

$$\frac{d}{dx} x^k = kx^{k-1}, \quad \frac{d}{dx} \ln x = \frac{1}{x}, \quad \frac{d}{dx} e^x = e^x.$$

- To compute derivatives of more complicated functions, we use these rules:

  1. *Constant multiple rule:* $[cf(x)]' = cf'(x)$.
  2. *Sum rule:* $[f(x) + g(x)]' = f'(x) + g'(x)$.
  3. *Product rule:* $\frac{d}{dx}[f(x) \cdot g(x)] = f'(x) \cdot g(x) + f(x) \cdot g'(x)$.
  4. *Quotient rule* $\frac{d}{dx}\left[\frac{f(x)}{g(x)}\right] = \frac{g(x)f'(x) - f(x)g'(x)}{[g(x)]^2}$.

5. *Chain rule:* $\dfrac{d}{dx} f[g(x)] = f'[g(x)] \cdot g'(x)$.

- The interpretation of the derivative as a *rate* of change makes it possible to model: motion, where **velocity** is the derivative of the position and **acceleration** is the derivative of the velocity; and economic quantities, where **marginal** cost, revenue, and profit are the derivatives of cost, revenue, and profit. Also using the derivative, we can model **exponential growth** and **decay**.

Chapter 3 Review Questions

- Explain why the derivative of a constant function is zero.

- Explain in your own words how you would determine the slope of a curved graph at one of its points.

- What is the main idea in linear approximation?

- Explain why a continuous function may not be differentiable. Give two examples.

- How you would estimate $\sqrt{224}$ without a calculator?

Chapter 3 Review Exercises

In Exercises 1–4, use the definition of the derivative and no other derivative formula to find $f'(x)$.

1. $f(x) = x^2 - 3x + 2$
2. $f(x) = 2x^3 - x$
3. $f(x) = \dfrac{2}{3x - 1}$
4. $f(x) = \sqrt{3x}$

In Exercises 5–12, find the slope of the graph and the equation of the tangent line at the given point. You may use any derivative rules you have learned.

5. $y = \sqrt{x}$ at $(9, 3)$

6. $y = \dfrac{1}{x^3}$ at $x = 2$

7. $y = x^2 + \dfrac{4}{x^2}$ at $(2, 5)$

8. $y = 1 + \ln x$ at $x = e$

9. $y = \dfrac{x}{x + 1}$ at $x = 1$

10. $y = \sqrt{x^2 + x + 2}$ at $(1, 2)$

11. $y = e^{x^2 - 1}$ at $x = 1$

12. $y = (2 - x^2)^{30}$ at $(1, 1)$

In Exercises 12–25, find $f'(x)$ using any rule you have learned.

13. $f(x) = x^5 - 3x^4 + 2x^3 + x^2 - 5x + 7$

14. $f(x) = \dfrac{x^2}{3} + \dfrac{3}{x^2}$

15. $f(x) = e^{x^2 - 3x + 1}$
16. $f(x) = x^5 + 5^x$

17. $f(x) = \sqrt[3]{3x + 1}$
18. $f(x) = \ln(4x^3)$

19. $f(x) = (\ln x)^4$
20. $f(x) = \dfrac{x + 2}{2x + 3}$

21. $f(x) = \dfrac{e^x + 1}{e^x}$
22. $f(x) = x\sqrt{x^2 + 1}$

23. $f(x) = x^3 \ln x$
24. $y = 3^{2x + 5}$

25. $f(x) = \sqrt{e^{3x} + 1}$
26. $f(x) = \dfrac{1}{(x^2 + x + 1)^2}$

27. For what x does the graph of $y = x^2 + 3x + 5$ have slope 9?

28. For what x does the graph of $y = x + \ln x$ have slope 4?

29. For what x is the line tangent to the graph of $y = x + e^x$ horizontal?

30. At what point is the line tangent to the graph of $y = x^2 + 3x + 1$ parallel to the line $x + y = 1$?

31. For what x does the graph of $y = x^{3/2} - 1$ have slope 1?

In Exercises 32–37, find $f''(x)$.

32. $f(x) = 2x^3 - 5x^2 + 7x - 1$

33. $f(x) = \ln(x^2)$

34. $f(x) = x^2 + e^{-x/2}$ **35.** $f(x) = \sqrt{2x + 1}$

36. $f(x) = \ln x - x^2$ **37.** $f(x) = xe^{2x}$

38. Suppose a ball is thrown upward from a height of 5 feet with an initial velocity of 30 feet per second, so that its height after t seconds is given by the formula $h = -16t^2 + 30t + 5$.
  (a) Find the instantaneous velocity of the ball at the end of 0.5 seconds and at the end of 1 second. What do the signs signify?
  (b) Find the average velocity of the ball during the time period from 1 to 1.5 seconds.
  (c) Find the average velocity of the ball during the time period between t and $t + \Delta t$ seconds. Simplify your answer as much as possible.
  (d) Find the instantaneous velocity of the ball at the end of t seconds.
  (e) What is the acceleration of the ball after 1 second? after $\frac{3}{2}$ seconds? after 1.762235 seconds?

39. The distance s in feet traveled by a car during the first 10 seconds after it starts from rest is given by the formula $s = \frac{5}{2}t^3 - \frac{4}{7}t^{7/2}$, where t is the time in seconds.
  (a) What is the instantaneous velocity (in feet per second) at the end of 1 second?
  (b) What is the average velocity of the car during the first second?
  (c) What is the acceleration of the car (in ft./sec²) at the end of 4 seconds?

40. Two cars are traveling along the same toll road. Let x be the position of car X and y be the position of car Y, both relative to the western entrance of the toll road. Translate the following statements into formulas involving x and y and their derivatives with respect to t. Take time in hours and east as the positive direction.
  (a) Both cars are heading east and car X is traveling 10 miles per hour faster than car Y.
  (b) The cars are heading in opposite directions.
  (c) Car X is heading east and is slowing down.
  (d) Car Y is heading west and speeding up.

41. Suppose that the demand for a certain product is given by $q = 5,000e^{-0.08p}$, where p is the price and q is the number sold at that price.
  (a) At what rate is the demand changing with respect to the price when the price is $10?

  (b) At what rate is the revenue changing with respect to the price when the price is $10?

42. A company finds that if it produces and sells x boxes of assorted chocolates per week, its profit (in dollars) is $0.02x^2 + 15x - 1,000$. What is its marginal profit (in dollars) at a production level of 100 boxes per week?

43. A manufacturing company determines that the revenue obtained from the sale of x items is given by $R(x) = 220x - 4x^2$ and the cost of producing these items by $C(x) = 900 + 40x$. For what output level x does the marginal revenue equal the marginal cost? Is that the same as the break-even point?

44. Suppose the cost of producing x pounds of fertilizer is given by the cost function $C(x) = 0.5x^2 + 1.5x + 8$.
  (a) Find the cost of producing 10 pounds.
  (b) Find the marginal cost for $x = 10$.
  (c) Use linear approximation to estimate the extra cost in going from 10 to 11 pounds.

45. Use linear approximation with the function $f(x) = \sqrt{x}$ to estimate:
  (a) $\sqrt{50}$ **(b)** $\sqrt{83}$ **(c)** $\sqrt{98}$

In Exercises 46–49, use linear approximation with an appropriate choice of function (and without a calculator) to estimate the given quantity.

46. $(8.1)^{1/3}$ **47.** $\dfrac{1}{2.108}$

48. $(1.027)^{12}$ **49.** $\ln(1.005)$

50. A hard-boiled egg at 98°C is put into a sink of water that is kept at a steady temperature of 18°C. At the end of 5 minutes, the egg's temperature is found to be 58°C. What is the egg's temperature 10 minutes after that?

51. Charcoal from a tree killed in the volcanic eruption that formed Crater Lake in Oregon contained 44.5% of the concentration of radioactive carbon found in a living sample of the same wood. How long ago did the volcanic eruption take place?

In Exercises 52–57, use the following table of values to find the derivative of the given function at the specified point:

x	$f(x)$	$g(x)$	$f'(x)$	$g'(x)$
1	2	3	$-1/2$	-2
2	5	$-1/3$	1	$3/5$
3	-1	2	$1/4$	$3/2$

52. $f(x) \cdot g(x)$ at $x = 3$ **53.** $f[g(x)]$ at $x = 3$

54. $g(x^3)$ at $x = 1$ **55.** $\dfrac{f(x)}{g(x)}$ at $x = 2$

56. $\ln[f(x) + g(x)]$ at $x = 3$

57. $\dfrac{1}{f(x)}$ at $x = 1$

58. If $f(0) = 4$ and $f'(0) = 2$, find the equation of the line tangent to the graph of $y = \sqrt{x + f(x)}$ at $(0, 2)$.

59. The graph of f is given in Figure 3.R.1.
(a) At what x is f discontinuous? Explain.
(b) At what x is f not differentiable? Explain.

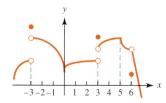

Figure 3.R.1

In Exercises 60–63, find dy/dx in terms of x and y.

60. $x^3 y + y^3 + x = 1$

61. $x^2 = \ln(1 + y^2)$

62. $x^2 + y^2 = xy + 1$

63. $xe^y - x^2 y = x^2 + 1$

64. Find the equation of the line tangent to the curve $5x^2 - y^2 = 1$ at the point $(1, 2)$.

65. Show that $y(t) = ce^{-t^2}$ is a solution of the differential equation $y' = -2ty$, where c is a given number.

66. Show that $y = 3/(1 + ce^{-3t})$ is a solution of the differential equation $y' = y(3 - y)$, where c is a given number.

67. Using implicit differentiation show that the equation $y^2 + t^2 = c^2$ implicitly defines solutions to the differential equation $yy' + t = 0$.

68. Show that $y(t) = 3e^t - 2$ is a solution to the differential equation $y' = y + 2$ satisfying the initial condition $y(0) = 1$.

69. Show that $y(t) = (1 + e^{-t})^{-1}$ is a solution to the differential equation $y' = y(1 - y)$ satisfying the initial condition $y(0) = \frac{1}{2}$.

70. Show that $A(t) = 50(e^{0.08t} - 1)$ is a solution to the differential equation $A' = 0.08A + 4$ satisfying the initial condition $A(0) = 0$.

71. Show that $y(t) = \frac{1}{3}t^2 + (5/3t)$ is a solution to the differential equation $ty' + y = t^2$ satisfying the initial condition $y(1) = 2$.

72. The displacement of an object moving on a straight line is given by $s(t) = 2 + 3t + \frac{1}{2}t^2$, where t is measured in seconds and s is measured in meters.
(a) Find its average velocity over the time interval $[2, 4]$.
(b) Find its velocity at $t = 2$.
(c) Find its acceleration at $t = 2$.

73. Sketch by hand the functions
$$f(x) = \begin{cases} x^2 & \text{if } x \geq 0 \\ x^3 & \text{if } x < 0 \end{cases} \quad \text{and} \quad g(x) = \begin{cases} x & \text{if } x \geq 0 \\ x^2 & \text{if } x < 0 \end{cases}$$
and decide if they are differentiable.

74. Determine m so that the line $y = mx - 1$ is tangent to the parabola $y = x^2$.

75. The volume and surface area of a sphere are both functions of the radius, given by the formulas $V = \frac{4}{3}\pi r^3$ and $S = 4\pi r^2$, where r is the radius, V is the volume, and S is the surface area. Suppose a spherical balloon is being inflated so that its volume is increasing at the rate of 1 cubic inch per second.
(a) How fast is the radius increasing at the moment when $r = 2$?
(b) How fast is the surface area increasing at that same moment?

Chapter 3 Practice Exam

1. If $f(x) = \sqrt{2x + 1}$, which (if any) of the following limits equals $f'(4)$?
(a) $\displaystyle\lim_{h \to 0} \dfrac{\sqrt{4 + h} - 2}{h}$

(b) $\displaystyle\lim_{h \to 0} \dfrac{1}{h}\left\{ \dfrac{1}{\sqrt{4 + h}} - \dfrac{1}{3} \right\}$

(c) $\displaystyle\lim_{h \to 0} \dfrac{\sqrt{9 + 2h} - 3}{h}$ (d) $\displaystyle\lim_{h \to 0} \dfrac{\sqrt{2h + 1} - 3}{h - 4}$

2. Consider the function $f(x) = 1/(2x + 5)$. Compute $f'(x)$ using the definition of the derivative and no other method.

3. Find the derivative of the function $f(x) = 5 + 2\sqrt{x} - (3/x^3)$.

4. Find the equation of the line tangent to the graph of $y = 1/\sqrt{x}$ at the point $(4, \frac{1}{2})$.

5. Find all the points where the graph of $y = 4x^3 - 5x + 1$ has slope 43.

6. A particle is moving in the positive direction along the x-axis so that its position on the axis at t seconds is given by the formula $s(t) = 1 + 1/(t + 1)$.
(a) Find its instantaneous velocity at $t = 0$ and $t = 2$.

(b) Find its average velocity between $t = 0$ and $t = 2$.
(c) Find its acceleration at $t = 1$.

7. A colony of bacteria is dying. It starts with an initial population of 10,000 organisms, and after t hours the population is given by the formula

$$P = 10,000e^{-0.04t}$$

How fast is the population changing at the end of 100 hours?

8. The monthly revenue of a small company is given by $R(x) = x^3 \ln x$, where x is the number of units sold. What is the marginal revenue during a month when they sell 10 units?

9. When linear approximation is used to estimate $\sqrt[3]{1,000 + h}$, what is the result?

10. A souvenir company sells 10,000 Beanie Babies per day and is operating at a net profit of $5,200 per day. Its marginal revenue is $1.40 and its marginal cost is $1.25. Estimate how much its profit per day will increase if it increases sales by 200 Beanie Babies per day.

11. Which of the following statements is true about the function below?

$$f(x) = |x - 2|$$

(a) It is continuous at $x = 2$ but has no derivative there.
(b) It has a derivative at $x = 2$ but is not continuous there.
(c) It is continuous at $x = 2$ and has a derivative there.
(d) It is not continuous at $x = 2$ and does not have a derivative there.

12. The graph of f is given in Figure 3.E.1.
(a) At what x is f discontinuous? Explain.
(b) At what x is f not differentiable? Explain.

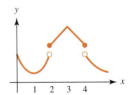

Figure 3.E.1

13. By computing the limits from the left and from the right of the difference quotient of the function $f(x) = x^{4/5}$, decide whether it is differentiable at $x = 0$.

14. Compute the derivative of $f(x) = e^{x^2+1}$.

15. Find the equation of the line tangent to the graph of $y = x^2 \ln x$ at $x = 1$.

16. Show that $y(t) = 1/(1 - t)$ is a solution to the differential equation $y' = y^2$.

17. Show that $y(t) = ce^{-t} + 2t - 2$ is a solution to the differential equation $y' = 2t - y$, where c is a constant.

18. Show that $y(x) = \ln(e^x + e)$ is a solution to the differential equation $y' = e^{x-y}$ satisfying $y(1) = 1 + \ln 2$.

19. A hamburger was left on a table in a room maintained at 20°C. At the end of 15 minutes, the temperature of the burger had dropped to 48 degrees and was decreasing at the rate of 3.5 degrees per minute. What was the initial temperature of the hamburger?

20. Find the slope of the tangent to the graph of $x^3 + y^3 = xy + 1$ at the point $(1, -1)$.

21. The volume of a sphere of radius r is given by the formula $V = \frac{4}{3}\pi r^3$. When helium is pumped into a spherical balloon, both the radius and the volume change with respect to time. Suppose that at a certain moment the volume is increasing at the rate of 8π cubic feet per minute, and the radius is increasing at the rate of $\frac{1}{2}$ feet per minute. What is the volume of the balloon at that moment?

22. Assume that the demand function for a product is

$$p = \frac{6}{0.01q^2 + 0.1q + 1},$$

where p is measured in dollars and q in thousands of units. Using implicit differentiation, find the rate of change of the quantity q with respect to price p when $q = 10$.

23. Show that the equation $x^3 + xy - y^3 = c$, where c is a constant, implicitly defines solutions to the differential equation

$$\frac{dy}{dx} = \frac{3x^2 + y}{3y^2 - x}.$$

24. An approximate value in billions of the world population t years after 1990 is given by

$$p(t) = \frac{10}{1 + e^{-0.03t}}.$$

(a) Show that $p(t)$ satisfies the **logistic** equation $dp/dt = 0.03p(1 - 0.1p)$.
(b) What was the population in 1990?
(c) What will the population be in 2100?
(d) Find the $\lim_{t \to \infty} p(t)$.

1. **Newton-Raphson method**   Finding an exact solution of an equation $f(x) = 0$ is often impossible, and we need to use approximation methods. The **Newton-Raphson** method, which is the one most widely used, is based on the idea of linear approximation—replacing a graph with its tangent line. The method is illustrated by the graph in Figure 3.P.1.

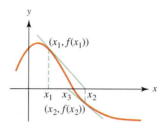

Figure 3.P.1

We start with a first guess at a solution and label it x_1. One way to choose it is to find an interval $[a, b]$ with $f(a) \cdot f(b) < 0$ and take x_1 to be a point on the interval.

Step 1   Write the equation of the tangent line to the graph of f at $x = x_1$ in the form $y = L(x)$, where $L(x)$ is a linear function. Linear approximation says that $f(x) \approx L(x)$ for x near x_1.

Step 2   Solve the equation $L(x) = 0$.

Let x_2 be the solution obtained above. If you completed the steps correctly, you should have

$$x_2 = x_1 - \frac{f(x_1)}{f'(x_1)}. \tag{61}$$

[In order for the method to work, we must have $f'(x_1) \neq 0$.] Because of the approximation $f(x) \approx L(x)$, we take x_1 as a first approximation to the solution of $f(x) = 0$.

To improve the approximation, we repeat steps 1 and 2, with x_2 in place of x_1 (as illustrated in Figure 3.P.1). Call the result x_3, so that

$$x_3 = x_2 - \frac{f(x_2)}{f'(x_2)}.$$

If we keep repeating this procedure, we generate a sequence of numbers $x_1, x_2, x_3, x_4, \ldots$ that tend toward a solution of $f(x) = 0$.

Questions   (a) Use the Newton-Raphson method with $x_1 = 1$ to find an approximate solution of

$$x^6 - 2x - 1 = 0.$$

Use a calculator and stop when you reach x_6. Check your answer by putting it into the left-hand side of the equation and seeing how close the result is to zero.

(b) Let $f(x) = e^x - 3x$. Find an interval $[a, b]$ with $f(a) > 0$ and $f(b) < 0$, and take x_1 to be the midpoint of the interval. Then use the Newton-Raphson method, stopping as soon as two successive iterations agree to six decimal places. Check your answer by evaluating f to see how close it comes to zero.

2. **Unmasking art forgeries**   White lead, also known as lead-210, is a radioactive substance that has been used by artists for many centuries as a pigment in painting. In the mid-twentieth century it was used to decide whether several paintings attributed to the Dutch master Jan Vermeer (1632–1675) were actually forgeries. Shortly after World War II, a minor Dutch painter named H. A. Van Meegeren was arrested for wartime collaboration with the Nazi occupation forces in Holland. He was accused of having supplied a number of priceless art works, among them a painting by Vermeer, to Hitler's second-in-command, Hermann Goering. Van Meegeren's surprising defense was that the painting was not genuine, but a forgery he had done himself! He even claimed to have forged other paintings attributed to Vermeer. Expert opinion was sharply divided on whether Van Meegeren was telling the truth, even after the paintings were subjected to x-ray and chemical examination. It was not until 1967 that conclusive proof was provided by the method of radioactive dating. Here is how it was done.

White lead, which has a half-life of 22 years, is created in the earth as a byproduct of the decay of radium-226, whose half-life is 1,600 years. While it is in the earth, it remains in a state of equilibrium—the amount that decays is replenished by the amount of radium decaying. When it is mined and manufactured as pigment, however, most of the radium is removed, and the lead decays faster than it can be replenished by the small amount of radium remaining. On that basis, scientists tried to date the paintings by the decay of their white lead pigment.

The amount of lead-210 in a sample t years after it was mined is the solution of the differential equation

$$\frac{dy}{dt} = -ky + r, \tag{62}$$

where k is the decay constant for lead-210, and r is the rate at which the small amount of radium-226 in the sample decays into lead-210. (Strictly speaking, r is a function of t, but we treat it as a constant because the half-life of radium is much longer than the time span we are checking.)

Task 1   Show that

$$y = \frac{r}{k}\left(1 - e^{-kt}\right) + Ae^{-kt} \tag{63}$$

is a solution of Eq. (62), with A being the amount of lead-210 in the sample at the time of manufacture.

Task 2   Solve Eq. (63) for A. Explain the basis for the following conclusion: The larger A is, the more recent the sample is.

It is not hard to measure y and r, but difficult to measure A. After a number of experiments with ore samples, experts concluded that a value of A greater than 10^6 would establish with certainty that a painting was less than 300 years old. The number was set very high to rule out experimental error and variation in different ore samples.

Task 3 The following table gives values for y and r measured in several paintings. Using these values with $t = 300$ and k equal to the decay constant of lead-210, compute A and determine whether each painting is a forgery.

Title of Painting	y	r
Disciples at Emmaus	270	0.8
Washing of Feet	400	0.26
Woman Reading Music	330	0.3
Woman Playing Mandolin	260	0.17

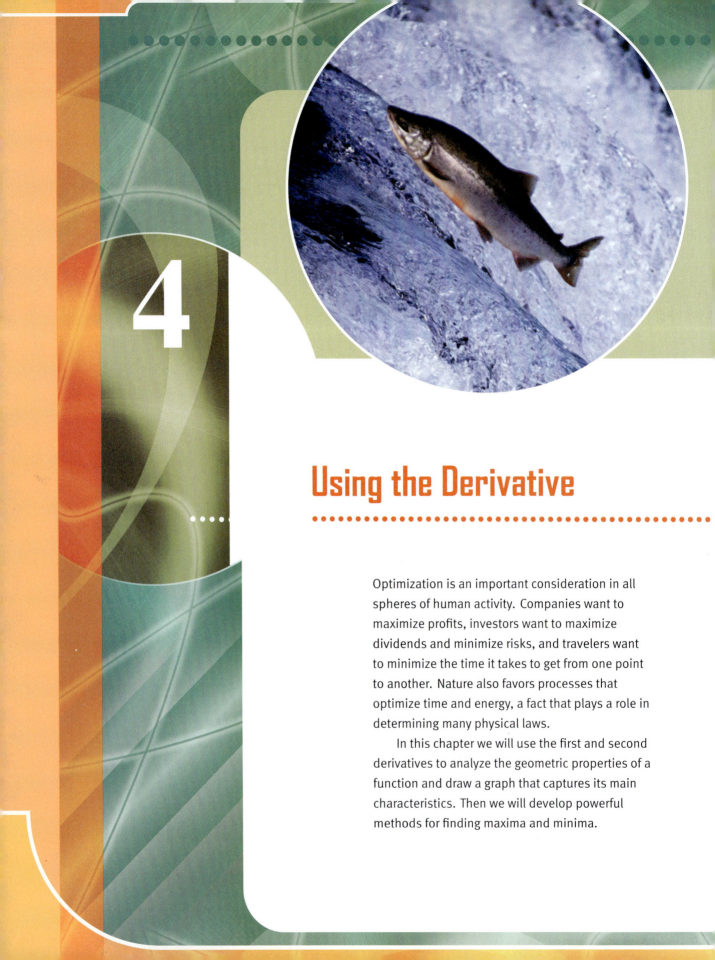

Using the Derivative

Optimization is an important consideration in all spheres of human activity. Companies want to maximize profits, investors want to maximize dividends and minimize risks, and travelers want to minimize the time it takes to get from one point to another. Nature also favors processes that optimize time and energy, a fact that plays a role in determining many physical laws.

In this chapter we will use the first and second derivatives to analyze the geometric properties of a function and draw a graph that captures its main characteristics. Then we will develop powerful methods for finding maxima and minima.

Here we shall use the information provided by the **first derivative** to find where a function is increasing and where it is decreasing, and to find the location of maxima and minima.

The derivative test for increasing and decreasing functions

The derivative measures the *slope* of the graph, and its sign indicates whether the graph is climbing or falling. Recall the case of a line:

- if the slope is positive, the line goes uphill to the right;
- if the slope is negative, the line goes downhill to the right.

The same is true for the graph of a nonlinear function. If such a function, say, $f(x)$, is differentiable in an interval, then near any point a of that interval the function is almost equal to its linear approximation. That is,

$$f(x) \approx L(x) = f(a) + f'(a)(x - a).$$

Therefore, $f(x)$ is increasing (or decreasing) near a if the slope $f'(a)$ is positive (or negative). This convincing intuitive argument can be turned into a rigorous proof of the following test.

Theorem 4.1.1

Derivative test for increasing and decreasing functions

- If $f'(x) > 0$ on an interval, then $f(x)$ is increasing on that interval;
- If $f'(x) < 0$ on an interval, then $f(x)$ is decreasing on that interval.

Figure 4.1.1 shows the graph of a function over four successive intervals, increasing on the first and third, decreasing on the second and fourth. The derivative (the slope of the graph) has the corresponding signs.

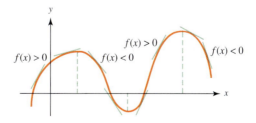

Figure 4.1.1

Next, we present a familiar example that illustrates the above test.

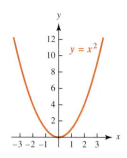

Figure 4.1.2

Determine where the function $f(x) = x^2$ is increasing and where it is decreasing.

SOLUTION   The derivative, $f'(x) = 2x$, is negative on the interval $(-\infty, 0)$ and positive on the interval $(0, \infty)$. Therefore, the function $f(x) = x^2$ is decreasing on $(-\infty, 0)$, and it is increasing on $(0, \infty)$, as shown in Figure 4.1.2.

Determining the sign of $f'(x)$

We now have a method for determining where a differentiable function is increasing and where it is decreasing: *Take its derivative and determine where it is positive and where it is negative.* Assuming that the derivative is continuous, we use the following simple procedure to determine the sign of $f'(x)$.

Method for determining the sign of $f'(x)$

1.   Find the points where $f'(x)$ equals zero; in other words, solve the equation $f'(x) = 0$.

2.   Mark off these points on a number axis. They divide the domain of f into a certain number of intervals, some of which may have infinite length.

3.   On each of these intervals the sign of $f'(x)$ stays the same. In other words, either $f'(x) > 0$ throughout the interval or else $f'(x) < 0$ throughout the interval. So choose one point inside each interval with which to test the sign of $f'(x)$ throughout that interval.

This procedure is based on the intermediate value theorem (see Theorem 1.3.2 in Chapter 1): *If a continuous function changes from positive to negative or vice versa on an interval, it must be zero at some point inside the interval.* Therefore, once we plot the points where the derivative is zero, we have marked off the only places where it can change sign (assuming that it changes continuously).

Determine where the function $f(x) = x^3 - 3x^2 + 3$ is increasing and where it is decreasing.

SOLUTION   First, we compute the derivative and set it equal to zero:

$$f'(x) = 3x^2 - 6x = 0.$$

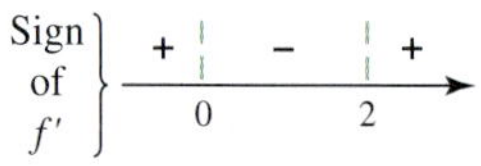

Figure 4.1.3

The solutions of this equation are $x = 0$ and $x = 2$. They are marked in Figure 4.1.3 and divide the x-axis into three intervals—namely, $(-\infty, 0)$, $(0, 2)$, and $(2, \infty)$—on each of which the sign of $f'(x)$ stays the same.

That means that we only have to test one point in each interval. For instance, we can use $x = -1$ as a test point in $(-\infty, 0)$. Plugging it into $f'(x)$ gives $f'(-1) = 9 > 0$, so that $f'(x) > 0$ on $(-\infty, 0)$.

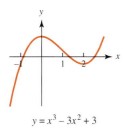

$y = x^3 - 3x^2 + 3$

Figure 4.1.4

Similarly, using $x = 1$ as a test point in $(0, 2)$ gives $f'(1) = -3 < 0$; and using 3 as a test point on $(2, \infty)$ gives $f'(3) = 9 > 0$. It follows that $f'(x) < 0$ throughout $(0, 2)$ and $f'(x) > 0$ throughout $(2, \infty)$.

The sign of $f'(x)$ is indicated in Figure 4.1.3. We conclude: $f(x) = x^3 - 3x^2 + 3$ is increasing on the intervals $(-\infty, 0)$ and $(2, \infty)$, and it is decreasing on $(0, 2)$. You can see these properties in the graph, shown in Figure 4.1.4.

To repeat: The places where $f'(x) = 0$ are the only ones where $f'(x)$ can change sign. However, that does not mean that $f'(x)$ *has* to change sign at those points—as the next example shows.

EXAMPLE 4.1.3

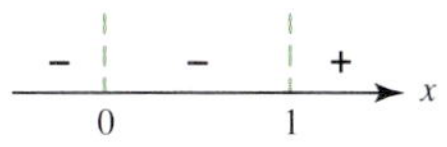

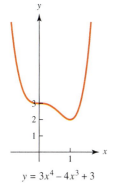

Figure 4.1.5

Determine where $f(x) = 3x^4 - 4x^3 + 3$ is increasing and where it is decreasing.

SOLUTION   Taking the derivative and setting it equal to zero give

$$12x^3 - 12x^2 = 12x^2(x - 1) = 0.$$

The solutions are $x = 0$ and $x = 1$, which are plotted on the x-axis in Figure 4.1.5. To determine the sign of $f'(x)$, we check one point on each interval. For instance,

- on $(-\infty, 0)$ we can use $x = -1$, which gives $f'(-1) = -24 < 0$;
- on $(0, 1)$ we can use $x = \frac{1}{2}$, which gives $f'(\frac{1}{2}) = -\frac{3}{2} < 0$;
- on $(1, \infty)$ we can use $x = 2$, which gives $f'(2) = 48 > 0$.

$y = 3x^4 - 4x^3 + 3$

Figure 4.1.6

The sign of $f'(x)$ is indicated in Figure 4.1.5. As you can see, it does not change at $x = 0$, even though $f'(0) = 0$.

In this case we conclude: $f(x)$ is increasing on $(1, \infty)$ and decreasing on $(-\infty, 1)$. The graph is shown in Figure 4.1.6.

What if the equation $f'(x) = 0$ has *no* solutions? In that case, the sign of $f'(x)$ cannot change—it is either always positive or always negative—and $f(x)$ is either increasing throughout or decreasing throughout.

EXAMPLE 4.1.4

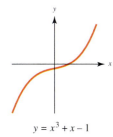

$y = x^3 + x - 1$

Figure 4.1.7

Determine where $f(x) = x^3 + x - 1$ is increasing and where it is decreasing.

SOLUTION   In this case, $f'(x) = 3x^2 + 1$, which is never zero. (Otherwise, there would be an x satisfying $3x^2 = -1$, which is impossible.) That means that $f'(x)$ never changes sign, and it is easy to see that it is always positive. (Test $x = 0$, for instance.) We conclude that $f(x)$ is always increasing. The graph is shown in Figure 4.1.7.

EXAMPLE 4.1.5

A company's cost for producing x calculators per week is given by the function

$$C(x) = -0.00001x^2 + x + 400, \quad 0 \le x \le 40{,}000,$$

where C is measured in hundreds of dollars. Show that its average cost, defined by $C(x)/x$, is a decreasing function.

SOLUTION   Let $A(x) = C(x)/x$ be the average cost. Then

$$A(x) = -0.00001x + 1 + \frac{400}{x}.$$

Taking its derivative gives

$$A'(x) = -0.00001 - \frac{400}{x^2}.$$

Since $A'(x) < 0$, we concude that the average cost $A(x)$ is a decreasing function. The graph is shown in Figure 4.1.8.

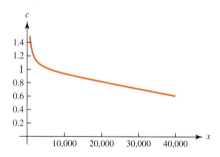

Figure 4.1.8

Critical points and vertical asymptotes

The procedure we have been using is based on the idea that the sign of $f'(x)$ can only change at a place where $f'(x)$ is zero. However, that principle is only valid if $f'(x)$ is continuous. That is true for most of the functions we will study in this course, but there are some exceptions.

EXAMPLE 4.1.6

Determine where $f(x) = x^{2/3}$ is increasing and where it is decreasing.

SOLUTION   The derivative is

$$f'(x) = \frac{2}{3}x^{-1/3} = \frac{2}{3x^{1/3}},$$

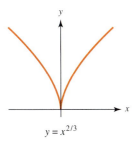

$y = x^{2/3}$

Figure 4.1.9

which is never zero, and we might be tempted to conclude that $f'(x)$ never changes sign. But we see that $f'(x)$ is negative for $x < 0$ and positive for $x > 0$. Therefore, $f(x)$ is decreasing on $(-\infty, 0)$ and increasing on $(0, \infty)$ as shown in Figure 4.1.9. The explanation is this: *The derivative is not defined at zero*, even though zero is in the domain of the function. The variable x has a positive exponent in the formula for $f(x)$, but a negative exponent in the formula for $f'(x)$. There is a discontinuity in the derivative at $x = 0$ that allows the sign to "jump" from negative to positive. In other words, we have a sharp corner (called a **cusp**) at the point $(0, 0)$.

This last example shows that we have to amend our procedure to take account of places where $f'(x)$ may not exist and to allow the possibility that the sign of $f'(x)$ may change sign at such points.

Definition 4.1.1

Critical point

A point c in the domain of f is a **critical point** if either

- $f'(c) = 0$ or
- $f'(c)$ does not exist.

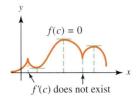

$f(c) = 0$

$f'(c)$ does not exist

Figure 4.1.10

The graph pictured in Figure 4.1.10 has five critical points—three at which the derivative is zero and two at which it does not exist. The sign of $f'(x)$ changes at each of them.

Critical points are the only places in the domain of f where the sign of $f'(x)$ can change. However, there is still one other case in which the sign of $f'(x)$ may change—at a point where the graph has a **vertical asymptote**. Such a point is *not* in the domain of f, but you sometimes have to be careful not to overlook it when studying the increasing or decreasing behavior of the graph.

EXAMPLE 4.1.7

Determine where $f(x) = x^{-2}$ is increasing and where it is decreasing.

SOLUTION   In this case, the derivative, $f'(x) = -2x^{-3}$, is positive if $x < 0$ and negative if $x > 0$. Therefore, $f(x) = x^{-2}$ is increasing on $(-\infty, 0)$ and decreasing on $(0, \infty)$. The graph is shown in Figure 4.1.11. As you can see, the point $x = 0$ is not in the domain of the function, and the vertical line through that point is a vertical asymptote.

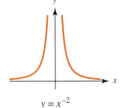

$y = x^{-2}$

Figure 4.1.11

Taking these examples into account, we get the final version of our method for deciding where a function is increasing and where it is decreasing.

Test for determining when a function is increasing or decreasing (final version)

1. Find all the critical points of the function—that is, all the points in the domain where the derivative is either zero or does not exist.

2. Find any points where the graph has a vertical asymptote.

3. Plot these points on a diagram of the x-axis. They divide the domain of the function into a collection of intervals (some of which may be infinite).

4. Test one point on each interval by plugging it into the derivative. The sign of the derivative at that point is its sign throughout the interval.

5. The function is increasing on the intervals where the derivative is positive and decreasing where the derivative is negative.

Figure 4.1.12

EXAMPLE 4.1.8

Determine where $f(x) = x + 1/x$ is increasing and where it is decreasing.

SOLUTION   The domain of this function consists of all $x \neq 0$. The derivative is $f'(x) = 1 - 1/x^2$, and it is zero at $x = 1$ and $x = -1$. These are the critical points. In addition, there is a vertical asymptote at $x = 0$.

The points $x = -1, 0$, and 1 divide the x-axis into four intervals. By testing a point in each interval, we get the diagram for the sign of $f'(x)$ shown in Figure 4.1.12.

We conclude: $f(x)$ is decreasing on $(-1, 0)$ and $(0, 1)$, and it is increasing on $(-\infty, -1)$ and $(1, \infty)$. A small point: We do not say $f(x)$ is decreasing on $(-1, 1)$ because that would suggest that the function is defined at $x = 0$, which it is not. The graph is shown in Figure 4.1.13.

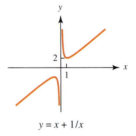

$y = x + 1/x$

Figure 4.1.13

First derivative test for maxima and minima

Definition 4.1.2

Let c be a point in the domain of a function $f(x)$. Then

• $f(x)$ has a **local maximum** (or relative maximum) at c if $f(c)$ is greater than or equal to the values $f(x)$ for all points x in an interval around c.

• $f(x)$ has a **local minimum** (or relative minimum) at c if $f(c)$ is less than or equal to the values $f(x)$ for all points x in an interval around c.

If $f(x)$ has either a local maximum or local minimum at c, then it is said to have a **local extremum** there.

A local maximum (or minimum) is not necessarily the point at which the function achieves its greatest (or smallest) value. Figure 4.1.14 shows the graph of a function $f(x)$, defined on the interval $(-2, 10)$, with three local maxima, at $x = 1$, 4, and 7,

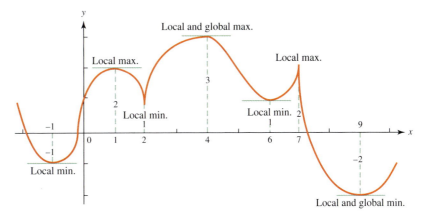

Figure 4.1.14

and four local minima, at $x = -1$, 2, 6, and 9. However, the graph reaches its highest point only at $x = 4$ and its lowest point only at $x = 9$.

We call a point at which the function achieves its greatest (or smallest) value a **global maximum** (or **minimum**). A more precise statement is given by the following:

Definition 4.1.3

Let c be a point in the domain of a function $f(x)$. Then

- $f(x)$ has a *global maximum* (or absolute maximum) at c if $f(c) \geq f(x)$ for all x in the domain of f.
- $f(x)$ has a *global minimum* (or absolute minimum) at c if $f(c) \leq f(x)$ for all x in the domain of f.

In the graph above, the global maximum is at $x = 4$, and the global minimum at $x = 9$. The illustration also indicates that a local extremum can only occur at a critical point, a fact that is important enough to be stated as a theorem.

Theorem 4.1.2

Critical points are the only candidates for extrema

If $f(x)$ has a local maximum or a local minimum at $x = c$, then c must be a critical point.

Here is the explanation. First, if $f(x)$ is not differentiable at c, then c is a critical point, by definition. On the other hand, if $f(x)$ *is* differentiable at c, then

$$f'(c) \approx \frac{f(x) - f(c)}{x - c}. \tag{1}$$

Now, suppose $f(x)$ has a local maximum at c. First, $f'(c)$ cannot be positive, for then the right-hand side of (1) would also be positive, and that would mean $f(x) < f(c)$ when $x < c$, which is not compatible with a local maximum. Second, $f'(c)$ cannot be negative, for then the right-hand side of (1) would be negative, and that would mean $f(x) < f(c)$ when $x > c$, which is also not compatible with a local maximum. Thus, the only possibility is $f'(c) = 0$. A similar argument shows that $f'(c) = 0$ if $f(x)$ has a local minimum at c.

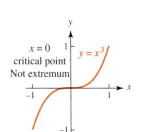

$x = 0$
critical point
Not extremum

Figure 4.1.15

Warning Critical points are **candidates** for extrema. A function may have a critical point without having an extremum there. The simplest example of such a function is $f(x) = x^3$. Since $f'(x) = 3x^2$, there is a critical point at $x = 0$. However, $f(x) = x^3$ has neither a local minimum nor a local maximum at zero, as you can clearly see in Figure 4.1.15. Indeed, the function is increasing throughout its domain.

Theorem 4.1.3

The first derivative test for maxima and minima

If $f(x)$ has a critical point at $x = c$, then

- there is a local maximum at $x = c$ if $f'(x)$ changes its sign from positive to negative, and

- there is a local minimum at $x = c$ if $f'(x)$ changes its sign from negative to positive.

These cases are shown in Figure 4.1.16.

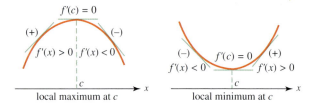

Figure 4.1.16

EXAMPLE 4.1.9

Find the local extrema of the function $f(x) = x^3 - 3x$.

SOLUTION We first take the derivative and set it equal to zero:

$$f'(x) = 3x^2 - 3 = 0.$$

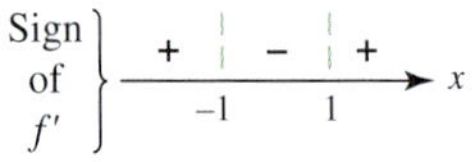

Sign
of
f'

Figure 4.1.17

Solving, we find that $f(x)$ has two critical points, at $x = -1$ and $x = 1$. These are marked on the diagram in Figure 4.1.17. They divide the x-axis into three intervals, $(-\infty, -1)$, $(-1, 1)$, and $(1, \infty)$, on each of which the sign of $f'(x)$ stays the same. That means that we only have to test one point in each interval. For instance,

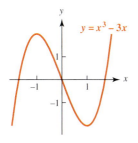

$y = x^3 - 3x$

Figure 4.1.18

- testing $x = -2$ in $(-\infty, -1)$ gives $f'(-2) = 9 > 0$,
- testing $x = 0$ in $(-1, 1)$ gives $f'(0) = -3 < 0$, and
- testing $x = 2$ in $(1, \infty)$ gives $f'(2) = 9 > 0$.

Therefore, the sign of $f'(x)$ is as shown in Figure 4.1.17.

Using the first derivative test, we conclude that our function $f(x) = x^3 - 3x$ has a local maximum at $x = -1$, with $f(-1) = 2$, and it has a local minimum at $x = 1$, with $f(1) = -2$. You can see these properties in the graph shown in Figure 4.1.18.

Before studying more examples of local minima and maxima, we will consider a rule for determining global maximum and minimum.

The single critical point principle

As we have already seen, a local extremum may not be a global maximum or minimum. There is one simple but useful criterion, however, that allows us to conclude that a local maximum (or minimum) is actually a global maximum (or minimum).

Theorem 4.1.4

The single critical point principle

Suppose the domain of $f(x)$ contains an interval containing *exactly one* critical point, say, at $x = c$. Then we have the following:

- If $f(x)$ has a local maximum at $x = c$, then $f(c)$ is the maximum value of $f(x)$ throughout the interval;
- If $f(x)$ has a local minimum at $x = c$, then $f(c)$ is the minimum value of $f(x)$ throughout the interval.

The reason is easy to see: If $x = c$ is the only critical point in (a, b), then c is the only point in the interval at which $f'(x)$ can change sign. If, say, the sign changes from positive to negative (a local maximum), then the graph is increasing from a to c and decreasing from c to b. Therefore, its highest point over the interval (a, b) is at $x = c$, and the maximum value is $f(c)$. Similarly, if the sign changes from negative to positive, the lowest point is at $x = c$ and the minimum value is $f(c)$.

EXAMPLE 4.1.10

Find the local extrema of the function $f(x) = x/(1 + x^2)$. Which are global extrema?

SOLUTION The domain of this function is the entire x-axis, and there is no vertical asymptote. (Notice, however, that there is a *horizontal* asymptote—namely, the x-axis.

That's because the denominator has a higher degree than the numerator.) To find the critical points, we take the derivative

$$f'(x) = \frac{1 - x^2}{(1 + x^2)^2}$$

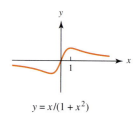

Figure 4.1.19

and find the places where it is zero. Using the basic principle that *a quotient is zero if and only if its numerator is zero*, we see that the critical points are at $x = -1$ and $x = 1$. These divide the x-axis into three intervals, $(-\infty, -1)$, $(-1, 1)$, and $(1, \infty)$. Evaluating $f'(x)$ at a test point in each of these intervals, we determine the sign of $f'(x)$, as shown in Figure 4.1.19. We conclude the following:

- $f(x)$ is decreasing on the intervals $(-\infty, -1)$ and $(1, \infty)$,
- $f(x)$ is increasing on $(-1, 1)$,
- $f(x)$ has a local minimum at $x = -1$, and
- $f(x)$ has a local maximum at $x = 1$.

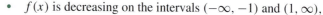

$y = x/(1 + x^2)$

Figure 4.1.20

The value of the function (and the height of the graph) at the local maximum is $f(1) = \frac{1}{2}$, and the value at the local minimum is $f(-1) = -\frac{1}{2}$.

The graph is shown in Figure 4.1.20. As you can see, the local maximum is, in fact, a global maximum. To see why, we observe that it is the only critical point in the interval $(0, \infty)$, so that $f(1)$ is the maximum value on $(0, \infty)$. Moreover, $f(x) \leq 0$ if $x \leq 0$, which means that $f(1)$ is the maximum value throughout the entire line. A similar argument shows that the local minimum is a global one.

In the next example, not all the local extreme points are global ones.

EXAMPLE 4.1.11

Find the local extrema of the function

$$f(x) = \frac{1}{4}x^4 + \frac{1}{3}x^3 - x^2 + 1.$$

What are the global extrema?

Figure 4.1.21

SOLUTION The derivative can be factored as follows:

$$f'(x) = x^3 + x^2 - 2x$$
$$= x(x^2 + x - 2) = x(x - 1)(x + 2).$$

From this we see that the critical points are at $x = -2, 0$, and 1. By using the usual method, we can get the diagram for the sign of $f'(x)$ shown in Figure 4.1.21.

We conclude: $f(x)$ has a local maximum at $x = 0$, and it has local minima at $x = -2$ and $x = 1$. Which of these, if any, are global?

First of all, if we compare the two local minimum values, we see that $f(-2) = -\frac{5}{3}$ and $f(1) = \frac{7}{12}$. Therefore, the height at $x = -2$ is lower than that at $x = 1$, and the latter cannot be a global minimum. On the other hand, there *is* a global minimum at

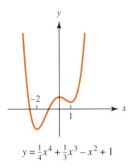

$y = \frac{1}{4}x^4 + \frac{1}{3}x^3 - x^2 + 1$

Figure 4.1.22

$x = -2$, as you can see in Figure 4.1.22. To explain why, we observe that there is only one critical point on $(-\infty, 0)$, so that $f(-2)$ is the smallest value on that interval. Since $f(-2) < 0$ and $f(x) > 0$ on $[0, \infty)$, we conclude that, in fact, $f(-2)$ is the smallest value over the entire domain of the function.

What about the maximum? There is only one local maximum, and the height there is $f(0)$, which equals 1. But, as the graph shows, it is not a global maximum. That is because the graph of an even-degree polynomial climbs higher and higher without bound as it moves either to the left or to the right (see Section 0.6). For instance, at $x = 2$ the height is given by $f(2) = \frac{11}{3}$, which is greater than the height at $x = 0$.

To summarize: The function in this example has one local maximum, at $x = 0$, but no global maximum. It has two local minima, one at $x = 1$ and the other at $x = -2$, the second of these being a global minimum as well.

Practice Exercises 4.1

1. Determine where the function $f(x) = 2x^3 + 7x^2 - 12x - 9$ is increasing and where it is decreasing, and find any local maxima or minima.

2. Find the local extrema of the function $f(x) = x^2 - \ln x$, $x > 0$. Are there any global extrema?

Exercises 4.1

In Exercises 1–20, find all critical points of the given function and use the derivative to determine where the function is increasing, where it is decreasing, and where it has a local maximum and minimum, if any.

1. $f(x) = 2x^2 - 7x + 3$ **2.** $f(x) = 1 - 4x - x^2$

3. $f(x) = 2x^3 + 3x^2 - 36x + 5$

4. $f(x) = 3x^4 - 4x^3 - 12x^2 + 6$

5. $f(x) = (x^2 - 4)^7$ **6.** $f(x) = x - \ln x$, $x > 0$

7. $f(x) = xe^x$ **8.** $f(x) = 2x^2 e^{5x} + 1$

9. $f(x) = x^{2/5}$ **10.** $f(x) = x^{1/5}$

11. $f(x) = xe^{-x}$ **12.** $f(x) = x^4 - 4x^3 + 1$

13. $f(x) = |x - 1|$

14. $f(x) = \sqrt{4 - x^2}$, $-2 < x < 2$

15. $f(x) = x^3 - 2x^2 + 3x - 4$

16. $f(x) = e^x - 2x$

17. $f(x) = \dfrac{1 + \ln x}{x}$, $x > 0$

18. $f(x) = \dfrac{x}{x^2 - 1}$, $x \ne \pm 1$

19. $f(x) = \dfrac{x^2}{x^2 + 1}$ **20.** $f(x) = x \ln x$, $x > 0$

21. A cross section of a tree trunk shows growth rings. The wider the ring, the more the tree grew that season, and this often correlates with the total amount of precipitation for the year. The trees of a region have their rings studied, and an attempt to model the precipitation gives the following formula:

$$P(y) = -y^2 + 20y + 80, \quad 0 \le y \le 20,$$

for the last 20 years, where y is in years and P is in centimeters.
(a) Where is $P' = 0$?
(b) Study the sign of P' for $0 < y < 20$.
(c) For what year(s) was there the greatest amount of rain? the least?

22. Powdery mildew is an interesting plant fungus in that it propagates when the weather is warm and dry, not wet. An organic method for fighting the mildew is to spray the infested plant with water, but it must be done when the air temperature is above 75°F. Suppose that the meteorological forecast for tomorrow predicts the following temperature:

$$T(h) = -\frac{1}{3}h^2 + 2h + 60, \quad 0 \le h \le 12,$$

where h is the number of hours past 9 A.M.
(a) Find the extrema for the 12-hour period from 9 A.M. to 9 P.M.

(b) If the forecast is correct, when can you spray your infested plants?

23. A salesman wishes to know when the purchases of his number-one-selling product will peak. To date, the data indicate that sales follow the pattern

$$S(t) = t^{1/3} + 100,$$

where t is the number of days since sales first started and S is the number of sales per day. What answer do you give him?

24. The temperature of an object is given by the formula

$$T(t) = 25e^{-0.5t}.$$

(a) Is it increasing or decreasing?
(b) Compare $T'(1)$, $T'(2)$, and $T'(3)$. What is happening to the slope of T?

25. A small business owner calculates that the cost of producing x hundreds of doll houses is modeled by the function

$$C(x) = (5x + 2)^{2/5} + 3,$$

where C is in thousands of dollars. Does the *marginal* cost increase or decrease with increased production?

26. For 5 weeks after a book on UFO encounters was first published, its sales grew slowly until a favorable review appeared in a nationwide magazine. Within a week it was on the best-seller list. It stayed there for 1 month, until another article was written proving the author falsified many of the facts, after which sales rapidly diminished to almost nothing. Figure 4.1.23 represents book sales s in millions per week versus time t in weeks.
(a) At what week did the sales achieve their global maximum? What is the value of the derivative there?
(b) What is the sign of the derivative during weeks 4 through 7?
(c) What is the sign of the derivative during weeks 12 through 15?
(d) True or false? Sales were increasing fastest in the first 5 weeks.

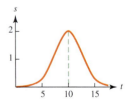

Figure 4.1.23

27. Wolf populations in the Midwest were drastically reduced with the expansion of American territory by pioneers. In fact, wolves were completely eliminated. But lately small groups have been slowly reintroduced into wilderness areas. These packs are considered successful because they are breeding, thus increasing their pack size.
(a) Sketch a smooth curve representing the trends of the wolf population in the Midwest.
(b) Label the parts where the derivative is positive and negative.
(c) Point out any extrema and identify them as maxima or minima.

Exercises 28–31 refer to the four graphs in Figure 4.1.24, each of which is the graph of $y = f(x)$ for some function. In each exercise identify the graph satisfying the given conditions.

28. $f'(x) < 0$ on $(-\infty, -1)$ and $(1, \infty)$; $f'(x) > 0$ on $(-1, 1)$

29. $f'(x) < 0$ on $(-\infty, \infty)$ **30.** $f'(x) > 0$ on $(-\infty, \infty)$

31. $f'(x) > 0$ on $(-\infty, -1)$ and $(1, \infty)$; $f'(x) < 0$ on $(-1, 1)$

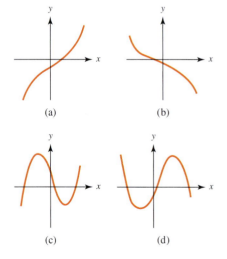

Figure 4.1.24

32. The graph of a function $f(x)$ is shown in Figure 4.1.25.
(a) On what intervals is $f'(x) > 0$?
(b) On what intervals is $f'(x) < 0$?
(c) At how many points is $f'(x) = 0$?

(d) At which x, if any, does the function have a local maximum?
(e) At which x, if any, does the function have a local minimum?

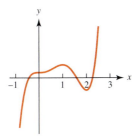

Figure 4.1.25

33. The graph of a function $f(x)$ is shown in Figure 4.1.26.
    (a) On what intervals is $f'(x) > 0$?
    (b) On what intervals is $f'(x) < 0$?
    (c) At how many points is $f'(x) = 0$?
    (d) At which x, if any, does the function have a local maximum?
    (e) At which x, if any, does the function have a local minimum?

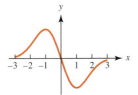

Figure 4.1.26

In Exercises 34–39, sketch the graph of a function $f(x)$ having the following properties:

34. $f'(x) < 0$ on $(-\infty, 0)$ and $(3, \infty)$; $f'(x) > 0$ on $(0, 3)$; $f'(0) = f'(3) = 0$

35. $f'(x) > 0$ on $(-\infty, 0)$; $f'(x) < 0$ on $(0, \infty)$; $f'(0) = 0$; $\lim_{x \to \pm\infty} f(x) = 0$

36. $f'(x) > 0$ on $(-\infty, 1)$ and $(1, \infty)$; $f'(1) = 0$

37. $f'(x) > 0$ on $(-\infty, -1)$ and $(0, 1)$; $f'(x) < 0$ on $(-1, 0)$ and $(1, \infty)$; $f(0) = 0$, $f(1) = 1$, $f(-1) = 1$; $f'(0) = 0$; $f'(-1)$ and $f'(1)$ do not exist

38. $f(x)$ has vertical asymptotes at $x = -2$ and $x = 2$; $f'(0) = 0$; $f'(x) > 0$ on $(-\infty, -2)$ and $(-2, 0)$; $f'(x) < 0$ on $(0, 2)$ and $(2, \infty)$

39. $f(x)$ has vertical asymptotes at $x = -2$ and $x = 2$; $f'(0) = 0$; $f'(x) > 0$ on $(-\infty, -2)$ and $(0, 2)$; $f'(x) < 0$ on $(-2, 0)$ and $(2, \infty)$

Exercises 40–44 refer to the graphs in Figure 4.1.27.

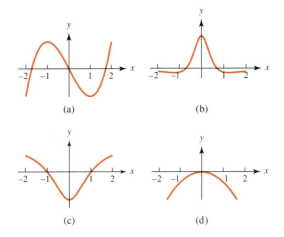

(a)   (b)

(c)   (d)

Figure 4.1.27

40. Figure 4.1.28 shows the graph of a function $f(x)$. Which of the graphs in Figure 4.1.27 is the graph of the derivative $f'(x)$?

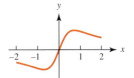

Figure 4.1.28

41. Figure 4.1.29 shows the graph of the derivative $f'(x)$ of a function $f(x)$. Which of the graphs in Figure 4.1.27 is the graph of $f(x)$?

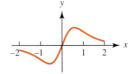

Figure 4.1.29

42. Figure 4.1.30 shows the graph of the derivative $f'(x)$ of a function $f(x)$. Which of the graphs in Figure 4.1.27 is the graph of $f(x)$?

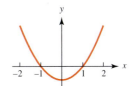

Figure 4.1.30

43. Figure 4.1.31 shows the graph of a function $f(x)$. Which of the graphs in Figure 4.1.27 is the graph of the derivative $f'(x)$?

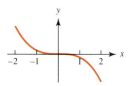

Figure 4.1.31

44. Let $f(x) = x^3 + x^2 - 5x - 2$.

~ **(a)** Graph both $f(x)$ and $f'(x)$ in the same $[-4, 4] \times [-6, 6]$ window.

(b) Verify that $f(x)$ is increasing where $f'(x) > 0$ and decreasing where $f'(x) < 0$.

(c) Use the calculator's zoom function to find those x where the graph of $f(x)$ changes from increasing to decreasing or vice versa.

(d) Use the calculator's zoom function to find the solutions of $f'(x) = 0$ and verify that they agree with the answers to part (c).

45. Suppose that $f(x)$ is a function with $f'(x) = 2x^4 - 3x^3 - 4x^2 - x + 2$. Graph $f'(x)$ and find any local minimum or maximum points of $f(x)$ on the interval $(-4, 4)$.

46. Let $f(x) = e^x - 2x^2$.

~ **(a)** Explain (without graphing) why $f'(x) = 0$ must have a solution in the interval $[0, 1]$ and another in $[1, 3]$.

(b) Graph $f'(x)$ in a window that clearly shows the two points where the graph crosses the x-axis. Then use the calculator's zoom function (or the equation solver) to find the solutions.

(c) Explain why $f'(x) = 0$ cannot have any other solutions. (*Hint:* Look at $f''(x)$.)

(d) Where is $f(x)$ increasing and where is it decreasing? Find any local minimum or maximum points.

(e) Graph $f(x)$ and verify your answer to part (d).

In Exercises 47–58, find the local and global extrema if any.

47. $f(x) = 5 + 2x - 3x^2$ **48.** $f(x) = (2x - 1)^4$

49. $f(x) = xe^{-x}$

50. $f(x) = \dfrac{1}{x^2 + 1}$

51. $f(x) = \ln(x^2 + 1)$

52. $f(x) = (x - 1)^{4/5}$

53. $f(x) = e^{-x^2}$

54. $f(x) = \dfrac{x^2}{x^2 - 1}$, $x \neq \pm 1$

55. $f(x) = (\ln x)^2$, $x > 0$

56. $f(x) = e^{x^2 - 3x}$

57. $f(x) = (x^2 - 1)^5$

58. $f(x) = (x^2 - 1)^4$

59. Let $f(x) = x^5 + x + 7$.

(a) Check that $f(-2) < 0$ and $f(-1) > 0$. Explain why the equation $x^5 + x + 7 = 0$ must have a solution between $x = -2$ and $x = -1$.

(b) Show that $f(x)$ has no critical points. Explain why the equation $x^5 + x + 7 = 0$ cannot have two solutions. (*Hint:* What is the sign of $f'(x)$ and what does it tell you?)

60. Suppose f is a function, with domain $(-\infty, \infty)$, such that $f(-1) = 4$, $f(2) = -3$, and $f'(x) < 0$ for all values of x. How many solutions does the equation $f(x) = 0$ have?

61. Show that the equation $e^x - x = 0$ does not have a solution. (*Hint:* Find the critical point and show that it is a global minimum. Then check its value.)

62. Suppose that f is a function, with domain $(-\infty, \infty)$, such that $f'(x) < 0$ for $x < 1$, $f'(x) > 0$ for $x > 1$, $f(1) = 2$, and $f'(1) = 0$. How many solutions does the equation $f(x) = 0$ have?

63. Census data for the world population $p(t)$ indicate that it satisfies the (logistic) differential equation

$$\frac{dp}{dt} = 0.03p \left(1 - \frac{p}{12} \right),$$

where p is measured in billions and t in years after 1990, when its size was about 5.23 billions. Assuming that it will always stay below 12 billion, show that $p(t)$ is an increasing function. Will the world population ever achieve a maximum value?

Solutions to practice exercises 4.1

1. We take the derivative and set it equal to zero:

$$f'(x) = 6x^2 + 14x - 12 = 2(3x^2 + 7x - 6) = 0.$$

Either by the quadratic formula or by factoring into $2(3x - 2)(x + 3)$, we see that the critical points occur at $x = -3$

and $x = \frac{2}{3}$. They divide the line into three intervals. By testing a point in each interval, such as

$$f'(-4) > 0, \quad f'(0) < 0, \quad f'(1) > 0,$$

we find the sign of $f'(x)$ as shown in Figure 4.1.32.

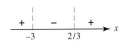

Figure 4.1.32

We conclude: $f(x)$ is increasing on $(-\infty, -3)$ and $(\frac{2}{3}, \infty)$, and decreasing on $(-3, \frac{2}{3})$, with a local maximum at $x = -3$ and a local minimum at $x = \frac{2}{3}$.

2. Taking the derivative and setting it equal to zero give

$$f'(x) = 2x - \frac{1}{x} = 0,$$

which simplifies to $x^2 = \frac{1}{2}$. The only positive solution is

the critical point $x = 1/\sqrt{2}$. By testing a point on each side of the critical point, say,

$$f'(1) = 1 > 0 \quad \text{and} \quad f'\left(\frac{1}{2}\right) = -1 < 0,$$

we see that $f'(x) > 0$ on $(1/\sqrt{2}, \infty)$ and $f'(x) < 0$ on $(0, 1/\sqrt{2})$. Therefore, by the first derivative test, there is a local minimum at $x = 1/\sqrt{2}$, with a value of $f(1/\sqrt{2}) = \frac{1}{2}(1 + \ln 2) \approx 0.8466$. Applying the single critical point principle, we conclude that it is also a global minimum. There is no global maximum, since $f(x)$ grows unboundedly large as $x \to 0^+$.

4.2 **Second Derivative Tests**

In the last section we saw how to find where a function is increasing or decreasing, and where it has maxima or minima by using the first derivative. Next, we shall see how to find where the *derivative* of a function is increasing or decreasing and how to use the *second* derivative to determine where the function has maxima and minima.

The second derivative test for concavity

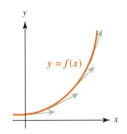

Figure 4.2.1

Our first task is to determine a method for deciding where the graph of a function is **bending upward** or **bending downward**. To see what that means, we consider the graphs of two different functions, drawn in Figures 4.2.1 and 4.2.2. As you can see, both functions are increasing—both graphs rise as they go to the right. But there is a significant difference between them. In Figure 4.2.1, the *slope increases* as the graph goes to the right. You can see it in the drawing, getting steeper to the right. On the other hand, the graph in Figure 4.2.2 flattens out as it goes to the right—in other words, the *slope decreases*.

The distinction is important. In many applications we not only want to know if the function is increasing or decreasing over a certain stretch, but we also want to know if its *derivative* is increasing or decreasing. The derivative of the function, as you know, is the slope of the graph. With that in mind, we can rephrase the difference as follows:

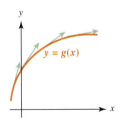

Figure 4.2.2

- In Figure 4.2.1 the function is increasing and the derivative is increasing;
- In Figure 4.2.2 the function is increasing and the derivative is decreasing.

It is customary to use geometric terminology in this context. We say that the graph in Figure 4.2.1 is **concave up**, and the one in Figure 4.2.2 is **concave down**. If you think of the graph as a drawing of a piece of a bowl, then

- concave up means the bowl is rightside up, and
- concave down means the bowl is upside down.

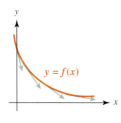

Figure 4.2.3

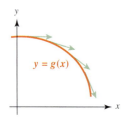

Figure 4.2.4

The same considerations apply to decreasing functions. Figures 4.2.3 and 4.2.4 show two graphs, both of them decreasing: The graph in Figure 4.2.3 is concave up (the bowl is rightside up). The one in Figure 4.2.4 is concave down (the bowl is upside down).

What about the slope? In Figure 4.2.3, the slope is negative, but it gets *less* negative as the graph goes to the right. That means the slope is *increasing*. So, once again, "concave up" means increasing slope. In Figure 4.2.4, the slope is also negative, but in this case it gets *more* negative as the graph goes to the right. So, once again, "concave down" means decreasing slope.

By using the derivative, we can define concavity precisely, as follows:

Definition 4.2.1

Concavity

Let $f(x)$ be a differentiable function on an interval. Then

- its graph is concave up if $f'(x)$ is increasing;
- its graph is concave down if $f'(x)$ is decreasing.

Notice that the definition says nothing about whether the *function* is increasing or decreasing, only the derivative. As we saw in Figures 4.2.1 through 4.2.4, a graph may be concave up whether it is increasing or decreasing, and the same for concave down.

How can we tell where a graph is concave up or down? To put it another way, how can we tell where $f'(x)$ is increasing and where it is decreasing? We simply use the same principle we applied to $f(x)$ in the last section, but this time we apply it to $f'(x)$. That is,

HISTORICAL PROFILE

Pierre de Fermat (1601–1665) was one of the first mathematicians to study maxima and minima by means of the tangent to the graph, even before calculus was a known subject. In fact, some eminent mathematicians and historians believe that he deserves credit as the true inventor of the subject.

In addition to his work in mathematics, Fermat was an eminent attorney, judge, and parliamentarian in Toulouse, France. He is considered one of the founders of probability theory and is also famous for his work in number theory. For

hundreds of years, the most celebrated problem in mathematics was the so-called "Fermat's last theorem," which states that for an integer $n \geq 3$, the equation $x^n + y^n = z^n$ has no solution in integers. Fermat wrote in the margin of one of his books, found after his death, that he had an elegant proof of that fact but there was no room to write it in the margin. Despite the efforts of some of the greatest mathematicians in history, the problem remained open until 1994, when it was solved by Andrew Wiles of Princeton University.

- $f'(x)$ is increasing if its derivative is positive, and
- $f'(x)$ is decreasing if its derivative is negative.

But the derivative of $f'(x)$ is simply the second derivative, $f''(x)$. Putting these ideas together, we obtain the following theorem:

Theorem 4.2.1

Second derivative test for concavity

Let $f(x)$ be a function that has a second derivative in an interval. Then

- its graph is concave up if $f''(x) > 0$ for all x;
- its graph is concave down if $f''(x) < 0$ for all x.

Figure 4.2.5 illustrates both cases. So, to study the concavity of the function $f(x)$, we can analyze the sign of $f''(x)$, just as we analyze the sign of $f'(x)$ to obtain information about where the function is increasing or decreasing.

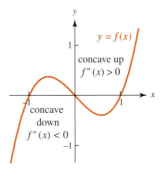

Figure 4.2.5

EXAMPLE 4.2.1

Let $f(x) = x^3 + x - 1$. Determine where the graph of this function is concave up or concave down.

SOLUTION   We already analyzed this function in Example 4.1.4 by using the sign of $f'(x)$ to determine where the function is increasing or decreasing. To study its concavity, we first compute the second derivative

$$f''(x) = 6x.$$

Clearly, $f''(x)$ is positive on the interval $(0, \infty)$ and negative on $(-\infty, 0)$. It follows that the graph is concave up on $(0, \infty)$ and concave down on $(-\infty, 0)$, as shown in Figure 4.2.6.

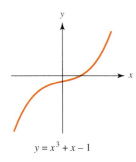

$y = x^3 + x - 1$

Figure 4.2.6

To find the sign of $f''(x)$ in more complicated cases, we follow a procedure similar to the one we used in the last section to determine the sign of $f'(x)$.

- Solve the equation $f''(x) = 0$ and plot the solutions on a diagram of the x-axis, dividing the domain of f into a number of intervals.
- Test a point in each interval.

Sign of f''

Figure 4.2.7

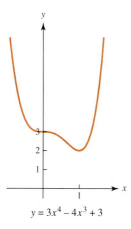

$y = 3x^4 - 4x^3 + 3$

Figure 4.2.8

EXAMPLE 4.2.2

Determine where the graph of $f(x) = 3x^4 - 4x^3 + 3$ is concave up or concave down.

SOLUTION   The second derivative is $f''(x) = 36x^2 - 24x = 12x(3x - 2)$. The places where $f''(x) = 0$ occur at $x = 0$ and $x = \frac{2}{3}$, which are marked in the diagram in Figure 4.2.7. On the interval $(-\infty, 0)$ we can test $x = -1$, which gives $f''(-1) = 60$. Testing $x = \frac{1}{3}$ gives $f''(\frac{1}{3}) = -4$, and testing $x = 1$ gives $f''(1) = 12$. This information is enough to obtain the sign chart in Figure 4.2.7.

We conclude the following: The graph is concave up on the intervals $(-\infty, 0)$ and $(\frac{2}{3}, \infty)$, and it is concave down on $(0, \frac{2}{3})$. You can see the concavity in the graph, shown in Figure 4.2.8.

Inflection points

A point where the concavity of a graph changes is called an **inflection point**. At such a point the sign of $f''(x)$ changes. That can only happen at a point where $f''(x)$ is either zero or undefined. Thus, inflection points are related to concavity in the same way that local extreme points (maxima or minima) are related to the question of where the graph is increasing or decreasing. Figure 4.2.9 shows two examples of inflection points.

Theorem 4.2.2

Inflection points

If $(c, f(c))$ is an inflection point, then either $f''(c) = 0$ or $f''(c)$ is undefined.

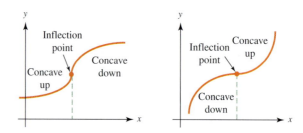

Figure 4.2.9

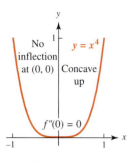

Figure 4.2.10

Warning Points c in the domain of $f(x)$ for which either $f''(c) = 0$ or $f''(c)$ is undefined are *candidates* for inflection points. A function may *not* have an inflection point at such a point c. The simplest example of such a function is $f(x) = x^4$, whose graph is shown in Figure 4.2.10. Since $f''(x) = 12x^2$, we have $f''(0) = 0$. However $f(x) = x^4$ does *not* have an inflection point at $(0, 0)$, as you can see in the graph on the right. Indeed, the function is concave up everywhere, since the sign of $f''(x)$ does *not* change at $x = 0$.

In Example 4.2.1 the graph has an inflection point at $x = 0$. More precisely: The inflection point occurs at $(0, -1)$. In Example 4.2.2 there are two inflection points: at $x = 0$ and $x = \frac{2}{3}$.

EXAMPLE 4.2.3

Determine the concavity of the graph of $f(x) = \dfrac{x}{1 + x^2}$ and find any inflection points.

SOLUTION In Example 4.1.10 we saw that

$$f'(x) = \frac{1 - x^2}{(1 + x^2)^2}.$$

To find $f''(x)$, we use the quotient and chain rules:

$$f''(x) = \frac{-2x(1 + x^2)^2 - 4x(1 + x^2)(1 - x^2)}{(1 + x^2)^4}.$$

Although that looks complicated, it can be simplified by factoring $(1 + x^2)$ out of both terms in the numerator and canceling it against the denominator. We get

$$f''(x) = \frac{-2x(1 + x^2) - 4x(1 - x^2)}{(1 + x^2)^3}$$

$$= \frac{-2x(3 - x^2)}{(1 + x^2)^3}.$$

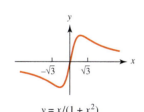

$y = x/(1 + x^2)$

Figure 4.2.11

There are three places where $f''(x) = 0$, at $x = 0$ and $x = \pm\sqrt{3}$. By testing points, we find that $f''(x) < 0$ on $(-\infty, -\sqrt{3})$ and on $(0, \sqrt{3})$, so that the graph is concave down on those intervals. On the other hand, $f''(x) > 0$ on $(-\sqrt{3}, 0)$ and on $(\sqrt{3}, \infty)$, so the graph is concave up on those intervals. We conclude that $f(x)$ has inflection points at $x = 0$ and at $x = \pm\sqrt{3} \approx 1.73$. The graph is shown in Figure 4.2.11.

The second derivative test for maxima and minima

The sign of $f''(x)$ can also be used to test for local maxima and minima.

Theorem 4.2.3

Second derivative test

Suppose that the first and second derivatives both exist in an interval around c and $f'(c) = 0$. Then,

- there is a local maximum at $x = c$ if $f''(c) < 0$, and
- there is a local minimum at $x = c$ if $f''(c) > 0$.

If $f''(c) = 0$, the test is inconclusive.

The graphs in Figure 4.2.12 suggest why the the two statements in the theorem are true. Here is the complete explanation of the first conclusion. From the definition of the derivative, we have

$$f''(c) = \lim_{x \to c} \frac{f'(x) - f'(c)}{x - c}.$$

If $f''(c) \neq 0$, the quotient on the right will have the same sign as $f''(c)$ for x close to c. If $f''(c) < 0$ and $f'(c) = 0$, that means

$$\frac{f'(x)}{x - c} < 0 \quad \text{for all } x \text{ near } c.$$

But then $f'(x)$ and $x - c$ must have opposite signs, which means that $f'(x)$ is positive for $x < c$ and negative for $x > c$. It follows from the first derivative test (Theorem 4.1.3) that $f(x)$ has a local maximum at $x = c$.

The explanation of the second conclusion is similar.

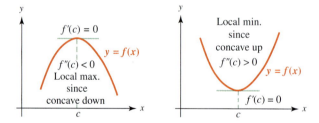

Figure 4.2.12

EXAMPLE 4.2.4

Find the local maximum and minimum points of

$$f(x) = x^3 - 3x^2 - 9x + 7.$$

SOLUTION   We start by taking the first derivative:

$$f'(x) = 3x^2 - 6x - 9 = 3(x - 3)(x + 1).$$

The solutions of $f'(x) = 0$ are $x = 3$ and $x = -1$. To test whether each of these is a

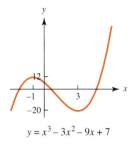

y = x³ − 3x² − 9x + 7

Figure 4.2.13

local maximum or minimum, we compute the second derivative

$$f''(x) = 6x - 6$$

and evaluate it at $x = 3$ and $x = -1$.

Since $f''(3) = 12 > 0$, we conclude that $f(x)$ has a local minimum at $x = 3$. The value (height of the graph) at that point is $f(3) = -20$.

Since $f''(-1) = -12 < 0$, we see that $f(x)$ has a local maximum at $x = -1$. The value at that point is $f(-1) = 12$. The graph is shown in Figure 4.2.13.

Warning   The second derivative test is only valid if $f'(c) = 0$ and $f''(c) \neq 0$. If both are zero, we cannot draw any conclusion. For example,

- $f(x) = x^4$ has a local minimum at $x = 0$,
- $f(x) = 1 - x^4$ has a local maximum at $x = 0$, and
- $f(x) = x^3$ has neither,

but all three functions have $f'(0) = 0$ and $f''(0) = 0$.

If $f'(c) = 0$ and $f''(c) = 0$, we have to *go back to the first derivative test*—that is, we must check the sign of $f'(x)$ on either side of $x = c$ to see if it changes.

Practice Exercises 4.2

1. Determine where the function $f(x) = e^{-x^2/2}$ is concave up or down, and find any inflection points.

2. Find the critical points of the function $f(x) = x^5 - 5x^4 + 5x^3 - 1$ and test each to see if it is a local maximum or min-

imum. Use the second derivative test if possible, otherwise the first derivative test.

Exercises 4.2

Exercises 1–4 refer to Figure 4.2.14, which is the graph of a function $f(x)$. Match each of the labeled points A through D with the conditions on $f(x)$ and $f'(x)$ given in the exercise.

1. $f(x)$ is increasing and $f'(x)$ is increasing.

2. $f(x)$ is increasing and $f'(x)$ is decreasing.

3. $f(x)$ is decreasing and $f'(x)$ is increasing.

4. $f(x)$ is decreasing and $f'(x)$ is decreasing.

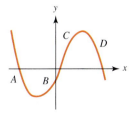

Figure 4.2.14

Exercises 5–18 refer to the 12 graphs in Figure 4.2.15, labeled (a) through (l), each of which is the graph of $y = f(x)$ for some function (see page 278). In each exercise identify any graphs satisfying the given conditions. There may be none or more than one.

5. $f'(x) > 0$ for all x

6. $f'(x) < 0$ for all x

7. $f''(x) > 0$ for all x

8. $f''(x) < 0$ for all x

9. $f'(x) > 0$ and $f''(x) > 0$ for all x

10. $f'(x) > 0$ and $f''(x) < 0$ for all x

11. $f'(x) < 0$ and $f''(x) > 0$ for all x

12. $f'(x) < 0$ and $f''(x) < 0$ for all x

13. $f''(x) > 0$ for $x < 0$ and $f''(x) < 0$ for $x > 0$, $f'(x) > 0$ for all x

14. $f''(x) > 0$ for $x < 0$ and $f''(x) < 0$ for $x > 0$, $f'(x) < 0$ for all x

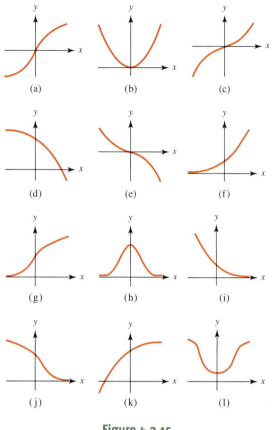

Figure 4.2.15

15. $f''(x) < 0$ for $x < 0$ and $f''(x) > 0$ for $x > 0$, $f'(x) > 0$ for all x

16. $f''(x) < 0$ for $x < 0$ and $f''(x) > 0$ for $x > 0$, $f'(x) < 0$ for all x

17. $f''(x) > 0$ on $(-\infty, -1)$ and $(1, \infty)$, $f''(x) < 0$ on $(-1, 1)$

18. $f''(x) < 0$ on $(-\infty, -1)$ and $(1, \infty)$, $f''(x) > 0$ on $(-1, 1)$

In Exercises 19–32, determine where the graph of the given function is concave up, where it is concave down, and where there are inflection points, if any.

19. $f(x) = 2x^3 + 6x^2 - 5x + 1$

20. $f(x) = (x - 1)^3 + 2$

21. $f(x) = \dfrac{1}{6}x^4 - x^3 - 4x^2 + 3x - 5$

22. $f(x) = \dfrac{1}{x+2}$ **23.** $f(x) = \dfrac{1}{(x-2)^2}$

24. $f(x) = x^4 + 4x^3 + 6x^2 + 2x - 3$

25. $f(x) = \dfrac{x}{x-1}$ **26.** $f(x) = \dfrac{x}{x^2-1}$

27. $f(x) = (\ln x)^2$, $x > 0$ **28.** $f(x) = xe^{-2x}$

29. $f(x) = \dfrac{x^2}{x^2+3}$ **30.** $f(x) = e^x - x^2$

31. $f(x) = \dfrac{1}{x^2+4}$ **32.** $f(x) = x^{5/3}$

In Exercises 33–44, find the critical points of the given function and use the second derivative test to determine whether each is a local maximum or minimum. Also, find the value of the function at each local extremum.

33. $f(x) = x^3 - 6x + 1$ **34.** $f(x) = \dfrac{x^2-1}{x^2+1}$

35. $f(x) = xe^{3x}$ **36.** $f(x) = \ln(x^2 + 1)$

37. $f(x) = x + \dfrac{4}{x}$, $x \neq 0$

38. $f(x) = x^2 - 2\ln x$, $x > 0$

39. $f(x) = e^{-x^2+2x}$ **40.** $f(x) = x^2 \ln x$, $x > 0$

41. $f(x) = x^4 - 2x^2 + 3$ **42.** $f(x) = x^2 e^{-x}$

43. $f(x) = x - \sqrt{x}$, $x > 0$ **44.** $f(x) = e^x - 2x$

45. A small dam controls the water that flows from several local creeks. Released water flows down a man-made channel and is diverted for irrigation. The depth of water in the channel was monitored for many days after heavy rains in the area and was modeled as

$$D(t) = \dfrac{8}{1 + e^{4-0.4t}},$$

where t is in days and D is in meters. Find the day on which the rate of depth increase was highest.

46. The harvesting period for peaches in a region lasts for about 5 weeks. The rate of harvesting for the first 4 weeks is modeled by the function

$$r(t) = \dfrac{800t}{t^4 + 48}, \quad 0 \leq t \leq 4,$$

where r is measured in thousands of pounds per week.
(a) When is the harvesting rate maximum?
(b) If $p(t)$ is the amount of peaches harvested by time t, then study the concavity of $p(t)$, find its inflection points, and sketch its graph.

In Exercises 47–50, find the critical points of the given function and determine whether each is a local maximum or minimum. Use the second derivative test if you can.

47. $f(x) = 3x^4 + 8x^3 + 6x^2 - 1$

48. $f(x) = 2x^5 - 5x^4 + 3$

49. $f(x) = (x^2 - 1)^4$ **50.** $f(x) = (x^2 - 1)^5$

51. Each year a school corporation's enrollment has been increasing. However, analysis of enrollment trends shows that the amount of increase is becoming smaller.

(a) Is the graph of total enrollment over time concave up or down?

(b) What one word change would you make in the above description to reverse the concavity?

52. Pennsylvania has been trying to reestablish its wild elk herd since 1913, when it brought in about 200 animals as an initial stock. However, the herd was not managed well, and it decreased faster and faster until around 1940, when the population bottomed out at a dozen animals. After that, the herd began to increase and kept increasing, but at a slower and slower rate, until 1970. That is when the Pennsylvania Game Commission and Bureau of Forestry began helping the herd, which grew at an ever-increasing rate, to 183 by 1992 and to more than 500 by 1999. Now biologists are transferring the excess stock to other states that wish to establish herds. Assume that the elk population is modeled as a smooth function.

(a) When did the herd achieve its global maximum? Minimum?

(b) When was the concavity up? Down?

(c) Could 1970 be described as an inflection point?

Source: Scott Weidensaul, *The Return of the Elk*, Washington, DC: Smithsonian, Dec. 1999, p. 88.

53. A company was doing well for 3 years, expanding its employee base at an ever-increasing rate as its product sales continued to grow. Then the collapse of the "dot.com" Internet industry drastically reduced company income, forcing layoffs. More and more employees were laid off each week. One year later, the board of directors brought in a new CEO who slowed down the layoff rate over a 6-month period, until no more people lost their jobs. After that, employment stayed constant, until 2 years year later, when the company began selling some innovative new items that allowed it to start rehiring at an ever-increasing rate.

(a) Sketch a graph representing the level of employment versus time.

(b) Identify the extrema and the sign of the derivative to the left and right of each one.

(c) Label the concavity and the signs of the second derivative.

(d) When do the inflection points occur?

(e) What words tell you this?

54. Some stars go through a sequence of events in their lifespans that cause them to explode, becoming supernovae. The steps are generally this: The star grows brighter and somewhat bigger, then it rapidly grows dimmer and considerably smaller. The core collapses under the weight of the outer layers, forcing the star to explode outward, sending most or all of its matter shooting out into space. After some time, the rate of expansion decreases.

(a) Which of the graphs of Figure 4.2.16 most accurately represents the radius of the star versus time?

(b) How many times does the concavity change during the sequence?

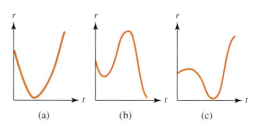

Figure 4.2.16

55. Suppose $f(x)$ is a function with $f''(x) = 2x^3 - 5x^2 - 11x + 9$. Graph $f''(x)$ in a $[-5, 5] \times [-20, 20]$ window and use the graph with the calculator's zoom function to determine where the graph of $f(x)$ is concave up, concave down, and for which x it has inflection points.

56. Suppose $f(x)$ is a function with $f'(x) = x^5 - \frac{1}{2}x^4 + x^3 + \frac{3}{2}x^2 - 12x + 5$. Find the inflection points and determine the concavity of $f(x)$ in the interval $(-2, 2)$.

57. Let $f(x) = (x^2 + x + 2)^{-1}$. Graph $f(x)$ and $f'(x)$ in the same $[-3, 3] \times [-0.5, 0.75]$ window.

(a) For what x is $f'(x) = 0$, and what happens to the graph of $f(x)$?

(b) For what x does $f'(x)$ have a maximum, and what happens to the graph of $f(x)$ at that x?

(c) For what x does $f'(x)$ have a minimum, and what happens to the graph of $f(x)$ at that x?

58. Data from 1945 to 1995 indicate that the capital stock (capital per worker) k in the U.S. economy is modeled by the differential equation

$$\frac{dk}{dt} = 0.2k^{0.3} - 0.1k,$$

where k is measured in millions and t in years. Assuming that the current level of capital stock is 1.54, that is, $k(0) = 1.54$, and k does not exceed 2, show that $k(t)$ is an increasing and concave down function.

59. A possible model for the U.S. population is given by the (logistic) differential equation

$$\frac{dp}{dt} = 0.02p(1-p),$$

where p is measured in billions and t in years. In 2000 the U.S. population was about 0.28 billion. Assuming that $p(t) < 1$ and $\lim_{t\to\infty} p(t) = 1$, show that $p(t)$ is increasing and concave up until it reaches the value 0.5 and after that it is increasing and concave down. Sketch $p(t)$.

60. Consider a lake that can sustain a maximum of 100 thousand fish of a certain species. If this species has a growth rate of 0.2 and is harvested at the rate of 10% of the population present, then a model governing the population $y(t)$ is given by the (logistic with constant effort of harvesting) equation

$$\frac{dy}{dt} = 0.2y\left(1 - \frac{y}{100}\right) - 0.1y,$$

with y in thousands. Suppose the initial population is 30 thousand and it can never exceed 50 thousand. Show that $y(t)$ is increasing and concave down.

Source: F. Brauer and C. Castillo-Chavez, *Mathematical Models in Population Biology and Epidemiology*, New York, Springer, 2001.

Solutions to practice exercises 4.2

1. The first derivative is $f'(x) = -xe^{-x^2/2}$, and the second derivative is obtained by the product rule:

$$f''(x) = -e^{-x^2/2} + x^2e^{-x^2/2} = e^{-x^2/2}(x^2 - 1).$$

Since $e^{-x^2/2}$ is always positive, $f''(x) = 0$ if and only if $x^2 = 1$, that is, at $x = \pm 1$. Moreover, the sign of $f''(x)$ is the same as that of $(x^2 - 1)$, which is shown in the sign diagram in Figure 4.2.17.

Figure 4.2.17

We conclude that the graph is concave up on $(-\infty, -1)$ and $(1, \infty)$ and down on $(-1, 1)$. Moreover, there are inflection points at $x = 1$ and $x = -1$.

2. Taking the derivative and simplifying gives

$$f'(x) = 5x^4 - 20x^3 + 15x^2 = 5x^2(x - 3)(x - 1).$$

Therefore, the critical points occur at $x = 0, 1$, and 3. The second derivative is

$$f''(x) = 20x^3 - 60x^2 + 30x = 10x(2x^2 - 6x + 3).$$

Since $f''(1) = -10$, there is a local maximum at $x = 1$, and since $f''(3) = 90$, there is a local minimum at $x = 3$. However, $f''(0) = 0$, so the second derivative test is inconclusive.

To apply the first derivative test: We test the sign of $f'(x)$ at a point in the interval $(-\infty, 0)$ and at one in the interval $(0, 1)$. For instance,

$$f'(-1) = 40 \quad \text{and} \quad f'\left(\frac{1}{2}\right) = \frac{25}{16}.$$

Since the sign does not change at $x = 0$, we conclude that there is neither a local maximum nor minimum there.

4.3 Sketching Graphs

In this section we combine all the ideas and techniques developed so far (in this chapter as well as in the previous ones) for the purpose of drawing a quick sketch of the graph of a function that captures its main characteristics. Seeing the graph of a function (its geometric shape) helps us understand it better. In general, the function is expressed by a formula. Drawing its graph is like animating a lifeless object.

To draw the graph, we need to examine the formula very carefully for information about the continuity of the graph, its asymptotic behavior, symmetries, points of intersection with the axes, places where it rises or falls, differentiability, maxima and minima, concavity, and other special properties. Even when using a graphing calculator or computer, that information is often necessary in making choices about the range of the graph and other parameters.

Guidelines for sketching graphs

1. Determine the natural domain of the function and locate any vertical asymptotes. For rational functions or other quotients, this usually means finding where the denominator is zero. But vertical asymptotes can sometimes arise in other ways—for instance, $\ln x \to -\infty$ as $x \to 0^+$. In cases like this, where a point is on the boundary of the domain, try to find the limit as x approaches that point.

2. In general, the functions we consider are continuous over their natural domain, so their graphs are continuous curves, breaking only at vertical asymptotes.

3. If possible, using sign charts, determine where the function is positive and where it is negative. This may be difficult if it involves solving a complicated equation, but in many cases—those involving exponentials or only even powers of x, for instance—it may be easy, and it gives you a better idea of where the graph "lives."

4. Use the first derivative to determine where the graph is increasing, where it is decreasing, and where there are local maxima or minima, if any. Sign charts are helpful here.

5. Use the second derivative to determine where the graph is concave up, where it is concave down, and where it has inflection points, if any. You can also use the second derivative test to help determine whether a critical point is a local maximum or minimum. Again, sign charts are helpful here.

6. Check the behavior of the graph as $x \to \pm\infty$. Does it climb higher and higher or fall lower and lower without bound, or does it level off toward a horizontal asymptote? We have already seen how to answer that for rational functions. Functions involving exponentials or logarithms may be a little trickier, but the following points should help:

   - e^x grows much faster than any power of x, so that for any exponent m, $e^x/x^m \to \infty$ and $x^m/e^x \to 0$ as $x \to \infty$.

   - $\ln x$ grows much slower than any power of x, so that for any positive m, $\ln x/x^m \to 0$ and $x^m/\ln x \to \infty$ as $x \to \infty$.

7. Check for symmetry, either about the y-axis or about the origin.

8. Plot a few points, if possible, that are easy to calculate. One that is usually easy is the y-intercept—that is, the point on the graph where $x = 0$ and $y = f(0)$.

We start with a function that plays a major role in probability and statistics. Its graph is related to the famous **bell-shaped curve**.

EXAMPLE 4.3.1

Sketch the graph of the function $f(x) = e^{-x^2}$.

SOLUTION We first observe that the natural domain of f is the entire x-axis, and there are no vertical asymptotes. To find out what the graph looks like, we start by computing the first derivative, which we get by using the chain rule:

$$f'(x) = -2xe^{-x^2}.$$

Since $e^{-x^2} > 0$ for all x, the sign of $f'(x)$ is the same as that of $-2x$, positive for $x < 0$ and negative for $x > 0$. The first derivative test says that

- $f(x)$ is increasing on $(-\infty, 0)$;
- $f(x)$ is decreasing on $(0, \infty)$;
- there is a local maximum at $x = 0$, with value $f(0) = 1$.

Applying the single critical point principle in the interval $(-\infty, \infty)$, we conclude that there is a global maximum at $x = 0$ with $f(0) = 1$.

We next observe a few key features of the graph. First, it lies entirely above the x-axis, since $e^{-x^2} > 0$ for every x. Second, it approaches the x-axis asymptotically to both the right and left, since

$$\lim_{x \to \pm\infty} e^{-x^2} = \lim_{x \to \pm\infty} \frac{1}{e^{x^2}} = 0.$$

Third, the graph is symmetric about the y-axis, since $f(-x) = f(x)$, that is,

$$e^{-(-x)^2} = e^{-x^2}.$$

Finally, we can use the second derivative to study concavity and inflection points. Using the product and chain rules, we get

$$f''(x) = 4x^2 e^{-x^2} - 2e^{-x^2} = 2e^{-x^2}(2x^2 - 1).$$

The solutions of $f''(x) = 0$ are $x = 1/\sqrt{2}$ and $x = -1/\sqrt{2}$.

By testing a point in each of the three intervals marked off by these solutions, we get the sign chart shown in Figure 4.3.1.

Here is a summary of what we know about the graph:

- It is increasing on $(-\infty, 0)$ and decreasing on $(0, \infty)$, and it has a global maximum of height 1 at $x = 0$.
- It is concave up on the intervals $(-\infty, -1/\sqrt{2})$ and $(1/\sqrt{2}, \infty)$, concave down on $(-1/\sqrt{2}, 1/\sqrt{2})$, and it has inflection points at $x = 1/\sqrt{2}$ and $x = -1/\sqrt{2}$.
- It lies entirely above the x-axis, and it has the x-axis as a horizontal asymptote as $x \to \pm\infty$.
- It is symmetrical about the y-axis.

This information leads to the sketch of the graph shown in Figure 4.3.2. (The numbers ± 0.707, marked on the x-axis, are the approximations of $\pm 1/\sqrt{2}$ to three decimal places.)

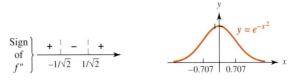

Figure 4.3.1   **Figure 4.3.2**

EXAMPLE 4.3.2

Sketch the graph of the function $f(x) = x - \ln x$.

SOLUTION In this case, the domain is $(0, \infty)$, since the logarithm is only defined for positive x. The first derivative is

$$f'(x) = 1 - \frac{1}{x},$$

which is negative on $(0, 1)$, positive on $(1, \infty)$ and zero at $x = 1$. The second derivative

$$f''(x) = \frac{1}{x^2},$$

is positive throughout $(0, \infty)$. From these facts we conclude that

- $f(x)$ is decreasing on $(0, 1)$ and increasing on $(1, \infty)$;
- it has a global minimum of height 1 at $x = 1$ (by the single critical point principle);
- it is concave up throughout $(0, \infty)$.

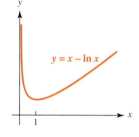

$y = x - \ln x$

Figure 4.3.3

There is no apparent symmetry in this graph, but there is an asymptote—namely, the y-axis. The graph climbs higher and higher as it approaches the y-axis, because x goes to zero and $\ln x$ goes to $-\infty$. Therefore, $x - \ln x \to \infty$ as $x \to 0^+$.

There is no horizontal asymptote, however. As $x \to \infty$, the graph climbs higher and higher, without bound. That is not altogether obvious. It is because x grows much faster than $\ln x$, so the difference, $x - \ln x$, gets unboundedly large.

Assembling all of our information results in the sketch shown in Figure 4.3.3.

EXAMPLE 4.3.3

Sketch the graph of the function $f(x) = \dfrac{3x}{x + 2}$.

SOLUTION The domain of this function consists of all $x \neq -2$, and the line $x = -2$ is a vertical asymptote. There is also a horizontal asymptote, the line $y = 3$. Also, $f(x) > 0$ on $(-\infty, -2)$ and $(0, \infty)$, and $f(x) < 0$ on $(-2, 0)$.

Using the quotient rule, we get

$$f'(x) = \frac{6}{(x + 2)^2},$$

which is positive for all x in the domain. Therefore, the function is increasing on both the intervals $(-\infty, -2)$ and $(-2, \infty)$.

To determine concavity, we first compute

$$f''(x) = \frac{-12}{(x + 2)^3}.$$

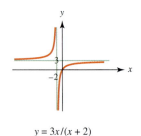

$y = 3x/(x+2)$

Figure 4.3.4

This quotient is positive on $(-\infty, -2)$ and negative on $(-2, \infty)$. That means the graph is concave up on the first of these intervals and concave down on the second. Since $f(0) = 0$, the graph goes through the point $(0, 0)$.

We now have enough information to obtain a sketch of the graph, which is shown in Figure 4.3.4.

EXAMPLE 4.3.4

Sketch the graph of the function $f(x) = x^2 e^{-x}$.

SOLUTION   The domain of this function is the entire x-axis, and there are no vertical asymptotes. It is also clear that $f(x) > 0$ for all x.

The first and second derivatives are

Figure 4.3.5

$$f'(x) = (2x - x^2)e^{-x} \quad \text{and} \quad f''(x) = e^{-x}(x^2 - 4x + 2).$$

The solutions of $f'(x) = 0$ are $x = 0$ and $x = 2$, and testing for signs results in the sign diagram of Figure 4.3.5.

We conclude the following: $f(x)$ is decreasing on the intervals $(-\infty, 0)$ and $(2, \infty)$ and increasing on $(0, 2)$, with a local minimum at $x = 0$ of height $f(0) = 0$ and a local maximum at $x = 2$ of height $f(2) = 4e^{-2} \approx 0.54$.

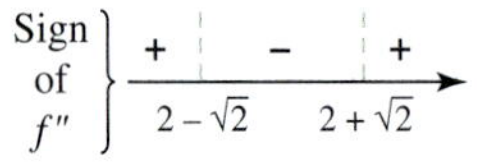

We can use the quadratic formula to solve $f''(x) = 0$. The solutions are $x = 2 \pm \sqrt{2}$, which are approximately 3.414 and 0.586. Testing for signs yields the diagram in Figure 4.3.6.

We conclude the following: The graph is concave up on $(-\infty, 2 - \sqrt{2})$ and $(2 + \sqrt{2}, \infty)$ and concave down on $(2 - \sqrt{2}, 2 + \sqrt{2})$, with inflection points at $x = 2 \pm \sqrt{2}$.

It is sometimes convenient to summarize the derivative information in a table such as the following:

Sign of f'': $+$ on $2 - \sqrt{2}$, $-$ between, $+$ on $2 + \sqrt{2}$

Figure 4.3.6

x	$f(x)$	$f'(x)$	$f''(x)$	graph characteristics
$-\infty < x < 0$	$+$	$-$	$+$	decreasing, concave up
$x = 0$	0	0	$+$	local minimum
$0 < x < 2 - \sqrt{2}$	$+$	$+$	$+$	increasing, concave up
$x = 2 - \sqrt{2}$	$+$	$+$	0	inflection point
$2 - \sqrt{2} < x < 2$	$+$	$+$	$-$	increasing, concave down
$x = 2$	$+$	0	$-$	local maximum
$2 < x < 2 + \sqrt{2}$	$+$	$-$	$-$	decreasing, concave down
$x = 2 + \sqrt{2}$	$+$	$-$	0	inflection point
$2 + \sqrt{2} < x < \infty$	$+$	$-$	$+$	decreasing, concave up

Neither of the criteria for symmetry—about the y-axis or about the origin—is satisfied. As for simple points to plot, we have already found a few. In particular, the graph goes through $(0, 0)$ and $(2, 4e^{-2}) \approx (2, 0.54)$.

What about the behavior as $x \to \pm\infty$? If we rewrite the function as a quotient, $f(x) = x^2/e^x$, we see that $f(x) \to 0$ as $x \to \infty$, which means that the x-axis is a horizontal asymptote on the right side.

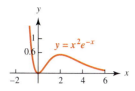

Figure 4.3.7

As $x \to -\infty$, the graph climbs higher and higher without bound. That is because both x^2 and e^{-x} get very large for large negative x. The graph is shown in Figure 4.3.7.

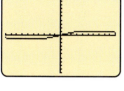

Figure 4.3.8

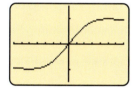

Figure 4.3.9

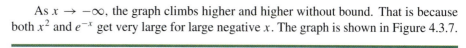

APPLYING TECHNOLOGY

Even with a computer or graphing calculator, it is important to know the key features of the graph, such as local maxima and minima, concavity, and asymptotes. If the graphing window is too small, it may exclude them. On the other hand, they may be impossible to see if the window is too large. In both cases the graph may be uninformative or misleading.

EXAMPLE 4.3.5

Use a graphing calculator to sketch the graph of the function $f(x) = \dfrac{10x}{x^2 + 25}$.

SOLUTION   We first try to sketch the graph using the default settings. On the TI-83 Plus, for instance, they produce a graphing window of $[-10, 10] \times [-10, 10]$, which results in the unsatisfactory graph shown in Figure 4.3.8.

One problem with that picture is that the vertical settings are too large for the range of the function. We might try to change the window setting by trial and error. Using $[-3, 3] \times [-2, 2]$, for example, produces the graph shown in Figure 4.3.9. However, that picture is misleading because it does not show the local (and global) maximum at $x = 5$ and the minimum at $x = -5$. We find those by applying the first derivative test with

$$f'(x) = \frac{10(25 - x^2)}{(x^2 + 25)^2}.$$

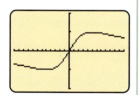

Figure 4.3.10

If we correct the setting, say, to $[-7, 7] \times [-2, 2]$, the picture improves, as shown in Figure 4.3.10, but it is still misleading.

We now see the maximum and minimum points, but we do not see that the graph has inflection points in addition to the one at $x = 0$. We find them by solving

$$f''(x) = \frac{20x(x^2 - 75)}{(x^2 + 25)^3} = 0,$$

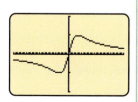

Figure 4.3.11

which gives $x = \pm\sqrt{75} \approx \pm 8.66$. By increasing the horizontal interval to $[-10, 10]$, we get Figure 4.3.11, which gives a little better sense of the inflection points but does not quite show the true features of the graph. For instance, the graph has the x-axis as a horizontal asymptote, which is suggested by Figure 4.3.11 but not convincingly displayed.

To show it, we increase the horizontal interval still more, to $[-30, 30]$, keeping the vertical set at $[-2, 2]$. The result, shown in Figure 4.3.12, shows the asymptote and the inflection points. We can see them even more clearly by changing the window setting

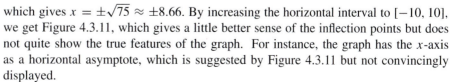
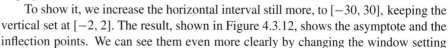

Figure 4.3.12

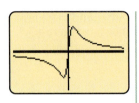

Figure 4.3.13

to $[-60, 60] \times [-1.5, 1.5]$, which results in the graph of Figure 4.3.13. However, we pay for it by sacrificing some clarity near $x = 0$.

Finally, we observe that the graph is *symmetric about the origin*, corresponding to the fact that $f(-x) = -f(x)$.

There are several morals to this story:

- First, and most important, the calculator cannot relieve you of all obligation to think. Without some understanding of the basic features of the graph, you may get an unintelligible or misleading picture.

- Second, even with calculus to guide you, a certain amount of trial and error is often necessary to get a clear picture of the graph.

- Third, there is no "right" picture of the graph (though there are certainly wrong ones). Emphasizing one feature may come at the expense of diminishing another. In some cases it may take two or three different graphs to get a clear insight into the properties of the function.

Another possible source of difficulty we have not yet touched on is that of solving the equations $f'(x) = 0$ and $f''(x) = 0$. Up until now, we have been able to do that by standard techniques of algebra. However, most equations do not yield so readily (if at all) to those techniques, and we have to use approximation methods—often with the aid of a calculator. Thus, not only can the calculator show us the graph, but it can, if used properly, help us choose the settings. We will explore this further in Section 4.6.

Practice Exercises 4.3

• •

Sketch the graphs of the following functions over their natural domains. Indicate any symmetry, asymptotes, local or global extreme points, and inflection points.

1. $f(x) = \dfrac{x^2}{x^2 - 1}$

2. $f(x) = xe^{x/2}$

Exercises 4.3

• •

In Exercises 1–7, sketch the graph of a function $f(x)$ satisfying the given conditions.

1. The domain consists of the entire x-axis;
$f(0) = 0$ and $f(1) = 1$;
$f'(x) > 0$ on $(-\infty, 1)$; $f'(x) < 0$ on $(1, \infty)$;
$f''(x) < 0$ on $(-\infty, 2)$; $f''(x) > 0$ on $(2, \infty)$;
$\lim\limits_{x \to \infty} f(x) = 0$ and $\lim\limits_{x \to -\infty} f(x) = -\infty$.

2. The domain consists of all $x \neq 0$; there is a vertical asymptote at $x = 0$; $f'(x) < 0$ on $(-\infty, 0)$ and on $(0, \infty)$;
$f''(x) < 0$ on $(-\infty, 0)$; $f''(x) > 0$ on $(0, \infty)$;
$\lim\limits_{x \to \infty} f(x) = 0$ and $\lim\limits_{x \to -\infty} f(x) = 0$.

3. The domain consists of the entire x-axis;
$f(-x) = f(x)$ for all x; $f(x) \geq 0$ for all x;
$f'(x) < 0$ on $(-\infty, 0)$; $f'(x) > 0$ on $(0, \infty)$;
$f''(x) > 0$ on $(-1, 1)$; $f''(x) < 0$ on $(-\infty, -1)$ and on $(1, \infty)$;
$\lim\limits_{x \to \infty} f(x) = 1$ and $\lim\limits_{x \to -\infty} f(x) = 1$.

4. The domain consists of the entire x-axis;
$f(0) = 0$; $f(-x) = -f(x)$ for all x;
$f'(x) > 0$ for all x; $f''(x) > 0$ on $(-\infty, 0)$;
$f''(x) < 0$ on $(0, \infty)$;
$\lim\limits_{x \to \infty} f(x) = 1$ and $\lim\limits_{x \to -\infty} f(x) = -1$.

5. The domain consists of all $x \neq \pm 1$; $\lim_{x\to\infty} f(x) = 0$ and $\lim_{x\to-\infty} f(x) = 0$; there are vertical asymptotes at $x = 1$ and $x = -1$;
$f(-x) = f(x)$ for all x; $f(0) = 0$;
$f'(x) > 0$ on $(-\infty, -1)$ and on $(-1, 0)$;
$f'(x) < 0$ on $(0, 1)$ and on $(1, \infty)$;
$f''(x) > 0$ on $(-\infty, -1)$ and on $(1, \infty)$;
$f''(x) < 0$ on $(-1, 1)$.

6. The domain consists of all $x > 0$; the y-axis is a vertical asymptote;
$f'(x) < 0$ on $(0, 2)$; $f'(x) > 0$ on $(2, \infty)$;
$f''(x) > 0$ for all x;
$f(2) = 4$; $\lim_{x\to\infty} f(x) = \infty$.

7. The domain consists of all $x \neq 0$; the y-axis is a vertical asymptote;
$f'(x) < 0$ on $(-\infty, 0)$ and $(2, \infty)$ and $f'(x) > 0$ on $(0, 2)$;
$f''(x) < 0$ on $(-\infty, 0)$ and $(0, 4)$ and $f''(x) > 0$ on $(4, \infty)$;
$f(2) = 3$; $\lim_{x\to\infty} f(x) = 1$ and $\lim_{x\to-\infty} f(x) = 1$.

8. A ship is surveying the ocean floor with sonar. Starting at a certain point, it travels in a straight line for a total of 1,000 feet and measures the depth below the surface. The computer returns the following information, where x is the ship's coordinate along its line of motion and $f(x)$ is the depth in feet.
$f(0) = -113$;
$f'(x) = 0$ and $f''(x) = 0$ for $0 \leq x \leq 337$;
$f'(x) < 0$ for $337 < x < 442$;
$f(442) = -202$, $f'(442) = 0$;
$f'(x) > 0$ for $442 < x < 546$;
$f''(x) > 0$ for $337 < x < 546$;
$f'(x) = 0$ and $f''(x) = 0$ for $546 \leq x \leq 1,000$;
$f(1,000) = -113$.
Sketch the floor of the ocean for that 1,000-foot stretch.

In Exercises 9–26, sketch the graph of the given function over its natural domain. Indicate any symmetry, asymptotes, local or global extreme points, and inflection points.

9. $f(x) = x^3 - 3x + 5$

10. $f(x) = x^3 - 3x^2 + 3x + 2$

11. $f(x) = x^4 - 8x^2 + 10$ 12. $f(x) = 2x^5 - 5x^4 + 8$

13. $f(x) = \dfrac{1}{x - 2}$ 14. $f(x) = \dfrac{1}{(x-2)^2}$

15. $f(x) = \dfrac{x}{2x + 1}$ 16. $f(x) = \dfrac{x}{2x - 3}$

17. $f(x) = x + \dfrac{4}{x}$ 18. $f(x) = x - \dfrac{4}{x}$

19. $f(x) = \dfrac{x}{x^2 + 4}$ 20. $f(x) = \dfrac{x^2}{x^2 + 4}$

21. $f(x) = xe^{-x}$

22. $f(x) = (\ln x)^2$

23. $f(x) = e^x + e^{-x}$

24. $f(x) = e^x - e^{-x}$

25. $f(x) = \ln(x^2 + 4)$ 26. $f(x) = \dfrac{1}{x^2 + 4}$

27. Let $f(x) = x^5 + 4x^3 - 12x^2 - 18x + 10$. Find $f'(x)$ and $f''(x)$ and then answer the following:
   (a) Find the inflection point of $f(x)$. Explain how you can be sure there is only one.
   (b) Find the critical points of $f(x)$. Explain how you can be sure there are exactly two.
   (c) What are $\lim_{x\to\infty} f(x)$ and $\lim_{x\to-\infty} f(x)$?
   (d) Using the previous answers, determine a suitable window and produce a graph of $f(x)$ that shows all the essential features.

28. Let $f(x) = e^x - 3x^2 + x$. Find $f'(x)$ and $f''(x)$ and then answer the following:
   (a) Find the inflection point of $f(x)$. Explain how you can be sure there is only one.
   (b) Find the critical points of $f(x)$. Explain how you can be sure there are exactly two.
   (c) What are $\lim_{x\to\infty} f(x)$ and $\lim_{x\to-\infty} f(x)$?
   (d) Using the previous answers, determine a suitable window and produce a graph of $f(x)$ that shows all the essential features.

29. Eucalyptus trees in Southern California are dying off faster than usual because of red gum lerp psyllids (a plant louse) and the larvae of the eucalyptus long-horned beetle borer. Let the rate of insect infection in a grove of 120 trees be given by

$$R = 0.00064x^2(120 - x),$$

where x is the number of infected trees and R is the rate at which the infection is spreading (in newly infected trees per month).

   Sketch the graph of this equation for $0 \leq x \leq 120$ to see how the rate of infection changes. At what stage is the infection spreading fastest?

Source: Renee Vogel, *G'day Mite*, Los Angeles Magazine, Oct. 1999.

30. Martin works in a factory where he assembles tiny electronic parts. Because the work causes eye strain and loss of concentration, his shift only lasts 6 hours. His productivity is modeled by

$$P(t) = -0.5t^3 + t^2 + 12t, \quad 0 \leq t \leq 6,$$

where t is the time in hours and $P(t)$ is the rate of output in parts per hour.

   Sketch the graph of $P(t)$ to see how Martin's productivity changes. When is he most productive?

31. Light penetrates the water in a lake according to the function

$$P(x) = e^{-rx},$$

where x is the depth in feet and $P(x)$ is the percentage of light that has reached that depth from the surface.

(a) Sketch the light penetration curve for a lake with $r = 0.1$.

(b) In order to survive, a particular aquatic plant needs to be reached by at least 50% of the light penetrating the surface. If $r = 0.1$, can i̇ ̇urvive at a depth of 10 feet? What is the greatest depth ... which it can survive?

32. Biologists call the graph of the equation $R = aSe^{-bS}$, $S \geq 0$, the **Ricker curve**, in honor of the Canadian biologist William Ricker, who used the equation to model the population of certain species of fish, among them Pacific salmon. In this model, S is the size of the parent population, R is the number of mature offspring, and a and b are positive constants. Sketch a graph of the Ricker curve for $a = 2$ and $b = 0.7$.

33. You have coordinated the administration of a district-wide 8th grade math test. Each student was given an adjusted score, obtained by subtracting the district-wide average score from the individual's score. The number of people receiving a given adjusted score, say, s, was found to fit the function

$$N(s) = 1,827e^{-0.005s^2}.$$

(a) You need to write a report on the results for the district superintendent. Make a sketch of the graph to include in your report.

(b) According to this model, what adjusted score was achieved by the largest number of students?

34. The use of wells affects the level and quality of ground water in a region. When the water level becomes low, salt water can be drawn into fresh water, rendering it unsuitable for drinking and agriculture. One large desert farm has noticed that the salt content of its well water is increasing, and samples indicate that the percentage of salt water fits the function

$$P(t) = 1 - e^{-0.362t},$$

where t is the time in years. Sketch the graph of $P(t)$ to help the farmer understand what the future of his ground water will be if there are no changes in well usage.

Solutions to practice exercises 4.3

1. Since $f(-x) = f(x)$, the graph is symmetric about the y-axis. The lines $x = \pm 1$ are vertical asymptotes, and $y = 1$ is a horizontal asymptote. We also observe that $f(0) = 0$ and $f(x) < 0$ for all $x \neq 0$ on $(-1, 1)$. Also, $f(x) > 0$ if $|x| > 1$. In fact, $f(x) > 1$ if $|x| > 1$, since the numerator is larger than the denominator and both are positive. Using the quotient rule, we get the first derivative:

$$f'(x) = -\frac{2x}{(x^2 - 1)^2}.$$

The denominator is positive for $x \neq \pm 1$, so that the sign of $f'(x)$ is the same as that of $-2x$, positive for $x < 0$ and negative for $x > 0$. We conclude that $f(x)$ is decreasing on both $(0, 1)$ and $(1, \infty)$, and increasing on both $(-\infty, -1)$ and $(-1, 0)$. The only critical point occurs at $x = 0$, and it is a local maximum, since the sign of $f'(x)$ changes from positive to negative.

Using the quotient rule again and simplifying the result, we get

$$f''(x) = \frac{2 + 6x^2}{(x^2 - 1)^3},$$

which is positive for $|x| > 1$ and negative for $|x| < 1$. Therefore, the graph is concave up on $(-\infty, -1)$ and $(1, \infty)$, and concave down on $(-1, 1)$. Putting together this information, we obtain the graph shown in Figure 4.3.14.

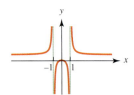

Figure 4.3.14

2. In this case, there is no apparent symmetry. The natural domain is the entire x-axis. The first derivative is

$$f'(x) = e^{x/2} + \frac{1}{2}xe^{x/2} = e^{x/2}\left(1 + \frac{x}{2}\right),$$

which is zero at $x = -2$. It is not hard to see that $f'(x) < 0$ for $x < -2$ and $f'(x) > 0$ for $x > -2$, so there is a local minimum at $x = -2$ with a value of $f(-2) = -2e^{-1} \approx -0.736$. By the single critical point principle, it is also a global minimum.

The second derivative is $f''(x) = e^{x/2}(1 + x/4)$, which is zero at $x = -4$, where it changes sign from negative to positive. Therefore, the graph is concave down on $(-\infty, -4)$, concave up on $(-4, \infty)$, and has an inflection point at $x = -4$.

We observe that $f(x) > 0$ on $(0, \infty)$ and $f(x) < 0$ on $(-\infty, 0)$. Moreover,

$$\lim_{x \to \infty} xe^{x/2} = \infty$$

because both factors become unboundedly large. On the left, we have

$$\lim_{x \to -\infty} x e^{x/2} = 0.$$

To see why, make the substitution $t = -x$. Then $t \to \infty$ as $x \to -\infty$, and $\lim_{x \to -\infty} x e^{x/2} = \lim_{t \to \infty} -t e^{-t/2} = \lim_{t \to \infty} -t/e^{t/2} = 0$. Putting together all the information gives the graph shown in Figure 4.3.15.

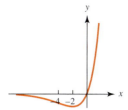

Figure 4.3.15

4.4 Optimization and Applications

Optimization is an important consideration in all spheres of human activity. Everyone wants to get the biggest bang for the buck. Companies want to maximize profit, investors want to maximize dividends and minimize risk, travelers want to minimize the time it takes to get from one point to another. Nature also favors processes that optimize time and energy, a fact that plays a role in determining many physical laws.

In this section we will consolidate and apply what we have learned about using the derivative to find maxima and minima. Let's start with a simple example.

EXAMPLE 4.4.1

A company manufacturing CD players has determined that the cost of producing x players a week is given by the function

$$C(x) = 6,600 + 70x,$$

and the revenue from selling x players a week is given by

$$R(x) = 180x - \frac{x^2}{4}, \quad 0 \le x \le 400.$$

Assuming that the company sells all the players it produces, how many should it make to maximize its profits? And what will its maximum profit be?

SOLUTION By subtracting cost from revenue, we obtain the profit function

$$P(x) = -\frac{x^2}{4} + 110x - 6,600, \quad 0 \le x \le 400,$$

whose derivative is

$$P'(x) = -\frac{x}{2} + 110.$$

Solving $P'(x) = 0$, we see that the only critical point occurs at $x = 220$, and it is not hard to check that the sign of $P'(x)$ changes from positive to negative at that point. Therefore, there is a local maximum at $x = 220$.

According to the single critical point principle, the local maximum is also a global maximum. Therefore, the company should make 220 players a week to maximize its profit. To find the maximum profit, we simply substitute $x = 220$ into the profit function, which gives $P(220) = 5,500$.

Next, we take up the general question of optimizing a continuous function $f(x)$ over an interval. There are two cases to consider.

Case 1: Optimizing $f(x)$ on a closed, bounded interval

By a **closed, bounded** interval we mean one of the form $[a, b]$, where both a and b are finite numbers. The following theorem is proved in advanced calculus:

Theorem 4.4.1

The extreme value theorem

If $f(x)$ is a continuous function on a closed, bounded interval $[a, b]$, then $f(x)$ has a global maximum and a global minimum on the interval.

If we think of the intuitive meaning of continuity—that we can draw the graph of $f(x)$ from one endpoint $(a, f(a))$ to the other endpoint $(b, f(b))$ without lifting our pencil from the paper—it seems reasonable to conclude that the graph must pass through a highest point (global maximum) and a lowest point (global minimum). A precise and convincing proof is based on the properties of the real number system and the rules of deductive reasoning, which we will not take up here. It is worth pointing out that the conclusion is not valid if the interval has infinite length or if either endpoint is excluded—as we shall see in some of the later examples.

In any case, by accepting the extreme value theorem on intuitive grounds, we obtain a procedure for finding maxima and minima on $[a, b]$.

Method for Finding Global Maxima and Minima

1. Find all critical points inside the interval—on (a, b), that is. These, plus the endpoints $x = a$ and $x = b$ are the only possible *candidates* for a global maximum or minimum on $[a, b]$.

2. *Evaluate* the function at each of the possible candidates—critical points inside the interval, plus endpoints—and *compare* the values. The highest value is the global maximum and the lowest is the global minimum.

EXAMPLE 4.4.2

Find the global maximum and minimum of each of the following functions on the interval specified:

(i) $f(x) = x^3 - 3x + 1$ on the interval $[0, 2]$
(ii) $f(x) = x^3 - 3x^2 + 3x + 2$ on the interval $[0, 2]$

SOLUTION   In each case, we begin by solving $f'(x) = 0$.

(i) In this case, $f'(x) = 3x^2 - 3$, and the solutions of $f'(x) = 0$ are $x = \pm 1$. However, only one of these, $x = 1$, is in the given interval. We construct a table of values for $f(x)$ at that critical point and the two endpoints:

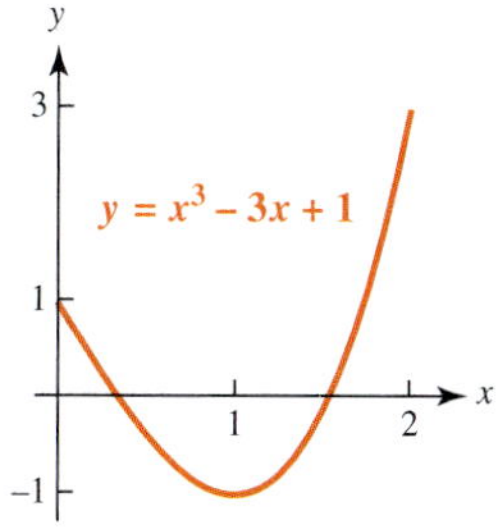

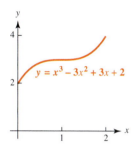

Figure 4.4.1

x	0	1	2
$f(x)$	1	-1	3

By comparing these values, we see that on the interval $[0, 2]$, the maximum value is 3, taken at $x = 2$, and the minimum value is -1, taken at $x = 1$. The graph is shown in Figure 4.4.1.

(ii) In this case, $f'(x) = 3x^2 - 6x + 3 = 3(x - 1)^2$, which has its only zero at $x = 1$. The values at the critical point and endpoints are given by the following table:

x	0	1	2
$f(x)$	2	3	4

We see that the minimum value is 2, occurring at the left-hand endpoint, $x = 0$. The maximum value is 4, occurring at the right-hand endpoint, $x = 2$.

The graph is shown in Figure 4.4.2. Observe that $f'(x)$ is positive on both sides of $x = 1$, so that $f(x)$ is increasing throughout $[0, 2]$.

Figure 4.4.2

Case 2: Optimizing $f(x)$ on an interval with one or both endpoints excluded

A continuous function always has a maximum and minimum on a closed, bounded interval. However, if one or both of the endpoints are excluded—for instance, if the interval is infinite in either direction—the function may not have a maximum or minimum, as you can see in the following example.

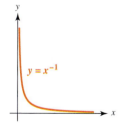

Figure 4.4.3

EXAMPLE 4.4.3

Show that the function $f(x) = x^{-1}$ does not have a maximum or minimum on the interval $(0, \infty)$.

SOLUTION   The derivative $f'(x) = -x^{-2}$ is negative throughout the interval, so the function is steadily decreasing. As $x \to 0^+$, the graph climbs higher and higher without bound, so that there is no maximum. And as $x \to \infty$, the graph levels off toward zero but never reaches it, so there is no minimum. The graph is shown in Figure 4.4.3.

We now consider a procedure for finding global extrema on an interval with one or both endpoints excluded—either a finite interval that is open or half-open or else an interval of infinite length. The method involves nothing more than the first derivative test, the single critical point principle, and a look at the behavior of the graph as it nears the ends of the interval. Use the graph as a guide whenever possible.

In the following we assume that f is defined and continuous throughout a given interval:

Steps for finding global maxima and minima, endpoints excluded

1. Find all critical points inside the interval

2. If there is exactly one critical point inside the interval and the sign of $f'(x)$ changes from negative to positive, then there is a global minimum there (first derivative test plus single critical point principle).

    In this case, there is no global maximum inside the interval. It can only occur at an endpoint if there is one. You must study the behavior of the graph as it moves toward both ends of the interval to determine if it achieves a maximum.

    A similar conclusion—with the words "maximum" and "minimum" switched—holds if the sign changes from positive to negative.

3. If there are no critical points inside the interval, there is no maximum and no minimum inside the interval. The only places they can occur are at the endpoints if there are any. (That's because the graph is either always rising or always falling throughout the interval.) Once again, you must examine the behavior of the graph as it moves toward the ends of the interval to determine whether it achieves a maximum or minimum at an endpoint.

    The same is true if there is a critical point but $f'(x)$ never changes sign.

4. If there is more than one critical point inside the interval,

    *   compute the values at the critical points and endpoints, if any;

    *   examine the behavior of the graph as it moves toward both ends of the interval.

EXAMPLE 4.4.4

Find the global maximum or minimum, if any, of the function $f(x) = xe^{-x}$ on the interval $(-\infty, \infty)$.

SOLUTION We first observe that $f(x)$ has the same sign as x, negative on $(-\infty, 0)$ and positive on $(0, \infty)$, and that $f(0) = 0$. There is no symmetry and no vertical asymptote.

The derivative is

$$f'(x) = (1 - x)e^{-x}.$$

It is zero at only one place, $x = 1$, where its sign changes from positive to negative. By applying the first derivative test and the single critical point principle, we conclude that the function has a global maximum at $x = 1$, with a value of $e^{-1} \approx 0.368$.

On the other hand, there is no minimum. As $x \to -\infty$, $f(x) \to -\infty$. To see why, make the substitution $x = -t$. Then $t \to \infty$ as $x \to -\infty$, so that

$$\lim_{x \to -\infty} xe^{-x} = \lim_{t \to \infty} -te^{t} = -\lim_{t \to \infty} te^{t} = -\infty.$$

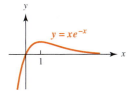

To the right, as $x \to \infty$, the graph has the x-axis as a horizontal asymptote, for

$$\lim_{x \to \infty} xe^{-x} = \lim_{x \to \infty} \frac{x}{e^x} = 0.$$

The graph is shown in Figure 4.4.4.

Figure 4.4.4

The problem of finding global extreme points depends on the interval in question as well as on the formula defining the function, as the next example shows.

EXAMPLE 4.4.5

Certain fish species, such as the Pacific salmon, breed only once in their lifetime. (Such species are called *semelparous*.) The average number of offspring produced by a female of the species is a function of her age. Let $r(x)$ be the average number of offspring produced by females of age x years, and suppose that for a certain semelparous species it is given by the formula

$$r(x) = \frac{4.6 \ln(2x)}{x}, \quad \text{for } 0.5 \le x.$$

(In this model, we assume that females under 0.5 years are not mature enough to breed.)

(i) Among all females, what age produces the maximum average number of off-spring?

(ii) For females between 2 and 4 years old, what age produces the maximum average number of offspring?

SOLUTION We begin by taking the derivative

$$r'(x) = 4.6 \left[\frac{x \cdot \frac{d}{dx} \ln(2x) - \ln(2x)}{x^2} \right] = 4.6 \left[\frac{(1 - \ln(2x))}{x^2} \right].$$

Solving $r'(x) = 0$ is equivalent to solving $1 - \ln(2x) = 0$, which leads to a single critical point at $x = e/2 \approx 1.36$. By checking the sign of $r'(x)$ on both sides of $e/2$, we can see that the critical point is a local maximum. For instance, $r'(1) = 4.6(1 - \ln 2) > 0$ and $r'(e) = -4.6(\ln 2)/e^2 < 0$.

(i) We want to find the point in the interval $[0.5, \infty)$ at which $r(x)$ is maximized. Since there is a local maximum inside this interval at $x = e/2$, the single critical point principle says that it is a global maximum. Therefore, we conclude that the maximum average number of offspring is produced by females of age $e/2 \approx 1.36$ years.

(ii) Here, we want to find the point in the interval $[2, 4]$ at which $r(x)$ is maximized. This time, the critical point at $x = e/2$ lies *outside* the interval, which means that the global maximum over $[2, 4]$ is attained at one of the endpoints. By testing each, we get

$$r(2) = 2.3 \ln 4 \approx 3.19 \quad \text{and} \quad r(4) = 1.15 \ln 8 \approx 2.39,$$

which shows that the maximum occurs at $x = 2$. Another way to see that is to observe that $r'(x)$ is negative throughout the interval $[2, 4]$ (why?), which means that $r(x)$ is decreasing over the interval.

We therefore conclude that females of age 2 years produce the maximum average number of offspring among those between 2 and 4 years old.

The next example involves a function with two critical points over an interval with no endpoints.

EXAMPLE 4.4.6

Find the global maximum and minimum, if any, of the function $f(x) = xe^{-x^2/2}$ on the interval $(-\infty, \infty)$.

SOLUTION   The derivative is $f'(x) = e^{-x^2/2}(1 - x^2)$, which is zero at $x = \pm 1$. Since $e^{-x^2} > 0$ for all x, the sign of $f'(x)$ is the same as that of $(1 - x^2)$, positive for $|x| < 1$ and negative for $|x| > 1$. Therefore, $f(x)$ has a local minimum at $x = -1$ and a local maximum at $x = 1$. The height of the graph at those points is

$$f(-1) = -e^{-1/2} \approx -0.61 \quad \text{and} \quad f(1) = e^{-1/2} \approx 0.61,$$

respectively.

To see whether the local extrema are global, we will study the behavior of the graph as $x \to \pm\infty$. To begin with, observe that $f(x)$ is negative if $x < 0$ and positive if $x > 0$. Moreover,

$$\lim_{x \to \pm\infty} xe^{-x^2/2} = \lim_{x \to \pm\infty} \frac{x}{e^{x^2/2}} = 0,$$

which means the x-axis is a horizontal asymptote in both directions.

As $x \to \infty$, the graph descends to the x-axis from above, and as $x \to -\infty$, it rises to it from below, as shown in Figure 4.4.5. That means the graph cannot get any higher than its height at $x = 1$ or any lower than its height at $x = -1$. We conclude that

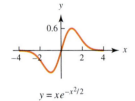

$y = xe^{-x^2/2}$

Figure 4.4.5

- $f(x)$ has a global maximum value of $e^{-1/2} \approx 0.61$, which occurs at $x = 1$, and
- $f(x)$ has a global minimum value of $-e^{1/2} \approx -0.61$, which occurs at $x = -1$.

Practice Exercises 4.4

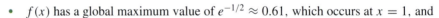

1. Find the global maximum and minimum of the function $f(x) = x^3 - x^2 + 1$ on the interval $[-1, 1]$.

2. Find the global maximum and minimum, if any, of the function $f(x) = e^x/x$ on the interval $(0, \infty)$.

Exercises 4.4

1. Let $f(x) = x^3 - 3x^2 + 1$. Find the global maximum and minimum values of $f(x)$ over each of the following intervals and state where they occur:
  (a) $[-1, 1]$ **(b)** $[1, 3]$ **(c)** $[-2, 3]$

2. Let $f(x) = x^3 - 12x$. Find the global maximum and minimum values of $f(x)$ over each of the following intervals and state where they occur:
  (a) $[0, 3]$ **(b)** $[-3, 0]$ **(c)** $[-1, 1]$

In Exercises 3–10, find the global maximum and minimum of the given function over the given interval.

3. $f(x) = x^2 + x + 3$, $[0, 2]$

4. $f(x) = x^2 + x + 3$, $[-1, 1]$

5. $f(x) = \dfrac{x}{x^2 + 4}$, $[1, 3]$

6. $f(x) = \dfrac{x}{x^2 + 4}$, $[-3, 3]$

7. $f(x) = (4x^2 - 9)^4$, $[-1, 2]$

8. $f(x) = 3x^4 - 4x^3 - 8$, $[-1, 2]$

9. $f(x) = \sqrt{x^2 - 2x + 2}$, $[0, 2]$

10. $f(x) = x + \dfrac{3}{x}$, $[1, 2]$

In Exercises 11–28, determine whether the function has a global maximum and/or minimum on the interval in question and, if so, where it occurs and what its value is.

11. $f(x) = x^2 - 2x - 3$ on $[0, 3]$

12. $f(x) = x^2 - 2x - 3$ on $[0, \infty)$

13. $f(x) = x + \dfrac{4}{x}$ on $(0, \infty)$

14. $f(x) = x - \dfrac{4}{x}$ on $(0, \infty)$

15. $f(x) = xe^{-2x}$ on $(-\infty, \infty)$

16. $f(x) = xe^{2x}$ on $(-\infty, \infty)$

17. $f(x) = x^3 + x - 1$ on $[-1, 1]$

18. $f(x) = x^3 + x - 1$ on $(-\infty, \infty)$

19. $f(x) = x^2 \ln x$ on $(0, \infty)$

20. $f(x) = x(\ln x)^2$ on $(0, \infty)$

21. $f(x) = \dfrac{e^x}{x^2}$ on $(0, \infty)$

22. $f(x) = \dfrac{x - 1}{x + 1}$ on $[0, \infty)$

23. $f(x) = x^4 - 6x^2$ on $[-2, 2]$

24. $f(x) = x^4 - 6x^2$ on $(-\infty, \infty)$

25. $f(x) = e^{2x} - 6e^x$ on $(-\infty, \infty)$

26. $f(x) = \dfrac{\ln x}{x^2}$ on $(0, \infty)$

27. $f(x) = \dfrac{x^2 - 1}{x^2 + 1}$ on $(-\infty, \infty)$

28. $f(x) = \dfrac{x^2 + 1}{x^2 - 1}$ on $(-1, 1)$

29. If a projectile is shot straight up from a 200-foot platform with an initial velocity of 160 feet per second, its height after t seconds is given by the formula $h = -16t^2 + 160t + 200$. After how many seconds will the projectile reach its highest point, and how high will it go?

30. A toy rocket is fired straight upward, and its height (in meters) is given by $h = t + 10 - \sqrt{2t^2 + 100}$, $0 \le t \le 20$, where t is the time in seconds. How high does the rocket go?

31. Suppose the average cost of producing x units of a certain product is given by the function

$$A(x) = 10x + \frac{4{,}000}{x + 1} + 300.$$

How many units should be produced to minimize the average cost, and what will the minimum be?

32. Some fish, such as Pacific salmon, produce many more mature offspring when their spawning grounds are less crowded. The Canadian biologist, William Ricker, developed a model of fish population dynamics based on the equation

$$R = aSe^{-bS}, \quad S \ge 0,$$

where S is the size of the parent population, R is the number of mature offspring, and a and b are positive constants. Biologists call the graph of this equation the **Ricker curve** in his honor. Show that R attains a global maximum for $S = 1/b$.

Solutions to practice exercises 4.4

1. We have $f'(x) = 3x^2 - 2x$. Solving $f'(x) = 0$ gives the critical points $x = \frac{2}{3}$ and $x = 0$. Both of these critical points are inside the interval. Next, we make a table of values for $f(x)$ at the two critical points as well as both endpoints, as follows:

x	-1	0	$2/3$	1
$f(x)$	-1	1	$23/27$	1

By comparing the values of $f(x)$, we conclude that on the interval $[-1, 1]$ the function $f(x) = x^3 - x^2 + 1$ has a minimum value of -1, occurring at $x = -1$, and a maximum value of 1, occurring at both $x = 0$ and $x = 1$. The graph is shown in Figure 4.4.6.

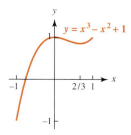

Figure 4.4.6

2. We first compute $f'(x) = (x-1)e^x/x^2$, and we see that $x = 1$ is the only solution of $f'(x) = 0$. Moreover, the sign of $f'(x)$ changes from negative to positive at $x = 1$, so the function has a local minimum there. By the single critical point principle, there is also a global minimum there, with value $f(1) = e$. There is no global maximum, since $\lim_{x\to\infty} f(x) = \infty$ and $\lim_{x\to 0+} f(x) = \infty$, which means the function increases without bound both as $x \to \infty$ and $x \to 0^+$. (Either would be enough to rule out a global maximum.)

4.5   Applied Optimization Problems

Applying optimization methods may require you to translate word problems into mathematical formulas. Here are a few suggestions for solving problems with the the help of calculus.

Solving word problems

1. **Read and think**   Before plunging into a problem, take a moment to read it carefully and think about it. What are its essential features? Have you seen such a problem before? What techniques seem appropriate? What is a reasonable estimate for the answer?

2. **Construct a mathematical model**   Be sure you understand what is given and what is asked. Assign symbols to all unknown quantities, both the primary one to be determined and any secondary ones that may be related to it. When possible, make a sketch and label it accurately. Write a set of equations relating the primary unknown in terms of given data and any secondary unknowns. In general, there should be as many equations as unknowns. Also, restrict the domains of the variables to those for which the problem makes sense.

3. **Apply reduction**   Reduce the mathematical model to a simpler, more familiar one. For example, eliminate any secondary unknowns by substitution, thereby reducing the set of equations to a single one involving only one unknown.

4. **Solve the reduced problem**   Solve the equation by familiar mathematical techniques, if possible, or by approximation methods. Avoid sloppiness and use complete sentences and mathematical expressions.

5. **Check your answer**   Once you have solved the problem, compare your answer with your original estimate and ask yourself if it makes sense. For example, a value that maximizes profit should not be smaller than a known value of the profit function.

EXAMPLE 4.5.1

A small independent magazine has been offering annual subscriptions at a price of $36. They currently have 6,000 subscribers. In an effort to increase its circulation, it has decided to lower the subscription rate, and it estimates that it will gain 200 new subscribers for every dollar it reduces the price. What price should the magazine set in order to maximize its revenue?

SOLUTION   Let p represent the price and q the number of subscribers. From the information given in the last paragraph we know that q is a linear function of p with slope -200. [For every \$1 that p decreases, q increases by 200. So the slope is $\Delta q / \Delta p = 200/(-1) = -200$.] And $q = 6{,}000$ when $p = 36$. Combining that information leads to the formula

$$q = 6{,}000 - 200(p - 36) = 13{,}200 - 200p.$$

To find the revenue, we multiply the price of a subscription by the number of subscribers:

$$R = pq = 13{,}200p - 200p^2. \tag{2}$$

We want to maximize this function over the interval $[0, 36]$. Taking the derivative gives

$$R'(p) = 13{,}200 - 400p.$$

By setting this equal to zero and solving, we get $p = 33$.

   Does this price actually give maximum revenue? Yes, because it is the only critical point inside the interval, and both the first and second derivative tests show it to be a local maximum. The single critical point principle then says that the revenue function has a global maximum there. Equation (2) tells us that at the price of \$33 the revenue will be \$217,800. By way of comparison, at the present price, with $p = 36$ and $q = 6{,}000$, the revenue is \$216,000 per year.

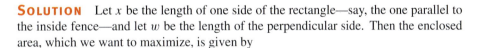

EXAMPLE 4.5.2

A livestock breeder has 1,200 feet of fencing, with which he plans to enclose a rectangular area and then divide it into two pens by erecting an inside fence parallel to one pair of sides (see Figure 4.5.1). What dimensions should he choose to maximize the enclosed area?

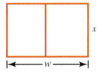

Figure 4.5.1

SOLUTION   Let x be the length of one side of the rectangle—say, the one parallel to the inside fence—and let w be the length of the perpendicular side. Then the enclosed area, which we want to maximize, is given by

$$A = xw. \tag{3}$$

As Figure 4.5.1 shows, there are three sides of length x—two on the boundary and one inside the pen—and two sides of length w. Therefore, $3x + 2w = 1{,}200$, the total length of fencing. Solving for w gives

$$w = 600 - \frac{3}{2}x,$$

and by substituting that into (3), we get

$$A = x\left(600 - \frac{3}{2}x\right) = 600x - \frac{3}{2}x^2.$$

Since there are three sides of length x and only 1,200 feet of fencing, we can restrict x to the interval $[0, 400]$.

Taking the derivative and setting it equal to zero give

$$A' = 600 - 3x = 0,$$

which means that the only critical point occurs at $x = 200$. There is a local maximum there, as we can see by using the second derivative test, and the single critical point principle says it is a global maximum. (Alternatively, we can compare the value of A at the critical point with its values at the endpoints, both of which are zero.) It follows that $w = 300$.

EXAMPLE 4.5.3

Find the point on the graph of $y = x^2$ in the first quadrant that is closest to the point $(0, 3)$ and find the minimum distance.

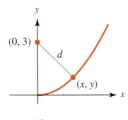

Figure 4.5.2

SOLUTION   The graph and the point $(0, 3)$ are shown in Figure 4.5.2. The distance d from any point (x, y) on the graph to the point $(0, 3)$ is given by the formula

$$d = \sqrt{x^2 + (y - 3)^2}.$$

However, any point (x, y) on the graph satisfies $y = x^2$, and by using that subsitution, we can write d as a function of y alone:

$$d = f(y) = \sqrt{y + (y - 3)^2},$$

which we must minimize for $y \geq 0$.

To do that, we first take the derivative:

$$f'(y) = \frac{1}{2}\left[y + (y - 3)^2\right]^{-1/2}\left[1 + 2(y - 3)\right]$$

$$= \frac{2y - 5}{2\sqrt{y + (y - 3)^2}}.$$

It follows that $f'(y) = 0$ if and only if $2y - 5 = 0$. Therefore, the only critical point of $f(y)$ on $[0, \infty)$ occurs at $y = \frac{5}{2}$.

How do we know that $f(y)$ has a global minimum there? First, a sign check of $f'(y)$ shows that $f(y)$ has a local minimum at $y = \frac{5}{2}$; for instance, $f'(1) < 0$ and $f'(3) > 0$. Then, by the single critical point principle, it must have a global minimum there.

We conclude that of all points on the graph of $y = x^2$ in the first quadrant, the point $(\sqrt{5/2}, 5/2)$ is closest to $(0, 3)$, and the distance between them is $f(5/2) = \sqrt{11/4}$.

Remark   We could have simplified our calculations a bit by observing that the point that minimizes the distance also minimizes the square of the distance. Therefore, we could have found the same point by minimizing $y + (y - 3)^2$. That way, we would have avoided dealing with the square root until the very end, when we had to compute the actual distance.

EXAMPLE 4.5.4

A graphic artist is designing a poster, which is to have margins of 2 inches at the top and along each side, and a 3-inch margin at the bottom. In order to save expenses, she wants the total area of the poster to be as small as possible, but the printed area (the part inside the margins) has to be 180 square inches. What dimensions will minimize the total area?

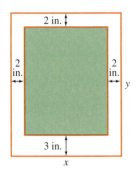

Figure 4.5.3

SOLUTION   In Figure 4.5.3, x is the width and y is the height of the total poster. We want to minimize $x \cdot y$, the total area.

By subtracting the top and bottom margins from the total height, we get $(y - 5)$ as the height of the printed area. Similarly, $(x - 4)$ is its width. Since the printed area (shaded in the picture) is 180 square inches, we get a relation between x and y,

$$(x - 4)(y - 5) = 180. \tag{4}$$

The idea is now to solve this equation for y in terms of x, in order to reduce the total area to a function of x alone. To do that, we first divide both sides of Eq. (4) by $(x - 4)$ and then add 5 to both sides. We get

$$y = 5 + \frac{180}{x - 4}.$$

Substituting this into the expression for the total area, we obtain the function

$$A(x) = x \cdot \left(5 + \frac{180}{x - 4}\right) = 5x + \frac{180x}{x - 4},$$

which we want to minimize over the interval $(4, \infty)$. (The total width x must be larger than 4 to allow for the side margins.) Taking the derivative and setting it equal to zero give

$$A'(x) = 5 - 180 \cdot \frac{4}{(x - 4)^2} = 0,$$

which simplifies to $(x - 4)^2 = 144$. The positive solution of this equation is $x = 16$. To see that it gives a local minimum, we use the second derivative test by checking that $A''(16) > 0$. (The details are left to you.)

Substituting $x = 16$ into (4) and solving for y gives $y = 20$. We conclude that the poster with the smallest total area has a height of 20 inches and a width of 16 inches.

EXAMPLE 4.5.5

A woman is in a rowboat 3 kilometers off shore. In order to get to point B on shore, 8 kilometers downstream, she first rows to point P and then walks from P to B (see

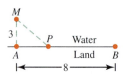

Figure 4.5.4

Figure 4.5.4). If she rows at the rate of 3 kilometers per hour and walks at the rate of 6 kilometers per hour, where should she choose P to minimize her time?

SOLUTION   Let x be the distance along the shore from A to P (as shown in Figure 4.5.5). Applying the Pythagorean theorem to the right triangle MAP, we see that the distance from M to P is given by $\sqrt{9 + x^2}$. At 3 kilometers per hour, the time required to travel that distance is

$$\frac{\sqrt{9 + x^2}}{3}$$

(recall that distance equals rate times time). The distance from P to B is $8 - x$, and, at 6 kilometers per hour, the time required to travel it is

$$\frac{8 - x}{6}.$$

By combining those times, we get the total time to travel from M to B as a function of x, given by

$$f(x) = \frac{\sqrt{9 + x^2}}{3} + \frac{8 - x}{6}.$$

To minimize $f(x)$, we first take the derivative

$$f'(x) = \frac{x}{3\sqrt{9 + x^2}} - \frac{1}{6}.$$

Solving $f'(x) = 0$ is done in the following steps:

$$\frac{x}{3\sqrt{9 + x^2}} = \frac{1}{6}$$

$$2x = \sqrt{9 + x^2} \quad \text{(multiply both sides by } 6\sqrt{9 + x^2}\text{)}$$

$$4x^2 = 9 + x^2 \quad \text{(square both sides)}$$

$$3x^2 = 9 \quad \text{(collect terms)}.$$

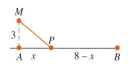

Figure 4.5.5

Therefore, $x = \sqrt{3}$. To see whether this gives a global minimum, we observe that $0 \le x \le 8$ (see Figure 4.5.5). Comparing the values of $f(x)$ at the critical point and the two endpoints leads to the following table:

x	0	$\sqrt{3}$	8
$f(x)$	2.33	2.20	2.85

Therefore, the minimum time is achieved for $x = \sqrt{3} \approx 1.73$ kilometers.

The next example belongs to a class of problems known as inventory problems.

EXAMPLE 4.5.6

A small service station sells 600 tires a year. It orders and receives r shipments of tires a year, with each shipment consisting of x tires. Each tire costs the station \$30. In addition, there are shipping and storage costs. The manufacturer charges the station \$25 per order for shipping, and the cost of storing the tires amounts to \$1.50 per tire per year (figured on the average number of tires per year).

The more shipments the station orders, the greater the shipping cost. On the other hand, if it orders only a few shipments a year, the station has to keep more tires in stock, which results in a greater storage cost. The question is: How many times a year should the station order tires to minimize the cost?

SOLUTION   How much is the annual shipping cost? If the station places r orders a year with a \$25 shipping cost per order, the total shipping cost for the year comes to $25r$. To compute the storage cost, we need to make a simplifying assumption—namely, that after each shipment is received, the stock declines at a steady rate and sells out totally just as the next shipment arrives.

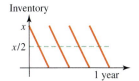

Figure 4.5.6

The saw-toothed graph in Figure 4.5.6 illustrates this property of the inventory. (This particular graph is for 4 shipments per year. At the beginning of each period there are x tires in stock, and the number steadily reduces to zero by the end of the period.) It is not hard to show that the average number in stock over the year is $x/2$. That number is represented by the dashed horizontal line at height $x/2$.

Since the storage cost is \$1.50 per tire per year, an average inventory of $x/2$ results in an annual storage cost of $0.75x$ dollars. Combining that with the shipping cost and the \$18,000 price of the 600 tires, we get the following annual cost function:

$$C(x) = 18,000 + 0.75x + 25r.$$

Since

$$rx = 600,$$

we can replace r by $600/x$, which gives

$$C(x) = 18,000 + 0.75x + \frac{15,000}{x},$$

where x is the number of tires in each shipment.

We want to find the x that minimizes $C(x)$ over the interval $x > 0$. Taking the derivative gives

$$C'(x) = 0.75 - \frac{15,000}{x^2}.$$

Setting this equal to zero and solving result in $x^2 = 20{,}000$, so that

$$x = \sqrt{20{,}000} \approx 141.42.$$

[Of course, $x = -\sqrt{20{,}000}$ is also a solution of $C'(x)$, but it's outside the interval under consideration, so we ignore it.] By using the first or second derivative tests, we see that

$C(x)$ has a local minimum at that point, and the single critical point principle says that it is actually a global minimum.

Combining this with equation $rx = 600$ gives

$$x = \sqrt{20{,}000} \approx 141.42 \quad \text{and} \quad r = \frac{600}{\sqrt{20{,}000}} \approx 4.24.$$

In other words, to minimize its cost, the company should order approximately 4.24 shipments of 141.42 tires each per year. That's not possible, of course, so they would have to round it off to the nearest reasonable numbers—say, 4 shipments a year of 150 tires each.

Practice Exercises 4.5

1. You want to build an open-topped cardboard box with a square base to contain a volume of 32 cubic inches. What dimensions should you choose to minimize the amount of cardboard?

2. Find the largest possible area of a rectangle inscribed in the graph

$$y = 1 - x^2 \quad \text{for} \; -1 \le x \le 1,$$

with its base on the x-axis and its two upper vertices on the graph, as shown in Figure 4.5.7.

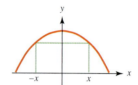

Figure 4.5.7

Exercises 4.5

1. Suppose the demand function for a certain product is given by the equation $q = 6{,}000 - 200\sqrt{p}$ for $0 \le p \le 900$, where p is the unit price and q is the number sold. What price maximizes the revenue?

2. Suppose the demand function for a certain product is given by the equation $q = 2{,}000e^{-0.005p}$. What price maximizes the revenue?

3. Recall that if $C(q)$ is the cost of producing q units of a certain product, then $C(q)/q$ is called the **average cost**. Suppose that $C(q) = 3{,}200 - 15q + 2q^2$. What choice of q minimizes the average cost?

4. An airport shuttle service carries an average of 1,200 passengers into town every day. The current fare is $15, but the company is thinking of raising it. It estimates that each dollar the fare increases will result in a loss of 50 passengers. What price should it charge to maximize its revenue?

5. Suppose that when a busy restaurant charges $9 for its octopus appetizer, an average of 48 people order the dish each night. When it raise the price of the appetizer to $12, the number ordering it drops to 42.

(a) Assuming that the demand q is a linear function of the price p, find an equation for q in terms of p.

(b) What price should the restaurant charge to maximize its revenue from the appetizer?

(c) Suppose each appetizer costs the restaurant $4 to make. What price should it charge to maximize its profit from the appetizer?

6. The manager of an almond orchard wants to harvest her nut crop at a time when it will bring the most money at the market. Currently almonds are bringing in 86 cents per pound, but the price will decrease by 2 cents per pound for each week she waits. Her orchard is bearing a harvest of 35 pounds per tree, which will increase by 1 pound per tree for each week she waits to harvest.

(a) Let t be the amount of time (in weeks) that she waits to harvest her crop. Find the income per pound as a function of t.

(b) Find the number of pounds per tree as a function of t.

(c) How long should she wait to harvest the nuts in order to maximize her revenue? If there are 4,000 trees, what is the maximum revenue?

7. A predatory animal expends energy in catching its prey (caloric expense) and gains energy from consuming it (caloric gain). Consider a small predator, a fox, and let the caloric expense be $E(w) = 0.1w^3$ and the caloric gain be $G(w) = 0.3w^2$, where w is the weight of the prey in pounds, with $0 \leq w \leq 2.5$. What is the optimum prey weight for the fox?

8. Hawks in North America feed their young almost entirely on small rodents. The successful raising of the chicks depends on timing the hatching to coincide with the maximum peak in the rodent population. Say the population model for rodents is given by $R = 4te^{-0.16t}$, where t is the time in weeks after the winter solstice and R is the number (in hundreds) of rodents per square mile. The chicks hatch 5 weeks after laying. What is the optimal time for the eggs to be laid?

9. Male deer grow antlers for a number of reasons, one of which is for advantage in competition for females. The animal with the bigger rack of antlers has a better chance of scaring off his opponent without risking a fight. However, there is a cost to having a large rack, for the deer with the bigger antlers is at greater risk of becoming entangled in brush or trees. Suppose that the success index of intimidating rivals is $G(w) = 1 - e^{-0.032w}$ (the gain), where w is the rack width in inches, and let $R(w) = 0.01w$ be the risk index of dying because of the encumbrance of antlers (the cost).
(a) Write the profit index (the gain minus the loss).
(b) Find the rack width for the deer that maximizes this index.

10. A medical research team is studying the human body's ability to metabolize a new drug used to prepare a patient for open heart surgery. By injecting specified dosages into its volunteers and taking blood samples every 30 minutes for analysis, the team concludes that the concentration of the drug in the bloodstream t hours after injection is given by

$$C(t) = \frac{3t}{t^2 + 4}.$$

The drug will be of the most help to patients if it is at its maximum concentration when the surgery begins. How many hours before surgery should the injection be given?

11. On the space shuttles, consistently warm conditions are ideal for the growth of bacteria. For that reason, astronauts carefully wipe their food trash with antibacterial cleaners before storing it away. Suppose that the cleansing is imperfect, so that after initially declining the bacteria population begins to grow again. If the number (in millions) of bacteria in the trash compartment after t hours is given by

$$B(t) = (t + 4)^2(t - 14) + 96t + 260,$$

find the minimum and maximum number of bacteria present during the first 8 hours.

12. Another (and slightly simpler) way to solve Example 4.5.4 is by letting x and y be the dimensions of the printed area, as shown in Figure 4.5.8. Write a formula for the total area of the poster and an equation that gives the printed area as 180 square inches. Then find the dimensions that minimize the total area.

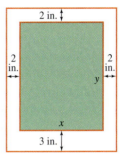

Figure 4.5.8

13. A book publisher is designing a book whose pages have 1-inch margins on the top and each side, and a 2-inch margin on the bottom. In this case, the publisher wants the *total page area* to be 150 square inches. Find the page dimensions that will maximize the *printed area.*

14. Referring to Example 4.5.2, suppose that the pen is to be built along a river, which will serve as one side. You can consider the river edge to be a straight line, with the inside fence running perpendicular to it, as in Figure 4.5.9. Find the dimensions that will maximize the enclosed area.

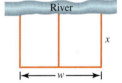

Figure 4.5.9

15. A kennel owner has 320 feet of fencing and wants to fence off a rectangular area and divide it into four adjacent sections, as shown in Figure 4.5.10. What dimensions will maximize the area?

Figure 4.5.10

16. Referring to the previous problem, suppose that there is no restriction on the amount of fencing, but the owner wants to enclose a total area of 5,000 square feet. What dimensions will minimize the amount of fencing used?

17. In modeling the spread of a contagious disease in a community, a working assumption is that the rate r at which the disease spreads is proportional to the product of the number of infected people, say, y, and the number of uninfected people. Suppose M is the total population of the community.

(a) Write an equation expressing r as a function of y, letting k be a constant of proportionality. (*Hint:* Express the number of uninfected people in the community in terms of M and y.)

(b) Show that the disease is spreading most rapidly when exactly half the community is infected.

18. You are hauling two Clydesdale horses in a trailer pulled by a large truck. The entire rig burns fuel at a rate of

$$G(s) = \frac{800 + s^2}{200s},$$

where s is the speed (in miles per hour) and $G(s)$ is the rate of fuel consumption (in gallons per mile).

(a) What speed will minimize the fuel consumption?

(b) If the trip lasts for 1,200 miles and the cost of fuel is $1.80 per gallon, what is the least possible fuel cost?

19. A trucking company has the following expenses in sending one of its trucks on a 1,600-mile round trip: $22.50 per hour for the driver's salary, 27 cents a mile for depreciation and wear on the truck, and fuel costs, which are estimated at $v/140$ dollars per mile if the truck is driven at a steady speed of v miles per hour. The speed limits for trucks on the route traveled are 45 mph minimum and 60 mph maximum. What speed should be used to minimize the cost of the trip?

20. A college bookstore orders 270 cartons of legal-size notepads per year, at a price of $9.60 per carton. In addition, it pays a shipping cost of $20 per shipment and storage costs of 75 cents per carton per year (based on the average number of cartons in stock over the year). Assume that the notepads are used at a constant rate throughout the year, all the shipments are of equal size, and each arrives just as the preceding shipment has been used up. How many orders a year should the bookstore place, and how many cartons should it order each time to minimize its costs?

21. A small company making chocolate sells 270,000 boxes of its five-pound assortment per year. The firm that supplies the boxes charges an order fee of $100 per shipment, in addition to the cost of the boxes. Annual storage costs amount to 2 cents per box, based on the average number of boxes stored per year. How many times a year should the company order boxes, and how many should it order at a time? Assume that the boxes are used at a steady rate throughout the year and that each shipment arrives just as the previous one is used up.

22. Find the point on the line $y = x$ that is closest to the point $(0, 2)$.

23. Find the point on the graph of $y = \sqrt{x}$ that is closest to the point $(1, 0)$.

24. You want to build a box with a square base and top to have a surface area of 600 square inches. What is the maximum possible volume?

25. A box with a square base and a top is to be built with a volume of 20 cubic feet. The material for the base costs $3 per square foot, the material for the top costs $2 per square foot. and the material for the sides costs $1 per square foot. What is the minimum cost of the box?

26. A company plans to make aluminum cans, each with a lid and containing a volume of 2,000 cubic centimeters.

(a) Find the dimensions of the can that will minimize the amount of aluminum used. (Recall that a cylinder with height h and base radius r has volume $V = \pi r^2 h$, and the area of its vertical surface—that is, excluding base and lid—is $2\pi r h$.)

(b) What is the relation between the height and the base diameter? Is this relation the same for a can of any volume if the surface area is minimized?

27. Find the area of the largest rectangle that can be inscribed in a semicircle of radius 1 with the base of the rectangle along a diameter of the semicircle, as shown in Figure 4.5.11.

Figure 4.5.11

28. A house is located at a point H in the woods, 2 miles from the nearest point A on a road, as shown in Figure 4.5.12. A telephone switching station is located at point B on the road, 5 miles from A. The homeowner wants to run a telephone cable through the woods from H to P and then along the road from P to B. The cost of laying the cable through the woods is three times as expensive per mile as it is along the road. Where should the point P be chosen to minimize the cost?

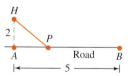

Figure 4.5.12

29. Given a square piece of cardboard whose sides are each 30 inches long, you want to make an open-topped box by cutting a square of side x out of each corner (as shown in Figure 4.5.13) and then folding up the sides (with the result shown in Figure 4.5.14).

(a) Write a formula for the volume of the box as a function of x. Find an interval in which x must fall. (*Hint:* Find an upper bound for $2x$.)

(b) Find the x that maximizes the volume of the box, and find the maximum volume.

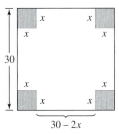

Figure 4.5.13

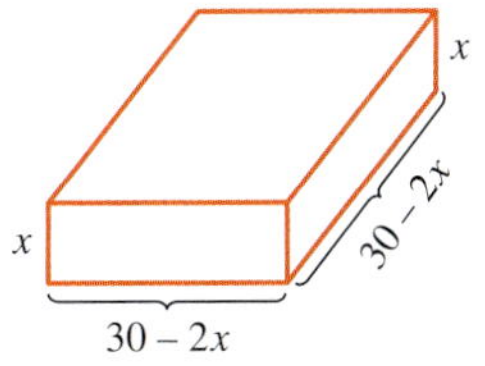

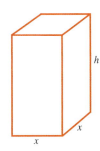

Figure 4.5.14

30. Two factories are located 8 miles apart at points A and B on a straight road (as shown in Figure 4.5.15). Both emit smoke, but the intensity coming from A is twice that of B, as measured by the particulate count. A contractor wants to build a house along that stretch of road at a point where the intensity is minimal.

Assume that the intensity of the pollutant at any point is inversely proportional to the distance from the source, with the same constant of proportionality for both factories. Where should the contractor put the house?

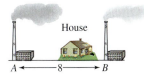

Figure 4.5.15

Solutions to practice exercises 4.5

1. Let h be the height of the box and x be the length of the base edge, as shown in Figure 4.5.16. The base has area x^2, and each of the four vertical faces has area xh. Therefore, the surface area (which we wish to minimize) is

$$S = x^2 + 4xh.$$

The volume of the box is 32 cubic inches, so that $x^2h = 32$, which we can write as $h = 32/x^2$. Substituting that into the formula for S gives

$$S = x^2 + \frac{128}{x}.$$

Taking the derivative and setting it equal to zero gives

$$2x - \frac{128}{x^2} = 0,$$

which reduces to the equation $x^3 = 64$. Therefore, the only critical point on the interval $(0, \infty)$ occurs at $x = 4$. It is a local minimum and therefore (by the single critical point principle) a global minimum. Since $h = 32/x^2$, we get $h = 2$.

Figure 4.5.16

2. The base of the rectangle is the interval $[-x, x]$, which has length $2x$. Its height is the height of the graph over x, which is $1 - x^2$. Therefore, the area is given by

$$A(x) = 2x(1 - x^2) = 2x - 2x^3,$$

for $0 \le x \le 1$. Setting the derivative equal to zero gives $2 = 6x^2$, which leads to $x = 1/\sqrt{3}$. Since $A(0) = A(1) = 0$ and $A(1/\sqrt{3}) = 4/(3\sqrt{3})$, we see that the area takes the maximum value of $4/(3\sqrt{3})$ at $x = 1/\sqrt{3}$.

4.6 Using a Graphing Calculator

APPLYING TECHNOLOGY

As we have already seen, a graphing calculator can give a poor or misleading picture if we rely on the default settings. We must take account of the key features of the graph, such as its domain and range and any local maxima and minima, inflection points, and asymptotes. In trying to do that, we may be faced with the problem of solving equations

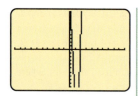

Figure 4.6.1

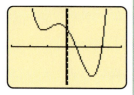

Figure 4.6.2

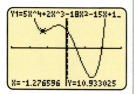

Figure 4.6.3

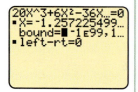

Figure 4.6.4

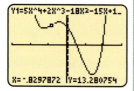

Figure 4.6.5

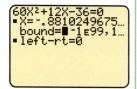

Figure 4.6.6

that are beyond the ordinary techniques of algebra. The calculator itself can help us to do that.

EXAMPLE 4.6.1

Graph the function $f(x) = 5x^4 + 2x^3 - 18x^2 - 15x + 12$ and find any intercepts, local maxima and minima, and inflection points.

SOLUTION   A first attempt, using the default window setting of $[-10, 10] \times [-10, 10]$ (on a TI-83 Plus), results in the diagram shown in Figure 4.6.1. It suffers from a number of glaring defects—the graph is only shown in the right half of the plane, although the domain is $(-\infty, \infty)$, and it is impossible to see any maxima, minima, or inflection points. It does suggest a few things, however. First, there are two x-intercepts in the picture. Second, the graph changes from decreasing to increasing somewhere between $x = 0$ and $x = 3$, which means there is a local minimum. Third, the y values get very large (in absolute value) for relatively small x.

The last observation suggests we might get a better graph by reducing the horizontal setting and increasing the vertical. A bit of trial and error yields the graph shown in Figure 4.6.2, with window $[-3, 3] \times [-25, 25]$. We can now see that there are at least two relative minima and one relative maximum. In addition, the graph changes from concave up to concave down and back to up, so there are at least two inflection points.

To find the relative extrema, we need to solve

$$f'(x) = 20x^3 + 6x^2 - 36x - 15 = 0.$$

We can use the equation solver of our calculator, but we must supply an initial guess. Figure 4.6.2 shows the leftmost minimum to be somewhere near $x = -1$. For a little more precision we might use the TRACE function to give approximate coordinates of the point. The result is shown in Figure 4.6.3. Next, using that approximation (rounded, say, to two decimal places) as our initial guess, we use the equation solver to find the x-coordinate of the relative minimum point, as shown in Figure 4.6.4. The y-coordinate can be found by evaluating $f(x)$.

The other local extrema, the x-intercepts, and the inflection points can be found in the same way (see the exercises). Figure 4.6.5 shows the TRACE cursor near the leftmost inflection point. With the given x-coordinate as our initial guess, we can use the equation solver to find a better approximation to the solution of $f''(x) = 60x^2 + 12x - 36 = 0$, as shown in Figure 4.6.6. (In this case, we could have solved "by hand," using the quadratic formula.)

How do we know we have found all of the local extrema and inflection points? Or, to put it another way, how do we know the graph does not turn around again, somewhere beyond the range of our graphing window? That is not always an easy question, but in this case a theorem of algebra (see Theorem 4.6.1 below) tells us that $f'(x)$ does not have more than three roots and $f''(x)$ does not have more than two. Therefore, we have found them all.

We can also conclude that there are no x intercepts other than those shown on the calculator screen. For, in order to cross the x axis again, the graph would have to change once more from increasing to decreasing (on the right side) or vice versa (on the left). But that would mean there is another critical point, which we know there is not.

In looking for critical points of polynomials and rational functions, the following theorem is often useful.

Theorem 4.6.1

A polynomial of degree n cannot have more than n roots.

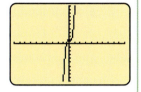

Figure 4.6.7

We used this theorem in Example 4.6.1 to conclude that we had found all of the critical and inflection points of the given function. However, it cannot be applied in every case, for a polynomial of degree n may have *less* than n real roots. In fact, the polynomial $f(x)$ of Example 4.6.1 has degree 4 but (as we just saw) only two roots.

EXAMPLE 4.6.2

Graph the function $f(x) = x^5 + 10x^3 - 2x^2 - x + 1$ and find any intercepts, local maxima and minima, and inflection points.

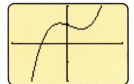

Figure 4.6.8

SOLUTION   As in the previous example, we will start by trying the default window setting, which gives the graph shown in Figure 4.6.7. The graph appears to be increasing, but the diagram is very unclear near $x = 0$. This time it seems both the horizontal and vertical settings are too large, and after some trial and error we obtain the graph shown in Figure 4.6.8, with window $[-1, 1] \times [-2, 2]$.

The picture shows one x-intercept, one inflection point, and two critical points—one a local maximum, the other a local minimum. We can find their coordinates as we did in the previous example. For instance, Figure 4.6.9 shows the TRACE screen, with the cursor positioned as close as possible to the local minimum. By using its x-coordinate as an initial guess, we get a more accurate approximation with the equation solver, as shown in Figure 4.6.10, which shows the (approximate) solution of $f'(x) = 0$. Figure 4.6.11 shows the solution of $f''(x) = 0$, using an initial guess of $x = 0$. (In fact, almost any guess will lead to that same answer.) The local maximum and the x-intercept can be found in the same way (see the exercises).

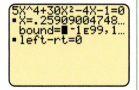

Figure 4.6.9

Once again, we ask whether we have found all the intercepts, critical points, and inflection points. This time the question is harder to answer. For instance, we have found two critical points, but $f'(x)$ has degree four, so it is possible there are more. In fact, however, there are not. To see why, we consider the second derivatve, $f''(x) = 20x^3 + 60x - 4$. Its derivative, the third derivative of f, is $60x^2 + 1$, which is positive for all x. Therefore, $f''(x)$ is always increasing and can have only only root. That means the graph of $f(x)$ can have only one inflection point, which we have found. And if there is only one inflection point, there cannot be more than two local extrema (why?). Combining that observation with the picture of the graph in Figure 4.6.9 leads to the conclusion that there is only one x-intercept.

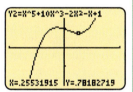

Figure 4.6.10

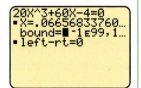

Figure 4.6.11

In general, the question of whether we have found all of the essential points of a graph may be difficult, and there is no sure recipe for answering it. We have to use a combination of algebra, reasoning, and experimentation with the calculator.

EXAMPLE 4.6.3

Graph the function $f(x) = e^x - e^{-x} - 3x$ and find any intercepts, local maxima and minima, and inflection points.

SOLUTION   Before using the calculator, we can get a certain amount of information by examining the function and its derivatives. First, we note that $f(0) = 0$, so that the graph goes through the origin. Second, we observe that $f(-x) = -f(x)$, which means the function is odd and the graph is symmetric about the origin. Taking the first and second derivatives, we obtain

$$f'(x) = e^x + e^{-x} - 3 \quad \text{and} \quad f''(x) = e^x - e^{-x}.$$

The equation $f''(x) = 0$ is equivalent to $e^x = e^{-x}$, which is the same as $e^{2x} = 1$. The only solution, and, therefore, the only candidate for an inflection point, occurs at $x = 0$. Since $f''(-x) = -f''(x)$, the sign of the second derivative is different on opposite sides of the origin. (A calculator check, for instance, shows that $f''(1) \approx 2.35$ and $f''(-1) \approx -2.35$.) Therefore, the graph does indeed have an inflection point at $(0, 0)$.

To find the roots of $f'(x)$, if any, we observe that $f'(0) = 1 + 1 - 3 < 0$. Since $f'(x) \to \infty$ as $x \to \infty$ (why?), $f'(x)$ must be positive for some positive x. Indeed, a bit of trial and error with the calculator shows that $f'(2) \approx 4.52$. Therefore, $f'(x)$ has a root somewhere in the interval $(0, 2)$. By symmetry, it also has one on the interval $(-2, 0)$. Moreover, these are the only critical points. If there were another, the graph would have to make another change of concavity, and there would be another inflection point.

With this information at hand, we use the calculator to graph the function. In this case, the default settings give a reasonable diagram (Figure 4.6.12), but by choosing a smaller window, we can zoom in on the key features. The inflection and critical points are all between $x = -2$ and $x = 2$, and the graph in Figure 4.6.12 shows that the intercepts are not far from there. Therefore, $[-3, 3] \times [-3, 3]$ seems a reasonable choice for a viewing window. The result is shown in Figure 4.6.13. The x-intercepts shown are the only solutions of $f(x) = 0$. Otherwise, the graph would have to turn around once more, but there are no other critical points.

We can now use the equation solver, as in the previous examples, to find the intercepts and critical points. The details are left as an exercise.

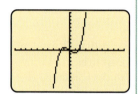

Figure 4.6.12

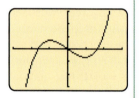

Figure 4.6.13

For many functions, it is difficult to find a single window setting that displays all the key features of the graph, but we can use the calculator to help us find them.

EXAMPLE 4.6.4

Let $f(x) = (x - 1)/(x^3 - 3x^2 + x + 4)$. Find any horizontal or vertical asymptotes, any intercepts, and any critical or inflection points. Draw the clearest diagram possible of the graph.

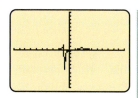

Figure 4.6.14

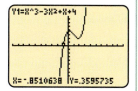

Figure 4.6.15

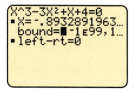

Figure 4.6.16

Figure 4.6.17

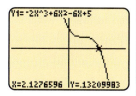

Figure 4.6.18

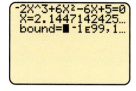

Figure 4.6.19

SOLUTION   We first attempt to graph the function using the default window, with the result shown in Figure 4.6.14. It is not very illuminating but does suggest there is a vertical asymptote in the negative half of the plane. To find the asymptote, we graph the denominator, as shown in Figure 4.6.15, where the TRACE cursor has been moved as close as possible to the x-intercept. We get a better approximation using the equation solver, as illustrated in Figure 4.6.16, which shows the root of the denominator to be $-0.893289\ldots$. We will denote that point by c. Figure 4.6.15 shows that the denominator is negative to the left of c and positive to the right.

Since the numerator is positive for $x > 1$ and negative for $x < 1$, we get the diagram shown in Figure 4.6.17 for the sign of $f(x)$.

Next, we note that $\lim_{x \to \pm\infty} f(x) = 0$, so that the x-axis is a horizontal asymptote. Since $f(1) = 0$ and $f(x) > 0$ for $x > 1$, we conclude that $f(x)$ has a local maximum on the interval $(1, \infty)$. To find it, we take the first derivative

$$f'(x) = \frac{-2x^3 + 6x^2 - 6x + 5}{(x^3 - 3x^2 + x + 4)^2}.$$

By graphing the numerator of this quotient in a $[-4, 4] \times [-10, 10]$ window (Figure 4.6.18) and then using the equation solver (Figure 4.6.19), we see that there is a point $b \approx 2.1447$ so that $f'(x)$ is positive to the left of b, negative to the right, and $f'(b) = 0$. Therefore, $f(x)$ has a local maximum at $x = b$. Since there are no other critical points, we conclude that

- $f(x)$ is increasing on the intervals $(-\infty, c)$ and (c, b),

- $f(x)$ is decreasing on (b, ∞),

- $f(x)$ has a vertical asymptote at $x = c$, where $b \approx 2.1447$ and $c \approx -0.8933$.

Combining all this information leads to the graph shown in Figure 4.6.20, drawn in a $[-2, 4] \times [-2, 2]$ window.

The inflection points are left as an exercise. One way to determine them is to graph $f'(x)$ (as shown in Figure 4.6.21, drawn in a $[-2, 4] \times [-1, 1]$ window). Then, use the TRACE and ZOOM functions, as illustrated by Figure 4.6.22, to find the points where $f'(x)$ has a relative extremum. (Recall that $f(x)$ has inflection points at precisely those x for which $f'(x)$ has a relative extremum.)

Figure 4.6.22 shows that one inflection point occurs at $x \approx 0.44548$. The approximate y-coordinate is $f(0.44548) \approx -0.14079$.

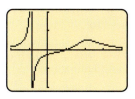

Figure 4.6.20

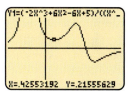

Figure 4.6.21

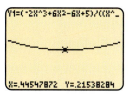

Figure 4.6.22

Exercises 4.6

1. Find all the x-intercepts, critical points, and inflection points of the graph of Example 4.6.1.

2. Find the x-intercept and local maximum point for the graph of Example 4.6.2.

3. Find the x-intercepts and local maximum and minimum points of the graph of Example 4.6.3.

4. Find the inflection points of the graph in Example 4.6.4.

In Exercises 5–18, use a calculator to graph the given function and find any intercepts, local maxima and minima, and inflection points.

5. $f(x) = x^3 - 8x^2 + 2x - 1$

6. $f(x) = x^3 - 3x^2 + 8x - 11$

7. $f(x) = x^4 + x^3 - 5x^2 + 4x + 3$

8. $f(x) = x^4 + x^3 - 5x^2 - 3x + 2$

9. $f(x) = x^5 - 8x^3 + 20x - 12$

10. $f(x) = x^5 - 8x^3 + 2x^2 + 20x$

11. $f(x) = e^x + e^{-x} - 4x$ 12. $f(x) = e^x - 20x$

13. $f(x) = \dfrac{x^2 + 2}{x^4 + 4}$ 14. $f(x) = \dfrac{x^3 - 1}{x^4 + 1}$

15. $f(x) = \dfrac{\ln(x^2 + 1)}{x^2 + 1}$ 16. $f(x) = e^x - 2x^2$

17. $f(x) = \dfrac{x^4 - 1}{e^x + 1}$ 18. $f(x) = e^{2x} - 4e^x + x$

In Exercises 19–22, find any vertical asymptotes and make a sign chart for $f(x)$. Then find any critical points and chart the sign of $f'(x)$ to determine any local maxima or minima. Check for horizontal asymptotes and inflection points. Use the information to make the clearest graph you can.

19. $f(x) = \dfrac{2x}{x^3 - x - 4}$ 20. $f(x) = \dfrac{2x}{x^3 - x^2 - 4}$

21. $f(x) = \dfrac{4x^3}{x^4 - 3x^2 - 4}$ 22. $f(x) = \dfrac{10x^4}{x^4 - 3x^2 - 4}$

23. Desertification is the process of changing the ecosystem of the land to desertlike conditions. This can occur from human modifications, such as deforestation, reduction of available groundwater, and poor farming practices. A region that is already a desert appears to be increasing in size. In 1950, it was predicted that the percentage increase in area per year would be

$$D_p(t) = \ln(e + 0.21t),$$

where t is the number of years after 1950. Efforts were made to reduce the level of desertification, and in 2000, the actual observed percentage increase per year since 1950 was calculated to be

$$D_a(t) = \ln(e + 0.05t).$$

(a) Graph both curves on the same set of axes. On the basis of the graphs, would you say the efforts were successful?

(b) Graph the difference $D_p(t) - D_a(t)$ between the predicted and the actual percentage increases. What is $\lim_{t \to \infty}[D_p(t) - D_a(t)]$?

Chapter 4 Summary

- If the derivative $f'(x)$ is *positive* (*negative*) on an interval, then the function $f(x)$ is *increasing* (*decreasing*) on that interval.

- A point c in the domain of $f(x)$ is a **critical point** if either $f'(c) = 0$ or $f'(c)$ does not exist.

- A function $f(x)$ has a **local maximum** (*minimum*) at a point c if $f(c)$ is *greater* (*less*) than or equal to the values $f(x)$ for all points x in an interval around c. A local maximum or minimum is called a **local extremum**. The *greatest* (*smallest*) value of a function throughout its domain is called a **global maximum** (*minimum*).

- **Critical points** are the only **candidates** for local extrema.

- **The first derivative test** for extrema says: If $f(x)$ has a critical point at $x = c$, then there is a local maximum (*minimum*) at $x = c$ if $f'(x)$ changes its sign from positive to negative (*negative to positive*).

- **The single critical point principle** says that if a function has **exactly one** critical point in an interval, and if the point is a **local** maximum (*minimum*) point, then it is a **global** maximum (*minimum*) point for that interval.

- The graph of a function $f(x)$ is **concave up** (*concave down*) in an interval if its second derivative $f''(x)$ is **positive** (*negative*) in that interval

- A point where the concavity of a graph changes is called an **inflection point**. Points where the *second derivative is zero or undefined* are candidates for inflection points.

- **The second derivative test** says that if $f'(c) = 0$, and if $f''(c)$ is negative (*positive*), then $f(x)$ has a local maximum (*minimum*) at c. To remember it, look at $f(x) = -x^2$ and $f(x) = x^2$ with $c = 0$.

- The information listed so far helps us to **sketch the graph** of a function. In addition we use its asymptotic behavior, symmetries, and a few points on the graph, such as x- and y-intercepts.

- The **extreme value theorem** states that if $f(x)$ is a continuous function on a closed, bounded interval $[a, b]$, then it has a global maximum and minimum on the interval. To compute them, we find all critical points, the values of the function at the critical points, and the two endpoints. The global maximum (*minimum*) is the biggest (*smallest*) of these values.

Chapter 4 Review Questions

- Find a function for which a critical point is *not* a local extremum.

- Define concavity and give an example for each case.

- Find a function whose second derivative is zero at a point that is *not* an inflection point.

- Find three functions for which $f'(c) = 0$ and $f''(c) = 0$, one having a local maximum at c, the second a local minimum, and the third having neither.

- How do we compute a possible global maximum or minimum of a continuous function on an interval with *one or both endpoints excluded*? Give examples.

Chapter 4 Review Exercises

In Exercises 1–12, determine where each of the functions is increasing and where it is decreasing, and find any local maxima or minima.

1. $f(x) = 3x^2 - 6x - 9$

2. $f(x) = 2x^3 + 3x^2 + 6x - 5$

3. $f(x) = x^3 - 4x^2 - 3x - 5$

4. $f(x) = x - \dfrac{4}{x}, \ x \neq 0$

5. $f(x) = \dfrac{x+1}{x^2}, \ x \neq 0$ **6.** $f(x) = x - 3x^{2/3}$

7. $f(x) = x^2 e^{-3x}$ **8.** $f(x) = x \ln x, \ x > 0$

9. $f(x) = e^x + e^{-2x}$ **10.** $f(x) = \sqrt{x^2 + x + 1}$

11. $f(x) = x(x^2 - 1)^{1/3}$ **12.** $f(x) = |x^2 - 1|$

In Exercises 13–22, find where the graph is concave up and where each is concave down, and find any inflection points.

13. $y = x^4 - 6x^2 - 3x - 5$

14. $f(x) = x^5 - 5x^4 + x - 2$

15. $y = e^{-2x}$ **16.** $y = \dfrac{x+1}{x^2}$

17. $y = x^2 - x \ln x, \ x > 0$ **18.** $y = xe^{-(2x+1)}$

19. $y = xe^{-x^2}$ **20.** $y = e^{2x} - e^x$

21. $y = (x^2 + x + 1)^{-1}$ **22.** $y = (\ln x)^3, \ x > 0$

23. The graph of $y = f(x)$ is shown in Figure 4.R.1. Which of the following statements is true?
  (a) $f(x)$ is increasing and $f'(x)$ is increasing.
  (b) $f(x)$ is increasing and $f'(x)$ is decreasing.
  (c) $f(x)$ is decreasing and $f'(x)$ is decreasing.
  (d) $f(x)$ is decreasing and $f'(x)$ is increasing.

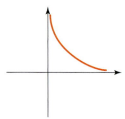

Figure 4.R.1

24. Figure 4.R.2 shows the graph of the derivative $f'(x)$ for some function $f(x)$.
  (a) Where is $f(x)$ increasing and where is it decreasing?
  (b) Where is it concave up and where is it concave down?

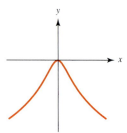

Figure 4.R.2

25. Figure 4.R.3 shows the graph of the derivative $f'(x)$ for some function $f(x)$.
  (a) Where is $f(x)$ increasing and where is it decreasing?
  (b) Where is it concave up and where is it concave down?

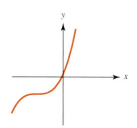

Figure 4.R.3

Exercises 26–29 refer to the six graphs, labeled (a) through (f), of Figure 4.R.4.

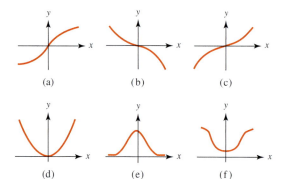

Figure 4.R.4

26. Which is the graph of a function with $f'(x) < 0$ for all x, $f''(x) > 0$ for $x < 0$, and $f''(x) < 0$ for $x > 0$?

27. Which is the graph of a function with $f''(x) > 0$ for all x?

28. Which is the graph of a function with $f''(x) < 0$ on $(-1, 1)$ and $f''(x) > 0$ on $(-\infty, -1)$ and $(1, \infty)$?

29. Which is the graph of a function with $f'(x) < 0$ for $x < 0$ and with two inflection points?

In Exercises 30–33, find the critical point(s) and use the second derivative test to see if each is a local maximum or minimum.

30. $f(x) = 4x^3 - x^2 - 2x + 1$

31. $f(x) = x - \ln x, \ x > 0$

32. $f(x) = e^{2x} + e^{-2x}$

33. $f(x) = \ln x - (\ln x)^2, \ x > 0$

In Exercises 34–36, sketch the graph of a function satisfying the given conditions.

34. The domain consists of the entire x-axis;
     $f(x) < 0$ for $x < 0$ and $f(x) > 0$ for $x > 0$;

$f'(x) < 0$ on $(-\infty, -1)$ and $(1, \infty)$; $f'(x) > 0$ on $(-1, 1)$;
$f''(x) < 0$ on $(-\infty, -2)$ and $(0, 2)$; $f''(x) > 0$ on $(-2, 0)$ and $(2, \infty)$

35. The domain consists of the entire x-axis;
$f(x) < 0$ for $x < 0$ and $f(x) > 0$ for $x > 0$;
$f'(x) > 0$ for all x;
$f''(x) < 0$ on $(-\infty, -1)$ and $f''(x) > 0$ on $(-1, \infty)$

36. The domain consists of all $x \neq -1$;
the line $x = -1$ is a vertical asymptote;
the x-axis is a horizontal asymptote;
$f(x) < 0$ for all $x \neq -1$ and $f''(x) < 0$ for all $x \neq -1$;
$f'(x) < 0$ for $x < -1$ and $f'(x) > 0$ for $x > -1$

In Exercises 37–42, find all the information you can about the function (domain, asymptotes, increasing/decreasing, maximum/minimum, concavity, inflection points, symmetry) and sketch the graph.

37. $f(x) = 9x^3 - 12x + 4$

38. $\dfrac{2x - 1}{2x + 1}$

39. $f(x) = \dfrac{x^2}{x^2 - 4}$

40. $f(x) = \dfrac{2x}{x^2 - 4}$

41. $f(x) = x^{5/3} - 15x$

42. $f(x) = x + e^{-x}$

In Exercises 43–46, determine whether the function has a global maximum and/or minimum on the interval in question and, if so, where it occurs and what its value is.

43. $f(x) = x^3 - 3x + 2$ on $(-\infty, \infty)$

44. $f(x) = x^3 - 3x + 2$ on $(0, \infty)$

45. $f(x) = x \ln x$ on $(0, \infty)$

46. $f(x) = x - e^x$ on $(-\infty, \infty)$

47. Let $f(x) = (x^3/3) - 2x^2 + 3x - 1$. Find where the global maximum and minimum of f occur over each of the following intervals:
(a) $[0, 2]$ **(b)** $[-1, 1]$ **(c)** $[2, 4]$

48. Suppose $C(x) = 40 + 6x + 200(3x + 1)^{-1}$, $x \geq 0$, is the cost function for a certain production process, where x is the production level in hundreds of units and $C(x)$ is the cost in thousands of dollars. What is the minimum possible cost and for what x is it achieved?

49. A baseball team plays in a stadium that holds 55,000 spectators. With ticket prices at \$10, the average attendance has been 27,000. When tickets were lowered to \$8, the average attendance rose to 33,000. Assuming that the demand function is linear, what price will maximize revenue?

50. The price per unit of a certain item is a function of the number of units produced and is given by

$$p = \frac{1,000}{250 + x} - 1, \quad 0 < x < 750.$$

Find the x that maximizes the revenue. How do you know the revenue is maximized?

51. A rancher wants to build a rectangular pen covering 2,400 square feet and divide it into four adjacent sections, as shown in Figure 4.R.5. He plans to use stronger fencing, costing \$12 a foot, for the outside of the pen. The fencing for the inside dividers costs \$4 a foot. What dimensions will minimize the cost?

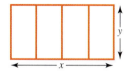

Figure 4.R.5

52. Find the line through the point $(2, 1)$ that cuts off the triangle of smallest area in the first quadrant. (*Hint:* Let t be the slope of the line and find the area of the triangle as a function of t. (See Figure 4.R.6.))

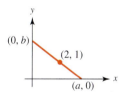

Figure 4.R.6

53. Find the point on the line $x + 3y = 1$ that is closest to the origin. (*Hint:* Write r as a function of x and minimize. (See Figure 4.R.7.))

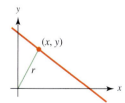

Figure 4.R.7

54. If a function $y(t)$ satisfies the differential equation $y' = 1 + y^2$, show that it is an increasing function.

55. If a function $y(t)$ is such that $y' = -e^{-y}$ in an interval, show that it is decreasing.

1. Find the critical points of the function $f(x) = (x^2 - 4)^{12}$ and state whether each is a local maximum or a local minimum or neither.

2. Find where each of the following functions is increasing and where it is decreasing:
   (a) $f(x) = x - e^{-4x}$ (b) $f(x) = x^2 e^{3x}$

3. Find any local extreme points of the function $f(x) = x^3 e^{-x}$. Determine if each is a local maximum or minimum and whether or not it is a global one.

4. Find where the function $f(x) = x^2 \ln x$, $x > 0$ is concave up and where it is concave down.

5. Suppose that f is a function with $f'(c) = 0$ and $f''(c) = 0$ at some point c in its domain. Which of the following statements is true?
   (a) f must have a local maximum at c.
   (b) f must have a local minimum at c.
   (c) f must have either a local maximum or local minimum at c.
   (d) f cannot have either a local maximum or minimum at c.
   (e) No conclusion can be drawn without more information.

6. Find all inflection points of the graph of $y = x^6 - 5x^4 + 2x - 1$.

7. Draw the graph of $y = f(x)$ satisfying the following properties: The domain consists of all $x \neq 0$, and the y-axis is a vertical asymptote.
   $f'(x) < 0$ on $(-\infty, 0)$ and $(0, 2)$; $f'(x) > 0$ on $(2, \infty)$;
   $f''(x) < 0$ for $x < 0$ and $f''(x) > 0$ for $x > 0$;
   $\lim_{x \to -\infty} f(x) = 0$.

8. Sketch the graph of $f(x) = x^3 - 6x^2 + 9x + 1$. Be sure to mark the relative maxima, relative minima, and points of inflection on your graph.

9. Find the minimum value of the function $f(x) = e^{2x}/x$ on the following intervals:
   (a) $[\frac{1}{4}, 1]$ (b) $[1, 4]$

10. Suppose f is a function, with domain $(-\infty, \infty)$, such that $f(-2) = 5$, $f(1) = -2$, and $f'(x) < 0$ for all values of x. How many solutions does the equation $f(x) = 0$ have? Explain.

11. Suppose the demand function for a certain product is given by the equation
$$q = 3,000 - 100\sqrt{p} \quad \text{for } 0 \leq p \leq 900,$$

where p is the unit price and q is the number sold. What price maximizes the revenue? How do you know it is a maximum?

12. Suppose you want to enclose 600 square yards with a rectangular fence and then divide the area in half with a fence parallel to one of the sides. What dimensions require the least amount of fencing?

13. If a function $y(t)$ satisfies the equation $y' = e^y$ in an interval, then show that it is increasing in that interval.

14. If a function $y(t)$ satisfies the equation $y' = -(1 + y^2)$ in an interval, then show that it is decreasing in that interval.

15. If a function $y(t)$ satisfies the equation $y' = e^y$ in an interval, then use the chain rule to show that $y(t)$ is concave up in that interval.

16. If a function $y(t)$ satisfies the equation $y' = -e^{-y}$ in an interval, then use implicit differentiation to show that $y(t)$ is concave down in that interval.

17. Assuming that the maximum population that the United States can sustain is about 800 million, its growth rate is 0.02, and its population in 2000 was about 280 million, then the logistic model says that t years after 2000 the U.S. population $p(t)$ (in millions) satisfies the equation

$$\frac{dp}{dt} = 0.02p\left(1 - \frac{p}{800}\right).$$

   (a) Show that the function

$$p(t) = \frac{22,400}{28 + 52e^{-0.02t}}$$

   is a solution and compute $p(0)$.
   (b) Find the U.S. population in the year 2050.
   (c) Find the $\lim_{t \to \infty} p(t)$.
   (d) Show that $p(t)$ is an increasing function for $t \geq 0$.
   (e) Show that $p(t)$ is concave up for $p < 400$ and concave down for $p > 400$.
   (f) Show that $p(t)$ has an inflection point at $p = 400$.
   (g) Find the rate of change of $p(t)$ when $p = 400$.
   (h) Sketch, by hand, the graph of the function $p(t)$ for $t \geq 0$.

18. The revenue function for a product is $R(q) = q(800 - q)$ and the cost function is $C(q) = q^3 - 5q^2 + 500q + 1,000$, where the quantity q is measured in hundreds of units and the profit is measured in thousands of dollars. Find the production level that maximizes the profit.

Chapter 4 Projects

1. **The Gaussian family**   Consider the famous Gaussian functions (useful in probability and statistics among other things)

$$f(x) = \frac{1}{\sigma\sqrt{2\pi}}\, e^{-\frac{1}{2}\left(\frac{x-\mu}{\sigma}\right)^2}, \quad -\infty < x < \infty,$$

where μ and σ are fixed numbers with $\sigma > 0$.

(a) Find where it increases or decreases, where it is concave up or down, its inflection points, extrema, asymptotes, symmetries, etc.

(b) Draw its graph when $\mu = 0.01, 1, 10$ and $\sigma = 0.01, 1, 10$.

(c) As you change μ and σ, make observations about the shape of the graph and explain them.

2. **Elasticity of demand**   is an important concept in economics, defined as follows. The *demand* for a certain manufactured product—that is, the number of units sold—is a function of the price, a relation we can symbolize by writing $q = f(p)$. The **elasticity** is defined by the formula

$$E(p) = -\frac{p}{q}\frac{dq}{dp}.$$

Verify the following facts about the elasticity.

(a) At the price that maximizes the revenue, $R = pq$, the elasticity satisfies $E(p) = 1$.

(b) If $E(p) > 1$, a small increase in price will result in a decrease in revenue. In that case, the demand is said to be **elastic**.

(c) If $E(p) < 1$, a small increase in price will result in an increase in revenue. In that case, the demand is said to be **inelastic**.

(d) Suppose the demand is elastic at a certain price. In that case, a small increase Δp in price will result in a decrease Δq in demand. The *percentage* changes will be

$$\text{Percentage increase in price} = 100 \cdot \frac{\Delta p}{p}$$

and

$$\text{Percentage decrease in demand} = -100 \cdot \frac{\Delta q}{q}.$$

(The minus sign is to make the right-hand side positive, since $\Delta q < 0$.) Show that

$$\text{Percentage decrease in demand} \approx E(p) \cdot (\text{percentage increase in price}).$$

Integration

In this chapter, we shall solved the **area problem** for a curved region by introducing the integral. The basic idea for solving this problem is to approximate the region by a set of rectangles, then refine the approximation over and over so that the rectangles fit the region better and better, and their total area approaches the area of the region as a limit. That leads us to the idea of the **definite integral**, which arises naturally in a variety of problems—computing total cost, revenue and profit, consumer and producer surplus, income streams, population predictions, and many others.

We will also take up a theorem that is at the heart of calculus—the **fundamental theorem of calculus**—which establishes a connection between the integral and the derivative.

In most of the examples and problems we have studied up until now, we started with a given function and computed its derivative in order to get information about the function. In many applications, however, the process is reversed—it is the derivative that is given and the function that we try to determine. That is because many of the laws of science, including social sciences, such as economics, are stated in terms of the *rate* at which a certain process changes, and our job is to find a formula to describe the process. For example, we may know the rate at which a population is increasing and want to make a prediction about its size at some future time. Or, we may have a marginal profit function and want to find the profit function from it, as in the following example.

EXAMPLE 5.1.1

Assume that a company's marginal profit function from producing and selling x units of a certain commodity per day is

$$MP(x) = -0.4x + 200, \tag{1}$$

in dollars per unit. If the profit is \$10,000 at the the production level of 100 units, determine the profit function.

SOLUTION Recalling that the marginal profit is the derivative of the profit function $P(x)$, we can rewrite Eq. (1) in the form

$$P'(x) = -0.4x + 200. \tag{2}$$

The information that the profit is \$10,000 at the the production level of 100 units can be written as $P(100) = 10,000$. Our task now is to find a function $P(x)$ that solves the equation $P'(x) = -0.4x + 200$ and satisfies the condition $P(100) = 10,000$. We will complete the solution after we discuss antiderivatives.

Antiderivatives

Definition 5.1.1

If $F'(x) = f(x)$, then we say that $F(x)$ is an **antiderivative** of $f(x)$.

Saying that $F(x)$ is an antiderivative of $f(x)$ is the same as saying that $f(x)$ is the derivative of $F(x)$, but with a change of emphasis. We think of $f(x)$ as the given function and $F(x)$ as the one to find. The process of finding an antiderivative is called **antidifferentiation**.

A significant difference between differentiation and antidifferentiation is that there is no single, unique antiderivative of a given function. *If a function has an antiderivative, it has infinitely many.* Here are a few simple examples.

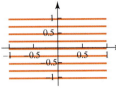

Figure 5.1.1

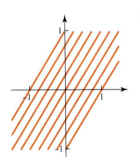

Figure 5.1.2

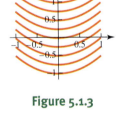

Figure 5.1.3

Figure 5.1.4

EXAMPLE 5.1.2

Find an antiderivative of each of the following functions:

$$\text{(i) } f(x) = 0, \quad \text{(ii) } f(x) = 1, \quad \text{(iii) } f(x) = x, \quad \text{(iv) } f(x) = x^2$$

SOLUTION (i) Since the derivative of any constant function is zero, any of the functions $F(x) = 1$, $F(x) = 2$, $F(x) = -\frac{3}{5}$, and $F(x) = \sqrt{2}$ are antiderivatives of the given function, $f(x) = 0$, as is $F(x) = c$ for any real number c. The graphs of these functions are all the lines parallel to the x-axis. There is a unique one through each point of the plane. A few of them are shown in Figure 5.1.1.

(ii) Since $\frac{d}{dx}(x) = 1$ and $\frac{d}{dx}(c) = 0$ for any constant c, we get an infinite family of antiderivatives of the form $F(x) = x + c$. The graphs of these functions are all the lines with slope 1. Again, there is a unique one through each point of the plane. Figure 5.1.2 shows the graphs corresponding to $c = -1, -0.75, \ldots, 0.75, 1$.

(iii) For any number c, we have

$$\frac{d}{dx}\left(\frac{1}{2}x^2 + c\right) = x.$$

Therefore, every function of the form $F(x) = \frac{1}{2}x^2 + c$ is an antiderivative of $f(x) = x$. Once again, there is an infinite family of antiderivatives. This time the graphs are parabolas that cover the whole plane without intersecting. Figure 5.1.3 shows those corresponding to $c = -1, -0.75, \ldots, 0.75, 1$.

(iv) Since

$$\frac{d}{dx}\left(\frac{1}{3}x^3 + c\right) = x^2,$$

the family of cubic functions $F(x) = \frac{1}{3}x^3 + c$ gives infinitely many antiderivatives of the given quadratic function $f(x) = x^2$. Again, the graphs cover the entire plane without intersecting. Figure 5.1.4 shows those corresponding to $c = -1, -0.75, \ldots, 0.75, 1$.

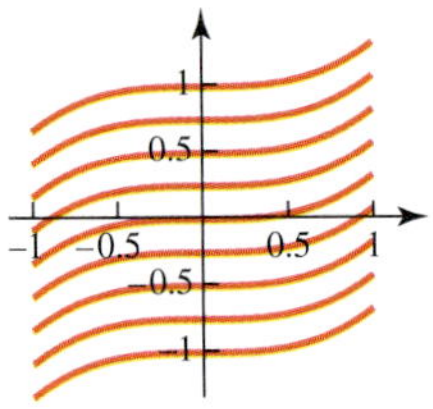

Not only is every constant function an antiderivative of $f(x) = 0$, but those are the *only* antiderivatives. The following theorem is a precise statement of that simple but useful fact:

Theorem 5.1.1

A function whose derivative is zero throughout an interval must be constant over that interval; that is, for $a < x < b$,

$$F'(x) = 0 \implies F(x) = c \quad \text{for some number } c.$$

We will not give a formal proof of this theorem, appealing instead to the intuitive idea that if the slope is zero at every point, the graph can neither rise nor fall, and must therefore be a horizontal line. As an immediate application, we have the following result:

Theorem 5.1.2

If $F_1(x)$ and $F_2(x)$ are antiderivatives of the same function throughout an interval, they differ by a constant c over that interval; that is, for $a < x < b$,

$$F_1'(x) = F_2'(x) \implies F_2(x) = F_1(x) + c.$$

If both $F_1(x)$ and $F_2(x)$ are antiderivatives of $f(x)$, so that

$$F_1'(x) = f(x) \quad \text{and} \quad F_2'(x) = f(x),$$

then $F_2'(x) - F_1'(x) = f(x) - f(x) = 0$, and, therefore,

$$\frac{d}{dx}[F_2(x) - F_1(x)] = 0.$$

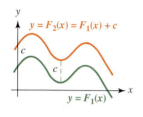

Figure 5.1.5

According to Theorem 5.1.1, that means $F_2(x) - F_1(x) = c$, for some number c, which is the same as $F_2(x) = F_1(x) + c$.

Figure 5.1.5 shows the graphs of two antiderivatives of the same function.

EXAMPLE 5.1.3

Find all antiderivatives of $f(x) = e^{2x}$.

SOLUTION The function $F_1(x) = \frac{1}{2}e^{2x}$ is one antiderivative, since

$$F_1'(x) = \frac{1}{2} \cdot 2e^{2x} = e^{2x}.$$

According to Theorem 5.1.2, any other antiderivative $F_2(x)$ will be of the form $F_2(x) = F_1(x) + c$ for some number c. Therefore, the formula

$$F(x) = \frac{1}{2}e^{2x} + c, \quad c \text{ any number},$$

gives all antiderivatives of $f(x) = e^{2x}$.

Now we are ready to complete the solution to Example (5.1.1).

SOLUTION OF EXAMPLE (5.1.1) CONTINUED According to Eq. (2), the profit function satisfies the equation $P'(x) = -0.4x + 200$, or, in other words, $P(x)$ is an antiderivative of $-0.4x + 200$. One antiderivative of this function is $-0.2x^2 + 200x$.

Therefore, $P(x) = -0.2x^2 + 200x + c$. Since the profit function satisfies the condition $P(100) = 10,000$, we have

$$P(100) = -0.2 \cdot 100^2 + 200 \cdot 100 + c = 10,000,$$

or $c = -8,000$. Therefore, $P(x) = -0.2x^2 + 200x - 8,000$.

The indefinite integral

In considering all the antiderivatives of a given function, we find it convenient to introduce some new teminology and notation, as follows:

Definition 5.1.2

Let $f(x)$ be a given function. The **indefinite integral** of $f(x)$ is the family of all antiderivatives of $f(x)$. It is denoted by $\int f(x)\,dx$. If $F(x)$ is any antiderivative of $f(x)$, then

$$\int f(x)\,dx = F(x) + c,$$

where c represents an arbitrary constant.

The symbol dx on the left-hand side of the formula serves as a reminder that x is the independent variable—rather like the d/dx in the derivative symbol.

With this notation we can write antiderivative formulas succinctly, as in the following example.

EXAMPLE 5.1.4

Find the indefinite integral of $f(x) = x^3$.

SOLUTION   An antiderivative is given by $F(x) = \frac{1}{4}x^4$, since

$$F'(x) = \frac{d}{dx}\left(\frac{1}{4}x^4\right) = \frac{1}{4} \cdot 4x^3 = x^3 = f(x).$$

Therefore, the indefinite integral is given by the formula

$$\int x^3\,dx = \frac{1}{4}x^4 + c,$$

where c is an arbitrary constant.

Basic indefinite integral formulas

Every known derivative formula can be converted into an indefinite integral formula. We now state three of the most basic ones.

Basic Indefinite Integrals

- Power Rule, $k \neq -1$: $\displaystyle\int x^k \, dx = \frac{1}{k+1} x^{k+1} + c.$ (3)

- Power Rule, $k = -1$: $\displaystyle\int \frac{1}{x} \, dx = \ln |x| + c.$ (4)

- Exponential Rule: $\displaystyle\int e^{kx} \, dx = \frac{1}{k} e^{kx} + c,$ for any $k \neq 0.$ (5)

To derive the first rule, which we will call the **power rule for integrals**, we observe that the power rule for derivatives tells us that

$$\frac{d}{dx} x^{k+1} = (k+1)x^k.$$

If $k \neq -1$ then $k + 1 \neq 0$, and we can divide both sides by it to get

$$\frac{d}{dx} \left(\frac{1}{k+1} x^{k+1} \right) = x^k.$$

Thus, $\frac{1}{k+1} x^{k+1}$ is an antiderivative of x^k if $k \neq -1$, which gives formula (3).

One way to remember the power rule for integrals is to notice that it reverses the power rule for derivatives. For instance, to find the derivative of x^4, we reduce the exponent by one and multiply by the old exponent. That gives $4x^3$. To find the indefinite integral of x^4, we *increase* the exponent by one and *divide* by the *new* exponent, which gives $\frac{1}{5} x^5$.

To derive the second rule, formula (4) we first recall that if $x > 0$,

$$\frac{d}{dx} \ln x = \frac{1}{x}.$$

Therefore, if $x > 0$,

$$\int \frac{1}{x} \, dx = \ln x + c = \ln |x| + c,$$ (6)

where c represents an arbitrary constant.

On the other hand, if $x < 0$, then $(-x) > 0$, and the chain rule gives

$$\frac{d}{dx} \ln(-x) = -\frac{1}{-x} = \frac{1}{x}.$$

Therefore, if $x < 0$,

$$\int \frac{1}{x}\,dx = \ln(-x) + c = \ln|x| + c, \tag{7}$$

where c again represents an arbitrary constant. Combining (6) and (7) gives (4).

The third rule, which we call the **exponential rule for integrals**, is a consequence of the derivative formula

$$\frac{d}{dx}\,e^{kx} = ke^{kx}.$$

If $k \neq 0$, we can divide both sides by k, which leads to

$$\frac{d}{dx}\,\frac{1}{k}e^{kx} = e^{kx}$$

and formula (5) follows. Notice that once again the integral formula is obtained by reversing the derivative formula and adding an arbitrary constant.

There are two other useful formulas that enable us to find the antiderivatives of many more functions. They both follow from the similar rules for derivatives.

Constant Multiple Rule for Indefinite Integrals

$$\int kf(x)\,dx = k \int f(x)\,dx, \quad \text{for any constant } k.$$

Sum Rule for Indefinite Integrals

$$\int [f(x) + g(x)]\,dx = \int f(x)\,dx + \int g(x)\,dx.$$

By combining these formulas, we can find antiderivatives of many functions.

EXAMPLE 5.1.5

Find $\int \left(5e^{-2x} + \frac{9}{x}\right) dx$.

SOLUTION   By applying the sum and constant multiple rules, we obtain

$$\int \left(5e^{-2x} + \frac{9}{x}\right) dx = 5 \int e^{-2x}\,dx + 9 \int \frac{1}{x}\,dx.$$

By the exponential and power rules, we know that

$$\int e^{-2x}\,dx = -\frac{1}{2}e^{-2x} + c_1 \quad \text{and} \quad \int \frac{1}{x}\,dx = \ln|x| + c_2,$$

where c_1 and c_2 represent arbitrary constants. Therefore,

$$\int \left(5e^{-2x} + \frac{9}{x}\right) dx = -\frac{5}{2}e^{-2x} + 9\ln|x| + c,$$

where $c = 5c_1 + 9c_2$.

Just as with the derivative, the sum rule may be applied to three or more functions.

EXAMPLE 5.1.6

Find $\displaystyle\int \left(u^{5/3} + 7u^{-3} - 4e^{9u} + 8\right) du$.

SOLUTION Applying the sum and constant multiple rules gives

$$\int \left(u^{5/3} + 7u^{-3} - 4e^{9u} + 8\right) du = \int u^{5/3} \, du + 7\int u^{-3} \, du - 4\int e^{9u} \, du + 8\int du$$

$$= \frac{3}{8}u^{8/3} - \frac{7}{2}u^{-2} - \frac{4}{9}e^{9u} + 8u + c.$$

Differential equations

Indefinite integral problems, such as those given in the last two examples, can also be stated in the language of **differential equations**. A differential equation is an equation involving an unknown function and its derivatives. A solution of such an equation is a function or a family of functions that satisfy the equation. Here is Example 5.1.5 restated as a differential equation.

EXAMPLE 5.1.7

Solve the differential equation

$$\frac{dy}{dx} = 5e^{-2x} + \frac{9}{x}.$$

SOLUTION The solution of this differential equation is the indefinite integral of Example 5.1.5, that is,

$$y(x) = -\frac{5}{2}e^{-2x} + 9\ln|x| + c.$$

A solution in this form—one that involves an arbitrary constant and includes all possible solutions—is called the **general solution** of the differential equation.

EXAMPLE 5.1.8

Find the general solution of the differential equation

$$y' = x^2 - 3x + 2.$$

SOLUTION   The general solution of this differential equation is simply the indefinite integral of the right-hand side, that is,

$$y(x) = \frac{1}{3}x^3 - \frac{3}{2}x^2 + 2x + c.$$

EXAMPLE 5.1.9

Solve the differential equation

$$y' = 140e^{-0.07t}.$$

SOLUTION   The solution $y = y(t)$ is the indefinite integral of the function $140e^{-0.07t}$. Thus,

$$y(t) = 140 \int e^{-0.07t}\, dt$$

$$= \frac{140}{-0.07}e^{-0.07t} + c = -2{,}000e^{-0.07t} + c.$$

Important note   One of the main reasons for studying indefinite integrals is to solve differential equations. Differential equations appear everywhere in applications of mathematics to science, engineering, economics, and business. In fact, the study of differential equations was one of the main motivations for the founders of calculus.

The initial value problem

According to Theorem 5.1.2, a function $y = y(x)$ is determined up to a constant once we know its derivative. To determine the constant, we need to know the value of $y(x)$ at a single point x_0. This value is often denoted by y_0, and the equation $y(x_0) = y_0$ is known as an **initial condition**.

Definition 5.1.3

Initial value problem

The differential equation $y' = f(x)$ together with the initial condition $y(x_0) = y_0$ is called an **initial value problem** and is often written as

$$\frac{dy}{dx} = f(x), \quad y(x_0) = y_0,$$

where $f(x)$ is a given function, and x_0 and y_0 are given numbers.

Solving an initial value problem

1. First, we find an antiderivative $F(x)$ of the given function $f(x)$. Once $F(x)$ is found, we write the general solution of the differential equation in the form

$$y(x) = F(x) + c.$$

2. Next, we use the initial condition $y(x_0) = y_0$ to compute the constant c from the equation

$$y_0 = F(x_0) + c.$$

We demonstrate these two steps in the following examples.

EXAMPLE 5.1.10

Find the solution of the initial value problem

$$\frac{dy}{dx} = 3x^2 - 2x - e^{-x}, \quad y(0) = \frac{1}{2}.$$

SOLUTION The general solution of the differential equation is given by the indefinite integral

$$y(x) = \int (3x^2 - 2x - e^{-x})\, dx = x^3 - x^2 + e^{-x} + c.$$

Since $y(0) = \frac{1}{2}$, we must have

$$\frac{1}{2} = y(0) = 0^3 - 0^2 + e^{-0} + c = 1 + c.$$

Therefore, $c = -\frac{1}{2}$ and the solution is $y(x) = x^3 - x^2 + e^{-x} - \frac{1}{2}$.

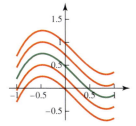

Figure 5.1.6

Remark Observe that in the last example the differential equation has infinitely many solutions obtained from the formula

$$y = x^3 - x^2 + e^{-x} + c$$

by giving c different values. In Figure 5.1.6, the curve passing though the point $(0, 0.5)$ is the graph of the solution of the initial value problem. The other curves are graphs of solutions of the differential equation *not* satisfying the given initial condition.

EXAMPLE 5.1.11

A company determines that the **marginal cost** (in dollars per unit) for producing x units of their product is given by the function

$$MC(x) = 3x^2 - 5x + 2.$$

If the fixed cost is \$1,000, then determine the **total cost** function and the cost of producing 100 units of the product. (Recall that the **fixed cost** is the constant amount of dollars required to start and maintain production, independent of the quantity of units produced. Another way to think of it is as the cost incurred if no units are produced. Usually it includes such things as rent, insurance, and so forth.)

SOLUTION   Since the marginal cost is the derivative of the cost function, we can write

$$C'(x) = 3x^2 - 5x + 2,$$

and taking the indefinite integral gives

$$C(x) = \int (3x^2 - 5x + 2)\, dx = x^3 - \frac{5}{2}x^2 + 2x + c, \tag{8}$$

where c is any fixed number.

To find c, we use the information that the fixed cost is 1,000, which means that $C(0) = 1,000$. Therefore, setting $x = 0$ in (8) gives $c = 1,000$, and the total cost function is

$$C(x) = x^3 - \frac{5}{2}x^2 + 2x + 1,000. \tag{9}$$

Now we can compute the cost of producing 100 units by setting $x = 100$ in formula (9), which gives

$$C(100) = 100^3 - \frac{5}{2}100^2 + 2 \cdot 100 + 1,000 = 976,200.$$

Objects moving under the influence of gravity

For another important example of an initial value problem, let's return to a formula we have already used in a number of examples throughout the text:

$$h = -16t^2 + v_0 t + h_0, \tag{10}$$

where h is the height (in feet) of an object moving under the influence of gravity, t is the time (in seconds), and v_0 and h_0 are constants representing the initial velocity (in feet per second) and height.

This formula can be derived from a more elementary principle, discovered by Galileo: *If an object is moving under the influence of gravity, with no other force acting on it, then its acceleration is constant.* It is customary to use the letter g for that constant. When the height is measured in feet, the value of g is approximately -32 feet per second squared.

Using v for the object's velocity and a for its acceleration, we have the relation $a = dv/dt$. Therefore, Galileo's principle can be stated in the form of the initial value problem:

$$\frac{dv}{dt} = g, \quad v(0) = v_0. \tag{11}$$

Solving the differential equation gives $v = gt + c$, and setting $t = 0$ and using the initial value yield $c = v_0$. Thus, from (11) we obtain the following equation for the velocity:

$$v = gt + v_0. \tag{12}$$

The velocity is the derivative of the height function, equation (12). This leads us to another initial value problem

$$\frac{dh}{dt} = gt + v_0, \quad h(0) = h_0.$$

This time, solving the differential equation gives

$$h = \int (gt + v_0)\, dt$$

$$= \frac{1}{2}gt^2 + v_0 t + b,$$

where b is a constant. Setting $t = 0$ and using the intitial value yield $h_0 = b$, so we get

$$h = \frac{1}{2}gt^2 + v_0 t + h_0.$$

With height in feet and $g = -32$ feet per second squared, this reduces to Eq. (10). Thus, we have derived the equation of motion from a much more basic principle.

APPLYING TECHNOLOGY

Estimating antiderivatives   Given a table of data representing the values of a continuous function, we can construct a table of approximate values for the antiderivative based on a single initial value.

EXAMPLE 5.1.12

Suppose that the table shown in Figure 5.1.7 gives the values $f(x)$ of a continuous function. Construct a table of approximate values of the antiderivative $F(x)$ using the initial value $F(0) = 0$.

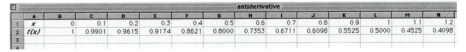

	A	B	C	D	E	F	G	H	I	J	K	L	M	N
1	*x*	0	0.1	0.2	0.3	0.4	0.5	0.6	0.7	0.8	0.9	1	1.1	1.2
2	*f(x)*	1	0.9901	0.9615	0.9174	0.8621	0.8000	0.7353	0.6711	0.6098	0.5525	0.5000	0.4525	0.4098
3														
4														

Figure 5.1.7

SOLUTION   The linear approximation formula of Section 3.4 applied to $F(x)$ can be written as

$$F(x + h) \approx F(x) + hF'(x).$$

Since $F'(x) = f(x)$, we can rewrite this as

$$F(x+h) \approx F(x) + hf(x).$$

Starting with $x = 0$ and using $h = 0.1$, we can successively approximate the values $F(0.1)$, $F(0.2)$, ..., $F(1.2)$. For instance,

$$F(0.1) \approx F(0) + (0.1)f(0) = 0 + (0.1) \cdot (1) = 0.1,$$

and

$$F(0.2) \approx F(0.1) + (0.1)f(0.1) \approx 0.1 + (0.1) \cdot (0.9901) = 0.19901.$$

The third row of Figure 5.1.8 shows the values computed by an Excel spreadsheet. The highlighted cell C3 contains the linear approximation formula, with B3 being the value of $F(0)$ and B2 the value of $f(0)$.

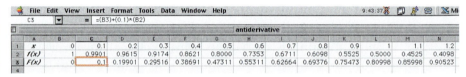

Figure 5.1.8

Practice Exercises 5.1

1. Find $\int \left(\dfrac{2}{t} + 6t^{-0.4} - 8e^{-0.1t} \right) dt$.

2. The marginal profit of a bike shop is $MP(x) = -\frac{3}{8}x^2 + 2x + 17$, where x is the number of bikes sold in a day. If

the shop sells only 2 bikes in a day, it loses \$100. Find the shop's profit for selling x bikes a day.

Exercises 5.1

1. The marginal profit function for a company producing and selling a certain product is

$$MP(x) = -0.8x + 2,000.$$

If 30 units are produced and sold, the company's profit is \$15,000. Determine the profit function $P(x)$.

2. The marginal cost for producing x units per month of a certain product is

$$MC(x) = 500 - 0.01x$$

for $0 \le x \le 10,000$. The company has a monthly fixed cost of \$4,000. Find the cost function $C(x)$.

In Exercises 3–10, find one antiderivative of the given function.

3. $f(x) = \sqrt{x}$

4. $g(x) = \dfrac{1}{\sqrt{x}}$

5. $h(s) = s - \dfrac{1}{s}$

6. $f(t) = e^{-2t}$

7. $f(x) = x^5 - \sqrt[3]{x} + x^{-4}$

8. $p(t) = 100e^{-0.02t} - \dfrac{50}{t}$

9. $A(r) = \pi r^2 - 2\pi r$

10. $q(s) = As^{0.7}$

In Exercises 11–14, verify that $F(x)$ is an antiderivative of $f(x)$.

11. $f(x) = 2xe^{x^2}$, $\quad F(x) = 1 + e^{x^2}$

12. $f(x) = xe^{x/2}$, $\quad F(x) = 2e^{x/2}(x - 2)$

13. $f(x) = \dfrac{2}{x^2 - 1}$, $\quad F(x) = \ln\left|\dfrac{x-1}{x+1}\right|$

14. $f(x) = \ln x$, $\quad F(x) = x \ln x - x$

In Exercises 15–18, find the value of the constant A that makes F(x) an antiderivative of f(x).

15. $f(x) = \dfrac{1}{5x + 1}$, $F(x) = A \ln |5x + 1|$

16. $f(x) = (2x - 3)^5$, $F(x) = A(2x - 3)^6$

17. $f(x) = \dfrac{x}{x^2 + 1}$, $F(x) = A \ln(x^2 + 1)$

18. $f(x) = xe^{-x}$, $F(x) = Ae^{-x}(x + 1)$

In Exercises 19–22, find the value of the constant A that makes the given integral formula true.

19. $\displaystyle\int (2t - 1)^5 \, dt = A(2t - 1)^6 + c$

20. $\displaystyle\int \sqrt{3x + 1} \, dx = A(3x + 1)^{3/2} + c$

21. $\displaystyle\int \dfrac{dt}{t^2 - t} = A \ln \left| \dfrac{t}{t - 1} \right| + c$

22. $\displaystyle\int te^{-t^2} \, dt = Ae^{-t^2} + c$

In Exercises 23–24, find the values of the constants A and B that make the given integral formula true.

23. $\displaystyle\int \dfrac{dx}{x^2 + x - 6} = A \ln |x - 2| + B \ln |x + 3| + c$

24. $\displaystyle\int te^{0.1t} \, dt = Ate^{0.1t} + Be^{0.1t} + c$

In Exercises 25–30, find five different antiderivatives of the given function and sketch all five on the same set of axes.

25. $f(x) = 2$ **26.** $f(x) = -2$ **27.** $f(x) = \dfrac{1}{2}x$

28. $f(x) = -\dfrac{1}{2}x$ **29.** $f(x) = 2e^{-x}$ **30.** $f(x) = \dfrac{1}{3x}$

31. The graph of a function f is shown in Figure 5.1.9. Sketch the graphs of an antiderivative F_1 with $F_1(0) = 2$ and of another antiderivative F_2 with $F_2(0) = 1$. Then compute $F_1(1)$, $F_2(1)$, and $F_1(3) - F_2(3)$.

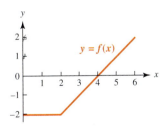

Figure 5.1.9

32. The graph of a function f is shown in Figure 5.1.10. Sketch the graphs of an antiderivative F_1 with $F_1(0) = -2$ and of another antiderivative F_2 with $F_2(0) = 0$. Then

compute $F_1(1)$, $F_2(1)$, $F_2(3) - F_1(3)$, and $F_1(4.8) - F_2(4.2)$.

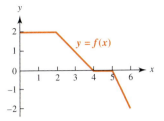

Figure 5.1.10

33. A study of a city's water reservoir provides the following information. During the first 5 months the water volume increased at a constant rate of 2 million cubic feet per month. For the next 6 months, volume remained unchanged; after that, it increased at a constant rate of 0.5 million cubic feet per month. Let $w(t)$ be the volume of water after t months.

(a) Graph the rate dw/dt as a function of t for $t \geq 0$. Be sure to label the parts.

(b) Is the rate graph continuous?

(c) Find the antiderivative for each part, given that $w = 900$ million cubic feet when $t = 0$.

(d) Sketch the graph of w as a function of t. Label the parts. Is this graph continuous?

In Exercises 34–42, find the given indefinite integral. Check your answer by differentiation.

34. $\displaystyle\int (x^2 - 3x + 1) \, dx$

35. $\displaystyle\int \dfrac{2}{x^3} \, dx$ **36.** $\displaystyle\int \sqrt{t} \, dt$

37. $\displaystyle\int e^{3t} \, dt$ **38.** $\displaystyle\int e^{-0.5t} \, dt$

39. $\displaystyle\int \dfrac{1}{2}(e^x + e^{-x}) \, dx$ **40.** $\displaystyle\int (1 - t)^2 \, dt$

41. $\displaystyle\int y^2 \left(5y^3 + \dfrac{1}{y} \right) dy$

42. $\displaystyle\int (e^{2x} + x^{3/2}) \, dx$

In Exercises 43–45, find an antiderivative $F(x)$, with $F(0) = 0$. What properties do you expect $F(x)$ to have when $f(x)$ is negative, when $f(x)$ is positive, when $f(x)$ is increasing, and when $f(x)$ is decreasing? Graph both functions and verify those properties.

43. Let $f(x) = 3x^2 - 12$. Use a $[-4, 4] \times [-40, 40]$ window.

44. Let $f(x) = e^x - 6x$. Use a $[-2, 4] \times [-8, 8]$ window.

45. Let $f(x) = x^{1/3} - x^3$. Use a $[-2, 2] \times [-2, 2]$ window.

In Exercises 46–50, find the general solution of the given differential equation.

46. $\dfrac{dy}{dx} = x + \dfrac{1}{x}$

47. $\dfrac{dy}{dx} = x^4 - 4x^{2/3}$

48. $\dfrac{dy}{dx} = 3e^{2x}$

49. $\dfrac{dy}{dt} = 100e^{-0.06t} + 350$

50. $\dfrac{dz}{dt} = \dfrac{t^3 - 1}{t}$

In Exercises 51–56, solve the given initial value problem and check your answer.

51. $\dfrac{dy}{dx} = 1 + 2x, \ \ y(0) = -1$

52. $\dfrac{dy}{dx} = 1 + 2x, \ \ y(1) = -1$

53. $\dfrac{dy}{dx} = x + e^{-x}, \ \ y(0) = 1$

54. $\dfrac{dy}{dx} = 1 - \dfrac{1}{x}, \ \ y(1) = 3$

55. $\dfrac{dQ}{dt} = e^{0.08t}, \ \ Q(0) = 80{,}000$

56. $\dfrac{dy}{dx} = \dfrac{1}{\sqrt{x}}, \ \ y(4) = 0$

In Exercises 57–59, solve the initial value problem for each of the given initial values. Then use a graphing calculator to graph the family of solutions on the same screen.

57. $\dfrac{dy}{dx} = 6x^2 - 4x + 1, \ \ y(0) = -2, -1, 0, 1, 2$

58. $\dfrac{dy}{dx} = e^{x/2} - e^{-x/2}, \ \ y(0) = -2, -1, 0, 1, 2$

59. $\dfrac{dy}{dx} = x + \dfrac{2}{x}, \ \ y(1) = 0, 1, 2, 3, 4$

60. Assume that the rate of production of an assembly line at time t is $80 + t - 0.25t^2$ units per hour, for $0 \le t \le 8$.
(a) Find the formula for the production function $P(t)$.
(b) When is the rate of production maximum during the 8-hour shift?
(c) When is the rate of production minimum during the 8-hour shift?

61. A small company that manufactures hiking boots determines that its marginal weekly profit is given by $P'(x) = 15 - 0.01x$, where x is the number of pairs of boots sold per week. If no boots are sold in a given week, the company loses $1,000.
(a) Find the profit function $P(x)$.
(b) What profit does the company make if it sells 1,000 pairs of boots in a given week?
(c) What is the maximum profit the company can earn in a single week?

62. A manufacturer's marginal cost function is $6q^2 - 4q + 8$ dollars per unit, where q is the number of units produced. The total cost of producing the first unit is $500. Find the total cost of producing 20 units.

63. A company installs a new computer system at a cost of $2,000,000, which is expected to generate savings at the rate $300{,}000\, e^{-0.03t}$ dollars per year after t years in operation. When will the system pay for itself?

64. Find the function whose tangent has slope $5x^4 - x + 5$ for each value of x, and whose graph passes through the point $(0, 8)$.

65. A function whose tangent has slope $2x - 3$ for each value of x passes through the point $(0, 4)$. What is the minimum height of its graph?

66. Find the function that passes through the point $(0, 0)$ and whose slope is twice that of the graph of $y = e^x + x + 1$ for every x.

67. Pipistrel bats, which roost in attics and old buildings and feed on insects, are declining in numbers in England because of home renovations, chemicals in the building materials, and insecticides. Biologists in one area believe that the decline is best modeled by $y' = -0.005t$, where y is the proportion of the bat population remaining after t years.
(a) Write y as an indefinite integral.
(b) What is the initial condition?
(c) How much of the population has died off after 14 years?
(d) Determine the year you would expect the local population to be completely eliminated.

68. Seismologists are monitoring the movement of a right lateral fault line. ("Lateral" means that two continental plates are sliding sideways past each other, and "right lateral" indicates that objects on the opposite side of the fault from you appear to be moving to the right.) Their research reports the rate of movement of the fault to be $y' = 0.6t + 3t^{-1/2}$ in centimeters per year. Find the function that represents the total number of centimeters of movement measured by the scientists. (*Hint:* Consider what the initial value must be.)

69. A funny joke begins to circulate by e-mail. Assume that the "infection" rate at which it spreads—measured by the number of new people receiving the joke per day—is given by $J'(t) = e^t - 0.9 + 15t$. If the joke was originally sent to 36 people, find $J(t)$, the number of people who have received the joke after t days.

70. Company A and company B manufacture similar products. Their marginal costs for producing x units are the same, but company B's start-up costs are \$2,300 higher per month than company A's. How much more does it cost company B to produce 9,500 units a month than company A?

71. The Book Nook and Adams Books are two competing book stores in the same town. The Book Nook's marginal profit is 92% that of Adams. If Adams Books does not sell any books, it loses \$2,400 per month, whereas the Book Nook only loses \$2,000 under the same circumstances. If Adams Books makes \$3,100 profit in November, how much is the Book Nook's profit that month?

72. An automobile is traveling along a straight road at 60 feet per second when the driver steps on the brake. At that point it starts decelerating at a constant rate of 12 feet per second squared until it stops.
(a) How long does it take to stop?
(b) How far does it travel in that time?

73. A car is moving along a straight road at 80 feet per second when the driver steps on the brake, causing it to decelerate at a constant rate. In the next 4 seconds it travels 240 feet. How fast is it going at the end of that time?

74. Suppose that the acceleration (in feet per second squared) of a falling hailstone is given by the formula

$$a = -30 + 9\sqrt{t} \quad \text{for } 0 \le t \le 10.$$

If it is falling at a rate of 20 feet per second when it is 1,200 feet above ground, what are its height and velocity 4 seconds later?

75. A farmer uses a large tank connected to his irrigation system to distribute 100 gallons of liquid fertilizer to his fields. When irrigation begins, he opens the tank's valve, allowing the fertilizer to leave the tank at the rate of $20e^{-0.2t}$ gallons per minute.

(a) Write the differential equation that represents this situation, with independent variable t being the time in minutes and the dependent variable V being the volume of fertilizer in the tank. Include the initial condition.
(b) Find the solution.

76. Exercising improves both physical and mental fitness. However, many people do not exercise due to inertia. Assume that available data show that the rate of change of a fitness index $F(t)$ due to exercising is

$$F'(t) = 7e^{-0.14t},$$

where t is the exercising time in hours per week.
(a) Find $F(t)$, assuming that $F = 50$ indicates no exercising at all.
(b) Find the exercise time needed for a fitness index of 90.
(c) Compute $F(30)$ and the limiting value of $F(t)$.

77. There are many variations of Fick's law of diffusion, including the one that describes the diffusion of a solute across a cell membrane:

$$c'(t) = \frac{kA}{V}[C - c(t)], \tag{13}$$

where $c(t)$ is the the solute concentration inside the cell at time t, A is the surface area of the membrane, V is the cell volume, C is the (constant) solute concentration outside the cell, and k is the permeability constant for the membrane. Verify that the solute concentration inside the cell

$$c(t) = (c_0 - C)e^{-kAt/V} + C$$

is a solution to this differential equation, given that $c(0) = c_0$. (*Hint:* Compute $c'(t)$, and show that $c(t)$ and $c'(t)$ satisfy (13).)

78. Assuming that the table in Figure 5.1.11 represents the values $f(x)$ of a continuous function, construct a table of approximate values of the antiderivative $F(x)$ under the initial condition $F(0) = 1$.

79. The table in Figure 5.1.12 gives the acceleration of a car during the first 3 seconds after starting from a position of rest. Find the approximate velocity at each of the given times.

x	0	0.1	0.2	0.3	0.4	0.5	0.6	0.7	0.8	0.9	1	1.1	1.2	1.3	1.4	1.5	1.6
$f(x)$	0.000	0.101	0.208	0.328	0.469	0.642	0.860	1.143	1.517	2.023	2.718	3.689	5.065	7.045	9.939	14.232	20.697

Figure 5.1.11

	A	B	C	D	E	F	G	H	I	J	K	L	M	N
								acceleration						
1	t	0	0.25	0.5	0.75	1	1.25	1.5	1.75	2	2.25	2.5	2.75	3
2	a	0.00	0.34	0.76	1.30	2.04	2.92	4.15	5.80	8.12	10.80	14.26	16.82	18.05
3	v													

Figure 5.1.12

In Exercises 80–82, construct a table of values for the antiderivative of the given function $f(x)$ in two ways—first, by finding the antiderivative $F(x)$ and computing its values, and, second, by constructing a table of values for $f(x)$ and using linear approximation to approximate the values of $F(x)$. Compare the results.

80. $f(x) = x^{1/2} - 1$, x from 1 to 5 in steps of 0.25, initial value $F(1) = 0$

81. $f(x) = e^{-x/3}$, x from 0 to 1 in steps of 0.1, initial value $F(0) = 0$

82. $f(x) = 1 - (1/x)$, x from 1 to 100 in steps of 1, initial value $F(1) = 0$

83. Complete the following problems by using a computer program, such as Maple or Mathematica:
   (a) Compute $\int 1/(24 - 10x - 3x^2 + x^3) \, dx$ and check your answer.
   (b) Solve $y' = 3x^2 + 1$ and draw the solutions for $-1 < x < 1$ and for $c = -4, -3, -2, -1, 0, 1, 2, 3, 4$.

Solutions to practice exercises 5.1

1. Applying the sum and constant multiple rules gives

$$\int \left(\frac{2}{t} + 6t^{-0.4} - 8e^{-0.1t} \right) dt$$

$$= 2 \int \frac{1}{t} \, dt + 6 \int t^{-0.4} \, dt - 8 \int e^{-0.1t} \, dt$$

$$= 2 \ln |t| + \frac{6}{0.6} t^{-0.6} - \frac{8}{(-0.1)} e^{-0.1t} + c$$

$$= 2 \ln |t| + 10t^{0.6} + 80e^{-0.1t} + c.$$

2. Since the marginal profit is the derivative of the profit function, the profit function satisfies the differential equation

$$\frac{dP}{dx} = -\frac{3}{8}x^2 + 2x + 17.$$

In addition, we have the initial condition $P(2) = -100$,

reflecting the fact that there is a \$100 loss when $x = 2$. The general solution of the differential equation is

$$P(x) = \int \left(-\frac{3}{8}x^2 + 2x + 17 \right) dx$$

$$= -\frac{1}{8}x^3 + x^2 + 17x + c.$$

To find the constant c, we use the initial condition, which gives

$$-100 = P(2) = -\frac{1}{8}2^3 + 2^2 + 17 \cdot 2 + c = 37 + c.$$

Therefore, $c = -137$, and the profit function is

$$P(x) = -\frac{1}{8}x^3 + x^2 + 17x - 137.$$

5.2 Integration by Substitution

In Section 5.1 we listed the integration formulas for the basic functions x^k and e^{kx}. We obtained these formulas simply by reversing the corresponding differentiation formulas. Now we will take up **integration by substitution**, an important method for reducing certain more complicated integrals to one of the standard forms. It is based on the chain rule for derivatives, which says that if F and g are two functions, then the derivative of the composite function $F(g(x))$ is given by the formula

$$\frac{d}{dx} [F(g(x))] = F'(g(x)) \cdot g'(x). \tag{14}$$

For instance, if $F(x) = x^7$ and $g(x) = x^2 + 1$, then $F(g(x)) = (x^2 + 1)^7$. Since $F'(x) = 7x^6$ and $g'(x) = 2x$, the chain rule gives

$$\frac{d}{dx} [F(g(x))] = \frac{d}{dx} \left[(x^2 + 1)^7 \right] = 7(x^2 + 1)^6 \cdot 2x. \tag{15}$$

Now assume that F is an antiderivative of a given function f; in other words, $F'(u) = f(u)$. Then we can rewrite (14) as

$$\frac{d}{dx}[F(g(x))] = f(g(x))g'(x) \quad \text{or} \quad f(g(x))g'(x) = \frac{d}{dx}[F(g(x))].$$

Taking the antiderivative of each side of this equation gives the following formula:

Integration by Substitution

$$\int f(g(x))g'(x)\,dx = F(g(x)) + c, \quad \text{where } F' = f. \qquad (16)$$

EXAMPLE 5.2.1

Compute $\displaystyle\int 7(x^2 + 1)^6\, 2x\, dx$.

SOLUTION   Referring to Eq. (15), we set $F(x) = x^7$, $f(x) = 7x^6$, and $g(x) = x^2 + 1$. We get

$$\int 7(x^2 + 1)^6\, 2x\, dx = \int \underbrace{7(x^2 + 1)^6}_{f(g(x))}\, \underbrace{(x^2 + 1)'}_{g'(x)}\, dx = \underbrace{(x^2 + 1)^7}_{F(g(x))} + c.$$

Formula (16) can be written in an equivalent way that is easier to apply. If we make the substitution

$$\boxed{u = g(x),}$$

then $F(u) = F(g(x))$. Since F is an antiderivative of f,

$$\int f(u)\,du = F(u) + c = F(g(x)) + c. \qquad (17)$$

Comparing formulas (16) and (17), we obtain the substitution formula in the form

$$\boxed{\int f(u)\frac{du}{dx}\,dx = \int f(u)\,du.} \qquad (18)$$

EXAMPLE 5.2.2

Compute the integral $\displaystyle\int \frac{2}{2x + 1}\, dx$.

SOLUTION If we let $u = 2x + 1$, then $du/dx = 2$. Therefore,

$$\int \frac{2}{2x+1}\, dx = \int \frac{1}{2x+1} 2\, dx$$

$$= \int \frac{1}{u} \frac{du}{dx}\, dx$$

$$= \int \frac{1}{u}\, du = \ln|u| + c = \ln|2x+1| + c.$$

To check, we differentiate the answer, using the chain rule. For simplicity, we assume that $2x + 1 > 0$, so that $\ln|2x + 1| = \ln(2x + 1)$. Then,

$$\frac{d}{dx}\left(\ln(2x+1) + c\right) = \frac{1}{2x+1} \frac{d}{dx}(2x+1) = \frac{2}{2x+1}.$$

You should convince yourself that the result is the same if $2x + 1 < 0$.

EXAMPLE 5.2.3

Compute the integral $\displaystyle\int (3x-2)^{-1/2}\, dx$. Check your answer.

SOLUTION If we substitute $u = 3x - 2$, we get $du/dx = 3$. This factor, which is a key ingredient in the substitution formula, is missing from the given integral. However, we can make an adjustment by multiplying and dividing by the same factor and then using formula (18), as follows:

$$\int (3x-2)^{-1/2}\, dx = \int (3x-2)^{-1/2} \cdot \frac{1}{3} \cdot 3\, dx$$

$$= \frac{1}{3} \int (3x-2)^{-1/2} \cdot 3\, dx \quad \text{(constant multiple rule)}$$

$$= \frac{1}{3} \int u^{-1/2} \frac{du}{dx}\, du$$

$$= \frac{1}{3}(2u^{1/2}) + c \quad \text{(substitution formula)}$$

$$= \frac{2}{3}(3x-2)^{1/2} + c.$$

We check the answer by taking its derivative, using the chain rule.

$$\frac{d}{dx}\left[\frac{2}{3}(3x-2)^{1/2} + c\right] = \frac{2}{3} \cdot \frac{1}{2}(3x-2)^{-1/2} \cdot 3 = (3x-2)^{-1/2}.$$

In making substitutions, a common and convenient practice is to write $du = g'(x)\, dx$ instead of $du/dx = g'(x)$. For instance, with $u = 3x - 2$, as in the last example, we have $du = 3\,dx$. If we write this in the equivalent form $dx = \frac{1}{3}\,du$, we get the substitution

$$\int (3x-2)^{-1/2}\,dx = \int \underbrace{(3x-2)^{-1/2}}_{u^{-1/2}}\,\underbrace{dx}_{\frac{1}{3}du} = \int \frac{1}{3}u^{-1/2}\,du = \frac{1}{3}\int u^{-1/2}\,du$$

EXAMPLE 5.2.4

Compute the integral $\displaystyle\int xe^{x^2}\,dx$. Check your answer.

PROBLEM-SOLVING TACTIC

Here are the keys to the substitution method:

- Look for a substitution $u = g(x)$ that simplifies the integral.
- Check that the derivative $g'(x)$ appears as a factor in the integral, at least up to a constant multiple.
- Use $u = g(x)$ and $du = g'(x)\,dx$ to eliminate x and reduce the integral to a known one in u.
- After computing the integral, replace u by $g(x)$ in the answer.

SOLUTION The substitution $u = x^2$ will simplify the exponential. Since $u' = 2x$, we can write $du = 2x\,dx$, which is equivalent to $\frac{1}{2}\,du = x\,dx$. Then

$$\int xe^{x^2}\,dx = \int \underbrace{e^{x^2}}_{e^u}\,\underbrace{x\,dx}_{\frac{1}{2}du}$$

$$= \int \frac{1}{2}e^u\,du$$

$$= \frac{1}{2}e^u + c = \frac{1}{2}e^{x^2} + c.$$

To check the answer, we differentiate it, using the chain rule:

$$\frac{d}{dx}\left(\frac{1}{2}e^{x^2} + c\right) = \frac{1}{2}e^{x^2}\cdot 2x = xe^{x^2}.$$

It is not always easy to see whether a substitution will work until you have tried it. Don't be afraid to make an educated guess and then do some calculations to see if it is right.

EXAMPLE 5.2.5

Find $\displaystyle\int \frac{x+1}{x^2+2x+3}\,dx$.

SOLUTION Here we try $u = x^2 + 2x + 3$. Then, $du = (2x+2)\,dx$, which we can rewrite as $\frac{1}{2}\,du = (x+1)\,dx$. By making these substitutions, we get

$$\int \frac{x+1}{x^2+2x+3}\,dx = \int \frac{1}{x^2+2x+3}(x+1)\,dx$$

$$= \int \frac{1}{u}\cdot\frac{1}{2}\,du$$

$$= \frac{1}{2}\ln|u| + c = \frac{1}{2}\ln|x^2+2x+3| + c.$$

EXAMPLE 5.2.6

At the moment a skydiver's parachute opens, she is falling at a rate of 20 feet per second. From then on, her acceleration is given by

$$a = \frac{6t}{(t^2 + 1)^{3/2}},$$

where t is the time in seconds (where the downward direction is taken to be positive). Find her velocity t seconds after her chute opens.

SOLUTION   Since $v = da/dt$, the velocity is given by

$$v = \int \frac{6t}{(t^2 + 1)^{3/2}} \, dt,$$

with $v(0) = 20$ feet per second.

To compute the integral, we make the substitution $u = t^2 + 1$, so that $du = 2t \, dt$. Then, $6t \, dt = 3 \, du$, and

$$v = \int 3u^{-3/2} \, du = -6u^{-1/2} + c = -\frac{6}{\sqrt{t^2 + 1}} + c.$$

Setting $t = 0$ gives $20 = -6 + c$, so that $c = 26$, and we obtain the following formula for the velocity:

$$v = 26 - \frac{6}{\sqrt{t^2 + 1}}.$$

Caution   The substitution method does not *always* work. As we have seen, we can compensate for a missing constant factor, but it is usually not possible to compensate for a missing nonconstant factor. For instance, in attempting to compute $\int e^{x^2} \, dx$, we might try the substitution $u = x^2$. But then $du = 2x \, dx$, and there is no way to adjust for the missing factor $2x$. In particular, the following attempt to multiply and divide by the same factor fails:

$$\int e^{x^2} \, dx = \int \underbrace{e^{x^2}}_{e^u} \frac{1}{2x} \, du \neq \frac{1}{2x} \int e^u \, du$$

because the extra factor $1/2x$ is not constant, and the constant multiple rule does not apply.

In certain cases we may be able to compensate for a missing nonconstant factor, but they are rare and require some ingenuity. For example, the following integral appears in one of the versions of the Solow model of economic growth, which we will come back to in a later chapter.

EXAMPLE 5.2.7

Find $\displaystyle\int \frac{1}{1+\sqrt{x}}\,dx$.

SOLUTION   One substitution that suggests itself is $u = 1 + \sqrt{x}$. If we try it, we get $du = 1/(2\sqrt{x})\,dx$. To compensate for the missing factor, we multiply and divide the integrand by $2\sqrt{x}$, as follows:

$$\int \frac{1}{1+\sqrt{x}}\,dx = \int \underbrace{\frac{1}{1+\sqrt{x}}}_{1/u} \cdot 2\sqrt{x} \cdot \underbrace{\frac{1}{2\sqrt{x}}\,dx}_{du}\,.$$

Once again, there is an extra factor $2\sqrt{x}$, which we cannot treat as a constant. However, looking back at the original substitution, we see that $\sqrt{x} = u - 1$. Therefore,

$$\int \frac{1}{1+\sqrt{x}}\,dx = \int \frac{1}{u} \cdot 2(u-1)\,du = \int \frac{2u-2}{u}\,du$$

$$= \int \left(2 - \frac{2}{u}\right)du = 2u - 2\ln|u| + c$$

$$= 2(1+\sqrt{x}) - 2\ln(1+\sqrt{x}) + c.$$

Practice Exercises 5.2
●●

1. Use the substitution $u = e^x$ to find $\int e^x/(e^x + e^{-x})\,dx$. Check your answer.

2. Compute the integral $\int (x+1)/(3x^2 + 6x + 8)\,dx$.

Exercises 5.2
●●

In Exercises 1–10, find the indefinite integral using the indicated substitution. Check your answer by differentiating.

1. $\displaystyle\int \frac{2x}{x^2+1}\,dx,\ u = x^2 + 1$

2. $\displaystyle\int \frac{\ln x}{x}\,dx,\ u = \ln x$

3. $\displaystyle\int e^{-t}\,dt,\ u = -t$

4. $\displaystyle\int \frac{e^t}{(2+e^t)^2}\,dt,\ u = 2 + e^t$

5. $\displaystyle\int \sqrt{3x+4}\,dx,\ u = 3x + 4$

6. $\displaystyle\int (x^2 - 1)e^{x^3 - 3x + 1}\,dx,\ u = x^3 - 3x + 1$

7. $\displaystyle\int \frac{x\ln(x^2+1)}{x^2+1}\,dx,\ u = \ln(x^2 + 1)$

8. $\displaystyle\int \frac{1}{e^x + e^{-x} + 2}\,dx,\ u = e^x \text{ or } e^x + 1$

9. $\displaystyle\int \frac{1-\sqrt{x}}{1+\sqrt{x}}\,dx,\ u = 1 + \sqrt{x}$

10. $\displaystyle\int \frac{e^{2x}}{e^x+1}\,dx,\ u = 1 + e^x$

11. (a) For the integral $\int 1/(1+\sqrt{x})\,dx$ of Example 5.2.7, show that the substitution $u = \sqrt{x}$ leads to the integral $\int 2u/(u+1)\,du$. Compute this integral by dividing $u+1$ into $2u$ to obtain a quotient and remainder. Then write the answer as a function of x.

(b) The answer to part (a) looks different from that of Example 5.2.7. Show that both answers are correct by taking their derivatives. How do you explain the fact that there are two different correct answers?

12. Given the integral $\int 1/(1 + x^2)\,dx$, will the substitution $u = (1 + x^2)$ reduce it to $\int (1/u)\,du$? Explain.

13. Given the integral $\int \sqrt{1 + e^t}\,dt$, will the substitution $u = 1 + e^t$ reduce it to $\int \sqrt{u}\,du$? Explain.

14. Waste water with a salinity of 45 parts salt per 1,000 parts water is released into the ocean by a manufacturing plant to mix with seawater. Measurements taken at 1-meter intervals away from the pipe show the rate of salinity decrease to be

$$S'(a) = -1.8ae^{-0.09a^2},$$

where a denotes the distance from the pipe in meters.
(a) What substitution $u = g(a)$ would make the integration of this function easier?
(b) Find the general solution $S(a)$.
(c) Use the initial condition $S(0) = 45$ to determine the specific solution.
(d) The limit as a goes to the infinity of S gives the average salinity of seawater. What is it?

Source: "Ocean and Oceanography (composition of seawater)," *Microsoft Encarta 98 Encyclopedia.*

In Exercises 15–30, find the indefinite integral, Check your answer by differentiating.

15. $\int \dfrac{2x}{\sqrt{x^2 + 4}}\,dx$

16. $\int (2x + 5)^7\,dx$

17. $\int (x^2 - 3)^5 x\,dx$

18. $\int t^3 \sqrt{t^4 + 1}\,dt$

19. $\int \dfrac{2x + 1}{(x^2 + x + 1)^3}\,dx$

20. $\int \dfrac{x + 1}{x^2 + 2x + 3}\,dx$

21. $\int t^3 e^{-0.1t^4}\,dt$

22. $\int e^{-4t}\,dt$

23. $\int \dfrac{3}{2x + 1}\,dx$

24. $\int \dfrac{(\ln t)^2}{t}\,dt$

25. $\int \dfrac{\ln(t^2)}{t}\,dt$

26. $\int \dfrac{e^t}{1 + e^t}\,dt$

27. $\int \dfrac{dt}{t \ln t}$

28. $\int e^{2x} \sqrt{1 + e^{2x}}\,dx$

29. $\int \dfrac{\sqrt{x}}{1 + \sqrt{x}}\,dx$

30. $\int \dfrac{1}{1 + x^{1/3}}\,dx$

In Exercises 31–34, solve the initial value problem.

31. $\dfrac{dy}{dt} = \sqrt{4t + 1}$, $y(0) = 3$

32. $\dfrac{dy}{dt} = \sqrt{4t + 1}$, $y(2) = 3$

33. $\dfrac{dr}{dx} = \dfrac{e^x - e^{-x}}{e^x + e^{-x}}$, $r(0) = 0$

34. $\dfrac{dy}{dx} = \dfrac{2x}{(x - 2)^2}$, $y(1) = 4$

35. A boat is moving in a straight line through an oil spill with an acceleration of $a = 5 - t\,(t^2 + 1)^{-1}$ miles per hour squared. Assuming that it started from a position of rest, what is its velocity at the end of 1 hour?

36. A company has fixed costs of $2,000 per day, and the marginal cost at a production level of x units is given (in dollars per unit per day) by

$$MC(x) = 100 - \dfrac{\log_{10}(10 + x)}{10 + x}.$$

Find the cost function.

37. The fish population in Bass Lake would grow exponentially if it were not for the fact that the number of fishermen is also growing exponentially. The Department of Natural Resources did a study and concluded that the future size of the population is modeled by the initial value problem

$$\dfrac{dy}{dt} = 70e^{0.14t} - 120e^{0.03t}, \quad y(0) = 3,700,$$

where y is the size of the fish population and t is the time (in years), starting from the date of its report. If this model is correct, how many fish will there be in the lake 10 years after the report?

38. A consulting firm, hired by a city to help develop its downtown renewal project, predicted that the number of jobs would increase at a rate of

$$\dfrac{dy}{dx} = 1,600 \left(1 - \dfrac{x}{\sqrt{1 + x^2}}\right),$$

where y is the number of jobs and x is the amount of money invested, in millions of dollars. If no money is invested, no new jobs will be created. If the prediction is correct, how many new jobs will be created by an investment of $6 million?

39. An advertisement appeared on television offering free concert tickets to anyone who responded to an Internet survey. The promotion company tabulated the number of responses and calculated the rate per hour to be

$$R'(h) = 37\dfrac{0.8h + 1}{0.4h^2 + h + 1},$$

where h is the number of hours after the ad appeared.
(a) Find the number of responses as a function of time (in hours), given that there were 168 immediate responses.
(b) About how many responses were received by $h = 1$?

40. In 1971, a person's social status was estimated to increase at the rate of

$$s'(t) = 0.462(t + 4)^{1.1},$$

where t is the number of years of education. Find $S(t)$, the social status level for t years of education, assuming that $S(0) = (0.22) \cdot 4^{2.1}$.

Source: "Mathematical Experimentation and Sociology Theory: A Critical Analysis," Robert L. Hamblin, *Sociometry*, 34: 423–452, 1971.

41. After a breakwater was built, the rate of erosion of a nearby beachside cliff was

$$E'(y) = -2.52y(0.1y^2 + 1)^{-4},$$

meters per year. A fence is 3.6 meters from the cliff's edge. Use integration and $E(0) = 3.6$ to find the distance from the fence to the edge of the cliff as a function of y. In how many years will the erosion reach the fence?

42. The **Gompertz growth functions** are important in mathematical biology. Among other things, they model the size of a growing tumor. These functions have the form $g(t) = e^{y(t)}$, where $y(t)$ is a solution of the initial value problem

$$\frac{dy}{dt} = kBe^{-kt}, \quad y(0) = \ln A - B,$$

for some given positive constants A, B, and k. Find an explicit solution of this initial value problem and use it to show that $g(t) = Ae^{-Be^{-kt}}$.

Source: D. S. Jones and B. D. Sleeman, *Differential Equations and Mathematical Biology*, London: George Allen and Unwin, 1983, p. 17: "All observations in animal tumors suggest that they obey the Gompertz growth law."

43. A drug is being intravenously administered to a patient at a constant rate of r units per hour. At the same time the patient's body is eliminating the drug at a rate proportional to the amount of drug present at any given time. Let k be the constant of proportionality. Under these circumstances,

the time at which there will be x units in the patient's system is an antiderivative of the function $f(x) = (r - kx)^{-1}$.

(a) Assuming that $x = 0$ when $t = 0$, find a formula giving t as a function of x.
(b) Find a formula giving x as a function of t.
(c) Find the limiting value of x as t becomes large.

In Exercises 44–46, find an antiderivative $F(x)$, with the given initial condition $F(0) = 0$. What properties do you expect $F(x)$ to have when $f(x)$ is negative, when $f(x)$ is positive, when $f(x)$ is increasing, and when $f(x)$ is decreasing? Graph both functions and verify those properties.

44. Let $f(x) = 10x(x^2 - 1)^4$ and $F(0) = 0$.
Use a $[-2.5, 2.5] \times [-2.5, 2.5]$ window.

45. Let $f(x) = \dfrac{e^x - e^{-x}}{(e^x + e^{-x})^2}$ and $F(0) = 0$.
Use a $[-3, 3] \times [-0.5, 0.5]$ window.

46. Let $f(x) = \dfrac{2\ln x}{x}$ and $F(1) = 0$.
Use a $[0, 3] \times [-1, 2]$ window.

In Exercises 47–49, solve the initial value problem for each of the given initial values. Then use a graphing calculator to graph the family of solutions on the same screen.

47. $\dfrac{dy}{dx} = \dfrac{x}{\sqrt{x^2 + 1}}, \quad y(0) = -1, 0, 1, 2, 3$

48. $\dfrac{dy}{dx} = 2xe^{-x^2}, \quad y(0) = -2, -1, 0, 1, 2$

49. $\dfrac{dy}{dx} = \dfrac{2\ln x}{x}, \quad y(1) = -2, -1, 0, 1, 2$

Solutions to practice exercises 5.2

1. If we let $u = e^x$, then $du = e^x \, dx$ and $e^{-x} = 1/u$. Therefore,

$$\int \frac{e^x}{e^x + e^{-x}} \, dx = \int \frac{1}{u + \dfrac{1}{u}} \, du = \int \frac{u}{u^2 + 1} \, du.$$

To evaluate this integral, we make another substitution, $w = u^2 + 1$, with $dw = 2u \, du$. Then

$$\int \frac{u}{u^2 + 1} \, du = \int \frac{1}{u^2 + 1} \cdot \frac{1}{2} \cdot 2u \, du$$

$$= \frac{1}{2} \int \frac{1}{w} \, dw = \frac{1}{2} \ln |w| + c$$

$$= \frac{1}{2} \ln(u^2 + 1) + c = \frac{1}{2} \ln(e^{2x} + 1) + c.$$

To check the answer, we differentiate, using the chain rule:

$$\frac{d}{dx} \frac{1}{2} \ln(e^{2x} + 1) = \frac{1}{2} \frac{2e^{2x}}{e^{2x} + 1} = \frac{e^{2x}}{e^{2x} + 1} = \frac{e^x}{e^x + e^{-x}},$$

where the last step was obtained by multiplying the numerator and denominator by e^{-x}.

2. If we let $u = 3x^2 + 6x + 8$, then $du = (6x + 6)\,dx = 6(x + 1)\,dx$. Dividing both sides by 6 gives $\frac{1}{6}\,du = (x + 1)\,dx$. Therefore,

$$\int \frac{x + 1}{3x^2 + 6x + 8} \, dx = \int \frac{1}{u} \cdot \frac{1}{6} \, du = \frac{1}{6} \ln |u| + c$$

$$= \frac{1}{6} \ln(3x^2 + 6x + 8) + c.$$

5.3   Integration by Parts and Partial Fractions

There are many other methods of integration, and most of them are applicable to special classes of functions. These methods have been used to derive the many integral formulas that can be found in books of integral tables and in software packages. Here we shall only introduce two of them: the method of **integration by parts**, and the method of **partial fractions**, which we will need later for the study of the logistic model of population growth.

Integration by parts

This method is based on the product rule for differentiation, which we can write in the form

$$(uv)' = uv' + u'v. \tag{19}$$

If we take antiderivatives on both sides of (19), we obtain

$$uv = \int uv' \, dx + \int u'v \, dx,$$

and rearranging terms gives the integration by parts formula:

$$\int uv' \, dx = uv - \int vu' \, dx. \tag{20}$$

The benefit is that it allows us to compute $\int uv' \, dx$ by reducing it to $\int vu' \, dx$, which may be a simpler integral. Using the notation

$$dv = v' \, dx \quad \text{and} \quad du = u' \, dx,$$

we can also write (20) in the abbreviated form

$$\int u \, dv = uv - \int v \, du. \tag{21}$$

EXAMPLE 5.3.1

Compute the indefinite integral $\int xe^x \, dx$.

SOLUTION   Let $u = x$ and $dv = e^x \, dx$, or, in other words, $dv/dx = e^x$. Then $v = e^x$ and $du = dx$, and formula (21) gives

$$\int xe^x \, dx = xe^x - \int e^x \, dx = xe^x - e^x + c.$$

We check the answer by taking its derivative:

$$\frac{d}{dx}(xe^x - e^x + c) = x\frac{d}{dx}e^x + e^x\frac{d}{dx}x - e^x = xe^x + e^x - e^x = xe^x.$$

Caution   Another choice here would be to let $u = e^x$, and $dv = x\,dx$. Then v would be an antiderivative of x, such as $v = \frac{1}{2}x^2$. By applying formula (21), we obtain

$$\int xe^x\,dx = \frac{1}{2}x^2e^x - \int \frac{1}{2}x^2\,d(e^x)$$

$$= \frac{1}{2}x^2e^x - \frac{1}{2}\int x^2e^x\,dx.$$

Here we have succeeded in expressing the integral $\int xe^x\,dx$ in terms of the more difficult integral $\int x^2e^x\,dx$, which, of course, is not desirable at all.

EXAMPLE 5.3.2

Compute $\displaystyle\int x^3\ln x\,dx$.

SOLUTION   Let $u = \ln x$ and $dv = x^3\,dx$. Then $v = \frac{1}{4}x^4$ and $du = (\ln x)'\,dx = (1/x)\,dx$. By applying the integration by parts formula, we obtain

$$\int x^3\ln x\,dx = \frac{1}{4}x^4\ln x - \frac{1}{4}\int x^4\frac{1}{x}\,dx$$

$$= \frac{1}{4}x^4\ln x - \frac{1}{4}\int x^3\,dx$$

$$= \frac{1}{4}x^4\ln x - \frac{1}{16}x^4 + c.$$

The wrong way   Another choice we have here is $u = x^3$ and $dv = \ln x\,dx$. Then $v = \int \ln x\,dx$, which is not a familiar integral. It can, in fact, be computed using integration by parts (see the exercises), but in the current example it leads to a more complicated integral instead of simplifying the one we have.

The moral   Integration by parts can make life easier, but if applied the wrong way, it may make things more complicated! And applying it in the correct way requires common sense and, of course, *experience*, like everything else in life. Once you have done a number of problems, such as those at the end of this section, you will get a better feeling for how to use it.

Integration by parts is particularly useful with integrals of the form $\int x^a e^{bx}\,dx$ and $\int x^a(\ln x)^b\,dx$, but it may have to be applied more than once.

EXAMPLE 5.3.3

Find $\int x^2 e^{3x}\, dx$ and check your answer.

SOLUTION   We let $u = x^2$ and $dv = e^{3x}\, dx$. Then $du = 2x\, dx$ and $v = \frac{1}{3}e^{3x}$ (by substitution). Therefore,

$$\int x^2 e^{3x}\, dx = \frac{1}{3}x^2 e^{3x} - \frac{2}{3}\int x e^{3x}\, dx. \qquad (22)$$

The integral on the right is not one of the basic ones, but it can be computed via integration by parts, with $u = x$ and $dv = e^{3x}\, dx$. Once again, $v = \frac{1}{3}e^{3x}$, but this time $du = dx$, and we get

$$\int x e^{3x}\, dx = \frac{1}{3}x e^{3x} - \frac{1}{3}\int e^{3x}\, dx = \frac{1}{3}x e^{3x} - \frac{1}{9}e^{3x}. \qquad (23)$$

Combining (22) and (23) gives the following answer:

$$\int x^2 e^{3x}\, dx = \frac{1}{3}x^2 e^{3x} - \frac{2}{9}x e^{3x} + \frac{2}{27}e^{3x} + c.$$

To check the answer, we take its derivative and verify that we get the original integrand:

$$\frac{d}{dx}\left(\frac{1}{3}x^2 e^{3x} - \frac{2}{9}x e^{3x} + \frac{2}{27}e^{3x} + c\right) = \frac{d}{dx}\left(\frac{1}{3}x^2 e^{3x}\right) - \frac{d}{dx}\left(\frac{2}{9}x e^{3x}\right) + \frac{d}{dx}\left(\frac{2}{27}e^{3x}\right)$$

$$= \left(\frac{2}{3}x e^{3x} + x^2 e^{3x}\right) - \left(\frac{2}{9}e^{3x} + \frac{2}{3}x e^{3x}\right) + \left(\frac{2}{9}e^{3x}\right)$$

$$= x^2 e^{3x}.$$

In some cases we can find the indefinite integral by more than one method. The solutions may look different but, if both are correct, they should only differ by a constant.

EXAMPLE 5.3.4

Find $\int x\sqrt{x+1}\, dx$ by

(i) substitution   and   (ii) integration by parts.

SOLUTION   (i) If we let $u = x + 1$, then $du = dx$, and we get

$$\int x\sqrt{x+1}\, dx = \int (u-1)\sqrt{u}\, du = \int (u^{3/2} - u^{1/2})\, du$$

$$= \frac{2}{5}u^{5/2} - \frac{2}{3}u^{3/2} + c = \frac{2}{5}(x+1)^{5/2} - \frac{2}{3}(x+1)^{3/2} + c.$$

(ii) To use integration by parts, we let $u = x$ and $v' = \sqrt{x+1}$. Then $du = dx$ and $v = \frac{2}{3}(x+1)^{3/2}$ (by substitution). Therefore,

$$\int x\sqrt{x+1}\,dx = \frac{2}{3}x(x+1)^{3/2} - \frac{2}{3}\int (x+1)^{3/2}\,dx$$

$$= \frac{2}{3}x(x+1)^{3/2} - \frac{4}{15}(x+1)^{5/2} + c.$$

The solutions to parts (i) and (ii) appear to be different, but we can check them both by differentiating. Therefore, they must differ by at most a constant. In fact, we can show with a bit of algebra that they are indeed the same. The details are left as an exercise.

The method of partial fractions

This method will be explained by means of an example.

EXAMPLE 5.3.5

Compute the integral $\displaystyle\int \frac{dx}{x^2 - 5x + 6}$.

SOLUTION First we factor the quadratic $x^2 - 5x + 6$ as follows:

$$x^2 - 5x + 6 = (x-2)(x-3).$$

Next, we look for numbers A and B such that

$$\frac{1}{(x-2)(x-3)} = \frac{A}{x-2} + \frac{B}{x-3}.$$

This decomposition of the given fraction into simpler fractions is called a **partial fraction decomposition**. To find A and B, we first multiply both sides of the equation by $(x-2)(x-3)$, which gives

$$1 = A(x-3) + B(x-2). \tag{24}$$

Since (24) must be true for all x, we can set $x = 2$, which reduces the equation to $1 = A \cdot (-1) + B \cdot 0$, or $A = -1$. Similarly, by setting $x = 3$, we obtain $1 = A \cdot 0 + B \cdot 1$, or $B = 1$. Therefore, we conclude that

$$\frac{1}{x^2 - 5x + 6} = \frac{1}{(x-2)(x-3)} = \frac{1}{x-3} - \frac{1}{x-2},$$

Now, our integral becomes

$$\int \frac{dx}{x^2 - 5x + 6} = \int \frac{dx}{x-3} - \int \frac{dx}{x-2}$$

$$= \ln|x-3| - \ln|x-2| + c = \ln\frac{|x-3|}{|x-2|} + c.$$

Remarks

1. Another way to determine A and B is to expand the right-hand side of (24) and collect terms, which gives

$$1 = (A + B)x - 3A - 2B.$$

For this equation to hold for all x, we must have

$$A + B = 0$$
$$-3A - 2B = 1.$$

Then by solving this system for A and B, we obtain $A = -1$ and $B = 1$.

2. The method of Example 5.3.5 applies to integrals of the form

$$\int \frac{dx}{(x - a)(x - b)} \quad \text{and} \quad \int \frac{x\,dx}{(x - a)(x - b)}$$

with $a \neq b$. It can also be extended to any integral of the form $\int P(x)/Q(x)\,dx$, where P and Q are polynomials, but we will omit the more complicated cases.

Many integral problems involve a combination of the various methods we have learned.

EXAMPLE 5.3.6

Find $\displaystyle\int \frac{e^x}{e^{2x} - 1}\,dx.$

SOLUTION   We begin with a substitution, letting $u = e^x$. Then $du = e^x\,dx$, and

$$\int \frac{e^x}{e^{2x} - 1}\,dx = \int \frac{1}{u^2 - 1}\,du.$$

To find this integral, we use the method of partial fractions. We first set

$$\frac{1}{u^2 - 1} = \frac{A}{u - 1} + \frac{B}{u + 1},$$

and then clear the denominators to get

$$1 = A(u + 1) + B(u - 1).$$

Setting $u = 1$ and solving for A gives $A = \frac{1}{2}$. Setting $u = -1$ and solving for B yields $B = -\frac{1}{2}$. Therefore,

$$\int \frac{1}{u^2 - 1}\,du = \int \frac{\frac{1}{2}}{u - 1}\,du - \int \frac{\frac{1}{2}}{u + 1}\,du = \frac{1}{2}\ln|u - 1| - \frac{1}{2}\ln|u + 1| + c.$$

Combining both methods gives

$$\int \frac{e^x}{e^{2x} - 1}\, dx = \frac{1}{2}\ln|e^x - 1| - \frac{1}{2}\ln|e^x + 1| + c = \frac{1}{2}\ln\left|\frac{e^x - 1}{e^x + 1}\right| + c.$$

APPLYING TECHNOLOGY

Integration by using computer programs There are many software programs dealing with symbolic integration. They can reproduce all known integral formulas and combine them by symbolic manipulation. **Mathematica** and **Maple** are examples of such software programs. For instance, using Mathematica, we can find an antiderivative of $1/(x^3 + 2x^2 - x - 2)$ with the command **Integrate[f(x), x]**.

In[1]: = **Integrate[1/(x^3 + 2x^2 − x − 2), x]**

Out[1]: $= \frac{1}{6}\text{Log}[-1 + x] - \frac{1}{2}\text{Log}[1 + x] + \frac{1}{3}\text{Log}[2 + x]$

Some graphing calculators (such as the TI89) can also do symbolic integration and differentiation.

Practice Exercises 5.3

1. Solve the initial value problem: $dy/dt = te^{-0.1t}$, $y(0) = 1$.

2. Find $\int 1/(x^2 - 4)\, dx$. Check your answer.

Exercises 5.3

In Exercises 1–2, use integration by parts with u and v′ as indicated.

1. $\displaystyle\int xe^{2x}\, dx; \quad u = x,\ v' = e^{2x}$

2. $\displaystyle\int \ln x\, dx; \quad u = \ln x,\ v' = 1$

In Exercises 3–12, use integration by parts to find the integral.

3. $\displaystyle\int te^{-t/2}\, dt$

4. $\displaystyle\int t \ln t\, dt$

5. $\displaystyle\int t^2 e^{-t}\, dt$

6. $\displaystyle\int x(\ln x)^2\, dx$

7. $\displaystyle\int x^2 \ln x\, dx$

8. $\displaystyle\int t^3 e^t\, dt$

9. $\displaystyle\int x\sqrt{3x + 2}\, dx$

10. $\displaystyle\int \frac{x^2}{\sqrt{x + 1}}\, dx$

11. $\displaystyle\int x^3\sqrt{x^2 + 1}\, dx$

12. $\displaystyle\int x^3 e^{x^2}\, dx$

In Exercises 13–14, find the partial fraction decomposition.

13. $\dfrac{1}{(x - 1)(x + 5)}$

14. $\dfrac{1}{x^2 - 9}$

In Exercises 15–20, use the method of partial fractions to find the indefinite integrals.

15. $\displaystyle\int \frac{dx}{x^2 - 9x + 20}$

16. $\displaystyle\int \frac{dt}{t^2 - t}$

17. $\displaystyle\int \frac{dt}{t^2 + t}$

18. $\displaystyle\int \frac{1}{t^2 + t - 6}\, dt$

19. $\displaystyle\int \frac{dx}{(x - a)(x - b)},\ (a \neq b)$

20. $\displaystyle\int \frac{dp}{p(1 - (p/k))},\ k$ constant

21. Find $\int x/[(x - 1)(x + 2)]\, dx$ by first finding a partial fraction decomposition:

$$\frac{x}{(x - 1)(x + 2)} = \frac{A}{x - 1} + \frac{B}{x + 2}.$$

(Clear denominators and solve for A and B.)

22. Find $\displaystyle\int \frac{3x}{x^2 - 5x + 6}\, dx.$

23. Find $\displaystyle\int \frac{x - 3}{x^2 - 3x - 4}\, dx.$

24. A psychologist studying the way in which students learn a foreign language proposes the following model. The number of new words an average student can learn per day is a function $r(t)$ of time (in days) that satisfies the differential equation

$$\frac{dr}{dt} = 0.1\, e^{-0.1t}(1 - t).$$

If the student begins by learning 20 words a day, how many new words per day can he learn after 10 days?

25. A toy rocket starts from ground level and moves vertically upward with velocity $v = (t + 1)/\sqrt{2t + 1}$ feet per second. How high does it go in the first 12 seconds?

26. Nematodes are a type of roundworm that can have significant impact on agricultural crops. Some nematodes spread plant diseases or damage the plants, leaving them weakened and susceptible to other problems. A farmer has a field infested with nematodes and wishes to reduce their numbers. He applies an experimental treatment and tests the soil regularly over several weeks. Testing results show the rate of change in the number of nematodes to be

$$N'(t) = -0.00043(0.7t + 1)e^{0.7t},$$

where t is in days and N is in thousands of nematodes. Use integration by parts to find N, given $N(0) = 23$.

27. Continuing the previous problem, make a table of values of N for $t = 0, 1, 2, \ldots, 12$ to evaluate the result of the experiment. Did the treatment work?

28. Immediately after a presidential debate, a TV network conducts a call-in poll. The rate at which votes arrive is

$$v'(t) = 18t^{-5/6}\left(1 + \frac{\ln t}{6}\right), \qquad 1 \le t \le 60,$$

where t is measured in minutes and v in thousands of votes. Assuming that $v(1) = 2$, use integration by parts to calculate the total number of votes arriving during the 60-minute duration of the poll.

29. Epidemiology is the study of contagious diseases. In one model of the spread of a disease through a community, the number of days it takes for x people to become infected is taken to be an antiderivative of the function $f(x) = a/[x(M - x)]$, where M is the size of the community's population and a is a positive constant (called a constant of proportionality). Suppose that $a = 10{,}000$ and $M = 4{,}000$ and that at the beginning of the epidemic there are 8 infected people. In how many days will there be 200 infected people? Use a calculator to approximate the answer to two decimal places.

30. At a certain moment a chemical plant begins emptying waste material into a river at a rate of

$$100\left[1 + \frac{\ln(t + 1)}{(t + 1)^2}\right]$$

gallons per day. How many gallons has it emptied into the river at the end of 5 days? Use a calculator to approximate the answer to two decimal places.

31. A group of communal rodents is studied for a year and found to have a home range of 17.7 acres. After the first year it is observed that the range changes at an annual rate of approximately

$$y' = \frac{9}{t^2 + 9t}, \qquad t \ge 1.$$

(a) Is the range expanding or contracting?
(b) Use the method of partial fractions and the fact that $y(1) = 17.7$ to find the size of the range at any time $t \ge 1$.
(c) Use the limit as t approaches infinity to estimate the final size of the range.

32. You have set up an ecosystem in a container with a population consisting of the single-celled organism you are studying and a number of competing organisms. Your hypothesis is that your study organism will eliminate the others and dominate the niche. Initially, the competing organisms outnumber the one you are studying by two to one, and you observe that the rate at which the ratio changes is

$$R'(t) = \frac{-8}{t^2 + 8t + 15},$$

where t is measured in days and $R(t)$ is the ratio of competing organisms to study organisms.
(a) Find $R(t)$.
(b) Find the value for t where the populations are about equal. (It happens within the first week.)

In Exercises 33–35, find an antiderivative $F(x)$, with the given initial condition. What properties do you expect $F(x)$ to have when $f(x)$ is negative, when $f(x)$ is positive, when $f(x)$ is increasing, and when $f(x)$ is decreasing? Graph both functions and verify those properties.

33. Let $f(x) = (1 - x^2)^{-1}$, $-1 < x < 1$, and $F(0) = 0$. Use a $[-1, 1] \times [-3, 3]$ window.

34. Let $f(x) = x \ln x$ and $F(1) = 0$. Use a $[0, 2] \times [-0.5, 0.5]$ window.

35. Let $f(x) = 1 + \ln x$ and $F(1) = 0$. Use a $[0, 3] \times [-1, 2]$ window.

In Exercises 36–38, solve the initial value problem for each of the given initial values. Then use a graphing calculator to graph the family of solutions on the same screen.

36. $\dfrac{dy}{dx} = \dfrac{1}{2x - x^2}$, $y(1) = -3, -2, -1, 0, 1, 2, 3$

37. $\dfrac{dy}{dx} = 2x + \ln x$, $y(1) = -3, -2, -1, 0, 1, 2, 3$

38. $\dfrac{dy}{dx} = 4xe^{2x}$, $y(0) = 0, 1, 2, 3, 4, 5$

Supplementary Exercises for Sections 5.2 and 5.3

Problem-Solving Tactic:

Here are some tips for deciding what method of integration to use:

- First, see if one of the basic formulas can be directly applied.
- The method of partial fractions applies only to a very specific type of integral.
- Otherwise, when in doubt, first try to find a substitution, then try integration by parts.

Find the following indefinite integrals. Check your answer by differentiation.

39. $\displaystyle\int (3x + 5)^9 \, dx$

40. $\displaystyle\int \dfrac{dx}{4x + 1}$

41. $\displaystyle\int \left(\dfrac{2}{x + 4} - \dfrac{3}{x - 1} \right) dx$

42. $\displaystyle\int \dfrac{t^3}{t^4 + 1} \, dt$

43. $\displaystyle\int t\sqrt{t^2 + 9} \, dt$

44. $\displaystyle\int x^6 \ln x \, dx$

45. $\displaystyle\int \dfrac{1}{t^2 - 1} \, dt$

46. $\displaystyle\int \dfrac{t}{t^2 - 1} \, dt$

47. $\displaystyle\int \dfrac{1}{(t - 1)^2} \, dt$

48. $\displaystyle\int \dfrac{t}{(t - 1)^2} \, dt$

49. $\displaystyle\int x^3(x^4 - 4)^5 \, dx$

50. $\displaystyle\int (2t + 1)(t^2 + t - 1)^8 \, dt$

51. $\displaystyle\int (t + e^{4t}) \, dt$

52. $\displaystyle\int te^{4t} \, dt$

53. $\displaystyle\int x(1 + \ln x) \, dx$

54. $\displaystyle\int \dfrac{1 + \ln x}{x} \, dx$

55. $\displaystyle\int te^{0.5t^2} \, dt$

56. $\displaystyle\int t^2 e^{0.5t} \, dt$

57. $\displaystyle\int \dfrac{dx}{x^2 - 3x + 2}$

58. $\displaystyle\int (x - 1)e^{2x} \, dx$

59. $\displaystyle\int (\ln x)^3 \, dx$

60. $\displaystyle\int \ln(x^3) \, dx$

61. $\displaystyle\int \dfrac{2t}{t^2 - 3t} \, dt$

62. $\displaystyle\int \dfrac{te^{-t}}{(t - 1)^2} \, dt$

63. $\displaystyle\int \dfrac{1 - e^x}{1 + e^x} \, dx$

64. $\displaystyle\int \dfrac{1}{1 + e^x} \, dx$

65. $\displaystyle\int \dfrac{1}{e^x - e^{-x}} \, dx$

66. $\displaystyle\int \dfrac{e^x}{e^{2x} - 4e^x + 3} \, dx$

67. $\displaystyle\int \dfrac{x}{1 + \sqrt{x}} \, dx$

68. $\displaystyle\int \dfrac{x}{\sqrt{1 + x}} \, dx$

69. $\displaystyle\int \dfrac{1}{\sqrt{x}(\sqrt{x} - 1)(\sqrt{x} + 2)} \, dx$

70. $\displaystyle\int \dfrac{1}{(\sqrt{x} - 2)(\sqrt{x} + 3)} \, dx$

In Exercises 71–76, solve the given initial value problem.

71. $\dfrac{dy}{dt} = \sqrt{2t + 1}$, $y(0) = 1$

72. $\dfrac{dy}{dt} = t - te^t$, $y(0) = -1$

73. $\dfrac{dy}{dt} = (\ln t)^2 - t$, $y(1) = 2$

74. $\dfrac{dy}{dt} = \dfrac{1}{t^2 + 3t + 2}$, $y(0) = 0$

75. $\dfrac{dy}{dt} = t\sqrt{3t + 1}$, $y(0) = 0$

76. $\dfrac{dy}{dt} = e^{2t}\sqrt{e^t + 2} \, dt$, $y(0) = 1$

Solutions to practice exercises 5.3

1. The solution of the differential equation is given by the indefinite integral $y(t) = \int te^{-0.1t} \, dt$. To compute it, use integration by parts with $u = t$ and $v' = e^{-0.1t}$. Then, $v = -10e^{-0.1t}$ and $du = dt$. Applying formula (21) gives

$$\int te^{-0.1t} \, dt = -10te^{-0.1t} + 10 \int e^{-0.1t} \, dt$$

$$= -10te^{-0.1t} - 100e^{-0.1t} + c.$$

Using the initial condition, we have $1 = y(0) = -100 + c$, or $c = 101$. Thus, the solution is $y(t) = -10te^{-0.1t} - 100e^{-0.1t} + 101$.

2. First, decompose the integrand into partial fractions

$$\frac{1}{x^2 - 4} = \frac{1}{(x-2)(x+2)} = \frac{A}{x-2} + \frac{B}{x+2}.$$

Multiplying both sides by $(x-2)(x+2)$ converts this to

$$1 = A(x+2) + B(x-2).$$

Setting $x = 2$ gives $1 = 4A$, which means that $A = \frac{1}{4}$. Setting $x = -2$ gives $1 = -4B$, which means that $B = -\frac{1}{4}$. Therefore,

$$\int \frac{1}{x^2 - 4}\, dx = \int \left(\frac{\frac{1}{4}}{x-2} - \frac{\frac{1}{4}}{x+2} \right) dx$$

$$= \frac{1}{4} \ln |x-2| - \frac{1}{4} \ln |x+2| + c$$

$$= \frac{1}{4} \ln \left| \frac{x-2}{x+2} \right| + c.$$

5.4 Area and the Definite Integral

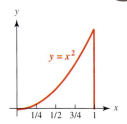

Figure 5.4.1

The problem of computing areas of regions in the plane is a very old one, and it arises in many important applications. The simplest and most basic case is the area of a rectangle, which, as you know, equals the base times the height. With this formula, it is not hard to compute the areas of other regions whose boundaries are made up of straight-line segments, such as triangles and parallelograms. The problem becomes harder, however, if part or all of the boundary is curved. One of the earliest recorded solutions of such a problem was done by the Greek mathematician Archimedes (287–212 B.C.), who computed the area under a parabolic curve. The curve shown in Figure 5.4.1 is a segment of the graph of $y = x^2$, and the problem is to compute the area between the graph and the x-axis for $0 \le x \le 1$.

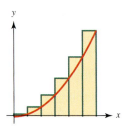

Figure 5.4.2

To compute the area under the parabola by the method of Archimedes, we cover the region under the graph with a collection of nonoverlapping rectangles, each with its base on the x-axis. The height of each rectangle is taken to be the value of the function at some point over the rectangle's base. In other words, we choose the height so that the topmost edge of the rectangle intersects the graph at some point. Figure 5.4.2 shows the region covered by six rectangles.

There are two key ideas. First, we can approximate the area of the region by adding the areas of these rectangles. Second, we can make the approximation very close by taking a lot of very thin rectangles, and the exact area can be found as a limit of this process. Figure 5.4.3 shows the same region covered by many thin rectangles. As you can see, the rectangles approximate the region better than those in Figure 5.4.2. (Actually, Archimedes used triangles rather than rectangles, but the basic method demonstrated here is essentially the same as his.)

The same idea works for *any* graph, $y = f(x)$, $a \le x \le b$, provided the function is continuous and nonnegative.

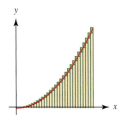

Figure 5.4.3

Riemann sums

Consider a region in the xy-plane whose base is an interval $[a, b]$ on the x-axis and that is bounded above by the graph of a nonnegative continuous function, $y = f(x)$ for $a \le x \le b$. Such a region is shown in Figure 5.4.4. We will refer to the area of this region as the **area under the graph of** $f(x)$ **for** $a \le x \le b$. We now proceed in three steps.

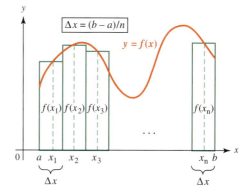

Figure 5.4.4

Step 1   For any given positive integer n, we cover the region with n nonoverlapping rectangles. To do that, we partition the interval $[a, b]$ into n subintervals of equal length. That length is called the **mesh** of the partition and symbolized by Δx. We have

$$\Delta x = \frac{b - a}{n}.$$

Next, we choose a point in each one of these n subintervals—a point x_1 in the first subinterval, a point x_2 in the second subinterval, and so on, up to a point x_n in the last subinterval. In principle these can be any points in the given subintervals, but in practice we usually choose either the left endpoint, the midpoint, or the right endpoint.

We now construct a rectangle over each subinterval, as pictured in Figure 5.4.4. Each of these rectangles has a base of length Δx. The first has height $f(x_1)$, the second has height $f(x_2)$, and so forth, up to the last rectangle, which has height $f(x_n)$.

HISTORICAL PROFILE

Archimedes (287–212 B.C.) is considered by many to be one of the three greatest mathematicians of all times. He was born and spent most of his life in Syracuse, Sicily, and devised a number of mechanical weapons for that country in its wars against the Romans. He invented the compound pulley as well as a type of pump, known as Archimedes's screw, which is still used in many parts of the world. He also discovered the principle—now called Archimedes's principle—that a body in water displaces its own weight.

His deepest love, however, was mathematics. In addition to devising the method for computing areas, he achieved a very accurate approximation of the number π, and he made significant contributions to the study of the centers of gravity of planar regions. According to legend, he was killed during the Roman invasion of Syracuse because he was too engrossed in his mathematical work to obey a soldier's order to move on.

Step 2   We compute the total area of the n rectangles constructed in Step 1, as follows:

$$\text{Area of the first rectangle} = (\text{height}) \cdot (\text{base}) = f(x_1) \cdot \Delta x$$
$$\text{Area of the second rectangle} = (\text{height}) \cdot (\text{base}) = f(x_2) \cdot \Delta x$$
$$\cdots$$
$$\text{Area of the } n\text{th rectangle} = (\text{height}) \cdot (\text{base}) = f(x_n) \cdot \Delta x.$$

By adding these, we get the total area of all the rectangles. The sum is called a **Riemann sum** for the function f, and we will denote it by $S_n(f)$. In other words, we have

Riemann Sum

$$S_n(f) = f(x_1)\Delta x + f(x_2)\Delta x + \cdots + f(x_n)\Delta x.$$

Step 3   We compute the limit of $S_n(f)$ as $n \to \infty$, if it exists.

In advanced calculus it is shown that if the function $f(x)$ is continuous on the interval $[a, b]$, the limit does indeed exist. In fact, $\lim_{n \to \infty} S_n(f)$ exists even in some cases when the function has discontinuities—for example, when $f(x)$ is continuous at all but a finite number of points and bounded throughout the interval. To summarize, we have the following:

The area under the graph of a nonnegative continuous function $f(x)$ can be approximated to any degree of accuracy by the Riemann sum $S_n(f)$ by choosing n large enough. Letting $n \to \infty$, we have

$$\text{The area under the graph of } f(x) \text{ for } a \leq x \leq b \; = \; \lim_{n \to \infty} S_n(f). \qquad (25)$$

In fact, the area under the graph is *defined* by formula (25).

EXAMPLE 5.4.1

Approximate the area under the graph of

$$f(x) = \frac{1}{x+1}, \quad 0 \leq x \leq 1,$$

using the Riemann sum with $n = 4$ and the midpoint of each subinterval.

SOLUTION   Here we have $\Delta x = \frac{1}{4}$, with $x_1 = \frac{1}{8}$, $x_2 = \frac{3}{8}$, $x_3 = \frac{5}{8}$, and $x_4 = \frac{7}{8}$, as shown in Figure 5.4.5. The area under the graph of $f(x)$ is approximated by the areas of the four rectangles, given by the Riemann sum

$$S_4(f) = [f(x_1) + f(x_2) + f(x_3) + f(x_4)]\Delta x$$

$$= \left[\frac{1}{\frac{1}{8} + 1} + \frac{1}{\frac{3}{8} + 1} + \frac{1}{\frac{5}{8} + 1} + \frac{1}{\frac{7}{8} + 1} \right] \cdot \frac{1}{4}$$

$$= 2 \left[\frac{1}{9} + \frac{1}{11} + \frac{1}{13} + \frac{1}{15} \right] \approx 0.69122.$$

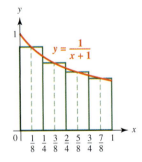

Figure 5.4.5

The definite integral: nonnegative case

The area under the graph is an intuitive notion, whereas the limit of the Riemann sums is a well-defined mathematical object. Thus, the Riemann sums and their limit give us a precise way of both discussing and computing the area under the graph. They also have many applications in addition to determining area. To emphasize the importance of this concept, we use a special name and symbol for the limit of the Riemann sums. We call it the **definite integral** of $f(x)$ for $a \leq x \leq b$ and denote it by the symbol $\int_a^b f(x)\,dx$.

Definition 5.4.1

Definition of the Definite Integral

$$\int_a^b f(x)\,dx = \lim_{n\to\infty}\,[f(x_1)\Delta x + f(x_2)\Delta x + \cdots + f(x_n)\Delta x],$$

provided the limit on the right exists.

In this definition it is understood that for each n the interval $[a, b]$ is partitioned into n equal subintervals, that x_i is a point in the ith subinterval, and that $\Delta x = (b - a)/n$. A function is called **integrable** on $[a, b]$ if the limit exists. As we have remarked, every function that is continuous on $[a, b]$ is integrable, but there are many others as well.

For emphasis, we rewrite Eq. (25) in terms of the definite integral. If $f(x)$ is a continuous, nonnegative function for $a \leq x \leq b$, then

$$\boxed{\text{The area under the graph of } f(x) \text{ for } a \leq x \leq b = \int_a^b f(x)\,dx.} \qquad (26)$$

EXAMPLE 5.4.2

Evaluate $\int_0^1 x\,dx$ in two ways:

(i) by finding the area under the graph by elementary geometry, and
(ii) by using Riemann sums.

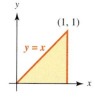

Figure 5.4.6

SOLUTION (i) The region under the graph is a right triangle with a base and height both equal to 1, as shown in Figure 5.4.6. Therefore, its area is $\frac{1}{2}$ (i.e., half the base times the height), and we conclude that $\int_0^1 x\,dx = \frac{1}{2}$.

(ii) To compute the integral by using Riemann sums, we first choose a positive integer n and divide the interval $[0, 1]$ into n equal subintervals, each of length $\Delta x = 1/n$. Next, let x_i be the left-hand endpoint of the ith subinterval, so that

$$x_1 = 0, \quad x_2 = \frac{1}{n}, \quad x_3 = \frac{2}{n}, \quad \ldots, \quad x_n = \frac{n-1}{n}.$$

Since $f(x) = x$, the Riemann sum is as follows:

$$S_n(f) = x_1 \Delta x + x_2 \Delta x + x_3 \Delta x + \cdots + x_n \Delta x$$

$$= 0 \cdot \frac{1}{n} + \frac{1}{n} \cdot \frac{1}{n} + \frac{2}{n} \cdot \frac{1}{n} + \frac{3}{n} \cdot \frac{1}{n} + \cdots + \frac{n-1}{n} \cdot \frac{1}{n}$$

$$= \frac{1}{n^2} + \frac{2}{n^2} + \frac{3}{n^2} + \cdots + \frac{n-1}{n^2}$$

$$= \frac{1}{n^2} \cdot [1 + 2 + 3 + \cdots + (n-1)].$$

This sum is the total area of a collection of rectangles, each constructed over a subinterval using the height of the graph over the left-hand endpoint as the height of the rectangle. The case of $n = 20$ is shown in Figure 5.4.7.

To compute $\lim_{n \to \infty} S_n(f)$, we need to use the following formula from algebra, whose proof we omit (but see the exercises). For any positive integer n,

$$1 + 2 + \cdots + (n-1) = \frac{n(n-1)}{2}. \tag{27}$$

Applying this formula, we get

$$S_n(f) = \frac{1}{n^2} \frac{n(n-1)}{2} = \frac{n-1}{2n} = \frac{1}{2}\left(1 - \frac{1}{n}\right).$$

Letting $n \to \infty$ gives

$$\lim_{n \to \infty} S_n(f) = \lim_{n \to \infty} \frac{1}{2}\left(1 - \frac{1}{n}\right) = \frac{1}{2}.$$

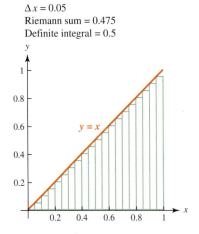

$\Delta x = 0.05$
Riemann sum $= 0.475$
Definite integral $= 0.5$

$y = x$

Figure 5.4.7

Therefore,

$$\int_0^1 x\,dx = \frac{1}{2},$$

which agrees with the result we obtained earlier by elementary geometry.

The next example illustrates Archimedes's method of computing the area under a parabola. Unlike the previous one, however, there is no way to find this integral by elementary geometry.

EXAMPLE 5.4.3

Compute $\int_0^1 x^2\,dx$.

SOLUTION Here we have $f(x) = x^2$, $a = 0$, and $b = 1$. Once again we choose a positive integer n and divide the interval $[0, 1]$ into n equal subintervals, each of length $\Delta x = 1/n$. This time we let x_i be the right endpoint of the ith subinterval, so that

$$x_1 = \frac{1}{n}, \quad x_2 = \frac{2}{n}, \quad \ldots, \quad x_n = \frac{n}{n} = 1.$$

Since $f(x) = x^2$, the Riemann sum is as follows:

$$S_n(f) = x_1^2 \Delta x + x_2^2 \Delta x + x_3^2 \Delta x + \cdots + x_n^2 \Delta x$$

$$= \left(\frac{1}{n}\right)^2 \cdot \frac{1}{n} + \left(\frac{2}{n}\right)^2 \cdot \frac{1}{n} + \cdots + \left(\frac{n}{n}\right)^2 \cdot \frac{1}{n}$$

$$= \frac{1}{n^3}(1^2 + 2^2 + \cdots + n^2).$$

This sum is the total area of a collection of rectangles, each constructed over a subinterval using the height of the graph over the right-hand endpoint. Figure 5.4.8 shows the case of $n = 20$.

To compute $\lim_{n\to\infty} S_n(f)$, we need another formula from algebra, whose proof we omit (but see the exercises). For any positive integer n,

$$1^2 + 2^2 + \cdots + n^2 = \frac{n(n+1)(2n+1)}{6}. \tag{28}$$

Using this formula, we get

$$S_n(f) = \frac{1}{n^3}\frac{n(n+1)(2n+1)}{6}$$

$$= \frac{(n+1)(2n+1)}{6n^2}$$

$$= \frac{2n^2 + 3n + 1}{6n^2} = \frac{1}{3} + \frac{1}{2n} + \frac{1}{6n^2}.$$

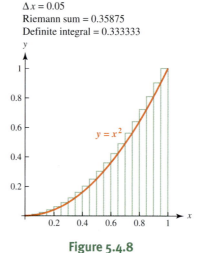

$\Delta x = 0.05$
Riemann sum = 0.35875
Definite integral = 0.333333

Figure 5.4.8

Therefore, $\lim_{n\to\infty} S_n(f) = \frac{1}{3}$, and we get

$$\int_0^1 x^2\,dx = \frac{1}{3}.$$

The definite integral: general case

The definite integral, as we have defined it, is not restricted to nonnegative functions, and it has important applications beyond finding the area under a graph. The method of forming the Riemann sums and taking their limit can be applied to functions that take both positive and negative values.

> ### Definition 5.4.2
>
> The *definite integral* of a function $f(x)$, $a \le x \le b$, is defined by
>
> $$\int_a^b f(x)\,dx = \lim_{n\to\infty} [f(x_1)\Delta x + f(x_2)\Delta x + \cdots + f(x_n)\Delta x],$$
>
> provided the limit on the right exists.

In this formula it is understood that for each n the interval $[a, b]$ is partitioned into n equal subintervals, each of length $\Delta x = (b-a)/n$, and that x_i is an arbitrarily chosen point in the ith subinterval for $i = 1, \ldots, n$. As in the nonnegative case, it can be shown by more advanced techniques that the limit exists if $f(x)$ is continuous on $[a, b]$, and that it also exists in many cases in which $f(x)$ has discontinuities.

We can obtain a geometric interpretation of the integral by observing that each term in the Riemann sum is related to the area of a rectangle. If $f(x_i)$ is nonnegative for a

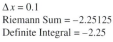

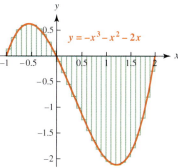

Figure 5.4.9

given i, the term $f(x_i)\Delta x$ has the same interpretation as before—the area of a rectangle whose base is the ith subinterval and whose height is $f(x_i)$. Such a rectangle extends vertically upward from the subinterval to the point $(x_i, f(x_i))$ on the graph. On the other hand, if $f(x_i)$ is negative, we construct a rectangle that extends vertically *downward* from the subinterval to the point $(x_i, f(x_i))$. The height of that rectangle is $|f(x_i)|$ and its area is $|f(x_i)\Delta x|$. Or, to put it another way, if $f(x_i) < 0$, then $f(x_i)\Delta x$ is the *negative* of the area of the rectangle extending downward from the subinterval to the point $(x_i, f(x_i))$. In Figure 5.4.9 we show the rectangles corresponding to the Riemann sum $S_n(x)$ of the function

$$f(x) = -x^3 - x^2 - 2x, \quad -1 \le x \le 2$$

for $n = 30$.

In general, then, we can interpret the Riemann sum as follows:

$S_n(f) = $ total area of the rectangles above the x-axis
$\quad\quad\quad - $ total area of the rectangles below the x-axis,

where the ith rectangle is above the x-axis if $f(x_i) \ge 0$ and below if $f(x_i) < 0$. Since

$$\int_a^b f(x)\,dx = \lim_{n\to\infty} S_n(f),$$

we are led to the following interpretation of the definite integral:

If R is the region between the graph of $f(x)$ and the x-axis for $a \le x \le b$, then

$\int_a^b f(x)\,dx = $ the area of that part of R lying above the x-axis
$\quad\quad\quad\quad - $ the area of that part of R lying below the x-axis.

For the function $f(x) = -x^3 - x^2 - 2x$, $-1 \le x \le 2$, whose graph is shown in Figure 5.4.9, the definite integral is equal to the area of the region above the interval $[-1, 0]$ minus the area of the region below the interval $[0, 2]$.

On the other hand, to find the *total* area enclosed between the graph of $f(x)$ and the x-axis for $a \le x \le b$, we would *add* the areas above and below the x-axis. For example, once again referring to the graph of

$$f(x) = -x^3 - x^2 - 2x, \quad -1 \le x \le 2,$$

pictured in Figure 5.4.9, we have

$$\int_{-1}^{0} f(x)\, dx = \text{area under the graph and above the } x\text{-axis}$$

and

$$\int_{0}^{2} f(x)\, dx = -\text{ area above the graph and below the } x\text{-axis},$$

so that the

$$\text{area between the } x\text{-axis and the graph of } f = \int_{-1}^{0} f(x)\, dx - \int_{0}^{2} f(x)\, dx.$$

EXAMPLE 5.4.4

The graph of $f(x) = x$, $-1 \le x \le 2$, is drawn in Figure 5.4.10. Use elementary geometry to determine

    (i) $\int_{-1}^{2} f(x)\, dx$,

    (ii) the total area enclosed between the graph and the x-axis.

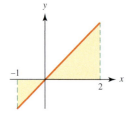

Figure 5.4.10

SOLUTION   The region between the graph and the x-axis consists of two isoceles right triangles—one below the interval $[-1, 0]$ and the other above the interval $[0, 2]$. The first has area $\frac{1}{2}$, and the second has area 2.

    (i) The definite integral is the difference of those two areas:

$$\int_{-1}^{2} x\, dx = 2 - \frac{1}{2} = \frac{3}{2}.$$

    (ii) The total area enclosed by the graph and the x-axis is the sum of those two areas, which equals $\frac{5}{2}$.

Other interpretations of the definite integral

The importance of the definite integral comes from the variety of ways it can be used to model phenomena in the physical, biological, and social sciences.

From velocity to displacement Suppose an object is moving in a straight line, with its velocity at time t given by a continuous function $v(t)$. In this case, its change of position (also called its **displacement**) between times a and b is given by a definite integral:

$$\text{Displacement between times } a \text{ and } b = \int_a^b v(t)\, dt. \tag{29}$$

To see why, think of dividing the time interval $[a, b]$ into n equal subintervals and choose a time t_i in each. The length of each time subinterval is given by $\Delta t = (b - a)/n$, and the object's velocity during that period is approximately $v(t_i)$. Therefore, its displacement during that time period is approximately $v(t_i) \cdot \Delta t$ (rate × time). By adding these, we approximate the displacement over the time period $[a, b]$,

$$\text{Displacement over } [a, b] \approx v(t_1)\Delta t + v(t_2)\Delta t + \cdots + v(t_n)\Delta t.$$

The sum on the right is a Riemann sum for the function $v(t)$ over the interval $[a, b]$. If we let $n \to \infty$,

- the approximations approach the actual displacement, and
- the Riemann sums approach the definite integral.

Thus, by letting $n \to \infty$, we get formula (29).

From acceleration to velocity In a similar way, if the object's acceleration at time t is given by a continuous function $a(t)$, the *change* in its velocity between times a and b is given by a definite integral:

$$\text{Change in velocity between times } a \text{ and } b = \int_a^b a(t)\, dt.$$

From rate of growth to total growth Suppose a living object—a plant, an animal, or a colony of cells—is growing at a rate of $r(t)$, where t denotes the time in suitable units. If we assume $r(t)$ is continuous, the total growth between two times a and b is also given as a definite integral:

$$\text{Total growth between times } a \text{ and } b = \int_a^b r(t)\, dt. \tag{30}$$

The rationale is similar to that in the previous cases. If we divide the time interval $[a, b]$ into n equal subintervals, each of size $\Delta t = (b - a)/n$, and choose a time t_i in the ith subinterval for each i, the growth rate is approximately $r(t_i)$ during that time period. Therefore, the total growth during the ith time subinterval is approximately $r(t_i) \cdot \Delta t$ (again, rate × time). By adding these, we get an approximation to the total growth over the entire time period:

$$\text{Total growth} \approx r(t_1)\Delta t + r(t_2)\Delta t + \cdots + r(t_n)\Delta t.$$

By letting $n \to \infty$, we get formula (30).

In each of these examples, the total change of a certain quantity (position, velocity, size) was expressed as the definite integral of its rate of change (velocity, acceleration, growth rate). As we know, the *derivative* of a function is its *rate of change*, which suggests the following general principle: *The total change in a function over an interval* [a, b] *is equal to the definite integral of its derivative from a to b.* That is indeed the case, provided the function has a continuous derivative. The formal statement of that fact is the **fundamental theorem of calculus**, which we will discuss in the next section.

In many applications, the rate is given by a finite set of values rather than a formula, especially when the information comes from experimental work or statistical analysis. If we assume the values come from a continuous function, we can use the Riemann sum to estimate the definite integral.

EXAMPLE 5.4.5

Suppose that the velocity (in meters per minute) of an object moving in a straight line is given by a continuous function $v(t)$, with the following known values:

$$v\left(\frac{1}{2}\right) = 1.4, \quad v(1) = 1.2, \quad v\left(\frac{3}{2}\right) = 1.8 \quad \text{and} \quad v(2) = 2.4.$$

Estimate its displacement during the time interval from 0 to 2 minutes.

SOLUTION   We want to estimate $\int_0^2 v(t)\,dt$. To do that, we divide the interval $[0, 2]$ into the following four equal subintervals:

$$\left[0, \frac{1}{2}\right], \quad \left[\frac{1}{2}, 1\right], \quad \left[1, \frac{3}{2}\right], \quad \left[\frac{3}{2}, 2\right], \quad \text{with } \Delta x = \frac{2}{4} = \frac{1}{2}.$$

Next, we form the Riemann sum using the right endpoint of each interval:

$$S_4(f) = v\left(\frac{1}{2}\right) \cdot \frac{1}{2} + v(1) \cdot \frac{1}{2} + v\left(\frac{3}{2}\right) \cdot \frac{1}{2} + v(2) \cdot \frac{1}{2}$$

$$= (1.4) \cdot \frac{1}{2} + (1.2) \cdot \frac{1}{2} + (1.8) \cdot \frac{1}{2} + (2.4) \cdot \frac{1}{2}$$

$$= 0.7 + 0.6 + 0.9 + 1.2 = 3.4.$$

Therefore,

$$\int_0^2 v(x)\,dx \approx 3.4 \text{ meters.}$$

Basic properties of the definite integral

Next, we list the basic properties of the definite integral that follow directly from the definition of the integral as a limit of Riemann sums. These properties are very useful in calculating integrals. The first two are similar to the sum and constant multiple rules for indefinite integrals.

If $f(x)$, $a \le x \le b$, and $g(x)$, $a \le x \le b$, are two continuous functions, then

$$\int_a^b [f(x) + g(x)]\,dx = \int_a^b f(x)\,dx + \int_a^b g(x)\,dx. \tag{31}$$

and if k is any constant, then

$$\int_a^b k f(x)\, dx = k \int_a^b f(x)\, dx. \tag{32}$$

EXAMPLE 5.4.6

Find each of the following definite integrals:

$$\text{(i)} \ \int_0^1 5x\, dx, \quad \text{(ii)} \ \int_0^1 -3x^2\, dx, \quad \text{(iii)} \ \int_0^1 (5x - 3x^2)\, dx.$$

SOLUTION   We already know from Examples 5.4.2 and 5.4.3 that

$$\int_0^1 x\, dx = \frac{1}{2} \quad \text{and} \quad \int_0^1 x^2\, dx = \frac{1}{3}.$$

Applying formulas (31) and (32) gives:

$$\text{(i)} \ \int_0^1 5x\, dx = 5 \int_0^1 x\, dx = \frac{5}{2},$$

$$\text{(ii)} \ \int_0^1 -3x^2\, dx = -3 \int_0^1 x^2\, dx = -3 \cdot \frac{1}{3} = -1, \text{ and}$$

$$\text{(iii)} \ \int_0^1 (5x - 3x^2)\, dx = \int_0^1 5x\, dx + \int_0^1 -3x^2\, dx = \frac{5}{2} - 1 = \frac{3}{2}.$$

The next property allows us to combine definite integrals over contiguous intervals:

$$\int_a^c f(x)\, dx + \int_c^b f(x)\, dx = \int_a^b f(x)\, dx, \quad a \le c \le b. \tag{33}$$

If $f(x) \ge 0$ for $a \le x \le b$, we can interpret this as saying that the area under the graph from a to b is the sum of the areas from a to c and from c to b.

In defining $\int_a^b f(x)\, dx$ as we have, we have assumed that $a < b$. It is sometimes useful to write the integral in the opposite way, from b to a. We define that as follows:

$$\int_b^a f(x)\, dx = - \int_a^b f(x)\, dx. \tag{34}$$

The minus sign is explained by the following fact. The Riemann sums for the first integral have $\Delta x = (a - b)/n$, which is the negative of $\Delta x = (b - a)/n$, used for the Riemann sums for the second integral. Integrals where a and b are reversed do appear in applications (e.g., in problems of motion).

Finally, if $a = b$, the length of the interval is zero, and the graph above it reduces to a single point. In that case, we have

$$\int_a^a f(x)\, dx = 0. \qquad (35)$$

APPLYING TECHNOLOGY

Computing Riemann sums Although the left- or right-hand endpoints are usually the easiest to use in computing Riemann sums, the midpoints generally give the best approximations of the definite integral. The following example illustrates the use of midpoints in approximations.

EXAMPLE 5.4.7

Use Riemann sums to estimate $\int_0^1 x^2\, dx$ by partitioning the interval $[0, 1]$ into n equal segments and using the midpoint of each segment for (i) $n = 10$ and (ii) $n = 20$. Compare these approximations with the actual value of $\int_0^1 x^2\, dx$ obtained in Example 5.4.3.

SOLUTION   (i) If $n = 10$, the length of each subinterval is $\Delta x = 0.1$ (which we will also call the step size). To compute the Riemann sum with a graphing calculator, we must first generate the sequence of numbers x_i^2, where x_i is the midpoint of the ith interval. The first subinterval, which is $[0, 0.1]$, has midpoint 0.05. The last subinterval, which is $[0.9, 1]$, has midpoint 0.95.

Most graphing calculators have a list editor that will generate a sequence of numbers and perform various operations on it. To produce the sequence, we generally need to enter the function, the independent variable, a starting point, an ending point, and a step

HISTORICAL PROFILE

Bernhard Riemann (1826–1866) was the son of a pastor who insisted he study theology. While at the university, however, Riemann developed a deep interest in mathematics. Despite the poor health and financial difficulties that afflicted him throughout his life, he became one of the most influential mathematicians of the nineteenth century, and his research laid the foundation for many twentieth-century developments.

Although his works are relatively small in number, they had a profound impact on several subjects, including real analysis, complex analysis, and geometry. He developed an important method of studying geometry with analytic tools, now called **Riemannian geometry**, that provided the mathematical framework for the general theory of relativity. His theory of the integral, using what are now called Riemann sums, was published in 1854.

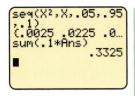

Figure 5.4.11

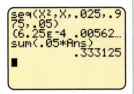

Figure 5.4.12

size. Figure 5.4.11 shows this information entered in a TI-83 Plus through the list editor's **seq** command. In response, the calculator returned the sequence $\{0.0025, 0.0225, \ldots\}$, which consists of the numbers x_i^2 for each midpoint x_i.

Next, by using the command sum(.1*Ans), we have instructed the calculator to multiply each term of the sequence by 0.1 (which equals Δx, the step size) and then take the sum to get the Riemann sum. (Ans refers to the previous output, in this case, the sequence generated in the previous step. Multiplying a sequence by a number, as in 0.1*Ans, means multiplying every term of the sequence by that number.)

The result, shown on the screen, is 0.3325. Since $\int_0^1 x^2\,dx = \frac{1}{3} = 0.\overline{3}$, we see that the approximation error (in absolute value) is $0.0008\overline{3}$.

(ii) If $n = 20$, the step size is $\Delta x = 0.05$. The first interval, which is $[0, 0.05]$, has midpoint 0.025. The last interval, which is $[0.95, 1]$, has midpoint 0.975. Figure 5.4.12 shows these entered as the starting and ending points in the list editor's seq command, preceded by the function x^2 and independent variable x, and followed by the step size. The sequence generated consists of the numbers x_i^2 for each midpoint x_i, beginning with 0.000625 (written in scientific notation as 6.25E^{-4}), which equals 0.025^2. Next, we use sum(0.05*Ans) to multiply every term of the sequence by the step size and then take the sum. The result, shown on the screen, is 0.333125. In this case, the approximation error (in absolute value) is 0.0002083.

Remark Notice that by doubling the number of intervals, we reduced the error by roughly one-fourth, from about 0.0008 to about 0.0002. That is related to a general property of midpoint approximations, one that we will come back to when we study numerical methods in Section 5.7.

Practice Exercises 5.4

1. Estimate the area under the curve $y = 1/(1+x^2)$, $0 \le x \le 1$, by using the Riemann sums with $\Delta x = 0.2$ and the left endpoints of the segments.

Estimate $\int_1^3 f(x)\,dx$ by dividing the interval $[1, 3]$ into eight equal segments and computing the Riemann sum, choosing the points $x_1, \ldots, x_8$ to be the right-hand endpoints of the segments.

2. Suppose you are given the following table of values for $f(x)$:

x	1.0	1.25	1.5	1.75	2.0	2.25	2.5	2.75	3.0
$f(x)$	-0.5	-0.28	-0.09	0.06	0.19	0.31	0.42	0.51	0.61

Exercises 5.4

1. Estimate the area under the graph of $y = 1/x$, $1 \le x \le 2$, by partitioning the interval $[1, 2]$ into four equal segments and computing the Riemann sum

$$S_4(f) = f(x_1)\Delta x + f(x_2)\Delta x + f(x_3)\Delta x + f(x_4)\Delta x,$$

where the points x_1, x_2, x_3, and x_4 are chosen to be:
(a) the left-hand endpoints of the segments,
(b) the right-hand endpoints of the segments,

(c) the mid-points of the segments.
Express each of your answers as the sum of four proper fractions and then convert it to a decimal by using a calculator.

2. Sketch the graph of $y = 1/x$, $1 \le x \le 2$ and draw each set of rectangles corresponding to parts (a), (b), and (c) of the last problem. Based on these sketches, which of the three estimates is closest to the actual integral?

3. Approximate the definite integral of the function whose graph is shown in Figure 5.4.13, using the left endpoints and $\Delta x = 0.2$.

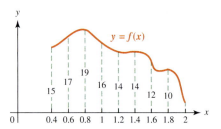

Figure 5.4.13

4. Figure 5.4.14 shows a garden plot with one curved and three straight sides. Based on the measurements shown, estimate the area.

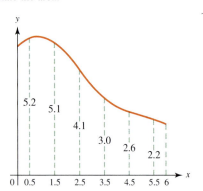

Figure 5.4.14

Exercises 5–6 refer to the plot of the function $f(x)$, $-1 \le x \le 2$, and the grid shown in Figure 5.4.15.

5. Estimate the following integrals:

(a) $\displaystyle\int_{-1}^{1} f(x)\,dx$ (b) $\displaystyle\int_{-1}^{0.25} f(x)\,dx$

(c) $\displaystyle\int_{0.25}^{2} f(x)\,dx$ (d) $\displaystyle\int_{-1}^{2} f(x)\,dx$

6. Estimate the area of the region between the graph of the function in Figure 5.4.15 and the x-axis, for $-1 \le x \le 2$.

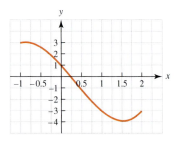

Figure 5.4.15

7. Park rangers want to know the approximate area in square meters of a lake inside a state park. Surveyors made shore-to-shore measurements and constructed the diagram in Figure 5.4.16. All measurements are in meters.
(a) Estimate the area using a mesh of 40 meters. Use the midpoint of each interval.
(b) Compare that result to one using a mesh of 20 meters and the left endpoint of each interval.

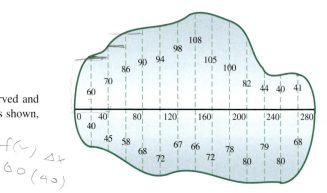

Figure 5.4.16

In Exercises 8–13, sketch the graph of the function and use formulas from elementary geometry (area of a triangle and area of a rectangle) to find the definite integral.

8. $\displaystyle\int_{0}^{1} 3x\,dx$ **9.** $\displaystyle\int_{-2}^{1} 2x\,dx$ **10.** $\displaystyle\int_{-3}^{-1} 5\,dx$

11. $\displaystyle\int_{0}^{3} (-2)\,dx$ **12.** $\displaystyle\int_{1}^{3} x\,dx$ **13.** $\displaystyle\int_{0}^{3} (1-x)\,dx$

14. Biologists studying pond turtles count the number of hatchlings they can find when they first arrive and then once a week for 5 more weeks. As determined by past research, the typical reproduction rate during those 5 weeks is shown in Figure 5.4.17. This year's reproduction rate is depicted in Figure 5.4.18.
(a) Find the area under each graph. What do those numbers represent?
(b) Construct the ratio of the area under this year's graph to that of the typical year. If the ratio is greater than 1, the

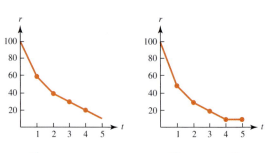

Figure 5.4.17   **Figure 5.4.18**

turtles were more successful in reproducing than usual. If it is less than 1, they were less successful. Evaluate the turtles' reproductive success.

15. Suppose you are given the following table of values for an integrable function $f(x)$:

x	1.0	1.25	1.5	1.75	2.0
$f(x)$	0	0.22	0.41	0.56	0.69

x	2.25	2.5	2.75	3.0
$f(x)$	0.81	0.92	1.01	1.10

Estimate $\int_1^3 f(x)\,dx$ by dividing the interval $[1, 3]$ into eight equal segments and computing the Riemann sum, choosing the points $x_1, \ldots, x_8$ to be
(a) the left-hand endpoints of the segments,
(b) the right-hand endpoints of the segments.

16. Let $f(x)$, $5 \leq x \leq 7.25$, be a continuous function whose known values are given by the table:

x	5	5.25	5.50	5.75	6
$f(x)$	0.8	−0.04	−0.8	−1.2	−2

x	6.25	6.50	6.75	7
$f(x)$	−1	0	1.2	1.6

Estimate $\int_5^{7.25} f(x)\,dx$ using the Riemann sum corresponding to nine subintervals and left-hand endpoints.

17. Every fall, thousands of sandhill cranes gather in northern Indiana en route from Canada to Texas. During 1 week of the migration season, an ornithologist observes the arrival rate of the cranes, in hundreds per day, and constructs the following table:

t = day	1	2	3	4	5	6	7
$r(t)$ = arrivals per day	37	43	48	56	62	68	76

(a) Write the definite integral that represents the total number of arrivals over the given time period.

(b) Identify Δt.
(c) Estimate the definite integral using Riemann sums.

18. Postal workers sort mail at a rate that was sampled every 2 days:

t = day	0	2	4	6	8	10	12
$s(t)$ = number of letters sorted per day (in thousands)	112	99	135	142	123	108	105

Estimate the total amount of mail that was sorted from days 4 through 12.

19. An object is falling vertically downward, and its velocity (in feet per second) is given by $v = -32t - 50$. Write a definite integral that gives the change in height in the first 2 seconds.

20. A bicycle starts from rest and moves in a straight line with acceleration given by

$$a(t) = 2 - \frac{t+2}{t^2+1} \text{ feet per second squared.}$$

Write a definite integral that gives its velocity at the end of 10 seconds.

21. The rate at which a certain tree grows is $0.05te^{-0.01t}$ centimeters per day. Write a definite integral that gives the total growth in the first 100 days.

22. Niagara Falls has a water volume flow rate of about 194,940 cubic feet per second. Sketch the flow rate graph and use elementary geometry to calculate the total volume flowing over the falls in 1 minute.

Source: "Niagara Falls (waterfall)," Microsoft Encarta 98 Encyclopedia.

23. The graph of a function $f(x)$, $-2 \leq x \leq 6$, is shown in Figure 5.4.19. Given that

$$\int_{-2}^1 f(x)\,dx = -2.8, \qquad \int_1^3 f(x)\,dx = 1.2,$$

and

$$\int_1^6 f(x)\,dx = -3.5,$$

find the area of the region enclosed by the graph of $f(x)$, $-2 \leq x \leq 6$, and the x-axis.

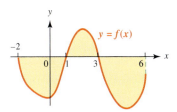

Figure 5.4.19

24. Given the areas of the regions indicated in Figure 5.4.20, compute $\int_{-4}^{5} f(x)\,dx$.

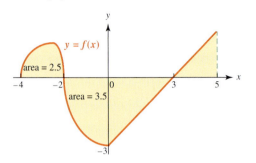

Figure 5.4.20

25. Suppose you know that $\int_{-1}^{5} f(x)\,dx = -6$ and $\int_{3}^{5} f(x)\,dx = 4$. Find $\int_{-1}^{3} 2f(x)\,dx$.

26. In the next section we will see that $\int_{0}^{1} x^3\,dx = \frac{1}{4}$. For now, estimate the integral with the help of a calculator, using Riemann sums obtained by partitioning the interval $[0, 1]$ into 20 equal segments and taking

(a) left-hand endpoints,
(b) right-hand endpoints, and
(c) midpoints.
Compare the errors.

27. In the next section we will see that $\int_{1}^{2} x^{-1}\,dx = \ln 2$. For now, estimate the integral with the help of a calculator, using Riemann sums obtained by partitioning the interval $[1, 2]$ in n equal segments for

(a) $n = 10$, (b) $n = 20$, and (c) $n = 40$
and taking the midpoints. Compare the results to the value of $\ln 2$ given by your calculator and compare the sizes of the error for the various choices of n.

28. It can be shown by more advanced techniques that
$$\int_{1}^{2} (x^2 - 2x + 2)^{-1}\,dx = \frac{\pi}{4}.$$

Estimate the integral with the help of a calculator, using Riemann sums obtained by partitioning the interval $[1, 2]$ in n equal segments for

(a) $n = 10$, (b) $n = 20$, and (c) $n = 100$
and taking the midpoints. Compare the results to the value of $\pi/4$ given by your calculator and compare the sizes of the error for the various choices of n.

 In Exercises 29–37, approximate the definite integral of the given function over the given interval by means of a Riemann sum. Use a calculator.

29. $f(t) = t^2 + t - 2$, $-1 \le t \le 1$; $n = 8$, midpoints

30. $f(x) = \dfrac{2 - x}{x}$, $1 \le x \le 3$; $\Delta x = 0.1$, midpoints

31. $f(t) = \dfrac{e^t}{t + 1}$, $0.5 \le t \le 1.5$; $n = 10$, left endpoints

32. $f(x) = \sqrt{x}$, $0 \le x \le 4$; $\Delta x = 0.5$, midpoints

33. $f(t) = 2^t$, $-1 \le t \le 1$; $\Delta t = 0.25$, left endpoints

34. $f(x) = x^x$, $0.5 \le x \le 1$; $n = 5$, right endpoints

35. $f(x) = x^3$, $0 \le x \le 2$; $n = 20$, midpoints

36. $f(x) = x^5 - 4x^2 + 5$, $-1 \le x \le 2$; $n = 20$, right endpoints

37. $f(x) = e^{-(1/2)x^2}$, $0 \le x \le 2.4$; $n = 30$, midpoints

38. The purpose of this exercise is to derive formula (27).

(a) Write the integers from 1 to n in two rows, the first in ascending order, the second in descending, as follows:

1	2	3	4	5	$\cdots$	n
n	$(n-1)$	$(n-2)$	$(n-3)$	$(n-4)$	$\cdots$	1

(b) The sum of each row is $1 + 2 + \cdots + n$, so that the sum of all the numbers in the table is $2(1 + 2 + \cdots + n)$.
(c) What is the sum of each vertical column? How many columns are there? On the basis of the answers to those questions, write another expression for the sum of all the numbers in the table.
(d) Equate the expressions you got in steps (b) and (c) and divide both sides by 2. You should have formula (27).

39. The purpose of this exercise is to derive formula (28). Let
$$S = 1^2 + 2^2 + 3^2 + \cdots + n^2$$
and
$$T = 1 + 2 + 3 + \cdots + n.$$

(a) First show that $2 \cdot 1 + 3 \cdot 2 + 4 \cdot 3 + \cdots + n \cdot (n - 1) = S - T$. (*Hint:* Write $2 \cdot 1 = 2 \cdot (2 - 1)$, $3 \cdot 2 = 3 \cdot (3 - 1)$, $4 \cdot 3 = 4 \cdot (4 - 1)$, etc.)
(b) Verify that $n^3 - 1 = (2^3 - 1^3) + (3^3 - 2^3) + (4^3 - 3^3) + \cdots + (n^3 - (n - 1)^3)$. Then use the identity $a^3 - b^3 = (a - b)(a^2 + ab + b^2)$ to show that $n^3 = 3S - T - n^2$.
(c) Finally, use formula (27) to substitute $T = n(n + 1)/2$ and solve for S.

Solutions to practice exercises 5.4

1. The area is approximately equal to the Riemann sum

$$S_5(f) = f(0) \cdot 0.2 + f(0.2) \cdot 0.2 + f(0.4) \cdot 0.2$$
$$+ f(0.6) \cdot 0.2 + f(0.8) \cdot 0.2$$

$$= \frac{0.2}{1 + 0^2} + \frac{0.2}{1 + 0.2^2} + \frac{0.2}{1 + 0.4^2}$$
$$+ \frac{0.2}{1 + 0.6^2} + \frac{0.2}{1 + 0.8^2} \approx 0.833732.$$

2. We have

$$\int_1^3 f(x)\, dx \approx f(1.25) \cdot 0.25 + f(1.5) \cdot 0.25$$
$$+ f(1.75) \cdot 0.25 + f(2) \cdot 0.25$$
$$+ f(2.25) \cdot 0.25 + f(2.5) \cdot 0.25$$
$$+ f(2.75) \cdot 0.25 + f(3.0) \cdot 0.25$$
$$= (-0.28 - 0.09 + 0.06 + 0.19$$
$$+ 0.31 + 0.42 + 0.51 + 0.61) \cdot 0.25$$
$$= 1.73 \cdot 0.25 = 0.4325.$$

5.5 The Fundamental Theorem of Calculus

There is a remarkable connection between the definite integral of a function and its indefinite integral. It was discovered in the late seventeenth century by two scientists working independently in different countries: the great English mathematician and physicist Isaac Newton and the celebrated German mathematician and philosopher Gottfried Wilhelm Leibniz. Their discovery was so important that it has come to be called the **fundamental theorem of calculus**.

Theorem 5.5.1

The fundamental theorem of calculus

Suppose $f(x)$ is a continuous function on the interval $[a, b]$, and $F(x)$ is an antiderivative of $f(x)$, that is, $F'(x) = f(x)$ for all $x \in [a, b]$. Then

$$\int_a^b f(x)\, dx = F(b) - F(a).$$

Before providing an explanation of this beautiful theorem, let us illustrate what it says by means of some examples.

EXAMPLE 5.5.1

Compute $\displaystyle\int_0^1 x\, dx$.

SOLUTION We already computed this integral in Example 5.4.2 in two different ways—first, by using the formula for the area of a right triangle and, second, by using

Riemann sums. Now let's compute it a third way—by using the fundamental theorem. We know that $F(x) = \frac{1}{2}x^2$ is an antiderivative of $f(x) = x$ and, therefore,

$$\int_0^1 x\,dx = F(1) - F(0) = \frac{1}{2}\cdot 1^2 - \frac{1}{2}\cdot 0^2 = \frac{1}{2}.$$

EXAMPLE 5.5.2

Compute $\displaystyle\int_0^1 x^2\,dx$.

SOLUTION   We computed this integral in Example 5.4.3 by using Riemann sums, in essentially the same way Archimedes did 23 centuries ago. Using the fundamental theorem, we obtain the answer much more easily. Recalling that $F(x) = x^3/3$ is an antiderivative of $f(x) = x^2$, we have

$$\int_0^1 x^2\,dx = \int_0^1 F'(x)\,dx = F(1) - F(0) = \frac{1}{3}\cdot 1^3 - \frac{1}{3}\cdot 0^3 = \frac{1}{3}.$$

A few remarks are in order. First, as we have already seen, a given function has more than one antiderivative, but any two of them just differ by a constant. In other words, if $F'(x) = f(x)$ and $G'(x) = f(x)$, then $G(x) = F(x) + C$ for some constant C. The question arises: What if we use a different antiderivative in applying the fundamental theorem? To take a specific example, suppose we had used $G(x) = (x^3/3) + 5$ as the antiderivative of $f(x) = x^2$ in the last example, instead of $F(x) = x^3/3$. How would that have changed the result? The answer is: Not at all! In that case, we would have had

$$\int_0^1 x^2\,dx = \int_0^1 G'(x)\,dx = G(1) - G(0) = \left(\frac{1}{3}\cdot 1^3 + 5\right) - \left(\frac{1}{3}\cdot 0^3 + 5\right) = \frac{1}{3}.$$

The point is that in subtracting $G(0)$ from $G(1)$, the extra 5 we added appears twice with opposite signs and therefore cancels. The same holds true in general. If $G(x) = F(x) + C$, then

$$G(b) - G(a) = [F(b) + C] - [F(a) + C] = F(b) - F(a).$$

For that reason, *any* antiderivative can be used in applying the fundamental theorem. The end result is the same.

A second remark concerns notation. We will have many occasions to write the difference $F(b) - F(a)$, and it is useful to have a shorthand way of denoting that. The customary notation is $F(x)\big|_a^b$, and we can write

$$\int_a^b F'(x)\,dx = F(x)\Big|_a^b = F(b) - F(a).$$

We will use this notation in the next few examples.

EXAMPLE 5.5.3

Find $\int_0^1 e^x\,dx$.

SOLUTION Since e^x is its own antiderivative, the fundamental theorem gives

$$\int_0^1 e^x\,dx = e^x\Big|_0^1 = e - 1 \approx 1.71828.$$

EXAMPLE 5.5.4

Find the area under the graph of $y = \dfrac{1}{x}$ from $x = 1$ to $x = 2$.

SOLUTION We know that $\dfrac{d}{dx}\ln x = \dfrac{1}{x}$, which means that

$$\int_1^2 \frac{1}{x}\,dx = \ln x\Big|_1^2 = \ln 2 - \ln 1 = \ln 2.$$

Therefore, the area under the graph is $\ln 2 \approx 0.693147$.

The fundamental theorem also applies to functions that take negative values.

EXAMPLE 5.5.5

Compute $\int_1^2 (5x^4 - 7)\,dx$.

SOLUTION Since $F(x) = x^5 - 7x$ is an antiderivative of $f(x) = 5x^4 - 7$,

$$\int_1^2 (5x^4 - 7)\,dx = (x^5 - 7x)\Big|_1^2 = (32 - 14) - (1 - 7) = 18 - (-6) = 24.$$

Explanation of the fundamental theorem of calculus

We can express the fundamental theorem in the following way: If $F'(x)$ exists and is continuous for $a \le x \le b$, then

$$\boxed{\int_a^b F'(x)\,dx = F(b) - F(a).}\tag{36}$$

The right-hand side of this equation is the total change in F over the interval $[a, b]$, and the equation says that the definite integral of the rate of change of F is equal to the total change in F.

Without giving a complete proof of that fact, we will explain the underlying idea. By definition, $\int_a^b F'(x)\,dx$ is the limit of Riemann sums, that is,

$$\int_a^b F'(x)\,dx \approx F'(x_1)\Delta x + F'(x_2)\Delta x + F'(x_3)\Delta x + \cdots + F'(x_n)\Delta x, \qquad (37)$$

where, for each n, we partition the interval into n equal subintervals and choose a point x_i in each subinterval. In particular, by choosing x_i to be the left endpoint of the ith subinterval, we get the sequence

$$a = x_1 < x_2 < \cdots < x_n < b,$$

and for each i from 2 to n,

$$x_i - x_{i-1} = \Delta x = \frac{b-a}{n}.$$

Now, if we apply linear approximation to $F(x)$ for $x_1 \le x \le x_2$, we get

$$F(x_2) \approx F(x_1) + F'(x_1) \cdot (x_2 - x_1).$$

Since $x_1 = a$ and $(x_2 - x_1) = \Delta x$, we can rewrite this as

$$F(x_2) - F(a) \approx F'(x_1) \cdot \Delta x.$$

Similarly, in the second subinterval, we obtain the relation

$$F(x_3) - F(x_2) \approx F'(x_2) \cdot \Delta x,$$

and in the third we get

$$F(x_4) - F(x_3) \approx F'(x_3) \cdot \Delta x.$$

We can continue this way in all subintervals until we reach the last subinterval, where we obtain the relation

$$F(b) - F(x_n) \approx F'(x_n) \cdot \Delta x.$$

Notice that the right-hand sides of these relations are precisely the terms in the Riemann sum of formula (37). Therefore, we can rewrite (37) as follows:

$$\int_a^b F'(x)\,dx \approx F'(x_1)\Delta x + F'(x_2)\Delta x + F'(x_3)\Delta x + \cdots + F'(x_n)\Delta x$$

$$\approx [F(x_2) - F(a)] + [F(x_3) - F(x_2)] + [F(x_4) - F(x_3)] + \cdots + [F(b) - F(x_n)].$$

Observe that the terms $F(x_2), F(x_3), \ldots, F(x_n)$ in this sum cancel because each appears twice with opposite signs. All that remains is $F(b) - F(a)$, so that

$$\int_a^b F'(x)\,dx \approx F'(x_1)\Delta x + F'(x_2)\Delta x + F'(x_3)\Delta x + \cdots + F'(x_n)\Delta x$$

$$\approx F(b) - F(a).$$

Letting $n \to \infty$ results in the equation $\int_a^b F'(x)\,dx = F(b) - F(a)$.

The area function as an antiderivative

Next, we will look at the fundamental theorem in a slightly different way. Suppose $f(x), a \leq x \leq b$, is a continuous function. We use it to define a new function $A(t)$ for $a \leq t \leq b$ as follows:

$$A(t) = \int_a^t f(x)\,dx, \quad a \leq t \leq b, \tag{38}$$

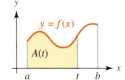

Figure 5.5.1

If $f(x)$ is nonnegative, $A(t)$ is the area under its graph from a to t, as shown in Figure 5.5.1. For that reason we call it the **area function**.

Now, suppose that $F(x)$ is any antiderivative of $f(x)$. By applying the fundamental theorem to the right-hand side of (38), we get

$$A(t) = F(t) - F(a),$$

and taking the derivative on both sides gives

$$A'(t) = F'(t) = f(t).$$

In other words, the area function is *also an antiderivative* of f. That observation leads to the following theorem.

Theorem 5.5.2

If $f(x), a \leq x \leq b$, is continuous, then the function

$$A(t) = \int_a^t f(x)\,dx, \quad a \leq t \leq b,$$

is an antiderivative of f; that is, $A'(t) = f(t)$.

EXAMPLE 5.5.6

Compute the derivative of the function

$$A(t) = \int_0^t e^{-x^2/2}\,dx.$$

SOLUTION $A(t)$ is the area function of $f(x) = e^{-x^2/2}$. Therefore by the last theorem, we have

$$A'(t) = e^{-t^2/2}.$$

Theorem 5.5.2 is actually another version of the fundamental theorem of calculus. It is equivalent to Theorem 5.5.1 in the sense that either theorem can be derived from the other. We have just seen how Theorem 5.5.2 follows from Theorem 5.5.1. To go the other way—that is, to deduce Theorem 5.5.1 from Theorem 5.5.2—suppose that F is any antiderivative of f. Since, according to 5.5.2, $A(x)$ is also an antiderivative and any two antiderivatives differ by a constant, we have

$$A(x) = F(x) + C, \quad a \leq x \leq b,$$

for some constant C. From the definition of the area function we see that

$$A(a) = 0 \quad \text{and} \quad A(b) = \int_a^b f(x)\, dx.$$

Therefore,

$$\int_a^b f(x)\, dx = A(b) - A(a)$$
$$= [F(b) + C] - [F(a) + C] = F(b) - F(a).$$

Thus, we have derived Theorem 5.5.1 from Theorem 5.5.2.

From marginal function to total function

As we just saw in formula (36), the fundamental theorem of calculus can be restated as the following **fundamental principle**:

> The definite integral of a rate of change equals the total change.

We saw a number of illustrations of this principle in the last section.

- The displacement (change in position) of an object moving in a straight line is the definite integral of its velocity.
- The change in its velocity is the definite integral of its acceleration.
- The total growth of a living object over a time period is the definite integral of its rate of growth.

Another important use of this principle is in economics, in which many decisions are based on marginal analysis of the cost and revenue. To see what that means, suppose a company's current operation entails a certain amount of cost and generates a certain amount of revenue. The company may want to know how the cost and revenue will change if it increases or decreases its production level. A key tool in analyzing that is the **marginal cost**, denoted by *MC*, which was introduced in Chapter 3. It is the change in cost per unit change in production—or, in other words, the *rate* at which the cost changes. As a working principle, we use the following definition:

Marginal cost

The marginal cost is the derivative of the cost function.

If production is measured in continuous units, such as pounds of coffee, economists use this as the definition of marginal cost. But if the production units are restricted to integers, such as the number of bicycles, the derivative is only a linear approximation to the actual change in cost. We will ignore this distinction and treat all quantities as continuous—a common and useful assumption when production is very large, so that a change of one unit is almost negligible and can be treated as a continuous change.

Similarly, the **marginal revenue**, denoted by *MR*, is the rate at which the revenue changes—in other words, the change in revenue per unit change in production—and we define it as follows:

Marginal revenue

The marginal revenue is the derivative of the revenue function.

If $C(x)$ denotes the cost of producing x units and $R(x)$ denotes the revenue generated by x units, then we can translate these statements into the following formulas:

$$MC(x) = C'(x) \quad \text{and} \quad MR(x) = R'(x),$$

where $MC(x)$ and $MR(x)$ are the marginal cost and marginal revenue.

Therefore, to find the additional cost involved in increasing production from a units to b units, we simply integrate the marginal cost—that is,

$$\text{Total change in cost} = C(b) - C(a) = \int_a^b C'(x)\,dx = \int_a^b MC(x)\,dx.$$

Similarly, the extra revenue resulting from this change in production is given by the formula:

$$\text{Total change in revenue} = R(b) - R(a) = \int_a^b R'(x)\,dx = \int_a^b MR(x)\,dx.$$

If the total change in revenue is greater than the total change in cost, an increase in the production level from a to b may be justified.

> **EXAMPLE 5.5.7**

Suppose the marginal cost involved in producing x units of a certain product is given by the function

$$MC(x) = \frac{1}{2}x + 3{,}000 \quad \text{when } x \geq 80.$$

Determine the increase in cost if production is increased from 80 to 100 units.

SOLUTION   Since $C'(x) = \frac{1}{2}x + 3{,}000$, the total increase in cost is given by

$$
\begin{aligned}
C(100) - C(80) &= \int_{80}^{100} C'(x)\,dx \\
&= \int_{80}^{100} \left(\frac{1}{2}x + 3{,}000 \right) dx \\
&= \left(\frac{1}{4}x^2 + 3{,}000x \right) \Big|_{80}^{100} \\
&= \left(\frac{1}{4} \cdot 100^2 + 3{,}000 \cdot 100 \right) - \left(\frac{1}{4} \cdot 80^2 + 3{,}000 \cdot 80 \right) \\
&= \frac{1}{4}(100^2 - 80^2) + 3{,}000(100 - 80) \\
&= 5 \cdot 180 + 60{,}000 = 60{,}900.
\end{aligned}
$$

> **EXAMPLE 5.5.8**

Suppose that when the production level of a certain product is at least 80 units, the marginal revenue is given by

$$MR(x) = -3x^2 + 200x + 10^4.$$

Find the change in total revenue if production is increased from 80 to 100 units.

SOLUTION   Since $R'(x) = MR(x)$, the total change in revenue is given by

$$
\begin{aligned}
R(100) - R(80) &= \int_{80}^{100} R'(x)\,dx \\
&= \int_{80}^{100} \left(-3x^2 + 200x + 10^4 \right) dx \\
&= \left(-x^3 + 100x^2 + 10^4 x \right) \Big|_{80}^{100} \\
&= \left(-100^3 + 100 \cdot 100^2 + 10^4 \cdot 100 \right) - \left(-80^3 + 100 \cdot 80^2 + 10^4 \cdot 80 \right)
\end{aligned}
$$

$$= \left(80^3 - 100^3\right) + 100 \left(100^2 - 80^2\right) + 10^4(100 - 80)$$

$$= -488{,}000 + 360{,}000 + 200{,}000 = 72{,}000.$$

EXAMPLE 5.5.9

A company has determined that its marginal cost is given by

$$MC(x) = \frac{1}{2}x + 3{,}000 \quad \text{when } x \geq 80,$$

and its marginal revenue by

$$MR(x) = -3x^2 + 200x + 10^4 \quad \text{when } x \geq 80.$$

If the company is currently operating at a production level of 80 units per day, is it profitable for it to increase its production to 100 units per day?

SOLUTION   In Example 5.5.7 we found that

$$\text{Increased cost} = 60{,}900,$$

while in Example 5.5.8 we found that

$$\text{Increased revenue} = 72{,}000.$$

Since the revenue increased by 11,100 more than the cost, we conclude that the increase in production from 80 to 100 units is profitable.

Here are some other **marginal functions** used in economics.

- The **marginal profit**, denoted by MP, which we define as the derivative of the profit function. It measures the change in profit per unit change in production. Since profit is the difference between revenue and cost, the marginal profit is the marginal revenue minus the marginal cost.

- The **marginal product of capital**, denoted by MPK, which is the change in production output per unit change in capital—in other words, the *rate* at which production changes as capital is increased or decreased.

- The **marginal product of labor**, denoted by MPL, which is the change in production output per unit change in labor.

Practice Exercises 5.5

1. Using the fundamental theorem of calculus, compute $\int_1^2 (16x^7 - (1/x))\, dx$.

2. The marginal profit function for a company producing and selling a certain product is $MP(x) = -0.8x + 1{,}000$. If 50 units are produced and sold, the company's profit is \$20,000. Determine the profit function.

Exercises 5.5

In Exercises 1–6, find the area under the graph of $f(x)$ from a to b by two methods: first, by elementary geometry and, second, by the fundamental theorem of calculus.

1. $f(x) = \frac{1}{2}$, $\quad a = 1$, $\quad b = 5$

2. $f(x) = 2$, $\quad a = 0$, $\quad b = 3$

3. $f(x) = \frac{x}{2}$, $\quad a = 0$, $\quad b = 4$

4. $f(x) = 3x$, $\quad a = 2$, $\quad b = 3$

5. $f(x) = 2x + 2$, $\quad a = 0$, $\quad b = 1$

6. $f(x) = -4x + 8$, $\quad a = 1$, $\quad b = 2$

In Exercises 7–16, sketch the graph of $f(x)$ from a to b and use the fundamental theorem of calculus to find the area under the graph.

7. $f(x) = x^3$, $\quad a = 0$, $\quad b = 2$

8. $f(x) = x^3 - 1$, $\quad a = 1$, $\quad b = 2$

9. $f(x) = 2x - x^2$, $\quad a = 0$, $\quad b = 2$

10. $f(x) = \sqrt{x}$, $\quad a = 1$, $\quad b = 4$

11. $f(x) = x + \frac{1}{x}$, $\quad a = 1$, $\quad b = 2$

12. $f(x) = 3 - 2x - x^2$, $\quad a = -3$, $\quad b = 0$

13. $f(x) = \frac{1}{x^2}$, $\quad a = 1$, $\quad b = 3$

14. $f(x) = x^3 - 3x^2 + 6$, $\quad a = 0$, $\quad b = 3$

15. $f(x) = e^x$, $\quad a = -1$, $\quad b = 1$

16. $f(x) = x^4 - 2x^2 + 1$, $\quad a = -1$, $\quad b = 1$

In Exercises 17–26, find the given definite integral by using the fundamental theorem of calculus.

17. $\int_0^1 (x - x^3)\,dx$

18. $\int_{-1}^1 (x - x^3)\,dx$

19. $\int_1^3 (1 - x)\,dx$

20. $\int_{-3}^1 x\,dx$

21. $\int_0^2 (x^3 - 4x^2 + 3x)\,dx$

22. $\int_{-1}^1 (e^x - 1)\,dx$

23. $\int_0^4 (x^4 - 4x^3)\,dx$

24. $\int_{-2}^{-1} \frac{1}{x}\,dx$

25. $\int_{-1}^1 x^{1/3}\,dx$

26. $\int_0^1 (3x - 2e^x)\,dx$

In Exercises 27–34, use the methods of Sections 5.2 and 5.3 to find an antiderivative and then apply the fundamental theorem of calculus to find the integral.

27. $\int_{-1}^1 e^{-2x}\,dx$

28. $\int_0^1 xe^x\,dx$

29. $\int_0^4 \sqrt{2x + 1}\,dx$

30. $\int_2^3 \frac{1}{x(x-1)}\,dx$

31. $\int_{-1}^0 (e^x - e^{-x})\,dx$

32. $\int_0^1 xe^{-x^2}\,dx$

33. $\int_1^2 x \ln x\,dx$

34. $\int_0^2 \frac{x^2}{\sqrt{x^3 + 1}}\,dx$

In Exercises 35–40, find both $\int_a^b f(x)\,dx$ and the total area between the graph of f and the x-axis from a to b.

35. $f(x) = x^2 + x - 2$, $\quad a = 0$, $\quad b = 2$

36. $f(x) = x^3$, $\quad a = -1$, $\quad b = 1$

37. $f(x) = \frac{x}{x^2 + 1}$, $\quad a = 0$, $\quad b = 1$

38. $f(x) = \frac{x}{x^2 + 1}$, $\quad a = -1$, $\quad b = 1$

39. $f(x) = \ln x$, $\quad a = 1$, $\quad b = 2$

40. $f(x) = \ln x$, $\quad a = \frac{1}{2}$, $\quad b = 2$

41. If $F(x)$ is an antiderivative of a function $f(x)$, and if $F(-1) = 5$ and $F(2) = 3$, find $\int_{-1}^2 f(x)\,dx$.

42. If $g(0) = -2$ and $g(3) = -1$, what is $\int_0^3 g'(x)\,dx$?

43. Let $f(x) = x, 0 \le x \le 1$. Use elementary geometry to find a formula for the area function $A(t)$ and verify that $A'(t) = f(t)$.

44. In the graph shown in Figure 5.5.2, the shaded area is equal to $\sqrt{t} - 1$ for any $t > 1$. What is $f(x)$?

45. In the graph shown in Figure 5.5.3, the shaded area is equal to $(t^3/3) + 16 \ln t - \frac{1}{3}$ for any $t > 1$. What is $f(x)$?

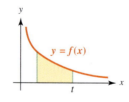

Figure 5.5.2

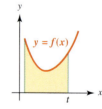

Figure 5.5.3

46. In the graph shown in Figure 5.5.4, the shaded area is equal to $t + e^{-t} - 1$ for any $t > 0$. What is $f(x)$?

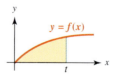

Figure 5.5.4

In Exercises 47–52, find $G'(t)$.

47. $G(t) = \displaystyle\int_0^t \sqrt{x^2 - 3x + 5}\, dx$

48. $G(t) = \displaystyle\int_1^t (1 + \ln x)^2\, dx$

49. $G(t) = t + \displaystyle\int_0^t x^2 e^{x^4}\, dx$

50. $G(t) = \displaystyle\int_t^1 e^{x^2}\, dx$

51. $G(t) = \displaystyle\int_1^{t^2} e^x \sqrt{x}\, dx$ (*Hint:* Use the chain rule.)

52. $G(t) = \displaystyle\int_0^{3t+1} \frac{1}{1 + x^2}\, dx$

53. A beaker of hot water is cooling at a rate of $-5.52e^{-0.07t}$ degrees Celsius per minute. Find the change in temperature during the first 5 minutes.

54. A continuous intravenous dosage of medicine is being absorbed into a patient's bloodstream at a rate of $0.9 - 0.27/\sqrt{1 + x^2}$ milligrams per hour. Write a definite integral that gives the amount absorbed into the patient's bloodstream between the 10th and 20th hours after the process begins.

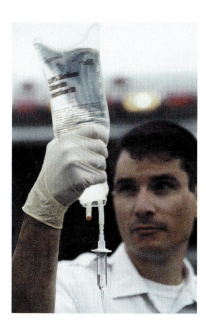

55. After the year 2000, the rate of growth, in millions of dollars per year, of sales of a new company is given by the function $r(t) = 1,000 - 10t^2$, where the time t is measured in years. If the company's sales were $800 million in 2000, find the projected sales of the company in the year 2010.

56. A company's marginal cost for producing x units per day of its product is given by the function $C'(x) = x^2 - 10x + 80$ dollars per unit. If the company operates at the production level of 20 units per day, find the increase in cost involved in raising the level to 30 units per day.

57. The marginal profit of a company producing and selling a certain product is given by $MP(x) = -0.2x + 50$, where x is the number of units sold per day. When 100 units are produced and sold, the company's profit is $1,500. Find the profit function.

58. Assume the marginal cost function in producing a certain product is $C'(q) = 3q^2 - 60q + 900$ for $0 \leq q \leq 30$ and that the fixed cost is $1,000. Find the total cost of producing 30 items.

59. The marginal profit function for producing and selling x units of a certain product is given by $MP(x) = -0.6x + 420$. The company is presently operating at a production level of 500 units and earning a profit of $150,000. Is it profitable for the company to increase production? If yes, find the production level that maximizes the profit, as well as the maximum profit.

60. A company has determined that when it produces at least 100 units of its product, its marginal cost is given by $MC(x) = 0.25x + 1,500$ and its marginal revenue by $MR(x) = -6x^2 + 500x + 20,000$. If the company is currently operating at a production level of 100 units per day, is it profitable for the company to increase production to 110 units per day? Justify your answer.

61. A company produces 40 computers a day. The company manager estimates that the marginal profit function for the computer production is given by $MP(x) = -0.006(x + 40)^2 + 500$. What will the change in profit be if the company increases production to 60 computers a day?

62. After instituting a new prevention program, a city has determined that its crime rate, in thousands of crimes per year, is modeled by

$$C' = \frac{-42}{(3 + t)^2}.$$

(a) Compute the projected drop in the number of crimes committed over the first 5 years of the program.
(b) Next, calculate the projected drop in the number of crimes committed over the second 5 years. What is your conclusion?

63. A man operating a small chicken ranch is testing out a new feed for his birds that is supposed to increase egg production. For an operation his size, the laying rate for chickens eating the new feed is claimed to be at least $E'(w) = 0.08w + 0.3$ in hundreds of eggs per week. He first started using the feed after a week in which the hens had produced 1,700 eggs. During the 5th week, he collected about 2,000 eggs. Did the feed perform as well as expected? Use a definite integral to answer this.

64. Water has been leaking out of an old water tower at a rate of

$$L'(t) = 0.00135e^{0.5t} + 0.3 \text{ gallons per day.}$$

(a) How many gallons leaked out in the first 2 days?
(b) How many leaked out during the 10th through 12th days?
(c) Would you conclude that the hole is staying the same size or widening?

65. One fish species, called Type A, grows at a rate of $0.072t^2$ centimeters per year. Another species, Type B, grows at $2.1t^{0.6}$ centimeters per year. Both are the same size when they are born.
(a) Which species has the faster growth rate at 5 years of age? at 12 years?
(b) Which species has achieved the most growth by age 5? age 12? age 20?

66. According to the fundamental theorem of calculus, $\ln 3 = \int_1^3 (1/x)\, dx$. Approximate $\ln 3$ by a Riemann sum obtained by dividing the interval $[1, 3]$ into 20 equal segments and using the midpoint of each segment.

67. The velocity of an object moving in a straight line was measured every quarter of a second for a period of 3 seconds, resulting in the following table, where t is the time in seconds and v is the velocity in meters per second:

t	0.25	0.5	0.75	1.0	1.25	1.5
v	2.11	2.18	2.24	2.29	2.33	2.37
t	1.75	2.0	2.25	2.5	2.75	3.0
v	2.40	2.42	2.45	2.47	2.48	2.5

Approximate the distance the object traveled during the 3 seconds.

Solutions to practice exercises 5.5

1. Since $F(x) = 2x^8 - \ln x$ is an antiderivative of $f(x) = 16x^7 - (1/x)$, we have

$$\int_1^2 \left(16x^7 - \frac{1}{x}\right) dx = (2x^8 - \ln x)\Big|_1^2 = (2 \cdot 2^8 - \ln 2)$$
$$- (2 \cdot 1^8 - \ln 1) = 510 - \ln 2.$$

2. Let $P(x)$ be the profit for producing and selling x units. Then for any production level, say, q, we have

$$P(q) - P(50) = \int_{50}^q P'(x)\, dx$$

$$= \int_{50}^q (-0.8x + 1{,}000)\, dx$$
$$= (-0.4x^2 + 1{,}000x)\Big|_{50}^q$$
$$= (-0.4q^2 + 1{,}000q) - (-0.4 \cdot 50^2 + 1{,}000 \cdot 50)$$
$$= -0.4q^2 + 1{,}000q - 49{,}000.$$

Since $P(50) = 20{,}000$, we have

$$P(q) = -0.4q^2 + 1{,}000q - 29{,}000.$$

5.6 Computing Definite Integrals, Areas, and Averages

Substitution in definite integrals

The method of substitution for indefinite integrals together with the fundamental theorem of calculus gives a method for evaluating definite integrals.

EXAMPLE 5.6.1

Find $\displaystyle\int_0^4 (1 + 2x)^{1/2}\, dx$.

SOLUTION We begin by finding an antiderivative. If we let $u = 1 + 2x$, then $du = 2\,dx$, or, equivalently, $dx = \frac{1}{2}\,du$. Therefore,

$$\int (1 + 2x)^{1/2}\,dx = \int u^{1/2} \frac{1}{2}\,du$$

$$= \frac{1}{3} u^{3/2} + c = \frac{1}{3}(1 + 2x)^{3/2} + c.$$

Since we can use *any* antiderivative in applying the fundamental theorem, we will ignore the constant c. Then

$$\int_0^4 (1 + 2x)^{1/2}\,dx = \frac{1}{3}(1 + 2x)^{3/2}\Big|_0^4 = \frac{1}{3} \cdot 9^{3/2} - \frac{1}{3} \cdot 1^{3/2} = \frac{26}{3}$$

An alternate method, which may sometimes be faster, is to change the limits of integration—that is the endpoints of the interval over which the integral is being taken—when the substitution is first made. More precisely, assume that we want to compute a definite integral that can be written in the form

$$\int_a^b f(g(x))g'(x)\,dx.$$

If F is an antiderivative of f, then $F(g(x))$ is an antiderivative of $f(g(x))g'(x)$ (chain rule). Therefore, by the Fundamental Theorem, we have

$$\int_a^b f(g(x))g'(x)\,dx = F(g(b)) - F(g(a)).$$

If we let $u = g(x)$, then we can write $du = g'(x)\,dx$, and this formula takes the form:

$$\int_a^b f(g(x))g'(x)\,dx \stackrel{u=g(x)}{=} \int_{g(a)}^{g(b)} f(u)\,du = F(g(b)) - F(g(a)) \tag{39}$$

EXAMPLE 5.6.2

Find $\displaystyle\int_0^2 \frac{1}{(x+1)^2}\,dx$.

SOLUTION If we let $g(x) = x + 1$, then $g(0) = 1$ and $g(2) = 3$. Substituting $u = g(x)$ gives $du = dx$, so that

$$\int_0^2 \frac{1}{(x+1)^2}\,dx = \int_1^3 \frac{1}{u^2}\,du = -\frac{1}{u}\Big|_1^3 = \frac{2}{3}.$$

EXAMPLE 5.6.3

Compute $\displaystyle\int_0^1 \frac{2x^3 + x}{(x^4 + x^2 + 2)^3}\, dx$ by both methods.

SOLUTION   We make the substitution $u = x^4 + x^2 + 2$, which gives

$$du = (4x^3 + 2x)\, dx.$$

Dividing both sides by 2 puts this in the form $\frac{1}{2}\, du = (2x^3 + x)\, dx$. Using this substitution, we find the indefinite integral, as follows:

$$\int \frac{2x^3 + x}{(x^4 + x^2 + 2)^3}\, dx = \int \underbrace{(x^4 + x^2 + 2)^{-3}}_{u^{-3}}\ \underbrace{(2x^3 + x)\, dx}_{\frac{1}{2}\, du}$$

$$= \frac{1}{2} \int u^{-3}\, du$$

$$= -\frac{1}{4} u^{-2} + c = -\frac{1}{4}(x^4 + x^2 + 2)^{-2} + c.$$

Therefore,

$$\int_0^1 \frac{2x^3 + x}{(x^4 + x^2 + 2)^3}\, dx = -\frac{1}{4}(x^4 + x^2 + 2)^{-2}\,\Big|_0^1 = \frac{1}{16} - \frac{1}{64} = \frac{3}{64}.$$

For the alternate method, we change the limits of integration, using the same substitution. If $x = 0$, then $u = 0^4 + 0^2 + 2 = 2$, and if $x = 1$, then $u = 1^4 + 1^2 + 2 = 4$. Therefore,

$$\int_0^1 \frac{2x^3 + x}{(x^4 + x^2 + 2)^3}\, dx = \int_2^4 \frac{1}{u^3}\frac{1}{2}\, du = \frac{1}{2}\int_2^4 u^{-3}\, du$$

$$= -\frac{1}{4} u^{-2}\,\Big|_2^4 = -\frac{1}{4}\left[\frac{1}{4^2} - \frac{1}{2^2}\right] = \frac{1}{16} - \frac{1}{64} = \frac{3}{64}.$$

Integration by parts in definite integrals

By using the fundamental theorem of calculus and the integration by parts formula for indefinite integrals, we obtain

$$\int_a^b u\, dv = uv\,\Big|_a^b - \int_a^b v\, du, \tag{40}$$

which is the integration by parts formula for definite integrals.

EXAMPLE 5.6.4

Compute the integral $\displaystyle\int_0^1 x e^x\, dx$.

SOLUTION Let $u = x$ and $dv = e^x \, dx$, which gives $v = e^x$ and $du = dx$. Applying formula (40) gives

$$\int_0^1 x e^x \, dx = x e^x \Big|_0^1 - \int_0^1 e^x \, dx = 1 \cdot e^1 - 0 \cdot e^0 - e^x \Big|_0^1 = 1.$$

Next, we give an example of an integral that can be computed in two different ways.

EXAMPLE 5.6.5

Compute $\displaystyle\int_1^e \frac{\ln x}{x} \, dx$ in two ways, by substituion and by integration by parts.

SOLUTION First, by substitution. Let $u = \ln x$, so that $du = (1/x)\,dx$. If $x = 1$ then $u = 0$, and if $x = e$ then $u = 1$. Therefore,

$$\int_1^e \frac{\ln x}{x} \, dx = \int_0^1 u \, du = \frac{1}{2} u^2 \Big|_0^1 = \frac{1}{2}.$$

Next, to use integration by parts, we let $u = \ln x$ and $dv = (1/x)\,dx$. Then $du = (1/x)\,dx$ and $v = \ln x$, and formula (40) gives

$$\int_1^e \frac{\ln x}{x} \, dx = (\ln x)^2 \Big|_1^e - \int_1^e \ln x \cdot \frac{1}{x} \, dx$$

$$= (\ln e)^2 - (\ln 1)^2 - \int_1^e \ln x \cdot \frac{1}{x} \, dx = 1 - \int_1^e \frac{\ln x}{x} \, dx.$$

Adding $\int_1^e (\ln x / x) \, dx$ to both sides, we obtain

$$2 \int_1^e \frac{\ln x}{x} \, dx = 1 \quad \text{or} \quad \int_1^e \frac{\ln x}{x} \, dx = \frac{1}{2}.$$

The area between two curves

If $f(x)$ and $g(x)$ are both continuous functions such that $f(x) \geq g(x)$ for $a \leq x \leq b$, then the graph of $f(x)$ lies above that of $g(x)$ throughout the interval $[a, b]$. In that case the area of the region bounded by the two curves from $x = a$ to $x = b$ is given by the formula

$$\text{Area between } f \text{ and } g = \int_a^b [f(x) - g(x)] \, dx, \quad \text{if } f(x) \geq g(x). \qquad (41)$$

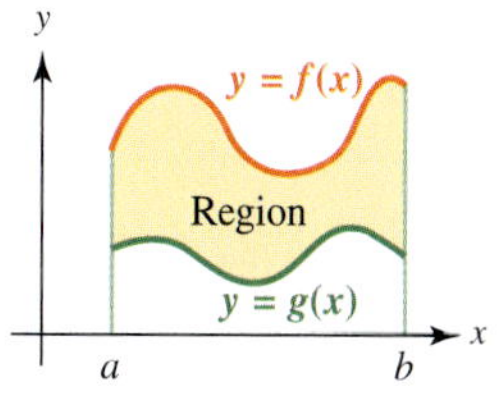

Figure 5.6.1

Figure 5.6.1 illustrates the case when $g(x) \geq 0$, so that both curves lie above the x-axis and the area between them is equal to the area under the upper curve minus the area under the lower curve.

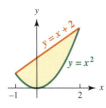

Figure 5.6.2

PROBLEM-SOLVING TACTIC

To compute the area between two curves, say, $y = f(x)$ and $y = g(x)$, over an interval $[a, b]$:

- Solve the equation $f(x) = g(x)$ to find the points on the x-axis over which the curves intersect. They divide the interval $[a, b]$ into subintervals.

- Over each subinterval, determine which function is above the other and compute the subarea they enclose by integrating the upper function minus the lower one.

- Find the total area by adding all these subareas. Keep in mind that the area is positive! not zero or negative!

EXAMPLE 5.6.6

Find the area between the curves $y = x^2$ and $y = x + 2$ for $-1 \leq x \leq 2$.

SOLUTION   The graphs are in Figure 5.6.2. As you can see, the line lies above the parabola throughout the interval $[-1, 2]$. To verify that analytically, observe that

$$x^2 - x - 2 = (x - 2)(x + 1) < 0,$$

if and only if $-1 < x < 2$. Therefore, $x + 2 \geq x^2$ if $-1 \leq x \leq 2$. We can now apply formula (41) to get

$$\text{area} = \int_{-1}^{2} (x + 2 - x^2)\, dx = \left(\frac{x^2}{2} + 2x - \frac{x^3}{3} \right) \Bigg|_{-1}^{2} = \frac{9}{2}.$$

If the curves cross, we have to find their points of intersection and determine where $f(x) \geq g(x)$ and vice versa.

EXAMPLE 5.6.7

Find the area of the region between the curves $y = 9 - x^2$ and $y = x^2 + 1$ from $x = 0$ to $x = 3$.

SOLUTION   To find the point of intersection of the two graphs, we solve the equation

$$x^2 + 1 = 9 - x^2. \tag{42}$$

Solving Eq. (42) gives

$$2x^2 = 8, \quad \text{or} \quad x^2 = 4, \quad \text{or} \quad x = \pm 2.$$

The graphs are shown in Figure 5.6.3. As you can see, the graph of $y = 9 - x^2$ is above $y = 1 + x^2$ for $0 \leq x \leq 2$, and the opposite is true for $2 \leq x \leq 3$.

To verify that analytically, we observe that $1 + x^2 \leq 9 - x^2$ reduces to $2x^2 \leq 8$, or, equivalently, $x^2 \leq 4$. This last inequality is the same as $-2 \leq x \leq 2$.

Thus, we have

$$\text{Area} = \int_{0}^{2} [(9 - x^2) - (x^2 + 1)]\, dx + \int_{2}^{3} [(x^2 + 1) - (9 - x^2)]\, dx$$

$$= \int_{0}^{2} (8 - 2x^2)\, dx + \int_{2}^{3} (2x^2 - 8)\, dx$$

$$= \left(8x - \frac{2}{3}x^3 \right) \Bigg|_{0}^{2} + \left(\frac{2}{3}x^3 - 8x \right) \Bigg|_{2}^{3}$$

$$= \left(16 - \frac{16}{3} \right) + (18 - 24) - \left(\frac{16}{3} - 16 \right) = \frac{46}{3}.$$

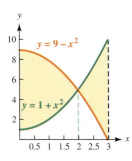

Figure 5.6.3

EXAMPLE 5.6.8

A small company making optical equipment has just leased a computer-operated robot arm for grinding and polishing lenses. It estimates that the resulting savings in personnel and material expenses will grow over the next 10 years at a rate of $\sqrt{4 + t}$ hundreds of thousands per year, where t is the time in years. On the other hand, the combined rental and maintenance costs will grow at a rate of $1 + \frac{1}{4}t$ hundred thousand dollars per year.

(i) Graph these functions over the interval [0, 10] and compute the area between the graphs.

(ii) What does the area represent?

SOLUTION (i) The graph of $1 + \frac{1}{4}t$ is a straight line of slope $\frac{1}{4}$ through the point $(0, 1)$. We can draw the graph of $\sqrt{4 + t}$ using the methods of Section 4.3. Its key properties are that it is positive, increasing, and concave down over its entire domain.

Both graphs are drawn in Figure 5.6.4, which shows that the savings function is greater than the rental and maintenance cost function throughout the interval [0, 10]. We can also check that algebraically by first finding the points where the graphs intersect. To do that, we solve

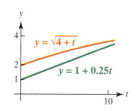

Figure 5.6.4

$$\sqrt{4 + t} = 1 + \frac{1}{4}t.$$

Squaring both sides and rearranging terms give

$$\frac{1}{16}t^2 - \frac{1}{2}t - 3 = 0,$$

whose solutions are -4 and 12. Since both of these are outside the interval [0, 10], the graphs do not cross inside that interval, and since the savings function is greater at zero, it is greater throughout the interval.

To find the area between the graphs, we integrate their difference:

$$\int_0^{10} \left(\sqrt{4 + t} - 1 - \frac{1}{4}t \right) dt = \left[\frac{2}{3}(4 + t)^{3/2} - t - \frac{t^2}{8} \right] \Bigg|_0^{10}$$

$$= \left[\frac{2}{3}(14)^{3/2} - 10 - \frac{100}{8} \right] - \frac{2}{3} \cdot 4^{3/2} \approx 7.0888.$$

(ii) Since $\sqrt{4 + t}$ is the *rate* at which the savings grow, its integral from 0 to 10 is the total amount saved over the first 10 years. Similarly, the integral of $1 + \frac{1}{4}t$ is the total rental and maintenance cost over that same period. Therefore, the area between the graphs, which is the integral of their difference, represents the total *net* savings over the period, with the rental and maintenance costs taken into account.

Average values of continuous quantities

Everyone knows what it means to average a finite set of numbers—say, the scores you received on four tests in a course last semester. Suppose you had grades of 91, 74, 82, and 86. To find the average, you simply add the grades and divide by the number of them—in this case, 4. So, the average is

$$\frac{91 + 74 + 82 + 86}{4} = 83.25.$$

In general, to find the average value of a set of n numbers, we add the numbers and divide by n.

But what if we are dealing with a continuous function rather than a finite set of numbers? For example, suppose you invest $10,000 in an account paying an annual interest rate of 4%, compounded continuously, and leave it for a year. What is the *average* amount in the account over that 1-year period? (You might need to know that for tax purposes, for instance.) As we already know, the amount A in the account at any time is given by the formula

$$A = 10{,}000e^{0.04t}, \tag{43}$$

where t is the time in years.

A crude way to approximate the average amount over the year would be to simply compute the account balance at the end of 6 months—in other words, midway through the year. To do that, we simply set $t = \frac{1}{2}$ in formula (43), which gives $10{,}000e^{0.02} \approx 10{,}202.01$. However, using that as the average ignores the fact that the amount is changing throughout the year. A better method might be to divide the year into two equal periods, compute the amount midway through each of those periods, and average those two numbers—in other words, average the amounts at the end of 3 months and 9 months. Setting $t = \frac{1}{4}$ and $\frac{3}{4}$ in (43) and averaging gives

$$\frac{10{,}000e^{0.01} + 10{,}000e^{0.03}}{2} \approx 10{,}202.52.$$

Or, to get an even more accurate average, we might divide the year into quarters, compute the amount midway through each quarter—compute it for $t = \frac{1}{8}, \frac{3}{8}, \frac{5}{8}$, and $\frac{7}{8}$, in other words—and then average those four numbers. If we did that, we would get

$$\frac{10{,}000e^{0.005} + 10{,}000e^{0.015} + 10{,}000e^{0.025} + 10{,}000e^{0.035}}{4} \approx 10{,}202.65.$$

We could continue in the same way—that is,

- divide the year up into n periods, for some number n,
- choose a particular time in each period—for instance, the midpoint of the period, as we just did,
- compute the balance at each of those times, and
- average those balances.

The larger we choose n, the more accurate our average is. And, by letting n get larger and larger—by letting $n \to \infty$, in other words—we get closer and closer to a definite integral, as you will see below. First, let's extend our method to a more general setting, replacing the compound interest formula by any function.

Let $f(x)$, $a \le x \le b$, be a continuous function. To find the "average" value of the function over the interval $[a, b]$, we first partition the interval into n equal subintervals, just as we did in defining the Riemann sums. Next, we choose a point in each subinterval—say, x_1 in the first interval, x_2 in the second, and so forth. If we evaluate the function at each of these points and average those values, we get

$$\frac{f(x_1) + f(x_2) + \cdots + f(x_n)}{n}, \tag{44}$$

which we can take as an approximate average of $f(x)$ over the interval $[a, b]$. The mesh of the partition—that is, the length of each of the subintervals—is given by

$$\Delta x = \frac{b - a}{n}.$$

Dividing both sides of this equation by $b - a$ gives

$$\frac{\Delta x}{b - a} = \frac{1}{n}. \tag{45}$$

If we replace $1/n$ in formula (44) by the left-hand side of (45), we get

$$[f(x_1) + f(x_2) + \cdots + f(x_n)] \cdot \frac{\Delta x}{b - a},$$

which we can rewrite in the form

$$\frac{1}{b - a}[f(x_1)\Delta x + f(x_2)\Delta x + \cdots + f(x_n)\Delta x].$$

Notice that the expression in brackets is a Riemann sum for the definite integral $\int_a^b f(x)\,dx$. Letting $n \to \infty$, we get the following formula for the **average value** of $f(x)$ over the interval $[a, b]$:

$$\text{Average value of } f \text{ over } [a, b] = \frac{1}{b - a}\int_a^b f(x)\,dx. \tag{46}$$

In the example we started with—an initial deposit of \$10,000 in an account paying 4% interest, compounded continuously—the function is given by $f(t) = 10{,}000e^{0.04t}$, and the average balance during the first year is

$$\int_0^1 10{,}000e^{0.04t}\,dt = \frac{10{,}000}{0.04}e^{0.04t}\Big|_0^1$$

$$= 250{,}000(e^{0.04} - 1) \approx 10{,}202.69.$$

Here is another example.

EXAMPLE 5.6.9

A sum of \$700 is deposited in a bank account that pays 7% interest, compounded continuously. Find the average value of the money during the first 10 years, assuming that no other deposits or withdrawals are made.

SOLUTION   Let $M(t)$ be the money in the bank after t years. Then,

$$M(t) = 700e^{0.07t}.$$

Therefore, its average value over the first 10 years is

$$\frac{1}{10}\int_0^{10} M(t)\,dt = \frac{1}{10}\int_0^{10} 700e^{0.07t}\,dt$$

$$= \frac{1}{10}\frac{700}{0.07}e^{0.07t}\Big|_0^{10}$$

$$= 1{,}000(e^{0.7} - 1) \approx \$1{,}013.75.$$

EXAMPLE 5.6.10

Suppose you are driving along a straight road.

(i) Show that your average velocity over a time interval $[a, b]$ is the same as the average value of the velocity over that interval.

(ii) After stopping at a stop sign, you move your car forward so that both the velocity *and* the acceleration keep increasing for the first 15 seconds. Show that your average velocity over that period is less than half your final velocity.

SOLUTION   (i) Recall that the *average velocity* over a time interval $[a, b]$ was defined in Section 3.5 by the formula $[s(b) - s(a)]/(b - a)$, where $s(t)$ is the distance traveled, relative to some fixed point on the road, at time t. On the other hand, formula (46) gives the average of the velocity function $v(t)$ over that time interval as

$$\frac{1}{b - a}\int_a^b v(t)\,dt.$$

Since the velocity is the derivative of the distance function, $v(t) = s'(t)$, we get

$$\frac{1}{b - a}\int_a^b v(t)\,dt = \frac{1}{b - a} s(t)\Big|_a^b = \frac{s(b) - s(a)}{b - a}.$$

Figure 5.6.5

(ii) Let $v(t)$ be the velocity as a function of time (in seconds). The acceleration is the derivative $v'(t)$ and the fact that it is increasing means that the graph of v is *concave up*, as shown in Figure 5.6.5. (Recall from Section 4.2 that the graph of a function is concave up if its derivative is increasing.) It follows that the graph lies entirely below the secant line between its endpoints (see Figure 5.6.6). Therefore, as we also see in Figure 5.6.6, the area under the graph of v is less than the area of the triangle determined by the points $(0, 0)$, $(15, 0)$, and $(15, v(15))$.

Thus, we have

Figure 5.6.6

$$\text{Area under the graph of } v = \int_0^{15} v(t)\,dt < \frac{15 \cdot v(15)}{2} = \text{Area of triangle.}$$

Dividing both sides by 15, we get

$$\frac{1}{15} \int_0^{15} v(t)\, dt \le \frac{v(15)}{2}.$$

The left-hand side of this inequality is the average velocity, and the right-hand side is half the final velocity.

Practice Exercises 5.6

1. Compute the average value of the function $f(t) = te^{-0.1t}$ on the interval $[0, 10]$.

2. Find the area between the curves $y = x/(x+1)$ and $y = x/(x^2+1)$ for $0 \le x \le 1$.

Exercises 5.6

Evaluate the following integrals:

1. $\displaystyle\int_0^{10} e^{-0.1t}\, dt$

2. $\displaystyle\int_1^2 (x-1)^{19}\, dx$

3. $\displaystyle\int_0^3 te^{t/3}\, dt$

4. $\displaystyle\int_1^e \frac{\ln x}{x^2}\, dx$

5. $\displaystyle\int_1^e \frac{(\ln x)^2}{x}\, dx$

6. $\displaystyle\int_1^e \ln x\, dx$

7. $\displaystyle\int_0^1 (3x^2 - 1)(x^3 - x)^4\, dx$

8. $\displaystyle\int_0^1 2^t\, dt$

9. $\displaystyle\int_0^7 \frac{t}{\sqrt{t+9}}\, dt$

10. $\displaystyle\int_1^2 (2x-5)e^{-x^2+5x-6}\, dx$

11. $\displaystyle\int_0^3 \frac{x+1}{2x^2+4x+1}\, dx$

12. $\displaystyle\int_0^1 t^2 e^{-t}\, dt$

13. $\displaystyle\int_4^5 x\sqrt{x^2-16}\, dx$

14. $\displaystyle\int_1^e t^4 \ln t\, dt$

15. $\displaystyle\int_2^3 \frac{dt}{t^2-t}$

16. $\displaystyle\int_0^1 \frac{6}{t^2-t-2}\, dt$

17. Express the area of the region enclosed by the curves $y = f(x)$ and $y = g(x)$, shown in Figure 5.6.7, in terms of integrals.

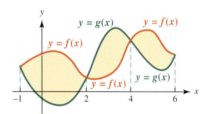

Figure 5.6.7

In Exercises 18–23, verify that $f(x) \ge g(x)$ over the given interval and find the area between the two graphs.

18. $f(x) = x,\ g(x) = x^2,\ 0 \le x \le 1$

19. $f(x) = 4 - x^2,\ g(x) = 3,\ 0 \le x \le 1$

20. $f(x) = 4 - x^2,\ g(x) = 2 - x,\ -1 \le x \le 2$

21. $f(x) = 3x^2 + 2,\ g(x) = 2x - 3,\ -1 \le x \le 2$

22. $f(x) = x,\ g(x) = \dfrac{1}{x},\ 1 \le x \le 2$

23. $f(x) = 2x^2,\ g(x) = x^3 - 3x,\ 0 \le x \le 3$

In Exercises 24–27, find the intersection points of the graphs and the area between the graphs over the given interval.

24. $f(x) = 1 - x^2,\ g(x) = x + 1,\ -1 \le x \le 1$

25. $f(x) = \dfrac{2}{x},\ g(x) = x - 1,\ 1 \le x \le 3.$

26. $f(x) = \sqrt{x},\ g(x) = x^2,\ 0 \le x \le 4$

27. $f(x) = e^x,\ g(x) = e^{-x},\ -1 \le x \le 1$

In Exercises 28–31, find the area between the graphs over the given interval.

28. $f(x) = \dfrac{8}{x^2}$, $g(x) = x$, $1 \le x \le 3$

29. $f(x) = x^2 - 1$, $g(x) = -\dfrac{x}{2}$, $0 \le x \le 1$

30. $f(x) = \dfrac{x}{1 + x^2}$, $g(x) = \dfrac{1}{x + 1}$, $0 \le x \le 1$

31. $f(x) = x^2$, $g(x) = x^3$, $-1 \le x \le 1$

32. Find the area of the region enclosed by the curves $y = 4 - x^2$ and $y = x^2 - 4x + 4$ between their points of intersection.

33. Compute the area of the region enclosed by the curves $y = x^2 + x + 1$ and $y = -2x^2 + 4x + 7$ between their points of intersection.

In Exercises 34–41, find the average value of the given function over the interval indicated.

34. $f(x) = 8$, $0 \le x \le 4$ **35.** $f(x) = x^2$, $0 \le x \le 3$

36. $f(t) = t^5$, $-1 \le t \le 2$

37. $h(t) = \dfrac{5}{t}$, $\dfrac{1}{2} \le t \le 2$

38. $M(t) = 10,000e^{-0.08t}$, $0 \le t \le 20$

39. $g(t) = te^{0.1t}$, $0 \le t \le 10$

40. $p(t) = t(t^2 + 1)^5$, $-1 \le t \le 1$

41. $f(x) = \dfrac{1}{x^2 - 5x + 4}$, $5 \le x \le 8$

42. If the average value of a function $f(x)$ over the interval $[2, 5]$ is -15, find $\int_2^5 f(x)\,dx$.

43. If $\int_{-1}^4 f(x)\,dx = 30$, find the average value of the function $f(x)$ over the interval $[-1, 4]$.

44. The average value of a function $f(x)$ over the interval $[1, 3]$ is 10, and its average over the interval $[3, 6]$ is 20. Find its average value over the interval $[1, 6]$.

45. Suppose an object is moving in a straight line, with its position at time t (relative to some coordinate axis) given by $s(t)$ and its velocity by $v(t)$. In Chapter 3, the *average* velocity over the time interval $[a, b]$ was defined to be $[s(b) - s(a)]/(b - a)$. Is that the same as the average value of the velocity function $v(t)$ over that same time interval? Explain.

46. A vehicle is moving on a straight road in such a way that its velocity (in miles per hour) is given by $v(t) = 60[1 - (1 + t)^{-1/2}]$, where t is the time in seconds. Express its average velocity over the first 10 seconds as an integral and compute it.

47. What is the average slope of the graph of $y = x^3$ over the interval $[0, 2]$? Is it the same as the average value of the slope at $x = 0$ and $x = 2$? Explain.

48. Suppose you deposit $10,000 in an account paying 8% annual interest, compounded continuously, and do not make any further deposits or withdrawals. Find the average amount of money in the account during the first 5 years.

49. The amount of medicine being continuously injected into a patient's bloodstream is given by the formula $q(t) = 30 - 20e^{-0.4t}$, where t is the time in hours and $q(t)$ is the dosage in milligrams. What is the average number of milligrams in the patient's bloodstream over the first 6 hours?

50. Linda works hard at her studies, especially in language and math. She has been studying a new foreign language, and she tries to memorize new words regularly. She calculates that the growth rate of her vocabulary, in words per day, is best modeled by

$$r = 203[0.08x^{-0.92}\ln(5x) + x^{-0.92}],$$

where x is the number of days since she started studying. What was the average number of words per day she learned in the time interval $[10, 30]$ days?

51. A population grows according to the formula $p(t) = 2,500e^{0.05t}$, where t is measured in months. Find the average value of the population over the first 2 years.

52. It is estimated that the world population (in billions) t years after 1980 is given by the formula

$$p(t) = \frac{48.6}{4.5 + 6.5e^{-0.29t}}.$$

Find the average population of the world from 1980 to 2030.

53. Biologists are attempting to reintroduce a species back into part of its historic range. Since it is not an endangered species, they choose to merely monitor the population, without assisting it. The rate of growth of the population in the first 6 years was modeled at

$$G'(y) = 10.736e^{-y}(1 - y), \quad 0 \le y \le 6,$$

where G' is in hundreds of animals per year.
(a) At what value y do you find a maximum or minimum? Which is it?
(b) Use that value of y to break up the given interval into two parts. Then compare the total growth during each part.
(c) Find the average rate of growth of the population in each of the two periods.
(d) Write a sentence that describes the behavior of the population during the entire study period.

54. The biologists reintroduce the species (see the previous problem) into a different part of its historical range, again only monitoring the population. During the 6-year study period, the population growth rate was

$$G'(y) = \frac{30e^{-y}}{(1 + 2e^{-y})^2} \quad \text{(in hundreds of animals/year).}$$

(a) Divide the study period into three equal parts. Write each interval in the form $[a, b]$.

(b) Compute the total growth for each interval.

(c) Find the average rate of growth of the population during each period.

(d) Write a sentence describing the population's behavior over the entire study period.

(e) Project the behavior of the population to 2 years beyond the study period. What is the total change?

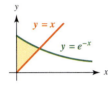

Figure 5.6.8

55. Estimate the area of the region enclosed between the y-axis and the curves $y = e^{-x}$, $y = x$, shaded in Figure 5.6.8. Here you will need to find an approximation of the solution to $e^{-x} = x$ by using a calculator or a computer.

56. Graph the functions $\ln x$ and $x - 2$ and use a calculator to approximate their two points of intersection to four decimal places. Using those approximations, estimate the area between the two graphs.

In Exercises 57–58, use a Riemann sum, partitioning the given interval into n equal segments and choosing the midpoint of each subinterval, to approximate the average value of the function.

57. $f(x) = \dfrac{1}{1 + x^2}, \quad -1 \leq x \leq 1, \ n = 8$

58. $f(x) = e^{-x^2}, \ 0 \leq x \leq 3, \ n = 12$

Solutions to practice exercises 5.6

1. By definition, the average value is $\frac{1}{10} \int_0^{10} te^{-0.1t} \, dt$. To compute the integral we use integration by parts with $u = t$ and $dv = e^{-0.1t} \, dt$, as follows:

$$\int_0^{10} te^{-0.1t} \, dt = (-10te^{-0.1t} - 100e^{-0.1t}) \Big|_0^{10} = 100 - 200e^{-1}.$$

To find the average, we divide by 10 and get $10 - 20e^{-1} \approx 2.6424$.

2. First, we find the intersection points of the two graphs by solving

$$\frac{x}{x + 1} = \frac{x}{x^2 + 1}.$$

Clearing denominators changes this to the form $x^3 + x = x^2 + x$, which reduces to $x^2(x - 1) = 0$, with solutions $x = 0$ and $x = 1$. On the interval $[0, 1]$, we have $x \geq x^2$, so that $x + 1 \geq x^2 + 1$. Taking the reciprocals reverses the inequality, and multiplying by x gives

$$\frac{x}{x + 1} \leq \frac{x}{x^2 + 1}$$

for $0 \leq x \leq 1$. The graphs are shown in Figure 5.6.9.

To find the area between the curves, we must compute

$$\int_0^1 \left(\frac{x}{x^2 + 1} - \frac{x}{x + 1} \right) dx.$$

The first integral can be handled by the substitution $u = x^2 + 1$, $du = 2x \, dx$, as follows:

$$\int_0^1 \frac{x}{x^2 + 1} \, dx = \frac{1}{2} \int_1^2 \frac{du}{u} = \frac{1}{2} \ln u \Big|_1^2 = \frac{1}{2} \ln 2.$$

For the second integral, we divide the denominator into the numerator with the remainder to get

$$\int_0^1 \frac{x}{x + 1} \, dx = \int_0^1 \left(1 - \frac{1}{x + 1} \right) dx$$

$$= (x - \ln |x + 1|) \Big|_0^1 = 1 - \ln 2.$$

Therefore, the area between the graphs is $\frac{3}{2} \ln 2 - 1 \approx 0.0397$.

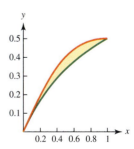

Figure 5.6.9

5.7 Numerical Methods

In order to compute an integral in closed form—that is, by using the fundamental theorem—we need to have an explicit formula for the given function and we also need to know an antiderivative. In many applications, there are two obstacles to this procedure. First, the function may only be known from a list of data—in other words, by its values at certain points—rather than by an explicit formula. Second, even if we know a formula, we may not be able to find an antiderivative. An example of this is $f(x) = e^{-x^2}$, whose antiderivative does not have a simple form.

In such cases, the only way to compute the definite integral is by approximation, and there are several methods for doing that.

The midpoint rule

We have already used this method in earlier approximation problems. By way of review, here are the steps.

Step 1 Choose a positive integer n and divide the interval $[a, b]$ into n equal subintervals, each of length $\Delta x = (b - a)/n$.

Step 2 Let x_i be the midpoint of the ith subinterval.

Step 3 Use the approximation

$$\int_a^b f(x)\, dx \approx f(x_1)\Delta x + f(x_2)\Delta x + \cdots + f(x_n)\Delta x$$

$$= [f(x_1) + f(x_2) + \cdots + f(x_n)] \cdot \Delta x.$$

EXAMPLE 5.7.1

Use the midpoint rule with $n = 4$ to approximate $\displaystyle\int_0^1 \frac{1}{x^2 + 1}\, dx$.

SOLUTION   We divide the interval into four equal segments. Each subinterval has length $\frac{1}{4}$, and

- the first subinterval $[0, \frac{1}{4}]$ has midpoint $x_1 = \frac{1}{8}$,
- the second subinterval $[\frac{1}{4}, \frac{1}{2}]$ has midpoint $x_2 = \frac{3}{8}$,
- the third subinterval $[\frac{1}{2}, \frac{3}{4}]$ has midpoint $x_3 = \frac{5}{8}$, and
- the fourth subinterval $[\frac{3}{4}, 1]$ has midpoint $x_4 = \frac{7}{8}$.

We next form the Riemann sum and use it to approximate the integral.

$$\int_0^1 \frac{1}{x^2+1}\,dx = \left(\frac{1}{x_1^2+1} + \frac{1}{x_2^2+1} + \frac{1}{x_3^2+1} + \frac{1}{x_4^2+1}\right) \cdot \frac{1}{4}$$

$$= \left(\frac{64}{65} + \frac{64}{73} + \frac{64}{89} + \frac{64}{113}\right) \cdot \frac{1}{4}$$

$$= \left(\frac{1}{65} + \frac{1}{73} + \frac{1}{89} + \frac{1}{113}\right) \cdot 16 \approx 0.7867.$$

By more advanced techniques, one can show that $\int_0^1 1/(x^2+1)\,dx = \pi/4 \approx 0.785$ (see Chapter 8).

The trapezoidal rule

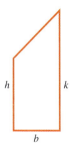

Figure 5.7.1

Another method of approximating definite integrals, and one that is often easier to use than the midpoint rule, is obtained by replacing the rectangles used in computing the Riemann sum by trapezoids, such as the one shown in Figure 5.7.1. Its area is given by a formula from elementary geometry:

$$\text{Area} = \frac{1}{2}(h+k) \cdot b. \tag{47}$$

(See the exercises at the end of this section for a derivation of this formula. An easy way to remember it is to observe that the right-hand side is simply the base times the *average* of the two heights.)

In our case, we construct trapezoids by first choosing a positive integer n and partitioning the interval $[a, b]$ into n equal segments, just as we did for Riemann sums. Let

$$a = x_0 < x_1 < \cdots < x_{n-1} < x_n = b$$

be the endpoints of the subintervals in increasing order. Over each subinterval we construct a trapezoid whose upper edge is the secant line connecting the points of the graph over the endpoints. A typical graph is shown in Figure 5.7.2. By applying

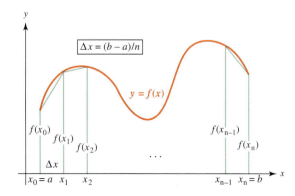

Figure 5.7.2

formula (47), we see that

$$\text{The area of the 1st trapezoid is } \frac{f(x_0) + f(x_1)}{2} \cdot \Delta x,$$

$$\text{The area of the 2nd trapezoid is } \frac{f(x_1) + f(x_2)}{2} \cdot \Delta x,$$

$$\cdots \qquad \cdots$$

$$\text{The area of the last trapezoid is } \frac{f(x_{n-1}) + f(x_n)}{2} \cdot \Delta x.$$

Adding these areas gives us the **trapezoidal rule**:

$$\int_a^b f(x)\,dx \approx [f(x_0) + 2f(x_1) + 2f(x_2) + \cdots + 2f(x_{n-1}) + f(x_n)] \cdot \frac{\Delta x}{2}.$$

This formula can also be written as

$$\int_a^b f(x)\,dx \approx \left[\frac{f(x_0)}{2} + f(x_1) + f(x_2) + \cdots + f(x_{n-1}) + \frac{f(x_n)}{2} \right] \cdot \Delta x.$$

EXAMPLE 5.7.2

Estimate $\int_0^2 \sqrt{1 + x^4}\,dx$ using the trapezoidal rule with four equal subintervals.

SOLUTION   We have $\Delta x = \frac{2-0}{4} = 0.5$, so that $x_0 = 0$, $x_1 = 0.5$, $x_2 = 1$, $x_3 = 1.5$, $x_4 = 2$. Thus,

$$\int_0^2 \sqrt{1 + x^4}\,dx \approx \left[\sqrt{1 + 0^4} + 2\sqrt{1 + 0.5^4} + 2\sqrt{1 + 1^4} + 2\sqrt{1 + 1.5^4} + \sqrt{1 + 2^4} \right] \cdot \frac{0.5}{2}$$

$$= \left[1 + 2\sqrt{1.0625} + 2\sqrt{2} + 2\sqrt{6.0625} + \sqrt{17} \right] 0.25 \approx 3.73.$$

EXAMPLE 5.7.3

Assume that we have the following table of values for a function $f(x)$, $1 \le x \le 3$:

x	1	1.25	1.50	1.75	2	2.25	2.50	2.75	3
$f(x)$	0.6	0.8	1.12	1.8	1.4	1.4	1	0.4	0.2

Approximate its integral from 1 to 3 by using the trapezoidal rule.

SOLUTION   Here we have $\Delta x = 0.25$, $x_0 = 1$, $x_1 = 1.25, \ldots, x_8 = 3$ and $f(x_0) = 0.6$, $f(x_1) = 0.8, \ldots, f(x_8) = 0.2$. By applying the trapezoidal rule, we obtain

$$\int_1^3 f(x)\,dx \approx [0.6 + 2 \cdot 0.8 + 2 \cdot 1.12 + 2 \cdot 1.8 + 2 \cdot 1.4 + 2 \cdot 1.4 + 2 \cdot 1$$
$$+ 2 \cdot 0.4 + 0.2] \cdot \frac{0.25}{2} = 2.08.$$

Estimating the error

In using any approximation method, it is very desirable to estimate the size of the error—where by *error* we mean the absolute value of the difference between the approximation and the actual value. In the midpoint rule, the error is defined by

$$E_M = \left| \int_a^b f(x)\,dx - [f(x_1)\Delta x + f(x_2)\Delta x + \cdots + f(x_n)\Delta x] \right|,$$

where x_i is the ith midpoint. The trapezoidal error is

$$E_T = \left| \int_a^b f(x)\,dx - [f(x_0) + 2f(x_1) + 2f(x_2) + \cdots + 2f(x_{n-1}) + f(x_n)] \cdot \frac{\Delta x}{2} \right|,$$

where $x_0, \ldots, x_n$ are the endpoints of the subintervals. If the function has continuous first and second derivatives throughout the interval $[a, b]$, the following theorem may be applied. Its proof can be found in any standard textbook on numerical analysis.

Theorem 5.7.1

Suppose that $f''(x)$ exists and is continuous for $a \le x \le b$ and let K be the maximum value of $|f''(x)|$ on $[a, b]$. Then

$$E_M \le \frac{K(b-a)^3}{24n^2} \quad \text{and} \quad E_T \le \frac{K(b-a)^3}{12n^2}.$$

One important consequence of this theorem is that if we increase n by a certain factor, the error (in either rule) decreases by roughly the *square* of that factor—doubling n reduces the error by about one-fourth, tripling n reduces it by about one-ninth, and so forth.

EXAMPLE 5.7.4

Approximate $\ln 2$ by applying the trapezoidal rule to the integral $\int_1^2 x^{-1}\,dx$, whose value is $\ln 2$, using

(i) $n = 10$,   (ii) $n = 20$,   (iii) $n = 100$.

Figure 5.7.3

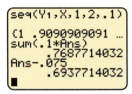

Figure 5.7.4

Estimate the error in all three cases and compare your approximation to the value of ln 2 given by your calculator.

SOLUTION   We will use a calculator (in this case a TI-83 Plus) to handle the calculations. Since we will be using the same function three times, it is convenient to enter it in the function editor as shown in Figure 5.7.3.

(i) For $n = 10$, each subinterval has length $\Delta x = 0.1$ (the step size). Using the list editor (shown in Figure 5.7.4), we create a sequence of values of $f(x) = x^{-1}$, from 1 to 2 in steps of 0.1. Next, we use the command sum(.1*Ans) to multiply each term of the sequence by 0.1 and take the sum. Comparing this with the trapezoidal rule, we see that we have to subtract

$$\frac{1}{2}[f(x_0) + f(x_{10})] \cdot \Delta x = \frac{1}{2}\left(1 + \frac{1}{2}\right) \cdot (0.1) = 0.075.$$

The result, shown in Figure 5.7.4, is 0.6937714032.

To estimate the error, we use Theorem 5.7.1. The second derivative, given by $f''(x) = 2x^{-3}$, is a decreasing function on $[1, 2]$. Therefore, its maximum value occurs at $x = 1$ and equals 2. Applying Theorem 5.7.1 with $K = 2$, $a = 1$, $b = 2$, and $n = 10$ gives

$$E_T \leq \frac{1}{600} = 0.001\overline{6}.$$

By comparing the trapezoidal approximation with the value of ln 2, we see that the error is actually

$$E_T = |\ln 2 - 0.6937714032| = 0.000624\ldots,$$

which is considerably better than estimated. (Remember: Theorem 5.7.1 only gives an *upper bound* for the error. The error is often smaller.)

(ii) For $n = 20$, we follow the same procedure, this time using $\Delta x = 0.05$ and subtracting $\frac{1}{2}(1 + \frac{1}{2}) \cdot (0.05) = 0.0375$ at the last step. The result, shown in Figure 5.7.5, is 0.6933033818.

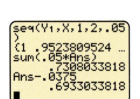

Figure 5.7.5

To estimate the error, we use the same values of K, a, and b with $n = 20$, which gives $E_T \leq \frac{1}{2400} \approx 0.00041\overline{6}$. The actual error is

$$E_T = |\ln 2 - 0.6933033818| = 0.000156\ldots.$$

Notice that increasing n from 10 to 20 reduced that error by roughly one-fourth.

(iii) For $n = 100$, we use $\Delta x = 0.01$, subtracting $\frac{1}{2}(1 + \frac{1}{2}) \cdot (0.01) = 0.0075$ in the last step. The result, shown in Figure 5.7.6, is 0.6931534305. The error estimate given by Theorem 5.7.1 for $n = 100$ is

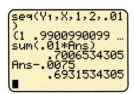

Figure 5.7.6

$$E_T \leq \frac{1}{60,000} = 0.00001\overline{6},$$

and the actual error is $E_T = |\ln 2 - 0.6931534305| = 0.0000062\ldots.$ Increasing n by a factor of 10, from 10 to 100, has decreased the error by roughly 0.01.

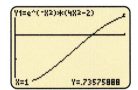

Figure 5.7.7

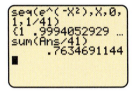

Figure 5.7.8

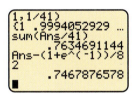

Figure 5.7.9

EXAMPLE 5.7.5

For what n can you be sure the trapezoid rule will approximate $\int_0^1 e^{-x^2}\,dx$ with an error less than 0.0001? Compute the approximation using that n

SOLUTION Letting $f(x) = e^{-x^2}$, we compute the second derivative:

$$f''(x) = e^{-x^2}(4x^2 - 2).$$

This function is increasing on the interval $[0, 1]$ [as we can see by computing $f^{(3)}(x)$ and checking its sign], which means that the minimum and maximum occur at the endpoints, $x = 0$ and 1, where

$$f''(0) = -2 \quad \text{and} \quad f''(1) = 2e^{-1} \approx 0.736.$$

Therefore, the maximum value of $|f''(x)|$ on $[0, 1]$ equals 2. A graph of $f''(x)$, drawn in a $[0, 1] \times [-2, 1]$ viewing window, is shown in Figure 5.7.7.

Applying Theorem 5.7.1, with $K = 2$, $a = 0$, and $b = 1$, we obtain the error estimate

$$E_T \leq \frac{K(b-a)^3}{12n^2} < \frac{2}{12n^2} = \frac{1}{6n^2}.$$

To make $E_T < 0.0001$, we need to find n satisfying:

$$\frac{1}{6n^2} \leq 0.0001 \quad \text{or} \quad n^2 \geq \frac{10{,}000}{6} = 1666.\overline{6}.$$

Thus, taking $n \geq \sqrt{1{,}667} \approx 40.8$ will give the desired accuracy. Using $n = 41$, we proceed to compute the terms of the sequence $f(x_i)$ for x_i from 0 to 1 in steps of $\Delta x = \frac{1}{41}$, multiply each term by $\frac{1}{41}$, take the sum, and subtract

$$\frac{1}{2}[f(x_0) + f(x_n)] \times \Delta x = \frac{(1 + e^{-1})}{82}.$$

The result is 0.7467876578, as shown in Figures 5.7.8 and 5.7.9.

Simpson's rule

This is the most widely used numerical method for approximating $\int_a^b f(x)\,dx$. It is based on approximating f by quadratic polynomials. We will simply state it here and give an example of its use, leaving its derivation as a project at the end of this chapter.

To begin with, we choose an *even* positive integer n and divide the interval $[a, b]$ into n equal segments, each of length $\Delta x = (b - a)/n$. Let

$$a = x_0 < x_1 < \cdots < x_{n-1} < x_n = b$$

be the endpoints of the subintervals in increasing order. Then we have:

Simpson's Rule

$$\int_a^b f(x)\,dx \approx [f(x_0) + 4f(x_1) + 2f(x_2) + 4f(x_3) + 2f(x_4) + \cdots$$
$$+ 2f(x_{n-2}) + 4f(x_{n-1}) + f(x_n)] \cdot \frac{\Delta x}{3}.$$

This approximation is generally much more accurate than the midpoint or trapezoidal rules, and the calculations are no harder. If we let E_S denote the approximation error (in absolute value) involved in using Simpson's rule, we have the following estimate:

$$E_S \leq \frac{K(b-a)^5}{180n^4},$$

where K is the maximum of $|f^{(4)}(x)|$ for $a \leq x \leq b$. The great advantage here is that increasing n by a factor reduces the error by roughly the *fourth power* of that factor. Thus, doubling n reduces the error by about one-sixteenth. Unfortunately, the error bound itself—involving, as it does, the optimization of $|f^{(4)}(x)|$—may be very difficult to compute. In practice, one often calculates the approximation for a sequence of n's, stopping when the difference between successive answers is less than some preset bound.

EXAMPLE 5.7.6

Approximate $\int_0^1 e^{-x^2}\,dx$ using Simpson's rule with $n = 2$ and with $n = 4$.

SOLUTION   If $n = 2$, $\Delta x = \frac{1}{2}$, and the division points are $x_0 = 0$, $x_1 = \frac{1}{2}$, and $x_2 = 1$. Applying Simpson's rule, we get

$$\int_0^1 e^{-x^2}dx \approx [f(x_0) + 4f(x_1) + f(x_2)] \cdot \frac{\Delta x}{3}$$

$$= [e^0 + 4e^{-1/4} + e^{-1}] \cdot \frac{1}{6} = 0.74718\ldots.$$

Comparing this with Example 5.7.5, we see that the approximations are about 0.0004 apart.

By doubling n, we expect to reduce the error by about one-sixteenth. If $n = 4$, $\Delta x = \frac{1}{4}$, and the division points are $x_0 = 0$, $x_1 = \frac{1}{4}$, $x_2 = \frac{1}{2}$, $x_3 = \frac{3}{4}$, and $x_4 = 1$. Applying Simpson's rule, we get

$$\int_0^1 e^{-x^2}dx \approx [f(x_0) + 4f(x_1) + 2f(x_2) + 4f(x_3) + f(x_2)] \cdot \frac{\Delta x}{3}$$

$$= [e^0 + 4e^{-1/16} + 2e^{-1/4} + 4e^{-9/16} + e^{-1}] \cdot \frac{1}{12}$$

$$= 0.7468553798\ldots.$$

APPLYING TECHNOLOGY

Graphing calculators have built-in procedures for approximating definite integrals. On the TI-83 Plus, for instance, one of the MATH operations is a function integral command fnInt shown in Figure 5.7.10. In using it, one must specify the function, the independent variable, and the lower and upper limits of integration. In Figure 5.7.11, for example, we have used fnInt to approximate $\int_0^1 1/\sqrt{1 + x^2}\, dx$.

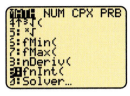

Figure 5.7.10   **Figure 5.7.11**

Practice Exercises 5.7

1. Approximate $\int_0^1 \sqrt{1 + x^2}\, dx$ by dividing the interval $[0, 1]$ into five equal segments and using
(**a**) the midpoint rule,   (**b**) the trapezoid rule.
Estimate the error in both cases.

2. Use Simpson's rule with $n = 2$ and $n = 4$ to estimate $\int_0^1 (1 + x^2)^{-1}\, dx$.

Exercises 5.7

1. (**a**) Estimate the errors involved in using the midpoint and trapezoidal rules with $n = 5$ to estimate $\int_0^1 x^2\, dx$.
(**b**) Compute the approximations and compare the answers to the actual value of the integral. What are the actual errors?

2. (**a**) Referring to the previous exercise, how much improvement in the error would you expect to see if you use $n = 20$?
(**b**) Use a calculator to compute the approximations and compare the answers to the actual value of the integral. How much have the errors actually improved?

3. (**a**) Use the midpoint rule with $n = 5$ to approximate $\int_0^2 (x + 1)^{-1}\, dx$. Estimate the error.
(**b**) For what n can you be sure the error is less than 0.001? Compute the approximation.
(**c**) Compare your answer to $\ln 3$, the actual value of the integral.

4. Scientists studying a type of antelope clock its speed when running for 8 minutes as

t minutes	0	1	2	3	4
$r(t)$ miles per hour	0	32	34	39	39
t minutes	5	6	7	8	
$r(t)$ miles per hour	38	36	32	26	

(**a**) Use the midpoint rule to estimate the total distance covered during that time.
(**b**) Estimate the antelope's average speed over that 8-minute period.

5. Assume that the north and south banks of a river have the shapes of curves represented by the equations

$$N(x) = \frac{1}{(x - 2)^{1/2}} + 1 \quad \text{and} \quad S(x) = \frac{1}{x^{1/2}},$$

respectively, where x is in hundreds of feet (see Figure 5.7.12 on page 396).

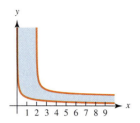

Figure 5.7.12

Use the midpoint rule and $n = 6$ to estimate the area of the river from $x = 3$ to $x = 6$.

6. (a) Use the trapezoidal rule with $n = 10$ to approximate $\int_0^3 e^{-x/3}\,dx$. Estimate the error.
   (b) For what n will the error be less than 0.001? Compute the approximation.
   (c) Compare your answer to the actual value of the integral.

7. Use the trapezoidal rule with $n = 10$ to approximate $\int_{-1}^{1} e^{-x^2/2}\,dx$. Estimate the error.

In the following exercises approximate the given integral by both the midpoint and trapezoidal rules.

8. $\displaystyle\int_0^1 e^{\sqrt{x}}\,dx; \quad n = 4$

9. $\displaystyle\int_{-1}^{2} \frac{dx}{1+x^2}; \quad n = 6$

10. $\displaystyle\int_{-1}^{1} \sqrt{1-x^2}\,dx; \quad n = 10$

11. $\displaystyle\int_{-1}^{0} 2^{-x}\,dx; \quad n = 10$

12. $\displaystyle\int_{-1}^{1} \ln(x^2+1)\,dx; \quad n = 20$

13. $\displaystyle\int_0^1 \frac{1}{e^x+1}\,dx; \quad n = 10$

14. $\displaystyle\int_{-1}^{1} \frac{1}{\sqrt{1+x^4}}\,dx; \quad n = 8$

15. $\displaystyle\int_1^2 \frac{\ln t}{t+1}\,dt; \quad n = 10$

16. Use the information in the following table to approximate $\int_0^4 f(x)\,dx$ by the trapezoid rule:

x	0	1	2	3	4
$f(x)$	0.1	0.4	0.5	1	−0.1

17. An object is shot straight upward from ground level. Its velocity v (in feet per seconds) is given by the the following table (with t in seconds):

t	0	1	2	3	4	5
v	0	9	15	22	31	44
t	6	7	8	9	10	
v	58	75	94	112	140	

Use the trapezoidal rule to approximate the height of the object at the end of 10 seconds.

18. Use the trapezoidal rule to estimate the average value of a continuous function $f(x)$, $-1.5 \le x \le 1.5$, which has the following table of values:

x	−1.5	−1.25	−1	−0.75	−0.5	−0.25	0
$f(x)$	4.8	4.5	3.9	3.1	2.1	2.3	2.7
x	0.25	0.5	0.75	1	1.25	1.5	
$f(x)$	3.0	3.2	3.5	4.2	4.9	5.6	

19. In Figure 5.7.13, assume that the vertical segments are spaced equally apart, and their lengths, from left to right, are as follows:

1.0, 1.38, 1.69, 2.04, 2.34, 2.4, 2.56, 2.52, 2.51, 2.48, 2.27, 2.0, 1.82, 1.73, 1.61, 1.65, 1.75.

Use the trapezoidal rule to estimate the area under the graph.

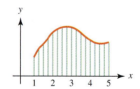

Figure 5.7.13

20. Some plants (such as watermelon) deplete the soil of some nutrients they need when they are planted in the same place for many years. Assume that such a plant was grown in the same location for 4 years and the following data were gathered for the rate of change of a nutrient. We assume that at $t = 0$ the amount of the nutrient in the soil is 1 (representing 100% of the natural level).

t	0	0.5	1	1.5	2
$r(t)$	−0.4	−0.327	−0.298	−0.219	−0.179
t	2.5	3	3.5	4	
$r(t)$	−0.147	−0.120	−0.098	−0.080	

(a) Use the trapezoidal rule to approximate the total change.

(b) What fraction of the natural value of the nutrient is in the soil at the end of 4 years?

21. Doctors in a city gathered together all the height measurements of 6-year-olds from their files. They calculated the mean (or average) height, subtracted it from each measurement, and obtained the following approximation for the height distribution:

$$N'(h) = \frac{1,274}{h^2 + 1},$$

where N is the number of children whose height differs from the mean by h inches.

Use the trapezoidal rule to calculate the approximate number of children whose heights are within 2 inches of the mean in either direction. Use $n = 8$.

22. Approximate $\int_1^2 x^{-1}\, dx$ by Simpson's rule, first using $n = 2$ and then $n = 4$. Compare your answer to $\ln 2$, the actual value of the integral. How did the error change in going from $n = 2$ to $n = 4$?

23. For what n will Simpson's rule approximate $\int_1^2 x^{-1}\, dx$ with error less than 0.00001? Calculate the approximation.

24. A seasonal pond is one that exists during wet or rainy periods and dries up after the wet season is over. One season, the change in the number of frogs in the pond was modeled by

$$f' = (3 - t)^{t^2} - 1, \quad 0 \le t \le 3,$$

in hundreds of frogs per month. Use Simpson's rule with $n = 16$ to estimate the total number of frogs in the first month, assuming there are no frogs at the beginning of the season.

25. The honey opossum (*Tarsipes rostratus*) is the smallest marsupial in Australia. It feeds on nectar and the babies weigh 2 to 6 milligrams at birth, which is about 4,000 times lighter than an average adult. One baby had its growth r rate measured, in milligrams per day, over the course of 2 weeks.

t	0	1	2	3	4	5	6	7
r	0	1.3	2	2.5	3.5	3.5	3.6	3.7

t	8	9	10	11	12	13	14
r	3	2.6	2.2	3.5	3.7	4	3.2

Use Simpson's rule to determine the total growth during this time.

26. To derive the formula for the area of a trapezoid with a horizontal base of length b and vertical sides of lengths h and k, divide the longer side into two segments as shown in Figure 5.7.14 (where we have assumed that $k > h$). Add the areas of the rectangle and triangle, and simplify.

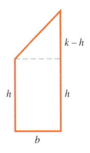

Figure 5.7.14

27. Use the function integration operation on a graphing calculator to approximate $\int_0^1 e^{-x^2}\, dx$. Compare the result with those obtained by using Simpson's rule in Example 5.7.6.

28. Use the function integration operation on a graphing calculator to approximate $\int_0^2 \sqrt{1 + x^4}\, dx$. Compare the result with that obtained by using the trapezoidal rule in Example 5.7.2.

29. Use the function integration operation on a graphing calculator to approximate $\int_0^1 (1 + x^2)^{-1}\, dx$. Then, use the calculator to compute $\pi/4$ and compare the results. (See Example 5.7.1.)

30. Figure 5.7.15 shows the graphs of $y = x^2$ and $y = (x^2 + 1)^{-1}$.

(a) Use a graphing calculator to find the intersection points of the two graphs.

(b) Use the calculator to approximate the area between the graphs from one intersection point to the other.

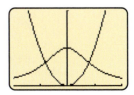

Figure 5.7.15

Solutions to practice exercises 5.7

1. In this case $\Delta x = 0.2$. The endpoints are 0, 0.2, 0.4, 0.6, 0.8, and 1. In addition, $f''(x) = (1 + x^2)^{-3/2}$, so that K, the maximum of $|f''(x)|$ on $[0, 1]$, equals 1.
(a) The midpoints are 0.1, 0.3, 0.5, 0.7, and 0.9, and the midpoint rule gives

$$\int_0^1 \sqrt{1 + x^2}\, dx \approx (0.2) \left(\sqrt{1.01} + \sqrt{1.09} + \sqrt{1.25} \right.$$
$$\left. + \sqrt{1.49} + \sqrt{1.81} \right) \approx 1.1466.$$

The error satisfies $|E_M| \le K(b-a)^3/(24 \cdot 25) = \frac{1}{600}$.
(b) The trapezoid rule gives

$$\int_0^1 \sqrt{1 + x^2}\, dx \approx (0.1) \left(\sqrt{1} + 2\sqrt{1.04} + 2\sqrt{1.16} \right.$$
$$\left. + 2\sqrt{1.36} + 2\sqrt{1.64} + \sqrt{2} \right) \approx 1.1501.$$

The error satisfies $|E_T| \le K(b-a)^3/(12 \cdot 25) = \frac{1}{300}$.

2. With $n = 2$, we obtain

$$\int_0^1 (1 + x^2)^{-1}\, dx \approx \left(\frac{1}{1 + 0^2} + \frac{4}{1 + \left(\frac{1}{2}\right)^2} + \frac{1}{1 + 1^2} \right) \cdot \frac{1}{6}$$
$$\approx 0.783.$$

With $n = 4$, we obtain

$$\int_0^1 (1 + x^2)^{-1}\, dx \approx \left(\frac{1}{1 + 0^2} + \frac{4}{1 + \left(\frac{1}{4}\right)^2} + \frac{2}{1 + \left(\frac{1}{2}\right)^2} \right.$$
$$\left. + \frac{4}{1 + \left(\frac{3}{4}\right)^2} + \frac{1}{1 + 1^2} \right) \cdot \frac{1}{12} \approx 0.7854.$$

Chapter 5 Summary

- An **antiderivative** of a given function $f(x)$ is a function $F(x)$ such that $F'(x) = f(x)$. In other words, it is a solution of the differential equation $dy/dx = f(x)$. The family of all antiderivatives of $f(x)$ is called the **indefinite integral** of $f(x)$ and is denoted by $\int f(x)\, dx$.

- If $F(x)$ is a known antiderivative of $f(x)$, then $\int f(x)\, dx = F(x) + c$, where c is any constant. To determine this constant, we need to know the value of the function $y(x)$ at a single point x_0. The problem $dy/dx = f(x)$, $y(x_0) = y_0$, is called an **initial value problem**.

- For finding antiderivatives, we use the **power, exponential, constant multiple**, and **sum rules for indefinite integrals** that are obtained by reversing the corresponding rules for derivatives.

- Using antiderivatives, we can determine a **total** function (such as **cost**, and **profit**) from its **marginal function** (marginal cost and marginal profit).

- For computing integrals of more complicated functions, we have the methods of **integration by substitution** and **integration by parts**. The first is based on the **chain rule** for differentiation and is expressed as

$$\int f(g(x))g'(x)\, dx \stackrel{u=g(x)}{=} F(g(x)) + c, \quad \text{where } F' = f.$$

The second is based on the product rule for differentiation and is expressed by the formula

$$\int u\, dv = uv - \int v\, du, \quad \text{where } dv = v'\, dx \text{ and } du = u'\, dx.$$

- The **definite integral** of a function on an interval $[a, b]$ is defined by

$$\int_a^b f(x)\,dx = \lim_{n \to \infty} [f(x_1)\Delta x + f(x_2)\Delta x + \cdots + f(x_n)\Delta x],$$

provided the limit exists. The sum on the right is called a **Riemann sum**. For each n, it is formed by partitioning the interval $[a, b]$ into n equal subintervals of length $\Delta x = (b - a)/n$ and by choosing a point x_i from the ith subinterval at which to evaluate f.

- If $f(x)$ is positive on the interval $[a, b]$, its Riemann sums provide an approximation of the **area** under its graph, while their limit (the definite integral) gives the actual value of the area.

- The **fundamental theorem of calculus** states that

$$\int_a^b f(x)\,dx = F(b) - F(a),$$

where $F(x)$ is an antiderivative of $f(x)$, or, equivalently, where $F'(x) = f(x)$. In other words: *The definite integral of a rate of change equals the total change.*

- With the fundamental theorem of calculus, we can also use the methods of substitution and integration by parts for evaluating definite integrals.

- If $f(x) \geq g(x)$ for $a \leq x \leq b$, the area between their graphs from a to b is given by $\int_a^b [f(x) - g(x)]\,dx$.

- The average value of a function $f(x)$ over an interval $[a, b]$ is given by $1/(b - a) \int_a^b f(x)\,dx$

- In addition to Riemann sums, there are other **numerical methods** for approximating definite integrals. One is the **trapezoidal rule**:

$$\int_a^b f(x)\,dx \approx [f(x_0) + 2f(x_1) + 2f(x_2) + \cdots + 2f(x_{n-1}) + f(x_n)] \cdot \frac{\Delta x}{2},$$

which is obtained by replacing the areas of rectangles in a Riemann sum by the areas of trapezoids whose upper edges are the secants connecting the endpoints of the graph over each subinterval.

Chapter 5 Review Questions

- State the power, exponential, constant multiple, and sum rules for indefinite integrals. Give concrete examples.

- State the substitution and integration by parts methods for evaluating definite integrals. Give concrete examples.

- Explain the relationship between the definite integral and the area between the graph of a function and the x-axis if part of the graph is below the axis. Give concrete examples.

- If you cannot find an antiderivative of a given function, or if you only know a table of values for the function, how can you estimate the definite integral? Give concrete examples.

Chapter 5 Review Exercises

1. Figure 5.R.1 shows the graphs of three functions f, g, and h. Decide which one of the functions g and h is an antiderivative of f. Explain.

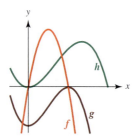

Figure 5.R.1

2. Figure 5.R.2 shows the graphs of four functions f, g, u, and v. The antiderivatives of three of them, f, u, and v, are also shown in the same diagram. Find those antiderivatives. Explain.

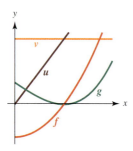

Figure 5.R.2

3. Find an antiderivative of

$$\frac{e^{3x} - e^{-3x}}{e^{3x} + e^{-3x}}.$$

Compute the following indefinite integrals:

4. $\displaystyle\int \frac{x^3 + 1}{x^4 + 4x + 8}\, dx$ 5. $\displaystyle\int x e^{x^2 + 1}\, dx$

6. $\displaystyle\int \sqrt{5x + 1}\, dx$

7. $\displaystyle\int t e^{-0.25t}\, dt$

8. $\displaystyle\int x^5 \ln x\, dx$

9. $\displaystyle\int \frac{1}{x^2 - 8x + 15}\, dx$

10. $\displaystyle\int \frac{1}{x^2 - 9x + 8}\, dx$

11. $\displaystyle\int \frac{t}{\sqrt{t - 4}}\, dt$

12. $\displaystyle\int (t + 1)^2 \ln(t + 1)\, dt$

13. Find a function $y = y(x)$ such that $y' = e^{-2x} - 4x^3 + 1$ and $y(0) = 12$.

14. Find the function whose graph has slope $x + \ln x$ for each positive value of x and passes through the point $(1, 8)$.

15. If $y = y(x)$ is the solution to the initial value problem: $y' + (1/x) = x^5$, $y(1) = 3$, find $y(e)$.

16. Solve the differential equation $y' = x^7 e^{x^8}$.

17. The marginal cost function for producing x units of a certain product is given by the function $MC(x) = \frac{1}{4}x + 5{,}000$. If the cost of producing 20 units is \$50,000, then determine the cost function $C(x)$.

18. The marginal profit function for a company producing and selling a certain product is $MP(x) = -0.6x + 3{,}000$. If 40 units are produced and sold, then the company's profit is \$25,000. Determine the profit function $P(x)$. Find the sales amount x that maximizes the profit. Explain in economic terms the meaning of the negative coefficient of x in $MP(x)$.

Exercises 19–22 refer to the function f whose graph is given in Figure 5.R.3.

19. Use four subintervals and their midpoints to estimate $\int_0^2 f(x)\, dx$.

20. Use eight subintervals and their right endpoints to estimate $\int_0^2 f(x)\, dx$.

21. Use eight subintervals and their left endpoints to estimate $\int_0^2 f(x)\, dx$.

22. Use three subintervals and their midpoints to estimate the area under the graph of $f(x)$, $0 \le x \le 1.5$.

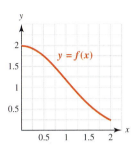

Figure 5.R.3

23. If the graph of a function $f(x)$, $-1 \le x \le 4$, is as shown in Figure 5.R.4, and if $\int_{-1}^{2} f(x)\,dx = 4.8$, $\int_{-1}^{4} f(x)\,dx = 3.6$, then find the area of the region enclosed by the graph of $f(x)$, $-1 \le x \le 4$, and the x-axis.

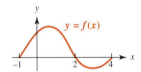

Figure 5.R.4

24. Let $f(x)$ be a continuous function on the interval $[2, 4]$, with $f(2) = 10$, $f(2.5) = 14$, $f(3) = 12$, and $f(3.5) = 8$. By computing the Riemann sum with four subintervals, using the value of the function at the left endpoints, estimate $\int_{2}^{4} f(x)\,dx$.

25. Given the following table of values, estimate the integral $\int_{1}^{3} f(x)\,dx$ by using a uniform partition of four subintervals and evaluating the function at the right-hand endpoint of each subinterval.

x	1	1.5	2	2.5	3
$f(x)$	0.8	0.5	-1	-2	-1.5

Compute the following definite integrals:

26. $\displaystyle \int_{-1}^{1} (x^4 - x^2)\,dx$

27. $\displaystyle \int_{-3}^{-2} \frac{6x^2 - 2}{x^3 - x}\,dx$

28. $\displaystyle \int_{-1}^{1} \frac{x^2 + 1}{x^3 + 3x + 9}\,dx$

29. $\displaystyle \int_{1}^{4} \frac{e^{\sqrt{x}}}{\sqrt{x}}\,dx$

30. $\displaystyle \int_{2}^{3} \frac{1}{x^2 - x}\,dx$

31. $\displaystyle \int_{0}^{1} t^2 e^{-0.1t}\,dt$

32. Find the area of the region between the curve $y = x^5$ and the x-axis, from $x = -1$ to $x = 1$.

33. Find the area between the curves $y = 2x^2 - 4x + 6$ and $y = -x^2 - 2x + 1$ from $x = 1$ to $x = 2$.

34. Compute the area of the region enclosed by the curves $y = 7 - x^2$ and $y = -1 + x^2$.

35. Calculate the average value of the function $f(x) = xe^{x/3}$, $0 \le x \le 3$.

36. Suppose $F(x)$ is an antiderivative of a continuous function $f(x)$, with $F(2) = 8$ and $F(4) = -2$. Compute $\int_{2}^{4} f(x)\,dx$.

37. Suppose $A(t)$ is the area under the graph of $y = 1/(3 + x^4)$ from 0 to t, for any positive number t. Compute $A'(t)$.

38. Consider the function $f(x) = 2x$. Compute the following quantities *using geometry* without the fundamental theorem of calculus:

   (a) The area between $y = f(x)$ and the x-axis from $x = -1$ to $x = 2$

   (b) $\displaystyle \int_{-1}^{2} f(x)\,dx$ **(c)** $\displaystyle \int_{0}^{2} f(x)\,dx$

In Exercises 39–40, find $A'(t)$.

39. $A(t) = \displaystyle \int_{0}^{t} \sqrt{x^2 - 2x + 8}\,dx$

40. $A(t) = t^2 + \displaystyle \int_{1}^{t} (1 + \ln x)^4\,dx$

41. If $A(t) = t^3 + \displaystyle \int_{1}^{t} e^x \sqrt{1 + x^3}\,dx$, compute $A'(2)$.

42. Write the Riemann sum for the function $f(x) = 1/x^3$ over the interval $[0, 1]$ if the interval is partitioned into 10 equal segments and the right-hand endpoint is used as the point x_j in the jth interval. Do not compute it.

43. Estimate the area of the region bounded by the two curves and the p-axis shown in Figure 5.R.5.

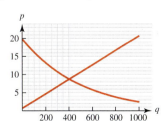

Figure 5.R.5

44. Suppose that the curves in Figure 5.R.5 are the graphs of $D(q) = 20e^{-0.002q}$ and by $S(q) = 0.02q + 1$. Compute the area of the region bounded by the two curves and the p-axis.

45. Estimate $\int_{0}^{1} \sqrt{1 + x^3}\,dx$ by dividing the interval $[0, 1]$ into five equal segments and using

   (a) the midpoint rule   **(b)** the trapezoid rule.

46. Use the trapezoid rule with $n = 10$ to estimate $\int_{-1}^{1} 3^x \, dx$.

~

47. After 2000 the rate of growth, in millions of dollars per year, of the revenue of a new company is given by the function $r(t) = 10e^{0.1t}$, where the time t is measured in years after 2000. If the company's revenue was 500 million in 2000, find its projected revenue for the year 2010.

48. A company operating at the production level of 30 units per day estimates that its marginal cost is given by the function $C'(x) = x^2 - 10x + 80$ dollars per unit. Find the additional cost for operating at the level of 40 units per day.

49. The value of a certain new car depreciates at the rate of $1000(t - 4)$ dollars per year, $0 \le t \le 3$. Compute the total depreciation in the first two years.

50. Compute the average value of $f(x) = x \ln x$ on $1 \le x \le 3$.

51. Figure 5.R.6 shows the graph of the velocity of a car accelerating, on a straight road, from rest to a speed of 60 miles per hour. Estimate the distance traveled during the first 40 seconds. (Pay attention to the units given.)

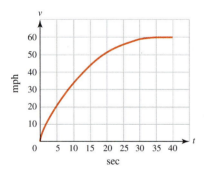

Figure 5.R.6

52. A company producing 40 units of a software package a day estimates that its marginal profit function is given by $MP(x) = -0.03(x + 50)^2 + 600$, $30 \le x \le 70$. How will the profit change if the company increases production to 50 units a day?

53. Figure 5.R.7 shows the shape of the ground of Notre Dame's football stadium with several measurments (all in yards). Estimate its area. For the curved corners use three subintervals and their midpoints to compute the corresponding Riemann sums.

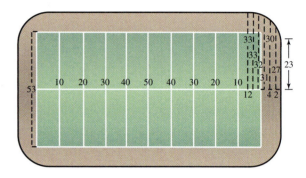

Figure 5.R.7

54. Using partial fractions and substitution, compute the integrals

(a) $\displaystyle \int \frac{dk}{(3\sqrt{k} + 1)(5\sqrt{k} + 1)}$

(b) $\displaystyle \int \frac{dk}{(M\sqrt{k} + 1)(N\sqrt{k} + 1)}$

where $M < N$.

Chapter 5   Practice Exam

1. Find the area of the region between the curve $y = x^3$ and the x-axis, from $x = -1$ to $x = 2$.

2. Compute the integral $\int_0^1 (8x^3 + (1/(x + 1)) + e^x) \, dx$.

3. Compute $\int 2xe^{x^2 - 5} \, dx$.

4. Consider the following differential equation $dy/dt = -\frac{2}{3}t^{3/2}$ with the initial condition $y(0) = 4$. Find $y(1)$.

5. Using integration by parts, find the definite integral $\int_1^e t^3 \ln t \, dt$.

6. Suppose f is the function whose graph is shown in Figure 5.E.1. Find $\int_0^5 f(x) \, dx$.

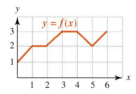

Figure 5.E.1

7. If the graph of a function $f(x)$, $-1 \leq x \leq 4$, is as shown in Figure 5.E.2, and if $\int_{-1}^{2} f(x)\,dx = -6.3$, $\int_{-1}^{4} f(x)\,dx = -4.2$, then find the area of the region enclosed by the graph of $f(x)$, $-1 \leq x \leq 4$, and the x-axis.

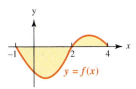

Figure 5.E.2

8. Using the trapezoid rule, estimate $\int_{1}^{2} \sqrt{4 - x^2}\,dx$ by dividing the interval $[1, 2]$ into four equal segments.

9. Estimate the integral $\int_{3}^{4} f(x)\,dx$ by using a Riemann sum with a uniform partition of five subintervals and by evaluating the function at the left side of each interval, when the following values of the function are given:

x	3	3.2	3.4	3.6	3.8	4
$f(x)$	0.8	0.2	-0.5	-0.9	-1.1	-1.3

10. If $g(t) = t^2 + \int_{1}^{t} \ln x \sqrt{1 + x^3}\,dx$, then compute $g'(1)$.

11. Compute the definite integral $\int_{1}^{2} (2x^3 + x)/(x^4 + x^2 + 2)\,dx$.

12. The value of a certain new computer changes at a rate of $10(t - 20)$ dollars per month, $0 \leq t \leq 36$. Compute the total change in the value of the computer over the first year.

13. Find the area enclosed by the curves $y = x^2$ and $y = -x^2 + 8$.

14. A company operating at a production level of 20 units per day estimates that its cost is \$10,000 and its marginal cost is $3x^2 - 6x + 100$ dollars per unit, $20 \leq x \leq 30$. Find the cost for operating at the level of 30 units per day.

15. After 2000 the rate of growth, in millions of dollars per year, of the revenue of a new company is given by the function $20\,e^{0.2t}$, where the time t is measured in years after 2000. If the company's revenue was 400 million in 2000, then find the revenue function.

16. Find the average value of $f(t) = 1/t^3$ over the interval $[1, 3]$.

17. Compute the integral $\int 1/(x^2 - 8x + 7)\,dx$.

18. Figure 5.E.3 shows the graph of the velocity of a car decelerating from a speed of 120 kilometers per hour to a speed of 20 kilometers per hour. Estimate the distance traveled during this period. (Pay attention to the units given.)

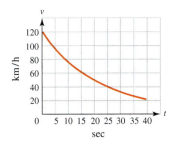

Figure 5.E.3

19. Figure 5.E.4 shows the shape of a lake together with several measurments of its width and length. Estimate the area of the lake using:
(a) the trapezoidal rule   **(b)** the midpoint rule.

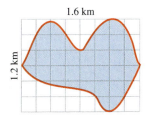

Figure 5.E.4

20. Compute the integral $\int 1/(4\sqrt{x} + 1)\,dx$.

21. If the average value of a function $f(x)$ over the interval $[2, 7]$ is 3, and its average value over the interval $[7, 10]$ is -3, then find its average value over the interval $[2, 10]$.

Chapter 5 Projects

1. In this project you are asked to derive Simpson's rule in three steps.

(a) If $x_0 < x_1 < x_2$ are three points such that $x_1 - x_0 = x_2 - x_1 = \Delta x$, and if $p(x) = Ax^2 + Bx + C$ is a quadratic polynomial, show that

$$\int_{x_0}^{x_2} p(x)\, dx = [p(x_0) + 4p(x_1) + p(x_2)] \cdot \frac{\Delta x}{3}.$$

(b) Suppose we are given a function $f(x)$, and we choose a quadratic polynomial $p(x) = Ax^2 + Bx + C$ for which $p(x_0) = f(x_0)$, $p(x_1) = f(x_1)$, and $p(x_2) = f(x_2)$. In other words, we choose the coefficients A, B, and C so that the graph of $p(x)$ passes through the points $(x_0, f(x_0))$, $(x_1, f(x_1))$, and $(x_2, f(x_2))$ as shown in Figure 5.P.1. Based on part (a), show that

$$\int_{x_0}^{x_2} f(x)\, dx \approx \int_{x_0}^{x_2} p(x)\, dx = [f(x_0) + 4f(x_1) + f(x_2)] \cdot \frac{\Delta x}{3}.$$

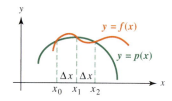

Figure 5.P.1

(c) Finally, suppose we divide the interval $[a, b]$ into an *even* number of subintervals, all of width Δx, and apply step (b) over the successive pairs of subintervals. In other words, for each three successive points x_{i-1}, x_i and x_{i+1}, we choose a quadratic polynomial as in step (b) and calculate its integral. Show that we get the following sequence of formulas:

$$\int_{x_0}^{x_2} f(x)\, dx \approx [f(x_0) + 4f(x_1) + f(x_2)] \cdot \frac{\Delta x}{3}$$

$$\int_{x_2}^{x_4} f(x)\, dx \approx [f(x_2) + 4f(x_3) + f(x_4)] \cdot \frac{\Delta x}{3}$$

$$\cdots$$

$$\int_{x_{n-2}}^{x_n} f(x)\, dx \approx [f(x_{n-2}) + 4f(x_{n-1}) + f(x_n)] \cdot \frac{\Delta x}{3}.$$

Since $x_0 = a$ and $x_n = b$, summing gives Simpson's rule:

$$\int_a^b f(x)\, dx \approx [f(x_0) + 4f(x_1) + 2f(x_2) + 4f(x_3) + 2f(x_4) + \cdots$$

$$+ 2f(x_{n-2}) + 4f(x_{n-1}) + f(x_n)] \cdot \frac{\Delta x}{3}.$$

2. The integral

$$A(b) = \frac{1}{\sqrt{2\pi}} \int_0^b e^{-x^2/2} \, dx$$

cannot be computed by using the fundamental theorem of calculus since the integrand has no antiderivative that can be expressed in terms of the known elementary functions. Nevertheless, it is important to estimate it because it plays an important role in many areas of mathematics, particularly in statistical analysis. (It is called the standard normal probability density function, and its graph is the famous "bell-shaped" curve of statistics.)

(a) Estimate $A(0.5)$ with $n = 10$, $A(1)$ with $n = 10$, and $A(2)$ with $n = 20$, by using both the trapezoidal and Simpson's rules.

(b) Considering the shape of the given function, which rule gives a better approximation of the integrals in part (a)?

6

Further Applications of the Integral

In this chapter, we present further applications of the integral, including the consumer and producer surplus, based on the fundamental law of supply supply and demand in economics, and the **present and future value** of a continuous income stream.

The second of these leads us to the study of **separable differential equations**, which have many interesting applications, including retirement plans, mortgage payments, population growth models, and various environmental situations. One fundamental population model called the **logistic equation** is studied in detail.

Finally, we introduce the notion of an **improper integral**, motivated by the present value of a perpetual income stream coming from an indestructible capital asset. These integrals also play an important role in computing probabilities.

One of the most fundamental economic models is the law of supply and demand for a certain product (milk, bread, beef, fuel, cars, etc.), or service (transportation, health care, education, etc.) in a free-market environment. In this model the quantity of a certain item produced and sold is described by two curves, called the supply and demand curves of the item. The **supply curve** gives the quantity that producers will supply at any given price, and the **demand curve** gives the quantity that consumers will demand at any given price.

To be more specific, consider the case of bread. In modeling the supply of bread, economists assume that the price of flour is fixed and that the quantity of bread supplied depends only on its price. In other words, the quantity supplied q is a function of the price per unit p.

Similarly, in modeling the demand, economists assume that the aggregate income (i.e., the total income in the economy) is fixed and that the quantity of bread demanded by consumers depends only on the price. In other words, the quantity of bread demanded by consumers, also denoted by q, is a function of the price p per unit.

As you might expect, the supply function is *increasing*—the higher the price, the more producers will supply, and the demand function is *decreasing*—the higher the price, the less consumers will buy.

Following a long-standing tradition in economics, we will reverse the roles of the independent and dependent variables, and write p as a function of q. What that amounts to is solving the supply and demand equations for p as a function of q. Thus, the supply curve is defined by the formula

$$p = S(q),$$

where p denotes the price, q denotes the quantity supplied, and S is a function obtained from data. Similarly, the demand curve is given by an equation of the form

$$p = D(q),$$

where D is a function obtained from data. The function S is increasing, and D is decreasing.

Typical supply and demand curves are sketched in Figure 6.1.1. The point (q_e, p_e) where the two curves cross is called the market equilibrium point, or simply the **equilibrium point**. In terms of economics its significance is this: The price of bread will keep changing until it reaches the **equilibrium price** p_e. At that point, the quantity supplied is equal to the quantity demanded, and we call that quantity the **equilibrium quantity** q_e.

In an ideal free market both consumers and producers gain by buying and selling at the equilibrium price. To see how that works, let's compute how much consumers gain by buying bread at the equilibrium price rather than at a higher price. (We can assume that every consumer pays at least the equilibrium price, for if the price falls below equilibrium, the demand exceeds the supply and forces up the price.)

First, we will compute the total amount spent by the consumers if everyone buys at the equilibrium price p_e. In that case, q_e units are supplied and bought, and the total amount spent is simply the number of units bought times the price per unit—that is,

Total amount spent at equilibrium price $= p_e q_e$.   (1)

407

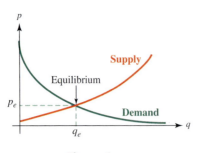

Figure 6.1.1

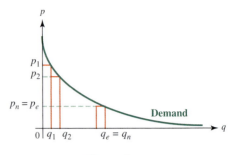

Figure 6.1.2

Next, we will suppose that instead of buying at equilibrium price those consumers pay the maximum price that each is willing to pay. What is the total expenditure in that case? The answer is this: *the integral of the demand function from 0 to q_e.* To see why, we will study the demand equation, $p = D(q)$, and to show things clearly, we will graph it by itself (without the supply curve). (See Figure 6.1.2.)

If we choose a large positive integer n and partition the interval $[0, q_e]$ into n equal segments, then the length of each of these subintervals is given by

$$\Delta q = \frac{q_e}{n}.$$

Let $q_1, q_2, \ldots, q_n$ be the right endpoints of the subintervals and $p_1, p_2, \ldots, p_n$ be the corresponding prices, as found on the demand curve. In particular, $q_n = q_e$ (the equilibrium quantity) and $p_n = p_e$ (the equilibrium price).

Now, we construct a rectangle over each segment (as shown in Figure 6.1.2) to the height of the graph over the right-hand endpoint of the subinterval. For instance, the first rectangle has height p_1 (the height of the graph over q_1). Its base has length Δq, which means its area is equal to $p_1 \cdot \Delta q$. But that is also the total amount paid for Δq units of bread at price p_1 per unit, which is approximately the total amount spent by consumers willing to pay at least p_1.

The next rectangle has area $p_2 \cdot \Delta q$, which is the total amount spent in buying Δq units at price p_2 per unit. That is approximately the total amount spent by consumers willing to pay at least p_2 but less than p_1. Continuing in this way up to n, we see that the total amount paid by consumers willing to pay at least p_e is approximately equal to

$$p_1\Delta q + p_2\Delta q + \cdots + p_n\Delta q. \tag{2}$$

The sum in (2) is a Riemann sum for the integral $\int_0^{q_e} D(q)\, dq$, and the sum gets closer and closer to the integral as we take n larger and larger. At the same time, the sum gives better and better approximations to the total expenditure. Therefore, we can conclude that the

$$\boxed{\text{Total amount paid at maximum prices} = \int_0^{q_e} D(q)\, dq.} \tag{3}$$

Quantity (3) is the area under the demand curve from $q = 0$ to $q = q_e$. As Figure 6.1.3 shows, it is greater than $p_e q_e$, which is the area of the rectangle with sides $[0, q_e]$ and $[0, p_e]$, and which, according to formula (1), represents the total amount spent by consumers at the equilibrium price. The difference between these two areas—quantity (3)

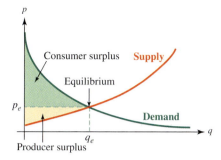

Figure 6.1.3

minus quantity (1)—represents the total that consumers save by buying at equilibrium price. For that reason it is called the **consumer surplus** (CS) for this product. In short,

---

Consumer Surplus

$$CS = \int_0^{q_e} D(q)\,dq - p_e q_e, \tag{4}$$

---

where $p = D(q)$ is the demand curve, p_e is the equilibrium price, and q_e is the equilibrium quantity. The consumer surplus is the benefit obtained by consumers engaged in free trade.

A similar analysis shows that the producers also gain by trading at the equilibrium price. Their gain, called the **producer surplus** (PS), is given by the following quantity.

Producer Surplus

$$PS = p_e q_e - \int_0^{q_e} S(q)\,dq, \tag{5}$$

---

where the supply curve is expressed in the form

$$p = S(q).$$

Figure 6.1.3 shows both the consumer and the producer surplus.

EXAMPLE 6.1.1

For a certain item the demand curve is

$$p = D(q) = \frac{20}{q+1},$$

and the supply curve is

$$p = S(q) = q + 2.$$

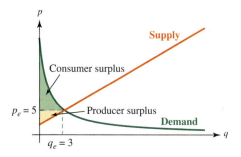

Figure 6.1.4

Find the equilibrium price and equilibrium quantity. Then compute the consumer and producer surplus.

SOLUTION   To find the equilibrium quantity, we let $D(q) = S(q)$ to obtain

$$\frac{20}{q+1} = q + 2.$$

Clearing the denominator gives $20 = (q + 1)(q + 2)$, which simplifies to $q^2 + 3q - 18 = 0$. The positive solution gives the equilibrium quantity, $q_e = 3$, and the equilibrium price is $p_e = 5$. The demand and supply curves are shown in Figure 6.1.4.

From formula (4), we have

$$
\begin{aligned}
\text{CS} &= \int_0^{q_e} D(q)\,dq - p_e q_e \\[4pt]
&= \int_0^3 \frac{20}{q+1}\,dq - 5 \cdot 3 \\[4pt]
&= 20 \ln(q + 1)\big|_0^3 - 15 \\[4pt]
&= 20(\ln 4 - \ln 1) - 15 \\[4pt]
&= 40 \ln 2 - 15 \approx 12.73.
\end{aligned}
$$

And from formula (5) we have

$$
\begin{aligned}
PS &= p_e q_e - \int_0^{q_e} S(q)\,dq \\[4pt]
&= 5 \cdot 3 - \int_0^3 (q + 2)\,dq \\[4pt]
&= 15 - \left(\frac{1}{2}q^2 + 2q\right)\bigg|_0^3 \\[4pt]
&= 15 - \left(\frac{9}{2} + 6\right) = 4.50.
\end{aligned}
$$

For more complicated functions, it may be necessary to use a graphing calculator to find the equilibrium point.

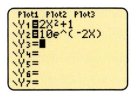

Figure 6.1.5

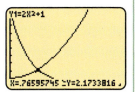

Figure 6.1.6

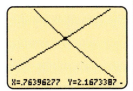

Figure 6.1.7

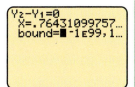

Figure 6.1.8

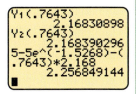

Figure 6.1.9

APPLYING TECHNOLOGY

EXAMPLE 6.1.2

Suppose the supply and demand functions are given by

$$S(q) = 2q^2 + 1 \quad \text{and} \quad D(q) = 10e^{-2q}.$$

Use a calculator to graph both curves on the same set of axes, find the equilibrium quantity and price, and estimate the consumer and producer surplus.

SOLUTION   After entering both functions in the function editor (shown in Figure 6.1.5), we have to choose an appropriate graphing window. We observe that $D(q)$ is decreasing, with $D(0) = 10$ and $0 \le D(q) \le 10$ for all q. In addition, $S(q)$ is increasing, with $S(0) = 1$ and $S(3) = 19$. Therefore, the curves cross at a point with $0 < q < 3$ and $D(q) \le 10$, so that a $[0, 3] \times [0, 10]$ window will contain the equilibrium point. The resulting graph is shown in Figure 6.1.6. The TRACE cursor has been placed as close as possible to the equilibrium point, giving the coordinates shown at the bottom of the screen. We can get a better estimate by zooming in, as shown in Figure 6.1.7, where we used the ZOOM command twice to get closer to the equilibrium point.

We can get a more accurate solution by using the equation solver, as shown in Figure 6.1.8. The solution depicted on the screen is the equilibrium quantity. To find the price, we evaluate either $S(q_e)$ or $D(q_e)$. Figure 6.1.9 shows the result of evaluating both functions, using the four-decimal approximation $q_e \approx 0.7643$. The values, rounded to three decimal places, agree, and we have $p_e \approx 2.168$.

Using these values, we calculate the consumer surplus:

$$\int_0^{0.7643} 10e^{-2q}\, dq - p_e q_e = -5e^{-2q} \Big|_0^{0.7643} - (0.7643) \cdot (2.168)$$

$$= 5 - 5e^{-1.5286} - (0.7643) \cdot (2.168) \approx 2.259,$$

(shown in Figure 6.1.9) and the producer surplus:

$$p_e q_e - \int_0^{0.7643} (2q^2 + 1)\, dq = (0.7643) \cdot (2.168) - \left(\frac{2q^3}{3} + q \right) \Big|_0^{0.7643}$$

$$= (0.7643) \cdot (2.168) - \frac{2 \cdot (0.7643)^3}{3} - 0.7643$$

$$\approx 0.595.$$

Another option, particularly useful in cases where the integrals cannot be computed in closed form, is to rewrite the consumer and producer surpluses in the forms

$$\int_0^{q_e} [D(q) - p_e]\, dq \quad \text{and} \quad \int_0^{q_e} [p_e - S(q)]\, dq$$

and use numerical methods to approximate the integrals.

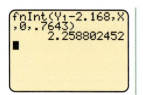

Figure 6.1.10

Every standard graphing calculator can perform numerical integration. Figure 6.1.10 shows the consumer surplus of the last example approximated using the numerical integration function fnInt on a TI-83 Plus.

Practice Exercises 6.1

1. Suppose the demand and supply functions are given by

$$D(q) = -q + 6 \quad \text{and} \quad S(q) = q^2 + 2q + 2, \quad q \geq 0.$$

Find the equilibrium price and quantity, and compute the consumer and producer surpluses.

Exercises 6.1

1. Figure 6.1.11 shows supply and demand curves, both of which are straight lines. Use elementary geometry to find the consumer and producer surplus.

2. Figure 6.1.12 shows supply and demand curves, both of which are straight lines. Use elementary geometry to find the consumer and producer surplus.

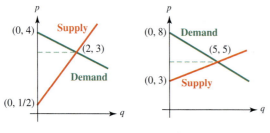

Figure 6.1.11   **Figure 6.1.12**

In each of the following systems the first equation defines a demand curve and the second equation defines a supply curve. Determine the equilibrium points, the consumer surplus, and the producer surplus. Also indicate these quantities as points and areas of regions in a coordinate plane.

3. $p = 10 - 0.4q$
   $p = 2 + 0.6q$

4. $p + 0.2q = 400$
   $p - 0.4q = 40$

5. $q = 5,000 - 50p$
   $q = 100p - 1,000$

6. $p = -0.1q + 200$
   $p = 0.2q + 20$

7. $75p + 45q = 2,250$
   $7.5p - 3q = 37.5$

8. $q + 250p = 60,000$
   $-q + 500p = 15,000$

Find the equilibrium quantity and price, the consumer surplus, and the producer surplus for each of the following demand and supply curves:

9. $D(q) = -0.4q + 23, \quad S(q) = 0.03q^2 + 3$

10. $D(q) = -0.2q + 60, \quad S(q) = 0.003q^2 + 0.02q + 8$

11. $D(q) = 0.3(q - 20)^2, \quad S(q) = 2q + 10, \ 0 \leq q \leq 20$

12. $D(q) = 0.005(q - 100)^2, \quad S(q) = 0.1q + 2,$
    $0 \leq q \leq 100$

13. $D(q) = \dfrac{25}{q + 2}, \quad S(q) = q + 2$

14. $D(q) = \dfrac{110}{q + 4}, \quad S(q) = q + 5$

15. $D(q) = (q - 5)^2, \quad S(q) = q^2 + q + 3, \ 0 \leq q \leq 5$

16. $D(q) = 0.03(q - 50)^2, \quad S(q) = 0.03q^2 + q + 3,$
    $0 \leq q \leq 50$

17. The demand and supply curves for a product are shown in Figure 6.1.13. Indicate the region whose area represents the consumer surplus and the one whose area represents the producer surplus. Then give your best estimate for the consumer and producer surplus.

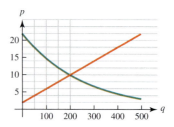

Figure 6.1.13

18. A mouse is gathering food for the winter. One particular food is preferred over all others by the mouse. Sometimes this food is plentiful and sometimes it's rare. When it is rare, the mouse will have to spend a lot of time looking for a few pieces. When plentiful, the mouse spends only a little time looking and acquires many pieces.
(a) Does this describe a supply or demand function?
(b) A mouse has no money. What price does he pay for a piece of the food?
(c) Draw a rough sketch of the function. Be sure to label the axes.

19. Biologists studying a certain animal noticed that the reproduction rate was low during drought years and high during wet years. They hypothesized that the amount of rainfall dictated the amount of food available per baby. If there were just a few calories available, only a few babies would be produced. When many calories were available, many babies would be produced.
(a) Does this describe a supply or demand function?
(b) Draw a rough sketch of this function. Label the axes.
(c) If c_0, the amount of calories available per baby, encourages females to produce b_0 babies, what is the total amount of calories available?

20. In some bird species, a male will mate with as many females as he can during breeding season. Sometimes the male has to fight to win the female away from a rival. The number of females he can expect to mate with is a function of the number x of fights per female he is willing to endure. Assume that function is given by

$$F(x) = 2x.$$

(a) Is this a supply or demand function?

(b) If he was willing to endure four fights per female, how many females would he expect to get?
(c) How many fights would he expect to fight?

21. Roger is breeding tropical fish to sell to hobbyists. He feels that his supply curve looks like

$$y_1 = 0.06x^2 + 5,$$

where x is the number of fish, in hundreds, sold at y_1 dollars each. Articles in his trade newsletters lead him to believe that the demand curve is

$$y_2 = 0.2x^2 - 4.8x + 31.8.$$

(a) The equilibrium point is very close to seven fish. Calculate the producer surplus if Roger sells at $y_1(7)$.
(b) What is the consumer surplus?

In Exercises 22–25, use a calculator to plot the demand and supply curves in the same window. Then find the equilibrium quantity to four decimal places, use it to find the equilibrium price to three decimal places, and calculate the consumer and producer surpluses. If necessary, use numerical integration.

22. $D(q) = 150e^{-0.004q}$, $S(q) = 5e^{0.003q} + 0.02q + 1$, $0 < q < 1{,}000$

23. $D(q) = 50e^{-0.002q}$, $S(q) = 0.0001q^2 + 0.004q + 1$, $0 < q < 1{,}000$

24. $D(q) = (1 + 0.003q^2)^{-1/2}$, $S(q) = 0.02(q + 1)$, $0 < q < 100$

25. $D(q) = \dfrac{10}{1 + 0.4q}$, $S(q) = \ln(10 + q^4)$, $0 < q < 10$

Solutions to practice exercises 6.1

1. Setting $D(q) = S(q)$, collecting terms, and simplifying give $q^2 + 3q - 4 = 0$. The only positive solution is $q = 1$, which is the equilibrium quantity. The equilibrium price is given by $D(1) = S(1) = 5$. The consumer surplus is given by

$$\int_0^{q_e} D(q)\,dq - p_e q_e = \int_0^1 (-q + 6)\,dq - 5$$

$$= \left(-\frac{q^2}{2} + 6q\right)\Bigg|_0^1 - 5 = \frac{1}{2}.$$

The producer surplus is given by

$$p_e q_e - \int_0^{q_e} S(q)\,dq = 5 - \int_0^1 (q^2 + 2q + 2)\,dq$$

$$= 5 - \left(\frac{q^3}{3} + q^2 + 2q\right)\Bigg|_0^1 = \frac{5}{3}.$$

6.2 Continuous Income Streams

Let's review some basic formulas from Chapter 2 involving the return on money deposited in a bank paying a given rate of interest. Suppose that an initial amount of M dollars is deposited in an account paying an interest rate of r per year. We will consider three cases—those in which the interest is compounded once a year, several times a year, and continuously—and list the formulas using the notation and terminology common in finance.

Interest compounded once a year   In this case, as we saw in Chapter 2, the amount at the end of t years is $M(1 + r)^t$. In the language of finance, the initial amount M is called the **present value** of the money, symbolized by PV, and the final amount is called the **future value**, denoted by FV. Using this terminology, we say that given a present value PV and an interest rate r, compounded annually, the future value FV after t years is given by the formula

$$FV = PV(1 + r)^t \quad \text{(interest compounded once a year at rate } r\text{).}$$ (6)

It is also useful to express PV in terms of FV, which we can do by solving Eq. (6) for PV, as follows:

$$PV = \frac{FV}{(1 + r)^t} \quad \text{(interest compounded once a year at rate } r\text{).}$$ (7)

EXAMPLE 6.2.1

If a bank account pays a 7% interest rate, compounded annually, find the present value of a future value of $1,000 after 5 years.

SOLUTION   Using formula (7), we get

$$PV = \frac{1,000}{(1 + 0.07)^5} \approx \$712.99.$$

Interest compounded n times a year   In this case, an initial amount of M dollars at annual interest rate r yields

$$M\left(1 + \frac{r}{n}\right)^{tn}$$

after t years, as we saw in Chapter 2. Recasting this in the language of finance, we say that at annual interest rate r, compounded n times a year, the future value FV after t years of a present value PV is given by the formula

$$FV = PV\left(1 + \frac{r}{n}\right)^{nt} \quad \left(\begin{array}{c}\text{interest compounded } n \text{ times}\\ \text{a year at rate } r\end{array}\right).$$ (8)

If we solve this equation for PV, then we obtain a formula for the present value in terms of the future value—that is,

$$
PV = \frac{FV}{\left(1 + \dfrac{r}{n}\right)^{nt}} \quad \left(\begin{array}{c}\text{interest compounded } n \text{ times} \\ \text{a year at rate } r\end{array}\right). \tag{9}
$$

EXAMPLE 6.2.2

Find the present value of a deposit that grows to $1,000 after 5 years in the bank with interest at an annual rate of 7%, compounded daily.

SOLUTION For simplicity, we will ignore leap years and interpret "daily" as meaning 365 times a year. To find the present value, we simply apply formula (9) with $FV = 1,000$, $r = 0.07$, $n = 365$, and $t = 5$. Then we obtain

$$
PV = \frac{1,000}{\left(1 + \frac{0.07}{365}\right)^{1,825}} \approx \$704.71.
$$

Interest compounded continuously In this case, the formulas for future and present value take the following forms:

$$
FV = PV e^{rt} \quad \text{(interest compounded continuously at rate } r\text{)} \tag{10}
$$

and

$$
PV = FV e^{-rt} \quad \text{(interest compounded continuously at rate } r\text{)}. \tag{11}
$$

These follow from (8) and (9) by letting $n \to \infty$, as we saw in Chapter 2.

EXAMPLE 6.2.3

Find the future value of $1,000 after 5 years with interest compounded continuously at an annual rate of 7%.

SOLUTION If we apply formula (10) with $PV = \$1,000$, $r = 0.07$, and $t = 5$, we obtain

$$
FV = 1,000 e^{0.35} \approx \$1,419.07.
$$

EXAMPLE 6.2.4

Find the present value of an amount that will become $1,000 in 5 years at an interest rate of 7%, compounded continuously.

SOLUTION   We apply formula (11) with $FV = 1,000$, $r = 0.07$, and $t = 5$ to find

$$PV = 1,000e^{-0.35} \approx \$704.69.$$

Future and present value of a continuous income stream

Up until now we have been calculating future value under the assumption that once the initial deposit is made, there are no future deposits or withdrawals. But there is no reason not to take future deposits into account, provided that they are regular enough to allow us to incorporate them into a formula. A common situation is that in which deposits are made at regular intervals over a long period of time. If the time between deposits is relatively short compared to the overall life of the account, we can think of the money as flowing continuously into the account in much the same way that water flows into a tank. Just as we ignore the individual molecules of water and think of them as merging into a stream, so we can think of the individual deposits merging into a continuous stream of money.

For example, suppose a person plans to retire in 30 years and arranges for money to be deposited continuously in the bank at a rate of $10,000 per year. (Think of this as a continuous stream of money being deposited.) If the interest is compounded continuously at a rate of 7%, what is the future value of the income in 30 years?

The new problem we face here is that we have a continuous stream of money instead of a single deposit. To deal with this problem, we choose a large integer n and partition the time interval $[0, 30]$ into n equal subintervals, each of length $\Delta t = 30/n$. Let

$$t_1 = \frac{30}{n} \quad \text{(the right-hand endpoint of the 1st subinterval),}$$

$$t_2 = 2 \cdot \frac{30}{n} \quad \text{(the right-hand endpoint of the 2nd subinterval),}$$

$$t_3 = 3 \cdot \frac{30}{n} \quad \text{(the right-hand endpoint of the 3rd subinterval),}$$

and so forth, up to

$$t_n = n \cdot \frac{30}{n} = 30 \quad \text{(the right-hand endpoint of the nth subinterval).}$$

Since the money is being continuously deposited at a steady rate of $10,000 per year, the amount that flows into the account during the time interval $[0, t_1]$ is equal to $10,000 \cdot \Delta t$, with $\Delta t = 30/n$. This money remains in the account for $30 - t_1$ more years, and its future value at the end of that time is given by formula (10) as

$$10,000 \cdot \Delta t \cdot e^{0.07(30-t_1)}.$$

(Actually, that is only an approximation of the future value because we have only computed it starting from time t_1 and ignored the interest accumulated during the time period $[0, t_1]$. However, if n is large, that period is very small relative to the total 30 years, and we get a good approximation.)

Next, we consider the time interval from $[t_1, t_2]$, which also has length Δt. The total amount deposited in the bank during that period is again equal to $10,000 \cdot \Delta t$. It remains in the account for $30 - t_2$ years, and its (approximate) future value is given by formula (10) as

$$10,000 \cdot \Delta t \cdot e^{0.07(30 - t_2)}.$$

Continuing in this way, we can find the (approximate) future value of the amount deposited in each one of the n subintervals, the last one being given by

$$10,000 \cdot \Delta t \cdot e^{0.07(30 - t_n)}.$$

If we add up the future values of the income in all the n subintervals, we obtain the amount

$$10,000 e^{0.07(30 - t_1)} \Delta t + 10,000 e^{0.07(30 - t_2)} \Delta t + \cdots + 10,000 e^{0.07(30 - t_n)} \Delta t.$$

(For convenience we have switched the positions of Δt and the exponential factor in each term, but that certainly does not change the value.) This is a Riemann sum for the definite integral of the function $10,000 e^{0.07(30 - t)}$ from $t = 0$ to $t = 30$, and as $n \to \infty$, it approaches the value of that integral. At the same time, the sum approaches the future value of the total income stream. Therefore, we can conclude that the

$$\text{Future value of the income stream} = \int_0^{30} 10,000 e^{0.07(30 - t)} \, dt$$

$$= 10,000 e^{2.1} \int_0^{30} e^{-0.07t} \, dt$$

$$= (10,000 e^{2.1}) \left(\frac{e^{-0.07t}}{-0.07} \right) \Big|_0^{30}$$

$$= -\frac{1,000,000}{7} e^{2.1} (e^{-2.1} - 1)$$

$$\approx \$1,023,739.$$

We can modify this example in two ways. First, by replacing the specific data by arbitrary positive constants, such as T for the number of years, S for the number of dollars per year deposited, and r for the interest rate. The analysis we just did remains valid, and we can conclude that *after T years the future value of an income stream deposited continuously at a rate of S dollars per year into an account paying interest rate r, compounded continuously, is given by*

$$\int_0^T S e^{r(T - t)} \, dt.$$

Second, we can get an even more general formula by dropping the assumption that the income stream is deposited at a constant rate. Let us suppose instead that the rate of deposit is given by some function, say, $S(t)$. In other words, we allow the rate of deposit to vary over time. Then we arrive at the following rule:

The Future Value of a Continuous Income Stream

The future value of a continuous income stream flowing at a rate of $S(t)$ dollars per year for T years, earning interest at an annual rate r, compounded continuously, is given by

$$FV = \int_0^T S(t)e^{r(T-t)}\, dt. \tag{12}$$

Similarly, we can compute the *present* value of the same income stream, with interest compounded continuously at an annual rate r. To do that, we again choose some large integer n and partition the interval $[0, T]$ into n subintervals, each of length $\Delta t = T/n$. If t_j is the right-hand endpoint of the jth subinterval, then the amount of money deposited during that time interval is approximately $S(t_j) \cdot \Delta t$. According to formula (11), the present value of that amount is approximately

$$S(t_j) \cdot \Delta t \cdot e^{-rt_j}.$$

By adding up the present values in each of the subintervals, we obtain the sum

$$S(t_1)e^{-rt_1}\Delta t + S(t_2)e^{-rt_2}\Delta t + \cdots + S(t_n)e^{-rt_n}\Delta t.$$

This sum approximates the present value of the total income over the time interval $[0, T]$, and we can make the approximation as good as we want by taking n large enough. In other words, the present value is the *limit* of this sum as $n \to \infty$. But this same sum is also a Riemann sum for the function $S(t)e^{-rt}$ over the interval $[0, T]$, and its limit as $n \to \infty$ is the integral of that function from 0 to T. Therefore, by letting $n \to \infty$, we get the following:

The Present Value of a Continuous Income Stream

The present value of a continuous income stream flowing at a rate of $S(t)$ dollars per year for T years, earning interest at an annual rate r, compounded continuously, is given by

$$PV = \int_0^T S(t)e^{-rt}\, dt. \tag{13}$$

EXAMPLE 6.2.5

Money is continuously deposited into an account at a steady rate of $10,000 per year for 30 years. If the account pays 7% annual interest, compounded continuously, find the present value of the income stream.

SOLUTION   We simply apply formula (13) with $S(t) = 10{,}000$, which gives

$$\text{Present value of income stream} = \int_0^{30} 10{,}000 e^{-0.07t} \, dt$$

$$= \frac{10{,}000}{-0.07} e^{-0.07t} \Big|_0^{30}$$

$$= \frac{1{,}000{,}000}{7} (1 - e^{-2.1}) \approx \$125{,}363.$$

Remark

1. Observe that formula (13) is simply the present value of the amount in formula (12), and formula (12) is simply the future value of the amount in formula (13). In short,

$$\boxed{FV = PV e^{rT}}$$

   This observation gives a way of deriving either one of the formulas from the other.

2. If $S(t) = S$, a constant, as in Example 6.2.5, then computing the integral

$$\int_0^T S e^{-rt} \, dt = -\frac{S}{r} e^{-rt} \Big|_0^T = \frac{S}{r}(1 - e^{-rT}),$$

   simplifies formulas (12) and (13) and gives

$$\boxed{PV = \frac{S}{r}(1 - e^{-rT}),} \tag{14}$$

$$\boxed{FV = \frac{S}{r}(e^{rT} - 1).} \tag{15}$$

EXAMPLE 6.2.6

Assume that a company manufactures a certain type of machine, and it estimates that each machine will generate a continuous income stream whose rate in the tth year of operation will be $15 - 2t$ million dollars per year. Assuming that the lifetime of a machine is about 7 years and that the money can be invested at an annual rate of 8%, compounded continuously, find the fair market price of each machine.

SOLUTION   In this situation it is reasonable to assume that the fair market price of each machine is the present value of the continuous income stream with $S(t) = 15 - 2t$,

$r = 0.08$, and $T = 7$. Using formula (13), we find that this price is

$$\int_0^7 (15 - 2t)e^{-0.08t}\, dt = 15 \int_0^7 e^{-0.08t}\, dt - 2 \int_0^7 te^{-0.08t}\, dt.$$

The first integral gives

$$\int_0^7 e^{-0.08t}\, dt = \frac{e^{-0.08t}}{-0.08}\bigg|_0^7 = \frac{1}{0.08}(1 - e^{-0.56}) \approx 5.35989.$$

To compute the second integral, we use integration by parts with $u = t$ and $dv = e^{-0.08t}\, dt$. Then we find $du = dt$, $v = e^{-0.08t}/(-0.08)$, and

$$\int_0^7 te^{-0.08t}\, dt = -\frac{1}{0.08}te^{-0.08t}\bigg|_0^7 + \frac{1}{0.08}\int_0^7 e^{-0.08t}\, dt$$

$$= -\frac{7}{0.08}e^{-0.56} - \frac{1}{(0.08)^2}e^{-0.08t}\bigg|_0^7$$

$$= -\frac{7}{0.08}e^{-0.56} - \frac{1}{(0.08)^2}(e^{-0.56} - 1) \approx 17.0178.$$

Putting it all together, we find that the fair market price of each machine is about \$46.363 million.

Formula (13) and the analysis leading up to it are also valid when money is being continuously *withdrawn* from an account.

EXAMPLE 6.2.7

A 60-year-old woman sets up a retirement fund by depositing an initial amount in an account paying 5%, compounded continuously. Her plan is to withdraw money from the account at an annual rate of \$50,000. Treating the withdrawals as a continuous stream, determine the amount she must deposit in order for the account to last for 30 years.

SOLUTION　In this case, money is flowing *out* of the account at a rate of \$50,000 per year. Applying formula (13), we see that the present value of the total amount flowing out of the account is

$$\int_0^{30} 50{,}000e^{-0.05t}\, dt = -1{,}000{,}000e^{-0.05t}\bigg|_0^{30}$$

$$= 1{,}000{,}000(1 - e^{-1.5}) = 776{,}869.84.$$

Therefore, she needs to deposit that amount initially to cover the total outflow.

In practice, interest is usually compounded using discrete time periods—such as months or days—rather than continuously. In such cases, the formulas we have developed using continuous compounding are only approximations to the actual present and future values. However, they are simpler to compute and give good estimates if the compounding period is relatively small. (See the exercises at the end of this section.)

Practice Exercises 6.2

1. Find the future value after 10 years of a continuous income stream of $1,200 per year deposited in an account paying 6% annual interest, compounded continuously.

2. Find the present value of a continuous income stream of $500 per month deposited for 20 years into an account paying 4% annual interest, compounded continuously.

Exercises 6.2

1. Find the future value of $5,000 after 4 years at a 6% annual interest rate, compounded monthly.

2. Find the present value of a deposit whose future value after 6 years at a 7% annual interest rate, compounded daily, is $20,000.

3. Perennial plants grow and produce seeds during a season, and then they die over the winter. When the next growing season arrives, they will regrow and, in addition, new seeds will sprout. One field has 1,000 members of a particular perennial at the beginning of a season. At the start of the next season, the number of seeds that sprout is 23% of the previous year's plant population. This goes on for 5 years.

(a) The number of plants each year (old and new) grows like a money account where interest is compounded n times a year. What is n?

(b) What are r and t?

(c) Calculate the number of plants in the field in 5 years. Assume that no plant dies or is otherwise removed during that time period.

4. Find the present value of a deposit whose future value after 8 years at an annual rate of 9%, compounded continuously, is $15,000.

5. Find the future value of $4,000 after 10 years at an annual rate of 8%, compounded continuously.

6. Suppose that an amount of money M is deposited in a bank that pays an annual interest rate r, compounded continuously.

(a) Find a formula expressing T, the time needed for M to double, as a function of r.

(b) Find T if r is 5%.

(c) If r is 10%, find the time needed for M to quadruple.

7. Find the interest rate that must be paid by a bank so that an initial deposit doubles in 7 years, assuming that the interest is compounded continuously.

8. Suppose that money is deposited steadily into a savings account at a constant rate of $20,000 per year. Find the balance at the end of 7 years if the account pays 9% interest, compounded continuously.

9. Suppose that money is deposited steadily into a savings account at a constant rate of $15,000 per year for 5 years. Find the present value of this income stream if the account pays 7.5% interest, compounded continuously.

10. If a continuous income stream flows into a savings account at a constant rate of $10,000 per year and earns 8% interest, compounded continuously, find the time required for the balance to become $2,000,000.

11. When the usual breeding tanks are full, a fish farm uses a lake on its property to store excess breeding fish. If the farm puts fish into the lake at the rate of 1,000 fish per year and if the fish increase their population exponentially at an annual rate of 33%, find the fish population in the lake after 6 years. Assume that no fish dies or is removed.

12. An employee of a company is offered a choice between two retirement plans. In the first plan, the company will deposit an initial amount of $5,000 in an IRA paying interest at an annual rate of 8%, compounded continuously, and it will then continue to deposit money at a constant rate of $15,000 per year for the next 25 years. In the second plan, the company will pay the employee $1,100,000 at the end of 25 years. Which one is the more beneficial plan for the employee?

13. A star player in the NBA is offered a 6-year contract by a team and two choices of compensation. In the first he is offered a lump sum of $40,000,000, paid at the beginning of his contract. In the second he is offered an initial payment of $6,000,000 and a 6-year continuous income stream at the rate of $7,500,000 per year deposited into a savings account paying 8% annual interest, compounded continuously. Assuming that the player can also invest his money with the same interest of 8%, determine which plan is better for the player.

17. The population of a fast-developing region is increasing in two ways. First, its intrinsic growth rate is 2% per year (birth rate minus death rate). Second, its annual net immigration rate (new arrivals minus departures) is given by $S(t) = 0.1t + 0.004$ million per year, where t is the time in years. If the initial population was 4 million, what is the population of the region after 10 years?

18. A self-employed software engineer estimates that her annual income over the next 10 years will steadily increase according to the formula $70,000(1.2)^t$, where t is the time in years. She decides to save 12% of her income in an account paying 6% annual interest, compounded continuously. Treating the savings as a continuous income stream over a 10-year period, find the present value.

19. A 30-year old man plans to invest money as a continuous income stream at a constant rate until he is 70. After that he will stop investing and start withdrawing at an annual rate of $60,000. If the account pays (and will always pay) 7% interest, compounded continuously, and he wants the fund to pay off for 20 years, at what rate should he invest?

14. Imagine yourself in the following situation. You have already gotten that great job, and you are thinking of buying that great house 7 years from now. You estimate that you will need $100,000 for a down payment and other expenses. Let us also assume that you have an account that pays 6.5% interest, compounded continuously.
(a) If you deposit money continuously into the account at a constant rate during the whole 7-year period, find the rate (dollars per year) you must use in order to meet your goal of $100,000.
(b) If you decide to make one lump sum deposit now, instead of continuous deposits, what is the amount you must deposit now in order to meet your goal of $100,000?

15. On the day their daughter is born, a couple begins saving for her college education by depositing money into an account that pays 5.5% interest, compounded continuously. If the deposits are made often enough so that they can be approximated by a continuous income stream flowing at the rate of $5000 + 500t$ dollars per year, calculate the amount that will be in the account on their daughter's eighteenth birthday.

16. Assume that estimates show that a growing company will be generating a continuous income stream for the next 5 years at the variable rate of $25 + 3t$ million dollars per year at any time t.
(a) Compute the future value of the income stream, assuming that money can be invested for that period at the annual rate of 9% interest, compounded continuously.
(b) Compute the present value of the income stream described in (a).
(c) Write an equation relating the numbers found in (a) and (b).

20. The Michigan Fish and Game Commission has built a man-made lake in one of its large state parks, and it plans to stock it with trout. It estimates that fishermen will remove trout from the lake at a steady rate of 6,000 per year. It also estimates that the fish will grow exponentially at a rate of 12% per year. If it does not plan to restock the lake for 5 years, what is the minimum number of trout it must put in the lake now?

21. Suppose you have $60,000 in a savings account paying 4.5% interest, compounded continuously. If you start withdrawing money continuously at the rate of $5,000 per year without making any additional deposits, how long will the account last?

22. A continuous income stream is entering an account at a variable rate of $10,000t/(1 + t)$ dollars per year, where t is the time in years. The account pays 5% annual interest, compounded continuously.
(a) Write an integral that gives the future value of the account after 10 years.
(b) Use the midpoint rule with $n = 10$ to estimate the future value.
(c) Compare the answer to (b) with the approximation you get with the numerical integrator on your calculator.

In Exercises 23–29 we consider a discrete, rather than continuous, income stream—that is, one consisting of a finite number of deposits made at regular intervals.

23. Suppose that a total of S dollars per year is deposited into an account paying interest rate r, compounded n times per year, with a deposit of S/n being made at the beginning of each period. Let $A(k)$ denote the amount in the account at the end of k periods.

(a) Show that

$$A(1) = \left(1 + \frac{r}{n}\right)\left(\frac{S}{n}\right);$$

$$A(k) = \left(1 + \frac{r}{n}\right)\left[A(k-1) + \frac{S}{n}\right], \quad k \geq 2. \tag{16}$$

(b) If $S = 1{,}000$, $r = 0.06$, and $n = 4$, compute the amount at the end of
(i) 1 year (ii) 2 years

24. Suppose $S = 1{,}200$ and $r = 0.05$. Use formula (16) with a spreadsheet to compute the amount in the account at the end of 1 year if
(a) $n = 12$ **(b)** $n = 365$
(The first 2 months for the case $n = 12$ are shown in Figure 6.2.1.) Compare your answers to the amount at the end of 1 year if the interest is compounded continuously.

	A	B	C	D
1	Month	Begin	End	
2	1	100	100.4167	
3	2	200.4167	201.2517	
4				
5				
6				

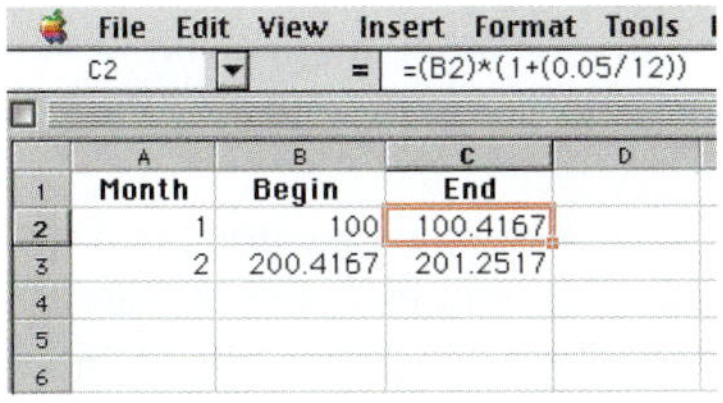

Figure 6.2.1

25. Suppose that a total of \$6,000 per year is deposited in four equal payments, one at the beginning of each quarter, into an account paying 4% annual interest, compounded quarterly. Use formula (16) to find the balance at the end of 10 years.

In Exercises 26–28, use formula (16) to find the balance at the end of Y years.

26. $S = 10{,}000,\ r = 0.025,\ n = 12,\ Y = 2.$

27. $S = 5{,}000,\ r = 0.075,\ n = 52,\ Y = 0.5$

28. $S = 20{,}000,\ r = 0.05,\ n = 12,\ Y = 5$

29. **(a)** Using formula (16), show that

$$A(2) = \frac{S}{n}\left[\left(1 + \frac{r}{n}\right)^2 + \left(1 + \frac{r}{n}\right)\right],$$

$$A(3) = \frac{S}{n}\left[\left(1 + \frac{r}{n}\right)^3 + \left(1 + \frac{r}{n}\right)^2 + \left(1 + \frac{r}{n}\right)\right],$$

and, in general,

$$A(k) = \frac{S}{n}\left[\left(1 + \frac{r}{n}\right)^k + \left(1 + \frac{r}{n}\right)^{k-1} + \cdots + \left(1 + \frac{r}{n}\right)\right].$$

(b) Use the geometric series formula: for $t \neq 1$,

$$1 + t + t^2 + \cdots + t^k = \frac{t^{k+1} - 1}{t - 1}$$

to show that

$$A(k) = \frac{S}{r}\left[\left(1 + \frac{r}{n}\right)^{k+1} - \left(1 + \frac{r}{n}\right)\right].$$

(If you are unfamiliar with the geometric series formula, see the next exercise.)
(c) Find the amount at the end of 10 years if $S = 1{,}200$, $r = 0.05$, and $n = 365$. Compare your answer to the amount at the end of 10 years if the interest is deposited and compounded continuously.

30. The purpose of this exercise is to derive the geometric series formula

$$1 + t + t^2 + \cdots + t^k = \frac{t^{k+1} - 1}{t - 1}. \tag{17}$$

Let

$$M = 1 + t + t^2 + t^3 + \cdots + t^k.$$

Multiply both sides of this equation by t to get

$$tM = t + t^2 + t^3 + \cdots + t^k + t^{k+1}.$$

Now subtract the first equation from the second and solve for M to get the formula.

Solutions to practice exercises 6.2

1. Using formula (12) with $T = 10$, $r = 0.06$, and $S(t) = 1{,}200$ gives

$$FV = \int_0^{10} 1{,}200e^{0.06(10-t)}\,dt = -\frac{1{,}200}{0.06}e^{0.06(10-t)}\Big|_0^{10}$$

$$= 20{,}000(e^{0.6} - 1) \approx 16{,}422.$$

2. A rate of \$500 per month is equal to \$6,000 per year, and using formula (13) with $T = 20$, $r = 0.04$, and $S(t) = 6{,}000$ gives

$$PV = \int_0^{20} 6{,}000e^{-0.04t}\,dt = -150{,}000e^{-0.04t}\Big|_0^{20}$$

$$= 150{,}000(1 - e^{-0.8}) \approx 82{,}600.$$

6.3 Separable Differential Equations

Continuous income streams can also be represented by initial value problems, which leads to an important type of differential equation with many applications. As we have already seen in Chapter 5, a differential equation is an equation involving an unknown function and its derivative, and an initial value problem consists of a differential equation and a pair of numbers that give the value of the function at one point. Differential equations play a major role in modeling phenomena in the physical, life, and social sciences, and solving them was a motivating factor in the development of calculus. In this section we will study the important class of **separable differential equations**.

Differential equation of a continuous income stream

Let's start with an example. Suppose you open a retirement account at age 30 with an initial amount of $1,000 and then make continuous deposits at the rate of $7,000 per year. If the account pays an interest rate of 7%, compounded continuously, what is the balance in the account at any given time? In particular, what is the balance at the retirement age of 65?

First, we will translate this problem into a differential equation (a procedure known as **modeling**). We start by letting $M(t)$ denote the amount of money in the account at any time t, where t is the number of years after age 30. This is the unknown function. What do we know about it? First, we know that the money at the beginning was $1,000. In mathematical terms that can be expressed by the equation

$$M(0) = 1,000. \tag{18}$$

This equation is called an **initial condition** for $M(t)$.

Next, we observe that $M(t)$ grows in two ways: first, because new money is continually being deposited and, second, because interest is continually being added. The first occurs at a rate of $7,000 per year, the second at a rate of 7% of the current balance. To translate this information into the form of a differential equation, we let Δt denote a very small amount of time, and we compare the balance at time t with the balance at time $t + \Delta t$. During that time interval the interest paid is approximately $0.07M(t)\Delta t$ (which represents money growing at a rate of 7% of the current balance $M(t)$ for a period of length Δt). The new money added is approximately $7,000\Delta t$ (which represents money being added at a constant rate of 7,000 for a period of length Δt). Therefore, the approximate change in the balance is given by the formula

$$M(t + \Delta t) - M(t) \approx 0.07M(t)\Delta t + 7,000\Delta t,$$

and we can make the approximation as good as we want by taking Δt small enough. Dividing both sides by Δt, we get

$$\frac{M(t + \Delta t) - M(t)}{\Delta t} \approx 0.07M(t) + 7,000,$$

and by letting $\Delta t \to 0$, we obtain the differential equation

$$\frac{dM}{dt} = 0.07M + 7,000. \tag{19}$$

Our money problem has now been *modeled* by Eqs. (18) and (19), which together are called an **initial value problem**. We can solve this problem by a method called **separation of variables**.

Separation of variables

A differential equation is called **separable** if it can be written in the form

$$f(y) \frac{dy}{dt} = g(t), \tag{20}$$

where f and g are known functions. Once it is written in that form, we say the variables are *separated*. To solve such an equation, we integrate both sides:

$$\int f(y) \frac{dy}{dt} \, dt = \int g(t) \, dt. \tag{21}$$

Using the substitution formula for indefinite integrals [formula (16) of Section 5.2], we can rewrite this in the form

$$\int f(y) \, dy = \int g(t) \, dt, \tag{22}$$

which explains the name "separation of variables," since the left-hand side involves only y and the right-hand side involves only t. By computing these integrals, we obtain an equation in y and t involving an arbitrary constant. If we can put that equation in the form $y = y(t)$, expressing y explicitly as a function of t, we say that we have an **explicit solution** of the differential equation; if not, the differential equation is said to have an **implicit solution**.

Separation of variables is the most fundamental method for solving differential equations, and many of the other techniques amount to reducing the equation to a separable form.

EXAMPLE 6.3.1

Solve the differential equation $dy/dt = 3t^2 y^2$. Then find the solution satisfying the initial condition $y(0) = 1$.

SOLUTION The first thing to observe is that the constant function $y = 0$ satisfies both sides of the equation. That is, $dy/dt = 0$ (because y is constant) and $3t^2 y^2 = 0$ (because $y = 0$). To find the other solutions, we separate variables. We first divide both sides of the equation by y^2 to get

$$\frac{1}{y^2} \frac{dy}{dt} = 3t^2,$$

which we can write in the form

$$\frac{1}{y^2}\, dy = 3t^2\, dt.$$

Then, we integrate both sides:

$$\int \frac{1}{y^2}\, dy = \int 3t^2\, dt,$$

which gives

$$-\frac{1}{y} = t^3 + c,$$

where c is an arbitrary constant. Solving for y, we get

$$y = -\frac{1}{t^3 + c}.$$

If we let $C = -c$, we can also write this solution as

$$y = \frac{1}{C - t^3}.$$

Finally, to evaluate C, we substitute the initial condition, $t = 0$ and $y = 1$, which gives $1 = 1/C$. Therefore, $C = 1$.

EXAMPLE 6.3.2

PROBLEM-SOLVING TACTIC

To apply separation of variables:

- Use algebra to put the equation into the form $f(y)\, dy = g(t)\, dt$.

- Integrate both sides to obtain an equation of the form $F(y) = G(t) + c$.

- If possible, express y as a function of t.

- Use the initial condition to evaluate c.

Solve the differential equation $dy/dt = y - y^2$, draw the graphs of several solutions for $t \geq 0$, and find their limiting value as $t \to \infty$.

SOLUTION   In this case, the functions $y = 0$ and $y = 1$ are both constant solutions of the differential equation. That is, $dy/dt = 0$ (because y is constant) and $y - y^2 = 0$. To find the other solutions, we separate variables, so that the differential equation takes the form $((1/y) - y^2)(dy/dt) = 1$, which we also write in the form

$$\frac{dy}{y - y^2} = dt.$$

Next, we integrate both sides. The integral of the right-hand side is equal to $t + c_1$, where c_1 is an arbitrary constant. To compute the integral of the left-hand side, we use the method of partial fractions to get

$$\frac{1}{y - y^2} = \frac{1}{y(1 - y)} = \frac{1}{y} + \frac{1}{1 - y}.$$

Therefore,

$$\int \frac{dy}{y - y^2} = \int \frac{dy}{y} + \int \frac{dy}{1 - y}$$

$$= \ln |y| - \ln |1 - y| + c_2 = \ln \left| \frac{y}{1 - y} \right| + c_2,$$

where c_2 is an arbitrary constant. Putting it all together, we obtain

$$\ln \left| \frac{y}{1 - y} \right| = t + c,$$

where we have written c in place of $c_1 - c_2$. By exponentiating, we obtain

$$\frac{y}{1 - y} = \pm e^c e^t.$$

If we now replace $\pm e^c$ by another constant, say, C, and solve for y, we obtain the formula

$$y(t) = \frac{Ce^t}{1 + Ce^t}.$$

Another form of this solution is obtained by multiplying numerator and denominator by e^{-t} to get

$$y(t) = \frac{C}{C + e^{-t}}. \tag{23}$$

The graphs of $y(t)$, $t \geq 0$ for $C = 0.1,\ 0.5,\ 1,\ 4,\ -5,\ -3.5,\ -2.5$, and -2 are shown in Figure 6.3.1, together with the constant solution $y = 1$. Notice that all of these

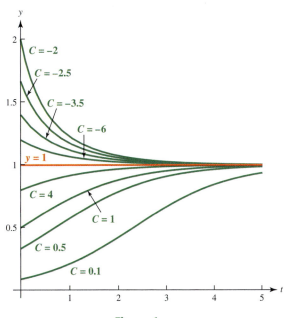

Figure 6.3.1

solutions seem to move asymptotically toward the constant solution $y = 1$ as $t \to \infty$. That is indeed the case for any solution other than the constant solution $y = 0$; referring to (23), we have $\lim_{t \to \infty} e^{-t} = 0$, which means that $\lim_{t \to \infty} y(t) = 1$.

Finally, we observe that for any $C \neq -1$, the solution satisfies the initial condition $y(0) = C/(C + 1)$. If $C = -1$, the y-axis is a vertical asymptote of the solution, and $y(0)$ is undefined. In general, if $C < 0$, the line $t = -\ln|C|$ is a vertical asymptote, situated to the left of the y-axis if $|C| > 1$ and to the right if $0 < |C| < 1$. Figures 6.3.3 and 6.3.4, further below, show the graphs corresponding to $C = -\frac{3}{2}$ and $C = -\frac{2}{3}$, respectively.

The solution given by formula (23) is called the **general solution** of the differential equation. It contains an arbitrary constant C, and any solution can be obtained from it by assigning an appropriate value to C. (Strictly speaking, the constant solution $y = 1$ is an exception, but we can think of it as the solution obtained by setting $C = \infty$.)

EXAMPLE 6.3.3

Solve the initial value problem $dy/dt = y - y^2$ for

$$\text{(i)}\ \ y(0) = \frac{1}{2}, \quad \text{(ii)}\ \ y(0) = 3, \quad \text{(iii)}\ \ y(0) = -2$$

Determine the domain of each solution and find any asymptotes.

SOLUTION In the last example we found the general solution of this differential equation to be

$$y(t) = \frac{C}{C + e^{-t}},$$

with $y(0) = C/(C + 1)$.

(i) Setting $t = 0$ and $y = \frac{1}{2}$, we get $\frac{1}{2} = C/(C + 1)$, and clearing denominators gives $2C = C + 1$. Therefore, $C = 1$, and the solution of our initial value problem is

$$y = \frac{1}{1 + e^{-t}}.$$

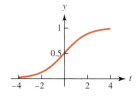

Figure 6.3.2

The domain of this function consists of the entire t-axis, and there is no vertical asymptote. The line $y = 1$ is a horizontal asymptote as $t \to \infty$, and the x-axis is a horizontal asymptote as $t \to -\infty$. The graph is shown in Figure 6.3.2.

(ii) Setting $t = 0$ and $y = 3$, we get $3 = C/(C + 1)$, and solving for C gives $C = -\frac{3}{2}$. Therefore,

$$y(t) = \frac{-\frac{3}{2}}{-\frac{3}{2} + e^{-t}} = \frac{-3}{-3 + 2e^{-t}} = \frac{3}{3 - 2e^{-t}}.$$

In this case, the denominator is zero if $e^{-t} = \frac{3}{2}$, which is equivalent to $t = \ln(\frac{2}{3}) \approx -0.405$. Therefore, the domain consists of all $t \neq \ln(\frac{2}{3})$, and the line $t = \ln(\frac{2}{3})$ is a

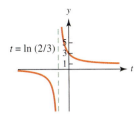

Figure 6.3.3

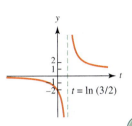

Figure 6.3.4

vertical asymptote. As before, the line $y = 1$ is a horizontal asymptote as $t \to \infty$, and the t-axis is a horizontal asymptote as $t \to -\infty$. The graph is shown in Figure 6.3.3.

(iii) Setting $t = 0$ and $y = -2$, we get $-2 = C/(C + 1)$, and solving for C gives $C = -\frac{2}{3}$. Therefore,

$$y(t) = \frac{-\frac{2}{3}}{-\frac{2}{3} + e^{-t}} = \frac{-2}{-2 + 3e^{-t}} = \frac{2}{2 - 3e^{-t}}.$$

The denominator is zero if $e^{-t} = \frac{2}{3}$, that is, $t = \ln(\frac{3}{2}) \approx 0.405$. Therefore, the domain consists of all $t \neq \ln(\frac{3}{2})$, and the line $t = \ln(\frac{3}{2})$ is a vertical asymptote. Once again, the line $y = 1$ is a horizontal asymptote as $t \to \infty$, and the t-axis is a horizontal asymptote as $t \to -\infty$. The graph is shown in Figure 6.3.4.

EXAMPLE 6.3.4

Solve the differential equation $y' = \dfrac{t^2}{y}$.

SOLUTION By clearing the denominator on the right-hand side of this equation, we separate the variables, as follows:

$$y \frac{dy}{dt} = t^2 \quad \text{or} \quad y\, dy = t^2\, dt.$$

Integrating both sides gives

$$\int y\, dy = \int t^2\, dt,$$

and the integrals are easily computed:

$$\frac{1}{2}y^2 = \frac{1}{3}t^3 + c_0,$$

where c_0 is an arbitrary constant. Multiplying both sides by 2, we get the implicit solution

$$y^2 = \frac{2}{3}t^3 + c,$$

where we have written c in place of $2c_0$.

Sometimes an implicit solution is the best we can do because the equation linking y and t is too hard to solve for y in terms of t. In the last example, we could solve for y by taking square roots, but there are two points to bear in mind. First, the square root only makes sense when the expression on the right is positive, and, second, a square root is only determined up to $\pm$, so there are actually two explicit solutions. To determine which one to use, we need some additional information, such as an initial condition. The following example illustrates that situation.

EXAMPLE 6.3.5

Solve the following initial value problems:

$$\text{(i)} \quad y' = \frac{t^2}{y}, \quad y(3) = 4 \quad \text{and} \quad \text{(ii)} \quad y' = \frac{t^2}{y}, \quad y(3) = -4$$

SOLUTION Both of these problems involve the differential equation of the last example, and we have already computed the general solution

$$y^2 = \frac{2}{3}t^3 + c.$$

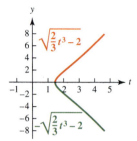

Figure 6.3.5

Substituting $y = \pm 4$ and $t = 3$ gives $16 = 18 + c$, so that $c = -2$. Therefore, we have two possible explicit solutions—namely,

$$y = \sqrt{\frac{2}{3}t^3 - 2} \quad \text{and} \quad y = -\sqrt{\frac{2}{3}t^3 - 2}.$$

The first of these is the solution to (i), the second to (ii). Figure 6.3.5 shows the graphs of both solutions on the same set of axes.

Now that we know how to solve separable differential equations, we are in a position to solve the income stream problem modeled by Eqs. (18) and (19).

EXAMPLE 6.3.6

Solve the initial value problem

$$\frac{dM}{dt} = 0.07M + 7,000, \quad M(0) = 1,000.$$

SOLUTION We first separate variables in the differential equation, which gives

$$\frac{dM}{0.07M + 7,000} = dt.$$

Next, we integrate both sides:

$$\int \frac{dM}{0.07M + 7,000} = \int dt. \tag{24}$$

The integral on the right is equal to $t + c$. To compute the one on the left, we make the substitution $u = 0.07M + 7,000$, so that $du = 0.07dM$, and we get

$$\frac{1}{0.07} \ln |0.07M + 7,000| = t + c,$$

where c is an arbitrary constant. By multiplying both sides by 0.07 and taking exponentials, we obtain

$$0.07M + 7,000 = \pm e^{0.07c} e^{0.07t}. \tag{25}$$

Solving for M and denoting the constant $\pm e^{0.07c}/0.07$ by the single letter C gives

$$M(t) = -100,000 + Ce^{0.07t}. \tag{26}$$

To find the particular solution to our specific income problem, we use the initial condition (18) to determine the constant C. Setting $t = 0$ on both sides of Equation (26) and using the fact that $M(0) = 1,000$, we get

$$1,000 = M(0) = -100,000 + C,$$

so that $C = 101,000$. Combining all these steps, we get

$$M(t) = -100,000 + 101,000e^{0.07t}, \tag{27}$$

which is the amount of money in the retirement account t years after the initial deposit at age 30. To find the balance in the account at age 65, we simply substitute $t = 35$ into (27):

$$M(35) = -100,000 + 101,000e^{2.45} \approx 1,070,423.$$

> **PROBLEM-SOLVING TACTIC**
>
> A point of confusion in solving differential equations is that *a constant may be written in many different forms.* However, the best form is the simplest. For example, in solving Eq. (25) for M, we get the constant $\pm e^{0.07c}/0.07$, which can be simply replaced by C. No matter how you decide to denote the constant, the initial condition [here $M(0) = 1,000$] will always give the same solution.

To end this section, here is an interesting problem concerning mortgages.

EXAMPLE 6.3.7

Suppose that a home buyer plans to take a 15-year mortgage at an interest rate of 6% but cannot spend more than $1,000 per month on payments. Here are two questions of obvious interest to the buyer.

(i) What is the maximum amount that she can afford to borrow?

(ii) If the buyer borrows this maximum amount, how much total interest will she pay?

For simplicity, we will assume that the interest is compounded continuously and that the payments are made continuously at the rate of $12,000 per year. In practice, these assumptions may not be precisely satisfied, but they greatly simplify the problem and give close approximations to the actual solution. The case in which interest is compounded using discrete time periods is taken up in the exercises at the end of this section.

SOLUTION   Let $M(t)$ denote the balance of the mortgage after t years, where t is any number in the interval $[0, 15]$. This balance is changing over time in two ways. First, it is increasing at a rate of $0.06M(t)$ (in other words, 6% of the current balance) because of the interest. At the same time, it is decreasing at a constant rate of 12,000 (the rate at which it is being paid off). The rate at which $M(t)$ is changing is the difference of these two rates (both in dollars per year). That is,

$$\frac{dM}{dt} = 0.06M - 12{,}000 = 0.06(M - 200{,}000).$$

(For simplicity, we have written M instead of $M(t)$ on the right.) The condition that the mortgage must be paid off in 15 years can be expressed by

$$M(15) = 0,$$

and we want to use that to find $M(0)$.

Separating the variables in the differential equation, we obtain

$$\frac{dM}{M - 200{,}000} = 0.06\, dt,$$

so that

$$\int \frac{dM}{M - 200{,}000} = \int 0.06\, dt,$$

Both sides are easy to integrate:

$$\ln |M - 200{,}000| = 0.06t + c.$$

Exponentiating gives

$$M - 200{,}000 = \pm e^c e^{0.06t} = C e^{0.06t},$$

where $C = \pm e^c$. Therefore, $M(t) = 200{,}000 + C e^{0.06t}$. To evaluate the constant C, we use the condition $M(15) = 0$. From that we get

$$0 = 200{,}000 + C e^{0.06 \cdot 15}$$

or $C = -200{,}000 e^{-0.9}$. Thus,

$$M(t) = 200{,}000(1 - e^{0.06t - 0.9}).$$

The maximum amount that she can afford to borrow is

$$M(0) = 200{,}000(1 - e^{-0.9}) \approx 118{,}686.$$

To answer question (ii), we simply compute the total amount to be paid off (15 years times \$12,000 per year) and subtract the amount borrowed, which gives $15 \cdot 12{,}000 - 118{,}686 = 61{,}314$. This is the amount the buyer will pay above the size of the loan, and it therefore equals the total interest.

Practice Exercises 6.3

1. Solve the initial value problem $dy/dt = y^2 - 2ty^2$, $y(2) = 1$.

2. At the time of his retirement, a man has accumulated $350,000 in a pension account that pays 6% annual interest, compounded continuously. He arranges to draw $24,000

year, to be calculated as a continuous stream, out of the account. Let $M(t)$ be the balance in the account t years after he retires. Find $M(t)$ by posing and solving a suitable initial value problem, and determine the amount in the account after 20 years.

Exercises 6.3

In Exercises 1–4, verify that the given function, with C as an arbitrary constant, is a solution of the differential equation.

1. $t \dfrac{dy}{dt} = (1 + t^2)y$, $y = Cte^{t^2/2}$

2. $\sqrt{x+2} \dfrac{dy}{dx} = y^2$, $y = \dfrac{1}{C - 2\sqrt{x+2}}$

3. $\dfrac{dz}{dx} = 3x^2 e^{-z}$, $z = \ln(C + x^3)$

4. $\dfrac{dx}{dt} = x \ln x$, $x = e^{Ce^t}$

In Exercises 5–13, solve the given differential equation.

5. $\dfrac{dy}{dt} = -y$

6. $\dfrac{dy}{dx} = xy$

7. $\dfrac{dM}{dt} = 0.1M + 10$

8. $\dfrac{dQ}{dt} = 2 - 0.5Q$

9. $\dfrac{dy}{dt} = ye^t - e^t$

10. $\dfrac{dy}{dx} = xy + y$

11. $\dfrac{dy}{dt} = (y-1)(y-2)$

12. $x \dfrac{dy}{dx} = y$

13. $\dfrac{dy}{dt} = r(y - a)$; $r, a \in \mathbb{R}$

In Exercises 14–21, solve the given initial value problem.

14. $\dfrac{dQ}{dt} = 0.02Q$, $Q(0) = 50$

15. $\dfrac{dy}{dt} = y^2$, $y(0) = 4$

16. $\dfrac{dy}{dx} = \dfrac{x-1}{y}$, $y(3) = 8$

17. $\dfrac{dy}{dt} = 9 - y$, $y(0) = 12$

18. $e^x \dfrac{dy}{dx} - xy^2 = 0$, $y(0) = \frac{1}{3}$

19. $\dfrac{dM}{dt} = 0.1M + 100$, $M(0) = 50$

20. $\dfrac{dy}{dt} = y^2 + y$, $y(0) = -2$

21. $\dfrac{dy}{dx} = xy^2 - x$, $y(0) = 2$

In Exercises 22–25, solve the given family of initial value problems and graph them on the same screen, using the specified window.

22. $y' = y + 1$, $y(0) = -4, -3, -2, -1, 0, 1, 2$; use $[-1, 2] \times [-10, 10]$.

23. $\dfrac{dy}{dx} = \dfrac{y}{x}$, $y(1) = -3, -2, -1, 0, 1, 2, 3$; use $[-5, 5] \times [-5, 5]$.

24. $\dfrac{dy}{dx} = \dfrac{x}{y}$, $y(0) = 1, 1.5, 2, 2.5, 3, 3.5, 4$; use $[0, 4] \times [0, 6]$.

25. $\dfrac{dy}{dt} = e^{-y}$, $y(0) = 0, 0.5, 1, 1.5, 2, 2.5, 3$; use $[0, 10] \times [0, 4]$.

26. Find an implicit solution of the equation $dy/dt = -ty^3$. Then write explicit solutions of the initial value problems consisting of this differential equation with each of the following initial conditions:
  (a) $y(0) = 1$ **(b)** $y(0) = -1$

27. Find an implicit solution of the equation $e^{x^2} y\, y' = -2x$. Then write explicit solutions of the initial value problems consisting of this differential equation with each of the following initial conditions:
  (a) $y(0) = 1$ **(b)** $y(0) = -2$

28. A home buyer plans to take a 30-year mortgage at an 8% annual interest rate, compounded continuously. Suppose she can only spend $1,200 a month on payments. What is the maximum amount she can borrow? Assume that the payments are made continuously at a fixed rate.

29. A retired person has $1,800,000 invested in an account paying an annual interest rate of 9%, compounded continuously. Over the next 30 years, he plans to withdraw money continuously at a fixed annual rate. If he wants to have $500,000 left in the account at the end of that time, at what rate should the withdrawals be made?

30. The net worth of a company is the difference between its revenue and expenses. Suppose that at any given time a

certain company's revenue is increasing at a rate equal to 6% of its net worth at that time. On the other hand, it has payroll obligations and other expenses amounting to $600 million per year, which it pays continuously. The company starts with an initial net worth of $10,500 million.
(a) Write an initial value problem describing the net worth of the company at any time.
(b) Solve the initial value problem to find a formula for its net worth at any time.
(c) Find its net worth 10 years later.
(d) Find the time required for the net worth to double its initial value.

31. It was reported in the news that the value of Michael Jordan's contract to play for the Bulls during the 1997–98 season was $33,000,000. Assume that in 1998 he invested this money so as to draw interest at an annual rate of 10%, compounded continuously, and that he arranged for continuous withdrawals at a rate of k dollars per year. Determine the withdrawal rate k at which the money $M(t)$ in his account will remain constant.

32. DNA has the shape of a double helix, whose two strands separate and reform at a rapid rate. The process of reforming, which is called **renaturation**, depends on the random collision of the complementary strands. Its rate is modeled by the differential equation

$$\frac{dC}{dt} = -kC^2,$$

where C is the concentration of DNA that remains single-stranded at time t, and k is a constant, called the **reassociation rate constant**.

Source: Benjamin Lewin, *Genes VII*, Oxford Univ. Press, 2000.

(a) If C_0 is the concentration of single-stranded DNA at $t = 0$, find a formula for C in terms of t, k, and C_0.
(b) Show that the reaction is half-complete at time $t = (kC_0)^{-1}$.

33. A drug is being intravenously fed to a patient at a constant rate of 100 milligrams per hour. At the same time, the drug is being eliminated from the patient's bloodstream at a rate that equals 12% of the amount present at any given time.
(a) Write a differential equation for $y(t)$, the amount in the patient's bloodstream at time t, with t in hours and y in milligrams.
(b) If there is an initial amount of 50 milligrams in the patient's bloodstream, how much will there be at the end of 10 hours?

34. Exercising improves both physical and mental fitness. However, many people do not exercise due to inertia. Let $F(t)$ be a fitness index, where t is the exercising time in hours per week, with $F = 100$ indicating perfect fitness and $F = 50$ indicating no exercising at all. Assume that available data show that the rate at which $F(t)$ changes is

proportional to its difference from 100, with a constant of proportionality equal to 0.12.
(a) Write a differential equation and an initial condition modeling the fitness index $F(t)$ for all t.
(b) Solve the initial value problem to find $F(t)$.
(c) Find the exercising time needed for a fitness index of 80 or more.

35. You are studying a new language. Already you know 50 words and you estimate that you can can learn new words at the rate of 200 per year. Unfortunately, you forget words at a rate of 1% a year.
(a) Write the differential equation that models $V(t)$, the size of your vocabulary at time t. Identify the initial value.
(b) Using separation of variables, find a solution.
(c) What is the size of your vocabulary in three years?

36. A pharmaceutical company makes a medication whose active ingredient starts at 10 milligrams but loses potency at a rate of

$$\frac{dP}{dy} = -0.35y \text{ milligrams per year.}$$

The company wants the expiration date on the package to represent the time when potency is less than 50%. Is 2 years a good date?

37. Environmental studies estimate that 40,000 pounds of chemical pollution are already in a lake. It is possible to clean up 2% of it annually, but an estimated 1,000 pounds are illegally dumped into the lake each year.
(a) Write and solve an initial value problem modeling the amount of chemical pollution in the lake.
(b) Compare the pollution levels at $t = 0, 1, 2, 3$. Is the clean-up effort successful?

38. A colony of wasps is building an underground nest. They start by finding a hole and begin constructing cells at a rate of

$$\frac{dy}{dt} = 0.8e^{0.04y} \quad \text{(in tens of cells per day)}.$$

How big is the colony after 10 days?

39. A donor would like to make a gift of M dollars to create a fund able to give out 20 annual scholarships of $25,000 each at his favorite university. Suppose the fund can earn interest at the rate of 9%, compounded continuously, and the scholarships are withdrawn continuously at a constant annual rate.
(a) Write a differential equation and an initial condition describing the fund's balance.
(b) Find the minimum amount M_0 needed to keep the fund paying out scholarships forever.

40. Referring to Example 6.3.7, suppose that the interest is compounded monthly rather than continuously. Use the following sequence of steps to determine the amount M_0 the buyer can afford to borrow.
(a) Letting M_k denote the balance owed at the end of the kth month, show that

$$M_{k+1} = (1.005)M_k - 1,000. \tag{28}$$

(Assume that a payment is made at the end of each month.)
(b) By using (28) repeatedly, show that the amount owed at the end of n months is given by

$$M_n = (1.005)^n M_0 - 1,000\{1 + (1.005)$$
$$+ (1.005)^2 + \cdots + (1.005)^{n-1}\}.$$

Then use the geometric series formula [see (17) in the last exercise of Section 6.2] to get

$$M_n = (1.005)^n M_0 - 1000\frac{(1.005)^n - 1}{0.005}.$$

(c) At the end of 180 months (15 years) the balance is zero. Setting $M_{180} = 0$, solve for M_0. Compare that amount to the one found in Example 6.3.7.

41. A person borrows $10,000 at an annual interest rate of 12% with the plan of repaying it in monthly payments of A dollars over a 5-year period.
(a) Suppose the interest is compounded continuously. Write an initial value problem similar to equations (18)

and (19) for the balance $M(t)$ at the end of t years. Then solve the initial value problem for $M(t)$ in terms of A and use the fact that the loan is paid off in 5 years to find A.
(b) Suppose the interest is compounded monthly. Write a formula similar to Eq. (28) for the balance M_k at the end of k months. Use that formula and the geometric series formula to find M_k in terms of A; then use the fact that the loan is paid off in 5 years to find A.

Modeling by Differential Equations

42. In the year 2000 about 1 million homes worldwide were getting their electricity from solar cell installations. Suppose that at any time t (in years), their number $h(t)$ is increasing at a rate of $0.2h(t)$ homes per year. Write a differential equation and an initial condition modeling $h(t)$.

Source: Lester R. Brown: *Eco-Economy*, Norton 2001

43. The wind energy generating capacity of the U.S. in 2000 was about 2,500 megawatts. Suppose that it keeps increasing at a rate (in megawatts per year) equal to 0.12 times the capacity present at any time. Write a differential equation and an initial condition modeling this situation.

Source: Lester R. Brown: *Eco-Economy*, Norton 2001

44. In 1996 coal consumption worldwide was about 2,250 million tons of oil equivalent. Since then it has declined at the annual rate of 1% per year. Write an initial value problem modeling coal's consumption.

Source: Lester R. Brown: *Eco-Economy*, Norton 2001

45. The population of a region increases at an annual rate of 2.5%, and, in addition, new residents arrive at the rate of 50,000 per year. Write an initial value problem modeling its population if there were 30 million people initially living in this region.

46. Initially, there were about 50,000 pounds of chemical pollutant in a lake. A clean-up installation removes 5 percent of it annually, but an estimated 500 pounds are illegally dumped into the lake each year. Write an initial value problem modeling the amount of chemical pollutant in the lake.

Solutions to practice exercises 6.3

1. Separating variables gives $(1/y^2)\,dy = (1 - 2t)\,dt$, and integrating both sides gives $-(1/y) = t - t^2 + C$. Therefore, $y = 1/(t^2 - t - C)$. Setting $t = 2$ and $y = 1$ and solving for C, we get $C = 1$, and the solution is

$$y = \frac{1}{t^2 - t - 1}.$$

2. The balance M is increasing at a rate of $0.06M$ and decreasing at a rate of 24,000 per year. The initial amount is 350,000. Therefore,

$$\frac{dM}{dt} = 0.06M - 24,000, \quad M(0) = 350,000.$$

We can solve this problem by separating variables to get

$$\frac{dM}{M - 400{,}000} = 0.06\,dt.$$

Integrating both sides, exponentiating, and solving for M, we obtain

$$M(t) = 400{,}000 + Ce^{0.06t}.$$

To find the constant C, we set $t = 0$ and $M(0) = 350{,}000$, and solve to get $C = -50{,}000$. Therefore,

$$M(t) = 400{,}000 - 50{,}000e^{0.06t}.$$

At the end of 20 years, the balance will be $M(20) = 400{,}000 - 50{,}000e^{1.2} \approx 233{,}994$.

6.4 The Logistic Growth Model

The simplest assumption about population growth is that the population grows at a rate proportional to its size at any given time—the larger the population, the faster it grows, and the ratio of growth rate to size stays fixed. As we saw in Section 3.5 of Chapter 3, this condition can be expressed mathematically by the equation

$$\frac{dp}{dt} = rp, \tag{29}$$

where p is the population at any given time t, and r is a constant. If the population is p_0 at time zero, then we also have the initial condition

$$p(0) = p_0. \tag{30}$$

We can compute the solution of this initial value problem by using separation of variables, which gives

$$p(t) = p_0 e^{rt}. \tag{31}$$

This model—sometimes called the **exponential growth model**—is good for certain very simple species over short periods of time. But it is based on an assumption of unrestricted growth, and it fails to take into account the limited supply of resources—such as food, water, and space—that will eventually force the growth rate to slow down or decline. For that reason, the exponential growth model is not good for long-term projections of the human population.

For example, it has been estimated that the population of the earth in 1990 was 5.2 billion and was growing at a rate of 1.6% per year. If Eq. (31) was a correct model, the population of the earth (in billions) t years after 1990 would be

$$p(t) = 5.2e^{0.016t}.$$

Now, the area of the earth is approximately 5,502,360 billion square feet, including seas and mountains. How long would it be, under the exponential growth assumption, before each person had only 1 square foot or less to stand (or float) on? We can answer that by solving the equation

$$5{,}502{,}360 = 5.2e^{0.016t}$$

for t. Dividing both sides by 5.2 and taking logarithms give

$$0.016t = \ln\left(\frac{5{,}502{,}360}{5.2}\right),$$

so that

$$t = \frac{1}{0.016}\ln\left(\frac{5{,}502{,}360}{5.2}\right) \approx 867.$$

Therefore, by the year 2857 (i.e., 867 years after 1990) people would be living so close to each other that human nature and earth's resources would not allow it. However, to our great relief, this model is unrealistic for such long-term predictions because it ignores so many important factors influencing population growth.

The logistic growth equation

In an attempt to take account of the competition resulting from limited space and resources, the Belgian mathematician P. F. Verhulst introduced a new population growth model in 1832. His work was largely overlooked until the early twentieth century, when it was rediscovered by the American biologist Raymond Pearl. This model is described by the following differential equation:

$$\frac{dp}{dt} = rp - kp^2, \tag{32}$$

which is sometimes called the Verhulst-Pearl equation, but is more commonly known as the **logistic equation**.

As you can see, the equation involves two constants, r and k. The first of these is called the **intrinsic growth rate**, and it is the relative rate at which the population would grow if there were no restrictions. In other words, it plays the same role as the constant r in the exponential model. The constant k, called the **damping factor**, reflects the damping effect on growth caused by competition for resources between members of the species. Verhulst's basic assumption was that the amount of competition is proportional to the number of encounters between organisms, which in turn is proportional to the square of the population. That accounts for kp^2, which appears in Eq. (32) with a minus sign because it decreases the growth rate.

If we also assume that the initial population is known, we get the initial condition

$$p(0) = p_0, \tag{33}$$

where p_0 is a given number.

The initial value problem posed by Eqs. (32) and (33) can be solved explicitly. Before we do that, however, let us find all constant solutions. These are functions $p(t)$ with $dp/dt = 0$ for all t. Therefore, equation (32) becomes $0 = rp - kp^2$, or $p(r - kp) = 0$, whose solutions are the constant functions

$$p = 0 \quad \text{and} \quad p = \frac{r}{k}.$$

If we let $K = r/k$, we can make the substitution $k = r/K$ on the right side of Eq. (32), which gives

$$\frac{dp}{dt} = rp - \frac{r}{K}p^2,$$

and factoring rp from both terms changes the equation to the form

$$\frac{dp}{dt} = rp\left(1 - \frac{p}{K}\right). \tag{34}$$

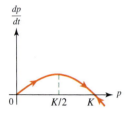

Figure 6.4.1

We can solve this equation by separation of variables. Before doing that, we mention that the two constant solutions, namely, $p = 0$ and $p = K$, are called **equilibrium solutions**, meaning that if the population starts at either of those levels, it never changes. That is obvious and not very interesting in the case of $p = 0$, for if there is no population to begin with, there is no growth, no change, and nothing to study. On the other hand, the constant solution $p = K$ has a very interesting interpretation. To explain it, we first graph dp/dt as a function of p, as given by Eq. (34), assuming that r and K are fixed. The graph is shown in Figure 6.4.1.

It shows that $dp/dt > 0$ if $0 < p < K$, which means that the population is increasing whenever it is less than K. On the other hand, $dp/dt < 0$ if $p > K$, which means the population is decreasing whenever it is greater than K. Putting together these two observations leads us to the following general principle: *No matter what its initial size, the population will move toward K as t becomes large.* For that reason, the solution $p = K$, in addition to being called the equilibrium solution, is also called the **environmental carrying capacity** or **saturation level**. This principle will become even more evident after we find an explicit solution of the initial value problem posed by Eqs. (34) and (33).

To solve Eq. (34), we start by separating variables:

$$\frac{1}{p\left(1 - \dfrac{p}{K}\right)} \, dp = r \, dt.$$

Then we integrate to obtain

$$\int \frac{1}{p\left(1 - \dfrac{p}{K}\right)} \, dp = \int r \, dt = rt + c.$$

By the method of partial fractions, we have

$$\frac{1}{p\left(1 - \dfrac{p}{K}\right)} = \frac{1}{p} + \frac{\dfrac{1}{K}}{1 - \dfrac{p}{K}},$$

and integrating gives

$$\int \frac{1}{p\left(1 - \dfrac{p}{K}\right)}\, dp = \int \frac{1}{p}\, dp + \int \frac{\dfrac{1}{K}}{1 - \dfrac{p}{K}}\, dp = \ln|p| - \ln\left|1 - \frac{p}{K}\right|.$$

Putting together these steps, we get

$$\ln\left|\frac{p}{1 - \dfrac{p}{K}}\right| = rt + c,$$

which is the same as

$$\frac{p}{1 - \dfrac{p}{K}} = \pm e^c e^{rt}.$$

Replacing the constant $\pm e^c$ by the single letter C gives

$$\frac{p}{1 - \dfrac{p}{K}} = C e^{rt}.$$

Next, by using the initial condition (33), we obtain

$$C = \frac{p_0}{1 - \dfrac{p_0}{K}},$$

and making this substitution in the previous equation gives

$$\frac{p}{1 - \dfrac{p}{K}} = \frac{p_0}{1 - \dfrac{p_0}{K}} e^{rt}.$$

By clearing denominators, we get

$$p\left(1 - \frac{p_0}{K}\right) = p_0 e^{rt} - p_0 \frac{p}{K} e^{rt},$$

and collecting all terms involving p together on the left side of the equation changes this to

$$p\left(1 - \frac{p_0}{K} + \frac{p_0}{K} e^{rt}\right) = p_0 e^{rt}.$$

Solving for p, we get

$$p = \frac{p_0 K e^{rt}}{K - p_0 + p_0 e^{rt}}.$$

Finally, we will change the appearance of the solution by multiplying both numerator and denominator on the right-hand side by e^{-rt}. In that way, we arrive at the following solution of the initial value problem:

$$p(t) = \frac{p_0 K}{p_0 + (K - p_0)e^{-rt}}. \tag{35}$$

The advantage of this form of the solution is that it makes it easy to see what happens as $t \to \infty$. The term involving e^{-rt} goes to zero, and then the p_0's in numerator and denominator cancel. Therefore, we get

$$\lim_{t \to \infty} p(t) = K = \frac{r}{k}. \tag{36}$$

That is, for any given initial value $p_0 > 0$, the solution tends toward the equilibrium solution $p = K = r/k$.

EXAMPLE 6.4.1

In 1980 the population of the world was approximately 4.45 billion, and in 1990 it was 5.23 billion. The intrinsic rate of growth for humans is estimated to be 0.029. Assuming that the logistic growth equation is an accurate model, find a formula for the world population and its equilibrium solution.

SOLUTION   Let t denote the number of years after 1980, and let $p(t)$ be the population in billions. By putting $p_0 = 4.45$ and $r = 0.029$ in formula (35), we obtain

$$p(t) = \frac{4.45K}{4.45 + (K - 4.45)e^{-0.029t}}. \tag{37}$$

To find the constant K, we use the 1990 population—that is, we substitute $t = 10$ and $p(10) = 5.23$, which gives

$$5.23 = \frac{4.45K}{4.45 + (K - 4.45)e^{-0.29}}.$$

Clearing the denominator, collecting terms, and solving for K, we get

$$K = \frac{(5.23)(4.45)(1 - e^{-0.29})}{4.45 - 5.23e^{-0.29}} \approx 10.92 \text{ billion.}$$

Substituting this into Eq. (37) and doing the necessary arithmetic give us the formula for the population of the world at any time:

$$p(t) \approx \frac{48.59}{4.45 + 6.47e^{-0.029t}}. \tag{38}$$

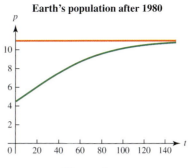

Figure 6.4.2

(We have written this as an approximation rather than an equality because the given data are only approximate and we have rounded our calculations to two places.)

Notice that $p(t) \to \frac{48.59}{4.45} \approx 10.92$, the equilibrium solution as $t \to \infty$. You can see that in Figure 6.4.2, which shows the graph of p as a function of t.

Suppose we had started with a different choice of initial population but with the same constants, $K = 10.92$ and $r = 0.029$. In other words, suppose we had solved the same differential equation but changed the initial condition. That would amount to replacing 4.45 by some other value in formula (37). Figure 6.4.3 shows several such graphs of the earth's population over time, each one corresponding to a different choice of initial population between 1 and 21 billion. As you can see, they all tend toward the same equilibrium solution.

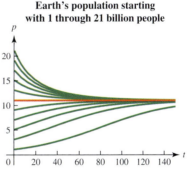

Figure 6.4.3

EXAMPLE 6.4.2

Data from the International Whaling Commission[1] indicate that there were about 26,500 gray whales in the North Pacific in the year 2000, with an average number of 174 caught per year. Suppose that the natural population growth, in the absence of whale fishing, would satisfy a logistic model with an intrinsic growth rate of 3% per year and an equilibrium solution of 28.1 thousand. Write an initial value problem modeling the

[1]http://www.iwcoffice.org/iwc.htm.

population $y(t)$ in thousands, assuming that the average annual catch remains at 174. What are the equilibrium solutions?

SOLUTION   Using the logistic model given by formula (34), we see that the population y would grow at a rate of

$$0.03y\left(1 - \frac{y}{28.1}\right)$$

thousands of whales per year. Taking the annual catch into account, we obtain the following formula for the rate of growth of the gray whale population:

$$\frac{dy}{dt} = 0.03y\left(1 - \frac{y}{28.1}\right) - 0.174, \tag{39}$$

with t in years and y in thousands of whales. Taking $t = 0$ to be the year 2000, we have

$$y(0) = 26.5 \tag{40}$$

Thus, the whale population is modeled by the initial value problem stated in equations (39) and (40).

To find the equilibrium solutions, we set the right-hand side equal to zero and solve. Expanding and rearranging terms gives

$$\frac{0.03}{28.1}y^2 - 0.03y + 0.174 = 0,$$

whose solutions are

$$y = \left(\frac{28.1}{0.06}\right)\left(0.03 \pm \sqrt{0.0009 - \frac{(0.12)(0.174)}{28.1}}\right) \approx 19.92 \text{ and } 8.18.$$

These are the equilibrium solutions.

HISTORICAL PROFILE

Pierre Verhulst (1804–1849) was born in Brussels and received a Ph.D. from the University of Ghent in 1825 after only 3 years of study. His first work was in the theory of numbers, but he soon developed a lifelong interest in social statistics, with particular attention to population growth. He spent his adult life as a professor in Brussels, where he gave courses in astronomy, celestial mechanics, calculus, geometry, and probability, and he was elected president of the Belgian Academy in 1835. His statistical analysis led him to estimate the upper limit of the Belgian population at 9,400,000. Although Verhulst neglected to allow for immigration as a factor in population increase, his prediction held up well, and in 1994 the population stood at 10,118,000.

Practice Exercises 6.4

•••

1. A certain region has an intrinsic growth rate of 4% per year. In 1990 its population was 2 million.
   (a) Assuming that the growth is exponential, write a formula for the population as a function of time and use it to predict the size of the population in 2010.

   (b) Write a formula for the population as a function of time, assuming that the growth is logistic and the population was 2.6 million in 2000. Use it to predict the size of the population in 2010.

Exercises 6.4

•••

1. You are reading a journal article describing a region in the United States as having a population function of the form

   $$P(t) = \frac{327K}{327 + (K - 327)e^{-0.015t}}$$

   where t is in years, P is number of people, and K is a positive constant. You immediately recognize this as a logistic growth solution. Identify p_0, r, and k.

2. Initial studies of a captive fruit fly (*Drosophila melanogaster*) indicate that the population increases exponentially at a rate of 0.7 per 2 weeks. There were originally 35 flies in the container.
   (a) Model this as an exponential growth problem. Write the differential equation, initial condition, and solution. Use 2 weeks as the unit of time.
   (b) What is the predicted population after 16 weeks?
   (c) Show what happens to the population level as $t \to \infty$.

3. **Continuation of the previous problem** Further studies of the fruit fly population suggest the existence of a damping effect with damping factor $k = 0.01$. [See (32) and the paragraph immediately following it.]
   (a) Model this as a logistic growth problem. Write the differential equation, initial condition, and solution. Use 2 weeks as the unit of time.
   (b) What is the carrying capacity?
   (c) What is the predicted population after 16 weeks?

4. The population of a fast-developing region is growing exponentially at a rate of 5% per year. In 1990 the population of the region was 1,500,000.
   (a) Write a differential equation describing the population as a function of time.
   (b) State an initial condition based on the given information.
   (c) Solve the initial value problem describing the population function.
   (d) Compute the population in the year 2010.
   (e) Find the time needed for the population to double its size.

5. The population of the United States was about 5 million in 1800, and it was about 280 million in the year 2000. Assume that the rate of growth of the population is proportional to its size at any given time.

   (a) Find the U.S. population in 1900 given by this model.
   (b) The actual population of the United States in 1900 was 76 million. How does that compare with your previous answer? Can you think of any reason why they are different?

6. The world population, after remaining stable for most of history, began to grow gradually after the agricultural revolution, and the rate of increase accelerated after the industrial revolution. In 1982 the world population was about 4.5 billion and in 1992 it was about 5.5 billion. Assume that the world population since 1982 has grown, and will continue to grow, at a rate proportional to its size at any given time.
   (a) Find the world population in 2032.
   (b) Find the time required for the population to triple.

   In Exercises 7–10, determine the intrinsic growth rate, the damping factor, and the equilibrium solution of the given logistic equation, and sketch dp/dt as a function of p.

7. $\dfrac{dp}{dt} = 0.04p - 0.001p^2$ 8. $\dfrac{dp}{dt} = 0.02p(1 - 0.002p)$

9. $\dfrac{dp}{dt} + 0.05p^2 = 0.25p$ 10. $\dfrac{dp}{dt} = -0.05p(p - 1)$

11. Write the logistic equation with intrinsic growth rate 0.03 and equilibrium solution $p = 9$.

12. The population of a region in 1975 was 25 million, and in 1995 it was 45 million. Assume that its intrinsic rate of growth is about 0.03 and that it follows the logistic growth model.
    (a) Write a differential equation and an initial condition describing the population of the region at any time.
    (b) Find the population of the region in the year 2035.

13. The logistic differential equation, in the form $y' = ay \cdot (M - y)$, is sometimes used to model the spread of an epidemic in a community, where M is the total population of the community, y is the number of infected people, and a is a constant of proportionality. The model is based on the assumption that the rate at which the epidemic spreads is proportional to the number of contacts between infected and noninfected members of the community, which, in turn, is proportional to the product of y (the number of infected people) and $(M - y)$ (the number of noninfected people).

(a) Taking $M = 3{,}000$, solve the differential equation. Your answer should involve the constant a and an arbitrary constant C.

(b) Suppose that 20 people are infected at the start and the number has risen to 314 after 3 days. Determine the constants a and C.

(c) If this model is correct, how many people will be infected after 5 days?

14. In 1900 the population of the world was approximately 1.63 billion, and in 1930 it was 2.07 billion. The intrinsic rate of growth for humans is estimated to be 0.03. Assume that the logistic growth equation models the world's population $p(t)$ and find a formula for $p(t)$ using these data. Then compare the actual world population in 2000 with the one predicted by the logistic equation.

15. In 1900 the U.S. population was approximately 76 million, and in 1910 it was 92 million. Assume that the intrinsic rate is 0.03, and that the U.S. populations follows the logistic growth model. Find a formula $p(t)$ for the U.S. population using the given data and then construct a table of the U.S. population from 1900 to 2000 in 4-year periods. Compare the table with the available census data since 1900.

16. Using formula (38), construct a table of the population of the earth predicted by the model of Example 6.4.1 for every 10 years from 2010 to 2100.

Solve the following initial value problems for the given logistic equations:

17. $\dfrac{dp}{dt} = 0.05p(1 - p), \quad p(0) = 0.8,$

18. $\dfrac{dp}{dt} = 0.04p\left(1 - \dfrac{1}{2}p\right), \quad p(0) = 1$

19. $\dfrac{dp}{dt} = 0.03p\left(1 - \dfrac{1}{5}p\right), \quad p(0) = 4$

20. $\dfrac{dp}{dt} = 0.01p - p^2, \quad p(0) = 50$

21. A corporation with 36,000 employees is connected by a computer network. Someone started a rumor that everyone would be receiving a large bonus at the end of the year. Initially, 50 people had heard the rumor. Two days later, 250 people had heard it.

(a) Assume that the spread of the rumor is modeled by $y' = ay(M - y)$, where M is the size of the community, y is the number of people who have heard the rumor, and a is a constant of proportionality. Determine the solution of this differential equation.

(b) How many people have heard the rumor by the end of the 10th day?

22. Consider a lake that can sustain a maximum of 100 thousand fish of a certain species. If this fish has a growth rate of 0.2 and is harvested at the rate of 10% of the population present, then the population $y(t)$ of this fish can be modeled by the differential equation

$$\frac{dy}{dt} = 0.2y\left(1 - \frac{y}{100}\right) - 0.1y$$

(called a **logistic equation with constant effort harvesting**). Find $y(t)$, assuming that the initial population is 30 thousand. What is $\lim_{t \to \infty} y(t)$?

23. Biologists have been evaluating the population dynamics of a deer herd. They calculate the annual growth rate to be 0.42 and the damping factor to be 0.02, in hundreds of deer per year. There is another wilderness area that matches the deer herd's location in many ways, and the biologists want to establish a successful deer population in that area as well. What is the maximum herd size they should start with in order to have a population that does not change over time?

24. An equation similar to the logistic differential equation is used to model a certain type of chemical reaction, such as the reaction of acetic acid (CH_3COOH) and ethanol (C_2H_5OH), which proceeds as follows: $CH_3COOH + C_2H_5OH \rightarrow CH_3COOC_2H_5 + H_2O$. Chemists call this type of reaction—in which there are two reactants and one molecule of the first reacts with one molecule of the second at a given fixed temperature—a **second-order reaction**. In 1862, the chemists Berthelot and St. Gilles, discovered the **law of mass action** for such reactions, which can be expressed by the differential equation

$$\frac{dx}{dt} = r(a - x)(b - x), \tag{41}$$

where a and b are the initial concentrations of the two reactants (such as acetic acid and ethanol), r is a constant of proportionality called the **specific reaction rate**, and x is the amount by which the substances decrease during the reaction. (Note that x is a function of t, and the second-order assumption implies that the amount and rate of decrease are the same for both reactants.) Assuming, for simplicity, that $a \neq b$, solve Eq. (41).

25. A population may have a variable intrinsic growth rate $r(t)$ and follow the more general logistic equation

$$\frac{dp}{dt} = r(t)p\left(1 - \frac{p}{K}\right).$$

(a) Show that the function

$$p(t) = \frac{Kp_0}{p_0 + (K - p_0)e^{-\int_0^t r(s)\,ds}}$$

is the solution to this differential equation satisfying the initial condition $p(0) = p_0$.

(b) Compute $p(t)$ if $K = 10$, $p_0 = 6$, and $r(t) = 0.2t/(t + 1)$.

(c) With $p(t)$ as in (b), evaluate $p(5)$ and find $\lim_{t \to \infty} p(t)$.

26. Solve the initial value problem posed in Eqs. (39) and (40).

27. In addition to the exponential and logistic growth models there are many other models of population growth in the literature that include the following.

(a) $\dfrac{dp}{dt} = rp \ln \dfrac{K}{p}$ (Gompertz, 1825)

(b) $\dfrac{dp}{dt} = rp \dfrac{K - p}{K + ap}$ (Smith, 1963)

(c) $\dfrac{dp}{dt} = rp \left[1 - \left(\dfrac{p}{K} \right)^{\theta} \right], \ 0 < \theta < 1$

(Ayala-Gilpin, Ehrenfeld, 1973)

(d) $\dfrac{dp}{dt} = p \left(re^{1-(p/K)} - d \right)$ (Nisbet-Gurney, 1982)

Set $dp/dt = 0$ in these models to find their constant (equilibrium) solutions.

Source: F. Brauer and C. Castillo-Chavez, *Mathematical Models in Population Biology and Epidemiology*, New York: Springer, 2001.

28. Estimates of the population of both the world and China (in millions) are listed in the following tables.

World population from 1000–2000

Year: t	World: $p(t)$	Year: t	World: $p(t)$
1000	275	1910	1755
1100	306	1920	1811
1200	348	1930	2070
1300	384	1940	2295
1400	373	1950	2490
1500	446	1960	2982
1650	553	1970	3632
1750	726	1980	4400
1850	1325	1990	5260
1900	1617	2000	6230

Population of China from 1110–2000

Year: t	China: $p(t)$	Year: t	China: $p(t)$
1110	46.7	1953	602
1303	60.5	1965	725
1757	190	1982	1032
1901	426	2000	1266
1928	474		

Investigate whether any of the models you have learned thus far in the theory and in the previous exercise can fit the given data. You may also want to produce your own models. You may discover, however, that models are never "correct," and all they do is to provide us with a "good" approximation over some time interval of the process we are studying.

Solutions to practice exercises 6.4

1. Let $p(t)$ be the population, in millions, t years after 1990.
(a) Our assumptions are that $dp/dt = 0.04p$ and $p(0) = 2$. The solution of this initial value problem is $p = 2e^{0.04t}$, and the population in 2010 is given by $p(20) = 2e^{0.8} \approx 4.45$ million.
(b) According to formula 34, the differential equation is

$$\frac{dp}{dt} = 0.04p \left(1 - \frac{p}{K} \right).$$

Its solution [using $p_0 = p(0) = 2$] is given by formula (35) as follows:

$$p(t) = \frac{2K}{2 + (K - 2)e^{-0.04t}},$$

where K is the saturation level. Setting $t = 10$ and $p = 2.6$ and solving for K give $K = (5.2)(1 - e^{-0.4})/(2 - 2.6e^{-0.4}) \approx 6.666$. Using this value yields $p(20) \approx 3.25$.

6.5   Improper Integrals and Applications

Earlier we discussed the present value PV of an income stream flowing at a uniform rate of S dollars per year with annual interest rate r, compounded continuously, for a total period of T years. The formula we found for PV is

$$PV = \int_0^T Se^{-rt}\, dt.$$

Now, if we assume that this cash flow persists forever—which may approximate the case, for instance, for income coming from an indestructible capital asset such as rent on a building or land—then the present value of such a **perpetual income stream** is given by the integral

$$PV = \int_0^\infty Se^{-rt}\, dt.$$

This integral is called an **improper integral**, since its upper limit is not finite, and it is defined as follows:

$$\int_0^\infty Se^{-rt}\, dt = \lim_{T \to \infty} \int_0^T Se^{-rt}\, dt.$$

EXAMPLE 6.5.1

Find the present value of a perpetual income stream flowing continuously at a rate $10,000 per year, with interest compounded continuously at an annual rate of 5%.

SOLUTION   In this case

$$PV = \int_0^\infty 10{,}000e^{-0.05t}\, dt$$

$$= \lim_{T \to \infty} \int_0^T 10{,}000e^{-0.05t}\, dt$$

$$= \lim_{T \to \infty} \frac{10{,}000}{-0.05} e^{-0.05t} \Big|_0^T$$

$$= \lim_{T \to \infty} -200{,}000(e^{-0.05T} - 1)$$

$$= -200{,}000 \left(\lim_{T \to \infty} e^{-0.05T} - 1 \right)$$

$$= -200{,}000 \cdot (0 - 1) = 200{,}000.$$

In general, an integral is called an improper integral if either

- one or both of the limits of integration is infinite or
- its integrand is an unbounded function.

The first of these cases covers integrals of the form $\int_a^\infty f(x)\,dx$, which are defined by

$$
\int_a^\infty f(x)\,dx = \lim_{b\to\infty} \int_a^b f(x)\,dx, \quad \text{provided that the limit exists,}
$$

and integrals of the form $\int_{-\infty}^b f(x)\,dx$, which are defined by

$$
\int_{-\infty}^b f(x)\,dx = \lim_{a\to-\infty} \int_a^b f(x)\,dx, \quad \text{provided that the limit exists.}
$$

EXAMPLE 6.5.2

Compute $\displaystyle\int_0^\infty e^{-x}\,dx$.

SOLUTION Let b be a positive number. We have

$$
\int_0^b e^{-x}\,dx = -e^{-x}\Big|_0^b = -(e^{-b} - e^0) = 1 - e^{-b}.
$$

Therefore,

$$
\int_0^\infty e^{-x}\,dx = \lim_{b\to\infty}(1 - e^{-b}) = 1 - \lim_{b\to\infty} e^{-b} = 1 - 0 = 1.
$$

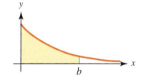

Figure 6.5.1

We can interpret this integral as the area under the graph of $y = e^{-x}$ for $x \geq 0$, as shown in Figure 6.5.1. The area of the shaded region is $\int_0^b e^{-x}\,dx$, and its limit as $b \to \infty$ is the area under the entire graph over the interval $[0, \infty)$. As we have just verified, the total area under this graph is finite, even though the graph extends infinitely far to the right.

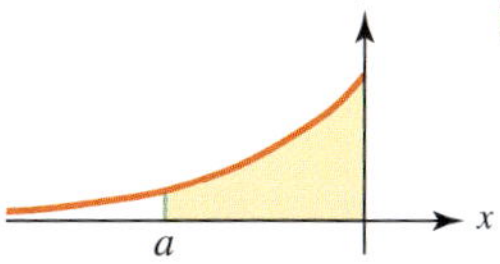

EXAMPLE 6.5.3

Find the area under the graph of $y = e^x$ over the interval $(-\infty, 0]$.

Figure 6.5.2

SOLUTION In this case, $\int_a^0 e^x\,dx$ is the area under the graph from $x = a$ to $x = 0$ (shaded in Figure 6.5.2), and the area under the entire graph over the interval $(-\infty, 0]$

is obtained by taking the limit as $a \to -\infty$, which gives

$$\int_{-\infty}^{0} e^x \, dx = \lim_{a \to -\infty} \int_{a}^{0} e^x \, dx = \lim_{a \to -\infty} e^x \Big|_{a}^{0} = \lim_{a \to -\infty} [e^0 - e^a] = 1.$$

EXAMPLE 6.5.4

You plan to start a business breeding white mice in a germ-free environment to sell to laboratories. You estimate that the mice will breed exponentially at an annual rate of 18%, and they will be sold at a continuous rate of 900 per year. How large an initial stock will keep your business going indefinitely without replenishment?

SOLUTION   We can treat this as an income stream problem, with the mice flowing out (like dollars from an account) at a rate of 900 per year and increasing at a (continuously compounded) rate of 0.18. (See Example 6.2.7.) The present value of the total outflow during the first T years is

$$\int_{0}^{T} 900 e^{-0.18t} \, dt = -\frac{900}{0.18} e^{-0.18t} \Big|_{0}^{T} = \frac{900}{0.18}(1 - e^{-0.18T}).$$

Letting $T \to \infty$, we obtain

$$\int_{0}^{\infty} 900 e^{-0.18t} \, dt = \lim_{T \to \infty} \frac{900}{0.18}(1 - e^{-0.18T}) = \frac{900}{0.18} = 5,000,$$

which is the present value of the perpetual total outflow. Therefore, an initial stock of 5,000 mice will last indefinitely without replenishment.

In many important statistical cases, the probability that a certain random number falls inside a particular interval (a, b) is given by the integral

$$\int_{a}^{b} \frac{1}{\sqrt{2\pi}} e^{-(1/2)x^2} \, dx.$$

Think, for instance, of a particle starting from the origin and moving randomly, back and forth along a straight line, for one unit of time. Motion of this kind, called **Brownian motion**, is observed in the up-and-down movement of a small particle of solid matter suspended in a liquid. The probability that the particle's coordinate falls inside the interval (a, b) is given by the integral above. If we take the interval to be the entire axis, or, in other words, $(-\infty, \infty)$, we get an improper integral with *both* limits of integration infinite, which we interpret as follows:

$$\int_{-\infty}^{\infty} \frac{1}{\sqrt{2\pi}} e^{-(1/2)x^2} \, dx = \lim_{a \to -\infty} \int_{a}^{0} \frac{1}{\sqrt{2\pi}} e^{-(1/2)x^2} \, dx + \lim_{b \to \infty} \int_{0}^{b} \frac{1}{\sqrt{2\pi}} e^{-(1/2)x^2} \, dx.$$

Since it is certain that the particle's coordinate falls in the interval $(-\infty, \infty)$, we should assign a value of 1 to that event, which suggests that

$$\int_{-\infty}^{\infty} \frac{1}{\sqrt{2\pi}} e^{-x^2/2} \, dx = 1.$$

This formula is indeed true and can be verified by more advanced methods. (But see the exercises at the end of this section for an approximation.) The function

$$f(x) = \frac{1}{\sqrt{2\pi}} e^{-x^2/2},$$

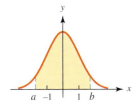

Figure 6.5.3

whose graph is shown in Figure 6.5.3, is called the **standard normal distribution**. It plays a major role in statistical analysis.

In general, improper integrals whose limits are both infinite are defined, as above, by the formula

$$\int_{-\infty}^{\infty} f(x) \, dx = \lim_{a \to -\infty} \int_{a}^{0} f(x) \, dx + \lim_{b \to \infty} \int_{0}^{b} f(x) \, dx, \quad \text{if both limits exist.}$$

Unbounded functions

An integral of the form $\int_{a}^{b} f(x) \, dx$ over a finite interval $[a, b]$ is also called *improper* if the function has a vertical asymptote at either end of the interval. If $\lim_{x \to a^+} f(x) = \pm\infty$, for example, the integral is defined as follows:

$$\int_{a}^{b} f(x) \, dx = \lim_{h \to a^+} \int_{h}^{b} f(x) \, dx, \quad \text{provided the limit exists.}$$

Similarly, if $\lim_{x \to b^-} f(x) = \pm\infty$, its integral over $[a, b]$ is improper and is defined by

$$\int_{a}^{b} f(x) \, dx = \lim_{h \to b^-} \int_{a}^{h} f(x) \, dx, \quad \text{provided the limit exists.}$$

EXAMPLE 6.5.5

Compute $\int_{0}^{1} \frac{1}{\sqrt{x}} \, dx$.

SOLUTION We have

$$\int_{0}^{1} \frac{1}{\sqrt{x}} \, dx = \lim_{h \to 0^+} \int_{h}^{1} x^{-1/2} \, dx = \lim_{h \to 0^+} 2x^{1/2} \Big|_{h}^{1} = \lim_{h \to 0^+} 2[1 - h^{1/2}] = 2.$$

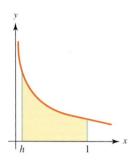

Figure 6.5.4

Since the integrand is positive for $x > 0$, the integral $\int_h^1 (1/\sqrt{x})\, dx$ is the area under the graph over $[h, 1]$ (shaded in Figure 6.5.4). Its limit as $h \to 0^+$ is the area under the graph from 0 to 1. Even though the graph is unbounded, the area it bounds is finite.

Convergent and divergent integrals

Not every improper integral has a finite value. If it does, we say the integral **converges**. If the limit is ∞ or $-\infty$, then we say that the improper integral **diverges**.

EXAMPLE 6.5.6

Show that the improper integral $\displaystyle\int_1^\infty \frac{dx}{x^{0.9}}$ diverges.

SOLUTION   This integral diverges since

$$\int_1^\infty \frac{dx}{x^{0.9}} = \lim_{b\to\infty} \int_1^b x^{-0.9}\, dx = \lim_{b\to\infty} \frac{x^{0.1}}{0.1}\bigg|_1^b = \lim_{b\to\infty} 10\,(b^{0.1} - 1) = \infty.$$

EXAMPLE 6.5.7

Determine whether each of the following integrals converges or diverges:

$$\text{(i)}\quad \int_1^\infty \frac{1}{x^2}\, dx \quad\text{and}\quad \text{(ii)}\quad \int_0^2 \frac{1}{x^2}\, dx.$$

SOLUTION   (i) We have

$$\int_1^h \frac{1}{x^2}\, dx = -\frac{1}{x}\bigg|_1^h = 1 - \frac{1}{h}.$$

Therefore, $\int_1^\infty (1/x^2)\, dx = \lim_{h\to\infty} (1 - 1/h) = 1$, and the integral converges.
(ii) In this case, the integrand is unbounded as $x \to 0^+$, and

$$\lim_{h\to 0^+} \int_h^2 \frac{1}{x^2}\, dx = \lim_{h\to 0^+} \left(\frac{1}{h} - \frac{1}{2}\right) = \infty.$$

Therefore, the integral diverges.

In evaluating improper integrals, the following limits are sometimes useful. (We omit the proofs, but see the exercises for the cases of $m = 1, 2, 3$.) For any positive number m,

$$\lim_{x\to\infty} x^m e^{-x} = 0 \quad\text{and}\quad \lim_{x\to 0^+} x(\ln x)^m = 0. \tag{42}$$

EXAMPLE 6.5.8

Determine whether the integral $\int_0^1 (\ln x)^2\, dx$ converges.

SOLUTION We first evaluate $\int_h^1 (\ln x)^2\, dx$, using integration by parts twice:

$$\int_h^1 (\ln x)^2\, dx = x(\ln x)^2\Big|_h^1 - 2\int_h^1 \ln x\, dx \quad (u = (\ln x)^2,\ dv = dx)$$

$$= -h(\ln h)^2 - 2\left\{ x\ln x\Big|_h^1 - \int_h^1 1\, dx \right\} \quad (u = \ln x,\ dv = dx)$$

$$= -h(\ln h)^2 + 2h\ln h + 2 - 2h.$$

Next, letting $h \to 0^+$ and using (42), we have

$$\lim_{h\to 0^+} \int_h^1 (\ln x)^2\, dx = \lim_{h\to 0^+} [-h(\ln h)^2 + 2h\ln h + 2 - 2h] = 2.$$

Therefore, $\int_0^1 (\ln x)^2\, dx = 2$, and we conclude that the integral converges.

It may happen that an improper integral neither converges nor diverges. In other words, the limit may not exist, either as a finite number or ∞ or $-\infty$. In such cases, we say that the improper integral *does not exist*. For instance, suppose that the "sawtooth" graph shown in Figure 6.5.5 extends indefinitely to the right, repeating the same pattern of alternating line segments of slope 1 and -1. If $f(x)$ is the corresponding function, then its integral can be computed using the areas of triangles. In particular,

$$\int_0^2 f(x)\, dx = 1, \quad \int_0^4 f(x)\, dx = 0, \quad \int_0^6 f(x)\, dx = 1, \quad \int_0^8 f(x)\, dx = 0, \quad \dots,$$

and, in general, $\int_0^h f(x)\, dx$ oscillates between 0 and 1 as $h \to \infty$. Therefore, $\lim_{h\to\infty} \int_0^h f(x)\, dx$ does not exist.

That type of behavior cannot happen if $f(x) \geq 0$ throughout the interval of integration. In that case, $\int_a^h f(x)\, dx$ is an increasing function of h, and it must either get unboundedly large as $h \to \infty$ (in which case the improper integral diverges) or approach a finite limit (in which case the integral converges). This observation is the basis of the following theorem, known as the **comparison test for improper integrals**:

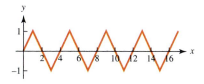

Figure 6.5.5

Theorem 6.5.1

Suppose $f(x)$ and $g(x)$ are continuous functions on $[a, \infty)$ with $0 \le f(x) \le g(x)$ for all $x \ge a$.

- If $\int_a^\infty g(x)\, dx$ converges, $\int_a^\infty f(x)\, dx$ also converges.
- If $\int_a^\infty f(x)\, dx$ diverges, $\int_a^\infty g(x)\, dx$ also diverges.

EXAMPLE 6.5.9

Determine whether each of the following integrals converges or diverges:

$$\text{(i)} \quad \int_0^\infty \frac{1}{x^2 + 1}\, dx \quad \text{and} \quad \text{(ii)} \quad \int_1^\infty \frac{1}{1 + \sqrt{x}}\, dx.$$

SOLUTION   (i) We first observe that $\int_0^\infty 1/(x^2 + 1)\, dx$ converges if and only if $\int_1^\infty 1/(x^2 + 1)\, dx$ converges. That is because

$$\int_0^\infty \frac{1}{x^2 + 1}\, dx = \int_0^1 \frac{1}{x^2 + 1}\, dx + \int_1^\infty \frac{1}{x^2 + 1}\, dx.$$

and the first integral on the right is finite. We have already seen in Example 6.5.7 that $\int_1^\infty (1/x^2)\, dx$ converges. Moreover,

$$\frac{1}{x^2 + 1} < \frac{1}{x^2} \quad \text{for all } x.$$

Therefore, we can apply Theorem 6.5.1 to conclude that $\int_1^\infty 1/(x^2 + 1)\, dx$ converges.

(ii) We first observe that if x is very large, $1 + \sqrt{x} \approx \sqrt{x}$. For that reason, we might try to compare the given integral with $\int_1^\infty (1/\sqrt{x})\, dx$. Since

$$\int_1^\infty \frac{1}{\sqrt{x}}\, dx = \lim_{b \to \infty} 2\sqrt{x}\, \Big|_1^b = \lim_{b \to \infty} (2\sqrt{b} - 2) = \infty,$$

we would then guess that the given integral diverges.

Unfortunately, a direct comparison of these two integrals does not work because

$$\int_1^\infty \frac{1}{1 + \sqrt{x}} < \int_1^\infty \frac{1}{\sqrt{x}}\, dx,$$

so that the divergence of the integral on the right does not imply the divergence of the one on the left. In other words, the inequality "goes the wrong way." Instead, we use the following comparison:

$$\frac{1}{1 + \sqrt{x}} \ge \frac{1}{2\sqrt{x}} \quad \text{for } x \ge 1.$$

Then, since

$$\int_1^\infty \frac{1}{2\sqrt{x}} = \lim_{b\to\infty} \int_1^b \frac{1}{2\sqrt{x}}\, dx = \lim_{b\to\infty}(\sqrt{b}-1) = \infty,$$

we conclude that the given integral diverges.

Remark   In applying Theorem 6.5.1, we may conclude that an integral converges, as we did in case (i) of the last example, *without being able to evaluate it.* Once we know it converges, however, we may apply numerical methods to approximate $\int_a^h f(x)\, dx$ for very large h as a way of estimating the improper integral (see the exercises).

There is also a comparison test for integrals of unbounded functions. In the following theorem, we consider the case of functions that are continuous on an interval $(a, b]$ with a vertical asymptote at a.

Theorem 6.5.2

Suppose $f(x)$ and $g(x)$ are continuous functions on $(a, b]$ with $0 \le f(x) \le g(x)$ for $a < x \le b$.

- If $\int_a^b g(x)\, dx$ converges, $\int_a^b f(x)\, dx$ also converges.
- If $\int_a^b f(x)\, dx$ diverges, $\int_a^b g(x)\, dx$ also diverges.

EXAMPLE 6.5.10

Determine whether $\displaystyle\int_0^1 \frac{1}{\sqrt{x^2+x}}\, dx$ converges.

SOLUTION   We first observe that $x + x^2 > x$ for $x > 0$, so that

$$\frac{1}{\sqrt{x^2+x}} < \frac{1}{\sqrt{x}} \quad \text{on } (0, 1].$$

Moreover,

$$\int_0^1 \frac{1}{\sqrt{x}}\, dx = \lim_{h\to 0^+} \int_h^1 \frac{1}{\sqrt{x}}\, dx = \lim_{h\to 0^+}(2 - 2\sqrt{h}) = 2.$$

Therefore, by the comparison test, $\int_0^1 (1/\sqrt{x^2+x})\, dx$ converges.

The following theorem is useful in applying the comparison test. Its proof is left as an exercise.

> ### Theorem 6.5.3
>
> Let p be any real number.
>
> - $\int_1^\infty (1/x^p)\, dx$ converges if $p > 1$ and diverges if $p \leq 1$.
> - $\int_0^1 (1/x^p)\, dx$ converges if $p < 1$ and diverges if $p \geq 1$.

Practice Exercises 6.5

1. Determine whether each of the following improper integrals converges, and, if so, compute it.

   (a) $\displaystyle\int_1^\infty \frac{1}{x^3}\, dx$ (b) $\displaystyle\int_0^1 \frac{1}{x}\, dx$

2. Use a comparison test to determine whether

   $$\int_2^\infty \frac{1}{x^4 - 1}\, dx$$

   converges.

Exercises 6.5

Compute the following improper integrals:

1. $\displaystyle\int_2^\infty \frac{dx}{x^5}$

2. $\displaystyle\int_3^\infty \frac{dx}{(x-2)^3}$

3. $\displaystyle\int_{-\infty}^0 e^{0.1t}\, dt$

4. $\displaystyle\int_0^1 x^{-1/3}\, dx$

5. $\displaystyle\int_0^\infty e^{-0.05t}\, dt$

6. $\displaystyle\int_{-2}^{-1} (x+2)^{-1/4}\, dx$

7. $\displaystyle\int_0^\infty \frac{x^2}{(x^3+1)^2}\, dx$

8. $\displaystyle\int_0^\infty t e^{-t^2/2}\, dt$

9. $\displaystyle\int_0^\infty t e^{-t}\, dt$

*A nonnegative function $f(x), a < x < b$, is called a **probability density function** if*

$$\int_a^b f(x)\, dx = 1.$$

(Such functions are used in various situations to determine probabilities.) In Exercises 10–12, show that the following functions are probability density functions:

10. $f(x) = 0.1 e^{-0.1x},\ 0 \leq x < \infty$

11. $f(x) = \lambda e^{-\lambda x},\ 0 \leq x < \infty$, where $\lambda > 0$

12. $f(x) = \lambda x^{-(\lambda+1)},\ 1 \leq x < \infty$, where $\lambda > 1$

13. Find the constant c so that $f(x) = c x^2 e^{-x}, 0 \leq x < \infty$, is a probability density function.

14. The improper integral $\int_0^\infty \lambda x e^{-\lambda x}\, dx$, where λ is a positive constant, arises in probability theory as the average waiting time between certain events, such as arrivals of customers at a checkout line or hits at a Website. Show that the integral converges and find its value.

15. Compute the improper integral $\int_0^1 \ln x\, dx$.

16. Find the area of the region between the curve $y = x^{-4}$, $1 \leq x < \infty$, and the x-axis.

17. Find the area of the region between the curve $y = x^{-1/3}$, $0 < x \leq 1$, and the x-axis.

In Exercises 18–29, determine whether the improper integral converges, and, if so, compute it.

18. $\displaystyle\int_1^\infty \frac{dx}{\sqrt{x}}$

19. $\displaystyle\int_0^1 \frac{dx}{x^2}$

20. $\int_0^\infty e^x\, dx$

21. $\int_0^1 x \ln x\, dx$

22. $\int_0^1 \frac{\ln x}{x}\, dx$

23. $\int_0^\infty \frac{x}{x^2+1}\, dx$

24. $\int_1^2 (x-1)^{-4/3}\, dx$

25. $\int_1^2 \frac{dx}{x^2-1}$

26. $\int_1^2 \frac{x}{\sqrt{x^2-1}}\, dx$

27. $\int_0^\infty \frac{1}{(x+1)(x+2)}\, dx$

28. $\int_0^\infty \frac{e^x}{e^{2x}+2e^x+1}\, dx$

29. $\int_1^2 \frac{1}{x \ln x}\, dx$

The next three exercises cover the proof of Theorem 6.5.3.

30. Show that $\int_1^\infty (1/x^p)\, dx$ converges if $p>1$ and diverges if $p<1$.

31. Show that $\int_0^1 (1/x^p)\, dx$ converges if $p<1$ and diverges if $p>1$.

32. Show that both $\int_0^1 (1/x)\, dx$ and $\int_1^\infty (1/x)\, dx$ diverge.

33. Find the present value of a perpetual income stream flowing continuously at a rate of $12,000 per year and with interest compounded continuously at the rate of 9%.

34. An apartment in Manhattan produces a perpetual income stream flowing continuously at a rate of $50,000 per year. This income is invested at the annual rate of 5%, compounded continuously. Find the present value of this stream.

35. What is the amount of money a donor needs to contribute in order to establish a scholarship fund generating an annual income of $100,000 forever if this fund earns an interest rate of 10%, compounded continuously?

36. A scholarship fund consists of $500,000 in an account paying 4% interest, compounded continuously. At what annual rate can the money be withdrawn if the fund is to last forever? Treat the withdrawals as a continuous income stream.

37. The rate of sales of a new book is estimated to be $s'(t) = 1,000,000e^{-0.5t}$ dollars per year, for all $t \geq 0$. Find the total revenue the book is expected to make.

38. Damage to the environment has caused a change in the snail population in an area. However, scientists have determined that it is growing again at the rate

$$g'(t) = 12te^{-0.3t}$$

in hundreds of snails per year. Find the total growth of the snail population if this growth rate persists forever.

39. A bacteria colony initially covered an area of 100 square millimeters. Since then the area has been growing at a rate of

$$A'(t) = \frac{180}{(t+1)^3}$$

square millimeters per day.
(a) Write an integral that gives the growth of the bacteria after h days.
(b) Does the colony's size grow indefinitely or is its area bounded?

40. The probability density function for the lifespan of a lightbulb with an average life of 1,000 hours is given by $f(x) = 0.001e^{-0.001x}$, $x \geq 0$. That means that the percentage of lightbulbs that will last more than h hours is equal to

$$L(h) = \int_h^\infty 0.001e^{-0.001x}\, dx.$$

Compute $L(500)$ and $L(1,500)$.

41. The manager of a game preserve wants to establish a wild boar population. The goal is to allow hunters to remove 75 animals a year but still maintain a viable population. The boars increase in numbers at a rate of 3% per year. Model this as a perpetual income stream problem with $S=75$. Find the number of animals the manager needs in the initial colony. Assume no catastrophic changes to the boar population happen over time.

In Exercises 42–50, use a comparison test to determine whether the integral converges or diverges.

42. $\int_1^\infty \frac{1}{x^4+2}\, dx$

43. $\int_0^\infty \frac{1}{\sqrt{x^3+1}}\, dx$

44. $\int_2^\infty \frac{1}{\sqrt{x^2-1}}\, dx$

45. $\int_0^1 \frac{1}{\sqrt{x^3(1-x)}}\, dx$

46. $\displaystyle\int_0^1 \frac{1}{\sqrt{x^3(1+x)}}\, dx$

47. $\displaystyle\int_0^1 \frac{1}{x\ln(x+1)}\, dx$

48. $\displaystyle\int_1^\infty e^{-x^2}\, dx$

49. $\displaystyle\int_{-\infty}^\infty e^{-x^2}\, dx$

50. $\displaystyle\int_0^\infty x^3 e^{-x^2}\, dx$

51. $\displaystyle\int_2^\infty \frac{1}{x^2-1}\, dx$

(*Hint:* Show that $x^2 - 1 \geq (x-1)^2$ if $x \geq 1$.)

52. $\displaystyle\int_0^1 \frac{1}{x+x^2}\, dx$ (*Hint:* $x^2 \leq x$ if $0 \leq x \leq 1$.)

53. In Example 6.5.9 we saw that $\int_0^\infty 1/(1+x^2)\, dx$ converges. Estimate this integral by approximating $\int_0^{1,000} (1/(1+x^2))\, dx$, using the numerical integration function on a graphing calculator.

54. Use the facts that

$$\int_0^\infty \frac{1}{1+x^2}\, dx = \int_0^{1,000} \frac{1}{1+x^2}\, dx + \int_{1,000}^\infty \frac{1}{1+x^2}\, dx$$

and

$$\frac{1}{x^2+1} \leq \frac{1}{x^2}$$

to estimate the approximation error in the previous problem.

55. The function

$$f(x) = \frac{1}{\sqrt{2\pi}} e^{-x^2/2}, \quad -\infty < x < \infty,$$

is known as the **standard normal distribution function**, and it plays a very important role in probability and statistics. Its graph, shown in Figure 6.5.6, is the famous "bell-shaped" curve. The total area under its graph is equal to 1, a fact whose proof will not be given here. However, to obtain some empirical evidence, use the numerical integration function on your calculator to find $\int_{-4}^4 e^{-x^2/2}$. Then, as a comparison, compute $\sqrt{2\pi}$. What do you conclude?

56. Continuation of the previous problem Estimate the error in approximating $1/\sqrt{2\pi}\int_{-\infty}^\infty e^{-x^2/2}$ by $1/\sqrt{2\pi}\int_{-4}^4 e^{-x^2/2}$ as follows:
(a) Show that if $x \geq 4$, then $e^{-x^2/2} \leq e^{-2x}$. Use that inequality to find an upper bound for $\int_4^\infty e^{-x^2/2}$.
(b) Explain why the same bound holds for $\int_{-\infty}^{-4} e^{-x^2/2}$.
(c) Use the previous steps to find a bound on the difference between $\int_{-\infty}^\infty e^{-x^2/2}$ and $\int_{-4}^4 e^{-x^2/2}$.

57. In Example 6.5.10 we saw that $\int_0^1 (1/\sqrt{x^2+x})\, dx$ converges. Approximate the improper integral by using the numerical integration function on a graphing calculator to approximate the integral over the interval $[0.0001, 1]$. Estimate the size of the error made in replacing 0 by 0.0001 as the lower bound of the integral.

In the following exercises we take up some cases of the formulas of (42).

58. Show that $\lim_{x\to\infty} xe^{-x} = 0$ in the following sequence of steps:
(a) Let $f_1(x) = e^x - x$. Show that $f_1(0) > 0$ and $f_1'(x) \geq 0$ for $x \geq 0$. Use these facts to conclude that $e^x > x$ for $x \geq 0$.
(b) Let $f_2(x) = e^x - x^2/2$. Show that $f_2(0) > 0$ and $f_2'(x) > 0$ for $x \geq 0$. Use these facts to conclude that $e^x > x^2/2$ for $x \geq 0$.
(c) Use that to show $\lim_{x\to\infty} e^x/x = \infty$ and, therefore, $\lim_{x\to\infty} xe^{-x} = 0$.

59. Continuation of the previous exercise Show that $\lim_{x\to\infty} x^2 e^{-x} = 0$ in the following sequence of steps:
(a) Let $f_3(x) = e^x - x^3/(3 \cdot 2)$. Show that $f_3(0) > 0$ and $f_3'(x) > 0$ for $x \geq 0$. Use these facts to conclude that $e^x > x^3/(3 \cdot 2)$ for $x \geq 0$.
(b) Use that to show $\lim_{x\to\infty} e^x/x^2 = \infty$ and, therefore, $\lim_{x\to\infty} x^2 e^{-x} = 0$.

60. Continuation of the previous exercises Use similar arguments to show that $\lim_{x\to\infty} x^m e^{-x} = 0$ for $m = 3$ and 4. Explain how you might continue this process up to any positive integer k to show that $\lim_{x\to\infty} x^k e^{-x} = 0$.

61. Continuation of the previous exercises Show that $\lim_{x\to 0} x(\ln x)^m = \lim_{u\to\infty}(-1)^m u^m e^{-u}$ after the substitution $u = -\ln x$. Combine that with the previous exercises to conclude that $\lim_{x\to 0} x(\ln x)^m = 0$ for $m = 1, 2, 3,$ and 4.

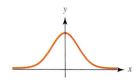

Figure 6.5.6

Solutions to practice exercises 6.5

1. (a) The integral converges and

$$\int_1^\infty \frac{dx}{x^3} = \lim_{b \to \infty} \int_1^b \frac{dx}{x^3} = \lim_{b \to \infty} \left(-\frac{x^{-2}}{2} \right) \Big|_1^b$$

$$= \lim_{b \to \infty} \left(\frac{1}{2} - \frac{1}{2b^2} \right) = \frac{1}{2}.$$

(b) The integral diverges because

$$\int_0^1 \frac{1}{x} \, dx = \lim_{h \to 0^+} \int_h^1 \frac{1}{x} \, dx = \lim_{h \to 0^+} (\ln x) \Big|_h^1$$

$$= \lim_{h \to 0^+} (-\ln h) = \infty.$$

2. We first observe that for large x, $x^4 - 1 \approx x^4$, so that we might try to compare the given integral with $\int_2^\infty (1/x^4) \, dx$, which converges. However, a direct comparison does not work because $1/(x^4 - 1) > 1/x^4$ on $[2, \infty)$. To get a comparison, we observe that $x^4 < 2x^4 - 2$ on $[2, \infty)$, and, therefore, $1/(x^4 - 1) < 2/x^4$. Moreover,

$$\int_2^\infty \frac{2}{x^4} \, dx = \lim_{b \to \infty} \int_2^b \frac{2}{x^4} \, dx = \lim_{b \to \infty} \left(\frac{1}{12} - \frac{2}{3b^3} \right) = \frac{1}{12}.$$

Therefore, the given integral converges.

· · · · | **Chapter** **6** **Summary** | ·

- If the market for a product is at **equilibrium**, that is, the **demand** $D(q)$ equals the **supply** $S(q)$, and if we denote the equilibrium point by (q_e, p_e), then the **consumer surplus** (CS) is the area of the region bounded above by the demand curve and below by the line $p = p_e$, while the **producer surplus** (PS) is the area of the region bounded above by the line $p = p_e$ and below by the supply curve. Thus,

$$\mathrm{CS} = \int_0^{q_e} D(q) \, dq - p_e q_e, \quad \mathrm{PS} = p_e q_e - \int_0^{q_e} S(q) \, dq.$$

- The **future value** after T years of an income stream deposited at a continuous rate of $S(t)$ dollars per year into an account paying interest rate r, compounded continuously, is given by $\int_0^T S(t) e^{r(T-t)} \, dt$, while the **present value** of such an income stream is given by $\int_0^T S(t) e^{-rt} \, dt$.

- A differential equation is called **separable** if it can be written in the form $f(y) \, dy = g(t) \, dt$.

- The **logistic growth equation** is $dp/dt = rp - kp^2$, where r denotes the **intrinsic growth constant** of the given population.

- An integral is called an **improper integral** if either *one or both of the limits of integration is infinite* or if *its integrand is an unbounded function*. In the first case, we have the definitions

$$\int_a^\infty f(x) \, dx = \lim_{b \to \infty} \int_a^b f(x) \, dx, \quad \int_{-\infty}^b f(x) \, dx = \lim_{a \to -\infty} \int_a^b f(x) \, dx,$$

provided that the limits exists.

Chapter 6 · Review Questions

- Explain why the demand curve must be decreasing and the supply curve increasing.

- Explain the difference between the present and future value of the same income stream. Give a concrete example.

- Give an example of a separable differential equation and also an example of a differential equation that is not separable.

- Let $K = r/k$ and write the logistic equation in the form $dp/dt = rp(1 - p/K)$. Then find its **equilibrium solutions**. What is the physical interpretation of K and what it is called? Give a concrete example.

- What is a **perpetual income stream** and how it is computed?

- Define the improper integral when the integrand is an **unbounded** function. Give concrete examples.

Chapter 6 · Review Exercises

1. The demand and supply curves for a product are as shown in Figure 6.R.1. Shade the region whose area represents the consumer surplus. Then make your best estimate of the consumer surplus.

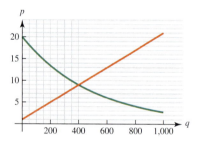

Figure 6.R.1

2. Suppose the supply and demand cuves are linear, as shown in Figure 6.R.2. Find the consumer and producer surpluses.

3. The demand curve of a certain item is $p = D(q) = (q - 5)^2$, $0 \le q \le 5$, and its supply curve is $p = S(q) = q^2 + q + 3$. Find the equilibrium price and equilibrium quantity. Then compute the consumer surplus.

4. The demand curve of a certain item is $p = D(q) = 56/(q + 2)$ and its supply curve is $p = S(q) = q + 3$. Find

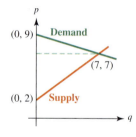

Figure 6.R.2

the equilibrium price and equilibrium quantity. Then compute the consumer surplus and the producer surplus.

5. The demand and supply curves for a certain alcoholic beverage are

$$p = -0.1q + 40 \quad \text{and} \quad p = 0.1q + 20.$$

(a) Find the equilibrium quantity and price, as well as the consumer and producer surplus.

(b) If the goverment imposes a \$4 tax per unit upon suppliers, then compute the new equilibrium, and the new consumer and producer surplus. (*Hint:* The supply cuve shifts four units upward.)

6. You have just discovered an investment that will pay 8% interest (continuously compounded) for at least 15 years.
(a) If you invest \$5,000 today, how many years do you need the investment to last in order to have \$15,000?

(b) Suppose, if instead, you deposit money continuously into the account at the rate of $2,000 per year for 4 years and then don't add any more. How many years after you start depositing will it take for the money to reach $15,000?

7. Assume that you deposit $300 a month in a bank paying 6% annual interest, compounded continuously. If you stop depositing money after 8 years and withdraw everything in the account to buy a new car, how much will you have available to spend? (You may assume that the deposits are made continuously.)

8. A person estimates that to retire comfortably in 30 years, she will need $2,000,000 in savings. If she can earn 10% interest, compounded continuously, how much should she deposit each year in order to meet this goal? Treat the deposits as a continuous income stream.

9. Suppose you win the lottery, and the officials tell you that you can take either (1) $50,000 per year for 20 years or (2) $500,000 in a lump sum. Assume that interest is compounded continuously and, in Option 1, the payments are made continuously.
(a) If interest rates are 10%, which is the better choice?
(b) What if interest rates are 5%?

10. Suppose that a small apartment building brings in revenue at a variable rate of $6,000(1.08^t)$ per year, where t is the time in years. Of the revenue 30% is deposited as a continuous income stream in an account paying 6% annual interest, compounded continuously. How much will be in the account at the end of 10 years?

11. As part of an incentive package, a firm agrees to pay a retirement bonus of $1,000(5 - t)/(t + 1)$ dollars per year during the first 5 years after an employee retires, where t is the time in years. Assume the money is continuously deposited into an account that pays 4% annual interest, compounded continuously.
(a) Write an integral that gives the present value of that continuous income stream.
(b) Find a numerical estimate.

12. A 30-year-old wins $1,000,000 in a lawsuit. He deposits it in an account earning interest of 7%, compounded continuously, and regularly makes withdrawals of $50,000 per year throughout his life. If $M(t)$ is the amount in his account after t years, find the differential equation satisfied by the function M and the initial value $M(0)$. Treat the withdrawals as a continuous income stream.

13. Find the constant k for which the function $y = 5x^2 - k$ is a solution of the differential equation $xy' - 2y = 8$.

14. Find the general solution of the differential equation $y' = (x - 1)(y - 2)$.

15. Find the solution of the differential equation $y' = xy + x$ satisfying $y(0) = 2$.

16. Let $y(x)$ be the solution of the initial value problem $y' = ye^x$, $y(0) = 1$. Find $y(1)$.

17. Find the solution of the initial value problem, $y' = xe^y$, $y(0) = \ln 2$.

18. Find the solution of the initial value problem $dy/dt = y^2 + y$, $y(0) = 1$.

19. A person opens a retirement account with an initial amount of $2,000. After that, money is continuously deposited in the account at the rate of $9,000 per year. Assume that the interest rate is 9%, compounded continuously. Model this problem as a differential equation and an initial condition describing the amount of money, $M(t)$, in the account at any time t. Then solve it to find $M(t)$ at any time t and the balance after 25 years.

20. A retired person has a sum of $1,500,000 invested so as to draw interest at an annual rate of 8%, compounded continuously. Withdrawals for living expenses are made at a rate of $75,000 per year. Assume that the withdrawals are made continuously.
(a) Model this problem as a differential equation and an initial condition describing the amount of money, $M(t)$, in the account at any time t.
(b) Solve this initial value problem to find $M(t)$.

21. A home buyer can afford to spend no more than $900 per month on mortgage payments. Suppose that the annual interest rate is 8%, compounded continuously, that the term of the mortgage is 30 years, and that payments are also made continuously.
(a) Determine the maximum amount that this buyer can afford to borrow.
(b) Determine the total interest paid during the term of the mortgage.

22. If a population grows exponentially at a rate of 4%, find (to the nearest year) how long it takes the population to double.

23. Solve the logistic equation $dp/dt = p - p^2$ with $p(0) = 5$ and compute $\lim_{t \to \infty} p(t)$.

24. Determine the intrinsic growth rate and the equilibrium solutions of the logistic equation $dp/dt = 0.02p - 0.001p^2$.

25. A population is modeled by the logistic equation $dp/dt = 0.015p - 0.0003p^2$, $p(0) = 10$. Find the environmental carrying capacity.

26. The U.S. population was 230 million in 1980 and 250 million in 1990. Assume that the U.S. population satisfies a logistic growth equation and has an intrinsic growth rate of about 0.02. Compute the U.S. population at any time t and its equilibrium value.

27. Find the equilibrium solutions of $dy/dt = 5y - y^2$.

28. Find the constant solutions of the differential equation $y' = y^2 - 5y + 6$.

29. Compute the improper integral $\int_0^\infty (x^2/(x^3 + 1)^4)\,dx$.

30. Compute the improper integral $\int_0^1 (x^2/\sqrt{1 - x^3})\,dx$.

31. Show that the function $f(x) = 0.5e^{-0.5x}$, $x > 0$ is a probability density function.

32. Find the constant c so that $f(x) = ce^{-x/8}$, $0 < x < \infty$ is a probability density function.

33. Find the present value of a perpetual income stream flowing continuously at a rate of $10,000 per year and with interest compounded continuously at the rate of 10%.

34. You are interested in purchasing an apartment complex. According to records, the complex brings in a continuous profit at the rate of $10,000 per year. You figure that you can safely invest this money simultaneously at 5% (compounded continuously).
(a) The current owner claims that the stream of money will last indefinitely. If that is true, what value would you place on this property?
(b) On the other hand, you are worried that changes in demographics will mean that the apartment complex has at most 15 good years left. As such, what value would you place on it?

In Exercises 35–40, determine whether the integral converges, and, if so, find its values.

35. $\int_0^1 \dfrac{e^t}{\sqrt{e^t - 1}}\,dt$

36. $\int_1^\infty \dfrac{e^t}{\sqrt{e^t + 1}}\,dt$

37. $\int_0^\infty \dfrac{e^x}{e^x + e^{-x}}\,dx$

38. $\int_0^1 \dfrac{1}{x^2 + x}\,dx$

39. $\int_0^1 (\ln x)^3\,dx$

40. $\int_2^\infty \dfrac{1}{x \ln x}\,dx$

In Exercises 41–46, use a comparison test to determine whether the integral converges.

41. $\int_1^\infty \dfrac{1}{x^3 + 1}\,dx$

42. $\int_1^\infty \dfrac{1}{1 + \ln x}\,dx$

43. $\int_2^\infty \dfrac{1}{\sqrt{x^3 - 1}}\,dx$

44. $\int_1^2 \dfrac{1}{\sqrt{x^2 - 1}}\,dx$

45. $\int_0^1 \dfrac{1}{\ln(x^2 + 1)}\,dx$

46. $\int_1^\infty e^{-x^2 + x}\,dx$

47. In macroeconomics, the initial value problem
$$\frac{dk}{dt} = s(A\sqrt{k} + 1)(B\sqrt{k} + 1), \quad k(0) = k_0,$$
where $A = a - \sqrt{b/s}$ and $B = a + \sqrt{b/s}$, with s, a, b, and k_0 being positive numbers, is a Solow model for capital stock. Using separation of variables with partial fractions, substitution, and algebraic manipulations, derive the implicit solution
$$\left(\frac{A\sqrt{k} + 1}{A\sqrt{k_0} + 1}\right)^{1/A}\left(\frac{B\sqrt{k} + 1}{B\sqrt{k_0} + 1}\right)^{-1/B} = e^{(\sqrt{bs})t}.$$

Source: "A Contribution to the Theory of Economic Growth," *Quarterly Journal of Economics*, (Feb., 1956), pp. 65–94.

••••• **Chapter 6 Practice Exam** ••••••••••••••••••••••••••••••

1. Figure 6.E.1 shows the graphs of the demand function $D(q)$ and the supply function $S(q)$. Which of the following is the best approximation for the producer surplus?
  (a) 9   (b) 24   (c) 15   (d) 5   (e) 18

2. Assume that for a certain commodity the demand curve is $D(q) = (q - 7)^2$, $0 \le q \le 7$, and the supply curve is $S(q) = q^2 + 6q + 9$. Find the equilibrium quantity and the equilibrium price.

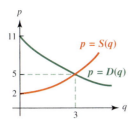

Figure 6.E.1

3. Assume that for a certain commodity the supply curve is $S(q) = q + 2$ and the demand curve is $D(q) = 16/(q + 2)$. Find the equilibrium quantity and price, and determine the consumer surplus.

4. The demand curve of some commodity is $D(q) = 49/(q + 3)$ and its supply curve is $S(q) = q + 3$. Find the producer surplus.

5. Suppose a person invests a lump sum of $50,000 at 8%, compounded continuously. How much will the investment be worth in 20 years?

6. If you deposit $1,400 each month into a retirement account that pays 7% interest, compounded continuously, how much money will be in your account after 30 years? Treat the deposits as a continuous income stream.

7. Suppose you win the lottery and are faced with two choices: a lump sum payment of $2,000,000 today, or an initial payment of $500,000 and a 5-year continuous income stream at the rate of $360,000 per year, deposited into an account paying 9% interest, compounded continuously.
(a) Find the present value of a 5-year continuous income stream of $360,000 per year at an interest rate of 9%, compounded continuously.
(b) Which is the better option? Justify your answer.

8. Suppose a continuous income stream is being deposited in an account at the variable rate of $1,000t$ dollars per year for a period of 10 years, where t is the time in years. If the account pays an annual interest rate of 6%, compounded continuously, what is its present value?

9. A woman retires on her 65th birthday and begins to withdraw $3,000 a month from her savings account, which earns 7%, compounded continuously. At that rate, she will run out of money on her 100th birthday. Let $M(t)$ be the amount in the account t years after her 65th birthday. Find an initial value problem satisfied by $M(t)$.

10. Find a constant k such that $y(t) = e^t + kt + 1$ is a solution of the differential equation $y' = y - 4t + 3$.

11. Solve the following initial value problem: $dy/dx = (x + 1)/y$ and $y(2) = 4$.

12. Solve the initial value problem $dQ/dt - 2Q = 6$ and $Q(0) = 5$.

13. What is the environmental carrying capacity K of the logistic equation given by the formula $dp/dt = 0.06p - 0.003p^2$?

14. The population of a region was 25 million in 1980 and 30 million in 2000. Assume that its intrinsic rate of growth is about 0.02 and that it follows the logistic growth model. Write a differential equation and an initial condition describing the population of the region at any time. Find the population of the region in the year 2020.

15. Find the present value of a perpetual income stream flowing continuously at a rate of $80,000 per year, with interest compounded continuously at 10%.

16. Compute the improper integral $\int_0^\infty 1/(x + 1)^{3/2}\, dx$.

17. Compute the improper integral $\int_2^3 (x - 2)^{-1/4}\, dx$.

18. Show that the improper integral $\int_1^\infty 1/(2 + x^4)\, dx$ converges.

Chapter 6 Projects

1. Assume that you are working for a construction company that is bidding for the right to construct a major new bridge. Because the local government is unable to finance the project, the winning bidder will have to do so in return for the right to collect tolls on the bridge for a certain number of years. Your proposal must specify that number, and the contract will be awarded to the bidder who specifies the lowest number. The government requires that the toll stay fixed at $5 per car throughout the entire period of the contract. Finally, your company has estimated that the construction of the bridge will cost $400 million and that cars will cross the bridge at the rate of 11,000,000 per year.

(a) If your company wins the contract, it plans to invest the revenue from the tolls as it comes in at an annual interest rate of 12%, compounded continuously. Compute the least number of years needed to make this arrangement at least as profitable as investing the entire $400 million right now at the same rate of interest.

(b) Suppose the number of years your company specifies is 1 greater than the minimum you computed and it wins the contract, but its estimate of the bridge traffic turns out to be wrong. What is the smallest number of cars per year that need to cross the bridge in order for the company to obtain as large a return on its investment as it would by investing the entire $400 million right now?

2. The use of the logistics equation as a model of population growth was pioneered in the early part of this century by the American biologist Raymond Pearl, who tested the model against various sets of experimental data. One such set, measuring the growth of a culture of yeast cells, is shown in Table 6.5.1. The left-hand column is the value of t (in hours). The second column is the average number of yeast cells observed over a number of repetitions of the experiment.

Pretend that it is the 1920s and you are Professor Pearl, and assume, as he did, that the data can be modeled by a logistic equation.

(a) On the basis of the given data, make a reasonable guess at the equilibrium solution (or saturation level). In other words, use the table to predict the maximum sustainable population.

(b) Using the cell count at time zero, write a formula for the logistic function that applies to this data. The only unknown constant in your formula should be the intrinsic growth rate r.

TABLE 6.5.1

No. of hours	Cell count
0	9.6
1	18.3
2	29.0
3	47.2
4	71.1
5	119.1
6	174.6
7	257.3
8	350.7
9	441.0
10	513.3
11	559.7
12	594.8
13	629.4
14	640.8
15	651.1
16	655.9
17	659.6
18	661.8

Source: **Raymond Pearl,** *The Quarterly Review of Biology,* **Vol. 2 (1927), p. 533.**

(c) Use $t = 10$ and the cell count at the end of 10 hours to obtain a value for r.

(d) Compute the number of cells predicted by the formula for all the values of t listed in the table and compare them with Pearl's cell count at each of those times.

(e) Repeat steps (c) and (d), using a different time and cell count from the table, and observe how much your predictions change.

HISTORICAL PROFILE

Raymond Pearl (1879–1940) was born in Farmington, New Hampshire, and received a Ph.D. from the University of Michigan in 1902. A fellowship at the Galton Laboratory of the University of London in 1905 and 1906 deepened his interest in biometrics and statistics, and he became a strong lifelong proponent of the use of statistical analysis in biology and medicine. After spending 2 years during World War I as the Chief of the Statistical Division of the U.S. Food Administration under Herbert Hoover, he accepted a position at Johns Hopkins University.

He was very prolific, and his bibliography contains over 700 items, including 17 books. He did pioneering studies in longevity, changes in world population, and genetics. In addition to being elected to the National Academy of Sciences and the American Academy of Arts and Sciences, he served as president of the American Statistical Association in 1939.

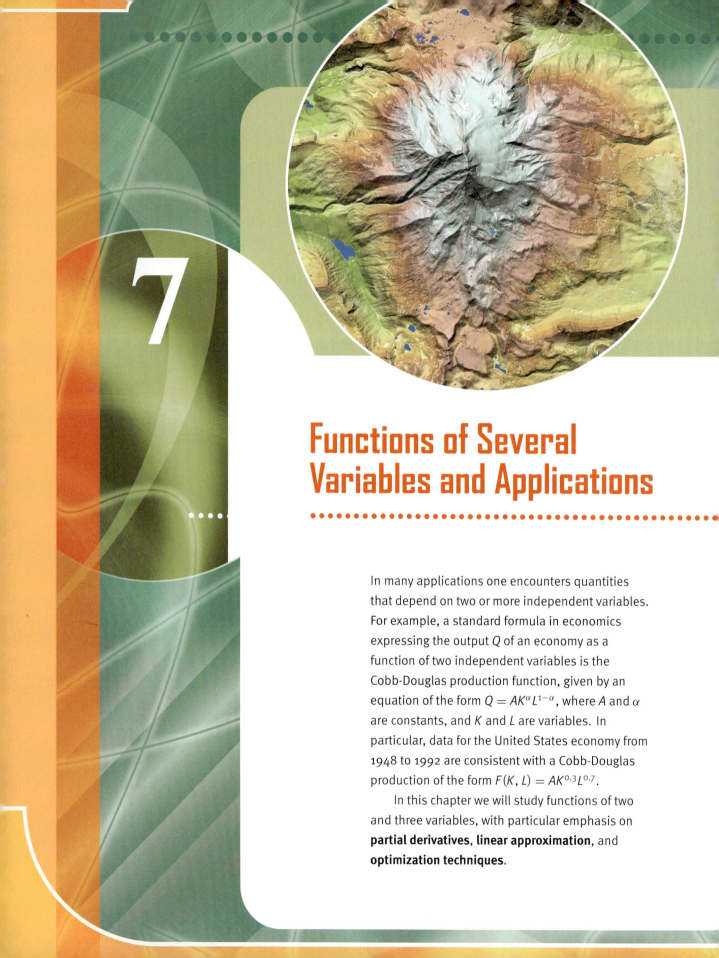

Functions of Several Variables and Applications

In many applications one encounters quantities that depend on two or more independent variables. For example, a standard formula in economics expressing the output Q of an economy as a function of two independent variables is the Cobb-Douglas production function, given by an equation of the form $Q = AK^{\alpha}L^{1-\alpha}$, where A and α are constants, and K and L are variables. In particular, data for the United States economy from 1948 to 1992 are consistent with a Cobb-Douglas production of the form $F(K, L) = AK^{0.3}L^{0.7}$.

In this chapter we will study functions of two and three variables, with particular emphasis on **partial derivatives**, **linear approximation**, and **optimization techniques**.

Recall that in Chapter 0 we introduced coordinates in a plane (called Cartesian coordinates). In a similar way, we can assign coordinates to the points of three-dimensional space. All we need is an additional axis. We start with a Cartesian plane, called the xy-plane, in horizontal position. Next, we draw an axis through the origin of the xy-plane and perpendicular to it. This vertical axis is called the **z-axis**. In that way we form a Cartesian system of coordinates in a three-dimensional space, as shown in Figure 7.1.1. The upward direction is the positive direction of the z-axis.

To any point Q in space we assign a unique ordered triple of numbers (x, y, z) as follows. We draw a vertical line through Q and let P be the point where it meets the xy-plane, as illustrated by Figure 7.1.1. The x- and y-coordinates of Q are the same as those of P. For the z-coordinate we take the length of the segment PQ if Q is above P and the negative of that length if Q is below P.

Conversely, we can assign a point Q in space to any ordered triple of real numbers (x, y, z) as follows. We start from the origin and move in the xy-plane to the point P with coordinates (x, y). Then we move vertically a distance of $|z|$ units, upward if $z > 0$ and downward if $z < 0$. Thus, we have established a one-to-one correspondence between all points of the three-dimensional space and all ordered triples of real numbers.

It is customary to denote the three-dimensional space with Cartesian coordinates by $\mathbb{R}^3$.

Distance in $\mathbb{R}^3$

By applying the Pythagorean theorem twice, as shown in the Figure 7.1.2, we see that the distance between the origin $(0, 0, 0)$ and a point (x, y, z) is $\sqrt{x^2 + y^2 + z^2}$. Similarly, the distance $d(P_1, P_2)$ between two points $P_1 = (x_1, y_1, z_1)$ and $P_2 = (x_2, y_2, z_2)$ is

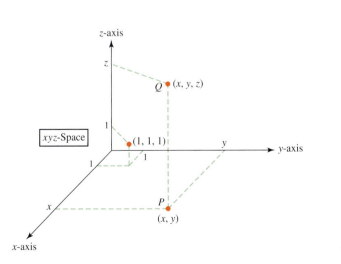

Figure 7.1.1

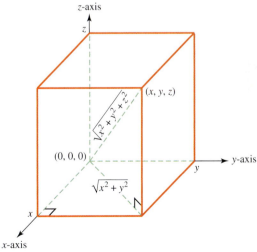

Figure 7.1.2

465

given by the formula

$$d(P_1, P_2) = \sqrt{(x_2 - x_1)^2 + (y_2 - y_1)^2 + (z_2 - z_1)^2}. \tag{1}$$

EXAMPLE 7.1.1

Find the distance between the points $P_1 = (-1, 3, 2)$ and $P_2 = (4, 1, -2)$.

SOLUTION   By applying formula (1), we get

$$d(P_1, P_2) = \sqrt{(4 + 1)^2 + (1 - 3)^2 + (-2 - 2)^2} = \sqrt{45}.$$

EXAMPLE 7.1.2

Show that the points $A = (2, 7, -1)$, $B = (1, 9, 4)$, and $C = (6, 4, 1)$ determine a right triangle in space.

SOLUTION   We need to use a fact from geometry: A triangle is a right triangle if and only if the lengths of its sides satisfy the Pythagorean formula—that is, $a^2 + b^2 = c^2$. In this case, the distance between the points A and B is

$$d(A, B) = \sqrt{(2 - 1)^2 + (7 - 9)^2 + (-1 - 4)^2} = \sqrt{30}.$$

The distance between A and C is

$$d(A, C) = \sqrt{(2 - 6)^2 + (7 - 4)^2 + (-1 - 1)^2} = \sqrt{29}$$

and the distance between B and C is

$$d(B, C) = \sqrt{(1 - 6)^2 + (9 - 4)^2 + (4 - 1)^2} = \sqrt{59}.$$

Since $(\sqrt{59})^2 = (\sqrt{30})^2 + (\sqrt{29})^2$, we conclude that the points A, B, and C form a right triangle.

Functions of two variables and their graphs

A function f of two variables consists of a rule or a formula that assigns a unique output $f(x, y)$ to each point (x, y) in some subset D of the xy-plane. The subset D is called the domain of the function. As we know, a function of one variable has a graph that is a curve in the plane. Similarly, a function of two variables has a graph that is a surface in three-dimensional space. If $f(x, y)$ is a function, then the set of points (x, y, z) with (x, y) in the domain of f and $z = f(x, y)$ forms a surface. That surface is the graph of the function, and it has the following characteristic property that held in the one-variable case:

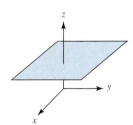

Figure 7.1.3

The simplest example of such a surface is the graph of a constant function, $f(x, y) = c$, where c is some fixed real number. The domain of this function consists of the entire xy-plane, and its graph is the set of all triples (x, y, z) in $\mathbb{R}^3$ with $z = c$. In geometric terms, it is a horizontal plane, parallel to the xy-plane, and $|c|$ units above it (if $c > 0$) or below it (if $c < 0$). An example of such a graph, with $c > 0$, is shown in Figure 7.1.3.

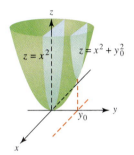

Figure 7.1.4

Visualizing a graph through x- and y-sections   One way to visualize the graph of a function $f(x, y)$ is to fix one of the variables and let the other vary freely. That way we obtain a function of one variable. For example, consider the function

$$f(x, y) = x^2 + y^2.$$

If we fix y at a given value y_0, we obtain the function

$$z(x) = f(x, y_0) = x^2 + y_0^2,$$

which is a function of the single variable x. If $y_0 = 0$, for instance, we obtain $z = x^2$, and if $y_0 = 3$, we obtain $z = x^2 + 9$. The function $z(x) = f(x, y_0)$ of the variable x that we obtain by fixing y at y_0 is called a **y-section**. Figure 7.1.4 shows the graphs of two y-sections, one of them corresponding to $y_0 = 0$. All y-sections of $f(x, y) = x^2 + y^2$ are parabolas. Their union, called a **paraboloid**, is the graph of $z = x^2 + y^2$.

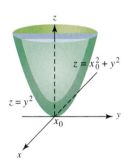

Figure 7.1.5

Similarly, if we fix $x = x_0$ in $f(x, y)$, we obtain a one-variable function of y, $z(y) = f(x_0, y)$, called an **x-section** of $f(x, y)$. For $f(x, y) = x^2 + y^2$, the x-sections are of the form $z(y) = x_0^2 + y^2$. Figure 7.1.5 shows the graphs of two x-sections, one of which corresponds to $x_0 = 0$. Again, all x-sections of $f(x, y) = x^2 + y^2$ are parabolas, and their union is the same paraboloid, the graph of $z = x^2 + y^2$. We can think of the x-sections as the curves that we get by slicing the surface with planes perpendicular to the x-axis, and the y-sections as the curves we get by slicing the surface with planes perpendicular to the y-axis.

Visualizing a graph by horizontal slicing   Another way to visualize the graph of a function $f(x, y)$ is to slice it by horizontal planes at different heights to obtain a family of horizontal curves. By projecting these curves down to the xy-plane, we partition the domain of f into disjoint sets, called level curves. For a given number z_0, the **level curve** of height z_0 is the set of all points in the domain of f satisfying the equation $f(x, y) = z_0$. Level curves are commonly seen in topographical maps, with each level curve representing the contour of points at a particular altitude. They are also seen in weather maps, where each level curve is the contour of points at a particular temperature (called an **isotherm**).

In general, the level curve given by $f(x, y) = z_0$ is actually a curve, but in certain cases it may consist of an isolated point (or points) or may even be empty.

EXAMPLE 7.1.3

Describe the level curves of the function $f(x, y) = x^2 + y^2$ and draw several of them.

SOLUTION   In this case the natural domain of f consists of the entire xy-plane. A level curve is given by an equation of the form $x^2 + y^2 = z_0$. There are three possibilities:

- If $z_0 < 0$, this equation has no solution and there is no level curve.
- If $z_0 = 0$, the only solution is $(x, y) = (0, 0)$, and the level curve reduces to a single point.
- If $z_0 > 0$, the level curve is the circle of radius $\sqrt{z_0}$ centered at $(0, 0)$.

In Figure 7.1.6 we show the level curves corresponding to $z_0 = \frac{1}{4}$, 1, and 4.

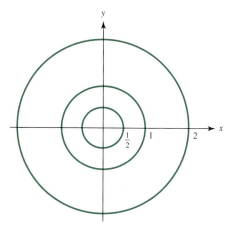

Figure 7.1.6

We can use level curves to help us visualize the graph of a function as a surface in the three-dimensional xyz-space by recalling that each level curve corresponds to a cross section of that surface by a horizontal plane. For instance, in Example 7.1.3 the level curve of height $z_0 > 0$ of the function $f(x, y) = x^2 + y^2$ is the circle of radius $\sqrt{z_0}$ centered at the origin. This circle, $z_0 = x^2 + y^2$, is the curve that results from slicing the graph of $z = x^2 + y^2$ with the horizontal plane $z = z_0$ and then projecting the resulting curve onto the xy-plane.

To visualize the surface, we reverse that process—that is, we think of lifting each level curve vertically upward to its height in $\mathbb{R}^3$. The result is a family of horizontal circles, whose radii expand as they get higher—with the circle at height z_0 having radius $\sqrt{z_0}$. In Figure 7.1.7 we have drawn the surface, two of its cross sections (corresponding to $z_0 = \frac{1}{4}$, 1), and the corresponding level curves in the xy-plane. To summarize: We can picture the surface as a collection of curves obtained by slicing it with a family of planes perpendicular to the z-axis.

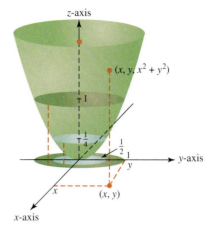

Figure 7.1.7

EXAMPLE 7.1.4

Let $f(x, y) = x^{1/3}y^{2/3}$ be a Cobb-Douglas production function, where $x > 0$ and $y > 0$.
  (i) Describe the level curves of this function.
  (ii) Graph the set of all inputs (x, y) that result in the production of 100 units.

SOLUTION   (i) The level curves are given by the system of equations

$$x^{1/3}y^{2/3} = z_0.$$

By cubing both sides, we can change this to the form $xy^2 = z_0^3$. Since we are restricting attention to positive x and y, the level curves only exist if $z_0 > 0$, and, in that case, solving for y gives $y = z_0^{3/2}/\sqrt{x}$. Several of these curves are shown in Figure 7.1.8.
  (ii) The inputs that result in the production of 100 units are simply the points on the level curve for $z_0 = 100$, that is, $y = 1{,}000/\sqrt{x}$. It is the middle curve in Figure 7.1.8.

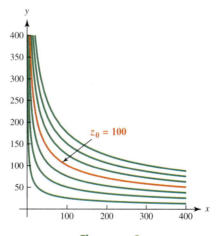

Figure 7.1.8

Below we show the graphs of a few functions together with some of their level curves, all drawn using Mathematica:

• $f(x, y) = x + y$, $-2 < x < 2, -2 < y < 2$ (Figures 7.1.9 and 7.1.10)
• $f(x, y) = 9 - x^2 - y^2$, $-2 < x < 2, -2 < y < 2$ (Figures 7.1.11 and 7.1.12)
• $f(x, y) = y^2 - x^2$, $-2 < x < 2, -2 < y < 2$ (Figures 7.1.13 and 7.1.14)

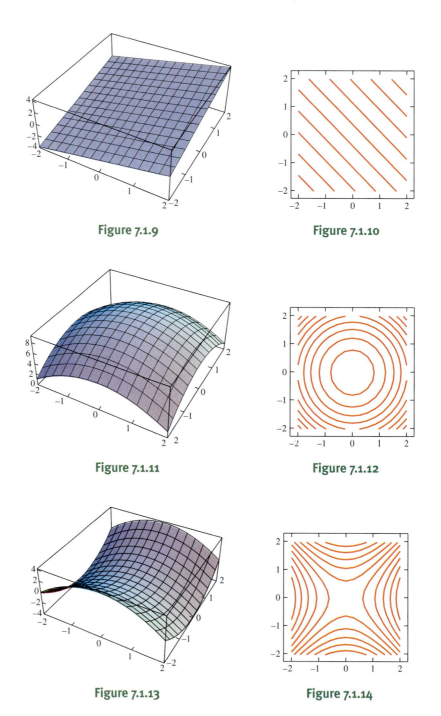

Figure 7.1.9   **Figure 7.1.10**

Figure 7.1.11   **Figure 7.1.12**

Figure 7.1.13   **Figure 7.1.14**

Planes and linear functions

Recall that a linear equation in two variables is represented geometrically by a straight line in the plane. Conversely, a straight line is described algebraically by a linear equation. As you already know, we can separate the lines and their equations into three types, as follows:

- A vertical line—one that is perpendicular to the x-axis—has an equation of the form $x = x_0$.

- A horizontal line—one that is perpendicular to the y-axis—has an equation of the form $y = y_0$.

- A line that is not perpendicular to either axis has an equation of the form $y = ax + b$ with $a \neq 0$, where a is the slope and b the y-intercept.

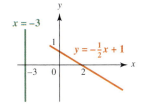

Figure 7.1.15

The second and third types are graphs of functions of x, whereas the first is not. Figure 7.1.15 shows two lines, one with equation $x = -3$ and the other with equation $y = -\frac{1}{2}x + 1$.

There is a similar correspondence between planes in $\mathbb{R}^3$ and linear equations in three variables. We have already observed that the graph of an equation of the form $z = z_0$ is a horizontal plane—in other words, a plane perpendicular to the z-axis—and it intersects the z-axis at the point $(0, 0, z_0)$. Similarly, the plane perpendicular to the x-axis at the point $(x_0, 0, 0)$ is described algebraically by the equation $x = x_0$, and the plane perpendicular to the y-axis at the point $(0, y_0, 0)$ is described by the equation $y = y_0$. (See Figure 7.1.16.)

A plane perpendicular to either the x- or y-axis is parallel to the z-axis. In general, a plane that is parallel to the z-axis is described by a linear equation in which the variable z does not appear. To see why, think of the line of intersection of such a plane with the xy-plane. Its equation in the xy-plane has the form $ax + by = c$, and the given plane, parallel to the z-axis, consists of all points in space directly above or below that line. In other words, if $(x_0, y_0, 0)$ is a point on the line, then every point of the form (x_0, y_0, z) is on the given plane, without restriction on z. And, conversely, if (x_0, y_0, z_0) is a point on the given plane, then so is $(x_0, y_0, 0)$. We conclude that

- an equation of the form $ax + by = c$ represents a plane parallel to the z-axis, and, conversely,

- any plane parallel to the z-axis is represented by an equation of that form.

Figure 7.1.16 shows the planes $x = 4$, $y = 3$, and $x + y = 2$.

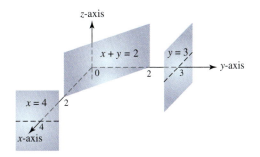

Figure 7.1.16

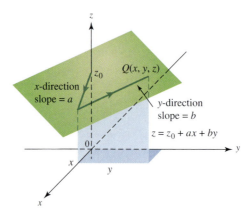

Figure 7.1.17

A plane that is parallel to the z-axis is not the graph of an equation of the form $z = f(x, y)$. In other words, it is not the graph of a function of x and y. (Think of the vertical line test.) On the other hand, any plane that is *not* parallel to the z-axis *is* the graph of a function of x and y (again, the vertical line test). In fact, it is the graph of a *linear* function—that is, of an equation of the form

$$z = z_0 + ax + by. \tag{2}$$

Without going into detail, we remark that the form of Eq. (2) can be derived from the fact that all the y-sections are lines of the same slope, and all the x-sections are lines of the same slope. Conversely, any equation of the form of (2) has a plane in $\mathbb{R}^3$ as its graph.

The constants on the right-hand side of Eq. (2) have the following interpretations, as illustrated by Figure 7.1.17.

- The coefficient z_0 is the *z-intercept*. That is, the plane intersects the z-axis at the point $(0, 0, z_0)$.
- The coefficient a is the *slope in the x-direction*. Any plane of the form $y = y_0$, perpendicular to the y-axis, intersects the plane of Eq. (2) in a line of slope $\Delta z / \Delta x = a$.
- The coefficient b is the *slope in the y-direction*. Any plane of the form $x = x_0$, perpendicular to the x-axis, intersects the plane of Eq. (2) in a line of slope $\Delta z / \Delta y = b$.

EXAMPLE 7.1.5

Find the equation of the plane through the points $(3, 0, 0)$, $(0, 4, 0)$ and $(0, 0, 5)$.

SOLUTION   Since the point $(0, 0, 5)$ is on the plane, we have $z_0 = 5$. To find the slope a in the x-direction, we use the points $(3, 0, 0)$ and $(0, 0, 5)$, which are both in the xz-plane, to obtain

$$a = \frac{\Delta z}{\Delta x} = \frac{5 - 0}{0 - 3} = -\frac{5}{3}.$$

To find the slope b in the y-direction, we use the points $(0, 4, 0)$ and $(0, 0, 5)$ in the yz-plane to obtain

$$b = \frac{\Delta z}{\Delta y} = \frac{5 - 0}{0 - 4} = -\frac{5}{4}.$$

Therefore, the equation of the given plane is

$$z = 5 - \frac{5}{3}x - \frac{5}{4}y.$$

The plane is shown in Figure 7.1.18.

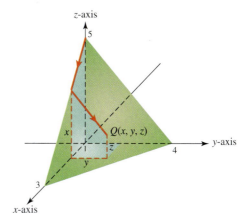

Figure 7.1.18

It should be emphasized that *a plane does not have a slope.* There is no single number that determines a plane up to parallel translation, the way the slope of a line does in $\mathbb{R}^2$. For a plane in $\mathbb{R}^3$ not parallel to the z-axis that is done by a *pair* of numbers, such as its slope in the x-direction (i.e., the slope of a y-section) and its slope in the y-direction (the slope of an x-section). To completely specify the plane, we need those two plus a third number, the z-intercept.

The equation of a plane not parallel to the z-axis can be determined from its slopes in the x- and y-directions and one point on the plane, as follows:

Equation of a Plane

The equation of the plane through an arbitrary point (x_1, y_1, z_1) having slope a in the x-direction and slope b in the y-direction is

$$z - z_1 = a(x - x_1) + b(y - y_1). \tag{3}$$

To derive it, suppose z_0 is the z-intercept. Then the equation of the plane has the form

$$z = z_0 + ax + by.$$

Since (x_1, y_1, z_1) is a point on the plane, the coordinates must satisfy

$$z_1 = z_0 + ax_1 + by_1.$$

Subtracting this equation from the previous one gives (3).

EXAMPLE 7.1.6

A company produces two products, X and Y. Assume that the cost (in hundreds of dollars) of producing x units of X and y units of Y is a linear function. Find the cost function $C(x, y)$, given the following data:

$$C(10, 20) = 120, \quad C(30, 15) = 210, \quad C(40, 50) = 330.$$

SOLUTION   The problem is equivalent to finding the equation of the plane through the points $(10, 20, 120)$, $(30, 15, 210)$, and $(40, 50, 330)$. Using Eq. (3) with the first of these points gives

$$z - 120 = a(x - 10) + b(y - 20),$$

where z is the cost function. To determine a and b, we first substitute $(x, y, z) = (30, 15, 210)$ into this equation, which gives

$$90 = 20a - 5b.$$

Next, we substitute $(x, y, z) = (40, 50, 330)$, which yields

$$210 = 30a + 30b.$$

We now have two linear equations in the unknowns a and b, which we can solve simultaneously to get $a = 5$ and $b = 2$. Therefore,

$$z - 120 = 5(x - 10) + 2(y - 20).$$

By expanding and collecting terms, we get

$$z = C(x, y) = 30 + 5x + 2y.$$

Practice Exercises 7.1

●●●

1. Compute the value of the function $f(x, y) = x^2/(x + y)$ at each of the points $(1, 2)$, $(0, 1)$, and $(-2, 1)$.

2. **(a)** Describe the level curves of the function $f(x, y) = 9 - x^2 - y^2$ corresponding to the values $z = 8, 5, 0, -7,$ and -16.
   (b) Describe the graph as a surface in three-dimensional space.
   (c) What kind of curve is formed by the intersection of the surface with a plane perpendicular to the x-axis? to the y-axis?

   (d) Are there any values of z for which there is no level curve? for which the curve reduces to a point?
   (e) What is the natural domain of the function?

3. Find the equation of the plane through the points $(-1, 3, 1)$, $(1, 2, 4)$, and $(2, 1, 5)$, and find its slope in both the x- and y-directions.

Exercises 7.1

Plot the following points using a system of Cartesian coordinates in three-dimensional space:

1. $(2, 0, 0)$, $(0, 3, 0)$, $(2, 3, 0)$, $(2, 3, 4)$, $(2, 3, -4)$

2. $(0, 0, 2)$, $(0, 0, -2)$, $(0, 1, 0)$, $(0, 1, 2)$, $(-1, 1, 2)$

3. $(0, -1, 0)$, $(1, 0, -2)$, $(1, 0, 2)$, $(3, 3, 3)$, $(3, 3, 0)$

4. You start at the point $(0, 2, 3)$, go a distance of five units in the positive direction along a line parallel to the y-axis, then move a distance of four units in the positive direction on a line parallel to x-axis and finally move vertically upward a distance of nine units. What are your new coordinates?

5. Starting at the point $(5, 4, 8)$, you move vertically downward a distance of seven units, next move two units in the negative x-direction, and then three units in the negative y-direction. Finally, you move 1 unit in the positive z-direction. Find the coordinates of your final position.

6. Find the coordinates of the vertices of a cube with its center at $(5, 8, 10)$, a side equal to four units, and each edge parallel to one of the coordinate axes.

In Exercises 7–10, plot the points and compute the distance between them:

7. $(0, 0, 0)$, $(1, 3, 4)$ **8.** $(2, 3, 1)$, $(3, 5, 4)$

9. $(2, 0, -2)$, $(1, 1, 1)$ **10.** $(-2, -3, 4)$, $(3, -1, -2)$

In Exercises 11–14, decide whether the given three points determine a right triangle.

11. $(0, 0, 0)$, $(3, 2, 0)$, $(1, 5, 1)$

12. $(2, 0, -1)$, $(3, 1, 3)$, $(2, 3, 0)$

13. $(0, 3, -5)$, $(4, -1, 5)$, $(10, -4, 2)$

14. $(4, 3, 6)$, $(0, 5, -6)$, $(-4, 9, -4)$

15. Scientists put a radio tag on a whale shark and began monitoring its activity. They labeled its original location as the origin, made north the direction of the positive y-axis, and then assigned three-dimensional coordinates to each contact point. They gathered the following data for location the first 5 days after tagging:

day	location
1	$(0.6, -3, -0.02)$
2	$(-1.3, -6, -0.3)$
3	$(-10.6, -4, 0)$
4	$(-21, 2, -0.4)$
5	$(-39, 7, -0.09)$

Each measurement is in hundreds of meters. Find the exact distance the whale shark was from the original location on each day. What does the z-axis represent?

16. Orienteering is the sport of using a compass and a very detailed topographic map to hike from checkpoint to checkpoint without getting lost. Ralph is new to the sport and is having trouble finding a checkpoint. His mentor tells him the checkpoint is at location $(50, 15, 6)$ as measured in yards from the last checkpoint. Ralph determines that he is at location $(10, 35, 1)$. How far away is he from the next checkpoint?

In Exercises 17–23, compute the value of the given function at the given points.

17. $f(x, y) = 8 - 2x + 4y$; $(1, 0)$, $(0, 1)$, $(-3, 2)$, $\left(-\frac{1}{2}, -\frac{1}{4}\right)$

18. $R(x, y) = 100 + 0.2x + 0.5y$; $(200, 100)$, $(1,000, 2,000)$

19. $f(x, y) = 2x^2 - 3xy + 5y^2 - x$; $(0, 0)$, $(0, 1)$, $(-2, 3)$, $(2, -3)$

20. $P(K, L) = 10K^{1/3} L^{2/3}$; $(0, 0)$, $(5, 0)$, $(8, 1)$, $(8, 27)$, $(125, 1,000)$

21. $r(s, t) = \dfrac{2st}{s^2 + t^2}$; $(1, 0)$, $(-3, 4)$, $(\sqrt{5}, \sqrt{5})$, $(8, 8)$, (a, a) for $a \neq 0$

22. $g(x, y) = e^{x^2 - y^2}$; $(0, 1)$, $(1, 0)$, $(-1, 1)$, $(2, -3)$, (a, a) for any a

23. $l(x, y) = \ln \sqrt{x^2 + y^2}$; $(1, 0)$, $(0, -1)$, $(-e, 0)$, $(-2, -3)$

Describe the natural domain of each of the following functions:

24. $f(x, y) = \sqrt{1 - x^2 - y^2}$

25. $g(x, y) = (y - x)^{-1/2}$

26. $f(x, y) = \dfrac{1}{1 + x^2 + y^2}$

27. $f(x, y) = \dfrac{1}{1 - x - y}$

28. $h(x, y) = y + \sqrt{x - 1}$

29. $f(x, y) = \dfrac{1}{\sqrt{xy - x}}$

For each function in Exercises 30–38 sketch the level curves in the xy-plane corresponding to the given heights z_0 and then sketch the graph as a surface in three-dimensional space. Finally, describe the x- and y-sections for each function when x and y are fixed at the given values.

30. $f(x, y) = 1 + x + y$; $z_0 = -3, -2, -1, 0, 1, 2,$
$x = 0,\ y = -1$

31. $f(x, y) = 3 + 2x + 3y$; $z_0 = -3, 0, 3, 6, 9, 12,$
$x = 1,\ y = 0$

32. $f(x, y) = y - 2x$; $z_0 = -4, -2, 0, 2, 4,\ x = 4,\ y = 2$

33. $f(x, y) = x - y^2$; $z_0 = -2, -1, 0, 1, 2,\ x = 0,\ y = -2$

34. $f(x, y) = y - x^2$; $z_0 = -2, -1, 0, 1, 2,\ x = 1,\ y = -1$

35. $f(x, y) = 1 + x^2 + y^2$; $z_0 = 2, 5, 10, 17,\ x = 2,\ y = -3$

36. $f(x, y) = (x - 1)^2 + y^2$; $z_0 = 0, \frac{1}{4}, 1, 4, 9,\ x = 2,\ y = 1$

37. $f(x, y) = xy$; $z_0 = -3, -2, -1, 0, 1, 2, 3,\ x = 1,$
$y = -1$

38. $f(x, y) = \sqrt{25 - x^2 - y^2}$; $z_0 = 0, 3, 4, 5,$
$x = 4,\ y = 0.$

39. In 1840, the French physician Jean-Louis-Marie Poiseuille discovered the following formula governing the flow of blood in capillaries, known as **Poiseuille's law:**

$$Q = \frac{\pi r^4 P}{8Lv},$$

where r is the radius and L is the length of the capillary, P is the difference in pressure between the capillary's ends, and v is the blood's viscosity. Assuming that P and v remain constant, Q becomes $Q(r, L)$, a function of the radius and length of the capillary.
(a) If we fix L at some value L_0 and allow r to vary, we obtain an L-section. What does the section look like? Sketch a typical one. Rewrite Q to show this.
(b) Next, fix r at some value r_0. What does the corresponding r-section look like? Sketch one.

Source: J. Keener and J. Sneyd, *Mathematical Physiology*, New York: Springer, 1998.

40. You are hiking in a canyon surrounded by steep hills, and you are at the point A, as shown in Figure 7.1.19. The topographic map shows you two trail choices that lead you up a hill and out of the canyon. Each line represents a level curve with z-axis increments of 100 feet. Which hill is steeper? Why?

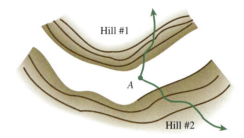

Figure 7.1.19

41. A company sells two products, X and Y. The unit price of X is \$2,000 and the unit price of Y is \$3,000. Suppose the company sells x units of X and y units of Y.
(a) Write a formula for the revenue function.
(b) Sketch the level curves corresponding to constant revenue of \$60,000 and \$120,000.

42. Let $Q(K, L) = 15K^{3/4}\, L^{1/4}$ be the Cobb-Douglas production function for a certain economy. Describe all inputs of capital and labor (K, L) that result in the production of 15,000 units. Sketch these inputs in a KL-coordinate plane. If the available capital is 100 units, find the labor needed to meet that production level.

43. The future value of a deposit of \$1,000, compounded continuously for t years with interest rate r, is a function of the two variables t and r, and is given by the formula $M(t, r) = 1,000e^{rt}$. Describe (by an equation) those inputs (r, t) that result in a future value of \$5,000. Sketch that set of inputs in an rt-coordinate plane. If the prevailing interest rate is 0.08, find the time needed for the future value to reach the level of \$5,000.

44. Suppose that the growth of a population is modeled by a logistic equation with an environmental carrying capacity (equilibrium solution) of 10 billion and intrinsic growth rate r. If the initial population is 5 billion, its population after t years is given by the formula

$$P(t, r) = \frac{10}{1 + e^{-rt}}.$$

Describe the inputs (r, t) that result in a population of 7.5 billion. Sketch those inputs in an rt-coordinate plane. If the intrinsic growth rate is 0.02, find the time needed for the population to reach the level of 7.5 billion.

In Exercises 45–57, sketch the plane with the given equation.

45. $x = 3$ **46.** $y = 8$ **47.** $z = 4$

48. $x - 1 = 0$ **49.** $y + 2 = 0$ **50.** $z + 5 = 0$

51. $x + y = 2$ **52.** $5x + 2y = 10$ **53.** $y + z = 1$

54. $x + y + z = 1$ **55.** $2x + 3y + 6z = 12$

56. $z = 12 - \frac{1}{2}x - \frac{1}{3}y$ **57.** $z = -6 + 2x + 3y$

In Exercises 58–61, find the equation of the plane passing through the given point and having the given slopes in the x- and y-directions.

58. $(0, 0, 6)$, slope in x-direction 1, slope in y-direction -2

59. $(4, -3, 3)$, slope in x-direction 2, slope in y-direction 5

60. $(2, 6, 4)$, slope in x-direction $\frac{1}{2}$, slope in y-direction $-\frac{1}{3}$

61. $(-4, -2, \frac{9}{2})$, slope in x-direction $-\frac{5}{2}$, slope in y-direction $\frac{7}{2}$

In Exercises 62–65, find the equation of the plane passing through the three given points.

62. $(2, 0, 0)$, $(0, 3, 0)$, $(0, 0, 6)$

63. $(1, 0, -1)$, $(0, 1, 3)$, $(1, -1, -2)$

64. $(1, 2, 1)$, $(-1, -3, -6)$, $(2, 8, -2)$

65. $(2, -3, -1)$, $(-2, -4, 4)$, $(1, \frac{1}{2}, 4)$

66. Assume that z is a linear function with slope 3 in the x-direction and slope 5 in the y-direction. Find the change in z when x changes by 0.1 and y changes by -0.2.

67. A private school has a capacity of 2,000 students, and its total enrollment is based on two factors: tuition and reputation. As tuition increases, enrollment decreases. As the school's reputation increases, so does enrollment. Even with a reputation level of zero, however, the school would be 100% full if it charged no tuition.
(a) If we assume the relationships are linear, this information can be visualized as a plane in $\mathbb{R}^3$. What is the sign of the slope in the tuition direction?
(b) What is the sign of the slope in the reputation direction?
(c) Write down the coordinates of the enrollment intercept.

In Exercises 68–71, decide whether the given planes have a common intersection and, if they do, find the intersection set.

68. $z = 5 + x - y$, $z = 3 + 2x + 3y$, $z = 10 - 5x + 2y$

69. $x - 2y + 2z = 4$, $x - y + z = 3$, $2x - 3y + 3z = 0$

70. $x + 2y - 3z = 1$, $x + 3y - 2z = 4$, $2x + 5y - z = 6$

71. $x + y - z = 4$, $x + 3y + z = 10$,
$x + 2y + z = 8$, $x + z = 4$

72. A company selling two items X and Y makes a profit of \$2 per unit for item X and \$5 per unit for item Y. When it sells only 100 units of X and 50 units of Y, then it loses \$500. Find the profit function of the company. Then compute the profit for selling 100 units of X and 500 units of Y.

73. Assume that the cost function of a company producing x units of one product and y units of another product is a linear function, and the following table of values is known:

$x\backslash y$	100	150	200
200	800	900	1000
300	850	950	1,050

Write a formula for the cost function.

74. Sociologists studying fashion trends have determined that participation in a trend is a function of both time (measured from the introduction of the trend) and expense. One trend in particular had a participation level of 50% of the target population 2 weeks after its introduction, at which time the cost of participating was \$20.
    For this trend, participation is estimated to be a linear function of time and expense, with slope -0.2 in the direction of time and -0.91 in the direction of expense. Find the equation of the plane that models this trend.

75. Using a software package, such as Maple or Mathematica, plot the graphs and level curves of the following functions:
(a) $f(x, y) = y^2 - x^2$, $-2 < x < 2$, $-2 < y < 2$
(b) $f(x, y) = xy$, $-2 < x < 2$, $-2 < y < 2$
(c) $f(x, y) = x^3 + 3xy^2 - 3x^2 - 3y^2$, $-5 < x < 5$, $-5 < y < 5$
(d) $f(x, y) = e^{-x^2} + e^{-9y^2}$, $-2 < x < 2$, $-2 < y < 2$

Solutions to practice exercises 7.1

1. $f(1, 2) = \dfrac{1^2}{1 + 2} = \dfrac{1}{3}$, $f(0, 1) = \dfrac{0^2}{0 + 1} = 0$,

$f(-2, 1) = \dfrac{(-2)^2}{-2 + 1} = -4$

2. (a) The level curves are circles centered at $(0, 0)$, with radii 1, 2, 3, 4, and 5, respectively.

(b) The graph is a paraboloid opening downward, with its vertex at $(0, 0, 9)$.
(c) Both intersections are parabolas.
(d) There are no level curves for $z > 9$. The level curve for $z = 9$ reduces to the single point $(0, 0)$.
(e) The natural domain is the entire xy-plane.

3. The equation has the form $z - z_0 = a(x - x_0) + b(y - y_0)$. Taking $(x_0, y_0, z_0) = (-1, 3, 1)$, we have $z - 1 = a(x + 1) + b(y - 3)$. Next, substituting $(1, 2, 4)$ for (x, y, z) gives $3 = 2a - b$. Similarly, substituting $(2, 1, 5)$ for (x, y, z) gives $4 = 3a - 2b$. By solving these equations, we get $a = 2$ and $b = 1$. Therefore, the plane has the equation $z - 1 = 2(x + 1) + (y - 3)$.

The coefficients give the slopes, so that 2 is the slope in the x-direction, and 1 is the slope in the y-direction.

7.2   Partial Derivatives

Given a function of two variables, $f(x, y)$, suppose we fix a value for y and let x vary freely (obtaining a y-section). Then f becomes a function of x alone, and we can take its derivative as a function of one variable in the usual way. This derivative is called the **partial derivative of f with respect to x**, and we denote it by

$$\frac{\partial f}{\partial x},$$

where the symbol ∂ is used in place of d to indicate the partial derivative. For example, if

$$f(x, y) = x^2 y^3 + y$$

and we fix $y = 5$, we obtain the function

$$f(x, 5) = x^2 5^3 + 5 \quad \text{or} \quad f(x, 5) = 125x^2 + 5.$$

This is a function of the variable x alone. Its derivative is the partial derivative of f with respect to x, which is written as

$$\frac{\partial f}{\partial x}(x, 5) = 2 \cdot 125x + 0 = 250x.$$

If we replace the value 5 of y with another constant y_0, then the function $f(x, y_0) = x^2 y_0^3 + y_0$ is again a function of the variable x alone, and we have

$$\frac{\partial f}{\partial x}(x, y_0) = 2x y_0^3.$$

In fact, we may think of y as being fixed without using a special notation such as y_0 to indicate it. We can simply write

$$\frac{\partial f}{\partial x}(x, y) = 2x y^3,$$

which means that we have kept y fixed and taken the derivative of f as a function of x alone.

Similarly, if we fix x and let y vary, we obtain a function of the variable y alone (x-section). It's derivative is called the **partial derivative of f with respect to y** and is denoted by

$$\frac{\partial f}{\partial y}.$$

For instance, setting $x = 4$ in the function $f(x, y) = x^2y^3 + y$ gives us the function $f(4, y) = 16y^3 + y$ with

$$\frac{\partial f}{\partial y}(4, y) = 48y^2 + 1.$$

More generally, keeping x fixed and taking the derivative with respect to y gives the partial derivative

$$\frac{\partial f}{\partial y}(x, y) = x^2 \cdot 3y^2 + 1 = 3x^2y^2 + 1.$$

EXAMPLE 7.2.1

Let $f(x, y) = x/y$, defined for $y \neq 0$. Find $\partial f/\partial x$ and $\partial f/\partial y$.

SOLUTION   If we fix y, we get a linear function of x with coefficient $1/y$, and the partial derivative of f with respect to x is given by

$$\frac{\partial f}{\partial x} = \frac{1}{y}.$$

On the other hand, if we fix x, we get y^{-1} multiplied by the coefficient x. Therefore, the partial derivative of f with respect to y is given by

$$\frac{\partial f}{\partial y} = -\frac{x}{y^2}.$$

Partial derivatives as limits and slopes

We can get a more precise definition of the partial derivative by using the basic limit formula defining the ordinary derivative. Specifically, if (x_0, y_0) is a point in the domain of f, the partial derivative with respect to x at (x_0, y_0) is defined by

$$\frac{\partial f}{\partial x}(x_0, y_0) = \lim_{h \to 0} \frac{f(x_0 + h, y_0) - f(x_0, y_0)}{h}, \tag{4}$$

provided, of course, that the limit exists.

We can interpret this geometrically by visualizing the graph of $z = f(x, y)$ as a surface in $\mathbb{R}^3$. Holding y fixed at y_0 amounts to slicing that surface by the plane $y = y_0$, which results in a curve defined by the equation $z = f(x, y_0)$, as shown in Figure 7.2.1. The slope of this curve—or, more precisely, the slope of its tangent line—is the derivative of $f(x, y_0)$ with respect to x at x_0. But that is exactly the same as the partial derivative $\partial f/\partial x$ at (x_0, y_0), as given by formula (4). As Figure 7.2.1 suggests, we can think of $(\partial f/\partial x)(x_0, y_0)$ as the slope of the surface $z = f(x, y)$ at (x_0, y_0) in the x-direction.

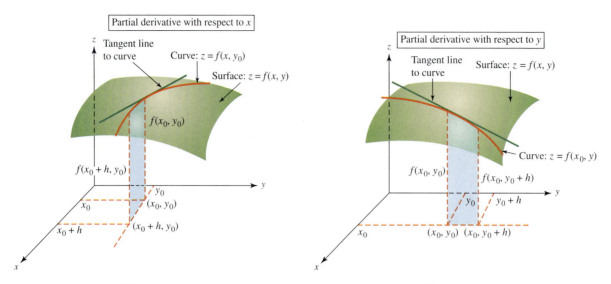

Figure 7.2.1   **Figure 7.2.2**

Similarly, the partial derivative of f with respect to y at (x_0, y_0) is defined by

$$\frac{\partial f}{\partial y}(x_0, y_0) = \lim_{h \to 0} \frac{f(x_0, y_0 + h) - f(x_0, y_0)}{h}. \tag{5}$$

Figure 7.2.2 illustrates the geometric interpretation of the partial derivative of f with respect to y at the point (x_0, y_0). It shows the curve defined by the equation $z = f(x_0, y)$, obtained by slicing the surface $z = f(x, y)$ in $\mathbb{R}^3$ by the plane $x = x_0$. The slope of this curve—or, more precisely, the slope of its tangent line—is the derivative of $f(x_0, y)$ with respect to y at y_0. But that is exactly the same as the partial derivative $(\partial f/\partial y)(x_0, y_0)$, as given by formula (5). Thus, we can think of $(\partial f/\partial y)(x_0, y_0)$ as the slope of the surface $z = f(x, y)$ at (x_0, y_0) in the y-direction.

The tangent plane

The two tangent lines—the one in the x-direction and the one in the y-direction—determine a plane tangent to the surface at the point $(x_0, y_0, f(x_0, y_0))$. It is shown in Figure 7.2.3. The partial derivative $(\partial f/\partial x)(x_0, y_0)$ is the slope of the tangent plane in the x-direction, and $(\partial f/\partial y)(x_0, y_0)$ is the slope of the tangent plane in the y-direction. Using the general equation of a plane given by formula (3) we obtain the following:

The Equation of the Tangent Plane

The equation of the tangent plane to the surface $z = f(x, y)$ at the point $(x_0, y_0, f(x_0, y_0))$ is

$$z = f(x_0, y_0) + \frac{\partial f}{\partial x}(x_0, y_0)(x - x_0) + \frac{\partial f}{\partial y}(x_0, y_0)(y - y_0). \tag{6}$$

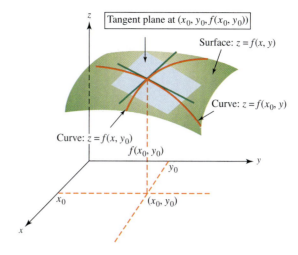

Figure 7.2.3

EXAMPLE 7.2.2

Find the equation of the tangent plane to the paraboloid $z = x^2 + y^2$ at the point where $x = -1$ and $y = 2$.

SOLUTION In this case $f(x, y) = x^2 + y^2$ and $f(-1, 2) = 5$. The partial derivatives are

$$\frac{\partial f}{\partial x} = 2x \quad \text{and} \quad \frac{\partial f}{\partial y} = 2y.$$

Applying formula (6) with $x_0 = -1$ and $y_0 = 2$ gives the equation of the tangent plane:

$$z = 5 - 2(x + 1) + 4(y - 2).$$

By expanding and collecting terms, we can also write the equation in the form

$$z = -5 - 2x + 4y.$$

Linear approximation

If a function has continuous partial derivatives, then its graph at a point $(x_0, y_0, f(x_0, y_0))$ is close to the tangent plane at that same point, at least for all (x, y) close to (x_0, y_0). In other words, we have the approximation

$$f(x, y) \approx f(x_0, y_0) + \frac{\partial f}{\partial x}(x_0, y_0)(x - x_0) + \frac{\partial f}{\partial y}(x_0, y_0)(y - y_0), \tag{7}$$

for all (x, y) that are sufficiently close to (x_0, y_0). This approximation is called the **linear approximation** of f at (x_0, y_0). It provides a good approximation to $f(x, y)$

near (x_0, y_0) by a linear function, and it is often used to simplify difficult computations at the expense of a small error. Notice the similarity between this formula and the one-variable linear approximation formula

$$f(x) \approx f(x_0) + f'(x_0)(x - x_0)$$

of Section 3.3.

EXAMPLE 7.2.3

Suppose $f(x, y) = 5 + e^{-2x+3y}$. Find
  (i) the partial derivatives $\partial f/\partial x$ and $\partial f/\partial y$,
  (ii) the equation of the tangent plane at the point $(0, 0)$, and
  (iii) the linear approximation of f at $(0,0)$.

SOLUTION   (i) First, by holding one variable fixed and differentiating with respect to the other, we get

$$\frac{\partial f}{\partial x} = -2e^{-2x+3y} \quad \text{and} \quad \frac{\partial f}{\partial y} = 3e^{-2x+3y}.$$

(ii) By substituting $(x, y) = (0, 0)$, we get

$$\frac{\partial f}{\partial x}(0, 0) = -2 \quad \text{and} \quad \frac{\partial f}{\partial y}(0, 0) = 3.$$

Since $f(0, 0) = 6$, the equation of the tangent plane at $(0, 0)$ is given by

$$z = 6 - 2(x - 0) + 3(y - 0) = 6 - 2x + 3y,$$

(iii) The linear approximation of f at $(0, 0)$ is given by

$$f(x, y) \approx 6 - 2x + 3y.$$

EXAMPLE 7.2.4

Suppose an object in the xy-plane moves from the point $(4, 3)$ to the point $(4.2, 2.9)$. Use linear approximation to estimate the change in its distance from the origin. Then compute the actual change in distance (to three decimal places) and compare it to the approximation.

SOLUTION   The distance function is $f(x, y) = \sqrt{x^2 + y^2}$, and its partial derivatives are

$$\frac{\partial f}{\partial x} = \frac{x}{\sqrt{x^2 + y^2}} \quad \text{and} \quad \frac{\partial f}{\partial y} = \frac{y}{\sqrt{x^2 + y^2}}.$$

In particular,

$$\frac{\partial f}{\partial x}(4, 3) = 0.8 \quad \text{and} \quad \frac{\partial f}{\partial y} = 0.6.$$

To estimate the change in distance, we apply formula (7) as follows:

$$f(4.2, 2.9) - f(4, 3) \approx \frac{\partial f}{\partial x}(4, 3)(4.2 - 4) + \frac{\partial f}{\partial y}(4, 3)(2.9 - 3)$$

$$= (0.8)(0.2) - (0.6)(0.1) = 0.1.$$

The actual change in distance (to three decimal places) is

$$\sqrt{4.2^2 + 2.9^2} - 5 \approx 0.104.$$

Marginal cost, revenue, and profit

In the language of business and economics, the partial derivatives are called **partial marginals** or simply **marginals**. At a given production level (x_0, y_0), the partial marginal cost with respect to x is the approximate increase in cost if y is held fixed at y_0 and x is increased by one unit, from x_0 to $(x_0 + 1)$. That follows from the linear approximation formula, for applying (7) with $x = x_0 + 1$ and $y = y_0$ gives

$$C(x_0 + 1, y_0) - C(x_0, y_0) \approx \frac{\partial C}{\partial x}(x_0, y_0) \cdot 1 + \frac{\partial C}{\partial y}(x_0, y_0) \cdot 0$$

$$= \frac{\partial C}{\partial x}(x_0, y_0).$$

Similarly, the partial marginal cost with respect to y is the approximate increase in cost if x is held fixed at x_0 and y is increased by one unit. The same interpretation applies to revenue and profit.

EXAMPLE 7.2.5

A company manufactures two products, X and Y. The total cost in dollars for producing x units of product X and y units of product Y is given by the function

$$C(x, y) = \frac{1}{4}x^2 + \frac{1}{5}y^2 + 6x + 7y + 800.$$

   (i) Find the cost of producing 20 units of X and 50 units of Y.
   (ii) Compute the partial marginal costs at $(20, 50)$.
   (iii) Find the approximate change in cost if the production level of X is increased to 21 units while keeping that of Y fixed at 50.
   (iv) Use linear approximation to estimate the cost of producing 22 units of X and 49 units of Y. Check the accuracy of the approximation by actually computing $C(22, 49)$ and comparing it to the estimated value.

SOLUTION   (i) The cost is given by

$$C(20, 50) = \frac{1}{4}(20)^2 + \frac{1}{5}(50)^2 + 6 \cdot (20) + 7 \cdot (50) + 800 = 1,870.$$

(ii) The marginal costs are

$$\frac{\partial C}{\partial x} = \frac{1}{2}x + 6 \quad \text{and} \quad \frac{\partial C}{\partial y} = \frac{2}{5}y + 7.$$

Therefore,

$$\frac{\partial C}{\partial x}(20, 50) = \frac{1}{2} \cdot 20 + 6 = 16,$$

and

$$\frac{\partial C}{\partial y}(20, 50) = \frac{2}{5} \cdot 50 + 7 = 27.$$

(iii) Because x changes by one unit, from 20 to 21, and y is held fixed the approximate change in cost is given by the partial marginal cost with respect to x and is therefore equal to \$16.

(iv) Set $(x_0, y_0) = (20, 50)$ and $(x, y) = (22, 49)$. Then $(x - x_0) = 2$ and $(y - y_0) = -1$. Therefore, the linear approximation formula gives

$$C(22, 49) \approx C(20, 50) + 2 \cdot \frac{\partial C}{\partial x}(20, 50) - 1 \cdot \frac{\partial C}{\partial y}(20, 50)$$

$$= 1,870 + 2 \cdot 16 - 27 = 1,875.$$

By way of comparison, a direct calculation gives

$$C(22, 49) = \frac{1}{4}(22)^2 + \frac{1}{5}(49)^2 + 6 \cdot (22) + 7 \cdot (49) + 800 = 1,876.20.$$

The Cobb-Douglas production function

As we have seen before, a Cobb-Douglas production function has the form

$$P(K, L) = AK^\alpha L^{1-\alpha},$$

where A and α are constants with $A > 0$ and $0 < \alpha < 1$. The partial derivative of P with respect to K is approximately equal to the **marginal product of capital** (MPK), which is defined to be the amount of extra output resulting from an increase of one unit in capital input. The partial derivative of P with respect to L is approximately equal to the **marginal product of labor** (MPL), which is defined as the amount of extra output resulting from an increase of one unit in labor input.

EXAMPLE 7.2.6

For the U.S. economy between the years 1899 and 1922, Cobb and Douglas estimated that

$$P(K, L) = 1.01 K^{1/4} L^{3/4}.$$

Use the partial derivatives to approximate the marginal productivity functions.

SOLUTION   Holding L fixed and taking the derivative of P as a function of K alone give us the partial derivative $\partial P / \partial K$, which approximates the marginal product of capital:

$$\frac{\partial P}{\partial K} = \frac{1.01}{4} K^{-3/4} L^{3/4}.$$

Similarly, to approximate the marginal product of labor by the partial derivative $\partial P / \partial L$, we hold K fixed and differentiate P as a function of L alone:

$$\frac{\partial P}{\partial L} = \frac{3.03}{4} K^{1/4} L^{-1/4}.$$

EXAMPLE 7.2.7

Suppose the production function of a certain country is given by a Cobb-Douglas function of the form

$$P(K, L) = 6 K^{1/3} L^{2/3},$$

where K is the amount of capital, P is the economy's output (both measured in billions of dollars), and L is the quantity of labor (measured in hundreds of thousands of workers). Estimate the MPK and MPL when $K = 64$ and $L = 125$.

SOLUTION   We have

$$\frac{\partial P}{\partial K} = 2 K^{-2/3} L^{2/3} \quad \text{and} \quad \frac{\partial P}{\partial L} = 4 K^{1/3} L^{-1/3}.$$

By substituting $K = 64$ and $L = 125$, we get the following approximations:

$$MPK \approx \frac{\partial P}{\partial K}(64, 125) = 2 \cdot 64^{-2/3} 125^{2/3} = \frac{50}{16} = 3.125$$

$$MPL \approx \frac{\partial P}{\partial L}(64, 125) = 4 \cdot 64^{1/3} 125^{-1/3} = \frac{16}{5} = 3.2.$$

We interpret these numbers as follows. If the quantity of labor is kept fixed at 125 units (1,250,000 workers), an increase of one unit of capital ($1 billion) will result in an approximate increase of 3.125 units of output ($3.125 billion). And if the amount of capital is kept fixed at $64 billion dollars (64 units), an increase of 100,000 workers (one unit of labor) will result in an approximate increase in output of $3.2 billion (3.2 units).

Higher-order partial derivatives

Just as in the one-variable case, differentiating a two-variable function more than once leads to higher-order derivatives. In the two-variable case there are four possible second derivatives:

$$\frac{\partial^2 f}{\partial x^2} = \frac{\partial}{\partial x}\left(\frac{\partial f}{\partial x}\right), \quad \frac{\partial^2 f}{\partial y^2} = \frac{\partial}{\partial y}\left(\frac{\partial f}{\partial y}\right),$$

$$\frac{\partial^2 f}{\partial y \partial x} = \frac{\partial}{\partial y}\left(\frac{\partial f}{\partial x}\right), \quad \frac{\partial^2 f}{\partial x \partial y} = \frac{\partial}{\partial x}\left(\frac{\partial f}{\partial y}\right).$$

As you can see above, the symbol $\partial^2/\partial y \partial x$ indicates that we first take the partial derivative of f with respect to x and then take the partial derivative with respect to y, whereas $\partial^2/\partial x \partial y$ indicates the same two operations performed in the opposite order. However, for well-behaved functions—including all that we will meet in this text—these two "mixed" second-order derivatives are actually equal. That is,

$$\frac{\partial^2 f}{\partial x \partial y} = \frac{\partial^2 f}{\partial y \partial x}.$$

EXAMPLE 7.2.8

Let $f(x, y) = x^3 y^4 + x^2 - 3y^2$. Compute all four second-order partial derivatives and verify that the mixed derivatives are equal

SOLUTION We begin by taking the first-order partial derivatives:

$$\frac{\partial f}{\partial x} = 3x^2 y^4 + 2x \quad \text{and} \quad \frac{\partial f}{\partial y} = 4x^3 y^3 - 6y.$$

Differentiating the first of these, first with respect to x and then with respect to y, gives

$$\frac{\partial^2 f}{\partial x^2} = \frac{\partial}{\partial x}(3x^2 y^4 + 2x) = 6xy^4 + 2 \quad \text{and} \quad \frac{\partial^2 f}{\partial y \partial x} = \frac{\partial}{\partial y}(3x^2 y^4 + 2x) = 12x^2 y^3.$$

Differentiating the second yields

$$\frac{\partial^2 f}{\partial x \partial y} = \frac{\partial}{\partial x}(4x^3 y^3 - 6y) = 12x^2 y^3 \quad \text{and} \quad \frac{\partial^2 f}{\partial y^2} = \frac{\partial}{\partial y}(4x^3 y^3 - 6y) = 12x^3 y^2 - 6.$$

As you can see,

$$\frac{\partial^2 f}{\partial y \partial x} = \frac{\partial^2 f}{\partial x \partial y} = 12x^2y^3.$$

An equation involving an (unknown) function of two or more independent variables and its partial derivatives is called a **partial differential equation**. These equations play a central role in many applications.

> **EXAMPLE 7.2.9**

Imagine an infinitely long cylindrical tube along the x-axis, filled with water. If, at time $t = 0$, we insert a drop of red coloring at the point $x = 0$, the coloring will slowly spread (diffuse) throughout the entire tube. Its concentration $c(x, t)$ at a point x at time t is described by the following equation, known as the **diffusion equation**:

$$\frac{\partial c}{\partial t} = D\frac{\partial^2 c}{\partial x^2}, \tag{8}$$

where D is called the **diffusion constant**. This equation models random movement and is used in biology to explain various phenomena, such as chemotaxis and pattern formation. In physics it models the diffusion of heat, and in economics it is used to model random changes, such as those in the Dow Jones index of the stock market. Show that if $D = 1$, a solution of Eq. (8) is given by the formula

$$c(x, t) = \frac{1}{\sqrt{4\pi t}}e^{-x^2/4t}.$$

SOLUTION Using the chain rule, we get

$$\frac{\partial c}{\partial x} = -\frac{x}{2t\sqrt{4\pi t}}e^{-x^2/4t}.$$

Next, using the product and chain rules, we get

$$\frac{\partial^2 c}{\partial x^2} = \frac{x^2}{4t^2\sqrt{4\pi t}}e^{-x^2/4t} - \frac{1}{2t\sqrt{4\pi t}}e^{-x^2/4t}$$

$$= \left(\frac{x^2}{4t^2\sqrt{4\pi t}} - \frac{1}{2t^{3/2}\sqrt{4\pi}}\right)e^{-x^2/4t}.$$

To compute $\partial c/\partial t$, we also use the product and chain rules:

$$\frac{\partial c}{\partial t} = \frac{1}{\sqrt{4\pi t}}e^{-x^2/4t}\left(\frac{x^2}{4t^2}\right) - \frac{1}{2(4\pi t)^{3/2}}4\pi\, e^{-x^2/4t}$$

$$= \left(\frac{x^2}{4t^2\sqrt{4\pi t}} - \frac{1}{2t^{3/2}\sqrt{4\pi}}\right)e^{-x^2/4t}.$$

Thus, we see that $\partial c/\partial t = \partial^2 c/\partial x^2$.

Practice Exercises 7.2

● ●

1. Find the first-order partial derivatives of the function

$$f(x, y) = x^3 + e^y + xy^2 + y \ln x.$$

2. Find the equation of the plane tangent to the graph of $f(x, y) = 2x/(1 + y^2)$ at the point $(2, -1, 2)$.

3. A company estimates that its daily profit (in thousands of dollars) from producing x units of its deluxe item and y units of its standard item is given by the function

$$P(x, y) = 60x + 70y - \tfrac{3}{10}x^2 - \tfrac{1}{5}y^2 - 100.$$

It is currently producing 50 deluxe and 200 standard items a day, and it is thinking of producing two more deluxe and five less standard items per day. Using linear approximation, estimate whether that will cause the profit to rise or fall, and by approximately how much.

Exercises 7.2

● ●

In Exercises 1–14, find the first-order partial derivatives of the given function. Assume that the variables are restricted to a domain in which the function is well defined.

1. $f(x, y) = 5 - 8x + 5y$ **2.** $f(x, y) = 3x^2 + xy - 2$

3. $f(x, y) = x^3 y^2 - 2x^2 y + y$

4. $f(x, y) = y^3 e^{2xy} - x^5$

5. $M(r, t) = 1,000 e^{rt}$ **6.** $f(x, y) = \sqrt{x^2 + y^2}$

7. $R(s, t) = \dfrac{5t^2}{s}$ **8.** $d(a, b) = (3a + b - 5)^2$

9. $l(x, y) = \ln(x^2 - 5y + 3)$

10. $f(x, y) = e^{-xy^2} \ln(x^2 + 1)$

11. $Q(K, L) = 10K^{0.4}L^{0.6}$ **12.** $f(x, y) = 16x^{1/8}y^{7/8}$

13. $M(t, r) = 1{,}000 \left(1 + \dfrac{r}{360}\right)^{360t}$

14. $P(t, K) = \dfrac{K}{1 + e^{-0.02t}}$

In Exercises 15–18, compute the partial derivatives of the given function at the given points, as indicated.

15. $\dfrac{\partial f}{\partial x}(2, 1)$ if $f(x, y) = x^2 + xy - 3y^2 + 4$

16. $\dfrac{\partial f}{\partial y}(1, 0)$ if $f(x, y) = x^3 e^{-y^2} - x$

17. $\dfrac{\partial f}{\partial x}(25, 16)$ if $f(x, y) = 100x^{1/2}y^{1/2}$

18. $\dfrac{\partial f}{\partial y}(0, 0)$ if $f(x, y) = \ln(x^2 + y^2 + 1)$

19. Coffee connoisseurs claim to taste a flavor difference between coffee made from beans that are of different quality or have different times between grinding and brewing. Suppose quality q is ranked from 1 to 10 and time t from grinding to brewing is measured in minutes. In a blind taste test, a panel of experts show that higher quality improves flavor and a longer time before brewing degrades flavor. One scientist attempts to quantify this information by modeling and derives the following equation:

$$F(t, q) = t^2 + 2tq + q^2.$$

Test the validity of the model by taking partial derivatives. Are the signs of the partial derivatives compatible with the experts' results?

20. Green plants manufacture their own food through photosynthesis but must first have the raw materials available. A herbaceous plant's growth based on the quantities of water w and nutrients n accessible to the roots is $G(w, n) = 0.006w^3 n^{0.7}$, where w is in milliliters, n is in micrograms, and G is the height in centimeters.
(a) Write the growth curve equation when there are 5 milliliters of water available.
(b) What is the rate of change of growth with respect to the amount of nutrients?

21. A sculptor is working with a rectangular metal plate. She applies a heat source to it and observes that after some time the temperature is distributed throughout the plate according to the formula

$$T = 5(x^2 - y^2) + 75,$$

where $(0, 0)$ is the center of the plate, and x and y are measured in centimeters.
(a) Compute $\partial T/dx$ and $\partial T/dy$.
(b) Sketch a typical x-section. For what y does it achieve a maximum or minimum?

(c) Sketch a typical y-section. For what x does it achieve a maximum or minimum?
(d) Use your previous conclusions to show that there is no minimum or maximum temperature inside the plate.

22. Centripetal force is a center-seeking force that causes an object in motion, which normally moves in a straight line, to move in a circle. One common example is an object tied to a string. Imagine holding one end of the string and twirling the object attached to the other end. The string provides tension, so the object moves in a circle. This centripetal force is measured by the formula

$$F = \frac{mv^2}{r},$$

where m is the mass of the object, v is its velocity, and r is the radius of the circle.
(a) Find the partial derivative with respect to the velocity and use linear approximation to estimate the change in force if the velocity changes from v to $v + 1$. Then compute the actual change and compare it to the estimate.
(b) Find the partial derivative with respect to the radius and use linear approximation to estimate the change in force if the radius changes from r to $r + 1$. Then compute the actual change and compare it to the estimate.

In Exercises 23–28, find the equation of the tangent plane to the graph of the given function at the given point and find the linear approximation to the function at the same point.

23. $f(x, y) = 3x^2 + 5y^2$; $(1, 2)$

24. $f(x, y) = x^2 - 3y^2$; $(-1, 2)$

25. $f(x, y) = e^{x^2 - y^4}$; $(1, -1)$

26. $f(x, y) = \ln(x^2 + 3y^2 + 2)$; $(0, 0)$

27. $f(x, y) = 8x^{1/2}y^2$; $(4, 1)$

28. $f(x, y) = x^2 e^{yx} + y^3$; $(0, -2)$

In Exercises 29–31, find the equation of the tangent plane and the linear approximation to $f(x, y)$ at the point in question, using the given information.

29. $f(0, 0) = 8$, $\quad \dfrac{\partial f}{\partial x}(0, 0) = 5$, $\quad \dfrac{\partial f}{\partial y}(0, 0) = -3$

30. $f(1, 2) = 5$, $\quad \dfrac{\partial f}{\partial x}(1, 2) = -\dfrac{1}{2}$, $\quad \dfrac{\partial f}{\partial y}(1, 2) = 4$

31. $f(-3, 5) = -2$, $\quad \dfrac{\partial f}{\partial x}(-3, 5) = 2$, $\quad \dfrac{\partial f}{\partial y}(-3, 5) = -\dfrac{8}{7}$

32. A rectangle whose base and height are 20 and 12 centimeters, respectively, has an area of 240 square centimeters. Use linear approximation to estimate the area of a rectangle whose base and height are 19 and 12.5 centimeters, respectively. Then compute the actual area and compare it with the approximation.

33. A rectangular box has a height of 6 inches. Its base is a square with a 4-inch side.
(a) Find its volume.
(b) Use linear approximation to estimate the change in volume if the height is increased to 6.02 inches and the side of the square is increased to 4.01 inches.

34. The distance from the origin to a point (x, y) in the plane is $\sqrt{x^2 + y^2}$. Use linear approximation to estimate the change in distance from the origin in moving from the point $(3, 4)$ to each of the following points:
(a) $(3.012, 4)$ (b) $(3, 3.995)$
(c) $(2.975, 4.002)$

35. You are designing a cylindrical can for a new beverage. Its volume is $V = \pi r^2 h$, where r is the radius of the base and h is the height, both in centimeters. Answer the following questions in two ways: first, by using linear approximation to get an estimate; second, by computing the actual change. For a fixed radius r_0 and height h_0, how much does the volume change with a one unit increase in radius?

36. The surface area S of a human body is a function of its weight W and height H. They are related by the formula

$$S = 0.0072W^{0.425}H^{0.725},$$

with S measured in square meters, W in kilograms, and H in centimeters. Compute the partial derivative $\partial S / \partial W$ and use it to answer the following:
(a) If a person who weighs 80 kilograms gains an additional 5 kilograms with no change in height, estimate the proportional change in the body's surface area. In other words, estimate the ratio of the increase in surface area to the original surface area.
(b) If a person's body weight increases by 10% with no change in height, estimate the percentage increase in the body's surface area.

37. The volume V, pressure P, and temperature T of a gas are related by a formula of the form

$$V = \frac{kT}{P},$$

where k is a constant depending on the particular gas and the units used. Suppose that $k = 0.5$ for a certain gas and choice of units.
(a) Compute $\partial V / \partial T$ and $\partial V / \partial P$.
(b) Estimate the change in V if P changes from 20 to 22 units and T changes from 400 to 396.

38. The number of camper vans sold per week by a certain company is a function of the form $f(x, y)$, where x is the average price of a van (in thousands of dollars) and y is the average price (in cents) of gasoline throughout the United States. Suppose that

$$\frac{\partial f}{\partial x}(23.5, 160) = -180 \quad \text{and} \quad \frac{\partial f}{\partial y}(23.5, 160) = -12.$$

Use this information to estimate the following:
(a) the change in the number of vans sold if the average price of a van goes down by \$500 while the price of gasoline remains steady
(b) the change in the number of vans sold if the average price of a van remains steady while the price of gasoline goes up by 5 cents a gallon
(c) the change in the number of vans sold if the average price of a van goes up by \$200 while the price of gasoline goes down by 10 cents a gallon

39. The production function of a certain country's economy is given by $P(K, L) = 20K^{0.3}L^{0.7}$. Use the partial derivatives to approximate the *MPK* and *MPL* when $K = 1{,}000$ and $L = 4{,}000$.

In Exercises 40–43, compute all four of the second-order partial derivatives and verify that the mixed partials are equal.

40. $f(x, y) = x^4 y + y^2 e^x$

41. $f(x, y) = \dfrac{x^2}{y}$

42. $f(x, y) = x^3 - 4x^2 y + 3xy^2 - 7y^3$

43. $f(x, y) = x \ln y$

Estimating derivatives *Formulas (4) and (5) imply that if h is small (positive or negative),*

$$\frac{\partial f}{\partial x}(x_0, y_0) \approx \frac{f(x_0 + h, y_0) - f(x_0, y_0)}{h}$$

and

$$\frac{\partial f}{\partial y}(x_0, y_0) \approx \frac{f(x_0, y_0 + h) - f(x_0, y_0)}{h}.$$

These formulas are called **forward difference formulas** *if h is positive and* **backward difference formulas** *if h is negative. In Exercises 44–47, use these formulas and the given list of values of $f(x, y)$ to estimate the first-order partial derivatives at the given point.*

44. At $(0, 0)$, use $f(0, 0) = 0$, $f(0.1, 0) = 0.3$, $f(0, 0.2) = -0.4$.

45. At $(1, 2)$, use $f(1, 2) = 8$, $f(1.5, 2) = 10$, $f(1, 2.6) = 8.3$.

46. At $(-4, 5)$, use $f(-4, 5) = -7$, $f(-3.8, 5) = -6$, $f(-4, 5.4) = -7.2$.

47. At $(5, 6)$, use $f(4.5, 6) = 20$, $f(5, 5.7) = 23$, $f(5, 6) = 21$.

More accurate formulas for approximating derivatives are the **central difference formulas**:

$$\frac{\partial f}{\partial x}(x_0, y_0) \approx \frac{f(x_0 + h, y_0) - f(x_0 - h, y_0)}{2h}$$

and

$$\frac{\partial f}{\partial y}(x_0, y_0) \approx \frac{f(x_0, y_0 + h) - f(x_0, y_0 - h)}{2h}.$$

In Exercises 48–49, use these formulas and the given list of values of $f(x, y)$ to estimate the first-order partial derivatives at $(0, 0)$.

48. $f(-0.1, 0) = 2.8$, $f(0.1, 0) = 3.6$, $f(0, -0.1) = -1$, $f(0, 0.1) = -0.4$

49. $f(-0.05, 0) = 8$, $f(0.05, 0) = 7.8$, $f(0, -0.05) = -1$, $f(0, 0.05) = 0$

50. Use the table of data to estimate the first partial derivatives $(\partial f / \partial x)(x_0, y_0)$ and $(\partial f / \partial y)(x_0, y_0)$ at the nine points (x_0, y_0) determined by setting $x_0 = 0, 0.1, 0.2$ and $y_0 = 1, 1.1, 1.2$. Use a central difference formula wherever possible; otherwise, use a forward or backward difference formula.

x	y	$f(x, y)$
0	1	4.2
0	1.1	4.5
0	1.2	4.6
0.1	1	5
0.1	1.1	4.9
0.1	1.2	4.6
0.2	1	4.1
0.2	1.1	4.2
0.2	1.2	4.4

51. The rising of a loaf of bread is due to yeast activity in the dough. The activity levels are dependent on the temperature and the quantity of sugar (in teaspoons). Based on his bread-making observations, a baker makes the following chart of rising times (in minutes) for a specific volume and a particular recipe:

Teaspoons\degrees Fahrenheit	85	87	89	91	93
1.5	71	69	68	66	65
1.75	67	66	63	62	60
2	60	59	57	56	53
2.25	55	53	51	49	48
2.5	51	50	49	47	47

(a) Use the central difference formula to estimate the change in rising time per unit change in temperature at 2 teaspoons of sugar and 89°F.

(b) Use the central difference formula to estimate the change in rising time per teaspoon of change in sugar content at 2 teaspoons of sugar and 89°F.

(c) Use linear approximation to estimate the rising time corresponding to $2\frac{1}{8}$ teaspoons of sugar and 90°F.

Functions of three or more variables *Partial derivatives and linear approximation may also be applied to functions of three or more variables. For instance, by keeping x and y fixed and letting z vary in a function $f(x, y, z)$ of three variables, we get the partial derivative $\partial f / \partial z$. The linear approximation formula in that case is*

$$f(x, y, z) \approx f(x_0, y_0, z_0) + \frac{\partial f}{\partial x}(x_0, y_0, z_0) \cdot (x - x_0)$$

$$+ \frac{\partial f}{\partial y}(x_0, y_0, z_0) \cdot (y - y_0)$$

$$+ \frac{\partial f}{\partial z}(x_0, y_0, z_0) \cdot (z - z_0).$$

In Exercises 52–57, find the first-order partial derivatives of the given function

52. $f(x, y, z) = x^2 yz + xz^3 + y^2 x^2$

53. $f(x, y, z) = \dfrac{z}{x + y^2}$ **54.** $f(x, y, z) = (x^2 + y^2)e^z$

55. $f(x, y, z) = xy^2 \ln z$ **56.** $f(x, y, z) = ze^{xy} + xe^{yz}$

57. $f(x, y, z) = \sqrt{x^2 + y^2 + z^2}$

58. Given the following data:

$$f(2, 0, -1) = 3, \quad \frac{\partial f}{\partial x}(2, 0, -1) = -0.5,$$

$$\frac{\partial f}{\partial y}(2, 0, -1) = 1.2, \quad \frac{\partial f}{\partial z}(2, 0, -1) = 0.3,$$

estimate

(a) $f(1.9, 0.2, -1.1)$ **(b)** $f(2.2, 0, -0.8)$

(c) $f(2, 0.25, -1)$

59. A rectangular box of length 10 inches, width 12 inches, and height 5 inches has a volume of 600 cubic inches. Use linear approximation to estimate the volume in each of the following cases:

(a) The height and width are kept the same, and the length is increased to 11 inches.

(b) The length is decreased to 9 inches, the width is increased to 13 inches, and the height is increased to 5.25 inches.

60. As one of its many applications, mathematics can be used to model musical phenomena. The partial differential equation

$$\frac{\partial^2 u}{\partial t^2} = c^2 \frac{\partial^2 u}{\partial x^2}, \qquad (9)$$

where t (time) and x (length) are independent variables and c is a constant, is known as the one-dimensional **wave equation**. It governs the motion of a vibrating string, such as a guitar string.

(a) Show that $u(x, t) = e^x(e^{ct} + e^{-ct})$ is one solution of this equation.

(b) Show that $u(x, t) = (x + ct)^2 + (x - ct)^2$ is another solution of this equation.

61. Suppose the ends of the string are clamped to the x-axis at the points 0 and L, as shown in Figure 7.2.4, and the string is raised to a position given by $y = f(x)$, for some twice-differentiable function $f(x)$. It can be shown that after the string is released, its position at any time t will satisfy $y = u(x, t)$, where $u(x, t)$ is a solution of Eq. (9) for a suitable constant c. Show that the function u defined by

$$u(x, t) = \frac{1}{2}(f(x + ct) + f(x - ct))$$

satisfies (9) and also the initial conditions $u(x, 0) = f(x)$ and $\partial u / \partial t(x, 0) = 0$.

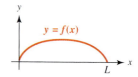

Figure 7.2.4

Solutions to practice exercises 7.2

1. $\dfrac{\partial f}{\partial x} = 3x^2 + y^2 + \dfrac{y}{x}$,

$\dfrac{\partial f}{\partial y} = e^y + 2xy + \ln x$

2. The partial derivatives are

$$\frac{\partial f}{\partial x} = \frac{2}{1 + y^2} \quad \text{and} \quad \frac{\partial f}{\partial y} = \frac{-4xy}{(1 + y^2)^2}.$$

HISTORICAL PROFILE

Cathleen S. Morawetz (1923–) is renowned for her work in partial differential equations. Among other things, she studied a higher-dimensional version of the wave equation and its relation to shock waves, which influenced the design of airplane wings. She is a member of the National Academy of Sciences and American Academy of Arts and Sciences, and a former president of the American Mathematical Society. In 1998 the President of the United States awarded her a National Medal of Science.

Therefore, $(\partial f/\partial x)(2, -1) = 1$ and $(\partial f/\partial y)(2, -1) = 2$, and the equation of the plane is

$$z - 2 = (x - 2) + 2(y + 1).$$

3. The partial derivatives are

$$\frac{\partial P}{\partial x} = 60 - \frac{3}{5}x \quad \text{and} \quad \frac{\partial P}{\partial y} = 70 - \frac{2}{5}y,$$

and the linear approximation formula tells us that

$$P(52,195) - P(50,200) \approx \frac{\partial P}{\partial x}(50,200) \cdot 2$$

$$+ \frac{\partial P}{\partial y}(50,200) \cdot (-5)$$

$$= 30 \cdot 2 + 10 \cdot 5 = 110.$$

Thus, the profit will rise by approximately \$110,000.

7.3 Maxima and Minima in Two Variables

As in the one-variable case, derivatives are an important tool in solving optimization problems. In fact, one of the main uses of partial derivatives is to find the minimum and maximum values of a given function. First, a few definitions.

We say a function $f(x, y)$ has a **local maximum** at a point (x_0, y_0) if

$$f(x, y) \leq f(x_0, y_0) \quad \text{for all } (x, y) \text{ close to } (x_0, y_0). \tag{10}$$

We say it has a **local minimum** at a point (x_0, y_0) if

$$f(x, y) \geq f(x_0, y_0) \quad \text{for all } (x, y) \text{ close to } (x_0, y_0). \tag{11}$$

Figure 7.3.1 shows surfaces in three-dimensional space in the neighborhood of a local maximum (a) and a local minimum (b). It is also common to use the phrase **relative maximum** (or **minimum**) in place of local maximum (or minimum).

If (10) holds true for *all* (x, y) throughout the domain of f, we say that f has a **global maximum** (also called an **absolute maximum**) at (x_0, y_0). If (11) is true on the whole domain of f, we say that f has a **global minimum** (or **absolute minimum**) at (x_0, y_0).

For example, the function $f(x, y) = x^2 + y^2$ has a local (and global) minimum at $(0, 0)$ since $f(0, 0) = 0$ and $x^2 + y^2 \geq 0$ for all (x, y). On the other hand, the function $f(x, y) = 1 - x^2 - y^2$ has a local (and global) maximum at $(0, 0)$ since $f(0, 0) = 1$ and $1 - x^2 - y^2 \leq 1$ for all (x, y).

Critical points and local extrema

The derivative is an important tool in finding the local maximum and minimum points of a function of several variables, just as it is in the one-variable case. In particular, the following theorem is the two-variable counterpart of a familiar criterion in one variable:

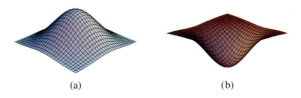

(a)   (b)

Figure 7.3.1

> ### Theorem 7.3.1
>
> **Necessary condition for local extrema**
>
> If $f(x, y)$ has a local maximum or minimum at a point (x_0, y_0) then
>
> $$\begin{cases} \dfrac{\partial f}{\partial x}(x_0, y_0) = 0 \\[2mm] \dfrac{\partial f}{\partial y}(x_0, y_0) = 0. \end{cases} \qquad (12)$$

This theorem is sometimes called the **first derivative test**. It is valid under the assumption that the partial derivatives exist at (x_0, y_0). We mention, however, that there are functions having a local maximum or minimum at a point where the partial derivatives do not exist. Such is the case with the distance function $f(x, y) = \sqrt{x^2 + y^2}$.

The idea behind Theorem 7.3.1 is simple. We reduce it to the one-variable case by considering each variable separately. To be specific, let's assume that $f(x, y)$ has a local extremum (i.e., maximum or minimum) at the point (x_0, y_0) and that both first-order partial derivatives exist. Then by keeping y_0 fixed and letting x vary, we get a one-variable function $f(x, y_0)$ with a local extremum at x_0. Therefore, its derivative must be zero (see Section 4.1), and we have $(\partial f/\partial x)(x_0, y_0) = 0$. Similarly, by keeping x_0 fixed and letting y vary, we get $f(x_0, y)$, which is a function of y only, with a local extremum at y_0. Therefore, the derivative must be zero at y_0, so that $(\partial f/\partial y)(x_0, y_0) = 0$.

Thus, the only possible candidates for local extrema are those at which both first-order partial derivatives are zero (assuming, as we are, that both partial derivatives exist throughout the domain of f). In other words, the search for local extema begins by solving the system (12). The solutions of that pair of equations are called the **critical points** of the function.

EXAMPLE 7.3.1

Find the critical points of the function

$$f(x, y) = \frac{1}{2}x^2 + 5y^2 - 3xy + x - 4y + 7.$$

SOLUTION   We have

$$\frac{\partial f}{\partial x} = x - 3y + 1 \quad \text{and} \quad \frac{\partial f}{\partial y} = 10y - 3x - 4.$$

The critical points are the solutions of the system of equations

$$x - 3y = -1$$
$$-3x + 10y = 4.$$

This system has only one solution, $x = 2$ and $y = 1$. Thus, $(x, y) = (2, 1)$ is the only critical point of f.

Finding the critical points is a necessary first step in locating maxima and minima, but it does not solve the problem. As in the one-variable case, a critical point may be a local maximum, local minimum, or neither. For example, consider a quadratic function of the form

$$f(x) = ax^2 + cy^2,$$

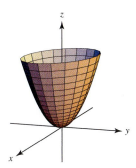

where a and c are nonzero constants. The partial derivatives are

$$\frac{\partial f}{\partial x} = 2ax \quad \text{and} \quad \frac{\partial f}{\partial y} = 2cy,$$

Figure 7.3.2

from which we conclude that $(0, 0)$ is a critical point (in fact, the only one), with $f(0, 0) = 0$. Now, consider the following three possibilities:

1. If a and c are both positive, then $f(x, y) > 0$ for all $(x, y) \neq (0, 0)$. Therefore, there is a local minimum at $(0, 0)$ (as shown in Figure 7.3.2).

2. If a and c are both negative, then $f(x, y) < 0$ for all $(x, y) \neq (0, 0)$. Therefore, there is a local maximum at $(0, 0)$ (as shown in Figure 7.3.3).

3. If $ac < 0$, there is neither a local maximum nor minimum at $(0, 0)$. For instance, if $a > 0$ and $c < 0$, then

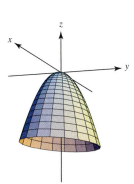

   - $f(x, 0) = ax^2 > 0$ for $x \neq 0$, and
   - $f(0, y) = cy^2 < 0$ for $y \neq 0$.

Figure 7.3.3

The case of $a > 0$ and $c < 0$, such as $f(x, y) = x^2 - y^2$, is shown in Figure 7.3.4. The curve of intersection with the plane $y = 0$ has a local minimum at the origin, whereas the curve of intersection with the plane $x = 0$ has a local maximum. This type of critical point is called a **saddle point**.

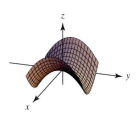

Quadratic functions

Figure 7.3.4

The key to understanding the max-min theory of functions of two variables is the analysis of quadratic functions—polynomials of degree two in x and y. We will now consider the special case in which

$$z = f(x, y) = ax^2 + 2bxy + cy^2, \tag{13}$$

where a, b, and c are constants, not all zero. Note that $f(0, 0) = 0$. The first partial derivatives are

$$\frac{\partial f}{\partial x} = 2ax + 2by \quad \text{and} \quad \frac{\partial f}{\partial y} = 2bx + 2cy,$$

which means the critical points are the solutions of the system of equations

$$ax + by = 0$$
$$bx + cy = 0.$$

If $ac - b^2 \neq 0$, the only solution is $(x, y) = (0, 0)$. We will assume that is the case, so that $(0, 0)$ is the unique critical point.

If both a and c are zero, Eq. (13) becomes $f(x, y) = 2bxy$, which has neither a maximum nor minimum at $(0, 0)$. To see why, suppose $b > 0$. Then $f(x, y) > 0$ if x and y have the same sign, and $f(x, y) < 0$ if x and y have opposite signs. The surface has a saddle point at $(0, 0)$, with a local minimum along the curve of intersection with the plane $y = x$ and a local maximum along the curve of intersection with $y = -x$. The surface is shown in Figure 7.3.5. Note that in this case $ac - b^2 < 0$, since both a and c are zero.

Now suppose that a and c are not both zero—say, $a \neq 0$. Referring back to (13), we will complete the square of $ax^2 + 2bxy$, treating y as a fixed number:

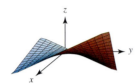

Figure 7.3.5

$$ax^2 + 2bxy = a\left(x + \frac{b}{a}y\right)^2 - \frac{b^2}{a}y^2.$$

Substituting this into (13) and simplifying, we get

$$f(x, y) = a\left(x + \frac{b}{a}y\right)^2 + \left(c - \frac{b^2}{a}\right)y^2$$
$$= \frac{1}{a}\left[(ax + by)^2 + (ac - b^2)y^2\right]. \tag{14}$$

We now consider the following cases:

1. If $ac - b^2 < 0$, let $u = ax + by$ and $v = y\sqrt{b^2 - ac}$. Then (14) becomes $(1/a)(u^2 - v^2)$, which has a saddle point at $x = 0$, $y = 0$.
2. If $ac - b^2 > 0$, let $u = ax + by$ and $v = y\sqrt{ac - b^2}$. Then (14) becomes $(1/a)(u^2 + v^2)$, which has

    - a minimum at $x = 0$, $y = 0$ if $a > 0$, and
    - a maximum at $x = 0$, $y = 0$ if $a < 0$.

Thus, we see that the behavior of $f(x, y)$ at the critical point $(0, 0)$ is governed by the signs of $ac - b^2$ and a. Finally, we observe that taking the second derivatives of f gives

$$\frac{\partial^2 f}{\partial x^2} = 2a, \quad \frac{\partial^2 f}{\partial x \partial y} = 2b, \quad \frac{\partial^2 f}{\partial y^2} = 2c.$$

Therefore, $\partial^2 f / \partial x^2$ has the same sign as a, and

$$\left(\frac{\partial^2 f}{\partial x^2}\right)\left(\frac{\partial^2 f}{\partial y^2}\right) - \left(\frac{\partial^2 f}{\partial x \partial y}\right)^2 = 4(ac - b^2)$$

has the same sign as $ac - b^2$.

We summarize our conclusions in a form that will be useful.

Second Derivative Test for Quadratic Functions

Given

$$f(x, y) = ax^2 + 2bxy + cy^2,$$

let

$$D = \left(\frac{\partial^2 f}{\partial x^2}\right)\left(\frac{\partial^2 f}{\partial y^2}\right) - \left(\frac{\partial^2 f}{\partial x \partial y}\right)^2.$$

Then

- $f(x, y)$ has a local minimum at $(0, 0)$ if $D > 0$ and $\partial^2 f/\partial x^2 > 0$.
- $f(x, y)$ has a local maximum at $(0, 0)$ if $D > 0$ and $\partial^2 f/\partial x^2 < 0$.
- $f(x, y)$ has neither a maximum nor minimum at $(0, 0)$ if $D < 0$. In this case $(0, 0)$ is called a *saddle point*.

EXAMPLE 7.3.2

Each of the following quadratics has a critical point at $(0, 0)$. Determine whether there is a local minimum there, a local maximum, or neither.

    (i) $f(x, y) = x^2 + xy + y^2$,
    (ii) $f(x, y) = -3x^2 + 3xy - 2y^2$,
    (iii) $f(x, y) = -x^2 + 3xy - 2y^2$.

SOLUTION (i) In this case, $\partial^2 f/\partial x^2 = \partial^2 f/\partial y^2 = 2$ and $\partial^2 f/\partial x \partial y = 1$. Therefore, $D = 3$, and there is a local minimum at $(0, 0)$.

(ii) Here, $\partial^2 f/\partial x^2 = -6$, $\partial^2 f/\partial y^2 = -4$, and $\partial^2 f/\partial x \partial y = 3$. Therefore, $D = 15$, and there is a local maximum at $(0, 0)$.

(iii) We have $\partial^2 f/\partial x^2 = -2$, $\partial^2 f/\partial y^2 = -4$, and $\partial^2 f/\partial x \partial y = 3$. Therefore, $D = -1$, and there is a saddle point at $(0, 0)$.

The second derivative test

The second derivative test for quadratic functions extends to a wide range of functions.

Theorem 7.3.2

Second Derivative Test

Let (x_0, y_0) be a critical point of the function $f(x, y)$, that is,

$$\frac{\partial f}{\partial x}(x_0, y_0) = 0 \quad \text{and} \quad \frac{\partial f}{\partial y}(x_0, y_0) = 0,$$

and assume that f has continuous first and second partial derivatives at all points near (x_0, y_0). Let

$$D(x, y) = \left(\frac{\partial^2 f}{\partial x^2}\right)\left(\frac{\partial^2 f}{\partial y^2}\right) - \left(\frac{\partial^2 f}{\partial x \partial y}\right)^2.$$

Then the following hold:

1. $f(x, y)$ has a local minimum at (x_0, y_0) if

$$D(x_0, y_0) > 0 \quad \text{and} \quad \frac{\partial^2 f}{\partial x^2}(x_0, y_0) > 0.$$

2. $f(x, y)$ has a local maximum at (x_0, y_0) if

$$D(x_0, y_0) > 0 \quad \text{and} \quad \frac{\partial^2 f}{\partial x^2}(x_0, y_0) < 0.$$

3. $f(x, y)$ has neither a local maximum nor minimum at (x_0, y_0) if

$$D(x_0, y_0) < 0.$$

In this case, (x_0, y_0) is called a **saddle point** for $f(x, y)$.

4. No conclusion about extrema of $f(x, y)$ at (x_0, y_0) can be drawn if

$$D(x_0, y_0) = 0.$$

We will discuss this test a little further on, but first we give some examples of its use.

EXAMPLE 7.3.3

Let $f(x, y) = x^3 - 6xy + 3y^2 - 9x + 7$. Find all the critical points and check whether each is a local maximum or minimum.

SOLUTION First, we compute the partial derivatives:

$$\frac{\partial f}{\partial x} = 3x^2 - 6y - 9 \quad \text{and} \quad \frac{\partial f}{\partial y} = -6x + 6y.$$

Setting them equal to zero and simplifying lead to the system of equations

$$x^2 - 2y - 3 = 0$$
$$y = x.$$

Since x and y are equal, we can substitute x for y in the first equation, which gives

$$x^2 - 2x - 3 = 0,$$

whose solutions are $x = -1$ and $x = 3$. Therefore, the critical points are $(-1, -1)$ and $(3, 3)$.

To check for local maximum or minimum, we first compute the second derivatives:

$$\frac{\partial^2 f}{\partial x^2} = 6x, \quad \frac{\partial^2 f}{\partial x \partial y} = -6, \quad \frac{\partial^2 f}{\partial y^2} = 6.$$

Therefore,

$$D(x, y) = \frac{\partial^2 f}{\partial x^2} \cdot \frac{\partial^2 f}{\partial y^2} - \left(\frac{\partial^2 f}{\partial x \partial y}\right)^2 = 36x - 36.$$

At the point $(-1, -1)$ we have $D(-1, -1) = -72$, which means there is a saddle point—neither a local minimum nor maximum. On the other hand, at $(3, 3)$, we have $D(3, 3) = 72$ and $(\partial^2 f/\partial x^2)(3, 3) = 18$. Therefore, there is a local minimum at $(3, 3)$.

PROBLEM-SOLVING TACTIC

To solve the nonlinear system (15):

- We exploit the fact that the second equation factors easily and we write it as $(x - 1)y = 0$.
- Setting each factor equal to zero, we obtain $x = 1$ or $y = 0$.
- Using $x = 1$ in the first equation, we obtain $y^2 - 1 = 0$ that gives $y = \pm 1$ and the system solutions $(1, 1)$ and $(1, -1)$.
- Using $y = 0$ in the first equation, we obtain $x^2 - 2x = 0$ that gives $x = 0$ or $x = 2$ and the system solutions $(0, 0)$ and $(2, 0)$.

EXAMPLE 7.3.4

Let $f(x, y) = x^3 + 3xy^2 - 3x^2 - 3y^2$. Find the critical points and use the second-derivative test to determine, if possible, the nature of $f(x, y)$ at each of these points.

SOLUTION As usual, we begin by computing the first derivatives:

$$\frac{\partial f}{\partial x} = 3x^2 + 3y^2 - 6x \quad \text{and} \quad \frac{\partial f}{\partial y} = 6xy - 6y.$$

Setting these equal to zero and simplifying lead to the system of equations

$$\begin{cases} x^2 + y^2 - 2x = 0 \\ xy - y = 0. \end{cases} \tag{15}$$

Solving system (15), we obtain four critical points: $(1, 1), (1, -1), (0, 0)$, and $(2, 0)$. To test these four points, we need to compute the second derivatives:

$$\frac{\partial^2 f}{\partial x^2} = 6x - 6, \quad \frac{\partial^2 f}{\partial x \partial y} = 6y, \quad \frac{\partial^2 f}{\partial y^2} = 6x - 6.$$

From these we get

$$D(x, y) = \frac{\partial^2 f}{\partial x^2} \cdot \frac{\partial^2 f}{\partial y^2} - \left(\frac{\partial^2 f}{\partial x \partial y}\right)^2 = (6x - 6)^2 - (6y)^2.$$

We then apply the second derivative test to each of the critical points as follows:

- At $(0, 0)$, $D(0, 0) = 36 > 0$ and $(\partial^2 f/\partial x^2)(0, 0) = -6 < 0$. Therefore, f has a local maximum at $(0, 0)$.
- At $(2, 0)$, $D(2, 0) = 36 > 0$, and $(\partial^2 f/\partial x^2)(2, 0) = 6 > 0$. Therefore, f has a local minimum at $(2, 0)$.
- At $(1, 1)$, $D(1, 1) = -36 < 0$. Thus, f has a saddle point at $(1, 1)$.
- At $(1, -1)$, $D(1, -1) = -36 < 0$. Thus, f has a saddle point at $(1, -1)$.

A Point to Remember   The second derivative test can only be applied in the case when $D(x_0, y_0) \neq 0$. That is similar to the one-variable case, in which no conclusion can be drawn if the second derivative is zero. To get an idea of what can happen when $D(x_0, y_0) = 0$, let's consider the function $f(x, y) = ax^4 + by^4$, where a and b are any nonzero constants. An easy calculation shows that

$$\frac{\partial f}{\partial x} = 4ax^3 \quad \text{and} \quad \frac{\partial f}{\partial y} = 4by^3,$$

and

$$D(x, y) = 144abx^2y^2.$$

It follows that $(0, 0)$ is a critical point with $D(0, 0) = 0$. Note that $f(0, 0) = 0$.

If a and b are both positive, then $f(x, y) > 0$ for every $(x, y) \neq (0, 0)$, and we conclude that the function has a local (and, in fact, global) minimum at $(0, 0)$.

If a and b are both negative, then $f(x, y) < 0$ for every $(x, y) \neq (0, 0)$, and we conclude that the function has a local (and, in fact, global) maximum at $(0, 0)$.

If $a > 0$ and $b < 0$, then $f(x, 0) > 0$ for every $x \neq 0$ and $f(0, y) < 0$ for every $y \neq 0$. It follows that $f(x, y)$ has neither a local maximum nor minimum at $(0, 0)$.

The basis for the second derivative test is a quadratic approximation formula, valid near a critical point of $f(x, y)$. For simplicity, let's assume that the critical point is at $(0, 0)$ and that f has continuous first and second partial derivatives at all nearby points. Then, for (x, y) near $(0, 0)$,

$$f(x, y) \approx ax^2 + 2bxy + cy^2,$$

where

$$a = \frac{1}{2}\frac{\partial^2 f}{\partial x^2}(0,0), \quad b = \frac{1}{2}\frac{\partial^2 f}{\partial x \partial y}(0,0), \quad c = \frac{1}{2}\frac{\partial^2 f}{\partial y^2}(0,0).$$

The behavior of f at $(0,0)$ can be determined by analyzing the quadratic, and that observation leads to the general form of the second derivative test. A precise justification can be found in advanced calculus texts.

Global maxima and minima

So far, we have only answered questions about *local* extrema. Finding the *global* maximum and minimum of a given function—or even determining if there are any—may be considerably more difficult, and the complications are substantially greater in two variables than in one. However, for quadratic functions we have the following simple situation.

> **Maxima and Minima for Quadratic Functions**
>
> If a quadratic function has a local maximum (minimum) at a point, then it also has global maximum (minimum) there.

EXAMPLE 7.3.5

Suppose the profit function of a company selling x units of one product and y units of another product is

$$P(x, y) = 12x + 9y - 450 - 0.01(4x^2 + xy + y^2).$$

Find the pair (x, y) that maximizes the company's profit and determine the maximum profit.

SOLUTION   Once again, we begin by computing the partial derivatives:

$$\frac{\partial P}{\partial x} = 12 - 0.01(8x + y) \quad \text{and} \quad \frac{\partial P}{\partial y} = 9 - 0.01(x + 2y).$$

Setting these equal to zero and simplifying lead to the system of equations

$$8x + y = 1{,}200$$
$$x + 2y = 900,$$

with the unique solution $x = 100$ and $y = 400$. Thus, the point $(100, 400)$ is the only critical point of P. To apply the second derivative test, we compute the second derivatives:

$$\frac{\partial^2 P}{\partial x^2} = -0.08, \quad \frac{\partial^2 P}{\partial x \partial y} = -0.01, \quad \frac{\partial^2 P}{\partial y^2} = -0.02.$$

Since $\partial^2 P / \partial x^2 < 0$ and

$$D(x, y) = \frac{\partial^2 P}{\partial x^2} \cdot \frac{\partial^2 P}{\partial y^2} - \left(\frac{\partial^2 P}{\partial x \partial y} \right)^2 = 0.0015,$$

there is a local maximum at the critical point. Since $P(x, y)$ is a quadratic function, the local maximum must also be a global maximum. Therefore, we find the maximum profit by evaluating the profit function at this point, which gives $P(100, 400) = 1,950$.

Practice Exercises 7.3

1. Find the point where the function $f(x, y) = 1 + 5x - 4y - xy - 2x^2 - y^2$ has a global maximum. Justify your answer and compute the maximum value.

2. Find all critical points of the function $f(x, y) = y^3 + x^2 y + 2x^2 - 3y^2 + 1$ and characterize each as a local maximum, local minimum, or saddle point.

Exercises 7.3

In Exercises 1–10, find the critical points of the given function.

1. $f(x, y) = x^2 + y^2 - 2x + 4y + 6$

2. $f(x, y) = 2y^2 + x^2 + xy + 3x - 2y + 1$

3. $f(x, y) = y^2 - 5xy - 2x^2 + 3$

4. $f(x, y) = x^2 + 3y^2 + 3y - 2x + 8$

5. $f(x, y) = 3xy - x^3 - y^3 + 10$

6. $f(x, y) = y^3 + 3x^2 - 6xy - 9y - 2$

7. $f(x, y) = xy - e^x$

8. $f(x, y) = \ln(1 + x^2 + y^2)$

9. $f(x, y) = x^2 - e^{y^2}$

10. $f(x, y) = y^2 - x \ln y, \ y > 0$

In Exercises 11–26, use the first and second derivative tests to find the local minima, maxima, and saddle points of the given function. If the second derivative test is inconclusive, state this.

11. $g(x, y) = x^2 - xy + y^2 + 3y - 1$

12. $f(x, y) = 2y^2 - xy + x^2 - 7x + 4$

13. $f(x, y) = x^2 - xy + 2y + x + 3$

14. $f(x, y) = 8 + 2y - x^2 - y^2$

15. $f(x, y) = x^3 + y^3 - 3xy + 4$

16. $f(x, y) = y^3 - x^3 - 3xy + 5$

17. $f(x, y) = 8 + 4x + 2y - x^2 - y^2$

18. $f(x, y) = xy - \frac{1}{2}x^2 - \frac{1}{3}y^3$

19. $f(x, y) = x^4 + y^4$

20. $f(x, y) = x^4 - y^4$

21. $f(x, y) = 1 - x^4 - y^4$

22. $f(x, y) = 1 + xy^3$

23. $f(x, y) = x \ln y, \ y > 0$

24. $f(x, y) = x^2 + x \ln y + 1, \ y > 0$

25. $f(x, y) = x^2 + e^{y^2}$

26. $f(x, y) = x^2 + e^{-y^2}$

27. A local science and mathematics children's museum has hired you, a craftsman, to build a plastic model of the surface defined by

$$f(x, y) = \frac{3}{2}x^2 + 6xy + \frac{2}{3}y^2$$

as part of a display on the geometry of surfaces. You notice that $f(0, 0) = 0$ and decide to start by modeling the surface near that point. Is it a critical point, and if so, is it a local maximum or minimum?

28. Your next assignment (continuing the previous problem) is to shape a thin metal plate to a graph of the form

$$f(x, y) = 1 - (x^2 + y^2 + Axy + Bx + Cy)$$

with a local maximum of 2 at $(1, -1)$. (The center of the plate is the origin.)
(a) Find the only A, B, and C that will give a critical point at $(1, -1)$ with $f(1, -1) = 2$.
(b) Show that there is indeed a local maximum at $(1, -1)$.

In Exercises 29–34, find the point where the given function has a global minimum or maximum. Justify your answer.

29. $f(x, y) = 2x^2 + 3y^2 - 12x - 6y + 9$

30. $f(x, y) = 3x^2 + y^2 + 2xy - 8x + 4y + 1$

31. $f(x, y) = 3 + 6x + 4y - 3x^2 - 4y^2$

32. $f(x, y) = 1 + 10x + 6y - 2xy - 3x^2 - 3y^2$

33. $f(x, y) = 5 - 2x - 6y - 3x^2 - 10xy - 9y^2$

34. $f(x, y) = 5x^2 + 13y^2 - 8xy - 22x - 2y + 35$

In Exercises 35–38, the function $P(x, y)$ is the profit function of a company selling x units of one product and y units of another product. In each case find the values of x and y that maximize the company's profit, as well as the maximum profit.

35. $P(x, y) = 8x + 11y - 900 - 0.01(x^2 + xy + 2y^2)$

36. $P(x, y) = 12x + 9y - 500 - 0.01(4x^2 + xy + y^2)$

37. $P(x, y) = 90x + 80y - 10{,}000 - 0.01(2x^2 + xy + 3y^2)$

38. $P(x, y) = 150x + 320y - 15{,}000$
$\qquad - 0.01(3x^2 + xy + 5y^2)$

39. A toy store carries two models of a certain board game—the standard and the deluxe. It has determined that if it charges x dollars for the standard and y dollars for the deluxe, then it will sell $(50 - 3x + 2y)$ standard and $(6 + \frac{1}{2}x - y)$ deluxe games per week.
(a) Write a formula for the weekly revenue as a function of x and y.
(b) Find the x and y that maximize the revenue. How do you know it gives a maximum and not a minimum or saddle point?
(c) Suppose the store pays the distributor $9 for each standard model and $14 for each deluxe. What price should it charge for each model in order to maximize the total profit? (You may use a calculator and round your answers to the nearest cent.)

40. A pharmaceutical company estimates that the demand function for one of its drugs in the United States is $p_1 = 25 - 0.1x$, while the demand function for the same drug in a foreign market is $p_2 = 20 - 0.05y$, where x and y are in thousands of units and p_1 and p_2 are in dollars per unit. Its cost function is $C(x, y) = 10 + 2(x + y)$ in thousands of dollars.
(a) Find the quantities x and y that maximize the company's profit.
(b) What are the corresponding prices?
(c) Explain the price discrimination.

41. The three most visited places in the kitchen during food preparation are the sink, the stove, and the refrigerator.

When designing floor plans, it is suggested that the total distance between them, the "work triangle," be minimized. The sink and the refrigerator are at $(0, 0)$ and $(3, 5)$, respectively. The stove is an electric countertop model that will be built into a work island in the kitchen.
(a) What coordinates should be chosen for the stove to minimize the total distance?
(b) Plot the sink, refrigerator, and stove on the xy-plane. Describe the location of the stove in terms of the refrigerator and sink.

42. A retail chain has three stores in a certain county. It is planning to build a new warehouse, and it wants to choose the location to minimize the sum of the squares of the distances from the three stores.
(a) If the stores are located at the points $(1, 0)$, $(-2, 2)$, and $(4, -5)$ relative to a set of coordinates, find the coordinates of the point where the warehouse should be located.
(b) Answer the same question if the stores are located at the points (x_1, y_1), (x_2, y_2), and (x_3, y_3).
(c) How do you know you have found the minimum and not a maximum or saddle point?

In Exercises 43–46, assume that $(2, 3)$ is a critical point of $f(x, y)$ and use the given information, if possible, to determine its nature. If not, say so.

43. $\dfrac{\partial^2 f}{\partial x^2}(2, 3) = -4, \quad \dfrac{\partial^2 f}{\partial y^2}(2, 3) = -7, \quad \dfrac{\partial^2 f}{\partial x \partial y}(2, 3) = 5$

44. $\dfrac{\partial^2 f}{\partial x^2}(2, 3) = 4, \quad \dfrac{\partial^2 f}{\partial y^2}(2, 3) = 7, \quad \dfrac{\partial^2 f}{\partial x \partial y}(2, 3) = 5$

45. $\dfrac{\partial^2 f}{\partial x^2}(2, 3) = 4, \quad \dfrac{\partial^2 f}{\partial y^2}(2, 3) = 7, \quad \dfrac{\partial^2 f}{\partial x \partial y}(2, 3) = 8$

46. $\dfrac{\partial^2 f}{\partial x^2}(2, 3) = 3, \quad \dfrac{\partial^2 f}{\partial y^2}(2, 3) = 12, \quad \dfrac{\partial^2 f}{\partial x \partial y}(2, 3) = -6$

In Exercises 47–48, find the point (a, b) where the function $E(a, b)$ has an absolute minimum. Note: In these exercises, a and b are the independent variables. Functions of this type will come up when we study the method of least squares in Section 7.4.

47. $E(a, b) = (a + b - 2)^2 + (2a + b - 2)^2 + (3a + b - 4)^2$

48. $E(a, b) = (b + 1)^2 + (2a + b)^2 + (a + b - 1)^2$
$\qquad + (a + b + 2)^2$

Solutions to practice exercises 7.3

1. The first derivatives are $\partial f/\partial x = 5 - y - 4x$ and $\partial f/\partial y = -4 - x - 2y$. Therefore, the critical points are the solutions of the equations

$$4x + y = 5 \quad \text{and} \quad x + 2y = -4.$$

The only solution of this system is $x = 2$ and $y = -3$.

To use the second derivative test, we compute the second-order partial derivatives: $\partial^2 f/\partial x^2 = -4$, $\partial^2 f/\partial x \partial y = -1$, and $\partial^2 f/\partial y^2 = -2$. Therefore,

$$D(2, -3) = \frac{\partial^2 f}{\partial x^2} \cdot \frac{\partial^2 f}{\partial y^2} - \left(\frac{\partial^2 f}{\partial x \partial y}\right)^2$$

$$= (-4) \cdot (-2) - (-1)^2 = 7 > 0,$$

and $(\partial^2 f/\partial x^2)(2, -3) < 0$, which means that there is a local maximum at $(2, -3)$. Since our function is quadratic, its local maximum is also a global maximum. Its maximum value is $f(2, -3) = 12$.

2. Setting the first-order partial derivatives equal to zero leads to the system of equations

$$2x(y + 2) = 0 \quad \text{and} \quad 3y^2 + x^2 - 6y = 0.$$

From the first equation we see that either $x = 0$ or $y = -2$.

Setting $x = 0$ in the second equation leads to $3y(y - 2) = 0$, whose solutions are $y = 0$ and $y = 2$. Therefore, the points $(0, 0)$ and $(0, 2)$ are both critical points.

Setting $y = -2$ in the second equation leads to $24 + x^2 = 0$, which has no solutions. Therefore, there are no other critical points.

The second-order partial derivatives are $\partial^2 f/\partial x^2 = 2y + 4$, $\partial^2 f/\partial y^2 = 6y - 6$, and $\partial^2 f/\partial x \partial y = 2x$. Therefore, $D(0, 0) = 4 \cdot (-6) < 0$, which means that there is a saddle point at $(0, 0)$. Since $D(0, 2) = 8 \cdot 6 > 0$ and $(\partial^2 f/\partial x^2)(0, 2) = 8 > 0$, there is a local minimum at $(0, 2)$.

7.4 The Method of Least Squares

An important application of the max-min theory for functions of two variables is the method of least squares, used for finding the line that best fits a given set of data points. This method is a mainstay of statistical analysis.

The least-squares line: an example

Suppose the demand (in billions of dollars) for the product of a certain company during the period 2000 through 2002 is given by the following table:

year	2000	2001	2002
demand	1	1	2

The company wants to find a formula—based on the data in the table—that approximates the demand as a function of time, which they can use to forecast the demand for the coming year. The simplest such formula is a linear equation of the form

$$y = at + b,$$

where t denotes the number of years after 1999 (so that $t = 1$ corresponds to 2000, and so forth), and y stands for the demand in billions of dollars. The problem is to choose the constants a and b to get the best fit between the linear function and the given data.

We first observe that the table gives us three data points: $(1, 1)$, $(2, 1)$, and $(3, 2)$. If we plot these points in the (t, y)-plane, we get a diagram known as a **scatter diagram**. It is shown in Figure 7.4.1.

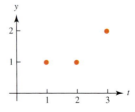

Figure 7.4.1

We want to find the line that is closest to these points—but in order to do that, we first have to decide what "closest" means. The interpretation most commonly used in statistics and other applications is this: For any given line

- take the vertical distance from each data point to the line,
- square that number, and
- add all the squares you get in this way.

The line for which that sum of squares is smallest is considered to be the closest to the set of data points. It is called the **least-squares line** or **regression line**.

Figure 7.4.2 shows the three data points from the table and a line with equation $y = at + b$. The numbers d_1, d_2, and d_3 are the vertical distances from the data points to the line. Of course, they depend on the coefficients a and b of the linear function, and our objective is to find the coefficients that minimize the sum of the squares of these distances, given by

$$E = d_1^2 + d_2^2 + d_3^2.$$

To do that, we have to write E as a function of a and b. To begin with,

$$d_1 = (a + b) - 1,$$

because the first data point is $(1, 1)$, which has height 1, whereas the line has height $(a + b)$ if $t = 1$. The second data point is $(2, 1)$, which has height 1, whereas the line has height $(2a + b)$ if $t = 2$. Therefore, $d_2 = (2a + b) - 1$. In the same way, we get $d_3 = (3a + b) - 2$. Squaring each of these and then adding give the formula

$$E(a, b) = [(a + b) - 1]^2 + [(2a + b) - 1]^2 + [(3a + b) - 2]^2.$$

Our task is to find the a and b that minimize this function. To do that, we first find the critical points. Taking partial derivatives and using the chain rule, we get

$$\frac{\partial E}{\partial a} = 2(a + b - 1) + 4(2a + b - 1) + 6(3a + b - 2)$$

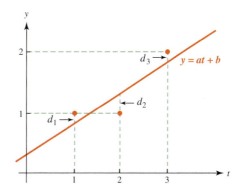

Figure 7.4.2

and

$$\frac{\partial E}{\partial b} = 2(a + b - 1) + 2(2a + b - 1) + 2(3a + b - 2).$$

Setting both of these equal to zero and simplifying give the system of equations

$$28a + 12b = 18$$
$$12a + 6b = 8.$$

To solve it, we multiply the second equation by -2 and add it to the first equation, which gives $4a = 2$ or $a = \frac{1}{2}$. Substituting this into either equation and solving for b give $b = \frac{1}{3}$. Therefore, the only critical point of $E(a, b)$ is given by

$$a = \frac{1}{2} \quad \text{and} \quad b = \frac{1}{3}.$$

To see that $E(a, b)$ actually has a minimum at its critical point, we can apply the second derivative test (Theorem 7.3.2). The second derivatives are

$$\frac{\partial^2 E}{\partial a^2} = 2 + 8 + 18 = 28$$

$$\frac{\partial^2 E}{\partial b^2} = 2 + 2 + 2 = 6$$

$$\frac{\partial^2 E}{\partial a \partial b} = 2 + 4 + 6 = 12,$$

so that

$$D = \frac{\partial^2 E}{\partial a^2} \cdot \frac{\partial^2 E}{\partial b^2} - \left(\frac{\partial^2 E}{\partial a \partial b} \right)^2 = 28 \cdot 6 - (12)^2 = 24.$$

Since $D > 0$ and $\partial^2 E / \partial a^2 > 0$, the error function $E(a, b)$ has a local minimum at the point $(\frac{1}{2}, \frac{1}{3})$. And, since $E(a, b)$ is quadratic, we conclude that it actually has a *global* minimum there. Therefore, the least-squares line—the line that best fits the given data—is given by the equation

$$y = \frac{1}{2}t + \frac{1}{3}.$$

If we use this equation to predict the production level for the 4th year, we get

$$y(4) = \frac{1}{2} \cdot 4 + \frac{1}{3} = \frac{7}{3}.$$

The general case

We now generalize the last example to the case in which we are given a set of n numerical data points, say,

$$(x_1, y_1), \quad (x_2, y_2), \quad \ldots, \quad (x_n, y_n)$$

and try to find the least-squares line (also called the regression line)—that is, the line

$$y = ax + b$$

that comes closest to those points, in the sense of minimizing the sum of the squares of the vertical distances between the line and the data points.

For x_1, the data point has height y_1, and the line has height $ax_1 + b$. Therefore, the vertical distance from the first data point to the line is given by

$$d_1 = (ax_1 + b) - y_1.$$

Similarly, the vertical distance from the second data point to the line is given by

$$d_2 = (ax_2 + b) - y_2,$$

and, in general, the distance from the ith data point to the line is given by

$$d_i = (ax_i + b) - y_i.$$

Squaring and adding all these distances give us the "error" function

$$E(a, b) = [(ax_1 + b) - y_1]^2 + [(ax_2 + b) - y_2]^2 + \cdots + [(ax_n + b) - y_n]^2,$$

and our job is to find the choice of a and b to minimize this function.

To do that, we first find the critical points. Taking partial derivatives and using the chain rule, we get

$$\frac{\partial E}{\partial a} = 2x_1(ax_1 + b - y_1) + 2x_2(ax_2 + b - y_2) + \cdots + 2x_n(ax_n + b - y_n)$$

$$\frac{\partial E}{\partial b} = 2(ax_1 + b - y_1) + 2(ax_2 + b - y_2) + \cdots + 2(ax_n + b - y_n).$$

Setting these equal to zero and simplifying give the system of equations

$$(x_1^2 + x_2^2 + \cdots + x_n^2)a + (x_1 + x_2 + \cdots + x_n)b = x_1y_1 + x_2y_2 + \cdots + x_ny_n$$

$$(x_1 + x_2 + \cdots + x_n)a + nb = y_1 + y_2 + \cdots + y_n,$$

which has the unique solution

$$a = \frac{n(x_1y_1 + \cdots + x_ny_n) - (x_1 + \cdots + x_n)(y_1 + \cdots + y_n)}{n(x_1^2 + \cdots + x_n^2) - (x_1 + \cdots + x_n)^2} \tag{16}$$

and

$$b = \frac{(y_1 + \cdots + y_n) - a(x_1 + \cdots + x_n)}{n}. \tag{17}$$

These are the coordinates of the critical point. As before, we can use the second derivative test to show that it is a local minimum point and therefore—because $E(a, b)$ is quadratic—a global minimum point.

EXAMPLE 7.4.1

Find the least-squares line $y = ax + b$ for the following data:

x	0.5	1	1.5	2
y	2.4	1.2	-0.7	-1.8

SOLUTION   In this case, we have $n = 4$ and

$$x_1 = 0.5, \quad x_2 = 1, \quad x_3 = 1.5, \quad x_4 = 2$$
$$y_1 = 2.4, \quad y_2 = 1.2, \quad y_3 = -0.7, \quad y_4 = -1.8.$$

Therefore,

$$x_1 + x_2 + x_3 + x_4 = 0.5 + 1 + 1.5 + 2 = 5$$
$$y_1 + y_2 + y_3 + y_4 = 2.4 + 1.2 - 0.7 - 1.8 = 1.1$$
$$x_1^2 + x_2^2 + x_3^2 + x_4^2 = (0.5)^2 + 1^2 + (1.5)^2 + 2^2 = 7.5$$
$$x_1 y_1 + x_2 y_2 + x_3 y_3 + x_4 y_4 = (0.5)(2.4) + 1(1.2) - (1.5)(0.7) - 2(1.8) = -2.25.$$

Substituting these into formulas (16) and (17), we get

$$a = \frac{-4(2.25) - 5(1.1)}{4(7.5) - 5^2} = -2.9$$

$$b = \frac{1.1 + (2.9)5}{4} = 3.9.$$

Thus, the least-squares line is $y = -2.9x + 3.9$.

EXAMPLE 7.4.2

During 1991–2000, a certain company has measured its profit per year in millions of dollars and the productivity of its workers (output per worker per week) in hundreds of dollars. The collected data are listed in Table 7.4.1, where x denotes productivity and y denotes profit.

   Find the least-squares line for these data and use it to make an approximate prediction of the profit the company will make if it can raise the workers' productivity to a level of 2.8.

TABLE 7.4.1

Year	x	y
1991	1.45	3.20
1992	1.60	3.50
1993	1.80	3.60
1994	1.90	3.65
1995	2.05	3.80
1996	2.20	4.25
1997	2.35	4.30
1998	2.40	4.35
1999	2.50	4.40
2000	2.75	4.55

SOLUTION   By using formulas (16) and (17) with $n = 10$, we find (with the help of a calculator) that, to five decimal places, $a = 1.09194$ and $b = 1.66694$. Therefore, the least-squares line is given by the formula

$$y = 1.09194x + 1.66694.$$

The given data and the least-squares line are graphed in Figure 7.4.3.

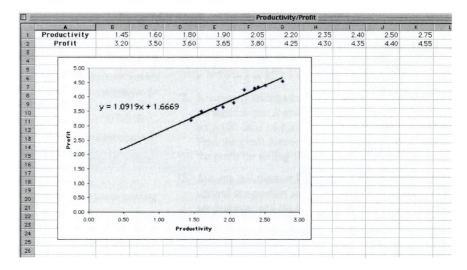

Figure 7.4.3

TABLE 7.4.2

x	y
0	2.18
0.2	1.75
0.4	1.5
0.6	1.16
0.8	1.24
1.0	0.85
1.2	0.7
1.4	1.12
1.6	1.34
1.8	1.42
2.0	2.07

Figure 7.4.4

Figure 7.4.5

By setting $x = 2.8$ in the least-squares equation, we get $y = 4.72$ (to two decimal places), which is an approximate prediction of the company's profit at that productivity level.

Both graphing calculators and spreadsheets have programs that calculate the coefficients of the regression line. Figure 7.4.3 shows an Excel spreadsheet containing the data of Example 7.4.2, a scatter diagram of the data points, and the equation and graph of the regression line.

Nonlinear regression

In Example 7.4.2 we implicitly assumed that the data in the table are linked by a relation that is approximately linear. But is that assumption warranted? How can we tell if the least-squares line actually fits the data? Unfortunately, there is no easy answer to that question. One simple method is to plot the scatter diagram and see if the points appear to cluster around the least-squares line. For instance, the scatter diagram shown in Figure 7.4.3 suggests that the least-squares line fits the data of Example 7.4.2 reasonably well. On the other hand, if we enter the data of Table 7.4.2 into the STAT editor of a graphing calculator (as shown in Figure 7.4.4) and create a scatter diagram based on that data, we get the diagram shown in Figure 7.4.5. The graph does not appear compatible with an approximately linear relation between the variables.

In fact, this scatter diagram is closer in shape to a parabola than a line, and we might try to use the least-squares principle to fit a quadratic function of the form

$$y = ax^2 + bx + c, \tag{18}$$

where a, b, and c are constants to be determined. This type of curve fitting is called

quadratic regression. Given data points

$$(x_1, y_1), \quad (x_2, y_2), \quad \ldots, \quad (x_n, y_n),$$

we determine the coefficients in Eq. (18) to minimize the following sum of squares:

$$E(a, b, c) = (ax_1^2 + bx_1 + c - y_1)^2 + (ax_2^2 + bx_2 + c - y_2)^2$$
$$+ \cdots + (ax_n^2 + bx_n + c - y_n)^2.$$

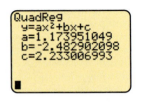

Figure 7.4.6

In principle, this is a simple optimization problem in three variables, but, as you might imagine, the formulas are rather complicated. Both spreadsheets and graphing calculators have built-in procedures that will calculate the quadratic regression coefficients and draw both the scatter diagram and the curve. Figure 7.4.6 shows the quadratic regression coefficients for the data of Table 7.4.1, and Figure 7.4.7 illustrates the quadratic regression curve.

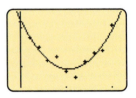

Figure 7.4.7

In addition to fitting linear and quadratic functions to sets of data pairs by the least-squares method, graphing calculators are programmed to fit other types of curves—third- and fourth-degree polynomials, power curves of the form $y = ax^b$, and exponential curves of the form $y = ae^{bx}$, where, in both cases, a and b are constants to be determined.

Practice Exercises 7.4

1. Given the points $(0, 2)$, $(1, 4)$, $(2, 3)$, and $(4, 4)$, find the coefficients a and b of the least squares line in the following sequence of steps:
   (a) Write the error function $E(a, b)$ giving the sum of the squares of the distances from the four points to the line $y = ax + b$.
   (b) Compute the first-order partial derivatives $\partial E/\partial a$ and $\partial E/\partial b$. Then, set them both equal to zero and solve the system of linear equations.

   (c) Explain how you can be sure that the solutions give the absolute minimum point.

2. Check your answer to the last problem by using formulas (16) and (17) to compute a and b.

Exercises 7.4

In Exercises 1–8, find the least-squares line for the given data.

1. $(1, 1)$, $(2, 2)$, $(3, 4)$
2. $(0, 3)$, $(1, 4)$, $(2, 4)$
3. $(0, 1)$, $(1, 2)$, $(2, 4)$
4. $(1, 1)$, $(2, 0)$, $(3, -3)$
5. $(0, 1)$, $(1, 2)$, $(2, \frac{5}{2})$
6. $(1, 9)$, $(2, 11)$, $(3, 24)$
7. $(0, 1)$, $(1, 2)$, $(2, 1)$, $(3, 2)$
8. $(1, 6)$, $(2, 5)$, $(3, 3)$, $(6, 1)$

In Exercises 9–14, you may need a calculator (or computer) to do the computations.

9. Suppose that the profit in billions of dollars of a large corporation for the last 5 years is given by the following table:

year	1	2	3	4	5
profit	2.3	2.7	2.8	3.0	3.5

Find the least-squares line to fit these data and then predict the profit for the 6th year.

10. The life expectancy at birth in the United States since 1950 is given by the following table:

year	1950	1960	1970	1980	1990
expectancy	68.2	69.7	70.8	73.7	75.4

Find the least-squares line to fit these data, letting t be the number of years after 1950 and y the life expectancy. Then use it to predict life expectancy in the year 2010.

11. The Gross National Product (GNP) of the United States in billions of dollars since 1960 is given by the following table:

year	1960	1970	1980	1990	2000
GNP	530.6	1046.08	2830.83	5832.23	9860.8

Find the least squares line to fit the given data and use it to predict the GNP in the year 2010.

Source: U.S. Department of Commerce, Bureau of Economic Analysis

12. A company's annual sales in billion of dollars for the past 9 years are shown in the following table:

year	1	2	3	4	5	6	7	8	9
sales	0.8	1.0	1.4	1.5	2.0	2.6	3.5	4.8	5.5

Find the equation of the least-squares line and then use it to predict the company's sales in the 10th year.

13. The U.S. population in millions from 1900 to 2000 is shown in the following table:

year	1900	1910	1920	1930	1940	1950
population	76	92	106	123	132	151

year	1960	1970	1980	1990	2000
population	179	203	227	249	281

Find the least-squares line to fit these data and use it to predict the U.S. population in the year 2010.

14. The cost in dollars of a first-class stamp since 1970 is given by the following table:

year	1971	1974	1975	1978	1981	1985
postage	0.08	0.10	0.13	0.15	0.20	0.22

year	1988	1991	1995	1999	2002
postage	0.25	0.29	0.32	0.33	0.34

Find the least-squares line to fit these data and use it to predict the price of a first-class stamp in the year 2010.

15. The National Audubon Society Christmas Bird Count is an early winter bird survey. Volunteers count all the birds they see in a 15-mile-diameter circle during a single day. It must be done within 2 weeks of December 25. A local report gives both a 6-year count of yellow-rumped warblers (*Dendroica coronata*) and the linear regression equation associated with it, as follows:

year	1980	1981	1982	1983	1984	1985
count	247	236	221	237	215	250

$$y = -6.63x + 257.3,$$

where x is the number of years after 1980.
(a) As you compare the function to the data, you should be able to see that a mistake has been made in determining the regression formula. What seems wrong?
(b) What is the correct function?

Source: J. R. Sauer, S. Schwartz, and B. Hoover, "The Christmas Bird Count Home Page," Version 95.1, Laurel, MD: Patuxent Wildlife Research Center, 1996. Go to www.mbr.nbs.gov/bbs/cbc.html.

16. The daily high temperatures recorded at Colfax, California, over 5 successive days beginning on November 16, 2000, were

day	16	17	18	19	20
temperature (in degrees Fahrenheit)	58	55	57	60	63

(a) Find the linear regression function for these data and then compute the average value of the function over this interval.
(b) Average the temperatures given in the table and compare that to your answer to part (a). What do you observe? Do you think it is a coincidence or a particular case of a general rule?

Source: Data from the National Oceanic and Atmospheric Administration.

17. Get data (from the library, for instance) on the yearly highs of the Dow Jones industrial average since 1980. Make a table, rounding the numbers to the nearest multiple of 10, and find the least-squares line that fits the data. Finally, use this line to predict how high the Dow Jones average will go in the year 2010.

18. The term "regression" was coined by Francis Galton, a British scientist of the nineteenth century. Galton, who

was a cousin of Charles Darwin and one of the founders of modern statistics, was interested in hereditary properties, and he developed the theory that, on the average, children would be closer to the mean in measurable traits, such as height, than were their parents. In other words, children whose parents were above average height might also be taller than average, but would generally be closer to the average than the parents. In a famous study, he tabulated the heights of children and their parents as follows:

x	64.5	65.5	66.5	67.5	68.5	69.5	70.5	71.5	72.5
y	65.8	66.7	67.2	67.6	68.2	68.9	69.5	69.9	72.2

where x is the average height of the mother and father, and y is the child's height, both in inches. (Actually, this is a somewhat simplified version of Galton's study, but the gist is there.) Use a graphing calculator to plot a scatter diagram of Galton's data, compute the equation of the least-squares line, and plot the line.

19. A study done in 1965 attempted to determine the effect on human health of radioactive waste that had leaked over the years into the Columbia River from the Hanford, Washington, nuclear reactor. Nine counties in Oregon that bordered the river were each assigned an exposure index (labeled as x in the table below) based on factors such as the distance from the reactor and the distance of the population centers from the river. The index was compared with the deaths from cancer per 100,000 (labeled as y in the table below), resulting in the following table:

x	8.34	6.41	3.41	3.83	2.57
y	210.3	177.9	129.9	162.3	130.1
x	11.64	1.25	2.49	1.62	
y	207.5	113.5	147.1	137.5	

(a) Use a graphing calculator to make a scatter diagram of these data points, then determine and graph the least-squares line.
(b) Using the least-squares line as a model, predict the approximate increase in the number of cancer deaths per 100,000 in a given county if the index increases by 1.
(c) Based on the least-squares model, approximately how many cancer deaths per 100,000 would you expect in a county with index zero?

In Exercises 20–23, use a graphing calculator to draw a scatter diagram of the given table of data. Make a judgment about what type of curve would best fit the data

points—linear, quadratic, or exponential—and find and graph it. (Note: All the data in these exercises are hypothetical.)

20. A study of the heights x and weights y of 15 5-year-old boys resulted in the following table:

x	42	46	39	35	36	38	40	36
y	38	49	32	37	31	36	42	37
x	41	34	37	41	43	37	33	
y	40	30	35	34	44	36	29	

21. A study compared the IQ scores x of a group of 5-year-old children with their attention span in minutes y in performing a repetitive task.

x	80	80	85	90	90	95	105	110
y	2	3	4.7	4.5	5.2	5	6	6.5
x	110	115	120	125	130	130	135	140
y	5.5	6.4	3.4	4.6	3.8	2.9	2	4.1

22. For a certain model used car, the average wholesale price y, in thousands of dollars, is tabulated as follows in terms of its age x.

x	1	2	3	4	5	6	7
y	8.2	7.0	6.1	5.25	4.3	3.5	2.9

23. A retail store that has been in business for 8 years made a table of its annual sales y, in thousands of dollars, in each given year x of operation.

x	1	2	3	4	5	6	7	8
y	24.3	28.6	32.2	41.1	49.4	69.5	85.6	110.4

24. Find and plot the least-squares line for the following set of data points: (1.45, 3.2), (1.8, 3.6), (1.9, 3.65), (2.05, 3.8), (2.2, 4.25), (2.35, 4.3), (2.4, 4.35), (2.5, 4.4), (2.75, 4.55), (2.9, 4.60), (3, 4.68), (3.2, 4.69), (3.5, 5), (3.65, 5.3), (4, 6).

Solutions to practice exercises 7.4

1. (a) The error function is

$$E(a, b) = (b-2)^2 + (a+b-4)^2$$
$$+ (2a+b-3)^2 + (4a+b-4)^2.$$

(b) The partial derivatives are

$$\frac{\partial E}{\partial a} = 2(a+b-4) + 4(2a+b-3) + 8(4a+b-4)$$

$$\frac{\partial E}{\partial b} = 2(b-2) + 2(a+b-4)$$
$$+ 2(2a+b-3) + 2(4a+b-4).$$

Setting them equal to zero and simplifying give the system of equations

$$21a + 7b = 26 \quad \text{and} \quad 7a + 4b = 13,$$

whose solution is $a = \frac{13}{35}$ and $b = \frac{13}{5}$.

(c) The second-order partials are $\partial^2 E/\partial a^2 = 42$, $\partial^2 E/\partial b^2 = 8$, and $\partial^2 E/\partial a \partial b = 14$. Therefore, $\partial^2 E/\partial a^2 > 0$ and $D > 0$, so there is a local minimum. Since the error function is quadratic, its local minimum is also a global minimum.

2. Formulas (16) and (17) give

$$a = \frac{4 \cdot 26 - 7 \cdot 13}{4 \cdot 21 - 49} = \frac{13}{35}, \quad b = \frac{13 - 7(13/35)}{4} = \frac{13}{5}.$$

7.5 Constrained Optimization and Lagrange Multipliers

Most optimization problems in business and economics are subject to one or more constraints. Suppose, for example, that a firm manufactures a certain product whose quantity is a function of the input of labor and of capital (in suitable units). Those inputs may be restricted by the total amount of money the firm has available, and the firm has to find the correct mix of labor and capital to maximize production without violating the restriction.

HISTORICAL PROFILE

Sir Francis Galton (1822–1911) was a cousin of Charles Darwin and was very influenced by his work. The son of a banker, he entered Trinity College, Cambridge, with the intention of studying medicine, but he soon became more interested in mathematics. He made notable contributions in a number of fields, first as an explorer of southwestern Africa, which earned him membership in the Royal Geographical Society, and also as a meteorologist, in which his pioneering work demonstrated previously unknown relationships between wind speed and direction and barometric pressure.

His overriding interest, however, was in using statistics to study how hereditary traits are transmitted, which led to his groundbreaking work in least-squares estimation. He named the least-squares line the **regression line** because he believed it gave evidence of his theory that genetic traits "regress" toward the average over succeeding generations—for instance, that, on average, unusually tall parents have children whose height is closer to the mean. He was made a Fellow of the Royal Society early in his career, and he was knighted in 1909.

In mathematical terms, a two-variable **constrained optimization problem** is one in which we try to maximize (or minimize) a function $f(x, y)$ whose input variables are subject to a side condition in the form of an equation, $g(x, y) = c$, called the **constraint equation**. In the situation described in the previous paragraph, $f(x, y)$ is a production function, and x and y represent the input of capital and labor. The constraint equation is a mathematical formulation of the restriction imposed on the variables. We shall discuss a problem of that type in Example 7.5.1. First, however, we will begin with a simpler example of a constrained optimization problem.

Problem   Find the minimum of the function

$$f(x, y) = x^2 + y^2, \tag{19}$$

subject to the constraint

$$x + y = 1.$$

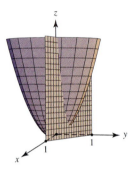

Figure 7.5.1

To begin with, let's make sure we understand the problem. We want to minimize the given function, but only over the set of points in the plane that satisfy the constraint equation $x + y = 1$. Without that extra condition, it is easy to find the minimum—it occurs at the point $(0, 0)$. But that point does not satisfy the constraint equation, so it is not a feasible solution to our constrained optimization problem. A geometric interpretation of the problem is pictured in Figure 7.5.1. The graph of $f(x, y)$ is a paraboloid in three-dimensional space, and the graph of the constraint equation is a plane parallel to the z-axis. The intersection of $z = x^2 + y^2$ and the plane $x + y = 1$ is a curve in three-dimensional space, and the constrained minimum is simply the minimum height of that curve. We will find the solution to this constrained minimization problem by two different methods.

The method of reduction to one variable

We first solve the constraint $x + y = 1$ for y to obtain $y = 1 - x$. Then we use that to eliminate y in $f(x, y)$ by substitution. We get

$$z = x^2 + (1 - x)^2,$$

which is a function of one variable only. To find its minimum we compute its critical points. We have $z' = 2x - 2(1 - x)$. Thus, $z' = 0$ if and only if $x = \frac{1}{2}$. Since $z'' = 4$, the function has a minimum at this critical point. To find the minimum value, we simply set $x = \frac{1}{2}$, to get

$$\left(\frac{1}{2}\right)^2 + \left(\frac{1}{2}\right)^2 = \frac{1}{2}.$$

Lagrange multipliers method

The key step in the previous method consisted of solving the constraint equation for one of the variables in terms of the other. That was easy in that case because the constraint was a linear equation. In most cases, however, the constraints are too complicated to solve. To deal with such situations, the French mathematician Joseph Louis Lagrange (1736–1813) devised a method that is now called the method of **Lagrange multipliers**. Here is how it works.

First, we observe that the level curves of our function $f(x, y) = x^2 + y^2$ are circles centered at zero. For example, the level curve with $z = \frac{1}{9}$ is given by

$$x^2 + y^2 = \frac{1}{9},$$

which is the circle of radius $\frac{1}{3}$ centered at zero. Several of these level curves are shown in Figure 7.5.2, together with the graph of the constraint equation

$$x + y = 1.$$

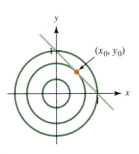

Figure 7.5.2

Notice—and this is a key point—that exactly one of the level circles is tangent to the constraint line $x + y = 1$. *The point of tangency, call it (x_0, y_0), is precisely the point where the function $f(x, y)$ takes a minimum subject to the constraint $x + y = 1$. To see why, observe that as the level circles expand—that is, as their radii become larger—the value of $f(x, y)$ increases. Second, we observe that every other circle that meets the constraint line has a larger radius than the tangent circle. Therefore, $f(x, y)$ has a larger value at those intersection points than at the tangency point (x_0, y_0). How can we find the point (x_0, y_0)? To do that, we use the following fact.

Lagrange Multiplier Principle

Suppose that two functions $f(x, y)$ and $g(x, y)$ have level curves that are tangent to one another at a point (x_0, y_0). Then the values of their first-order partial derivatives are proportional at that point—that is, there is a constant λ so that

$$\begin{cases} \dfrac{\partial f}{\partial x} = \lambda \dfrac{\partial g}{\partial x} \\[2mm] \dfrac{\partial f}{\partial y} = \lambda \dfrac{\partial g}{\partial y} \end{cases} \qquad (20)$$

at the point (x_0, y_0). The proportionality constant λ is called a Lagrange multiplier.

The geometric idea behind this principle is simply that if the two level curves have the same tangent line, their slopes are the same. But that means that the two functions must have first-order partial derivatives that are proportional.

In the present case $f(x, y) = x^2 + y^2$, and we can take $g(x, y) = x + y - 1$. In other words, we can think of the constraint line $x + y = 1$ as a level curve of $g(x, y)$ with height zero. The first-order partial derivatives are as follows:

$$\frac{\partial f}{\partial x} = 2x, \quad \frac{\partial g}{\partial x} = 1$$

$$\frac{\partial f}{\partial y} = 2y, \quad \frac{\partial g}{\partial y} = 1.$$

Thus, (20) takes the form

$$2x = \lambda$$
$$2y = \lambda.$$

From this we see that $x = y$, and combining this with the constraint equation $x + y = 1$ yields the condition $2x = 1$, or $x = \frac{1}{2}$. It follows that $y = \frac{1}{2}$ and $\lambda = 1$. Therefore, we conclude that the minimum of $f(x, y)$ subject to the constraint $x + y = 1$ occurs at the point $(\frac{1}{2}, \frac{1}{2})$, and the minimum value is given by

$$f\left(\frac{1}{2}, \frac{1}{2}\right) = \left(\frac{1}{2}\right)^2 + \left(\frac{1}{2}\right)^2 = \frac{1}{2}.$$

Two remarks are in order. First, as was to be expected, this result—the minimum value and the point where it occurs—is the same as the one we obtained under the method of reduction to one variable. Second, the actual value of λ is irrelevant—the Lagrange multiplier is simply a tool for finding the point (x_0, y_0).

Here is a summary of the Lagrange multiplier method in two variables.

Theorem 7.5.1

Lagrange multiplier in two variables

In order to solve the constrained optimization problem

$$\begin{cases} \text{Maximize or minimize} & f(x, y) \\ \text{subject to the constraint} & g(x, y) = 0, \end{cases} \tag{21}$$

we first solve the following system of equations:

$$\begin{cases} \dfrac{\partial f}{\partial x} = \lambda \dfrac{\partial g}{\partial x} \\[2mm] \dfrac{\partial f}{\partial y} = \lambda \dfrac{\partial g}{\partial y} \\[2mm] g(x, y) = 0, \end{cases} \tag{22}$$

for x, y, and λ. If the optimization problem (21) has a solution (x_0, y_0) then (x_0, y_0, λ_0) must be one of the solutions of the system (22) for some λ_0.

Remark   We can think of the solutions of the system (22) as the critical points of a function of three variables—the "**Lagrangian function**"

$$F(x, y, \lambda) = f(x, y) - \lambda g(x, y).$$

EXAMPLE 7.5.1

Suppose that the production function of a firm manufacturing a certain product is the Cobb-Douglas production function

$$f(x, y) = x^{2/3}y^{1/3},$$

where x is the number of units of labor and y is the number of units of capital. Also, assume that the firm has a total budget of \$120,000 and that a unit of labor costs \$80 and a unit of capital costs \$20. What combination of inputs (x, y) will maximize the output $f(x, y)$?

SOLUTION If the input of labor is x units and the input of capital is y units, then we must have $80x + 20y = 120,000$. Therefore, we need to solve the following constrained optimization problem:

$$\begin{cases} \text{Maximize } f(x, y) = x^{2/3}y^{1/3} \\ \text{subject to } g(x, y) = 80x + 20y - 120{,}000 = 0. \end{cases}$$

We have

$$\frac{\partial f}{\partial x} = \frac{2}{3}x^{-1/3}y^{1/3}, \quad \frac{\partial g}{\partial x} = 80,$$

$$\frac{\partial f}{\partial y} = \frac{1}{3}x^{2/3}y^{-2/3}, \quad \frac{\partial g}{\partial y} = 20.$$

Thus, we need to solve the system

$$\begin{cases} \frac{2}{3}x^{-1/3}y^{1/3} = 80\lambda \\ \frac{1}{3}x^{2/3}y^{-2/3} = 20\lambda \\ 80x + 20y = 120{,}000. \end{cases} \tag{23}$$

PROBLEM-SOLVING TACTIC

The key to solving system (23) is the observation that the right sides of the first two equations are "almost the same." So, dividing the first equation by 80 and the second by 20, and equating the resulting left sides lead to the simple equation $y = 2x$. We can then substitute that into the third equation and solve for x, and the rest follows. Needless to say, to solve a system there is more than one way.

From the first two equations, we obtain

$$\frac{1}{120}x^{-1/3}y^{1/3} = \frac{1}{60}x^{2/3}y^{-2/3},$$

which reduces to $60y = 120x$, or $y = 2x$. By substituting this into the third equation of our system, we get $120x = 120{,}000$, or $x = 1{,}000$. And it follows that $y = 2{,}000$. Therefore, the only candidate for a maximum point is

$$(x_0, y_0) = (1{,}000, 2{,}000),$$

and the value of the function at that point is given by

$$f(1{,}000, 2{,}000) = 1{,}000^{2/3} \cdot 2{,}000^{1/3} \approx 1{,}260.$$

One way to test whether $f(1,000, 2,000)$ is a maximum is to compare it with the function values at other points of the constraint line $80x + 20y = 120,000$. Two such points are $(0, 6,000)$ and $(1,500, 0)$. Evaluating the function at these points gives

$$f(1,500, 0) = 0 \quad \text{and} \quad f(0, 6,000) = 0.$$

Since the function must take a maximum value on the closed segment joining these points and $(1,000, 2,000)$ is the only candidate, we conclude that we have indeed found the maximum.

Testing for maxima and minima

Theorem 7.5.1 gives a method for finding the **constrained critical points** (x_0, y_0), which are the possible places of constrained maxima and minima. Next, we shall discuss two methods for deciding the nature of a critical point.

The first way to test the nature of a critical point (x_0, y_0) is to consider small perturbations of the form $(x_0 + \Delta x, y_0 + \Delta y)$, where Δx and Δy denote small (positive or negative) quantities. Of course, these perturbations must satisfy the constraint equation—that is, we must choose Δx and Δy so that

$$g(x_0 + \Delta x, y_0 + \Delta y) = 0.$$

Then, by computing $f(x_0 + \Delta x, y_0 + \Delta y)$ and comparing it with $f(x_0, y_0)$, we may be able to draw conclusions about the nature of (x_0, y_0).

For example, we have already seen that $(x_0, y_0) = (0.5, 0.5)$ is a critical point of $f(x, y) = x^2 + y^2$ under the constraint $x + y = 1$. Now, we make the following perturbations:

- Choosing $\Delta x = 0.1$ and $\Delta y = -0.1$ gives the point $(0.6, 0.4)$.
- Choosing $\Delta x = -0.1$ and $\Delta y = 0.1$ gives the point $(0.4, 0.6)$.

It is easy to see that both of these perturbations satisfy the constraint equation. An easy calculation shows that $f(0.5, 0.5) = 0.5$, $f(0.6, 0.4) = 0.52$, and $f(0.4, 0.6) = 0.52$, which strongly suggests that there is a minimum at $(0.5, 0.5)$.

The second method is to use the following second derivative test for constrained optimization. It is very similar to the test for unconstrained optimization in two variables.

Theorem 7.5.2

Second derivative test for constrained optimization

Let (x_0, y_0, λ_0) be a critical point of the Lagrangian function

$$F(x, y, \lambda) = f(x, y) - \lambda g(x, y)$$

and let

$$D(x, y, \lambda) = \left(\frac{\partial^2 F}{\partial x^2}\right)\left(\frac{\partial^2 F}{\partial y^2}\right) - \left(\frac{\partial^2 F}{\partial x \partial y}\right)^2.$$

Then the following hold:

1. If $D(x_0, y_0, \lambda_0) > 0$ and $(\partial^2 F / \partial x^2)(x_0, y_0, \lambda_0) > 0$, then $f(x, y)$ has a local minimum at (x_0, y_0) under the constraint $g(x, y) = 0$.
2. If $D(x_0, y_0, \lambda_0) > 0$ and $(\partial^2 F / \partial x^2)(x_0, y_0, \lambda_0) < 0$, then $f(x, y)$ has a local maximum at (x_0, y_0) under the constraint $g(x, y) = 0$.

EXAMPLE 7.5.2

Optimize the function

$$f(x, y) = 5x^2 + 3y^2 - 2xy - 6x - 10y + 1$$

under the constraint

$$g(x, y) = x + y - 8 = 0.$$

SOLUTION   Here the Lagrangian function is

$$F(x, y, \lambda) = 5x^2 + 3y^2 - 2xy - 6x - 10y + 1 - \lambda(x + y - 8).$$

The first-order partial derivatives are as follows:

$$\frac{\partial F}{\partial x} = 10x - 2y - 6 - \lambda, \quad \frac{\partial F}{\partial y} = 6y - 2x - 10 - \lambda, \quad \frac{\partial F}{\partial \lambda} = -(x + y - 8).$$

Setting these equal to zero gives us the system of equations

$$10x - 2y - 6 - \lambda = 0$$
$$6y - 2x - 10 - \lambda = 0$$
$$x + y - 8 = 0,$$

whose solutions are the critical points of the Lagrangian function $F(x, y, \lambda)$. From the first two equations we see that

$$10x - 2y - 6 = \lambda = 6y - 2x - 10, \tag{24}$$

which simplifies to $3x - 2y = -1$. Our system is now reduced to the pair of equations

$$3x - 2y = -1$$
$$x + y = 8,$$

whose solution is $x = 3$, $y = 5$. Substituting these into (24) gives $\lambda = 14$. Therefore, $(x_0, y_0, \lambda_0) = (3, 5, 14)$ is the only critical point of the Lagrangian function.

Next, we compute the second derivatives:

$$\frac{\partial^2 F}{\partial x^2} = 10, \quad \frac{\partial^2 F}{\partial x \partial y} = -2, \quad \frac{\partial^2 F}{\partial y^2} = 6,$$

so that $D(3, 5, 14) = 10 \cdot 6 - (-2)^2 = 56$. We are now in a position to apply Theorem 7.5.2. Since $D > 0$ and $\partial^2 F / \partial x^2 > 0$ at $(3, 5, 14)$, we conclude that $f(x, y)$ has a local minimum at the point $(3, 5)$. We can also conclude that the function has a global minimum, subject to the constraint, at that same point (see the remark immediately following). The minimum value is given by $f(3, 5) = 23$.

Remark Optimization—that is, finding global minima or maxima—is a complex and interesting subject, into which we have only taken a few initial steps. In particular, determining whether a local extremum is also a global one can be complicated. In the previous example, the constraint is a line in the xy-plane, which essentially reduces the problem to a one-dimensional max-min problem. In that case, we can modify the single critical point principle—valid in one variable but not in two—to fit the situation. (We omit the details.)

Lagrange multipliers in three variables

The method of Lagrange multipliers can be applied to functions of more than two variables. In three variables, for example, we have the following:

Theorem 7.5.3

Lagrange multipliers in three variables

If the optimization problem

$$\begin{cases} \text{Maximize or minimize} & f(x, y, z) \\ \text{subject to constraint} & g(x, y, z) = 0, \end{cases} \tag{25}$$

has a solution, then the optimum value is taken at a point (x_0, y_0, z_0) that is a solution to the system of equations

$$\begin{cases} \dfrac{\partial f}{\partial x} = \lambda \dfrac{\partial g}{\partial x} \\[2mm] \dfrac{\partial f}{\partial y} = \lambda \dfrac{\partial g}{\partial y} \\[2mm] \dfrac{\partial f}{\partial z} = \lambda \dfrac{\partial g}{\partial z} \\[2mm] g(x, y, z) = 0 \end{cases} \tag{26}$$

for some number λ.

As in the case of two variables, we can think of the solutions (x, y, z, λ) of the system (26) as the critical points of the Lagrangian function

$$F(x, y, z, \lambda) = f(x, y, z) - \lambda g(x, y, z).$$

EXAMPLE 7.5.3

The function $f(x, y, z) = 2x^2 + y^2 + z^2$ has a minimum subject to the constraint $x + 3y - z = 7$. Use the method of Lagrange multipliers to find it.

SOLUTION　In this case, $g(x, y, z) = x + 3y - z - 7$, so that

$$\frac{\partial g}{\partial x} = 1, \quad \frac{\partial g}{\partial y} = 3, \quad \text{and} \quad \frac{\partial g}{\partial z} = -1.$$

In addition, we have

$$\frac{\partial f}{\partial x} = 4x, \quad \frac{\partial f}{\partial y} = 2y, \quad \text{and} \quad \frac{\partial f}{\partial z} = 2z.$$

Applying Theorem 7.5.3 leads to the system of equations

$$4x = \lambda$$
$$2y = 3\lambda$$
$$2z = -\lambda$$
$$x + 3y - z - 7 = 0.$$

From the first two equations we see that $y = 6x$, and from the first and third we get $z = -2x$. Substituting these into the fourth equation gives $21x = 7$. Therefore, $x = \frac{1}{3}$, from which it follows that $y = 2$ and $z = -\frac{2}{3}$. The minimum, subject to the constraint, occurs at the point $(\frac{1}{3}, 2, -\frac{2}{3})$, and the minimum value is $f(\frac{1}{3}, 2, -\frac{2}{3}) = \frac{14}{3}$.

The method may also be applied to problems in which there is more than one constraint. In three variables, for instance, we may have two constraints. Then the problem is

$$\begin{cases} \text{Maximize or minimize} & f(x, y, z) \\ \text{subject to constraints:} & g(x, y, z) = 0 \\ & h(x, y, z) = 0. \end{cases} \tag{27}$$

If the optimization problem (27) has a solution, then it is achieved at a point (x_0, y_0, z_0) where the Lagrangian function

$$F(x, y, z, \lambda, \mu) = f(x, y, z) - \lambda g(x, y, z) - \mu h(x, y, z)$$

has a critical point at $(x_0, y_0, z_0, \lambda_0, \mu_0)$ for some numbers λ_0 and μ_0.

EXAMPLE 7.5.4

Find the point where the curve formed by the intersection of the sphere $x^2 + y^2 + z^2 = 4$ with the plane $x + y + z = 1$ has maximum height above the xy-plane.

SOLUTION The height of a point above the xy-plane is the same as its z-coordinate. Therefore, we can rephrase the problem as a constrained optimization problem, namely, to maximize the function $f(x, y, z) = z$ subject to the constraints

$$x^2 + y^2 + z^2 - 4 = 0$$
$$x + y + z - 1 = 0.$$

The Lagrangian function is

$$F(x, y, z, \lambda, \mu) = z - \lambda(x^2 + y^2 + z^2 - 4) - \mu(x + y + z - 1).$$

Taking the partial derivatives and setting them equal to zero lead to the following system of equations:

$$\frac{\partial F}{\partial x} = -2\lambda x - \mu = 0 \tag{28}$$

$$\frac{\partial F}{\partial y} = -2\lambda y - \mu = 0 \tag{29}$$

$$\frac{\partial F}{\partial z} = 1 - 2\lambda z - \mu = 0 \tag{30}$$

$$\frac{\partial F}{\partial \lambda} = -(x^2 + y^2 + z^2 - 4) = 0 \tag{31}$$

$$\frac{\partial F}{\partial \mu} = -(x + y + z - 1) = 0. \tag{32}$$

We first observe that $\lambda \neq 0$; otherwise, it would follow from (28) that $\mu = 0$, and (30) would then become $1 = 0$. Therefore, we can solve (28) for x and (29) for y, as follows:

$$x = -\frac{\mu}{2\lambda} \quad \text{and} \quad y = -\frac{\mu}{2\lambda}.$$

In particular, $y = x$, and making that substitution in (32) gives $z = 1 - 2x$. Finally, substituting for both y and z in (31) gives

$$2x^2 + (1 - 2x)^2 - 4 = 0,$$

which, after expanding and collecting terms, becomes $6x^2 - 4x - 3 = 0$. The quadratic formula gives two solutions: $x = (2 \pm \sqrt{22})/6$, which leads to two critical points:

$$x = \frac{2 + \sqrt{22}}{6}, \quad y = x = \frac{2 + \sqrt{22}}{6}, \quad z = 1 - 2x = \frac{1 - \sqrt{22}}{3}$$

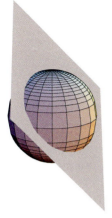

and

$$x = \frac{2 - \sqrt{22}}{6}, \quad y = x = \frac{2 - \sqrt{22}}{6}, \quad z = 1 - 2x = \frac{1 + \sqrt{22}}{3}.$$

As Figure 7.5.3 shows, there is a unique point of maximum height on the curve, and it occurs at the larger of the two possible z-values. Therefore, the maximum occurs at

$$\left(\frac{2 - \sqrt{22}}{6}, \frac{2 - \sqrt{22}}{6}, \frac{1 + \sqrt{22}}{3} \right).$$

Figure 7.5.3

Practice Exercises 7.5

1. Given that the function $f(x, y) = x^2 y$ has a maximum in the first quadrant subject to the constraint $x^2 + y^2 = 3$, use the method of Lagrange multipliers to find it.

2. Find the point on the line $2x + y = 1$ that is closest to the origin, and compute the minimum distance.

Exercises 7.5

In Exercises 1–10, use the method of Lagrange multipliers to find the maximum or minimum points, if any, of $f(x, y)$ subject to the given constraint.

1. $f(x, y) = x^2 + y^2, \ 2x + 6y = 2,000$

2. $f(x, y) = xy, \ x + y = 100$

3. $f(x, y) = x^2 + 3xy + y^2, \ x + y = 1,000$

4. $f(x, y) = x^2 - xy + y^2, \ x + y = 1,000$

5. $f(x, y) = (x - 1)^2 + (y - 2)^2 - 4, \ 3x + 5y = 47$

6. $f(x, y) = x + 2y, \ x^2 + y^2 = 1$

HISTORICAL PROFILE

Joseph-Louis Lagrange (1736–1813) was born in the city of Turin, which changed hands a number of times between France and Sardinia, giving both the French and Italians grounds to claim him as a native son. Although he was the son of an important public official, his family was not well off because of his father's unwise speculations. Lagrange was a largely self-taught mathematician who published his first article at the age of 18. His work caught the attention of the leading figures of his day, and he was appointed as a professor at the Royal Artillery School of Turin at the age of 19. At 30, he accepted a post as

Director of Mathematics at the Berlin Academy of Science, where he remained until 1787, when he moved to Paris as a member of the Academy of Sciences. His work in celestial mechanics won him the Academy's prize three times, in 1772, 1774, and 1780. He also did important work in number theory and probability. His invention of the Lagrange multiplier method is only part of his work in the calculus of variations, a subject concerned with finding paths that optimize certain physical quantities, such as energy, for particles moving along them. In 1808 Napoleon honored Lagrange with the Legion of Honor award.

7. $f(x, y) = xy$, $x^2 + y^2 = 1$

8. $f(x, y) = x^2 + (y-1)^2$, $x^2 - y^2 = 1$

9. $f(x, y) = x^2 + y^2$, $xy = 1$

10. $f(x, y) = 3x^{1/3}y^{2/3}$, $40x + 20y = 1,500,000$.

11. You are hiking across a mountain trail *ABCDE* as shown in Figure 7.5.4. The topographic map displays several level curves of the mountain's height above sea level, in increments of 200 feet.
    (a) At what point does the trail reach the highest elevation? What is its value?
    (b) How does the trail curve meet the level curve at that point?
    (c) If the equation of the path is $g(x, y) = 0$ and the equation of the mountain's surface is $z = f(x, y)$, use partial derivatives to express what happens at the point (x_0, y_0) where the trail reaches its highest elevation.

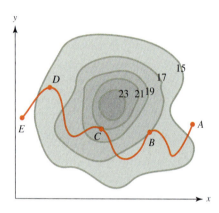

Figure 7.5.4

12. Among all pairs of positive numbers whose product equals 25, find the pair with the smallest sum. Formulate this as a two-variable optimization problem with one constraint and solve it in two ways—first, by reduction to one variable and, second, by Lagrange multipliers.

13. Find the point on the line $3x + y = 1$ that is closest to the origin. (*Hint:* Minimize the function $x^2 + y^2$ subject to the constraint $3x + y = 1$.)

14. Use Lagrange multipliers to find the maximum and minimum values of the function $f(x, y) = xy$ on the curve $x^2 + 2y^2 = 1$.

15. The function $f(x, y, z) = x^2 + y^2 + z^2$ has a minimum subject to the constraint $2x - 2y + z = 9$. Find it by the method of Lagrange multipliers.

16. The function $f(x, y, z) = x + y + z$ has a maximum subject to the constraint $z = 1 - x^2 - y^2$. Find it in two ways.
    (a) Reduce it to a two-variable problem by substituting for z.

(b) Apply the Lagrange multiplier method in three variables.

17. Among all triples (x, y, z) of positive numbers whose product is 8, there is a unique triple of minimum sum. Find it by means of Lagrange multipliers.

18. The function $f(x, y, z) = 4x - 3y$ has both a maximum and a minimum subject to the constraint $x^2 + y^2 + z^2 = 1$. Find the maximum and minimum values and the points where they occur. (*Hint:* Use the method of Lagrange multipliers in three variables to find the possible maximum and minimum points, and then compare the function values.)

19. Find the point on the plane $x + y - 2z = 5$ that is closest to the origin.

20. Find the point where the curve formed by the intersection of the paraboloid $z = x^2 + y^2$ with the plane $x + y = 2$ has minimum height above the xy-plane. Solve the problem in two ways: (a) by Lagrange multipliers and (b) by reducing it to a one-variable problem. In (b), use the second derivative test to show that the point is indeed a minimum.

21. The intersection of the surfaces $x^2 + y^2 = 8$ and $x + y + z = 5$ is a closed, oval-shaped curve known as an ellipse. It has a unique point of maximum height and a unique point of minimum height above the xy-plane. Formulate the problem of finding the points as an optimization problem with two constraints and solve it by means of Lagrange multipliers.

22. The planes $x + y + z = 1$ and $y = 2x$ intersect in a line in three-dimensional space. Find the point on that line that is closest to the origin.

23. Assume that a company sells two products *X* and *Y*. Its profit from selling x units of *X* and y units of *Y* is given by the function

$$P(x, y) = 10x + 20y - 0.1(x^2 + y^2).$$

If the company can produce a total of 100 units of the two products together, find the combination that will maximize its profit.

24. A company's output is given by the Cobb-Douglas production function $P(K, L) = 100K^{1/3}L^{2/3}$, where K and L denote the units of capital and labor, respectively. Suppose that the company's total budget for capital and labor is $1,800,000 and that each unit of capital costs $60 and each unit of labor costs $20. Find the combination of capital and labor that gives the maximum output.

25. The quantity Q of a product manufactured by a company is given by $Q(K, L) = 40K^{0.35}L^{0.65}$, where K is the quantity of capital and L is the quantity of labor used. Capital costs are $70 per unit, labor costs are $65 per unit, and the company wants to keep costs for capital and labor combined equal to $4,000,000. Find the combination of capital and labor that gives the maximum output.

26. The quantity Q of a product manufactured by a company is given by $Q(K, L) = 80K^{0.2}L^{0.8}$, where K is the quantity of capital and L is the quantity of labor used. Capital costs are \$50 per unit, labor costs are \$25 per unit, and the company wants to keep costs for capital and labor combined equal to \$500,000. Suppose you are asked to consult for the company and learn that 3,000 units of capital and 14,000 of labor are being used. What do you advise? Should the plant use more or less labor? more or less capital? If so, by how much? Remember that your objective is to maximize the quantity produced with costs fixed at \$500,000.

27. Postal regulations require that a package whose length plus girth exceed 108 inches must be mailed at an oversized rate. (For a rectangular package, as shown in Figure 7.5.5, the length is z, and the girth is $2x + 2y$.) What dimensions will maximize the volume of a rectangular package that is within the size limit?

28. **Continuation of the previous exercise** Answer the same question for a cylindrical package, as shown in Figure 7.5.6. In this case, the girth is the circumference.

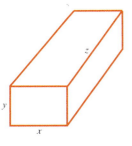

Figure 7.5.5

Figure 7.5.6

29. Students in a middle-school classroom are studying Native-American history. They learned that Plains Indians often built teepees, cone-shaped structures that were very portable. The students want to build one with the largest volume possible, but the teacher limits the combined height and radius to 15 feet. The volume of a cone of radius r and height h is given by

$$V = \frac{1}{3}\pi r^2 h.$$

(a) What dimensions should the students use for their teepee?
(b) What is the largest volume they can achieve?
(c) Show that the volume is a maximum by comparing it to nearby values along the constraint line.

30. An animal that is most active at dawn and dusk is called **crepuscular**. A particular crepuscular mammal spends 4 hours at dawn looking for food and defending its territory. Its social status depends on its ability to keep a prime territory and to eat enough food to stay in top condition. Biologists quantified the species' status ranking as

$$S(x, y) = x^{3/4}y^{1/4},$$

where x is the time spent foraging and y is the time spent defending.
(a) Optimize the status function given the time constraint.
(b) Did you find a maximum or minimum?
(c) What proportion of the dawn time period should be spent on foraging? on defending?

31. Find the dimensions of a rectangular tank with an open top of volume 32,000 cubic centimeters that minimizes the surface area.

32. Find the rectangular box that has the sum of length, width, and height equal to 120 centimeters and a maximum volume.

33. Find the point that is closest to the origin on the line formed by the intersection of the planes $x + 2y + z = 1$ and $x - y + z = 3$.

34. Find the dimensions of a closed rectangular box with a volume of 27,000 cubic centimeters and the minimum surface area. Can you formulate a general proposition about the dimensions of a rectangular box with a given volume and minimum surface area?

35. What are the relative dimensions of a closed rectangular box with given surface area and maximum volume? Justify your assertion.

36. Find the dimensions of the rectangular box of largest volume in the first octant with three faces on the coordinate planes, one of its corners at the origin, and the opposite vertex on the plane $2x + 3y + 4z = 12$.

Solutions to practice exercises 7.5

1. In this case $g(x, y) = x^2 + y^2 - 3$. Taking the first-order partial derivatives of f and g and applying Theorem 7.5.1 leads to the system of equations

$$2xy = 2\lambda x$$
$$x^2 = 2\lambda y$$
$$x^2 + y^2 = 3.$$

To solve this system, we first note that if $x = 0$, we must have $y = \sqrt{3}$ and $f(0, \sqrt{3}) = 0$. If $x \neq 0$, we can divide both sides of the first equation by $2x$ to get $\lambda = y$. Substituting this into the second equation gives $x^2 = 2y^2$, and substituting that into the third equation gives $3y^2 = 3$. Since we are looking for a point in the first quadrant, we must have $y = 1$. It follows that $x = \sqrt{2}$, and we conclude that the critical point is $(\sqrt{2}, 1)$, which is the only candidate for a maximum point. The maximum value, subject to the constraint, is $f(\sqrt{2}, 1) = 2$.

2. The distance from a point (x, y) to the origin is $\sqrt{x^2 + y^2}$, and we want to minimize that function, subject to $2x + y - 1 = 0$. However, the point that minimizes the distance also minimizes the *square* of the distance, so we can change the problem to that of minimizing the function $f(x, y) = x^2 + y^2$, subject to the given constraint. Taking the first-order partial derivatives and applying The-

orem 7.5.1 lead to the system of equations

$$2x = 2\lambda$$
$$2y = \lambda$$
$$2x + y = 1.$$

The first two equations imply that $x = 2y$, and substituting that into the third equation gives $5y = 1$. Therefore, the minimum is taken at the point $(\frac{2}{5}, \frac{1}{5})$. The minimum distance is $\sqrt{(4/25) + (1/25)} = \sqrt{1/5}$.

Chapter 7 Summary

- **Vertical line test**: A surface in $\mathbb{R}^3$ is the graph of a function of the form $z = f(x, y)$ if and only if no vertical line meets the surface more than once.

- If $f(x, y)$ is a function of two variables and z_0 is a given number, then the **level curve of height** z_0 is the set of all points in the domain of f satisfying the equation $f(x, y) = z_0$.

- The equation of the plane through an arbitrary point (x_0, y_0, z_0) having slope a in the x-direction and slope b in the y-direction is

$$z - z_0 = a(x - x_0) + b(y - y_0).$$

- The **partial derivative** of a function of two variables $f(x, y)$ with respect to x, denoted by $\partial f/\partial x$, is the ordinary x-derivative of the function when the variable y is held fixed at a given value. $\partial f/\partial y$ is defined similarly. In applications to economics partial derivatives are sometimes called **partial marginals**.

- The **linear approximation** of a continuously differentiable function $f(x, y)$ at a point $(x_0, y_0, f(x_0, y_0))$ is

$$f(x, y) \approx f(x_0, y_0) + \frac{\partial f}{\partial x}(x_0, y_0)(x - x_0) + \frac{\partial f}{\partial y}(x_0, y_0)(y - y_0).$$

- If $f(x, y)$ has a local **maximum** or **minimum** at a point (x_0, y_0), then its first-order partial derivatives must be zero there. Thus, the solutions of the system $(\partial f/\partial x)(x, y) = 0$, $(\partial f/\partial y)(x, y) = 0$ are the only candidates for maxima and minima and are called **critical points**. The **second derivative test** decides whether there is a maximum, minimum, or neither at a critical point.

- Given a set of n numerical data points, $(x_1, y_1), (x_2, y_2), \ldots, (x_n, y_n)$, the **least-squares line**, or **regression line**, is the line $y = ax + b$ that minimizes the sum of the squares of the vertical distances between the line and the data points. That is, a and b are chosen so that they minimize the error function

$$E(a, b) = [(ax_1 + b) - y_1]^2 + [(ax_2 + b) - y_2]^2 + \cdots + [(ax_n + b) - y_n]^2.$$

Applying the first and second derivative tests, we obtain explicit formulas for both a and b.

• If the **constrained optimization problem** to maximize or minimize $f(x, y)$, subject to the constraint $g(x, y) = 0$, has a solution (x_0, y_0), then it must be a solution of the following system of three equations:

$$\frac{\partial f}{\partial x} = \lambda \frac{\partial g}{\partial x}, \quad \frac{\partial f}{\partial y} = \lambda \frac{\partial g}{\partial y}, \quad g(x, y) = 0.$$

for some λ. This is the **Lagrange multiplier method** in two variables.

Chapter 7 Review Questions

• What is the formula of a linear funtion in two variables? What does its graph look like? Give concrete examples.

• State the definitions of the first-order partial derivatives of $f(x, y)$ as limits of difference quotients. Compute the difference quotients in concrete examples with $\Delta x = 0.1$ and $\Delta y = 0.2$.

• Write the equation of the tangent plane to the surface $z = f(x, y)$ at the point $(x_0, y_0, f(x_0, y_0))$. Give a concrete example.

• Define the second-order partial derivatives of a function $f(x, y)$. How many are there, and are there any relationships between them?

• State the second derivative test and give concrete examples for each case in the statement.

• Give an example of linear regression and an example of nonlinear regression.

• State the second derivative test for the Lagrange multiplier method.

• State the Lagrange multiplier method in three variables, first with one constraint and then with two constraints.

Chapter 7 Review Exercises

1. Which of the points $A = (3, 2, 8)$, $B = (-6, 7, 0)$, and $C = (6, 1, -9)$ is closest to the yz-plane? Which one lies on the xy-plane?

2. A cube having a volume of 64 cubic units is situated in 3-space so that each edge is parallel to an axis. If the center of the cube is at the point $(3, 5, 7)$, find the coordinates of its vertices.

3. Write the equation of the plane through the point $(1, 2, 3)$ and having slope -5 in the x-direction and slope $\frac{2}{7}$ in the y-direction.

4. Find the distance between the points $(2, -1, 4)$ and $(-1, -2, 4)$. Which of these points is closer to the xz-plane?

5. State whether or not each of the following functions is linear.
- **(a)** $f(x, y) = 1,998x - 12y + 74$
- **(b)** $f(x, y) = 12x^{1/2} - 7y + 10$
- **(c)** $f(x, y) = 24x - 2xy + 7$

6. Write the equation of the plane passing through the points
- **(a)** $(2, 0, 0)$, $(0, 3, 0)$, and $(0, 0, 12)$
- **(b)** $(0, 2, 3)$, $(0, -7, 2)$, and $(0, 1, 1)$
- **(c)** $(2, 0, -1)$, $(1, 1, 3)$, and $(2, -1, -2)$

7. Assume that a function $f(x, y)$ is linear and that $f(1, 2) = 12$, $f(1, 3) = 14$, $f(4, 5) = 33$. Write a formula for $f(x, y)$.

8. Assume that the cost function of a company producing x units of a product and y units of another product is a linear function and the following table of values is known:

$x \backslash y$	100	150	200
200	8,000	9,000	10,000
300	8,500	9,500	10,500

Write a formula for the cost function.

9. Describe the level curves of the function $f(x, y) = -x^2 - y^2$ with heights $z = -9, -4, 0, 4,$ and 9 and draw each one that is nonempty.

10. Draw the level curves of the function $f(x, y) = y/x^2$ with heights $z = -1, 0, 1,$ and 4.

In Exercises 11–16, find both first-order partial derivatives of the given function.

11. $f(x, y) = x^2y + 3y^2 - 2xy^3 + x - 5y$

12. $f(s, t) = \dfrac{s}{t^2 + 1}$ **13.** $g(x, y) = \ln(x^2 + xy)$

14. $h(u, v) = e^{u^2v - v^2u}$

15. $f(x, y) = x^2 \ln(1 + y^2)$

16. $g(s, t) = e^{s^2/t}$

17. Find $\dfrac{\partial f}{\partial x}(1, -2)$ and $\dfrac{\partial f}{\partial y}(1, -2)$ if $f(x, y) = xe^{y+2} + y/x^2$.

18. Find
$$\lim_{h \to 0} \frac{\sqrt{(x+h)^2 + y^2} - \sqrt{x^2 + y^2}}{h}.$$
(*Hint:* Think derivative.)

19. Find $\lim_{h \to 0} (e^{x/(y+h)} - e^{x/y})/h$.

20. A manufacturer of a certain product determines that its production function is given by the Cobb-Douglas function, $Q(L, K) = \sqrt{LK}$, where L is the number of labor

units and K is the number of capital units. Determine the marginal productivity of labor and the marginal productivity of capital. Then, compute the sum of the two marginal productivity functions for $L = 25$ and $K = 64$.

21. If $f(x, y) = x^4 + 3y^2 + 1$, find the equation of the plane tangent to the graph of f at the point $(x_0, y_0) = (1, 2)$.

22. Let $f(x, y) = x^2 + 5y^2$. Find the equation of the tangent plane at $(x, y) = (3, -2)$ and the linear approximation of f at $(3, -2)$.

23. Let $f(x, y) = ye^{-x^2 + 0.5x}$. Estimate $f(0.2, 3.1)$ by using the linear approximation of f at $(0, 3)$.

24. Suppose that $f(x, y)$ is a function satisfying $f(5, 4) = 10$, $(\partial f/\partial x)(5, 4) = -2$, and $(\partial f/\partial y)(5, 4) = 1$. Estimate the following quantities:
- **(a)** $f(5.5, 4)$ **(b)** $f(5, 3.8)$ **(c)** $f(4.9, 4.5)$

25. Suppose $f(x, y)$ is a function such that $f(10, 20) = 5$, $(\partial f/\partial x)(10, 20) = 2$, and $(\partial f/\partial y)(10, 20) = -1$. Find the equation of the tangent plane to $f(x, y)$ at the point $(10, 20)$ and use the linear approximation of f to estimate $f(9.9, 20.2)$.

26. Find all six first- and second-order partial derivatives of $f(x, y) = x^2 - y^3 + x \ln(y)$.

27. If $f(x, y) = \ln(4 + x^2 + y^2)$, compute all four second derivatives and verify that $(\partial^2 f/\partial x \partial y) = (\partial^2 f/\partial y \partial x)$.

28. If $f(x, y) = xe^{x+y^2}$, compute all four second derivatives and verify that $(\partial^2 f/\partial x \partial y) = (\partial^2 f/\partial y \partial x)$.

29. Find the critical points of $f(x, y) = 3x^2 + 3y^2 - 4xy + 10y - 20x + 5$.

30. Find all points (x, y) where the function $f(x, y) = x^3 - 6xy + 3y^2 - 9x + 7$ has a possible local maximum or minimum.

31. Find the critical points of $f(x, y) = 2x^2 - x^4 - y^2$ and determine the nature of each of these points.

32. Find the critical points of the function $f(x, y) = y + \frac{1}{2}x^2 - \ln(xy^2)$ and determine the nature of each.

33. A function $f(x, y)$ satisfies $\partial f/\partial x = 0 = \partial f/\partial y$, $\partial^2 f/\partial x^2 = -2$, $\partial^2 f/\partial y^2 = -1$, and $\partial^2 f/\partial x \partial y = 1$ at $(1, 2)$. Which of these does the function have there: a local minimum, a local maximum, or a saddle point?

34. Find the critical points of the function $f(x, y) = 2x^2y - 6x^2 + 2y^3 - 9y^2 - 24y + 1$ and use the second derivative test to determine, if possible, the nature of each of these points. If the second derivative test is inconclusive, say so.

35. Let $f(x, y) = x^3 + 3xy + y^3$. Find any local maximum, local minimum, or saddle points.

36. The profit of a company selling x units of a product X and y units of another product Y is $P(x, y) = 60x + 100y - 150 - 0.1x^2 - 0.2xy - 0.2y^2$, in thousands of dollars. Find the production levels x and y that maximize the company's profit. Justify your answer.

37. A firm produces two kinds of products: X, which sells for $3 each, and Y, which sells for $5 each. The company determines that the total cost, in thousands of dollars, of producing x thousand of X and y thousand of Y is given by $C(x, y) = 2x^2 - 2xy + y^2 - 3x + y + 7$. Find the amount of each type of product that must be produced and sold in order to maximize profit.

38. Find the least-squares line for a data set consisting of the three points $(0, 0)$, $(1, 2)$, and $(2, 1)$. Work this out directly from the definition *without* using formulas (16) and (17).

39. Find the least-squares line for the points $(-1, 2)$, $(0, 1)$, $(2, 3)$.

40. Find the least-squares line for the points $(1, 3)$, $(2, 3)$, $(3, 5)$, $(5, 6)$.

41. Find the function $E = E(a, b)$ that must be minimized in order to find the least-squares line $y = ax + b$ for the six points $(0, 2)$, $(1, 1)$, $(2, 1)$, $(4, 2)$, $(5, 5)$, $(6, 4)$.

42. Show that if a set of data points (x_1, y_1), (x_2, y_2), $\dots$, (x_n, y_n) all lie on a straight line, then that line is the least-squares line for the points.

43. On 4 different days, a service station compared the price of a gallon of high-test gasoline with the number of gallons it sold, resulting in the following table:

price	1.26	1.64	1.80	2.10
gallons (in thousands)	9.4	6.1	5.4	4.1

Assuming that the relationship between price and demand is approximately linear, use the least-squares line to estimate the number of gallons sold when the price is $1.50 per gallon.

44. Find the maximum and minimum of the function $f(x, y) = 2xy$ subject to constraint $g(x, y) = x^2 + y^2 - 2 = 0$.

45. Use the method of Lagrange multipliers to optimize the function $f(x, y) = 4x^2 + y^2$ subject to the constraint $xy = 2$.

46. Find the highest point on the curve $x^2 - 6x + 2y^2 - 8y + 1 = 0$.

47. Find the point on the plane $z = 2x - 3y + 7$ that is closest to the origin.

48. A company's output is given by the Cobb-Douglas production function $P = 600L^{3/4}K^{1/4}$, where L and K are the number of units of labor and capital. Each unit of labor costs the company $30, and each unit of capital costs $150. If the company has a total of $3 million for labor and capital, how much of each should it use to maximize production?

Chapter 7 Practice Exam

1. Find the distance between the points $(1, 0, 2)$ and $(-1, -2, 1)$.

2. Sketch the level curve of height 3 of $f(x, y) = x - y^2 + 1$.

3. Find an equation of the plane through the point $(2, 6, 4)$ with slope 1 in the x-direction and slope -2 in the y-direction.

4. Given that $f(1.1, 3) = 0.2$, $f(1, 3) = -0.1$, and $f(1, 2.9) = 0.1$, estimate $(\partial f / \partial y)(1, 3)$.

5. If $f(x, y) = x^3 + x^2y + e^{xy}$, what is $(\partial f / \partial x)(2, 3)$?

6. Find the equation of the tangent plane to the graph of the function $f(x, y) = -xy$ at the point $(1, -1)$.

7. Find the linear approximation of $f(x, y) = x^2 + y + 3xy$ at $(1, 2)$.

8. Find all critical points of the function $f(x, y) = 2x^2 + 3xy - 6y$.

9. If $f(x, y) = x^3 + 2xy - 6y^2$, then $(0, 0)$ is a critical point. Determine whether there is a local minimum, a local maximum, or a saddle point at $(0, 0)$.

10. Suppose that $(2, 1)$ is a critical point of a function $f(x, y)$ satisfying

$$\frac{\partial^2 f}{\partial x^2}(2, 1) = 8,$$

$$\frac{\partial^2 f}{\partial y^2}(2, 1) = 4, \qquad \frac{\partial^2 f}{\partial x \partial y}(2, 1) = -5.$$

Use the given information about $f(x, y)$ to determine the nature of the critical point $(2, 1)$.

11. The prices of Dotcomco stock at the end of the first three quarters are shown in the following table:

quarter	x	1	2	3
stock price	y	18	14	12

(a) Find the least-squares line that fits this data.
(b) Use your answer to part (a) to predict the stock price at the end of the fourth quarter.

12. Find the maximum value of $f(x, y) = 6x - 8y$ subject to the constraint $3x^2 + 4y^2 = 7$.

13. A company's output is given by the Cobb-Douglas production function $P(K, L) = 100K^{0.36}L^{0.64}$, where K and L denote the units of capital and labor, respectively. Each unit of capital costs \$72 and each unit of labor costs \$32. If the company's total budget for capital and labor is \$360,000, find the combination of capital and labor that gives the maximum output.

Chapter 7 Projects

1. Exponential regression The following table gives the population of the United States from 1950 to 1990 in millions:

year	1950	1960	1970	1980	1990
population	151	179	203	227	249

(a) Draw a scatter diagram of the points (t, p) for $t = 0, 10, 20, 30,$ and 40, where t is the number of years after 1950 and p is the population in millions. Observe that the diagram suggests the data are best fitted by an exponential curve of the form $p = Be^{at}$.

(b) Take the natural logarithm of both sides of the exponential equation, which changes it to the form $y = at + b$, where

$$y = \ln p \quad \text{and} \quad b = \ln B. \tag{33}$$

Next, do the same with the population data, which changes the table as follows:

t	0	10	20	30	40
$y = \ln p$	$\ln 151$	$\ln 179$	$\ln 203$	$\ln 227$	$\ln 249$

(c) Use linear regression, as discussed in Section 7.4, to compute the coefficents a and b that give the least-squares line $y = at + b$ for these data.

(d) Use (33) to convert this to the form $p = Be^{at}$.

(e) Predict the U.S. population for the year 2010.

2. The Coca-Cola can In this project, we will investigate whether a Coca-Cola can is designed to minimize the amount of aluminum used for the volume of soda it contains.

(a) First, use the method of Lagrange multipliers to solve the following problem. For a cylindrical can, closed at the top and bottom, with given volume V, find the ratio h/d of height to diameter that minimizes the total surface area A.

(b) Second, by measuring the height and the diameter of the base of a Coca-Cola can, determine whether it minimizes the surface for the volume it contains.

3. **Monopoly versus duopoly**   Here we consider a theory put forth by Augustin Cournot, a nineteenth-century French economist.[1] He considered a situation known as a **duopoly**, in which there are two firms producing the same product. Cournot assumed that the demand curve is linear and is known by both firms, and also that the two firms have identical cost functions. Each firm assumes that the other will fix its output at a value maximizing its profit, and each adjusts it own production accordingly. This game plan results in a joint output that maximizes the profit of both companies, and because the duopoly price is lower than that resulting from a monopoly, it also benefits consumers. The resulting production level is called the **Nash equilibrium** after the American mathematician John Nash, who won the 1994 Nobel prize for economics. The work for which he won the prize is in an area of mathematics known as game theory, which deals with questions of competition and strategy in economics and other areas.

As an example, suppose the demand curve for milk in a region is

$$p = 5 - 0.0005x, \quad 0 \le x \le 8,000,$$

where the quantity x is in gallons (per day) and the price p is in dollars. And suppose the cost for producing x gallons is $C(x) = 500 + x$ dollars.

(a) Assumming that there is only one company X producing milk in the region, find the production level x that maximizes its profit (resulting in a **monopoly**).

(b) Now assume that a second company Y is established in the region producing milk at the same cost as company X. If the two companies are in competition for the milk market of the region, then find the corresponding levels of production x and y that will maximize their individual profit (resulting in a duopoly).

(c) If the companies X and Y are in **collusion**, they will act as a single company (a monopoly) rather than competing, and they will split the profits between them. Find each firm's maximum profit in that case, and compare it and the consumer price with the duopoly case.

[1] Edwin Mansfield, *Managerial Economics*, New York: Norton, 1996, p. 442.

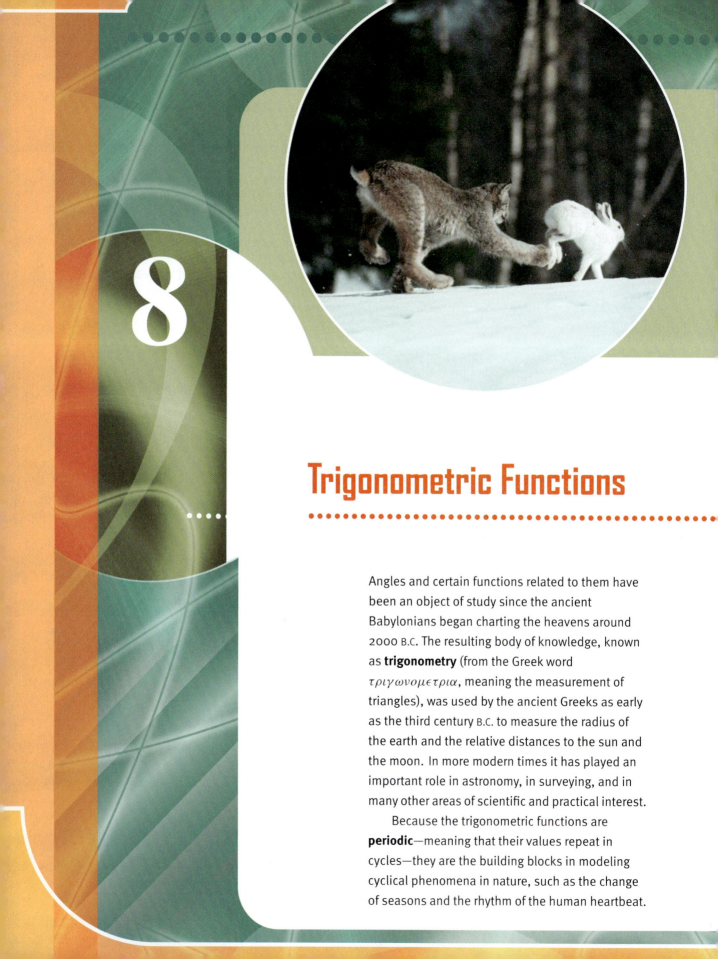

8

Trigonometric Functions

Angles and certain functions related to them have been an object of study since the ancient Babylonians began charting the heavens around 2000 B.C. The resulting body of knowledge, known as **trigonometry** (from the Greek word $\tau\rho\iota\gamma\omega\nu\omega\mu\epsilon\tau\rho\iota\alpha$, meaning the measurement of triangles), was used by the ancient Greeks as early as the third century B.C. to measure the radius of the earth and the relative distances to the sun and the moon. In more modern times it has played an important role in astronomy, in surveying, and in many other areas of scientific and practical interest.

Because the trigonometric functions are **periodic**—meaning that their values repeat in cycles—they are the building blocks in modeling cyclical phenomena in nature, such as the change of seasons and the rhythm of the human heartbeat.

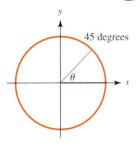

45 degrees

Figure 8.1.1

There are two methods of measuring angles—by *degrees* and by *radians*. The first is familiar from common usage, whereas the second is more prevalent in scientific work, particularly in applications involving calculus. In both methods, we put the vertex of the angle at the center of a circle and measure the size of the arc cut off by the two sides of the angle. Customarily, we place one side along the horizontal axis, as shown in Figures 8.1.1 and 8.1.2.

In measuring by degrees, we arbitrarily divide the full circle into 360 equal segments, each equal to 1 degree, and assign to an angle the number of degrees it cuts off on the circle. The angle θ shown in Figure 8.1.1 cuts off $\frac{1}{8}$ of the circle, and we assign it 45 degrees, which is $\frac{1}{8}$ of 360.

The use of 360 degrees for the full circle dates back to the number system used by the ancient Babylonians. A more natural method is to use the *length* of the arc cut off by the angle on a circle of radius 1. In Figure 8.1.2, s is the length of arc cut off by the angle θ, and we assign that number as the radian measure of the angle.

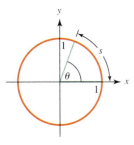

Figure 8.1.2

Definition 8.1.1

The **radian measure** of an angle is the length of the arc cut off on a circle of radius 1 by the sides of the angle with its vertex at the center of the circle.

Since the circumference of a circle of radius 1 equals 2π, we have the following formula relating degrees and radians:

$$360 \text{ degrees } = 2\pi \text{ radians,}$$

from which we get

$$1 \text{ degree } = \frac{\pi}{180} \text{ radians} \quad \text{and} \quad 1 \text{ radian } = \frac{180}{\pi} \text{ degrees.} \tag{1}$$

EXAMPLE 8.1.1

Convert 12 degrees to radians.

SOLUTION   Using (1), we get

$$12 \text{ degrees} = \frac{12\pi}{180} = \frac{\pi}{15} \text{ radians.}$$

The following table gives the correspondence between degrees and radians for some frequently occurring angles:

degrees	30	45	60	90	180	270
radians	$\pi/6$	$\pi/4$	$\pi/3$	$\pi/2$	π	$3\pi/2$

EXAMPLE 8.1.2

Find the angles of an isoceles right triangle in radians.

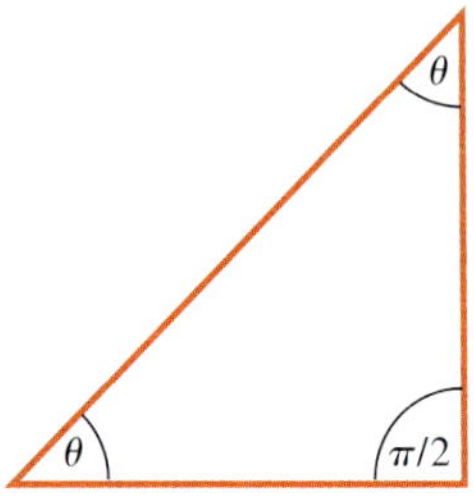

SOLUTION   From elementary geometry, we know that

- the angles opposite the equal sides of an isoceles triangle are equal, and
- the angles of a triangle add up to 180 degrees, which equals π radians.

Let θ be the radian measure of each of the equal angles, as shown in Figure 8.1.3. Since the right angle has $\pi/2$ radians, we must have $2\theta + \pi/2 = \pi$. Therefore, $\theta = \pi/4$.

Figure 8.1.3

In addition to measuring the size of an angle, we can also specify its *orientation* by assigning a *sign* to the radian measure. Think of one side of the angle as the *initial* side, which we place along the positive horizontal axis. Then place the other side by moving its tip along the circumference of the circle—much as you might move the minute hand of a clock—until it reaches its terminal position. If the movement is in the **counterclockwise** direction, we assign a **positive** sign to the radian measure. If it is in the **clockwise** direction, we assign a **negative** sign. Figure 8.1.4 shows the cases of $\pi/4$ and $-\pi/4$ radians.

This method of measuring angles allows for values that are greater than 2π or less than -2π. In fact, *any* number t can be the radian measure of an angle. We simply start at the tip of the initial side of the angle and move along the circumference of the circle—counterclockwise if t is positive, clockwise if t is negative—until we have traveled a total of $|t|$ units. We make at least one full revolution if $|t| \geq 2\pi$, at least two full revolutions if $|t| \geq 4\pi$, and so forth. The angles shown in Figure 8.1.5 illustrate a few simple examples.

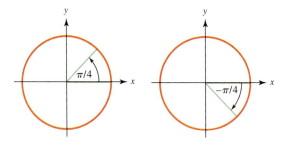

Figure 8.1.4

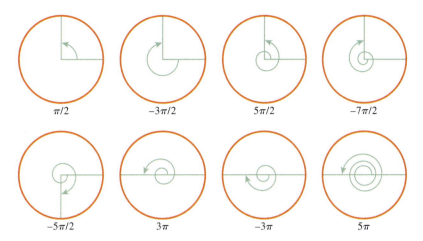

Figure 8.1.5

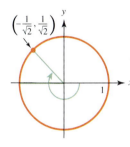

Figure 8.1.6

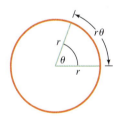

Figure 8.1.7

> **EXAMPLE 8.1.3**

Draw an angle of $-5\pi/4$ radians, with its initial side along the positive horizontal axis.

SOLUTION   Noting that $5\pi/4 = \pi + \pi/4$, we conclude that the angle cuts off $\frac{5}{8}$ of the circumference of the circle in which it is inscribed—half the circle (π) plus half of a right angle ($\pi/4$). Since the radian measure is negative, the terminal side of the angle is moved clockwise until it marks off $\frac{5}{8}$ of the circle. The angle is shown in Figure 8.1.6. The terminal point is $(-1/\sqrt{2}, 1/\sqrt{2})$.

Finally, we remark that if an angle of θ radians is placed with its vertex at the center of a circle of radius r, the arc that is cut off has length $r\theta$ (see Figure 8.1.7). To see why, let s denote that length of arc. Then, $s/2\pi r$ (the ratio of the length of arc to the full circumference) is equal to $\theta/2\pi$ (the ratio of the radian measures of the angle to that of the complete circle). Equating them and solving for s give $s = r\theta$.

Practice Exercises 8.1

1. Convert the following degrees to radians:
   (a) 15 degrees   (b) 210 degrees

2. Convert $3\pi/4$ radians to degrees.

3. Find the radian measure of each of the following angles. In both cases the terminal point is $(1/\sqrt{2}, 1/\sqrt{2})$.

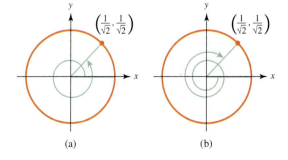

Exercises 8.1

Convert the following to radians:

1. 360 degrees, 450 degrees, 540 degrees

2. 120 degrees, 150 degrees, 240 degrees

3. 135 degrees, 225 degrees

4. 20 degrees, 110 degrees, 160 degrees

Convert the following to degrees:

5. $5\pi/4, 7\pi/4$ **6.** $\pi/8, 5\pi/8, 7\pi/8$

7. $5\pi/6, 7\pi/6, 11\pi/6$ **8.** $\pi/12, 11\pi/12, 13\pi/12$

9. Find the radian measure of the angles of an equilateral triangle.

10. If one of the acute angles of a right triangle is $\pi/6$ radians, what is the other?

Sketch each of the following angles (in radian measure) inscribed in a circle with the initial side along the positive horizontal axis.

11. $-\pi/4$ **12.** $7\pi/4$ **13.** -4π

14. $\pi/3$ **15.** $-2\pi/3$ **16.** $4\pi/3$

17. $13\pi/6$ **18.** $-7\pi/6$

In Exercises 19–27, find the radian measure of the angle pictured.

19.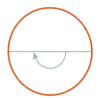

20.

21.

22.

23.

24.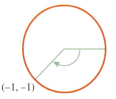
(−1, −1)

25.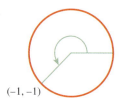
(−1, −1)

26.
(1, 1)

27.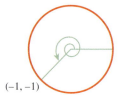
(−1, −1)

28. What is the length of the arc cut off by an angle of $\pi/10$ radians inscribed in a circle of radius 2?

29. What is the length of the arc cut off by an angle of $4\pi/3$ radians inscribed in a circle of radius 5?

30. Several members of the basketball team reported to the first day of practice out of shape, so the coach took them into the circular field house and had them run laps, moving counterclockwise and following the walls. The first player to drop out made it $2\frac{3}{4}$ times around. Most of the rest did $5\frac{1}{3}$ laps. The coach stopped the few remaining runners after 10 laps and had them cool down by running in the other direction. The last one stopped running after $2\frac{5}{8}$ clockwise laps.
(a) Rewrite the descriptions of the players' accomplishments using radian measurements.
(b) If the diameter of the fieldhouse is 200 feet, what was the longest total distance run by a player?

31. Margie set up her base camp on the top of a hill. Then she hiked 1 kilometer due east and set out her #1 flag marker. Keeping the same distance from base, she hiked 30 degrees counterclockwise to place flag #2. She continued in the same direction, keeping the same distance, to place flag

#3 at a point 70 degrees away from flag #2. The last flag was erected 120 degrees away from #1. Now the stations for the upcoming competition were set.

(a) Draw an above-view diagram of Margie's hike. Label every location she visited.

(b) Label the arc length in kilometers between consecutive flags.

32. A regular polygon has sides that are all the same length. This makes all the vertex angles equal in measure. The sum of these interior angles for a polygon with n sides is given by the formula $S = 180(n - 2)$ degrees.

(a) Convert the formula to radians.

(b) Use the formula to find the sum (in radians) of the interior angles of an equilateral triangle and also of a square. Do these agree with what you know from elementary geometry?

(c) Make a table for $n = 4, 5, \ldots, 10$ listing S in radians and also listing A, the size of each interior angle.

Solutions to practice exercises 8.1

1. (a) Recalling that 90 degrees $= \pi/2$ radians, we have 15 degrees $= \frac{1}{6} \cdot 90$ degrees $= \pi/12$.

(b) Recalling that 180 degrees $= \pi$ radians, we have

$$210 \text{ degrees} = 180 \text{ degrees} + 30 \text{ degrees} = \pi + \frac{\pi}{6} = \frac{7\pi}{6}.$$

2. $\dfrac{3\pi}{4} = \dfrac{\pi}{2} + \dfrac{\pi}{4} = 90 \text{ degrees} + 45 \text{ degrees} = 135 \text{ degrees}.$

3. (a) The terminal point is obtained by moving one full revolution and an additional 45 degrees counterclockwise. Therefore, the radian measure of the angle is $2\pi + \pi/4 = 9\pi/4$.

(b) The terminal point is obtained by moving one complete revolution clockwise and an additional $\frac{7}{8}$ of a complete revolution clockwise. Therefore, the radian measure is $-2\pi - 7\pi/4 = -15\pi/4$.

8.2 The Sine, Cosine, and Tangent

The sine, cosine, and tangent functions are defined in elementary trigonometry by means of right triangles. We place an acute angle θ at the base of a right triangle, as shown in Figure 8.2.1, letting O be the length of the vertical (or *opposite*) leg, A the length of the horizontal (or *adjacent*) leg, and H the length of the *hypotenuse*. Then we define the **sine**, **cosine**, and **tangent** of θ as follows:

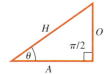

Figure 8.2.1

> **The Sine Function**
> $$\sin \theta = \frac{O}{H}.$$
>
> **The Cosine Function**
> $$\cos \theta = \frac{A}{H}.$$
>
> **The Tangent Function**
> $$\tan \theta = \frac{O}{A}.$$

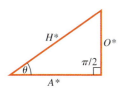

Figure 8.2.2

Any two right triangles having θ as an acute angle, such as those shown in Figures 8.2.1 and 8.2.2, are similar, which means that

$$\frac{O}{H} = \frac{O^*}{H^*}, \quad \frac{A}{H} = \frac{A^*}{H^*}, \quad \frac{O}{A} = \frac{O^*}{A^*}.$$

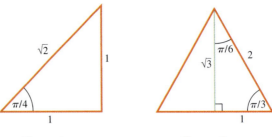

Figure 8.2.3   **Figure 8.2.4**

From that we see that the values of the sine, cosine, and tangent do not depend on the particular right triangle used.

We can use right triangles to find the values of the trigonometric functions for certain special angles. For instance, an isoceles right triangle with $O = A = 1$ (as shown in Figure 8.2.3) has a base angle of $\pi/4$ radians. By the Pythagorean theorem, $H = \sqrt{2}$. Therefore,

$$\sin\frac{\pi}{4} = \frac{1}{\sqrt{2}}, \quad \cos\frac{\pi}{4} = \frac{1}{\sqrt{2}}, \quad \tan\frac{\pi}{4} = 1. \tag{2}$$

As another example, suppose we construct an equilateral triangle with every side having length 2. The angles are all $\pi/3$ radians. If we draw a line from the top vertex perpendicular to the base, as shown in Figure 8.2.4, the line divides the original triangle into two right triangles, each with acute angles $\pi/3$ and $\pi/6$. The base of each of these triangles has length 1, the hypotenuse has length 2, and the Pythagorean theorem shows that the vertical leg has length $\sqrt{3}$. Using either of these right triangles, we see that

$$\sin\frac{\pi}{3} = \frac{\sqrt{3}}{2}, \quad \cos\frac{\pi}{3} = \frac{1}{2}, \quad \tan\frac{\pi}{3} = \sqrt{3}. \tag{3}$$

$$\sin\frac{\pi}{6} = \frac{1}{2}, \quad \cos\frac{\pi}{6} = \frac{\sqrt{3}}{2}, \quad \tan\frac{\pi}{6} = \frac{1}{\sqrt{3}}. \tag{4}$$

The sine and cosine as coordinates of points on the circle

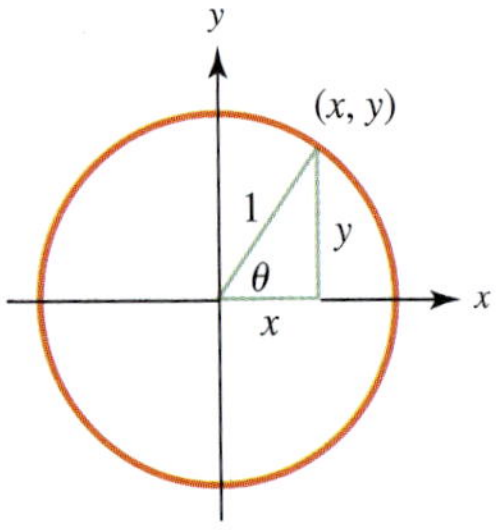

A more general way to define the trigonometric functions is to use a circle of radius 1, just as we did in defining radian measure. Take the circle in the xy-plane, with its center at the origin, and place the initial side of the angle along the x-axis. If (x, y) are the coordinates of the tip of the terminal side, as shown in Figure 8.2.5, then

$$\sin\theta = y, \quad \cos\theta = x, \quad \tan\theta = \frac{y}{x}. \tag{5}$$

Figure 8.2.5

In other words, $(\cos\theta, \sin\theta)$ are the coordinates of the terminal point of the angle, and

$$\boxed{\tan\theta = \frac{\sin\theta}{\cos\theta}.} \qquad (6)$$

Defining the sine and cosine in this way allows us to extend their domains to the entire set of real numbers.

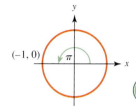

Figure 8.2.6

EXAMPLE 8.2.1

Find $\sin\theta$, $\cos\theta$, and $\tan\theta$ for $\theta = 0$, π, $\pi/2$, and $3\pi/2$.

SOLUTION   (i) If $\theta = 0$, the terminal side of the angle coincides with its initial side, so that the terminal point has coordinates $(1, 0)$. Therefore $\sin 0 = 0$, $\cos 0 = 1$, and $\tan 0 = 0$.

(ii) If $\theta = \pi$, the terminal point is obtained by completing half of a complete revolution counterclockwise, as shown in Figure 8.2.6. Therefore, its coordinates are $(-1, 0)$, so that $\sin\pi = 0$, $\cos\pi = -1$, and $\tan\pi = 0$.

(iii) If $\theta = \pi/2$, the terminal point is obtained by completing $\frac{1}{4}$ of a complete revolution counterclockwise, as shown in Figure 8.2.7. Therefore, its coordinates are $(0, 1)$, so that $\sin(\pi/2) = 1$ and $\cos(\pi/2) = 0$. However, $\tan(\pi/2)$ is undefined because the denominator of (6) is zero. In fact,

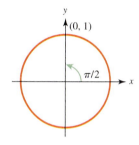

Figure 8.2.7

$$\lim_{\theta\to\pi/2^-}\tan\theta = \infty \quad \text{and} \quad \lim_{\theta\to\pi/2^+}\tan\theta = -\infty.$$

(iv) If $\theta = 3\pi/2$, the terminal point is obtained by completing $\frac{3}{4}$ of a complete revolution counterclockwise, as shown in Figure 8.2.8. Its coordinates are $(0, -1)$, so that $\sin(3\pi/2) = -1$ and $\cos(3\pi/2) = 0$. Once again, $\tan(3\pi/2)$ is undefined, and

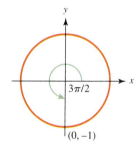

Figure 8.2.8

$$\lim_{\theta\to 3\pi/2^-}\tan\theta = \infty \quad \text{and} \quad \lim_{\theta\to 3\pi/2^+}\tan\theta = -\infty.$$

By using formulas (5) and referring to Figure 8.2.5, we can determine the signs of the trigonometric functions in each of the four quadrants of the plane, as given by Table 8.2.1 and shown in Figure 8.2.9.

$$
\begin{array}{c|c}
\begin{array}{l}\sin\theta > 0 \\ \cos\theta < 0 \\ \tan\theta < 0\end{array} & \begin{array}{l}\sin\theta > 0 \\ \cos\theta > 0 \\ \tan\theta > 0\end{array} \\
\hline
\begin{array}{l}\sin\theta < 0 \\ \cos\theta < 0 \\ \tan\theta > 0\end{array} & \begin{array}{l}\sin\theta < 0 \\ \cos\theta > 0 \\ \tan\theta < 0\end{array}
\end{array}
$$

Figure 8.2.9

TABLE 8.2.1

θ	$(0, \pi/2)$	$(\pi/2, \pi)$	$(\pi, 3\pi/2)$	$(3\pi/2, 2\pi)$
$\sin\theta$	+	+	−	−
$\cos\theta$	+	−	−	+
$\tan\theta$	+	−	+	−

EXAMPLE 8.2.2

Find

$$\text{(i)} \ \cos\left(\frac{7\pi}{3}\right), \quad \text{(ii)} \ \sin\left(-\frac{\pi}{4}\right), \quad \text{(iii)} \ \tan\left(\frac{7\pi}{6}\right).$$

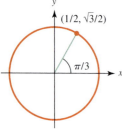

Figure 8.2.10

SOLUTION We will use formulas (2), (3), and (4).

(i) Since $7\pi/3 = 2\pi + \pi/3$, the terminal point of the angle is obtained by completing one full revolution counterclockwise and then adding an additional $\pi/3$ (as shown in Figure 8.2.10). Therefore,

$$\cos\frac{7\pi}{3} = \cos\frac{\pi}{3} = \frac{1}{2}.$$

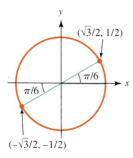

Figure 8.2.11

(ii) The terminal point of this angle is obtained by moving $\pi/4$ units along the circumference of the circle in the clockwise direction. Therefore, it is the reflection in the horizontal axis of the terminal point of $\pi/4$ (as shown in Figure 8.2.11), whose coordinates are $(\cos(\pi/4), \sin(\pi/4)) = (1/\sqrt{2}, 1/\sqrt{2})$. Therefore,

$$\sin\left(-\frac{\pi}{4}\right) = -\frac{1}{\sqrt{2}}.$$

(iii) Since $7\pi/6 = \pi + \pi/6$, the terminal point of this angle is obtained by completing half of a complete revolution plus an additional $\pi/6$ in the counterclockwise direction (as shown in Figure 8.2.12). Thus, we see that it is the reflection across the origin of the terminal point of $\pi/6$. The latter has coordinates $(\cos(\pi/6), \sin(\pi/6)) = (\sqrt{3}/2, 1/2)$. Thus, $\cos(7\pi/6) = -\sqrt{3}/2$, $\sin(7\pi/6) = -1/2$, and $\tan(7\pi/6) = 1/\sqrt{3}$.

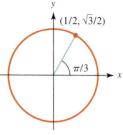

Figure 8.2.12

In general, if an angle has radian measure θ, the angle $2\pi + \theta$ determines the same (x, y) coordinates on the unit circle, just as we saw in case (i) of the previous example. Therefore, the trigonometric functions are **periodic**, with period 2π. In other words,

$$\sin(2\pi + \theta) = \sin\theta \quad \cos(2\pi + \theta) = \cos\theta \quad \tan(2\pi + \theta) = \tan\theta. \qquad (7)$$

Actually, $\tan\theta$ has period π (see Exercise 42).

By generalizing part (ii) of the same example, we see that if two angles have radian measures θ and $-\theta$, respectively, their terminal points are reflections of one another across the horizontal axis. Therefore, their horizontal coordinates are the same, while their vertical coordinates differ by a sign. From that we conclude:

$$\sin(-\theta) = -\sin\theta, \quad \cos(-\theta) = \cos\theta, \quad \tan(-\theta) = -\tan\theta. \qquad (8)$$

Another important property of the sine and cosine is the following: For any θ,

$$\sin^2 \theta + \cos^2 \theta = 1. \tag{9}$$

That follows from the fact that $(\cos \theta, \sin \theta)$ is a point on the unit circle.

The addition formulas

There are two other basic identities that are very useful. For any numbers u and v,

$$\sin(u + v) = \sin u \, \cos v + \cos u \, \sin v \tag{10}$$

and

$$\cos(u + v) = \cos u \, \cos v - \sin u \, \sin v. \tag{11}$$

The proofs can be found in any book on trigonometry, and we will omit them here.

EXAMPLE 8.2.3

Show that $\sin\left(\dfrac{\pi}{2} + \theta\right) = \cos \theta$.

SOLUTION Using (10), we have

$$\sin\left(\frac{\pi}{2} + \theta\right) = \sin\frac{\pi}{2}\,\cos\theta + \cos\frac{\pi}{2}\,\sin\theta$$
$$= 1 \cdot \cos\theta + 0 \cdot \sin\theta = \cos\theta.$$

The graphs of sine, cosine, and tangent

With the information we have gathered so far, including Example 8.2.3, we can make a table of values for the sine:

x	0	$\pi/6$	$\pi/4$	$\pi/3$	$\pi/2$	$2\pi/3$	$3\pi/4$	$5\pi/6$	π
$\sin x$	0	$1/2$	$1/\sqrt{2}$	$\sqrt{3}/2$	1	$\sqrt{3}/2$	$1/\sqrt{2}$	$1/2$	0

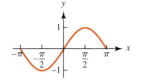

Figure 8.2.13

By plotting the points of this table and interpolating a smooth curve between them, we can sketch the graph of $y = \sin x$ over the interval $(0, \pi)$. And since, according to (8), the sine function is symmetric about the origin, we can extend the graph to cover the interval $(-\pi, \pi)$. The result is shown in Figure 8.2.13.

Next, we use the periodicity of $\sin x$, as stated in (7), to extend the graph over the entire x-axis, as shown in Figure 8.2.14.

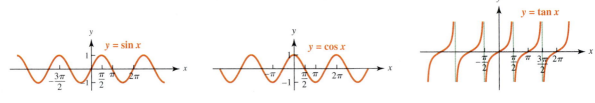

Figure 8.2.14   **Figure 8.2.15**   **Figure 8.2.16**

We can draw the graph of $y = \cos x$ by a similar procedure. However, a simpler method is to use the formula $\cos x = \sin(x + \pi/2)$, obtained in Example 8.2.3, to conclude that the graph of $\cos x$ is obtained by shifting the graph of $\sin x$ horizontally by $\pi/2$ units to the left. The result is shown in Figure 8.2.15.

For the graph of the tangent function, we use the results obtained in Example 8.2.1, Eq. (8), and Table 8.2.1 to draw the following conclusions:

• There are vertical asymptotes at $x = \pm\pi/2$.

• $\tan x > 0$ on the intervals $(-\pi, -\pi/2)$ and $(0, \pi/2)$.

• $\tan x < 0$ on the intervals $(-\pi/2, 0)$ and $(\pi/2, \pi)$.

• $\tan(-x) = -\tan x$.

By using these facts and constructing a small table of values (which is left as an exercise for you), we obtain the graph of $y = \tan x$ over the interval $(-\pi, \pi)$. We then extend it over the entire x-axis by periodicity to get the graph of $\tan x$, as shown in Figure 8.2.16.

In constructing their graphs, we assumed that the sine, cosine, and tangent are continuous throughout their domains. That is indeed the case, and we now state it formally.

Theorem 8.2.1

The functions $\sin t$ and $\cos t$ are continuous for all t. The function $\tan t$ is continuous at every t for which it is defined.

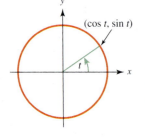

Figure 8.2.17

We will not prove this theorem, simply appealing to the definition of $\cos t$ and $\sin t$ as coordinates of a point on the unit circle as illustrated in Figure 8.2.17. As t varies, the coordinates $(\cos t, \sin t)$ change continuously, and so does their quotient $\tan t$, except at the points such as $t = \pm\pi/2, \pm3\pi/2$, and so forth, where it is not defined.

APPLYING TECHNOLOGY

Inverse trigonometric functions Any scientific or graphing calculator will compute sines, cosines, and tangents. Generally, in default of other instructions, the angles are assumed to be in radians. For instance, entering $\tan(45)$ on a TI-83 Plus will return 1.619775191, which is the tangent of 45 radians to nine decimal places. To find the tangent of 45 degrees, we can either

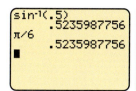

Figure 8.2.18

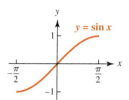

Figure 8.2.19

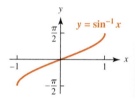

Figure 8.2.20

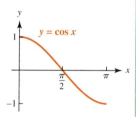

Figure 8.2.21

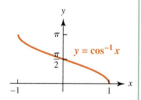

Figure 8.2.22

- convert to $\pi/4$ radians and enter $\tan(\pi/4)$, or
- specify that degrees are to be used. On the TI-83 Plus this can be done using the ANGLE key.

In some applications we want to find an angle having a given sine (or cosine or tangent—see the last exercises at the end of this section). To do that, there is a key usually marked $\sin^{-1}$ and placed above the $\boxed{\sin}$ key. For instance, to find an angle whose sine equals $\frac{1}{2}$, we enter $\sin^{-1}(0.5)$. The answer is in radians to a certain number of decimal places (in this case, nine). Since we already know an answer to this one, $\pi/6$, we can compare it to $\sin^{-1}(0.5)$ by entering it on the same screen and finding its decimal approximation. The result is shown in Figure 8.2.18. As you can see, the values agree. This is not the only angle, however, whose sine equals 0.5. Others are $\pi/6 \pm 2\pi$, $\pi/6 \pm 4\pi$, ... In general, given a number c between -1 and 1, there are infinitely many solutions to the equation $c = \sin x$.

In order to obtain a well-defined inverse for the sine function, we must restrict the domain of $\sin x$. Recall that a function has an inverse if and only if it satisfies the horizontal line test: *No horizontal line meets the graph in more than one point* (see Section 0.2). Looking at the graph of $y = \sin x$ shown in Figure 8.2.13, we see that this criterion is satisfied by the part of the graph over the interval $[-\pi/2, \pi/2]$. In other words, the function defined by

$$f(x) = \sin x, \quad -\frac{\pi}{2} \le x \le \frac{\pi}{2}$$

has an inverse, which we call the **inverse sine** (or **arcsine**) and denote by $\sin^{-1} x$. The graph of $f(x)$ is shown in Figure 8.2.19 and that of its inverse, $y = \sin^{-1} x$, in Figure 8.2.20. Each is obtained from the other by reflection in the line $y = x$ (see Section 0.2). We observe that

- the domain of $f(x)$ is $[-\pi/2, \pi/2]$ and the range is $[-1, 1]$, and
- the domain of $\sin^{-1} x$ is $[-1, 1]$, and the range is $[-\pi/2, \pi/2]$.

Thus, only numbers between -1 and 1 can be entered using the $\boxed{\sin^{-1}}$ command, and the value returned by the calculator is the radian measure of an angle between $-\pi/2$ and $\pi/2$.

Note In this case, the exponent -1 does *not* mean the reciprocal $1/\sin x$.

To find inverse functions for the other trigonometric functions, we must similarly restrict their domains. For $\cos x$, the general practice is to restrict the domain to the interval $[0, \pi]$. Figure 8.2.21 shows the part of the graph over that interval, and Figure 8.2.22 displays the inverse, denoted by $\cos^{-1} x$.

We observe that the domain of $\cos^{-1} x$ is $[-1, 1]$, and its range is $[0, \pi]$. Thus, only numbers between -1 and 1 can be entered using the $\boxed{\cos^{-1}}$ command, and the value returned by the calculator is the radian measure of an angle between 0 and π. Figure 8.2.23 shows several input-output pairs for $\cos^{-1} x$. (You should be able to rewrite them as multiples of π.)

For $\tan x$, Figure 8.2.16 shows that the part of the graph over the interval $(-\pi/2, \pi/2)$ satisfies the horizontal line test. Its inverse is shown in Figure 8.2.25. Its domain is

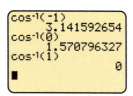

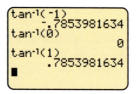

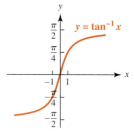

Figure 8.2.23   **Figure 8.2.24**   **Figure 8.2.25**

the entire x-axis, and its range is $(-\pi/2, \pi/2)$. Thus, any number can be entered using the $\boxed{\tan^{-1}}$ command, and the value returned by the calculator is the radian measure of an angle between $-\pi/2$ and $\pi/2$. Figure 8.2.24 shows several input-output pairs for $\tan^{-1} x$. (Once again, you should be able to rewrite them as multiples of π.)

HISTORICAL PROFILE

Aristarchus of Samos (c. 310 B.C.– 230 B.C.) was an astronomer and mathematician in ancient Greece. In addition to his pioneering attempt to determine the relative distances to the sun and the moon, he is also credited as being the first person to propose the idea of a heliocentric universe, many centuries before Copernicus. In fact, Archimedes (considered one of the greatest mathematicians of all time) wrote of him: *His hypothesis is that the fixed stars and the sun remain unmoved, that the earth revolves about the sun on the circumference of a circle, the sun lying in the middle of the orbit.*

Aristarchus also accepted the fact that the earth rotates on its axis, and that it is the earth's rotation that gives the illusion that the fixed stars are rotating daily. In addition, he is believed to have invented a sundial in the shape of a hemispheric bowl, with the time being indicated by the shadow cast inside the bowl by a rod.

HISTORICAL PROFILE

Eratosthenes (276 B.C.–194 B.C.) was born in the North African city of Cyrene. He studied in Athens and became director of the famous library in Alexandria.

His measurement of the radius of the earth was surprisingly accurate considering the period in which it was made and the technology available to him. In addition, he reputedly made an accurate estimate of the tilt of the earth's axis, developed a calendar that included leap years, made an accurate map of the Nile River, and compiled a star catalogue containing 675 stars. He also studied prime numbers and invented a method called the **sieve of Eratosthenes**, which, in modified form, is used to this day in number theory.

Practice Exercises 8.2

● ●

1. Find $\cos(5\pi/6)$ and $\tan(5\pi/3)$ without using a calculator.

2. If θ is between 0 and $\pi/2$ and $\tan\theta = 2$, what are $\sin\theta$ and $\cos\theta$?

3. Use the addition formula (11) to find $\cos(\pi/12)$ without using a calculator.

Exercises 8.2

● ●

In Exercises 1–4, find $\sin\theta$, $\cos\theta$, *and* $\tan\theta$.

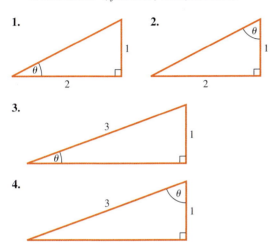

1.

2.

3.

4.

In Exercises 5–6, use a calculator to find the value of an appropriate trigonometric function.

5. In attempting to measure the height of a building (see Figure 8.2.26), you stand 20 feet away from it and measure the angle that the line of sight to the top of the building makes with the ground. If the angle is 67 degrees, how high is the building?

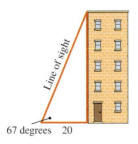

Figure 8.2.26

Figure 8.2.27

6. A taut rope runs from the top of a tree to a point on the ground 6 feet away from the base of the tree (see Figure 8.2.27). The angle between the ground and the rope is $4\pi/9$ radians. How long is the rope?

7. A group of high-school students are planning to paint a huge angle on the school gym floor for Math Appreciation

Week. Their design on paper has one leg that is 1 inch long and a second of 2 inches, with the angle between them measuring $\pi/3$ radians. The finished design calls for each leg to be 900 times the length in the drawing.
(a) By what constant do you multiply the angle measure in order to correctly scale the drawing?
(b) What is the cosine of the angle in the drawing? on the gym floor?

8. Aristarchus of Samos, who lived from about 310–230 B.C., attempted to measure the relative distances to the sun and the moon from the earth. To do that, he chose a time when there was a half-moon visible, which meant that the earth, moon, and sun were at the vertices of a right triangle (as in Figure 8.2.28). Then he measured the angle θ between his lines of sight to the sun and the moon. In terms of θ, what is the ratio of the distances of the sun and the moon from the earth?
(a) Aristarchus estimated the angle θ to be 87 degrees. If that were correct, how much further would the sun be from the earth than the moon?
(b) A better estimate of θ is 89.85 degrees. If that is correct, how much further is the sun from the earth than the moon?

Figure 8.2.28

9. Eratosthenes (275–195 B.C.) was the first person on record to compute the radius of the earth. His method depended on measuring the angle the sun's rays made with a vertical stick placed in the earth in the city of Alexandria, Egypt (labeled A in Figure 8.2.29). In order to find the angle, Eratosthenes measured both the length of the stick and its shadow (see Figure 8.2.30).
(a) What is the relation between those two lengths and the angle θ?
(b) Eratosthenes measured the angle θ at the moment when the sun was directly overhead in another Egyptian city, Syene (labeled S in Figure 8.2.29). He knew the distance between Alexandria and Syene. Can you explain

how he used that information and his determination of the angle θ to compute the radius of the earth?

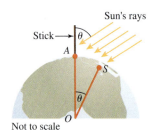

Sun's rays

Stick→

Not to scale

Figure 8.2.29

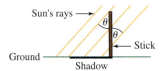

Sun's rays →

Ground

Shadow

Stick

Figure 8.2.30

10. The minute hand of a clock is set at 12 and the clock is plugged in. The hand is 1 inch long. Without using a calculator, track the coordinates of the tip of the hand as it travels around for 1 hour by completing the table. Start by drawing the path followed by the tip of the minute hand over the course of an hour and then draw the position hand at each 5-minute interval.

Min.	θ	$x = \cos\theta$	$y = \sin\theta$
5	$\pi/6$		
10	$\pi/3$		
15			
20			
25			
30			
40			
40			
45			
50			
55			
60			

11. Find $\sin(5\pi/4)$, $\cos(5\pi/4)$, and $\tan(5\pi/4)$ without using a calculator.

12. Find $\sin(2\pi/3)$, $\cos(2\pi/3)$, and $\tan(2\pi/3)$ without using a calculator.

13. Find $\sin(-\pi/3)$, $\cos(-\pi/3)$, and $\tan(-\pi/3)$ without using a calculator.

In Exercises 14–21, find the values without using a calculator.

14. $\sin(-\pi/2)$ **15.** $\tan(3\pi/4)$ **16.** $\cos(-\pi)$

17. $\sin(-\pi/6)$ **18.** $\sin(4\pi/3)$ **19.** $\cos(-2\pi/3)$

20. $\tan(5\pi/6)$ **21.** $\cos(3\pi)$

22. If θ is between 0 and $\pi/2$ and $\cos\theta = \frac{3}{5}$, what is $\tan\theta$?

23. If θ is between 0 and $\pi/2$ and $\tan\theta = \frac{2}{3}$, what is $\sin\theta$?

24. If θ is between $\pi/2$ and π and $\sin\theta = \sqrt{2}/3$, what are $\cos\theta$ and $\tan\theta$?

25. If θ is between $3\pi/2$ and 2π and $\tan\theta = -\frac{4}{3}$, what are $\sin\theta$ and $\cos\theta$?

26. If θ is between π and $3\pi/2$ and $\cos\theta = -\frac{1}{3}$, what are $\sin\theta$ and $\tan\theta$?

27. Find formulas similar to the addition formulas (10) and (11) for $\sin(u - v)$ and $\cos(u - v)$.

28. Show that $\cos(\pi/2 - \theta) = \sin\theta$. What is $\cos(\pi/2 + \theta)$?

29. Suppose that on a typical June day in La Jolla, California, the temperature oscillates sinusoidally between a low of 65°F at midnight and a high of 75°F at noon. Find a formula of the form $T(t) = a + b\sin(kt + d)$ that gives the temperature T (in degrees Fahrenheit) as a function of t (the time in hours after midnight).

30. Suppose the deer population in southern Michigan oscillates between a low of 20,000 and a high of 25,000 in a period of 4 years. Model it by a function of the form $p(t) = a + b\cos(ct)$, choosing $t = 0$ to be a time when the population is at its maximum level.

31. Normally, the heart beats between 60 and 80 times per minute (although it may reach 200 or more beats during intense exercising). Assume that a normal heartbeat of 70 times a minute results in a sinusoidal oscillation of arterial pressure, varying between 80 to 120 (measured in millimeters of mercury). Letting t be measured in seconds, with the pressure being at a maximum when $t = 0$, find a formula of the form $p(t) = a + b\sin(ct + d)$ for the pressure at any time t. Then sketch the graph of $p(t)$.

32. The rate of growth of an economy oscillates from -1% to 3% and back to -1% in a 5-year cycle. Find a formula of the form $r(t) = a + b\cos(ct + d)$ describing it, with t being the time in years.

33. Graph $\sin(\pi/2 + x)$ and $\cos x$ on the same screen, using a $[-3\pi, 3\pi] \times [-1, 1]$ window. What do you conclude? Explain.

34. Find $\sin(\pi - \theta)$ and $\cos(\pi - \theta)$.

35. Graph $\sin(\pi + x)$ and $\sin x$ on the same screen, using a $[-3\pi, 3\pi] \times [-1, 1]$ window. What do you conclude? Explain.

36. Graph $\sin(\pi/4 + x)$ and $\sin x + \cos x$ on the same screen, using a $[-3\pi, 3\pi] \times [-1.5, 1.5]$ window. What do you conclude? Explain.

37. Use the addition formula (10) to show that $\sin(2x) = 2 \sin x \cos x$.

38. (a) Use the addition formula (11) to show that $\cos(2x) = \cos^2 x - \sin^2 x$.
(b) Derive the formulas $\cos(2x) = 1 - 2\sin^2 x$ and $\cos(2x) = 2\cos^2 x - 1$.

39. Derive formulas for $\sin(3x)$ and $\cos(3x)$ in terms of $\sin x$ and $\cos x$, using the fact that $3x = 2x + x$.

40. Use the addition formulas to show that $\tan(u + v) = (\tan u + \tan v)/(1 - \tan u \tan v)$.

41. Graph $\tan(\pi + x)$ and $\tan x$ on the same screen, using a $[-\pi, \pi] \times [-5, 5]$ window. What do you conclude? Explain.

42. Show that $\tan(\theta + \pi) = \tan(\theta)$.

In Exercises 43-50, find the given values without using a calculator. (Hint: $\frac{1}{3} - \frac{1}{4} = \frac{1}{12}$ and $\frac{1}{8} + \frac{1}{8} = \frac{1}{4}$.)

43. $\sin(-\pi/12)$ **44.** $\cos(5\pi/12)$ **45.** $\tan(7\pi/12)$

46. $\cos(\pi/8)$ **47.** $\sin(\pi/8)$ **48.** $\cos(5\pi/8)$

49. $\cos(\pi/24)$ **50.** $\sin(-\pi/24)$

In Exercises 51–54, use a graphing calculator to find the value of an appropriate inverse function. Your answers should be in radians to 4 decimal places.

51. Find an angle between $-\pi/2$ and $\pi/2$ whose tangent equals 3.

52. Find an angle between 0 and π whose cosine equals 0.7. Find one between π and 2π with the same cosine.

53. If a right triangle has sides of lengths 3, 4, and 5, what are its angles?

54. If one leg of a right triangle is $\frac{2}{3}$ as long as the other, what are its angles?

55. A carpenter needs to cut several boards at angles for his project. Use your calculator to determine at which angle in radians the saw must be set to get the necessary cut. Round your answer to 3 decimal places.
(a) The board is $\frac{1}{2}$ inch wide. The right side needs to be $\frac{1}{8}$ inch lower than the left. (See Figure 8.2.31.)

Figure 8.2.31

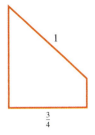

Figure 8.2.32

(b) Another is $\frac{3}{4}$ inch wide and the length of the finished cut must be 1 inch. (See Figure 8.2.32.)

In Exercises 56–58, make a table of values of the function for t near zero (on both sides) and try to guess the limit as $t \to 0$.

56. $f(t) = \dfrac{\sin t}{t}$ **57.** $g(t) = \dfrac{\cos t - 1}{t}$

58. $f(t) = \dfrac{\tan t}{t}$

*Exercises 59–62, arise from **robotics**, the design and operation of industrial robots. The model is simpler than those that arise in practice, which are generally three-dimensional with many links, but it gives some idea of the mathematics involved.*

59. Figure 8.2.33 shows a two-dimensional, two-link robot arm, fixed at point O, with an "elbow" at B. The lengths of the two links are fixed numbers r_1 and r_2. The angles they make with the horizontal, θ_1 and θ_2, are variables controlled by the operator or program. The point E, called the **end effector**, is the part that does the work. (Imagine, for instance, a pen or drill, or stapler attached to it.) Show that the coordinates (x, y) of E satisfy the equations

$$x = r_1 \cos\theta_1 + r_2 \cos\theta_2 \qquad (12)$$
$$y = r_1 \sin\theta_1 + r_2 \sin\theta_2. \qquad (13)$$

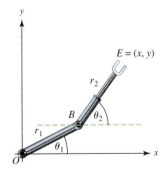

Figure 8.2.33

60. Suppose that $r_1 = 4$ inches and $r_2 = 5$ inches. Find (x, y) for the following choices of angles:
   (a) $\theta_1 = 2\pi/5, \theta_2 = \pi/10$ radians,
   (b) $\theta_1 = 22, \theta_2 = 39$ degrees

61. In practice, the point (x, y) may be specified, and the angles θ_1 and θ_2 must be found by solving Eqs. (12) and (13). Show that by squaring and adding the equations, we get

$$x^2 + y^2 = r_1^2 + r_2^2 + 2r_1 r_2 \cos(\theta_2 - \theta_1). \quad (14)$$

62. Suppose that $r_1 = 4$ inches and $r_2 = 5$ inches, and $(x, y) = (3, 7)$.

(a) Use (14) to find $\theta_2 - \theta_1$ (in radians to three decimal places).
(b) In Eqs. (12) and (13), write $\theta_2 = \theta_1 + (\theta_2 - \theta_1)$ and substitute the value for $\theta_2 - \theta_1$ found in step (a). Then use the addition formulas to put the equations in the form

$$3 = a \cos \theta_1 - b \sin \theta_1$$
$$7 = b \cos \theta_1 + a \sin \theta_1,$$

where a and b are constants. Next, solve the system for $\cos \theta_1$ and $\sin \theta_1$.
(c) Find θ_1 and then use it and step(a) to find θ_2.

Solutions to practice exercises 8.2

1. Using the addition formula, we get

$$\cos \frac{5\pi}{6} = \cos\left(\pi - \frac{\pi}{6}\right)$$

$$= \cos \pi \, \cos\left(-\frac{\pi}{6}\right) - \sin \pi \, \sin\left(-\frac{\pi}{6}\right)$$

$$= (-1) \cdot \cos \frac{\pi}{6} + 0 \cdot \sin \frac{\pi}{6} = -\frac{\sqrt{3}}{2}.$$

Or, simply observe that $\pi - \pi/6$ and $\pi/6$ are reflections of one another in the y-axis (see Figure 8.2.34). Therefore, their horizontal coordinates differ only by a sign, so that

$$\cos \frac{5\pi}{6} = -\cos \frac{\pi}{6}.$$

For $\tan(5\pi/3)$, we have $5\pi/3 = 2\pi - \pi/3$, which is the reflection of $\pi/3$ in the x-axis (see Figure 8.2.35). Therefore, $\cos(5\pi/3) = \cos(\pi/3) = 1/2$ and $\sin(5\pi/3) = -\sin(\pi/3) = -\sqrt{3}/2$, from which we get $\tan(5\pi/3) = -\sqrt{3}$.

2. Consider a right triangle in which θ is one of the angles, with the opposite side of length 2 and the adjacent side of length 1, so that $\tan \theta = 2$, as in Figure 8.2.36. We can compute the length of the hypotenuse by the Pythagorean formula: $h = \sqrt{5}$. From that, we see that

$$\sin \theta = \frac{2}{\sqrt{5}} \quad \text{and} \quad \cos \theta = \frac{1}{\sqrt{5}}.$$

3. We first observe that $\pi/12 = \pi/3 - \pi/4$. Then, using the addition formula, we get

$$\cos \frac{\pi}{12} = \cos\left(\frac{\pi}{3} - \frac{\pi}{4}\right)$$

$$= \cos \frac{\pi}{3} \cos\left(-\frac{\pi}{4}\right) - \sin \frac{\pi}{3} \sin\left(-\frac{\pi}{4}\right)$$

$$= \frac{1}{2} \cdot \frac{1}{\sqrt{2}} - \frac{\sqrt{3}}{2} \cdot \left(-\frac{1}{\sqrt{2}}\right) = \frac{1 + \sqrt{3}}{2\sqrt{2}}.$$

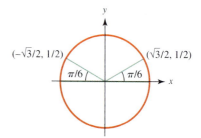

Figure 8.2.34

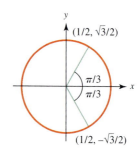

Figure 8.2.35

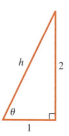

Figure 8.2.36

8.3 Derivatives of Trigonometric Functions

The derivatives of $\sin x$ and $\cos x$ are given by the following formulas:

$$\frac{d}{dx} \sin x = \cos x \tag{15}$$

and

$$\frac{d}{dx} \cos x = -\sin x. \tag{16}$$

An explanation of these rules will be given at the end of this section, but first we will consider some examples.

EXAMPLE 8.3.1

Find $\dfrac{d}{dx} (x \sin x)$.

SOLUTION   Using the product rule, we have

$$\frac{d}{dx} (x \sin x) = x \frac{d}{dx} \sin x + 1 \cdot \sin x = x \cos x + \sin x.$$

EXAMPLE 8.3.2

Find the maximum value of $\sin x + \cos x$ for $0 \le x \le 2\pi$.

SOLUTION   Let $f(x) = \sin x + \cos x$. We first look for critical points in the open interval $(0, 2\pi)$, by solving

$$0 = f'(x) = \cos x - \sin x. \tag{17}$$

We can assume that $\cos x \ne 0$; otherwise, $x = 3\pi/2$ or $\pi/2$, and $f'(x) = \pm 1$. Therefore, (17) is equivalent to $\tan x = 1$, which has the solutions $x = \pi/4$ and $x = 5\pi/4$ in $(0, 2\pi)$. Those are the critical points, and their values are as follows:

$$f\left(\frac{\pi}{4}\right) = \sin \frac{\pi}{4} + \cos \frac{\pi}{4} = \sqrt{2} \quad \text{and} \quad f\left(\frac{5\pi}{4}\right) = \sin \frac{5\pi}{4} + \cos \frac{5\pi}{4} = -\sqrt{2}.$$

At the endpoints we have $f(0) = f(2\pi) = 1$. Comparing these to the values at the critical points, we see that the maximum is $\sqrt{2} \approx 1.414$, attained at $x = \pi/4 \approx 0.7854$. The graph of $f(x)$ is shown in Figure 8.3.1, as drawn with a TI-83 Plus using a viewing window of $[0, 2\pi] \times [-2, 2]$, with the cursor positioned near the maximum point.

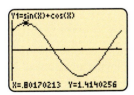

Figure 8.3.1

EXAMPLE 8.3.3

Verify the formula

$$\frac{d}{dx}\tan x = \frac{1}{\cos^2 x}. \tag{18}$$

SOLUTION   By the quotient rule,

$$\frac{d}{dx}\tan x = \frac{d}{dx}\frac{\sin x}{\cos x}$$

$$= \frac{\cos x \dfrac{d}{dx}\sin x - \sin x \dfrac{d}{dx}\cos x}{\cos^2 x}$$

$$= \frac{\cos^2 x + \sin^2 x}{\cos^2 x} = \frac{1}{\cos^2 x}.$$

We can also combine formulas (15) and (16) with the chain rule.

EXAMPLE 8.3.4

Find $\dfrac{dy}{dx}$ if

(i) $y = \cos^2 x$ and   (ii) $y = \sin(x^2 + 3x)$.

SOLUTION   (i) Using the chain rule with $u = \cos x$ and $y = u^2$, we obtain

$$\frac{dy}{dx} = \frac{dy}{du}\frac{du}{dx} = -2u \sin x = -2 \cos x \sin x.$$

(ii) Using the chain rule with $u = x^2 + 3x$ and $y = \sin u$, we get

$$\frac{dy}{dx} = \frac{dy}{du}\frac{du}{dx} = (2x + 3)\cos u = (2x + 3)\cos(x^2 + 3x).$$

EXAMPLE 8.3.5

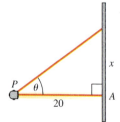

Figure 8.3.2

A spotlight (at point P of Figure 8.3.2) is rotating counterclockwise at a constant rate, making one full rotation every 3 minutes. How fast is its beam moving along a wall 20 feet away

(i) when it passes point A, and
(ii) $\frac{1}{2}$ minute later?

SOLUTION   Let t be the time in minutes, with $x = 0$ when $t = 0$. From the triangle in Figure 8.3.2, we get $\tan \theta = x/20$, or

$$x = 20 \tan \theta. \tag{19}$$

Both x and θ are functions of t, and taking derivatives on both sides of (19) gives

$$\frac{dx}{dt} = \frac{20}{\cos^2 \theta} \frac{d\theta}{dt}. \tag{20}$$

Since the angle θ is changing at a constant rate of $2\pi/3$ radians per minute, $d\theta/dt = 2\pi/3$. Therefore,

$$\frac{dx}{dt} = \frac{40\pi}{3\cos^2 \theta}. \tag{21}$$

(i) The beam hits point A when $\theta = 0, 2\pi, 4\pi, \ldots$. Therefore, $\cos \theta = 1$, and

$$\frac{dx}{dt} = \frac{40\pi}{3} \approx 41.9 \text{ feet per minute,}$$

which is the velocity of the point of light on the wall when it passes A.

(ii) If $t = \frac{1}{2}$, $\theta = \pi/3 + 2k\pi$, where $k = 0, 1, 2, \ldots$. Therefore, $\cos \theta = \frac{1}{2}$, and

$$\frac{dx}{dt} = \frac{40\pi}{\frac{3}{4}} \approx 167.55 \text{ feet per minute.}$$

The other trigonometric functions

There are three additional trigonometric functions, the **secant**, **cosecant**, and **cotangent**, defined as follows:

The Secant Function

$$\sec x = \frac{1}{\cos x}.$$

The Cosecant Function

$$\csc x = \frac{1}{\sin x}.$$

The Cotangent Function

$$\cot x = \frac{1}{\tan x} = \frac{\cos x}{\sin x}.$$

Because they are the reciprocals of the functions we have already studied, it is not hard to find their derivatives.

EXAMPLE 8.3.6

Find $\dfrac{d}{dx}\sec x$.

SOLUTION   Using the chain rule, we obtain

$$\frac{d}{dx}\sec x = \frac{d}{dx}\left(\frac{1}{\cos x}\right) = -\frac{1}{\cos^2 x}\frac{d}{dx}\cos x = \frac{\sin x}{\cos^2 x}.$$

Since

$$\frac{\sin x}{\cos^2 x} = \frac{1}{\cos x}\cdot\frac{\sin x}{\cos x} = \sec x \, \tan x,$$

we can rewrite the last formula in the form

$$\frac{d}{dx}\sec x = \sec x \, \tan x. \tag{22}$$

The derivative of $\csc x$ is derived similarly:

$$\frac{d}{dx}\csc x = -\csc x \, \cot x. \tag{23}$$

We can use the secant to modify formula (18) to get another, more customary, expression for the derivative of $\tan x$, as follows:

$$\frac{d}{dx}\tan x = \sec^2 x. \tag{24}$$

A similar formula holds for the derivative of $\cot x$:

$$\frac{d}{dx}\cot x = -\csc^2 x. \tag{25}$$

A predator-prey model

Trigonometric functions are very useful in modeling cyclical phenomena. Imagine, for instance, an ecological system in which one species, the **prey**, is the food source for a second species, the **predators**. In parts of Canada, for example, lynxes are the predators of snowshoe hares. The populations of both species go up and down in cycles of about 9 to 10 years. An indicator of the population trend of Canadian lynxes is shown in Figure 8.3.3, which is a graph illustrating the number of lynx pelts turned in to the Hudson Bay Company over a 100-year period. As you can see, the population oscillated during that time, peaking about every 9 to 10 years.

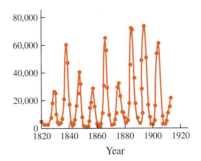

Figure 8.3.3

An explanation for the cyclical behavior of these populations is this. When hares are plentiful, the lynxes have plenty of food, and their population increases. The larger it grows, however, the more hares are consumed, and the hare population begins to diminish. That causes the lynx population to diminish as well, since its food supply is decreasing. After a time, the size of the lynx population has sunk to a level that is safer for the hares, and their population begins to increase. That, in turn, leads to an increase in lynxes, and the cycle continues.

A simple first step in modeling such an ecosystem is given by the following pair of differential equations:

$$\frac{dP}{dt} = a^2(H - H_e), \tag{26}$$

$$\frac{dH}{dt} = -b^2(P - P_e). \tag{27}$$

Here, t denotes time (in suitable units), and P and H are the numbers of predators and prey, respectively, at any given time. The symbols P_e and H_e represent constants. They are the equilibrium sizes of the respective populations—the numbers around which the populations oscillate. The numbers a^2 and b^2 represent constants of proportionality.

If $P > P_e$, the right-hand side of (27) is negative. Therefore, $dH/dt < 0$, and the prey population decreases. On the other hand, $dH/dt > 0$ when $P < P_e$, and the number of prey increases. Similarly, the predator population increases when $H > H_e$ (lots of prey) and decreases when $H < H_e$ (fewer prey).

To solve Eqs. (26) and (27), we begin by making the following substitutions. Let

$$u = P - P_e \quad \text{and} \quad v = H - H_e.$$

Then (26) and (27) are changed to

$$\frac{du}{dt} = a^2 v \quad \text{and} \quad \frac{dv}{dt} = -b^2 u. \tag{28}$$

These equations bear a close resemblance to formulas (15) and (16) for the derivatives of the sine and cosine, which suggests that the solution might be obtained by modifying the trigonometric functions (say, by introducing a chain-rule factor). That is indeed the case, as the next example shows.

EXAMPLE 8.3.7

Show that a solution of Eqs. (28) is given by

$$u = a \sin(abt) \quad \text{and} \quad v = b \cos(abt).$$

SOLUTION   Using the chain rule, we get

$$\frac{du}{dt} = a^2 b \cos(abt) = a^2 [b \cos(abt)] = a^2 v$$

and

$$\frac{dv}{dt} = -ab^2 \sin(abt) = -b^2 [a \sin(abt)] = -b^2 u.$$

These are not the only solutions of Eqs. (28). In fact, for any constants M and N, the functions defined by

$$u = a[M \sin(abt) + N \cos(abt)] \tag{29}$$
$$v = b[M \cos(abt) - N \sin(abt)] \tag{30}$$

are also solutions of (28). The verification is left for the exercises.

EXAMPLE 8.3.8

Suppose that the population sizes of a certain predator-prey pair of species in a particular region satisfy the differential equations

$$\frac{dP}{dt} = 0.25(H - 12) \quad \text{and} \quad \frac{dH}{dt} = -0.16(P - 2), \tag{31}$$

where time is measured in years and P and H in thousands of animals. Suppose that at $t = 0$, there are 1.7 thousand predators and 12.5 thousand prey. Find P and H as functions of t.

SOLUTION   We first make the substitutions $u = P - 2$ and $v = H - 12$, which changes Eqs. (31) into the form of (28), with $a = 0.5$ and $b = 0.4$. The solutions, given by (29) and (30), are

$$u = 0.5[M \sin(0.2t) + N \cos(0.2t)] \quad \text{and} \quad v = 0.4[M \cos(0.2t) - N \sin(0.2t)],$$

for some constants M and N. Therefore,

$$P = 2 + 0.5[M \sin(0.2t) + N \cos(0.2t)] \tag{32}$$
$$H = 12 + 0.4[M \cos(0.2t) - N \sin(0.2t)]. \tag{33}$$

To find M and N, we set $t = 0$ in (32) and (33) and use the initial conditions $P(0) = 1.7$ and $H(0) = 12.5$, which give

$$1.7 = 2 + 0.5N \quad \text{and} \quad 12.5 = 12 + 0.4M.$$

Therefore, $N = -0.6$, $M = 1.25$, and the solutions are

$$P = 2 + 0.5[1.25 \sin(0.02t) - 0.6 \cos(0.02t)] \tag{34}$$
$$H = 12 + 0.4[1.25 \cos(0.02t) + 0.6 \sin(0.02t)]. \tag{35}$$

Equations (26) and (27) are only a small first step in modeling very complicated predator-prey relations. They take no account, for instance, of the intrinsic growth rate of the prey. More sophisticated models can be found in a number of books on differential equations and mathematical ecology. (See the projects at the end of this chapter.)

Deriving the basic formulas

To compute the derivatives of $\sin x$ and $\cos x$, we start with the definition of the derivative in the form

$$f'(x) = \lim_{h \to 0} \frac{f(x+h) - f(x)}{h}.$$

Applying this to the sine function and using the addition formula (10) for the sine, we get

$$\begin{aligned}
\frac{d}{dx} \sin x &= \lim_{h \to 0} \frac{\sin(x + h) - \sin x}{h} \\
&= \lim_{h \to 0} \frac{\sin x \, \cos h + \cos x \, \sin h - \sin x}{h} \\
&= \lim_{h \to 0} \frac{\sin x (\cos h - 1) + \cos x \, \sin h}{h} \\
&= \lim_{h \to 0} \left(\sin x \, \frac{\cos h - 1}{h} \right) + \lim_{h \to 0} \left(\cos x \, \frac{\sin h}{h} \right) \\
&= \sin x \left(\lim_{h \to 0} \frac{\cos h - 1}{h} \right) + \cos x \left(\lim_{h \to 0} \frac{\sin h}{h} \right).
\end{aligned} \tag{36}$$

Thus, the question of finding the derivative of $\sin x$ has been reduced to evaluating two limits:

$$\lim_{h \to 0} \frac{\cos h - 1}{h} \quad \text{and} \quad \lim_{h \to 0} \frac{\sin h}{h}.$$

Both of these limits are indeterminate. They are, respectively, the derivatives of $\cos x$

and $\sin x$ at $x = 0$. A look at the cosine and sine graphs suggests that

$$\lim_{h \to 0} \frac{\sin h}{h} = 1 \tag{37}$$

and

$$\lim_{h \to 0} \frac{\cos h - 1}{h} = 0. \tag{38}$$

These are, in fact, true, and applying them to (36) gives

$$\frac{d}{dx} \sin x = (\sin x) \cdot 0 + (\cos x) \cdot 1 = \cos x.$$

A similar argument—applying the definition of the derivative to $\cos x$, using the addition formula (11) for the cosine, and taking (37) and (38) into account—shows that

$$\frac{d}{dx} \cos x = -\sin x.$$

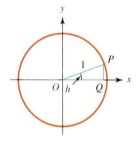

Figure 8.3.4

Although we will omit a precise proof of (37), the intuitive idea behind it can be seen in Figure 8.3.4. The point P has coordinates $(\cos h, \sin h)$ and Q has coordinates $(\cos h, 0)$. Therefore, the area of the right triangle OQP equals $\frac{1}{2} \sin h \cos h$.

On the other hand, the sector of the circle marked off by OQP has area $h/2$. To see why, recall that a circle of radius 1 has area π, and the proportion of the area inside the sector OQP is $h/2\pi$ (the ratio of the radian measure of angle QOP to that of the full circle). Therefore, the area of the sector is $(h/2\pi) \cdot \pi$, which equals $h/2$.

If h is very small, the area of the sector is close to that of the triangle, so

$$\frac{1}{2} \sin h \cos h \approx \frac{h}{2}.$$

Multiplying both sides by 2, dividing both sides by $h \cos h$, and taking the limit as $h \to 0$ give

$$\lim_{h \to 0} \frac{\sin h}{h} = \lim_{h \to 0} \frac{1}{\cos h} = 1.$$

(The last step follows from the fact that $\cos h$ is continuous and $\cos 0 = 1$.)

We can use (37) to derive (38) and then use (38) to find the derivative of $\cos x$. The details are omitted (but see the exercises below).

Practice Exercises 8.3

● ●

1. Find the derivatives of

   (a) $\dfrac{\sin x}{1 + \cos x}$

   (b) $\tan^2(3x)$

2. Sketch the graph of $y = x + \cos x$ over the interval $[0, 2\pi]$. State where the graph is increasing and where it is decreasing; discuss its concavity; list any vertical asymptotes, local extreme points, and inflection points.

Exercises 8.3

In Exercises 1–18, find the derivative.

1. $2\sin x - 3\cos x$

2. $\tan(2x)$

3. $\dfrac{1}{1 + \cos x}$

4. $e^t \cos t$

5. $\sin^3 x - \cos^3 x$

6. $x^3 \sin(x^2)$

7. $\tan(1 + x^4)$

8. $(\sin x) \cdot (\sin 2x)$

9. $\cos^2(3t) - 2\sin^2(3t)$

10. $\dfrac{\cos(2t)}{\sin(3t)}$

11. $\dfrac{1}{\tan(2x + 1)}$

12. $\ln(2 + \cos x)$

13. $e^{2t}[\cos(3t) + \sin(3t)]$

14. $\sec^3 x$

15. $x \csc x$

16. $\csc x \cot x$

17. $e^{\tan x}$

18. $\ln(1 + \cot^2 x)$

19. A small sailboat is out on the ocean, at rest in one area while its crew fishes. An observer on shore notices that the boat is moving up and down with the regular swells that pass by. The vertical displacement of the boat at any time t is given by $D(t) = 5\cos t \sin^2 t + 1$, where t is in minutes and D is in feet.
   (a) Find the vertical velocity at any time t.
   (b) Compute the vertical velocity when $t = k\pi$ for $k = 0, 1, \ldots$.

20. Find the equation of the line tangent to the graph of $f(x) = x + \cos x$ at $x = 0$. Verify your answer visually by graphing the function and the tangent line in the same window.

21. Find the equation of the line tangent to the graph of $f(x) = \cos^2(x + \pi/4)$ at $x = 0$. Verify your answer visually by graphing the function and the tangent line in the same window.

22. Derive formula (23). **23.** Derive formula (25).

24. Find the critical points of the function $f(x) = (1 + \cos x)^{-1}$ on the interval $(-\pi/2, \pi/2)$ and determine whether the function has a local maximum or minimum at each.

25. Find the critical points of the function $f(x) = \sin^2 x$ on the interval $(-\pi, \pi)$ and determine whether the function has a local maximum or minimum at each.

26. Find the absolute maximum of the function $f(x) = 2\sin x + \cos(2x)$ on the interval $[0, \pi]$.

27. Find where the function $f(x) = \sin x - \cos x$ is increasing and where it is decreasing on the interval $[0, 2\pi]$.

28. Find where the function $f(x) = (2 + \sin x)^{-1}$ is increasing and where it is decreasing on the interval $[0, 2\pi]$.

29. Find where the function $f(x) = x - 2\sin x$ is increasing and where it is decreasing on the interval $[0, 2\pi]$.

30. Use derivatives to show that the graph of $y = \tan x$ is concave down on $(-\pi/2, 0)$ and concave up on $(0, \pi/2)$.

31. Find all inflection points of the graph of $y = x + \cos x$ on the interval $(0, 2\pi)$.

32. Mark has mood swings that affect his outlook on life and his ability to interact well with others. His doctor asked him to keep a daily journal of his moods and found that they followed the pattern of $M(t) = \sin(\pi t/14)$, where t is the time in days and M is an index of the mood level.
   (a) Assuming that $M > 0$ is happy and $M < 0$ is sad, use derivatives to determine when Mark is most happy and when Mark is most sad.
   (b) What is the range for M?

33. Continuation of the previous problem Mark's doctor has prescribed daily medication for him. The journal now shows that his mood levels are best modeled by $M(t) = 1.2\sin^2(\pi t/14) - 0.25$.
   (a) Using derivatives, determine when Mark is most happy and most sad.
   (b) What values does M take on those days?
   (c) Compare the frequency and range of his mood levels when using the medicine to those without any medicine. How has the medicine affected him?

34. Find the minimum value of the function $f(x) = \tan x - 2x$ on the interval $[0, \pi/2)$ in two ways:
   (a) by using the calculator to solve the equation $f'(x) = 0$, and
   (b) by graphing $f(x)$ in a suitable window and using the calculator's ZOOM function.

35. A patient's blood pressure is given as a function of time by the formula

$$p = 100 + 18\cos(7t) + 12\sin(7t),$$

where t is the time in seconds. Find the maximum pressure (called **systolic**) and minimum pressure (called **diastolic**) in two ways:
   (a) by graphing the function and using the ZOOM function on your calculator, and
   (b) by solving the equation $dp/dt = 0$, using the equation solver.

36. Continuing the previous problem, find the times during the first 8 seconds when the patient's blood pressure is increasing and when it is decreasing.

In Exercises 37–40, graph $f(x)$ and $f'(x)$ on the same screen, using the specified window. What properties do you expect $f(x)$ to have when $f'(x)$ is negative, when $f'(x)$ is positive, when $f'(x)$ is increasing, and when $f'(x)$ is decreasing? Graph both functions and verify those properties.

37. $f(x) = \sin(2x) - \sin x$, $[-\pi, \pi] \times [-2.5, 2.5]$

38. $f(x) = x \cos x$, $[-3\pi/2, 3\pi/2] \times [-3.5, 3.5]$

39. $f(x) = x + 2 \cos x$, $[0, 3\pi] \times [-2, 11]$

40. $f(x) = \tan(2x)$, $[-\pi/4, \pi/4] \times [-5, 5]$

41. A lighthouse is located in Lake Michigan, 300 feet from the nearest point on shore, as shown in Figure 8.3.5. The light rotates at a constant rate, making k complete revolutions per hour. At the moment that the beam hits a point on the shore 500 feet from the lighthouse, the point of light is traveling along the shoreline at a rate of 2,500 feet per minute. Find k.

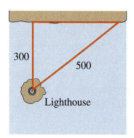

Figure 8.3.5

42. An airplane is flying due east on a horizontal course 3 miles above ground level, as shown in Figure 8.3.6. On the ground at point P, a revolving camera is tracking the movement of the plane.
(a) Write a formula relating the velocity of the plane and the rate $d\theta/dt$ at which the angle is changing. *Hint:* Use a trigonometric function to write an equation linking θ and x. Then, considering both variables as functions of t (time), take the derivative of both sides of your equation with respect to t, using the chain rule.
(b) If the plane is moving at a constant rate of 4 miles per minute, at what rate is the angle changing at the moment when $\theta = \pi/4$?

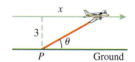

Figure 8.3.6

43. In order to videotape animals without being seen, a field biologist has constructed a small, camouflaged enclosure—called a **blind**—in a tree overlooking a meadow. From 20 feet up, he can tape deer as they use a salt lick. A buck

using the lick is startled and moves away at a speed of $\frac{22}{3}$ feet per second, with the biologist keeping the camera trained on him as he moves. How fast (in radians per second) is the camera angle changing at the moment when the deer is 15 feet away from the base of the tree? Include a sketch of the situation.

In Exercises 44–49, sketch the graph over the given interval. State where the graph is increasing and where it is decreasing; discuss its concavity; list any vertical asymptotes, local extreme points, and inflection points.

44. $y = \cos(2x)$, $[0, 2\pi]$ **45.** $y = \cos^2 x$, $[0, 2\pi]$

46. $y = \sec x$, $(-\pi, \pi)$ **47.** $y = \cot x$, $(-\pi, \pi)$

48. $y = \csc x$, $(-\pi, \pi)$ **49.** $y = x + \sin x$, $[0, 2\pi]$

50. Use a calculator to graph $f(x) = \sin(2x) + 2\sin(3x)$ over the interval $[0, 2\pi]$. Find the maximum and minimum values correct to three decimal places.

51. Graph $y = x \sin x$ and $y = x$, $0 \leq x \leq 3\pi$, in the same window. What do you observe about the graphs? In particular, answer the following:
(a) How do their heights compare?
(b) Where do they intersect, and what is the slope of the graph of $x \sin x$ at those points? Does that pattern continue beyond 3π?
(c) Use the equation solver to find all local maxima of $x \sin x$ in $[0, 3\pi]$. Where do they fall in relation to the line $y = x$?
(d) Now add the graph of $y = -x$ and increase the domain to 5π. Answer similar questions, comparing the graph of $y = x \sin x$ to that of $y = -x$.

52. Graph $y = x \cos x$ and $y = x$, $0 \leq x \leq 5\pi$, in the same window. What do you observe about the graphs? In particular, answer the following:
(a) How do their heights compare?
(b) Where do they intersect, and what is the slope of the graph of $x \cos x$ at those points? Does that pattern continue beyond 5π?
(c) Use the equation solver to find all local maxima of $y = \cos x$ in $[0, 5\pi]$. Where do they fall in relation to the line $y = x$?
(d) Now add the graph of $y = -x$. Answer similar questions, comparing the graph of $y = x \cos x$ to that of $y = -x$.

53. Suppose $f(x)$ is a positive function that is differentiable on $(0, \infty)$.
(a) Show that the graph of $y = f(x) \sin x$ is always between that of $y = -f(x)$ and $y = f(x)$.
(b) Find the points where the graph of $y = f(x) \sin x$ meets that of $y = f(x)$, and show that the two graphs have the same tangent at those points.
(c) Do the same with the graphs of $y = f(x) \sin x$ and $y = -f(x)$.

54. Show that the functions defined by formulas (29) and (30) are solutions of the system of differential equations (28).

55. Solve the system of differential equations

$$\frac{du}{dt} = 4v \quad \text{and} \quad \frac{dv}{dt} = -u$$

with initial conditions $u(0) = -2$, and $v(0) = 1$.

56. Suppose that the population sizes of a certain predator-prey pair of species in a particular region satisfy the differential equations

$$\frac{dP}{dt} = 0.09(H - 8) \quad \text{and} \quad \frac{dH}{dt} = -0.16(P - 3), \quad (39)$$

where time t is measured in years and P and H in thousands of animals. Suppose that at $t = 0$, there are 3.2 thousand predators and 7.6 thousand prey. Find P and H as functions of t.

57. Suppose a two-link robot arm, as shown in Figure 8.3.7, is programmed to move along a vertical wall at a constant velocity of 1 unit per second. Show that the angles θ_1 and θ_2 satisfy the following system of differential equations [see Eqs. (12) and (13) in the exercises of Section 8.2.]:

$$r_1 \sin\theta_1 \frac{d\theta_1}{dt} + r_2 \sin\theta_2 \frac{d\theta_2}{dt} = 0$$

$$r_1 \cos\theta_1 \frac{d\theta_1}{dt} + r_2 \cos\theta_2 \frac{d\theta_2}{dt} = 1.$$

58. In this exercise you will show that

$$\lim_{h \to 0} \frac{\cos h - 1}{h} = 0$$

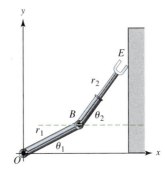

Figure 8.3.7

in the following sequence of steps:

(a) By multiplying the numerator and denominator by $\cos h + 1$, demonstrate that

$$\frac{\cos h - 1}{h} = \frac{-\sin^2 h}{h(\cos h + 1)}.$$

(b) Use the previous step to show that

$$\lim_{h \to 0} \frac{\cos h - 1}{h} = -\lim_{h \to 0} \sin h \cdot \lim_{h \to 0} \frac{\sin h}{h} \cdot \lim_{h \to 0} \frac{1}{\cos h + 1}.$$

(c) Evaluate each of the three limits on the right-hand side and take the product.

59. Use formulas (37) and (38) and the addition formula to show that

$$\frac{d}{dx} \cos x = -\sin x.$$

Solutions to practice exercises 8.3

1. (a) Using the quotient rule, we get

$$\frac{d}{dx} \frac{\sin x}{1 + \cos x} = \frac{(1 + \cos x)\dfrac{d}{dx}\sin x - \sin x \dfrac{d}{dx}(1 + \cos x)}{(1 + \cos x)^2}$$

$$= \frac{(1 + \cos x) \cdot \cos x - \sin x \cdot (-\sin x)}{(1 + \cos x)^2}$$

$$= \frac{\cos x + \cos^2 x + \sin^2 x}{(1 + \cos)^2}$$

$$= \frac{\cos x + 1}{(1 + \cos x)^2} = \frac{1}{1 + \cos x}$$

(b) Using the chain rule twice, we obtain

$$\frac{d}{dx} \tan^2(3x) = 2\tan(3x)\frac{d}{dx}\tan(3x)$$

$$= 2\tan(3x)\frac{1}{\cos^2(3x)}\frac{d}{dx}(3x) = \frac{6\tan(3x)}{\cos^2(3x)}.$$

2. Since $y' = 1 - \sin x \geq 0$ for all x, the function is increasing throughout the entire interval. It has a critical point at $x = \pi/2$, where the tangent is horizontal. The second derivative $y'' = -\cos x$ is negative on $[0, \pi/2)$ and $(3\pi/2, 2\pi]$, and the graph is concave down on those intervals. It is positive on $(\pi/2, 3\pi/2)$, and the graph is concave up on those intervals. There are inflection points at $x = \pi/2$ and $x = 3\pi/2$. There are no vertical asymptotes. The graph is shown in Figure 8.3.8.

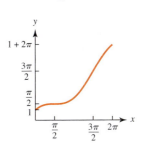

Figure 8.3.8

8.4 **Integrals of Trigonometric Functions**

From the derivative formulas (15) and (16) we get the following antiderivative formulas:

$$\int \sin x \, dx = -\cos x + c \tag{40}$$

and

$$\int \cos x \, dx = \sin x + c. \tag{41}$$

EXAMPLE 8.4.1

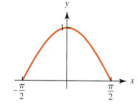

Figure 8.4.1

Find the area under the cosine curve between $-\pi/2$ and $\pi/2$ (shown in Figure 8.4.1).

SOLUTION Using the antiderivative formula (41) and the fundamental theorem of calculus, we get

$$\int_{-\pi/2}^{\pi/2} \cos x \, dx = \sin x \Big|_{-\pi/2}^{\pi/2} = \sin \frac{\pi}{2} - \sin \left(-\frac{\pi}{2} \right) = 2.$$

EXAMPLE 8.4.2

A spring is hanging vertically downward with its top end anchored to a ceiling. In its equilibrium state, the bottom tip of the spring is 4 feet below the ceiling. Suppose it is compressed by pushing it 1 foot upward, then released, and its velocity at any time t (in seconds) is given by

$$v = \sin t,$$

with the positive direction taken downward. Find a formula expressing the distance from the ceiling to the bottom of the spring as a function of time.

SOLUTION Figure 8.4.2 shows the spring in two positions: at equilibrium and at the moment of being released. If we let y be the distance of the bottom of the spring from the ceiling, the given information says that $dy/dt = \sin t$. Therefore,

$$y = \int \sin t \, dt = -\cos t + c. \tag{42}$$

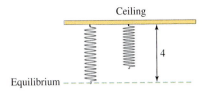

Figure 8.4.2

To find the constant c, we use the fact that $y = 3$ at the moment of release. Setting $t = 0$ and $y = 3$ in (42) and solving for c give $c = 4$, and we get the solution

$$y = 4 - \cos t.$$

We can use formulas (40) and (41) together with any of our known integration methods, such as substitution or integration by parts.

EXAMPLE 8.4.3

Find

$$\text{(i) } \int \sin^3 t \, \cos t \, dt, \quad \text{(ii) } \int t \cos(t^2 + 1) \, dt, \quad \text{(iii) } \int_0^{\pi/2} t \cos t \, dt.$$

SOLUTION (i) Using the substitution $u = \sin t$, so that $du/dt = \cos t$, we get

$$\int \sin^3 t \, \cos t \, dt = \int u^3 \, du = \frac{u^4}{4} + c = \frac{\sin^4 t}{4} + c.$$

(ii) Using the substitution $u = t^2 + 1$, we have $du = 2t \, dt$. Therefore,

$$\int t \cos(t^2 + 1) \, dt = \frac{1}{2} \int \cos u \, du = \frac{1}{2} \sin u + c = \frac{1}{2} \sin(t^2 + 1) + c.$$

(iii) We use integration by parts, with $u = t$ and $dv/dt = \cos t$. Then, $du/dt = 1$ and $v = \sin t$, so that

$$\int_0^{\pi/2} t \cos t \, dt = t \sin t \Big|_0^{\pi/2} - \int_0^{\pi/2} \sin t \, dt$$

$$= \frac{\pi}{2} + \cos t \Big|_0^{\pi/2} = \frac{\pi}{2} - 1.$$

EXAMPLE 8.4.4

Find $\int \tan x \, dx$.

SOLUTION Writing

$$\int \tan x \, dx = \int \frac{\sin x}{\cos x} \, dx,$$

we make the substitution $u = \cos x$, so that $du/dx = -\sin x$. Therefore,

$$\int \tan x \, dx = -\int \frac{1}{u} \, du = -\ln |u| + c = -\ln |\cos x| + c.$$

We obtain another way of expressing the antiderivative by observing that

$$-\ln |\cos x| = \ln \left(\frac{1}{|\cos x|} \right) = \ln |\sec x|,$$

so that

$$\int \tan x \, dx = \ln |\sec x| + c. \tag{43}$$

Similarly (see the exercises),

$$\int \cot x \, dx = -\ln |\csc x| + c. \tag{44}$$

EXAMPLE 8.4.5

Find $\int \sec x \, dx$.

SOLUTION We first change the form of the integrand by multiplying and dividing by the same expression:

$$\int \sec x \, dx = \int \sec x \, \frac{\sec x + \tan x}{\sec x + \tan x} \, dx = \int \frac{\sec^2 x + \sec x \tan x}{\sec x + \tan x} \, dx.$$

Although we seem to have made the problem more complicated, we have actually simplified it. Letting $u = \tan x + \sec x$ and $du = (\sec^2 x + \sec x \tan x) \, dx$, we obtain

$$\int \sec x \, dx = \int \frac{du}{u} = \ln |u| + c.$$

Therefore,

$$\int \sec x \, dx = \ln |\tan x + \sec x| + c. \tag{45}$$

Similarly (see the exercises),

$$\int \csc x \, dx = -\ln|\cot x + \csc x| + c. \tag{46}$$

EXAMPLE 8.4.6

Find $\displaystyle\int e^x \sin x \, dx$.

SOLUTION   We will try integration by parts, with $u = \sin x$ and $v' = e^x$. Then $u' = \cos x$ and $v = e^x$, and we obtain

$$\int e^x \sin x \, dx = e^x \sin x - \int e^x \cos x \, dx.$$

We next try integration by parts again, this time with $u = \cos x$ and $v' = e^x$, so that $u' = -\sin x$ and $v = e^x$, with the following result:

$$\int e^x \sin x \, dx = e^x \sin x - \left(e^x \cos x + \int e^x \sin x \, dx \right)$$

$$= e^x \sin x - e^x \cos x - \int e^x \sin x \, dx.$$

We now have the unknown integral on both sides of the equation, but with opposite signs. Adding it to both sides gives

$$2 \int e^x \sin x \, dx = e^x \sin x - e^x \cos x.$$

Therefore, dividing by 2 and adding an arbitrary constant gives

$$\int e^x \sin x \, dx = \frac{1}{2}(e^x \sin x - e^x \cos x) + c.$$

We can sometimes make use of trigonometric identities to help simplify an integral.

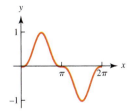

Figure 8.4.3

EXAMPLE 8.4.7

Find the area between the x-axis and the graph of $y = \sin^3 x$ from 0 to 2π.

SOLUTION   Since $\sin^3 x$ has the same sign as $\sin x$, it is positive on $(0, \pi)$ and negative on $(\pi, 2\pi)$. (The graph is shown in Figure 8.4.3.) Therefore, the area between the graph and the x-axis is given by

$$\int_0^\pi \sin^3 x\, dx - \int_\pi^{2\pi} \sin^3 x\, dx.$$

To find $\int \sin^3 x\, dx$, we use the identity $\sin^2 x + \cos^2 x = 1$, as follows:

$$\begin{aligned}
\int \sin^3 x\, dx &= \int \sin^2 x\, \sin x\, dx \\
&= \int (1 - \cos^2 x)\, \sin x\, dx \\
&= \int \sin x\, dx - \int \cos^2 x\, \sin x\, dx \\
&= -\cos x + \int u^2\, du \quad (\text{with } u = \cos x,\ du = -\sin x\, dx) \\
&= -\cos x + \frac{u^3}{3} + c = -\cos x + \frac{\cos^3 x}{3} + c.
\end{aligned}$$

(*Exercise:* Verify that this is indeed an antiderivative of $\sin^3 x$ by taking its derivative. That requires another use of the same trigonometric identity.) Therefore,

$$\int_0^\pi \sin^3 x\, dx = \left(-\cos x + \frac{\cos^3 x}{3} \right)\Bigg|_0^\pi = \frac{4}{3}$$

and

$$\int_\pi^{2\pi} \sin^3 x\, dx = \left(-\cos x + \frac{\cos^3 x}{3} \right)\Bigg|_\pi^{2\pi} = -\frac{4}{3}.$$

We conclude that the area between the graph and the x-axis equals $\frac{8}{3}$.

EXAMPLE 8.4.8

Find $\displaystyle\int_0^\pi \sin^2 x\, dx$.

SOLUTION   This time we use the addition formula for the cosine in the following form (see Exercise 38 of Section 8.2):

$$\cos(2x) = \cos^2 x - \sin^2 x = (1 - \sin^2 x) - \sin^2 x = 1 - 2\sin^2 x.$$

Therefore, $\sin^2 x = \frac{1}{2}[1 - \cos(2x)]$, and

$$\int \sin^2 x \, dx = \frac{1}{2} \int 1 \, dx - \frac{1}{2} \int \cos(2x) \, dx$$

$$= \frac{1}{2}x - \frac{1}{4}\sin(2x) + c \quad \text{(using } u = 2x, du = 2\,dx\text{)}.$$

Thus, we have

$$\int_0^\pi \sin^2 x \, dx = \left[\frac{1}{2}x - \frac{1}{4}\sin(2x)\right]\Bigg|_0^\pi = \frac{\pi}{2}.$$

Practice Exercises 8.4

1. Find $\int \frac{\sin t}{\cos^2 t} \, dt$.

2. Find the area between the graphs of $y = \sin x$ and $y = \sin(2x)$ from zero to $\pi/2$ (shown in Figure 8.4.4).

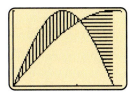

Figure 8.4.4

Exercises 8.4

In Exercises 1–12, find the indefinite integral.

1. $\int \sin(3x) \, dx$

2. $\int \sin x \, \cos x \, dx$

3. $\int \sec^2(x + 1) \, dx$

4. $\int \tan x \, \sec^2 x \, dx$

5. $\int t \, \sin t \, dt$

6. $\int t \, \sin(t^2) \, dt$

7. $\int t^2 \, \cos t \, dt$

8. $\int e^{-x} \cos(2x) \, dx$

9. $\int \sin(2x) \, \cos x \, dx$

10. $\int \ln(\sin x) \, \cot x \, dx$

11. $\int \frac{\cot^2 x}{\sin^2 x} \, dx$

12. $\int \frac{\sin^2 t}{\cos^2 t} \, dt$

13. Find the area under the sine curve from zero to π.

14. Find the area between the graphs of $y = \sin x$ and $y = \cos x$ from zero to $\pi/4$.

15. Find the area between the graph of $y = x \sin x$ and the x-axis from zero to π.

16. An architect with a passion for mathematics designs a courtyard in the center of an office complex in the shape of a curvilinear triangle. His sketch shows the base of the triangle (the south side of the courtyard) to be a straight line of length $\pi/2$ inches, placed along the interval $[0, \pi/2]$ on the x-axis. The other boundaries lie along the graphs of $y = \sin x$ on the left and $y = \cos x$ on the right, with their intersection being the northern apex of the triangle. The scale of the drawing is 1 inch per 100 feet.
   (a) Sketch the shape of the courtyard.
   (b) What is its area?

Find the following definite integrals:

17. $\int_{-\pi/2}^{\pi/2} \sin(3x) \, dx$

18. $\int_{-\pi/2}^{\pi/2} \cos(2x) \, dx$

19. $\int_0^1 t \, \cos(\pi t^2) \, dt$

20. $\int_{-\pi/2}^{\pi/2} \sin^2 x \, \cos x \, dx$

21. $\int_{-\pi/4}^{\pi/4} \frac{1}{\cos^2 x} \, dx$

22. $\int_{-\pi/4}^{\pi/4} \frac{\sin t}{\cos^2 t} \, dt$

23. $\int_0^{\pi/4} \tan x \, dx$

24. $\int_0^\pi x^2 \cos x \, dx$

25. $\int_0^{\pi/3} \sec^3 x \tan x \, dx$ **26.** $\int_{-\pi/2}^{\pi/2} \cos^2 x \, dx$

27. Referring to Example 8.3.8 of Section 8.3, find the average number of predators per year in the region over the 10-year period, $0 \le t \le 10$.

28. The water level in a channel fluctuates with the tide according to the function

$$h = 50 + 6 \sin\left(\frac{\pi t}{12}\right) + 4 \cos\left(\frac{\pi t}{12}\right),$$

where h is the water depth in feet and t is the time in hours after midnight. Find the average depth of the water
(a) during the period from midnight to noon, and
(b) over a 24-hour period.

29. The birth rate of baby robins that hatched in a private aviary over a 2-year period followed the model $R(y) = \sec(1 - y)$, where R is in hundreds of birds per year and $0 \le y \le 2$.
(a) The lowest birth rate coincided with a viral infection. When did that occur?
(b) What is the total number of birds that hatched during the 2 years?

30. By reviewing historical records, biologists conjectured that the rate at which a certain rare species of wild cat was being sighted was modeled by

$$C_1(y) = 0.3 \sin\left(\frac{\pi y}{2}\right) + 0.4 \cos\left(\frac{\pi y}{2}\right), \quad 0 \le y \le 1,$$

where y is the time in years and C_1 is in hundreds of sightings per year. However, sightings were fewer throughout all of last year and these fit the model

$$C_2(y) = 1.1 \sin^2\left(\frac{\pi y}{2}\right) \cos\left(\frac{\pi y}{2}\right).$$

What is the difference between the total number of conjectured sightings and the total number of actual sightings?

In Exercises 31–34, evaluate the given integral in two ways: first, by using the fundamental theorem of calculus and, second, by using the numerical integration function of your calculator. Compare the answers.

31. $\int_0^{\pi/2} \frac{\cos t}{1 + \sin t} \, dt$ **32.** $\int_0^{\pi/4} (x + \sec^2 x) \, dx$

33. $\int_0^{\pi} x \sin x \, dx$ **34.** $\int_0^{\pi} \sin t \, \cos^3 t \, dt$

35. Physicians are studying the change in the level of a particular chemical in the blood, and they observe a rather unusual case in their study group. During the first 4 weeks, one subject's level changed at a rate given by $y' = e^x \cos(e^x)$, where x is the time in weeks and y' is in milligrams per week.

(a) Use substitution to find the indefinite integral that represents the level of chemical in the subject's blood at any given time x.
(b) At the start of the study, the subject's chemical level was at 4 milligrams. Solve this initial value problem.
(c) Graph this function in a $[0, 4] \times [0, 5]$ window to see why the subject caught the attention of the doctors. Describe the behavior of the function.

36. Find the area under the graph of $y = \sec^3 x$ for $0 \le x \le \pi/4$.

37. Use a calculator to find the intersection of the graphs of $y = x$ and $y = \cos x$ over the interval $(0, \pi/2)$. Then find the area between the graphs from zero to $\pi/2$. (See Figure 8.4.5.)

38. Find the area between the graphs of $y = \sin x$ and $y = \sin(3x)$ for zero to $\pi/3$. (See Figure 8.4.6.)

39. Find the area above the graph of $y = \sec x$ and below the line $y = 3$. (See Figure 8.4.7.)

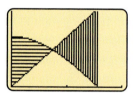

Figure 8.4.5

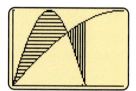

Figure 8.4.6

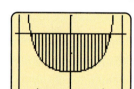

Figure 8.4.7

40. Derive formula (44) for $\int \cot x \, dx$.

41. Verify formula (46) for $\int \csc x \, dx$.

42. Use the identity $\cos^2 x = 1 - \sin^2 x$ to find $\int \cos^3 x \, dx$.

43. Show that for any integer $k \ne 0$, $\int_0^{\pi} \cos(kt) \, dt = 0$.

44. **(a)** Use the addition formula for the cosine to show that if m and n are integers with $m \ne \pm n$,

$$\sin(mt) \sin(nt) = \frac{1}{2}\{\cos[(m - n)t] - \cos[(m + n)t]\}.$$

(b) Use this identity and the preceding exercise to show that if m and n are integers with $m \ne \pm n$, $\int_0^{\pi} \sin(mt) \sin(nt) \, dt = 0$.

45. Show that for any integer $k \neq 0$, $\int_0^\pi \sin^2(kt)\,dt = \pi/2$. (See Example 8.4.8.)

46. Radio and television signals have waveforms that are combinations of functions of the form $\cos(kt)$ and $\sin(kt)$. A simple example is

$$f(t) = a_1 \sin t + a_2 \sin(2t) + a_3 \sin(3t),$$

where a_1, a_2, and a_3 are constants. In processing the signals, it is necessary to determine these coefficients from the function values. That can be done by integration, as follows:
(a) Using the previous two exercises, show that $a_1 = (2/\pi) \int_0^\pi f(t) \sin t\, dt$.
(b) Find similar formulas for a_2 and a_3.

47. Suppose a signal waveform is given by a function of the form $f(t) = c_1 \sin(2t) + c_2 \sin(4t) + c_3 \sin(6t)$, and

$$\int_0^\pi f(t) \sin(2t)\,dt = 1, \quad \int_0^\pi f(t) \sin(4t)\,dt = -1,$$

$$\int_0^\pi f(t) \sin(6t)\,dt = 2.$$

Find c_1, c_2, and c_3.

Solve the following initial value problems:

48. $y' = y^2 \sin t$, $y(0) = 1$ **49.** $y' = \dfrac{\cos t}{y}$, $y(0) = 2$

50. $y' = \dfrac{y}{\cos^2 t}$, $y(0) = 1$ **51.** $y' = \dfrac{\cos t}{\cos y}$, $y(0) = 2\pi$

52. For the spring of Example 8.4.2, show that y satisfies the differential equation $y'' + y = 4$. (This is an example of a **second-order** differential equation.)

53. Second-order differential equations of the form $y'' + a^2 y = 0$, where a is a constant, play an important role in analyzing electric circuits, mechanical vibrations, and many other oscillatory phenomena. Show that for any

constants M and N, the function

$$y = M \sin(at) + N \cos(at)$$

is a solution of this differential equation.

54. Second-order initial value problems involve a second-order differential equation and two specified values, one for y and one for y'. Solve the following second-order initial value problems:
(a) $y'' + y = 0$, $y(0) = 2$, $y'(0) = -1$
(b) $y'' + 4y = 0$, $y(0) = 1$, $y'(0) = -2$

In the following two exercises, we will use the function $\tan^{-1} x$ *(introduced in Section 8.2) to evaluate* $\int (1 + x^2)^{-1}\,dx$ *and approximate* π.

55. **(a)** Suppose $y = \tan^{-1} x$ with $-\pi/2 < y < \pi/2$. By definition of the inverse tangent, this is equivalent to $x = \tan y$. Applying implicit differentiation to this last equation, show that

$$\frac{dy}{dx} = \frac{1}{\sec^2 y}. \tag{47}$$

(b) Verify the identity $\sec^2 y = 1 + \tan^2 y$ (using $\tan y = \sin y / \cos y$). Use it to change (47) to the form

$$\frac{dy}{dx} = \frac{1}{1 + \tan^2 y} = \frac{1}{1 + x^2}.$$

(c) Find $\displaystyle \int \frac{1}{1 + x^2}\,dx$.

(d) Show that $\displaystyle \int_0^1 \frac{1}{1 + x^2}\,dx = \pi/4$.

56. Verify that $(\pi/4) = \int_0^1 1/(1 + x^2)\,dx$, as follows:
(a) Use the midpoint rule with $n = 4$ to approximate the integral and compare it to the value of $\pi/4$ given by your calculator.
(b) Use the numerical integration function on your calculator to get a better approximation of the integral.

Solutions to practice exercises 8.4

1. Using the substitution $u = \cos t$, with $du/dt = -\sin t$, we get

$$\int \frac{\sin t}{\cos^2 t}\,dt = -\int \frac{1}{u^2}\,du = \frac{1}{u} + c = \frac{1}{\cos t} + c.$$

2. To find the point of intersection, we set $\sin x = \sin(2x) = 2 \sin x \cos x$. If $0 < x \leq \pi/2$, we can divide both sides by $\sin x > 0$ (since it is not zero), and we get $\cos x = \frac{1}{2}$. Therefore, $x = \pi/3$. By testing a point in each of these intervals, we see that $\sin(2x) > \sin x$ on $(0, \pi/3)$, whereas

$\sin x > \sin(2x)$ on $(\pi/3, \pi/2)$. Therefore, the area between the graphs is given by

$$\int_0^{\pi/3} [\sin(2x) - \sin x]\,dx + \int_{\pi/3}^{\pi/2} [\sin x - \sin(2x)]\,dx$$

$$= \left[-\frac{\cos(2x)}{2} + \cos x \right]\Big|_0^{\pi/3} + \left[-\cos x + \frac{\cos(2x)}{2} \right]\Big|_{\pi/3}^{\pi/2}$$

$$= \frac{1}{2}.$$

Chapter 8 Summary

- The **radian measure** of an angle is the length of the arc cut off on a circle of radius 1 by the sides of the angle with vertex at the center of the circle. A right angle is $\pi/2$ radians, which corresponds to 90 degrees.

- To define the **trigonometric functions**, we use a circle of radius 1 in the xy-plane with its center at the origin. If θ is an angle with its vertex at the origin, initial side along the positive x-axis, and terminal side intersecting the unit circle at the point (x, y), then we define

$$\sin \theta = y, \quad \cos \theta = x, \quad \tan \theta = \frac{y}{x} = \frac{\sin \theta}{\cos \theta}.$$

  It follows from the Pythagorean theorem that

$$\sin^2 \theta + \cos^2 \theta = 1.$$

  The functions $\sin \theta$ and $\cos \theta$ are continuous for all θ, while $\tan \theta$ is continuous at every θ for which it is defined. They are all periodic with period 2π, that is,[1]

$$\sin(2\pi + \theta) = \sin \theta, \quad \cos(2\pi + \theta) = \cos \theta, \quad \tan(2\pi + \theta) = \tan \theta.$$

- The derivatives are given by the formulas

$$\frac{d}{dx} \sin x = \cos x, \quad \frac{d}{dx} \cos x = -\sin x, \quad \frac{d}{dx} \tan x = \frac{1}{\cos^2 x},$$

  which, when reversed, give the integral formulas:

$$\int \cos x \, dx = \sin x + c, \quad \int \sin x \, dx = -\cos x + c, \quad \int \frac{dx}{\cos^2 x} = \tan x + c.$$

- The reciprocal trigonometric functions and their derivatives are

$$\sec x = \frac{1}{\cos x}, \quad \csc x = \frac{1}{\sin x}, \quad \cot x = \frac{1}{\tan x}.$$

$$\frac{d}{dx} \sec x = \sec x \tan x, \quad \frac{d}{dx} \csc x = -\csc x \cot x, \quad \frac{d}{dx} \cot x = -\csc^2 x.$$

Chapter 8 Review Questions

- Define the sine, cosine, and tangent of an acute angle of a right triangle.

- Write the formulas for $\cos(u \pm v)$ and $\sin(u \pm v)$.

---

[1] Actually, $\tan \theta$ has period π

- There are infinitely many values of t for which $\sin t = 0$. What are they? Is $\tan t$ defined for these values of t?

- Find all the zeros of the function $\cos t$.

- What are the maximum and minimum values of the functions $\sin t$ and $\cos t$?

- In the interval $(-\pi, \pi)$ find where each of the functions $\sin t$ and $\cos t$ is increasing or decreasing.

- Is the sum of two periodic functions periodic?

- Draw the graph of the function $\sin(2t)$. What is its period?

- Draw the graph of the function $\sin(t/2)$. What is its period?

Chapter 8 Review Exercises

1. Convert to radians:
 (a) 6 degrees   (b) 150 degrees
 (c) 345 degrees

2. Convert to degrees:
 (a) $2\pi/3$ radians   (b) $\pi/9$ radians
 (c) $7\pi/12$ radians

3. Draw each of the following angles (in radian measure) inscribed in a circle, with its initial side along the positive horizontal axis:
 (a) $7\pi/3$ (b) $-5\pi/4$

4. Find the radian measures of the following angles:

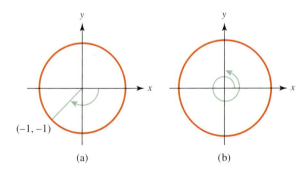

 (a)   (b)

5. For the triangle shown in Figure 8.R.1, find $\sin\theta$, $\cos\theta$, and $\tan\theta$.

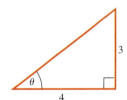

Figure 8.R.1

6. Without using a calculator, find $\sin\theta$, $\cos\theta$, and $\tan\theta$ for θ equal to
 (a) $5\pi/3$ (b) $-5\pi/6$
 (c) $13\pi/12$ (d) $3\pi/8$

7. Without using a calculator, find
 (a) $\sec(\pi/6)$ (b) $\csc(5\pi/4)$
 (c) $\cot(3\pi/2)$ (d) $\sec(2\pi/3)$

8. Without using a calculator, find
 (a) $\sin^{-1}(1/2)$ (b) $\cos^{-1}(-1/2)$
 (c) $\tan^{-1}(\sqrt{3})$ (d) $\sin^{-1}(-1/\sqrt{2})$

9. Show that $1 + \cot^2 t = \csc^2 t$.

10. What is the domain of $\cot x$?

11. Show that $\cot(u+v) = \dfrac{\cot u \; \cot v - 1}{\cot u + \cot v}$.

In Exercises 12–15, find dy/dx.

12. $y = \sin^3(2x)$ **13.** $y = x\tan x$

14. $y = \ln(\sec x)$

15. $y = \dfrac{1}{1 + \cos^2 x}$

16. Find the equation of the line tangent to the graph of $y = \tan x$ at $x = \pi/3$.

17. Find $f''(x)$ if $f(x) = e^x \sin x + e^x \cos x$.

18. Find all critical points of the function $f(x) = \sin x \cos x$ on the interval $(0, 2\pi)$ and determine which are local maxima and which are local minima.

19. Find the inflection points, if any, of the function $f(x) = \sin(2x) - 4\sin x$ on the interval $(0, \pi/2)$.

20. Find where the function $f(x) = x - \cos x$ is increasing and where it is decreasing over the interval $(0, 2\pi)$.

21. Sketch the graph of $y = 2\sin x - x$ over the interval $(-\pi, \pi)$. State where the graph is increasing and where it is decreasing; discuss its concavity; list any vertical asymptotes, local extreme points, and inflection points.

22. Sketch the graph of $y = x - \tan x$ over the interval $(-\pi/2, \pi/2)$. State where the graph is increasing and where it is decreasing; discuss its concavity; list any vertical asymptotes, local extreme points, and inflection points.

23. A spotlight (at point P of Figure 8.R.2) is rotating counterclockwise at a constant rate, making one full rotation every minute. Show that the velocity of the point of light on the wall is at a minimum when it passes point A.

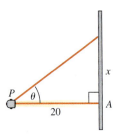

Figure 8.R.2

24. Solve the system of "predator-prey" differential equations

$$\frac{dP}{dt} = \frac{1}{4}(H - 200) \quad \text{and} \quad \frac{dH}{dt} = -\frac{1}{9}(P - 40)$$

with initial conditions $P(0) = 45$ and $H(0) = 190$.

In Exercises 25–27, find the indefinite integral.

25. $\displaystyle\int \sin(3x)\cos(3x)\,dx$

26. $\displaystyle\int \sin(2x)\cos(3x)\,dx$

27. $\displaystyle\int x\sec^2 x\,dx$

In Exercises 28–30, find the definite integral.

28. $\displaystyle\int_0^1 x\sin(2\pi x)\,dx$

29. $\displaystyle\int_0^{\pi/2} \frac{\cos t}{1 + \sin t}\,dt$

30. $\displaystyle\int_0^{\pi/2} \tan\theta\,d\theta$

31. Find the area under the graph of $y = \cos^3 x$ from $-\pi/2$ to $\pi/2$.

32. Without using a calculator, find the area between the graphs of $y = \sin x$ and $y = \cos x$ from zero to $\pi/2$.

33. Without using a calculator, find the area above the graph of $y = \sec x$ and below the line $y = 2$ over the interval $(-\pi/3, \pi/3)$.

34. Show that if m and n are integers with $m \neq \pm n$, $\int_0^\pi \cos(mt)\cos(nt)\,dt = 0$.

35. Find $\displaystyle\int_0^\pi \cos^2 t\,dt$.

In Exercises 36–39, find the limits. Hint: Use formulas (37) and (38).

36. $\displaystyle\lim_{h\to 0} \frac{\tan h}{h}$

37. $\displaystyle\lim_{t\to 0} \frac{\sin(2t)}{t}$

38. $\displaystyle\lim_{x\to 0} \frac{\sin(x/3)}{x}$

39. $\displaystyle\lim_{h\to 0} \frac{\cos h - 1}{\sin h}$

Chapter 8 Practice Exam

1. If one of the acute angles of a right triangle is $2\pi/5$ radians, what is the other?

2. What is the length of the arc cut off by an angle of $5\pi/6$ in a circle of radius 4?

3. If $3\pi/2 < \theta < 2\pi$ and $\cos\theta = \frac{2}{3}$, find $\sin\theta$, $\tan\theta$, and $\sec\theta$.

4. Find $\sin\left(\dfrac{5\pi}{6} - \dfrac{\pi}{4}\right)$.

5. Find $\cos^{-1}(-\sqrt{3}/2)$.

6. Identify the graphs of $\sin x$, $\cos x$, $\tan x$, and $\sec x$, $-2\pi \le x \le 2\pi$, from the four graphs shown in Figure 8.E.1.

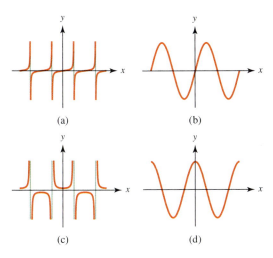

(a)

(b)

(c)

(d)

Figure 8.E.1

7. Find $d/dt\ \tan(\cos t)$.

8. Find the equation of the line tangent to the graph of $y = \sin(3x) + \cos(2x)$ at $x = \pi/6$.

9. Find the points where the function $f(x) = x - \sin(2x)$ achieves a global maximum and minimum on the interval $[0, 2\pi]$.

10. Sketch the graph of $y = \tan x - 2x$ over the interval $(-\pi/2, \pi/2)$. State where the graph is increasing and where it is decreasing; discuss its concavity; list any vertical asymptotes, local extreme points, and inflection points.

11. Solve the system of differential equations

$$\frac{du}{dt} = v \quad \text{and} \quad \frac{dv}{dt} = -\frac{1}{4}u$$

with initial conditions $u(0) = 1$, and $v(0) = -2$.

12. Find the area under the graph of $y = \tan x$ from zero to $\pi/3$.

13. The predicted size of the fish population in Bass Lake during the next 20 years is given by the function

$$P(t) = 200 + 30 \sin\left(\frac{\pi t}{10}\right) - 12 \cos\left(\frac{\pi t}{10}\right),$$

where t is the time in years. Assuming this model is correct, what is the average fish population over the next 5 years?

14. Find $\displaystyle\int e^{2x} \cos x\, dx$.

15. Evaluate $\displaystyle\lim_{h \to 0} \frac{\sin^2 h}{h}$.

Chapter 8 Projects

A predator-prey model The predator-prey model that we studied in Section 8.3 is very simplified. A more sophisticated model, proposed by J. Maynard Smith,[2] is given by the system of differential equations

$$\frac{dP}{dt} = c(H - H_e) \tag{48}$$

$$\frac{dH}{dt} = -a(H - H_e) - b(P - P_e), \tag{49}$$

where a, b, and c are positive constants. This model takes account of the intrinsic growth (if $H < H_e$) or decline (if $H > H_e$) of the prey population, independent of the change in the number of predators. In what follows, we discuss the solutions of this system of differential equations.

(a) Let $u = H - H_e$ and $v = P - P_e$, which changes (48) and (49) to

$$\frac{dv}{dt} = cu. \tag{50}$$

$$\frac{du}{dt} = -au - bv \tag{51}$$

[2]J. Maynard Smith, *Mathematical Ideas in Biology*, New York: Cambridge University Press, 1968–1969, pp. 44–46.

By differentiating both sides of (50) with respect to t, we get $d^2v/dt^2 = c\,(du/dt)$. Show that by combining this with (51) and (50), we can eliminate u and get the second-order differential equation

$$\frac{d^2v}{dt^2} + a\,\frac{dv}{dt} + bcv = 0. \tag{52}$$

(b) To solve Eq. (52), we consider the quadratic equation

$$x^2 + ax + bc = 0 \tag{53}$$

with the same coefficients as (52). Suppose that $a^2 - 4bc > 0$; show the following:

1. Equation (53) has two real roots, r_1 and r_2, which are both negative.

2. Both $v = e^{r_1 t}$ and $v = e^{r_2 t}$ are solutions of (52).

3. More generally, if M and N are any constants, a solution of (52) is given by $v = Me^{r_1 t} + Ne^{r_2 t}$.

4. For any choices of M and N, $\lim_{t \to \infty} v = 0$. In other words, the population of predators moves toward its equilibrium size over time. Show that the prey population also moves toward equilibrium.

(c) Suppose that $a^2 - 4bc < 0$. Show the following:

1. Equation (53) has no real roots.

2. If $q = \frac{1}{2}\sqrt{4bc - a^2}$, then both

$$v = e^{-at/2}\cos(qt) \quad \text{and} \quad v = e^{-at/2}\sin(qt)$$

are solutions of (52).

3. More generally, if M and N are any constants, a solution of (52) is given by

$$v = e^{-at/2}[M\cos(qt) + N\sin(qt)].$$

4. In this case, both populations oscillate around their equilibrium sizes, but both tend toward equilibrium as $t \to \infty$.

(d) In J. Maynard Smith's formulation of this model,

$$a = \frac{R_H - 1}{R_P}, \quad b = \frac{KH_e^*}{R_P}, \quad c = \frac{(R_H - 1)(R_P - 1)}{KH_e^*},$$

where R_P and R_H are the maximum reproductive rates of the predator and prey species, respectively, H_e^* is the equilibrium population of the prey population in the absence of predators, and K is a constant of proportionality. Assuming that R_p and R_H are both greater than 1, show that

$$a^2 - 4bc > 0 \iff R_H > (2R_P - 1)^2$$

and

$$a^2 - 4bc < 0 \iff R_H < (2R_P - 1)^2.$$

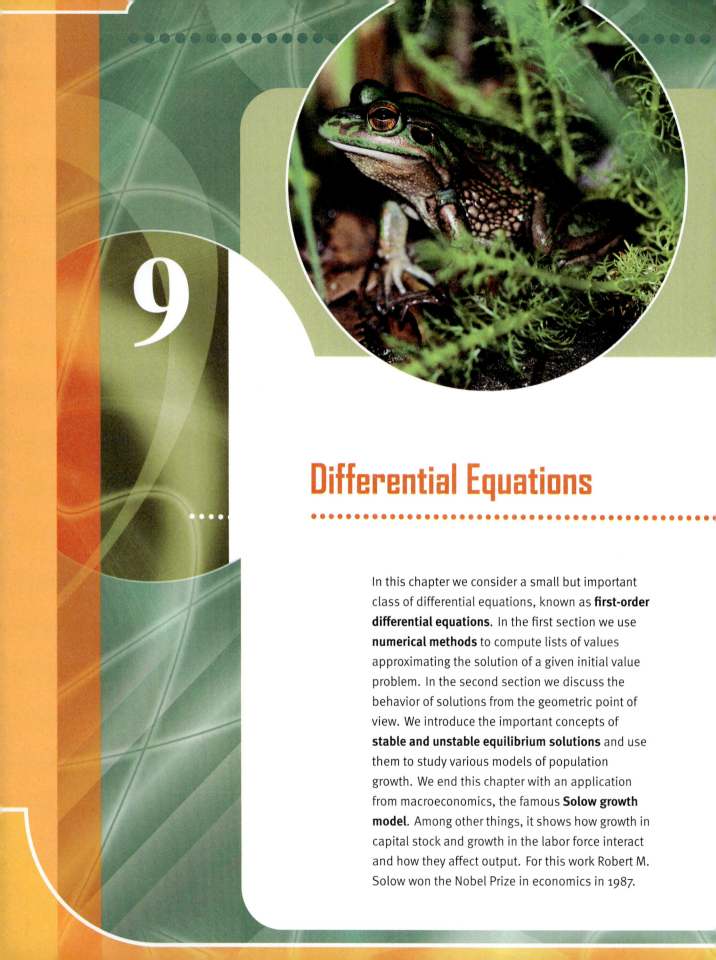

Differential Equations

In this chapter we consider a small but important class of differential equations, known as **first-order differential equations**. In the first section we use **numerical methods** to compute lists of values approximating the solution of a given initial value problem. In the second section we discuss the behavior of solutions from the geometric point of view. We introduce the important concepts of **stable and unstable equilibrium solutions** and use them to study various models of population growth. We end this chapter with an application from macroeconomics, the famous **Solow growth model**. Among other things, it shows how growth in capital stock and growth in the labor force interact and how they affect output. For this work Robert M. Solow won the Nobel Prize in economics in 1987.

In Sections 5.1, 5.2, and 5.3 we solved the simplest type of first-order differential equation, $y' = f(t)$, using indefinite integrals. For example, the solutions of the equation $y' = 2t$ are given by

$$y(t) = \int 2t \, dt = t^2 + c,$$

where c is any number. In Section 6.3 we went a step further. We learned how to solve a first-order differential equation

$$\frac{dy}{dt} = f(t, y),$$

provided we can separate the variables—that is, provided we can write $f(t, y) = h(t)g(y)$. For example if $f(t, y) = (1 - 2t)y^2$, then separating variables gives

$$y^{-2} \, dy = (1 - 2t) \, dt,$$

and by integrating both sides, we get

$$-y^{-1} = t - t^2 + c,$$

which we can solve for y to obtain $y(t) = 1/(t^2 - t - c)$.

    Not every differential equation is separable, however, and although there are other, more advanced analytic techniques, there are many important equations for which they fail to provide a solution. In those cases, we have to use approximation methods, just as we did earlier in evaluating definite integrals for which an antiderivative could not be found. The simplest such method for differential equations, which we shall discuss in this section, is due to the great Swiss mathematician, Leonhard Euler (1707–1783), and is based on linear approximation (which we studied in Section 3.4) and on the concept of direction fields.

Direction fields

We will introduce this concept by using a concrete example. Suppose we know that the initial value problem

$$\frac{dy}{dt} = y - t, \quad y(0) = \frac{3}{4} \tag{1}$$

has a solution, but we do not know how to find it with the methods we have learned (observe that the equation is not separable). In that case, our recourse is to extract enough information from the differential equation itself to construct the graph of its solution, called a **solution curve**.

    How can we accomplish this? Simply by reading the equation carefully. The differential equation $y' = y - t$ tells us that *the slope of the solution curve at any point*

(t, y) *is equal to* $y - t$. For example the slope at the point $(0, \frac{3}{4})$ is $\frac{3}{4} - 0 = \frac{3}{4}$, and the slope at the point $(\frac{1}{4}, \frac{3}{4})$ is $\frac{3}{4} - \frac{1}{4} = \frac{1}{2}$. To visualize this key information, we draw a short line segment, or arrow, with slope $y - t$ through the point (t, y). This segment is tangent to the solution curve passing through that point, and the collection of all these segments is called a **direction field** of the differential equation. If we could draw one through every point of the ty-plane, the segments would merge into a family of curves— the solution curves of the differential equation. Of course, it is not possible to draw a segment through *every* point in the plane, but we can draw enough of them to get a pretty good idea of the shapes of the solution curves. Usually, we choose a rectangular grid of points and draw a segment (or arrow) through each point of the grid.

In the present case, we are interested in the solution curve passing through $(0, \frac{3}{4})$, and we choose the grid to include that point. Figures 9.1.1 and 9.1.2 (generated by the computer program Mathematica^{TM}) show the direction field for the rectangular grid of points (t, y) formed by taking t from 0 to 4 and y from -1 to 3, both in steps of size $\frac{1}{4}$. Figure 9.1.1 gives an idea of the shapes of the solution curves in that part of the plane. In Figure 9.1.2 we have included the actual solution curve through the point $(0, \frac{3}{4})$.

In fact, the solution of the differential equation $y' = y - t$ is given by the formula

$$y(t) = 1 - e^t + ce^t + t, \tag{2}$$

where c is any constant. [You should check that (2) does indeed satisfy the differential equation.] By setting $c = \frac{3}{4}$, we obtain the solution to our initial value problem (1).

Figures 9.1.3 and 9.1.4 show the family of solution curves obtained from (2) by setting $c = -1, -0.75, -0.5, \ldots, 2.5$. In Figure 9.1.3 we see the curves by themselves; in Figure 9.1.4 they are superimposed on the direction field.

The preceding discussion and graphs suggest the following geometric method for solving first-order differential equations graphically:

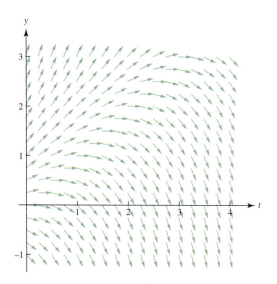

Figure 9.1.1

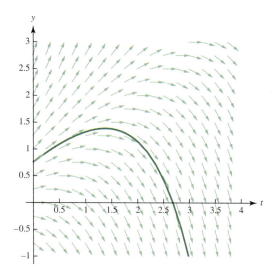

Figure 9.1.2

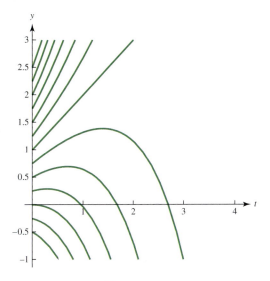

Figure 9.1.3

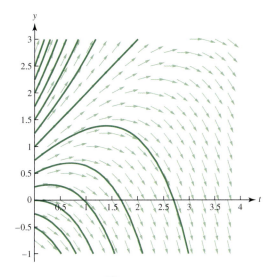

Figure 9.1.4

Method of the Direction Fields

To sketch the solution curve of the initial value problem

$$\frac{dy}{dt} = f(t, y), \quad y(t_0) = y_0,$$

we proceed as follows:

- Choose a rectangular grid of points containing the initial point (t_0, y_0), and through each grid point (t, y) draw a short arrow (called a **direction element**) whose slope equals $f(t, y)$.

- Next, starting from the point (t_0, y_0), draw a curve that follows (fits) the direction elements you meet as you move in the ty-plane. Think of the direction elements as signposts showing you the way across the plane.

The smaller the distance between the grid points, the closer the curve obtained this way will approximate the solution curve.

Euler's method

The geometric procedure we just described for sketching solution curves by following the direction field suggests a numerical method for approximating solutions of first-order initial value problems. Starting from the initial point (t_0, y_0), we move, for a small time interval Δt, along the line of slope $f(t_0, y_0)$ passing through that point. Since the equation of that line is

$$y = y_0 + f(t_0, y_0)(t - t_0),$$

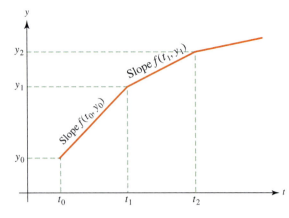

Figure 9.1.5

we reach a new point (t_1, y_1) with $t_1 = t_0 + \Delta t$ and $y_1 = y_0 + f(t_0, y_0)\Delta t$. (See Figure 9.1.5.) If the time interval is *small*, the linear approximation formula (Theorem 3.4.1 of Section 3.4) tells us that $y_1 \approx y(t_1)$. Having reached (t_1, y_1), we compute the slope $f(t_1, y_1)$ of the direction field there.

Heading in this new direction, once again using the time interval Δt, we reach a second point (t_2, y_2), whose coordinates are

$$t_2 = t_1 + \Delta t, \quad y_2 = y_1 + f(t_1, y_1)\Delta t.$$

By repeating this process, we obtain a piecewise linear curve that gives a numerical approximation of the solution. This approximation technique is known as **Euler's method**. Next, we describe it in more detail.

Euler's Method

To approximate the solution of an initial value problem

$$\frac{dy}{dt} = f(t, y), \quad y(t_0) = y_0,$$

where $f(t, y)$, t_0, and y_0 are all given, we proceed as follows:

- We choose a nonzero number Δt to use as a **step size** along the t-axis, and we let

$$t_1 = t_0 + \Delta t, \quad t_2 = t_1 + \Delta t, \quad \ldots, \quad t_n = t_{n-1} + \Delta t.$$

- Using the given function $f(t, y)$, we successively compute

$$y_1 = y_0 + f(t_0, y_0)\Delta t$$
$$y_2 = y_1 + f(t_1, y_1)\Delta t$$
$$\ldots$$
$$y_n = y_{n-1} + f(t_{n-1}, y_{n-1})\Delta t.$$

- We use the numbers $y_1, y_2, \ldots, y_n$ above as the approximations of the actual solution values $y(t_1), y(t_2), \ldots, y(t_n)$.

Remark For good approximations the step size Δt must be small. In real situations it is determined by analyzing the maximum size of the error involved. In the problems encountered here, Δt will either be given or will be computed from given data. Notice that t_0, t_n, and Δt are related by the formulas

$$\Delta t = \frac{t_n - t_0}{n} \quad \text{and} \quad n = \frac{t_n - t_0}{\Delta t}.$$

EXAMPLE 9.1.1

Use Euler's method with step size 0.2 to make a table of approximate values for the solution of the initial value problem

$$\frac{dy}{dt} = y - t, \quad y(0) = \frac{3}{4}$$

over the interval $0 \le t \le 2$.

SOLUTION We are given $f(t, y) = y - t$, $t_0 = 0$, $y_0 = 0.75$, and $\Delta t = 0.2$. Therefore,

$$n = \frac{2 - 0}{0.2} = 10$$

and

$$t_1 = 0.2, \quad t_2 = 0.4, \quad t_3 = 0.6, \quad \dots, \quad t_{10} = 2.$$

The first three approximations are

$$y_1 = y_0 + f(t_0, y_0)\Delta t = 0.75 + (0.75 - 0)(0.2) = 0.9$$
$$y_2 = y_1 + f(t_1, y_1)\Delta t = 0.9 + (0.9 - 0.2)(0.2) = 1.04$$
$$y_3 = y_2 + f(t_2, y_2)\Delta t = 1.04 + (1.04 - 0.4)(0.2) = 1.168.$$

Proceeding in this way, we obtain the following table of approximations to the solution $y(t)$:

$n - 1$	t_{n-1}	y_{n-1}	$f(t_{n-1}, y_{n-1})\Delta t$	$y_n = y_{n-1} + f(t_{n-1}, y_{n-1})\Delta t$
0	0	0.75	0.15	0.9
1	0.2	0.9	0.14	1.04
2	0.4	1.04	0.128	1.168
3	0.6	1.168	0.1136	1.2816
4	0.8	1.2816	0.09632	1.37792
5	1	1.37792	0.07558	1.4535
6	1.2	1.4535	0.0507	1.5042
7	1.4	1.5042	0.02084	1.52504
8	1.6	1.52504	-0.015	1.51004
9	1.8	1.51004	-0.05799	1.45205

Error in Euler's method   The solution to the initial value problem in Example 9.1.1 is given by the formula

$$y(t) = 1 - 0.25e^t + t.$$

(Check it.) Figure 9.1.6 shows the graph of this solution and the 10 points (t_n, y_n) from the table above, obtained using Euler's method.

Observe, that at the beginning the error is very small. However, as we move further away from the initial point $(0, \frac{3}{4})$, the error becomes larger. We can make the error smaller by decreasing the step size. For instance, using Euler's method with $\Delta t = 0.05$, we obtain 40 points (t_n, y_n) that are much closer to the graph of the solution, as shown in Figure 9.1.7. In Figure 9.1.8 we see both the approximate and actual values as computed using an Excel spreadsheet, with $\Delta t = 0.05$ and $n = 20$.

Figure 9.1.6

Figure 9.1.7

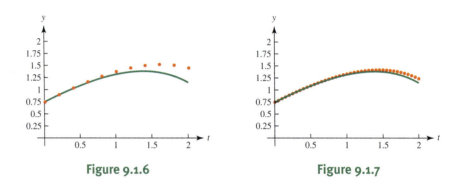

	A	B	C	D	E	F
1	t	Approximate value	Approximate value	Actual value	Error	
2	0	0.75	0.7875	0.75	0	
3	0.05	0.7875	0.824375	0.787182226	0.00031777	
4	0.1	0.824375	0.86059375	0.82370727	0.00066773	
5	0.15	0.86059375	0.896123438	0.859541439	0.00105231	
6	0.2	0.896123438	0.930929609	0.89464931	0.00147413	
7	0.25	0.930929609	0.96497609	0.928993646	0.00193596	
8	0.3	0.96497609	0.998224894	0.962535298	0.00244079	
9	0.35	0.998224894	1.030636139	0.995233113	0.00299178	
10	0.4	1.030636139	1.062167946	1.027043826	0.00359231	
11	0.45	1.062167946	1.092776343	1.057921954	0.00424599	
12	0.5	1.092776343	1.12241516	1.087819682	0.00495666	
13	0.55	1.12241516	1.151035918	1.116686746	0.00572841	
14	0.6	1.151035918	1.178587714	1.1444703	0.00656562	
15	0.65	1.178587714	1.2050171	1.171114793	0.00747292	
16	0.7	1.2050171	1.230267955	1.196561823	0.00845528	
17	0.75	1.230267955	1.254281353	1.220749996	0.00951796	
18	0.8	1.254281353	1.276995421	1.243614768	0.01066659	
19	0.85	1.276995421	1.298345192	1.265088287	0.01190713	
20	0.9	1.298345192	1.318262451	1.285099222	0.01324597	
21	0.95	1.318262451	1.336675574	1.303572585	0.01468987	
22	1	1.336675574		1.320429543	0.01624603	
23						

Euler Method Error

Sheet1 / Sheet2 / Sheet3

Figure 9.1.8

We will not go into a detailed analysis of the error in Euler's method, which you can find in books on differential equations and numerical analysis. However, the general rule is that the approximation error is proportional to the step size.

EXAMPLE 9.1.2

Let $p(t)$ be the U.S. population in millions, where t is the number of years after 2000. Assume that the population is modeled by the logistic differential equation

$$\frac{dp}{dt} = 0.015p(1 - 0.002p),$$

and that in the year 2000 it was about 280 million. Use Euler's method with five steps to estimate the U.S. population in the year 2010.

SOLUTION Here $f(t, p) = 0.015p(1 - 0.002p)$, $t_0 = 0$, $p_0 = 280$, and we are asked to estimate $p(10)$. Since $n = 5$, the step size is $\Delta t = (10 - 0)/5 = 2$, and the t values are

$$t_1 = 2, \quad t_2 = 4, \quad t_3 = 6, \quad t_4 = 8, \quad t_5 = 10.$$

To estimate $p(10)$, we must compute p_5. We have

$$p_1 = p_0 + f(t_0, p_0)\Delta t = 280 + [0.015 \cdot 280(1 - 0.002 \cdot 280)] \cdot 2 \approx 283.7$$
$$p_2 = p_1 + f(t_1, p_1)\Delta t \approx 283.7 + [0.015 \cdot 283.7(1 - 0.002 \cdot 283.7)] \cdot 2 \approx 287.4$$
$$p_3 = p_2 + f(t_2, p_2)\Delta t \approx 287.4 + [0.015 \cdot 287.4(1 - 0.002 \cdot 287.4)] \cdot 2 \approx 291$$
$$p_4 = p_3 + f(t_3, p_3)\Delta t \approx 291 + [0.015 \cdot 291(1 - 0.002 \cdot 291)] \cdot 2 \approx 294.7$$
$$p_5 = p_4 + f(t_4, p_4)\Delta t \approx 294.6 + [0.015 \cdot 294.6(1 - 0.002 \cdot 294.6)] \cdot 2 \approx 298.2.$$

Therefore, the population of the United States in the year 2010 will be approximately equal to 298.2 million.

Remark With the power of modern computing equipment, it is not difficult to implement Euler's method with a very large n and very small Δt, and one can even get reasonably good results with a programmable calculator or a simple spreadsheet. Nevertheless, the method is slow and rudimentary, and more sophisticated techniques—based on refinements of Euler's method—are generally used in practical applications. You can learn more about them from any book on numerical analysis.

Practice Exercises 9.1

● ●

1. Consider the initial value problem $y' = t + y$, $y(1.5) = -0.5$.

(a) Apply Euler's method with step size 0.1 to calculate approximate values of the solution for $1.5 \leq t \leq 2$.

(b) Check that $y(t) = 2e^{t-1.5} - t - 1$ is the solution to the given initial value problem.

(c) Compare the approximate values in (a) with the actual values of the solution in (b).

Exercises 9.1

1. A direction field of the differential equation $y' = y + t$ is shown in Figure 9.1.9. Sketch the graphs of the solutions that satisfy the given initial conditions:
   (a) $y(0) = -1$ **(b)** $y(0) = 0$
   (c) $y(0) = 1$ **(d)** $y(3) = -2$
   Does the direction field indicate that one of these solutions is a linear function? If yes, which one?

Does the direction field indicate that one of these solutions is a quadratic function? If yes, which one?

In Exercises 3–6, match the differential equation with one of the four direction fields given in Figure 9.1.11 Explain your answer.

3. $y' = -x$ 4. $y' = -y$

5. $y' = 1 - y$ 6. $y' = \dfrac{1}{y}$

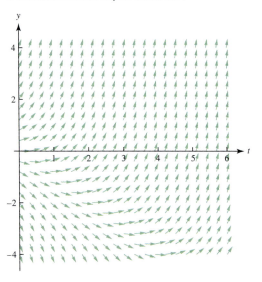

Figure 9.1.9

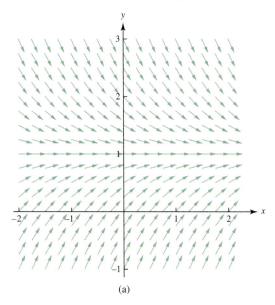

(a)

2. Figure 9.1.10 shows a direction field of the differential equation $y' = x^2 - y$. Sketch the graph of the solutions that satisfy the give initial conditions:
   (a) $y(0) = 2$ **(b)** $y(0) = 1$
   (c) $y(-1) = 0$ **(d)** $y(-1) = -1$

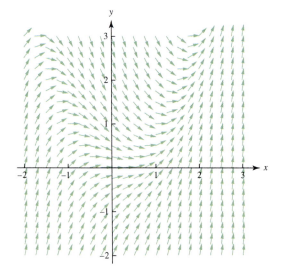

Figure 9.1.10

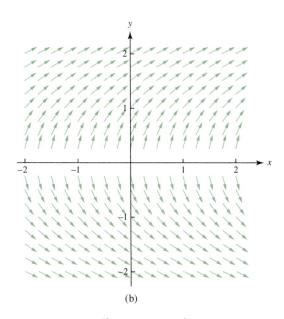

(b)

Figure 9.1.11 a, b

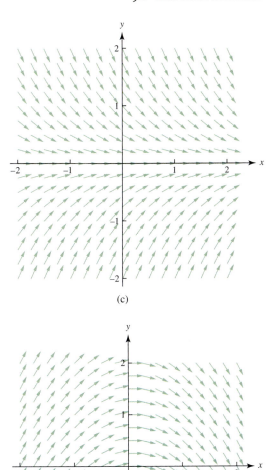

(c)

(d)

Figure 9.1.11 c, d

In Exercises 7–10, draw the direction field for the given differential equation over the rectangle $-2 \leq x \leq 2$, $-2 \leq y \leq 2$, using enough steps to give you an idea of the shape of the solution curves. Then, without solving the equation, sketch four of the solution curves using the direction field as a guide.

7. $y' = x$ **8.** $y' = y$

9. $y' = xy$ **10.** $y' = y^2$

11. School district records show the rate of change of the percentage of students absent throughout the district over a 6-year period. The data are given in the following table, where t is the time in years and y is the absentee percentage. (Note that in this case y' depends only on t and not y.)

(t, y)	$(1, y)$	$(2, y)$	$(3, y)$
$y' = f(t, y)$	0	0.49	0.77
(t, y)	$(4, y)$	$(5, y)$	$(6, y)$
$y' = f(t, y)$	0.97	1.13	1.25

(a) Is the function y increasing or decreasing? Why?
(b) Create a 6×8 rectangular (t, y) grid and sketch a direction element at each integer value of (t, y).
(c) The first year had a 3% rate of total absenteeism. Draw the curve that fits the direction elements.

12. The number of immigrants moving into a city's suburbs the year the city started recording such a statistic was 1.75 thousand. Every 6 months another count was made, and the changes were best described by $dI/dt = -0.4tI$, where t is in years and I in thousands of people.
(a) Taking t from 0 to 3 in steps of 0.5 and I from 0 to 2 in steps of 0.25, create a table of values of dI/dt.
(b) Construct a direction field from the table.
(c) Sketch a solution curve. About how many immigrants moved to the city's suburbs during the third year?

13. Consider the initial value problem $y' = 1 + y$, $y(0) = 0$.
(a) Use Euler's method with $n = 4$ to estimate $y(1)$.
(b) Check your estimate by solving the equation and computing $y(1)$.

14. Consider the initial value problem $y' = 2ty$, $y(0) = 1$.
(a) Use Euler's method with $\Delta t = 0.1$ to estimate $y(0.3)$.
(b) Check your estimate by solving the equation and computing $y(0.3)$.

15. Consider the initial value problem $y' = x/y$, $y(0) = 2$.
(a) Use Euler's method with $\Delta x = 0.25$ to estimate $y(1)$.
(b) Check your estimate by solving the equation and computing $y(1)$.

16. For the initial value problem $y' = 2y - x$, $y(0) = 1$, use Euler's method with $\Delta x = 0.1$ to estimate $y(0.5)$.

17. Assume that the current balance of a mortgage is $120,000, the interest is 7%, compounded continuously, and payments are made continuously at the rate of $14,000 per year. The balance $M(t)$ is modeled by the initial value problem

$$\frac{dM}{dt} = 0.07M - 14{,}000, \quad M(0) = 120{,}000,$$

where t is the time in years. Use Euler's method with $\Delta t = 1$ to estimate the balance of the mortgage 5 years later.

18. Biologists were concerned about the frog population in a polluted marshland. They found that its rate of change is

modeled by the equation

$$\frac{dy}{dt} = -0.1\sqrt{t} - 0.3y,$$

where t is measured in years and y is the size of the frog population, measured as a proportion of the original number. (That is, the number of frogs present when the biologists first arrived is designated as 1.)

(a) What are t_0 and y_0?
(b) Given a step size of 0.25, how many steps are needed to approximate the frog population in the first year?
(c) Use Euler's method to approximate the population in the first year with $\Delta t = 0.25$.
(d) Estimate the percentage of the original frog population that survived the first year.
(e) What is happening to the frogs?

19. Assume that the U.S. population $p(t)$, in billions, is approximately modeled by the logistic differential equation

$$\frac{dp}{dt} = 0.02p(1 - p)$$

and was 0.28 billion in the year 2000. Using Euler's method with $\Delta t = 4$, estimate the U.S. population in the year 2020.

20. The world population $p(t)$, in billions, is approximately modeled by the logistic differential equation

$$\frac{dp}{dt} = 0.03\, p\left(1 - \frac{1}{12}p\right)$$

and was 6 billion in the year 2000. Using Euler's method with $\Delta t = 2$, estimate the world population in the year 2010.

21. A company that produces 1970s-style clothing experienced an increase in sales when a popular television series outfitted all its characters in that type of clothing. Before the series aired, the company's sales rate was $250,000 per year. During the series' 2-year run, the changes in the sales rate were modeled by

$$\frac{dr}{dt} = 3t^{0.2} - 2tr,$$

where r is in millions of dollars per year.
(a) Use Euler's method with eight steps to construct a table estimating the sales rate at the end of each quarter over the 2-year run of the series.
(b) Use the table to estimate the total sales over the 2-year period. (*Hint:* Total sales equal $\int_0^2 r(t)\, dt$.)

22. You are assigned the task of learning how many new pine trees grow in a 100-square-mile forest each year. Since it is not practical to count every new tree in the forest, you divide the forest into small blocks and do your count in a number of them. After several years, you calculate the rate of increase in new trees to be given by

$$\frac{dT}{dt} = 1 + 2t - 0.1T^2,$$

with $T(0) = 2$, where T is in thousands of trees and t in years. You feel that your blocks adequately represent the forest as a whole, so the function can represent the entire forest. However, if you could have counted the whole forest exactly, you might have learned that the actual rate was

$$\frac{dT_a}{dt} = 1.2 + 2.7t - 0.08T_a^2,$$

with $T_a(0) = 2.5$. Using Euler's method with $\Delta t = 1$, estimate the solutions of both equations for $t = 1, 2, \ldots, 5$ and include the approximation error $|T_a(t) - T(t)|$ for each year.

23. Assume that the deer population of a park is governed by the differential equation

$$\frac{dy}{dt} = 0.2y(e^{1-(y/5)} - 1),$$

where y is in hundreds of deers and t is in years. Using Euler's method with 20 steps, estimate the population 20 years later if the initial population is 150.

Solutions to practice exercises 9.1

1. (a) We are given $f(t, y) = t + y$, $t_0 = 1.5$, $y_0 = -0.5$, and $\Delta t = 0.1$. Therefore, the t-values are

$$t_1 = 1.6, \quad t_2 = 1.7, \quad t_3 = 1.8, \quad t_4 = 1.9, \quad t_5 = 2,$$

and the approximate values are:

$$y_1 = y_0 + f(t_0, y_0)\Delta t = -0.5 + (-0.5 + 1.5)(0.1) = -0.4$$
$$y_2 = y_1 + f(t_1, y_1)\Delta t$$
$$= -0.4 + (-0.4 + 1.6)(0.1) = -0.28$$
$$y_3 = y_2 + f(t_2, y_2)\Delta t$$
$$= -0.28 + (-0.28 + 1.7)(0.1) = -0.138$$

$$y_4 = y_3 + f(t_3, y_3)\Delta t$$
$$= -0.138 + (-0.138 + 1.8)(0.1) = 0.0282$$

$$y_5 = y_4 + f(t_4, y_4)\Delta t$$
$$= 0.0282 + (0.0282 + 1.9)(0.1) = 0.22102.$$

(b) Since $y' = 2e^{t-1.5} - 1 - 0$ and $t + y = t + 2e^{t-1.5} - t - 1 = 2e^{t-1.5} - 1$, we see that the given function solves the differential equation $y' = t + y$. Moreover, $y(1.5) = 2e^{1.5-1.5} - 1.5 - 1 = -0.5$. Thus, $y(t)$ is the solution to the given initial value problem.

(c) Using a calculator and rounding to the first five decimal places, we find the actual values of the solution:

$$y(1.6) = -0.38966, \quad y(1.7) = -0.25719,$$
$$y(1.8) = -0.10028, \quad y(1.9) = 0.083649,$$
$$y(2) = 0.29744,$$

and observe that they agree with the approximate values produced by Euler's method to the first decimal place. Also, notice that the error grows with time.

9.2 Graphical Solutions

In this section we shall consider differential equations of the form

$$\frac{dy}{dt} = g(y), \tag{3}$$

where g is a given function. A differential equation of the form (3) is called **autonomous**[1] because the right-hand side $g(y)$ is independent of the variable t. One such example is the logistic equation

$$\frac{dy}{dt} = ry\left(1 - \frac{y}{K}\right),$$

modeling a population of intrinsic growth rate r moving toward its steady-state level K. Although Eq. (3) is separable and the formula

$$\int \frac{dy}{g(y)} = t + c \tag{4}$$

gives the family of its solutions, computing the integral on the left-hand side is difficult even for reasonably simple functions $g(y)$. In this section we shall use a graphical approach to study the solutions of (3). We start with the definition of an equilibrium solution.

Definition 9.2.1

An **equilibrium solution** (or **steady-state solution**) of a differential equation is a constant solution, one for which $dy/dt = 0$. Therefore, the equilibrium solutions of the differential equation

$$\frac{dy}{dt} = g(y)$$

are all the solutions $y_1, y_2, \ldots,$ of the equation

$$0 = g(y).$$

Note The graph of an equilibrium solution is a **horizontal line**.

[1] From the Greek word $\alpha\upsilon\tau\grave{o}\nu o\mu o\varsigma$, pronounced "aftonomous," meaning *independent*.

EXAMPLE 9.2.1

Find the equilibrium solutions of the differential equations

$$\text{(i)} \ \frac{dy}{dt} = -y, \quad \text{(ii)} \ \frac{dy}{dt} = y, \quad \text{(iii)} \ \frac{dy}{dt} = y(2 - y).$$

SOLUTION   (i) In this case, $g(y) = -y$. Solving $0 = -y$, we find that $y = 0$ is the only equilibrium solution.

(ii) Here, $g(y) = y$, and we again find that $y = 0$ is the only equilibrium solution.

(iii) In this case, $g(y) = y(2 - y)$. Solving $0 = y(2 - y)$ gives $y_1 = 0$ and $y_2 = 2$, which are the two equilibrium solutions of the given differential equation.

Next, we shall determine the behavior of the other (nonconstant) solutions, using the equilibrium solutions as a guide. Let us begin by drawing a direction field near the equilibrium solution for each of the differential equations in Example 9.2.1. Such direction fields are shown in Figure 9.2.1. Observe that the direction fields (a) and (b) indicate that the equilibrium solution $y = 0$ divides the ty-plane into two horizontal strips, and each nonconstant solution lies completely inside one of these two strips.

In (a) each nonconstant solution increases if it lies in the strip $y < 0$ (since $dy/dt > 0$ there) and decreases if it lies in the strip $y > 0$ (since $dy/dt < 0$ there). Therefore, all solutions move closer and closer to the equilibrium solution as t gets larger; or, in other words, the equilibrium solution *attracts* all the other solutions.

In (b), on the other hand, each nonconstant solution decreases if it lies in the strip $y < 0$ and increases if it lies in the strip $y > 0$, so that it moves farther and farther away from the equilibrium solution as t gets larger. In other words, the equilibrium solution *repels* all the other solutions.

Finally, in (c) we have both situations. The equilibrium solutions $y = 0$ and $y = 2$ divide the ty-plane into three horizontal strips, and each nonconstant solution lies completely inside one of these three strips. A solution decreases if it lies in the strip $y < 0$, increases if it lies in the strip $0 < y < 2$, and decreases if it lies in the strip $y > 2$. In other words, the equilibrium solution $y_1 = 0$ repels all nearby solutions, while the equilibrium solution $y_2 = 2$ attracts all nearby solutions.

The behavior of the solutions indicated by the direction fields (a), (b), and (c) can be verified by the formulas giving the solutions explicitly. They are

$$\text{(a)} \ y(t) = y_0 e^{-t}, \quad \text{(b)} \ y(t) = y_0 e^t, \quad \text{(c)} \ y(t) = \frac{2y_0}{y_0 + (2 - y_0)e^{-2t}},$$

where y_0 is the initial value at $t = 0$. Notice that in each case the equilibrium solution $y = 0$ is obtained by setting $y_0 = 0$, while in (c) the equilibrium solution $y = 2$ is obtained by setting $y_0 = 2$. Drawing several of the solution curves in each case, we obtain the graphs in Figure 9.2.2.

Definition 9.2.2

An equilibrium solution is called **asymptotically stable** if small changes in the initial condition yield solutions that approach the equilibrium value as t tends to infinity.

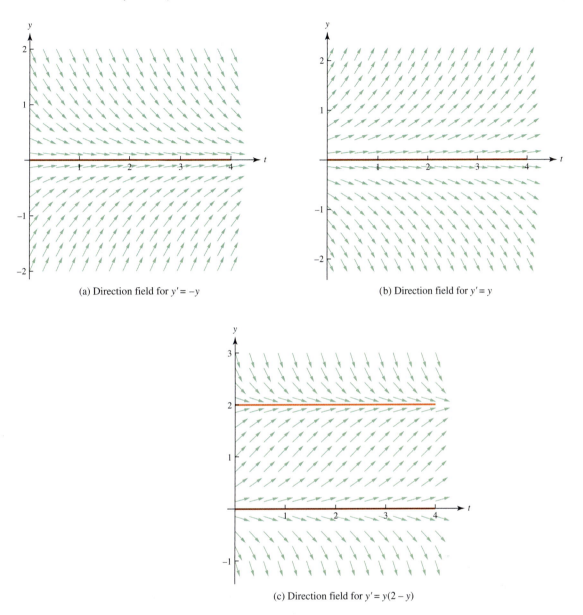

(a) Direction field for $y' = -y$

(b) Direction field for $y' = y$

(c) Direction field for $y' = y(2 - y)$

Figure 9.2.1

The graphs of the solution curves (shown in Figure 9.2.2), the solution formulas (a), (b) and (c), and the discussion preceding them all lead to the following conclusions.

(a) The equilibrium solution $y = 0$ of the differential equation $y' = -y$ is asymptotically stable since for any $y_0 \neq 0$ we have $\lim_{t \to \infty} y(t) = 0$.

(b) The equilibrium solution $y = 0$ of the differential equation $y' = y$ is not asymptotically stable since for any $y_0 > 0$ we have $\lim_{t \to \infty} y(t) = \infty$, while for any $y_0 < 0$ we have $\lim_{t \to \infty} y(t) = -\infty$.

(c) For the differential equation $y' = y(2 - y)$, the equilibrium solution $y_1 = 0$ is not asymptotically stable since $\lim_{t \to \infty} y(t) = 2$ if $y_0 > 0$. The equilibrium solution $y_2 = 2$ is asymptotically stable.

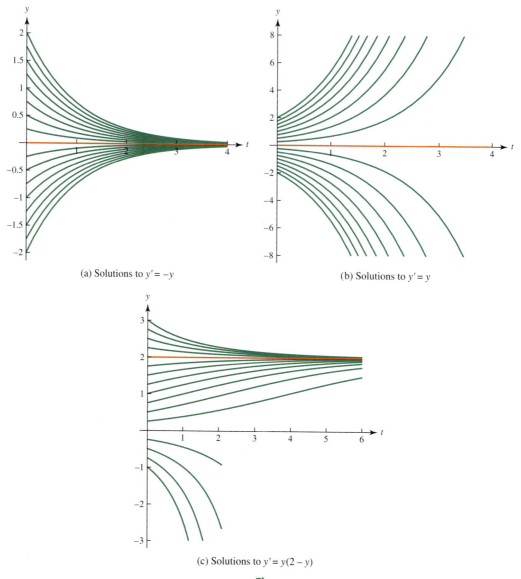

(a) Solutions to $y' = -y$

(b) Solutions to $y' = y$

(c) Solutions to $y' = y(2 - y)$

Figure 9.2.2

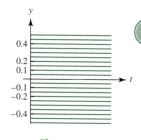

Figure 9.2.3

EXAMPLE 9.2.2

Determine whether $y = 0$ is an asymptotically stable equilibrium solution to $dy/dt = 0$.

SOLUTION   It is clear that $y = 0$ is an equilibrium solution. However, it is not asymptotically stable because all the solutions are constant functions, $y = c$, which do not approach 0 as $t \to 0$ unless $c = 0$. Figure 9.2.3 shows this clearly.

Remark The equilibrium solution of this last example has another type of stability, roughly defined by saying that small changes in the initial condition yield solutions that stay close to the equilibrium solution. In Example 9.2.2, the initial condition for the equilibrium solution is $y_0 = 0$. If we change it by a small amount, say, to $y_0 = c$ for a very small c, the solution becomes the constant function $y = c$, which is close to the equilibrium solution.

In what follows, we will consider only asymptotic stability. In the interest of brevity, we will use the word **unstable** to mean **not asymptotically stable**.

A stability criterion

We next describe a criterion for deciding the stabililty of equlibrium solutions for an equation of the form

$$\frac{dy}{dt} = g(y),$$

We shall always assume that $g(y)$ has a continuous derivative throughout its natural domain. To motivate this stability criterion, let us reexamine the differential equations of Example 9.2.1. We begin with the following observations:

1. If $g(y) = -y$, then $g'(0) = -1 < 0$ and (as we have already seen) the equlibrium solution $y = 0$ is asymptotically stable.
2. If $g(y) = y$, then $g'(0) = 1 > 0$ and the equilibrium solution $y = 0$ is unstable.
3. If $g(y) = y(2 - y)$, then

    - $g'(0) = 2 > 0$ and $y = 0$ is an unstable equilibrium solution, while
    - $g'(2) = -2 < 0$ and $y = 2$ is an asymptotically stable equilibrium solution.

These observations suggests the following result:

Theorem 9.2.1

Stability criterion

Suppose that the function $g(y)$ is differentiable in an interval around y_0 and consider the initial value problem

$$\frac{dy}{dt} = g(y), \quad y(0) = y_0.$$

If $g(y_0) = 0$, then $y = y_0$ is an equilibrium solution, which

- is asymptotically stable if $g'(y_0) < 0$, and
- is unstable if $g'(y_0) > 0$.

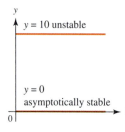

Figure 9.2.4

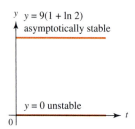

Figure 9.2.5

EXAMPLE 9.2.3

Find the equilibrium solutions of each of the differential equations

$$\text{(i)} \quad \frac{dy}{dt} = y(y - 10) \quad \text{and} \quad \text{(ii)} \quad \frac{dy}{dt} = 0.6ye^{1-(y/9)} - 0.3y$$

and determine their stability.

SOLUTION   (i) Here we have $g(y) = y(y - 10)$, which is differentiable and is zero if $y = 0$ and $y = 10$. These equilibrium solutions are shown in Figure 9.2.4. The derivative is $g'(y) = 2y - 10$. Using Theorem 9.2.1, we conclude that the equilibrium solution $y = 0$ is asymptotically stable, since $g'(0) = -10 < 0$, and the equilibrium solution $y = 10$ is unstable, since $g'(10) = 10 > 0$.

   (ii) In this case, $g(y) = 0.6ye^{1-(y/9)} - 0.3y = 0$ if either $y = 0$ or $0.6e^{1-(y/9)} = 0.3$. The second of these conditions reduces to $e^{1-(y/9)} = \frac{1}{2}$, whose solution is $y = 9(1 + \ln 2)$. Thus, we obtain two equilibrium solutions, $y = 0$ and $y = 9(1 + \ln 2)$, shown in Figure 9.2.5.

   To test the stability, we first take the derivative,

$$g'(y) = 0.6e^{1-(y/9)} - \frac{0.6}{9}ye^{1-(y/9)} - 0.3.$$

Since $g'(0) = 0.6e - 0.3 > 0$, the solution $y = 0$ is unstable. On the other hand,

$$g'(9(1 + \ln 2)) = 0.6e^{-\ln 2} - 0.6(1 + \ln 2)e^{-\ln 2} - 0.3 = -0.3(1 + \ln 2) < 0,$$

which means that the solution $y = 9(1 + \ln 2)$ is asymptotically stable.

   To explain the idea behind Theorem 9.2.1, we will use linear approximation to show the importance of the sign of $g'(y_0)$ at an equilibrium value y_0 of the differential equation

$$\frac{dy}{dt} = g(y).$$

Let $y = y(t)$ be a solution of the differential equation that starts near y_0, and let $h = y - y_0$. Note that both y and h are functions of t and

$$\frac{dh}{dt} = \frac{dy}{dt} = g(y) = g(y_0 + h) \approx g(y_0) + g'(y_0)h, \tag{5}$$

where the last step is linear approximation. Since y_0 is an equilibrium value, $g(y_0) = 0$, and (5) becomes

$$\frac{dh}{dt} \approx g'(y_0)h,$$

whose solution is

$$h(t) \approx h(0)e^{g'(y_0)t}.$$

If $g'(y_0) < 0$, then $h(t)$ goes to 0 as $t \to \infty$, which means that $y(t) \to y_0$, and the equilibrium solution y_0 is asymptotically stable. On the other hand, if $g'(y_0) > 0$, then

$h(t)$ moves away from 0 as $t \to \infty$, which means that $y(t)$ moves away from y_0, and the equilibrium solution y_0 is unstable.

Graphical method

The equilibrium solutions and their stability, together with the information we can extract from a careful reading of the differential equation

$$\frac{dy}{dt} = g(y),$$

can help us draw an accurate sketch of the graph of any solution once we know its initial value y_0 at $t = 0$. Here are the basic steps of this graphical method.

Step 1 Determine the zeros $y_1, y_2, \ldots,$ of the function $g(y)$ and mark them on the horizontal y-axis of a yz-coordinate system and on the vertical y-axis of a ty-coordinate system.

Step 2 Sketch the graph of the function $z = g(y)$ on the yz-coordinate system, using its zeros, its sign in each interval formed by the zeros, the sign of its derivative $g'(y)$, and its local extrema.

Step 3 The constant functions $y = y_1, y = y_2, \ldots,$ give the equilibrium solutions. Their graphs are horizontal lines in the ty-coordinate system. These lines separate the ty-plane into horizontal strips. By computing $g'(y)$, or by examining the shape of the graph of $z = g(y)$, decide on the stability of the equilibrium solutions.

Step 4 To sketch the graph of the solution to the initial value problem with $y(0) = y_0$ when y_0 is not a zero of $g(y)$, begin by placing y_0 on the y-axis of both coordinate systems and find the sign of $g(y_0)$.

Step 5 If the initial value y_0 is between two equilibrium values, then the solution $y(t)$ exists for all t, remains inside the same strip, and approaches one of the equilibrium values as $t \to \infty$. In particular,

- if $g(y_0) > 0$, then $y(t)$ is always increasing and approaches the upper equilibrium value as $t \to \infty$, and
- if $g(y_0) < 0$, then $y(t)$ is always decreasing and approaches the lower equilibrium value as $t \to \infty$.

The signs of $g(y)$ and $g'(y)$ determine the convexity of $y = y(t)$ and the extremum points of $g(y)$ give possible inflection values of the solution.

Step 6 If y_0 is above all equilibrium values and $g(y_0) > 0$, then the solution $y(t) \to \infty$ as t gets larger, but it may not exist for all t. (That is, it may approach a vertical asymptote.) On the other hand, if $g(y_0) < 0$, then $y(t)$ exists for all t, and it decreases and approaches the largest equilibrium value as $t \to \infty$.

Step 7 Similarly, if y_0 is below all equilibrium values and $g(y_0) < 0$, then the solution $y(t) \to -\infty$ as t gets larger, but it may not exist for all t. On the other hand, if $g(y_0) > 0$, then $y(t)$ exists for all t, and it increases and approaches the smallest equilibrium value as $t \to \infty$.

The next example demonstrates all these steps.

EXAMPLE 9.2.4

For the differential equation

$$\frac{dy}{dt} = -(y - 1)(y - 3)(y - 5),$$

(i) Find the equilibrium solutions and draw them on a ty-coordinate system.
(ii) Sketch the graph of $z = g(y)$ in a yz-coordinate system.
(iii) Determine the stability of the equilibrium solutions.
(iv) Sketch the graphs of the solutions corresponding to the initial values $y_0 = 0.7$, 1.6, 2.8, 3.2, 4.2, and 5.6.

SOLUTION   (i) Here the function $g(y)$ is

$$g(y) = -(y - 1)(y - 3)(y - 5).$$

We find its zeros by solving the equation $-(y - 1)(y - 3)(y - 5) = 0$ and obtain the equilibrium solutions

$$y_1 = 1, \quad y_2 = 3, \quad y_3 = 5.$$

They are drawn in Figure 9.2.6. Observe that they divide the ty-plane into four horizontal strips. The graph of any other solution must lie totally inside one of these strips.

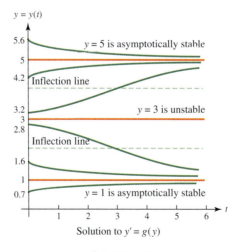

Figure 9.2.6

(ii) Expanding the product gives $g(y) = -y^3 + 9y^2 - 23y + 15$. Therefore, $g'(y) = -3y^2 + 18y - 23$. Solving $g'(y) = 0$, we find the critical points for $g(y)$ to be

$$y = \frac{1}{3}(9 - 2\sqrt{3}) \approx 1.845 \quad \text{and} \quad y = \frac{1}{3}(9 + 2\sqrt{3}) \approx 4.155.$$

Checking the sign of $g'(y)$, we find that $g(y)$ has a local minimum at the first of these and a local maximum at the second. Therefore, a solution $y(t)$ has an inflection point where it takes on either of these values. We sketch the corresponding horizontal lines (inflection lines) in the ty-plane (see Figure 9.2.6).

With this information and taking the sign of $g(y)$ into account, we sketch $z = g(y)$ in Figure 9.2.7.

(iii) Using the stability criterion of Theorem 9.2.1, we arrive at the following conclusions:

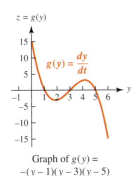

Graph of $g(y) = -(y-1)(y-3)(y-5)$

Figure 9.2.7

- The equilibrium solution $y = 1$ is asymptotically stable, since

$$g'(1) = -8 < 0.$$

- The equilibrium solution $y = 3$ is unstable since $g'(3) = 4 > 0$.
- The equilibrium solution $y = 5$ is asymptotically stable since

$$g'(5) = -8 < 0.$$

(iv) Next, using the information displayed in the graph of $z = g(y)$, we sketch the solutions with the given initial values, as shown in Figure 9.2.6.

- Since $g(0.7) > 0$, the solution $y(t)$ with initial value $y(0) = 0.7$ increases and approaches the equilibrium solution $y = 1$ from below as $t \to \infty$. Its graph is concave down, since $y(t)$ increases as t increases, which means that $g(y)$ (which equals dy/dt) is decreasing as t increases.
- Since $g(1.6) < 0$, the solution $y(t)$ with initial value $y(0) = 1.6$ decreases and approaches the equilibrium solution $y = 1$ from above as $t \to \infty$. Its graph is concave up, since $y(t)$ decreases as t increases, and $g(y)$ increases as y decreases. Therefore, $g(y)$ (which equals dy/dt) is increasing as a function of t.
- Since $g(2.8) < 0$, the solution $y(t)$ with initial value $y(0) = 2.8$ decreases for all t and approaches the equilibrium solution $y = 1$ from above as $t \to \infty$. Until it intersects the inflection line $y = (\frac{1}{3})(9 - 2\sqrt{3})$, $y = y(t)$ is a decreasing function of t and $g(y)$ is an increasing function of y. Therefore, $g(y) = dy/dt$ is a decreasing function of t, and the graph of $y(t)$ is concave down. After the graph crosses the inflection line, $y = y(t)$ continues to decrease as a function of t, and $g(y)$ decreases as a function of y. Therefore, $g(y) = dy/dt$ is increasing as a function of t, and the graph is concave up.
- Since $g(3.2) > 0$, the solution $y(t)$ with initial value $y(0) = 3.2$ increases for all t and approaches the equilibrium solution $y = 5$ from below as $t \to \infty$. Its graph is concave up until it intersects the inflection line $y = (\frac{1}{3})(9 + 2\sqrt{3})$ since both $y = y(t)$ and $g(y)$ are increasing as functions of t. After crossing the line, $y = y(t)$ increases and $g(y)$ decreases as t increases. Therefore, the graph is concave down.

- Since $g(4.2) > 0$, the solution $y(t)$ with initial value $y(0) = 4.2$ increases for all t and approaches the equilibrium solution $y = 5$ from below as $t \to \infty$. Its graph is concave down since $y = y(t)$ increases and $g(y)$ decreases as t increases.

- Since $g(5.6) < 0$, the solution $y(t)$ with initial value $y(0) = 5.6$ decreases for all t and approaches the equilibrium solution $y = 5$ from above as $t \to \infty$. Its graph is concave up since $y = y(t)$ is decreasing as t increases and $g(y)$ is increasing as y decreases. Therefore, $g(y) = dy/dt$ is increasing as a function of t.

In the next example, we combine graphical and numerical methods.

EXAMPLE 9.2.5

Data from the International Whaling Commission[2] indicate that there were about 26,500 gray whales in the North Pacific in the year 2000. Assume that the population $y(t)$ grows at the rate of $0.02y \ln(45/y)$ thousand whales per year and that an average catch of 174 whales per year is permitted.

    (i) Write the initial value problem modeling $y(t)$, where t measures years after 2000.

    (ii) Find all possible equilibrium solutions and determine their stability.

    (iii) Sketch the graph of $y(t)$.

    (iv) Use Euler's method with three steps to estimate the whales' population in 2030.

SOLUTION   (i) The problem reads as follows: t years after 2000 the rate of change of the whales' population $y(t)$ equals the rate of growth $0.02y \ln(45/y)$ minus the catch rate of 0.174 thousand whales per year. That is,

$$\frac{dy}{dt} = 0.02y \ln \frac{45}{y} - 0.174.$$

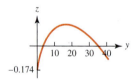

Figure 9.2.8

The initial condition is is $y(0) = 26.5$.

    (ii) Here we have $g(y) = 0.02y \ln(45/y) - 0.174$. To find the equilibria, we need to solve $0 = g(y)$, or $0.02y \ln(45/y) - 0.174 = 0$. However, this equation cannot be solved explicitly, and we must use numerical methods. Sketching the graph of $z = 0.02y \ln(45/y) - 0.174$ (see Figure 9.2.8), we observe that it crosses the y-axis at two points, one between 0 and 10, and the other between 30 and 40. We can approximate the solution by one of the standard numerical methods, such as the bisection method (see Section 1.3) or Newton's method (see the first project of Chapter 3), or, most easily, with a calculator. With an initial guess of 30, we find the equilibrium solution $y = 35.128$, and with an initial guess of 10 we find the equilibrium solution $y = 3.349$. It is easy to check, as the graph of $g(y)$ indicates, that $g'(35.128) < 0$ and $g'(3.349) > 0$. Therefore, we see by the stability criterion that $y = 35.128$ is asymptotically stable, while $y = 3.349$ is unstable.

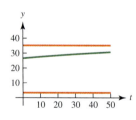

Figure 9.2.9

    (iii) Since $g(26.5) > 0$, our solution $y(t)$ increases, approaching the equilibrium $y = 35.128$ from below as $t \to \infty$. And since $g(y)$ (which is equal to dy/dt) decreases as y increases, we see that $y(t)$ is concave down. Thus, the graph of $y(t)$ resembles that of Figure 9.2.9.

[2]Go to www.iwcoffice.org/iwc.htm.

PROBLEM-SOLVING TACTIC

In finding equilibrium solutions and determining stability, we must solve $g(y) = 0$ and check the sign of $g'(y)$. In simple cases that can be done by direct computation. In more involved applications the graph of $g(y)$ is of great help both in locating the roots and determining the slope.

(iv) We are given $f(t, y) = 0.02y \ln(45/y) - 0.174$, $t_0 = 0$, $y_0 = 26.5$, and we are asked to estimate $y(30)$. Since $n = 3$, the step size is $\Delta t = 10$, and the t-values are $t_1 = 10$, $t_2 = 20$, and $t_3 = 30$. To estimate $y(30)$, we must compute y_3. We have

$$y_1 = y_0 + f(t_0, y_0)\Delta t = 26.5 + \left[0.02 \cdot 26.5 \ln\left(\frac{45}{26.5}\right) - 0.174\right] \cdot 10 \approx 27.566$$

$$y_2 = 27.564 + \left[0.02 \cdot 27.564 \ln\left(\frac{45}{27.564}\right) - 0.174\right] \cdot 10 \approx 28.528$$

$$y_3 = 28.528 + \left[0.02 \cdot 28.528 \ln\left(\frac{45}{28.528}\right) - 0.174\right] \cdot 10 \approx 29.389.$$

Therefore, the whales' population in 2030 will be approximately 29,389.

Remark If the initial number of whales is less than the unstable equilibrium value, the solution $y(t)$ decreases toward zero, and the population approaches extinction. Furthermore, increasing the size of the permitted annual catch results in an increase in the unstable equilibrium value. Thus, it is necessary that the size of the catch be regulated.

Logistic growth model with threshold

We have seen that the logistic model for population growth,

$$\frac{dp}{dt} = rp\left(1 - \frac{p}{K}\right),$$

has two equilibrium solutions: $p = 0$, which is unstable, and $p = K$ which is asymptotically stable and represents the environmental carrying capacity. According to this model, if the initial population $p_0 = p(0)$ is positive, no matter how small, it eventually will approach the equilibrium value K. However, that is not true for many species, and the model must be modified. To do that, we must take account of a critical number T called the **threshold level**, which varies from species to species. If the population of a certain species is under the threshold level, it will steadily decrease and move toward zero (the extinction of the species), as we observed in the remark following Example 9.2.5.

In the case of logistic growth, the following differential equation models that situation:

$$\frac{dp}{dt} = -rp\left(1 - \frac{p}{T}\right)\left(1 - \frac{p}{K}\right), \quad \text{where } 0 < T < K. \tag{6}$$

The reader can check that this equation has three equilibrium solutions. Two of them, $p = 0$ and $p = K$, are asymptotically stable, and a third, $p = T$, is unstable.

Practice Exercise 9.2

1. Consider the following population model with threshold:

$$\frac{dp}{dt} = 0.01p(1 - 2p)\left(1 - \frac{p}{6}\right).$$

Find the equilibrium solutions and discuss their stability.

Exercises 9.2

In Exercises 1–9, find the equilibrium solutions and discuss their stability. Then find the general solution by separation of variables and draw the graphs of four solutions with nearby initial values on both sides of the equilibrium value.

1. $\dfrac{dy}{dt} = -0.1y$ **2.** $\dfrac{dy}{dt} = 0.2y$ **3.** $\dfrac{dy}{dt} = 1 - y$

4. $\dfrac{dy}{dt} = 6 - 0.3y$ **5.** $\dfrac{dy}{dt} = y - 3$ **6.** $\dfrac{dy}{dt} = 0.5y - 2$

7. $\dfrac{dp}{dt} = 0.1p\left(1 - \dfrac{p}{10}\right)$ **8.** $\dfrac{dy}{dt} = y^2$

9. $\dfrac{dy}{dt} = y^3$

In Exercises 10–21, find the equilibrium solutions of the given differential equations and discuss their stability.

10. $\dfrac{dy}{dt} = 1 - e^y$ **11.** $\dfrac{dy}{dt} = 1 - e^{2-y}$

12. $\dfrac{dy}{dt} = y^2 - 9y$ **13.** $\dfrac{dy}{dt} = 4y - y^2$

14. $\dfrac{dy}{dt} = y^2 - 5y + 6$ **15.** $\dfrac{dy}{dt} = -y^2 + y + 2$

16. $\dfrac{dy}{dt} = y(1 - y^2)$ **17.** $\dfrac{dy}{dt} = -y^3 + 6y^2 - 8y$

18. $\dfrac{dy}{dt} = y^4 - 5y^2 + 4$ **19.** $\dfrac{dk}{dt} = 0.4\sqrt{k} - 0.1k$

20. $\dfrac{dp}{dt} = -rp\left(1 - \dfrac{p}{K}\right)$, where $r > 0$

21. $\dfrac{dp}{dt} = -rp\left(1 - \dfrac{p}{T}\right)\left(1 - \dfrac{p}{K}\right)$, where $r > 0$ and $0 < T < K$

⌇ *In each of the following exercises, you are given a differential equation of the form $dy/dt = g(y)$, and one or more initial conditions.*

(a) Find the equilibrium solutions and draw their graphs in a ty-coordinate system.
(b) Sketch the graph of the function $z = g(y)$ in a yz-coordinate system and use it to decide the stability of the equilibria.
(c) Use the graphical method to sketch the graphs of the solution to the given initial value problem(s). In all cases provide explanations for the shape of the graph: increasing/decreasing, concavity, inflection values, limiting value. (A graphing calculator, or computer, will be helpful here, in particular when completing Exercises 34–37.)

22. $\dfrac{dy}{dt} = 3 - \dfrac{1}{2}y$, $y(0) = 2$, $y(0) = 8$

23. $\dfrac{dy}{dt} = 1 - \dfrac{2}{3}y$, $y(0) = -1$, $y(0) = 3$

24. $\dfrac{dy}{dt} = \dfrac{2}{3}y - 2$, $y(0) = 0$, $y(0) = \dfrac{9}{2}$

25. $\dfrac{dy}{dt} = 0.2y - 0.1$, $y(0) = -\dfrac{1}{2}$, $y(0) = \dfrac{3}{2}$

26. $\dfrac{dy}{dt} = y(4 - y)$, $y(0) = -1$, $y(0) = 1$, $y(0) = 3$, $y(0) = 5$

27. $\dfrac{dy}{dt} = 0.1y(6 - y)$, $y(0) = -\dfrac{1}{2}$, $y(0) = 2$, $y(0) = 4$, $y(0) = 8$

28. $\dfrac{dy}{dt} = y(y - 3)$, $y(0) = -1$, $y(0) = \dfrac{1}{2}$, $y(0) = 2$, $y(0) = 5$

29. $\dfrac{dy}{dt} = 0.2y(y - 5)$, $y(0) = -\dfrac{1}{2}$, $y(0) = 1$, $y(0) = 3$, $y(0) = 7$

30. $\dfrac{dy}{dt} = 4 - y^2$, $y(0) = -3$, $y(0) = -1$, $y(0) = 1$, $y(0) = 3$

31. $\dfrac{dy}{dt} = 4 + 3y - y^2$, $y(0) = -2$, $y(0) = 0$, $y(0) = 2$, $y(0) = 5$

32. $\dfrac{dy}{dt} = y^2 - 9$, $y(0) = -4$, $y(0) = -1$, $y(0) = 2$, $y(0) = 4$

33. $\dfrac{dy}{dt} = y^2 - 8y + 12$, $y(0) = \dfrac{1}{2}$, $y(0) = 3$, $y(0) = 5$, $y(0) = 7$

34. $\dfrac{dy}{dt} = y(y - 1)(8 - y)$, $y(0) = -1$, $y(0) = 0.9$, $y(0) = 2$, $y(0) = 6$, $y(0) = 9$

35. $\dfrac{dy}{dt} = -0.2y(1 - 2y)(2 - y)$, $y(0) = 0.2$, $y(0) = 0.75$, $y(0) = 1.5$, $y(0) = 3$

36. $\dfrac{dy}{dt} = y^2(3 - y)$, $y(0) = -1$, $y(0) = 0.1$, $y(0) = 2.5$, $y(0) = 5$

37. $\dfrac{dy}{dt} = y^2(16 - y^2)$, $y(0) = -5$, $y(0) = -3$, $y(0) = 1$, $y(0) = 3$, $y(0) = 5$

38. Sketch the solutions to the differential equation $dy/dt = g(y)$ corresponding to the initial values $y(0) = 0$ and $y(0) = 4$, where $g(y)$ is as shown in Figure 9.2.10. In both cases compute $\lim_{t \to \infty} y(t)$.

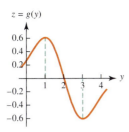

Figure 9.2.10

39. Sketch the solution to the initial value problem $dy/dt = g(y)$, $y(0) = -1.5$, where $g(y)$ is as shown in Figure 9.2.11. How many inflection points does it have? Does it intersect the t-axis? Explain your answer.

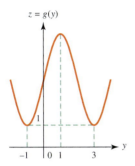

Figure 9.2.11

40. Assume that the U.S. population in 2000 was about 0.28 billion, its intrinsic growth rate is 0.01, and its future size $p(t)$ follows a logistic differential equation with a carrying capacity of 0.7 billion. Sketch the graph of $p(t)$ using the graphical method. Make sure that it has the correct shape, taking the concavity and inflection point into account.

41. Plants that are too crowded compete for light, water, and nutrients. Individuals that are not fit to survive such competition die off, decreasing the overall density. On the other hand, if the overall density is too low, plants that require cross-pollination do not reproduce and the population dies off. Thus, there is a range of overall densities that allows the plant community to grow and thrive. And if there are no plants in an area, the density remains unchanged.
(a) What variables would a differential equation modeling this problem involve?
(b) How many equilibrium solutions are described here?
(c) Write a differential equation modeling this problem.
(d) Identify stability, threshold, and carrying capacity, and sketch one possible solution in each strip.

42. A city planner is studying the "carrying capacity" of his city. With an idea of how many people the city can sustain, he can plan the total number of residences to be built in the area. Records show that the annual rate at which the city's population changes is given by

$$\frac{dp}{dt} = 0.01 p^{-1/2}(1{,}600 - p^2),$$

where p is in thousands of people.
(a) Determine the carrying capacity by finding the equilibrium solutions and checking their stability.
(b) Assuming an average of four people per family, how many total residences does the city need?

43. Rangers are attempting to introduce ring-necked pheasants (*Phasianus colchicus*) into a private game park. They need to know how many birds to release in order to achieve a stable population. The number of pheasants changes at the rate

$$\frac{dy}{dt} = 0.4 y^{1/3}(15.3 y - 36.9 - y^2),$$

where y is in tens of birds and t is in years.
(a) What is the largest sustainable population?
(b) What is the fewest number of birds the rangers can release in order to achieve it?

44. Sonja, being mathematically inclined, studied the rate of change $w'(t)$ of her daily average weight and decided that it can be modeled approximately by the equation

$$\frac{dw}{dt} = -4 \ln\left(\frac{w}{150}\right),$$

where t is in days and $w'(t)$ is in pounds per day.
(a) What is the average daily weight that Sonja tends to maintain once she achieves it?
(b) Sketch the solutions $w(t)$ corresponding to the initial weights of 140 and 160 pounds, respectively.
(c) If Sonja's initial weight is 160 pounds, estimate her weight 10 days later by using Euler's method with five steps.

45. The size $p(t)$ of a certain population is modeled by the initial value problem

$$\frac{dp}{dt} = 0.01 p(p - 1)\left(1 - \frac{p}{20}\right), \quad p(0) = 8.$$

(a) Sketch the graph of $p(t)$ and find the value at which its rate of growth is maximum.
(b) Compute its limiting value as $t \to \infty$.
(c) Find the threshold level in this model and the environmental carrying capacity.

46. The population size $p(t)$ of a certain species follows a logistic growth model with an intrinsic growth rate of 0.02, environmental carrying capacity of 100 million, and threshold level of 10 million.
(a) Sketch the graph of $p(t)$ if its value is 40 million at $t = 0$ and find the value at which its rate of growth is maximum.

(b) Sketch the graph of $p(t)$ if its value is 9 million at $t = 0$ and find the value at which its rate of decay is maximum.

47. With reference to Example 9.2.5 and the remark following it, increasing the size of the annual fish catch moves the equilibrium points closer together, until they merge into a single point. In that case, the whale population, no matter what its initial size, is doomed to extinction. Find the annual whale catch size at which that happens.

Solution to practice exercise 9.2

1. We have $g(p) = -0.01 p(1 - 2p)(1 - (p/6))$. The equilibrium solutions are the roots of the equation $0 = -0.01 \times p(1 - 2p)(1 - (p/6))$. Solving gives $p = 0$, $p = \frac{1}{2}$, and $p = 6$. Expanding the product in the expression of $g(p)$ gives $g(p) = -0.01(\frac{1}{3}p^3 - \frac{13}{6}p^2 + p)$, so that $g'(p) = -0.01(p^2 - \frac{13}{3}p + 1)$. To decide the stability of the equilibrium solutions, we shall use Theorem 9.2.1. Since $g'(0) = -0.01 < 0$, we see that $p = 0$ is asymptotically stable. Since $g'(\frac{1}{2}) = (0.01)(\frac{11}{12}) > 0$, the equilibrium solution $p = \frac{1}{2}$ (the threshold level) is unstable. Finally, we have $g'(6) = -(0.01) \cdot 11 < 0$, and therefore the equilibrium solution $p = 6$ (the environmental carrying capacity) is asymptotically stable.

9.3 The Solow Growth Model

The Solow growth model was developed by the economist Robert M. Solow in "A Contribution to the Theory of Economic Growth."[3] It shows how growth in capital stock, growth in the labor force, and advances in technology interact and how they affect output. For this work Solow won the Nobel Prize in economics in 1987. Here we shall discuss the fundamental equation of the Solow growth model. Our description follows Solow's original work and Mankiw's textbook on macroeconomics.[4]

The fundamental object in the Solow model is the **production function**, which expresses the output of goods produced as a function of capital and labor, each measured in suitable units. If Q is the annual quantity of goods produced by K units of capital and L units of labor, the production function takes the general form

$$Q = F(K, L), \quad \text{with } K, L > 0, \tag{7}$$

where F is an appropriate function whose precise form depends on certain assumptions about the economy in question. We will study one such function, the Cobb-Douglas production function, later in this section.

First, however, we will modify Eq. (7), based on a standard assumption concerning the production function. Experience has shown that if the amount of labor and the amount of capital both increase or decrease by a common factor, the output will change by that same factor. In other words, if K and L are both multiplied by the same constant, say, m, then the output is also multiplied by m. This property can be expressed quite concisely in terms of the production function by the following equation:

$$F(mK, mL) = mF(K, L).$$

[3] *Journal of Economics*, Feb., 1956, pp. 65–94.

[4] N. Gregory Mankiw, *Macroeconomics*, New York: 1994, Worth Publishers.

In the language of economics we express this fact by saying that the production function F has **constant returns to scale**, while in mathematical language F is said to be **homogeneous of degree 1**. In terms of Q it can also be written as

$$mQ = F(mK, mL). \tag{8}$$

By using property (8) with $m = 1/L$, we can rewrite Eq. (7) in the form

$$\frac{Q}{L} = F\left(\frac{K}{L}, 1\right).$$

Next, we replace the variables K and Q by new variables as follows. Let

$$k = \frac{K}{L} \quad \text{and} \quad q = \frac{Q}{L}.$$

k is called the **capital per worker** (or sometimes the **capital stock**), and q is called the annual **ouput per worker**. We can now rewrite Eq. (7) in the form

$$\boxed{q = f(k),} \tag{9}$$

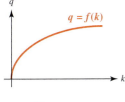

Figure 9.3.1

where f is defined by setting $f(k) = F(k, 1)$. The function f is also called a production function. The graph of a typical production function is shown in Figure 9.3.1.

As you can see, the graph is increasing and concave down. In other words, $f(k)$ is increasing but $f'(k)$ is decreasing. Moreover, the graph becomes more and more horizontal as it moves to the right and more and more vertical as it moves to the left. In mathematical terms, $f'(k) \to 0$ as $k \to \infty$, and $f'(k) \to \infty$ as $k \to 0^+$.

The linear approximation formula [formula (3.29) of Section 3.3] tells us that

$$f'(k) \approx f(k + 1) - f(k). \tag{10}$$

The right-hand side of this formula is the increase in annual output per worker resulting from the input of one extra unit of capital stock. In economics that is called the **marginal product of capital**, and Eq. (10) shows that it is approximated by the derivative of the production function. Since $f'(k)$ is decreasing, it follows that the marginal product of capital is diminishing. To put it another way, each additional unit of capital stock produces less output than the preceding one. This makes good sense, for when there is only a little capital, each extra unit makes a relatively big difference, resulting in a large increase of output. When there is plenty of capital, on the other hand, one additional unit makes relatively little difference and results in much less additional output.

Consumption, investment, and depreciation

In the Solow model, as in other macroeconomic models, the total output of an economy is the same as its total expenditure and also the same as its total income. What that means is that all the goods and services produced (the total output) are purchased by consumers, and the money they spend (the total expenditure) is realized as income by the companies and workers—either in the form of wages to the workers or profits to the owners (which together are the total income). Of course, the workers and owners are also consumers, and the economy moves in a circular flow. These three equal quantities—the total output, the total expenditure, and the total income—go under the single label of the **gross domestic product**.

Economists divide the gross domestic product into two components:[5]

- **consumption**, which is the amount spent on goods and services such as food, clothing, cars, movie tickets, and so forth; and

- **investment**, which is the amount spent on goods bought for future production use, including manufacturing equipment, real estate, and inventory.

Using Q for the total annual output, C for the total annual consumption, and I for the total annual investment, we can represent this decomposition by the equation $Q = C + I$. But, as we have already seen, the Solow model deals with the output *per worker*—that is, the total output Q divided by the number of units of labor L. If we divide both sides of the previous equation by L, we get

$$q = c + i, \tag{11}$$

where

$$q = \frac{Q}{L} \quad \text{(the \textbf{annual output per worker}),}$$

$$c = \frac{C}{L} \quad \text{(the \textbf{annual consumption per worker}),}$$

$$i = \frac{I}{L} \quad \text{(the \textbf{annual investment per worker}).}$$

From now on, in referring to these quantities, we will omit the terms *annual* and *per worker* so that output will be understood to mean annual output per worker, with similar interpretations of consumption and investment.

The Solow model makes an additional assumption—that *investment and consumption are proportional to output*. In mathematical terms, we can write

$$i = sq = sf(k), \tag{12}$$

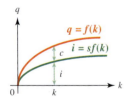

Figure 9.3.2

where s is a constant between 0 and 1, called the **rate of savings**. If we substitute this into Eq. (11) and solve for c, we get

$$c = (1 - s)q = (1 - s)f(k). \tag{13}$$

The graphs of these quantities are shown in Figure 9.3.2.

Finally, the Solow model assumes that capital depreciates at a fixed rate. In terms of the capital stock k, the amount of depreciation in 1 year is given by

$$\text{Annual depreciation } = \delta k, \tag{14}$$

where δ is a proportionality constant, called the **depreciation rate**.

[5]There are actually two additional components—government purchases and net exports—that economists generally consider, but they are not included in our description of the Solow model.

Solow's fundamental differential equation

The capital stock $k(t)$ varies over time, increasing as a result of investment and decreasing as a result of depreciation. Therefore, we have

$$\begin{Bmatrix} Rate\ of\ change \\ of\ capital\ stock \end{Bmatrix} = \begin{Bmatrix} rate\ at\ which \\ investment\ is\ made \end{Bmatrix} - \begin{Bmatrix} rate\ at\ which \\ depreciation\ occurs \end{Bmatrix}.$$

Since the basic assumptions in the Solow model are that investment is made at the rate $sf(k)$ and depreciation occurs at the rate δk, we obtain the differential equation

$$\frac{dk}{dt} = sf(k) - \delta k. \tag{15}$$

In addition, by letting k_0 represent the capital stock at the time we begin studying the economy, we get the initial condition

$$k(0) = k_0. \tag{16}$$

The initial value problem described by Eqs. (15) and (16) is the **Solow model** for an economy in which technology and the supply of labor do not change. Its solution $k = k(t)$ describes the evolution of the capital stock as time goes on. Putting it into Eq. (9) gives the output, putting it into Eq. (13) gives the consumption, putting it into Eq. (12) gives the investment, and putting it into Eq. (14) gives the depreciation at any

HISTORICAL PROFILE

Robert M. Solow (1924–) who was awarded the Nobel Prize for economics in 1987, was born in Brooklyn, New York, the grandchild of immigrants. He was educated in New York City public schools and entered Harvard University on a scholarship in September 1940. With the outbreak of World War II, he left Harvard in 1942 to join the U.S. Army, in which he served until 1945. About that period, which he mostly spent in the North-African and Italian campaigns, he wrote, "I think that those three years as a soldier formed my character. I found myself part of a tight-knit group, doing a hard job with skill and mutual loyalty." After the war he returned to Harvard, where he earned a Ph.D. under the direction of Wassily Leontieff (also a future Nobel laureate, who was to win the prize for economics in 1973). After spending a fellowship year at Columbia University, Solow accepted a position in the economics department of M.I.T., where he is currently a Professor Emeritus. In addition, he served on President Kennedy's Council of Economic Advisers in the early 1960s and was president of the American Economic Asociation in 1979.

Solow stressed the importance of technological advancement in producing and sustaining economic growth, and his work was an important factor in influencing governments to invest in technological research and development.

time t. A variation of this model, which we will not go into, describes the situation when the supply of labor changes, and another takes account of advances in technology.

Qualitative behavior of solutions

We can get information about the solutions to Eq. (15) using the few assumptions we have already made about $f(k)$ earlier in this section—specifically, that $f'(k)$ is decreasing with $\lim_{k\to\infty} f'(k) = 0$ and $\lim_{k\to 0^+} f'(k) = \infty$.

First, to find the equilibrium solution of (15), we set the right-hand side

$$g(k) = sf(k) - \delta k$$

equal to zero and solve for k. That is, we solve the equation

$$sf(k) = \delta k. \tag{17}$$

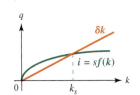

q

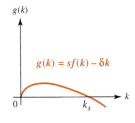

Figure 9.3.3

A solution of this equation is simply a point where the line of slope δ intersects the graph of the function $sf(k)$. As Figure 9.3.3 shows, there are two such points: zero, which is not of economic interest, and a positive one, denoted by k_s. We call k_s the **steady-state level** of capital stock. It satisfies the equation

$$\boxed{\frac{k_s}{f(k_s)} = \frac{s}{\delta},} \tag{18}$$

Figure 9.3.4

which is just another way of writing Eq. (17).

What about the stability of the equilibrium solution k_s? To answer that, we will graph $g(k)$ (which equals the derivative dk/dt) as a function of k. It is drawn in Figure 9.3.4. As you can see, $dk/dt > 0$ if k is in the interval $(0, k_s)$. Therefore, if we start with an initial condition k_0 satisfying $0 < k_0 < k_s$, the solution $k(t)$ will keep increasing toward the equilibrium level k_s, as shown in Figure 9.3.5.

On the other hand, $dk/dt < 0$ when $k > k_s$. That means that if $k_0 > k_s$, the solution $k(t)$ will keep decreasing toward k_s. In both cases, then, the solution moves toward the steady-state level k_s, which means that the constant solution $k = k_s$ is asymptotically stable (also shown in Figure 9.3.5).

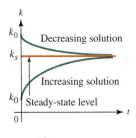

Figure 9.3.5

The Cobb-Douglas production function

As we have already noted, the solution of the Solow differential equation (15) depends on the choice of production function. One such function that is used extensively in economic analysis is the **Cobb-Douglas production function**, named for its discoverers—Charles Cobb, a mathematician, and Paul Douglas, an economist who also served with distinction as a U.S. Senator from Illinois from 1949 to 1966. As a young economics professor in 1927, Douglas observed that the distribution of national income between capital and labor had not changed for a long time. In other words, the output Q was divided into two portions—one part αQ going to capital, and the other part $(1 - \alpha)Q$ going to labor—and the proportionality constant α was fixed over time. (In fact, it was about 0.3.) He asked Cobb what kind of function would produce this result, and Cobb showed that the production function would take the form

$$Q = F(K, L) = AK^\alpha L^{1-\alpha}. \tag{19}$$

In this equation α is a constant between 0 and 1, and A is another positive constant. If we divide both sides of this equation by L, we get

$$\frac{Q}{L} = \frac{AK^\alpha L^{1-\alpha}}{L} = A\frac{K^\alpha L^{1-\alpha}}{L^\alpha L^{1-\alpha}} = A\left(\frac{K}{L}\right)^\alpha.$$

As we have done before, we will let $q = Q/L$ (the output per worker) and $k = K/L$ (the capital per worker or capital stock). In terms of these variables we can rewrite the last equation in the form $q = Ak^\alpha$; in other words, the production function is given by $f(k) = Ak^\alpha$. Finally, we set $A = 1$ for simplicity. The production equation now takes the form

$$q = f(k) = k^\alpha, \quad 0 < \alpha < 1. \tag{20}$$

With this choice of production function, Solow's differential equation (15) now takes the form

$$\frac{dk}{dt} = sk^\alpha - \delta k. \tag{21}$$

Graphical solving   To study (21), we follow Section 9.2. Here we have $g(k) = sk^\alpha - \delta k$. Solving $0 = sk^\alpha - \delta k = k^\alpha(s - \delta k^{1-\alpha})$, we find that the equilibrium solutions are $k = 0$ and and $k = k_s$, where

$$k_s = \left(\frac{s}{\delta}\right)^{1/(1-\alpha)}. \tag{22}$$

The derivative of g is $g'(k) = s\alpha k^{\alpha-1} - \delta$. Computing it at k_s, we find

$$g'(k_s) = \alpha\delta - \delta < 0,$$

since $\alpha < 1$. Therefore, the equilibrium solution k_s is asymptotically stable. It is called the steady-state level of capital stock. Next, solving the equation $g'(k) = 0$, we find $k = k_i$, where

$$k_i = \left(\frac{\alpha s}{\delta}\right)^{1/(1-\alpha)} = \alpha^{1/(1-\alpha)} k_s. \tag{23}$$

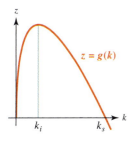

Figure 9.3.6

Checking the sign of $g'(k)$, we find that it is positive for $0 < k < k_i$ and negative for $k > k_i$. Thus, $g(k)$ has a maximum at k_i and its graph looks like that shown in Figure 9.3.6.

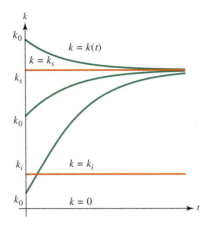

Figure 9.3.7

By studying the graph of $z = g(k)$, we conclude that every solution with initial value k_0 in the interval $(0, k_s)$ is increasing for all times and is concave up until its graph meets the horizontal line $k = k_i$. There it has an **inflection** point, after which it is concave down and approaches k_s as $t \to \infty$, with its graph always staying between the lines $k = 0$ and $k = k_s$. On the other hand, if $k_0 > k_s$, then the corresponding solution curve lies above the line $k = k_s$, it is concave up, decreasing, and approaches k_s as $t \to \infty$. Both cases are shown in Figure 9.3.7. Note that at the inflection value k_i the rate of change of the capital stock is **maximum**.

EXAMPLE 9.3.1

Assuming that $\alpha = 0.5$, $s = 0.3$, and $\delta = 0.1$, compute the steady-state level of capital stock and the value at which the rate of change of the capital stock is maximum.

SOLUTION As we have seen above, the steady-state level of capital stock is given by

$$k_s = \left(\frac{s}{\delta}\right)^{1/(1-\alpha)} = \left(\frac{0.3}{0.1}\right)^{1/(1-0.5)} = 3^2 = 9,$$

while the value at which the rate of change of the capital stock is maximum is given by

$$k_i = \alpha^{1/(1-\alpha)} k_s = (0.5)^{1/(1-0.5)} 9 = (0.5)^2 9 = 2.25.$$

Analytical solving As it happens, Eq. (21) is a classical differential equation, studied long ago by the great Swiss mathematician Jakob Bernoulli (1654–1705). The method for solving it is to make a change of variables by defining

$$y = k^{1-\alpha}. \tag{24}$$

From the chain rule, we have

$$\frac{dy}{dt} = (1 - \alpha)k^{-\alpha}\frac{dk}{dt},$$

which can be rewritten as

$$\frac{dk}{dt} = \frac{1}{1-\alpha} k^{\alpha} \frac{dy}{dt}.$$

Substituting this into the left-hand side of (21) changes the equation to

$$\frac{1}{1-\alpha} k^{\alpha} \frac{dy}{dt} = sk^{\alpha} - \delta k.$$

Next, we divide both sides by k^{α} and use (24), which gives

$$\frac{1}{1-\alpha} \frac{dy}{dt} = s - \delta k^{1-\alpha} = s - \delta y.$$

Multiplying both sides by $(1 - \alpha)$, we obtain

$$\frac{dy}{dt} = (1-\alpha)(s - \delta y),$$

which can be solved by separation of variables. The solution (which you should work out as an exercise) is

$$y = \frac{s}{\delta} + Ce^{-\delta(1-\alpha)t},$$

where C is an arbitrary constant. Finally, using (24), we replace y by $k^{1-\alpha}$ and get

$$k^{1-\alpha} = \frac{s}{\delta} + Ce^{-\delta(1-\alpha)t}.$$

If we have an initial condition of the form $k(0) = k_0$, we can eliminate the arbitrary constant by substituting $t = 0$ and $k = k_0$ and solving for C. We get

$$C = k_0^{1-\alpha} - \frac{s}{\delta}.$$

Substituting this into the previous equation and raising both of its sides to the power $1/(1 - \alpha)$ give us the formula for the capital stock k at any time t:

$$k = \left[\frac{s}{\delta} + \left(k_0^{1-\alpha} - \frac{s}{\delta} \right) e^{-\delta(1-\alpha)t} \right]^{1/(1-\alpha)}. \tag{25}$$

Notice that as $t \to \infty$, the second term on the right-hand side goes to zero, so that $k \to k_s$, where k_s is given by (22). In Figure 9.3.8 we show several solutions when $\alpha = 0.5$, $s = 0.3$, and $\delta = 0.1$.

As Eq. (22) shows, the steady-state level of capital stock k_s depends on two things: the rate of savings s and the depreciation rate δ. Observe that the steady-state level increases when the savings rate increases and decreases when the depreciation rate increases, which is consistent with common economic experience. In addition, this formula indicates one reason why the Federal Reserve Bank and the U.S. government frequently adjust interest rates. Of course, there are additional reasons, such as inflation.

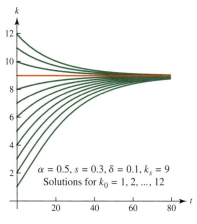

$\alpha = 0.5, s = 0.3, \delta = 0.1, k_s = 9$
Solutions for $k_0 = 1, 2, \dots, 12$

Figure 9.3.8

Numerical solving The Solow differential equation can be solved numerically. In fact, it provides an ideal situation for numerical experimentation. In the following example we apply Euler's method to estimate the solution of a Solow differential equation.

EXAMPLE 9.3.2

Assume that the capital stock for an economy is modeled by the Solow differential equation (21) with $s = 0.2$, $\delta = 0.1$, and $\alpha = 0.3$. That is,

$$\frac{dk}{dt} = 0.2k^{0.3} - 0.1k.$$

(i) Find the stable steady-state level of capital stock.

(ii) Assuming an initial condition $k(0) = 1$, use Euler's method with $\Delta t = 2$ to numerically solve this differential equation over the next 60 years.

SOLUTION   (i) The steady-state level of capital stock is given by

$$k_s = \left(\frac{s}{\delta}\right)^{1/(1-\alpha)} = \left(\frac{0.2}{0.1}\right)^{1/(1-0.3)} = 2^{1/0.7} \approx 2.6918.$$

(ii) Using Euler's method with $\Delta t = 2$ gives

$$k_n = k_{n-1} + (0.2\, k_{n-1}^{0.3} - 0.1\, k_{n-1}) \cdot 2.$$

Then starting with $n = 1$ and $k_0 = 1$ and using a calculator, or a computer, we obtain the following pairs $[t, k(t)]$:

$(0, 1)$, $(2, 1.2)$, $(4, 1.38249)$, $(6, 1.54681)$, $(8, 1.69337)$, $(10, 1.82317)$,
$(12, 1.9375)$, $(14, 2.03779)$, $(16, 2.12547)$, $(18, 2.2019)$, $(20, 2.26839)$,
$(22, 2.32613)$, $(24, 2.3762)$, $(26, 2.41955)$, $(28, 2.45705)$, $(30, 2.48946)$,

$(32, 2.51745)$, $(34, 2.54162)$, $(36, 2.56246)$, $(38, 2.58043)$, $(40, 2.59593)$,
$(42, 2.60928)$, $(44, 2.62078)$, $(46, 2.63068)$, $(48, 2.63921)$, $(50, 2.64655)$,
$(52, 2.65287)$, $(54, 2.65831)$, $(56, 2.66299)$, $(58, 2.66701)$, $(60, 2.67048)$.

By plotting these as points in the plane and connecting each successive pair by a line segment, we get the graph of the approximate solution illustrated in Figure 9.3.9. Observe that as t gets large, $k(t)$ approaches the steady-state level of capital stock 2.6918.

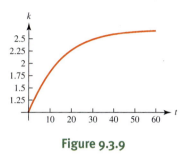

Figure 9.3.9

Further exploration   The differential equation (15) can be modified to include growth in the labor force (due to population growth) and improvements in technology. In that case, the steady-state level is the one at which the output per worker is fixed. But, if we assume that the labor force keeps growing and technology advances, that means the total output of the economy must also grow, which is what really happens in good economic systems. Moreover, a high standard of living means a high rate of consumption, and there is one steady-state level among all those possible that achieves the highest consumption rate. It is called the **golden rule level of capital accumulation**, and it is another very important reason why the Federal Reserve and the U.S. government change interest rates. All this and much more are explained in great detail in books on macroeconomics (see, e.g., Mankiw).

Exercises 9.3

A graphing calculator or a computer is needed for most of the following exercises.

1. Data from 1945 to 1995 (see Mankiw) indicate that the U.S. economy follows a Cobb-Douglas production function with $\alpha = 0.3$, and that the investment rate is about $s = 0.22$. Assuming that the depreciation rate is 0.1,
   (a) Write the corresponding Solow differential equation.
   (b) Compute its steady-state level of capital.
   (c) Compute the value at which the rate of change of the capital stock is maximum.
   (d) Use the stability criterion to show that the equilibrium solution is asymptotically stable.
   (e) Sketch the graphs of the solutions corresponding to initial capital 0.2, 2.2 and 4.2 units, and explain their shape (increasing/decreasing, concavity, inflection points, limiting value).

(f) Use formula (25) and a graphing calculator, or a computer, to draw the solutions corresponding to $k_0 = 0.5, 1, 1.5, \ldots, 6$.

2. Data from the German economy indicate that the investment rate is 0.3 (see Mankiw). Assume that it follows a Cobb-Douglas production function with $\alpha = 0.3$, and that the depreciation rate is $\delta = 0.1$. Redo parts (a–f) of Exercise 1 but in the context of the German economy, using $k_0 = 0.5, 3$, and 6 in part (e) and $k_0 = 0.6$, $1.2, 1.8, \ldots, 9.6$ in part (f).

3. Data from the Japanese economy indicate that the investment rate is 0.34 (see Mankiw). Assume that it follows a Cobb-Douglas production function with $\alpha = 0.3$, and that the depreciation rate is $\delta = 0.1$. Redo parts (a–f) of Exercise 1 but in the context of the Japanese

economy, using $k_0 = 0.5, 3,$ and 8 in part (e) and $k_0 = 0.7,$ $1.7, 2.7, \ldots, 11.7$ in part (f).

4. Consider the Solow differential equation $dk/dt = 0.3\sqrt{k} - 0.1k$ with initial condition $k(0) = 1$.
   (a) Find the nonzero equilibrium solution and show that it is asymptotically stable.
   (b) Use Euler's method with $n = 4$ to approximate $k(1)$ to three decimal places.
   (c) Find an explicit solution of the initial value problem [see formula (25)] and use it to compute $k(1)$ to three decimal places.

5. Consider the Solow differential equation $dk/dt = 0.25k^{0.45} - 0.15k$ with initial condition $k(0) = 2$.
   (a) Use Euler's method with $\Delta t = 1$ to approximate $k(5)$ to four decimal places.
   (b) Find an explicit solution of the initial value problem [see formula (25)] and use it to compute $k(5)$ to four decimal places.
   (c) Find the nonzero equilibrium solution and show that it is asymptotically stable.

6. Consider the Solow differential equation $dk/dt = 0.5k^{0.31} - 0.12k$ with initial condition $k(0) = 2$.
   (a) Use Euler's method with $n = 4$ to approximate $k(4)$.
   (b) Use the data from part (a) to draw a polygonal curve that approximates the solution graph for $0 \le t \le 4$.

7. Follow the analytical method described in this section to solve the Solow differential equation $dk/dt = 0.3\sqrt{k} -$

$0.1k$ with initial condition $k(0) = 25$ and draw a sketch of the solution graph for $0 \le t < 50$.

8. The steady-state level of capital with the highest consumption is called the golden rule level of accumulation and is denoted by k_{gold}.
   (a) Assuming that the depriciation rate δ is fixed and the savings rate s is variable, show that in a Solow model with Cobb-Douglas production function $f(k) = k^\alpha$, the golden rule level k_{gold} is achieved when $s = \alpha$.
   (b) Then using the information given in Exercises 1–3, compute k_{gold} for the U.S., German, and Japanese economies.
   (c) Finally, decide which economy's steady-state level of capital is closest to its k_{gold}.

9. Another example of a production function is given by
   $\boxed{\text{C}}$ $f(k) = (0.6\sqrt{k} + 1)^2$.
   (a) Write the Solow differential equation for this production function, assuming that the savings rate is $s = 0.2$ and the depreciation rate is $\delta = 0.1$.
   (b) Find the equilibrium solution and show that it is asymptotically stable.
   (c) Use Euler's method with $n = 3$ to approximate $k(15)$ to three decimal places if $k(0) = 20$.
   (d) Use Euler's method with many steps (or any other numerical solver, such as Mathematica's NDSolve) to estimate $k(200)$, if $k(0) = 20$, and to sketch the graph of $k(t)$.

Chapter 9 Summary

- If for a given function $f(t, y)$, we choose a rectangular grid of points and from each grid point (t, y) draw an arrow of slope $f(t, y)$ and of unit length, then we obtain a **direction field** for the differential equation $dy/dt = f(t, y)$. If we start from a point (t_0, y_0) of the grid and draw a curve following the arrows, then this curve is close to the graph of the solution to our differential equation with initial condition $y(t_0) = y_0$.

- Starting from (t_0, y_0), we move along the line segment of slope $f(t_0, y_0)$ for time Δt to reach (t_1, y_1) with $t_1 = t_0 + \Delta t$ and $y_1 = y_0 + f(t_0, y_0)\Delta t$. There we recompute the slope $f(t_1, y_1)$ and repeat the first step again and again to obtain the formula $y_n = y_{n-1} + f(t_{n-1}, y_{n-1})\Delta t$, which is **Euler's method** for approximating $y(t_n)$, the value of the solution to the initial value problem $dy/dt = f(t, y)$, $y(t_0) = y_0$.

- A constant solution to $dy/dt = g(y)$ is called an **equilibrium solution**. To find the equilibrium solutions, we solve $0 = g(y)$.

- An equilibrium solution is called **asymptotically stable** if small changes in the initial condition yield solutions that approach the equilibrium value as t tends to infinity.

An equilibrium solution $y = y_0$ is asymptotically stable if $g'(y_0) < 0$ and is *not* asymptotically stable if $g'(y_0) > 0$.

- The graphs of the equilibrium solutions divide the ty-plane into horizontal strips. Using them as guides, we can sketch the graph of any other solution without knowing a formula for it. Each nonconstant solution lies entirely in one of these horizontal strips, always increasing (decreasing) toward the upper equilibrium solution or ∞ (toward the lower equilibrium solution or $-\infty$) if $g(y_0) > 0$ ($g(y_0) < 0$), where y_0 is the initial value of the solution. Solving $g'(y) = 0$, we find the values where solutions have inflection points.

- The **Solow differential equation** is $dk/dt = sf(k) - \delta k$, where k is the capital stock, $f(k)$ is the production function, s is the savings rate, and δ is the depreciation rate. A typical production function is given by the formula $q = k^\alpha$, $0 < \alpha < 1$. It is called a **Cobb-Douglas production function**. The U.S. economy is described by the Cobb-Douglas production function with $\alpha = 0.3$. The Solow differential equation can be studied numerically, graphically, and in some cases (Cobb-Douglas production function) analytically.

Chapter 9 Review Questions

- Describe the first three steps of Euler's method.

- What is an autonomous differential equation? Give two examples of such equations.

- Write differential equations with one, two, and three equilibrium solutions.

- Write a differential equation that has $y = 8$ as an asymptotically stable solution and $y = 1$ as an unstable solution.

- Write a Solow model for the U.S. economy and compute its steady-state capital stock.

Chapter 9 Review Exercises

1. If $y(t)$ satisfies the differential equation $dy/dt = 9y - 5t$ and its graph passes though the point $(3, 2)$, find the slope of the graph at that point.

2. A solution curve of the differential equation $dy/dt = t^3 y + y^2$ passes through the point $(1, 2)$. Find its slope at that point.

3. Let $y(t)$ be the solution of the initial value problem $y' = e^{3-y} - 1$, $y(0) = 2$. Find $y'(0)$.

4. If $y(t)$ solves the initial value problem $y' = \ln(y/8)$, $y(3) = 16$, find $y'(3)$.

5. Let $y(t)$ be the solution of the initial value problem $y' = 4y - y^2$, $y(0) = 1$.
   (a) Find $y'(0)$ and the linear approximation $L(t)$ of the solution at $(0, 1)$.
   (b) Estimate $y(0.1)$ by computing $L(0.1)$.

6. Consider the initial value problem $dy/dt = 3 - y$, $y(0) = 1$.
   (a) Use Euler's method with five steps to estimate $y(1)$.
   (b) Solve the initial value problem by separation of variables and compute $y(1)$.

7. Consider the initial value problem $y' = 2y - 10t$, $y(0) = 3$.
 (a) Use Euler's method with five steps to estimate the solution in the interval $[0, 1]$.
 (b) Check that $y(t) = \frac{1}{2}(5 + e^{2t} + 10t)$ is the solution to the given initial value problem.
 (c) Compute the error at each step. Where is it largest?

8. Use Euler's method with $\Delta t = 0.25$ to approximate the solution of the initial value problem $y' = t + 2y - 8$, $y(0) = 4$, for $0 \le t \le 1$.

9. The rate of change of the population of a city is modeled by the equation $dp/dt = 0.04 p \ln(2/p)$, where p is in millions and t is in years. Use Euler's method with $\Delta t = 2$ to estimate the size of its population 10 years later if its initial size is 0.8 million.

10. There are 10 million pine trees in a forest and their number $y = y(t)$ is growing at the rate of $0.2y[e^{1-(y/8)} - 1]$, where y is measured in millions and t in years. They are also harvested at the rate of $0.1y$.
 (a) Model this situation as an initial value problem.
 (b) Use Euler's method with five steps to estimate the number of pine trees 20 years later.

In Exercises 11–13, find the equilibrium solutions of the given differential equation and determine whether they are asymptotically stable.

11. $\dfrac{dy}{dt} = 0.1(10y - 16 - y^2)$

12. $\dfrac{dy}{dt} = 0.4y \ln \dfrac{9}{y}$

13. $\dfrac{dy}{dt} = 0.2ye^{1-(y/8)} - 0.3y$

14. Explain why any solution $y(t)$ of the differential equation $y' = e^{1-y}$ must be an increasing function.

15. Explain why any solution $y(t)$ of the differential equation $y' = -1 - y^2$ must be a decreasing function.

16. Sketch the solution of the initial value problem $y' = e^{-(y-1)^2}$, $y(0) = 1$.

17. Sketch the solution of the initial value problem $y' = -y^2$, $y(0) = -1$.

18. Sketch the solution of the initial value problem $y' = 3 - y$, $y(0) = 2$. Then do the same if $y(0) = 4$.

19. Sketch the solution of the initial value problem $y' = 3 + y$, $y(0) = -2$. Then do the same if $y(0) = -4$.

20. Sketch the solution $y(t)$ of the initial value problem $y' = y(4 - y)$, $y(0) = 1$. Explain the shape of the graph: increasing/decreasing, concavity, inflection value, and limiting value. Then do the same work if $y(0) = 5$.

21. According to the 2000 census, the U.S. population then was 0.28 billion. Assume that the population is governed by the logistic model with an intrinsic growth rate of 0.02 and environmental carrying capacity of 0.5 billion. Also assume that after the year 2000 immigrants arrive in the United States at the rate of 1 million per year.
 (a) Write the initial value problem that models the time evolution of the U.S. population $p(t)$, t years after 2000.
 (b) Find any equilibrium value of the population and decide its stability.
 (c) Sketch the graph of $p(t)$ and explain its shape.

22. Biologists studying an endangered species know from historical accounts that the local environment's carrying capacity for this animal is about 6.7 thousand. They also feel that if the population falls below 0.36 thousand, genetic diversity will not be great enough to maintain any population.
 (a) What are the three equilibrium solutions for the population? (*Hint:* One is trivial.)
 (b) For each solution, explain its stability.
 (c) Write a differential equation $dy/dt = g(y)$ for this situation, where y is the population in thousands.
 (d) Verify that your equation's equilibrium solutions have the appropriate stability.

23. There are 500,000 fish in a lake and the population $y(t)$ is growing at the rate of $0.2ye^{1-(y/5)} - 0.1y$, where y is measured in hundreds of thousands and t in years.
 (a) Write the initial value problem modeling the fish population, compute its asymptotically stable equilibrium value, and sketch the graph of the solution $y(t)$.
 (b) Suppose harvesting is allowed at the annual rate of $0.3y$. Write the initial value problem modeling the fish population in this situation, compute its new asymptotically stable equilibrium value, and sketch the graph of the solution $y(t)$.
 (c) Suppose harvesting is allowed at the annual rate of ky instead of $0.3y$. Find the value of k that will drive the fish population into extinction.

Chapter 9 Practice Exam

Exercises 1–6 refer to Figure 9.E.1, which displays the direction field for a differential equation of the form $y' = f(t, y)$, for $0 \le t \le 4$ and $0 \le y \le 4$.

1. What is the sign of the derivative of the solution passing through the point $(1, 3)$? Explain.

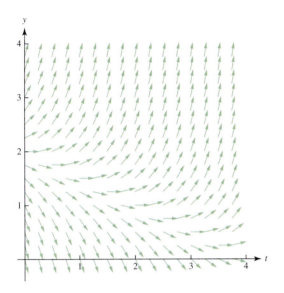

Figure 9.E.1

2. What is the sign of the derivative of the solution corresponding to the initial condition $y(\frac{1}{2}) = 1$? Explain.

3. Is there an equilibrium solution with a value between 0 and 4? Explain.

4. This differential equation has a solution of the form $y = at + b$. Find it.

5. Does the solution corresponding to the initial condition $y(0) = 2$ have a maximum for $0 \le t \le 1$? Explain.

6. Sketch the graph of the solutions corresponding to the initial conditions $y(0) = 2$ and $y(0) = 1$.

Exercises 7–10 refer to Figure 9.E.2, which displays the direction field for a differential equation of the form $y' = f(t, y)$, for $0 \le t \le 4$ and $-0.5 \le y \le 4.5$.

7. Find the three equilibrium solutions and decide their stability. Explain.

8. Which of the solutions corresponding to the initial conditions $y(0) = -0.25$, $y(0) = 1$, $y(0) = 3$, and $y(0) = 4.25$, are increasing functions of t and which are decreasing? Explain.

9. Write an initial condition whose corresponding solution goes from concave up to concave down, and one whose corresponding solution goes from concave down to concave up. Explain. Sketch the graphs of both solutions.

10. Find the limit as $t \to \infty$ of the solutions corresponding to the initial conditions $y(0) = -0.25$, $y(0) = 0.25$, and $y(0) = 3.25$. Explain. Sketch the graphs of these solutions.

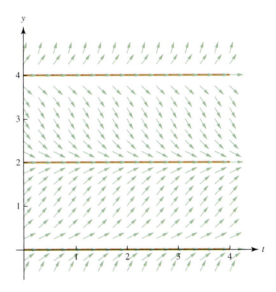

Figure 9.E.2

11. Use Euler's method with $\Delta t = 0.1$ to estimate $y(0.2)$, where $y(t)$ is the solution to the initial value problem $y' = 2y + t$, $y(0) = 1$.

12. Isaac, recovering from an illness, determines that his daily average weight changes according to the differential equation

$$\frac{dw}{dt} = -5 \ln \frac{w}{180},$$

where t is in days and w in pounds. At the beginning of his recovery, Isaac was 165 pounds.
(a) Is there an average daily weight that he tends not to deviate from once he achieves it?
(b) Sketch the graph of his weight $w(t)$ and explain its shape.
(c) Use Euler's method with $\Delta t = 5$ to estimate his weight 15 days later.

13. The population of a region is governed by the logistic model with an intrinsic growth rate of 0.02, environmental carrying capacity of 8 million, and an initial population of 6 million. Since its unemployment level is much higher than most parts of the country, this region's population is leaving at the rate of 1% of the population per year.

(a) Write the initial value problem that models the time evolution of the population $p(t)$, treating it a continuously changing quantity.

(b) Find the equilibrium solutions and then determine their stability.

(c) Sketch the graph of $p(t)$ and explain its shape.

14. Repeat the previous exercise, this time assuming that the population leaves the region at the rate of 14,400 people per year.

Chapter 9 Projects

1. **The error in Euler's method**   If the function $f(t, y)$ is well behaved (e.g., if its partial derivatives are continuous), then the initial value problem

$$\frac{dy}{dt} = f(t, y), \quad y(t_0) = y_0,$$

has a unique solution in an interval [a, b] containing t_0. Moreover, if $y_1, y_2, \ldots, y_n$ are the approximate values of the solution, computed using Euler's method with step size Δt, and $y(t_1), y(t_2), \ldots, y(t_n)$ are the actual solution values, then the **global error**

$$E_n = y_n - y(t_n)$$

satisfies the estimate

$$|y_n - y(t_n)| \leq C \cdot \Delta t \tag{26}$$

for small enough step size Δt, where C is a constant that depends on $f(t, y)$. That is, the global error in Euler's method is proportional to the step size Δt. For an explanation of these facts, we refer the reader to a book on differential equations[6] and numerical methods. Here you will verify estimate (26) numerically in the following two special cases:

(a) $\dfrac{dy}{dt} = te^{-t} - y, y(0) = 2,$ with solution $y(t) = \dfrac{1}{2}e^{-t}(4 + t^2)$

(b) $\dfrac{dy}{dt} = t^2 - y, y(0) = 1,$ with solution $y(t) = 2 - e^{-t} - 2t + t^2.$

First, use Euler's method with step size $\Delta t = 0.1$ to compute y_{10}, the approximate value of $y(1)$, and compute the global error $|y_{10}(1) - y(1)|$ and the ratio $|y_{10}(1) - y(1)|/\Delta t$. Second, repeat the process, but changing the step size from $\Delta t = 0.1$ to $(\Delta t)/2 = 0.05$. Third, repeat it again with step size $(\Delta)/4$, and then again and again, halving the step size each time until it becomes $(\Delta t)/(2^{10}) = (0.1)/(2^{10})$. According to (26), the global error should decrease each time, approximately by a factor of 2.

Finally, find an approximate value for the constant C that makes inequality (26) true. A spreadsheet program may help you to organize these computations.

---

[6]W. Boyce and R. Diprima, *Elementary Differential Equations and Boundary Value Problems*, Seventh Edition New York: John Wiley & Sons, 2001.

2. **Another family of production functions**[7] This is given by

$$Q = (aK^\alpha + L^\alpha)^{1/\alpha}, \quad 0 < \alpha < 1,$$

where a is a positive constant. Consider the particular case $\alpha = \frac{1}{2}$ and do the following:

    (a) Write the production function in the form $q = f(k)$ and sketch its graph.

    (b) Write the Solow differential equation for this production function.

    (c) If $sa^2 > \delta$, show that there is no equilibrium solution.

    (d) If $sa^2 < \delta$, show that there is an equilibrium solution and determine its stability.

    (e) For $a = 0.4$, savings rate $s = 0.3$, and depreciation rate $\delta = 0.1$, compute the steady-state level of capital stock. Then sketch the graphs of the solutions corresponding to the initial conditions $k_0 = 1, 15$, and 40. Make sure to explain their shape (increasing/decreasing, concavity, inflection points, limiting value).

    (f) Consider the Solow differential equation with a, s, and δ as in part (e). Assuming that at $t = 0$ the capital stock is 20, use Euler's method with $\Delta t = 2$ to approximate the values of $k(t)$ over the next 40 years. Then plot them.

    (g) Find Solow's original paper and locate all examples of production functions discussed there. Describe one that is not considered in this chapter.

[7] Robert M. Solow, "A Contribution to the Theory of Economic Growth," *Journal of Economics*, Feb. 1956, pp. 65–94.

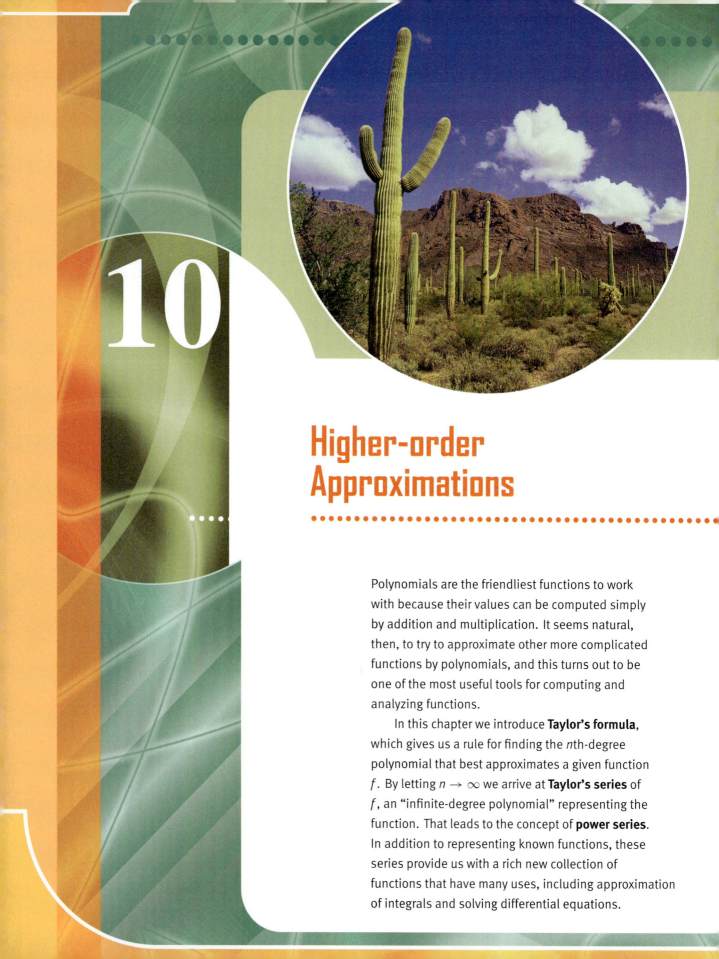

10

Higher-order Approximations

Polynomials are the friendliest functions to work with because their values can be computed simply by addition and multiplication. It seems natural, then, to try to approximate other more complicated functions by polynomials, and this turns out to be one of the most useful tools for computing and analyzing functions.

In this chapter we introduce **Taylor's formula**, which gives us a rule for finding the nth-degree polynomial that best approximates a given function f. By letting $n \to \infty$ we arrive at **Taylor's series** of f, an "infinite-degree polynomial" representing the function. That leads to the concept of **power series**. In addition to representing known functions, these series provide us with a rich new collection of functions that have many uses, including approximation of integrals and solving differential equations.

The simplest example of approximating a function by a polynomial is *linear approximation*, which we studied in Section 3.4. In that case, we use the first-degree polynomial

$$f(a) + f'(a)(x - a) \tag{1}$$

to approximate $f(x)$ for x near a. The guiding idea is that of replacing the graph by its tangent line, as shown in Figure 10.1.1. In other words, we approximate $f(x)$ near a by the linear function through the point $(a, f(a))$ whose derivative (i.e., slope) is the same as that of f at a.

That suggests the following question as the starting point for our study of polynomial approximation: Given $f(x)$, how do we find a polynomial of degree n, say $P(x)$, satisfying the condition

$$P^{(k)}(a) = f^{(k)}(a), \tag{2}$$

for $k = 0, 1, 2, \ldots, n$? (Recall that $f^{(k)}$ stands for the kth derivative, with $f^{(0)}$ denoting f itself.)

Taylor polynomials about zero

For simplicity, we will begin by taking $a = 0$. Write the polynomial in the form (assuming $n \geq 5$)

$$P(x) = c_0 + c_1 x + c_2 x^2 + c_3 x^3 + c_4 x^4 + c_5 x^5 + \cdots + c_n x^n,$$

with the coefficients to be determined. Taking the first three derivatives and evaluating at zero lead to the following table:

k	$P^{(k)}(x)$	$P^{(k)}(0)$
0	$c_0 + c_1 x + c_2 x^2 + c_3 x^3 + c_4 x^4 + c_5 x^5 + \cdots + c_n x^n$	c_0
1	$1 c_1 + 2 c_2 x + 3 c_3 x^2 + 4 c_4 x^3 + 5 c_5 x^4 + \cdots + n c_n x^{n-1}$	$1 c_1$
2	$2 \cdot 1 c_2 + 3 \cdot 2 c_3 x + 4 \cdot 3 c_4 x^2 + 5 \cdot 4 c_5 x^3 + \cdots + n(n-1) c_n x^{n-2}$	$2 \cdot 1 c_2$
3	$3 \cdot 2 \cdot 1 c_3 + 4 \cdot 3 \cdot 2 c_4 x + 5 \cdot 4 \cdot 3 c_5 x^2 + \cdots + n(n-1)(n-2) c_n x^{n-3}$	$3 \cdot 2 \cdot 1 c_3$

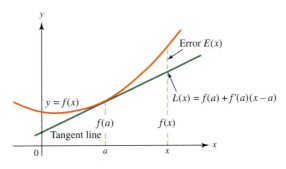

Figure 10.1.1

Examining the right-hand column, we can see a pattern that might lead us to guess that $P^{(4)}(0) = 4 \cdot 3 \cdot 2 \cdot 1 c_4$. That is indeed the case, as we see by taking the derivative of $P^{(3)}(x)$ and evaluating at zero:

$$P^{(4)}(x) = 4 \cdot 3 \cdot 2 \cdot 1 c_4 + 5 \cdot 4 \cdot 3 \cdot 2 c_5 x + \cdots + n(n-1)(n-2)(n-3)x^{n-4}$$
$$P^{(4)}(0) = 4 \cdot 3 \cdot 2 \cdot 1 c_4.$$

Continuing in this way leads to the formula

$$P^{(k)}(0) = k(k-1)(k-2)\cdots 1 c_k. \tag{3}$$

A more concise way of writing this formula is to use the symbol $k!$ (read as k *factorial*) to represent the product of the first k positive integers, that is, for any positive integer k,

$$k! = k(k-1)(k-2)\cdots 1,$$

so that $1! = 1$, $2! = 2$, $3! = 6$, $4! = 24$, $5! = 120$, and so forth. For convenience in later formulas, we also define $0! = 1$.

Solving Eq. (3) for c_k leads to the following expression for the coefficients of a polynomial in terms of its derivatives.

Theorem 10.1.1

If $P(x) = c_0 + c_1 x + c_2 x^2 + \cdots + c_n x^n$, then

$$c_k = \frac{P^{(k)}(0)}{k!}. \tag{4}$$

Example 10.1.1

Verify formula (4) if

$$P(x) = 1 - 7x + 3x^2 + \left(\frac{1}{2}\right)x^3 - x^4 + \left(\frac{2}{3}\right)x^5.$$

SOLUTION   Taking successive derivatives, evaluating at zero, and dividing by the appropriate factorial lead to the following table:

k	$P^{(k)}(x)$	$P^{(k)}(0)$	$P^{(k)}(0)/k!$
0	$1 - 7x + 3x^2 + (1/2)x^3 - x^4 + (2/3)x^5$	1	1
1	$-7 + 6x + (3/2)x^2 - 4x^3 + (10/3)x^4$	-7	-7
2	$6 + 3x - 12x^2 + (40/3)x^3$	6	3
3	$3 - 24x + 40x^2$	3	$1/2$
4	$-24 + 80x$	-24	-1
5	80	80	$2/3$

As you can see, the entries in the right-hand column are equal to the coefficients of $P(x)$.

We are now in a position to answer the question posed at the beginning of the section, at least for the case $a = 0$.

Theorem 10.1.2

Suppose $f(x)$ has derivatives at least through order n at zero. Then there is a unique polynomial

$$P(x) = c_0 + c_1 x + c_2 x^2 + \cdots + c_n x^n$$

of degree n such that $P^{(k)}(0) = f^{(k)}(0)$ for $k = 0, 1, 2, \ldots, n$. The coefficients are given by

$$c_k = \frac{f^k(0)}{k!}. \tag{5}$$

PROOF We already saw in Theorem 10.1.1 that $c_k = P^k(0)/k!$. It follows that $P^{(k)}(0) = f^{(k)}(0)$ if and only if $c_k = f^k(0)/k!$.

The nth-degree poynomial whose coefficients satsify (5) is called the nth-degree **Taylor polynomial** of f about zero. Its coefficents are called the **Taylor coefficients** of f at zero.

EXAMPLE 10.1.2

Find the Taylor polynomials of degrees 1 through 4 of the function $f(x) = (1 - x)^{-2}$ about zero.

SOLUTION   We compute the derivatives through order four and evaluate them at zero. The results are displayed in the following table, with the Taylor coefficients in the right-hand column:

k	$f^{(k)}(x)$	$f^{(k)}(0)$	$f^{(k)}(0)/k!$
0	$(1-x)^{-2}$	1	1
1	$2(1-x)^{-3}$	2	2
2	$3\cdot 2(1-x)^{-4}$	$3\cdot 2$	3
3	$4!(1-x)^{-5}$	$4!$	4
4	$5!(1-x)^{-6}$	$5!$	5

It follows that the Taylor polynomials of degrees 1 through 4 are

$$P_1(x) = 1 + 2x$$
$$P_2(x) = 1 + 2x + 3x^2$$
$$P_3(x) = 1 + 2x + 3x^2 + 4x^3$$
$$P_4(x) = 1 + 2x + 3x^2 + 4x^3 + 5x^4.$$

Figure 10.1.2 shows the graphs of $(1-x)^{-2}$ and its Taylor polynomials of degrees 1 through 4. As you can see, the polynomials approximate the function near zero, and $P_4(x)$ gives a better approximation than do the lower-degree polynomials.

EXAMPLE 10.1.3

Find a formula for the Taylor coefficients of $f(x) = e^x$ and determine the nth-degree Taylor polynomial about zero.

SOLUTION   Since $\dfrac{d}{dx}e^x = e^x$, the successive derivatives are all the same—that is,

$$f^{(k)}(x) = e^x,$$

for every k. Therefore, $f^{(k)}(0) = 1$, and the kth Taylor coefficient is simply $1/k!$. It follows that the nth-degree Taylor polynomial about zero is

$$P_n(x) = 1 + x + \frac{x^2}{2!} + \frac{x^3}{3!} + \cdots + \frac{x^n}{n!}.$$

Figure 10.1.3 on the next page shows the graph of e^x and its Taylor polynomials through degree 4.

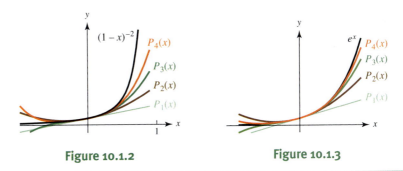

Figure 10.1.2   **Figure 10.1.3**

Taylor polynomials about other points

Next, we derive a formula for the nth-degree Taylor polynomial of a given function $f(x)$ about a, where a is a given number in the domain of f.

Theorem 10.1.3

Suppose that $f(x)$ has derivatives through order n at a. Then the polynomial

$$P(x) = f(a) + \frac{f'(a)}{1!}(x - a) + \frac{f^{(2)}(a)}{2!}(x - a)^2$$

$$+ \cdots + \frac{f^{(n)}(a)}{n!}(x - a)^n$$

(6)

is the unique polynomial of degree n with $P^{(k)}(a) = f^{(k)}(a)$ for $k = 0, 1, 2, \ldots, n$.

PROOF   We make a change of variables by setting $t = x - a$ and let

$$g(t) = f(t + a).$$

By the chain rule, $g(t)$ has derivatives through order n at $t = 0$, and

$$g^{(k)}(0) = f^{(k)}(a),$$

(7)

for k from zero to n. Letting $Q(t)$ denote the Taylor polynomial of g about zero, we have

$$Q(t) = g(0) + \frac{g'(0)}{1!}t + \frac{g^{(2)}(0)}{2!}t^2 + \frac{g^{(3)}(0)}{3!}t^3 + \cdots + \frac{g^{(n)}(0)}{n!}t^n.$$

(8)

Replacing t by $(x - a)$ gives us another polynomial $P(x) = Q(x - a)$, and, taking (7) into account, we see that

$$P(x) = f(a) + \frac{f'(a)}{1!}(x - a) + \frac{f^{(2)}(a)}{2!}(x - a)^2 + \cdots + \frac{f^{(n)}(a)}{n!}(x - a)^n.$$

By the chain rule,

$$P^{(k)}(a) = Q^{(k)}(0) = g^{(k)}(0) = f^{(k)}(a).$$

To see that $P(x)$ is the only polynomial of degree n with this property, suppose there were another, say, $\widehat{P}(x)$. Then, by defining $\widehat{Q}(t) = \widehat{P}(t+a)$, we would obtain a second Taylor polynomial of degree n for $g(t)$, in contradiction to Theorem 10.1.2.

The polynomial defined by formula (6) is called the nth-degree Taylor polynomial of f about a, and its coefficients are called the Taylor coefficients of f at a.

EXAMPLE 10.1.4

Find the nth-degree Taylor polynomial of $f(x) = 1/x$ about $x = 1$.

SOLUTION Taking the first five derivatives and evaluating them at $x = 1$ yield the following table:

k	$f^{(k)}(x)$	$f^{(k)}(1)$	$f^{(k)}(1)/k!$
0	x^{-1}	1	1
1	$-x^{-2}$	-1	-1
2	$2x^{-3}$	2	1
3	$-3!x^{-4}$	$-3!$	-1
4	$4!x^{-5}$	$4!$	1
5	$-5!x^{-6}$	$-5!$	-1

There is a clear pattern for the Taylor coefficients displayed in the right-hand column—specifically, they equal 1 for even k and -1 for odd k, which we can express concisely in the form $(-1)^k$. To see that the pattern continues in that way, suppose we arrive at the kth stage and have

$$f^{(k)}(x) = (-1)^k k! \, x^{-(k+1)},$$

as we have in the table up to $k = 5$. Then taking one more derivative gives

$$f^{(k+1)}(x) = (-1)^k k! \, \frac{d}{dx} x^{-(k+1)} = (-1)^{k+1}(k+1)! \, x^{-(k+2)},$$

and, by setting $x = 1$ and dividing by $(k+1)!$, we see that the $(k+1)$st coefficient equals $(-1)^{k+1}$. Therefore, the coefficients continue to follow the same pattern, and the Taylor polynomial of degree n about $x = 1$ is

$$P(x) = 1 - (x-1) + (x-1)^2 - (x-1)^3 + \cdots + (-1)^n (x-1)^n.$$

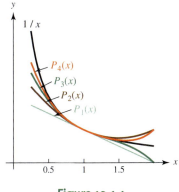

Figure 10.1.4

Figure 10.1.4 shows $f(x)$ and the Taylor polynomials $P_n(x)$ for $n = 1$ through 4 near $x = 1$.

Approximation by Taylor polynomials

How well do the Taylor polynomials approximate the given function? If we let

$$R_n(x) = f(x) - P_n(x),$$

where $f(x)$ is a given function and $P_n(x)$ is its nth-degree Taylor polynomial about a, then $R_n(x)$ is the *error* involved in replacing $f(x)$ by $P_n(x)$. We also call it the **remainder term** to indicate that it is the difference between the actual function and the Taylor polynomial.

Finding a simple, exact expression for $R_n(x)$, one that we can use in calculating, is usually very difficult. The best we can do is *estimate* the size of the remainder as a way of determining the approximation error involved in using $P_n(x)$. The following theorem, known as **Taylor's theorem**, gives just such an estimate.

Theorem 10.1.4

Suppose $f(x)$ is defined and has continuous derivatives through order $(n + 1)$ in an interval around a. Then

$$f(x) = P_n(x) + R_n(x), \tag{9}$$

where $P_n(x)$ is the nth-degree Taylor polynomial about a, and $R_n(x)$ satisfies the inequality

$$\left| R_n(x) \right| \leq \frac{M |x - a|^{n+1}}{(n + 1)!}, \tag{10}$$

with M being the maximum value of $|f^{(n+1)}|$ over the interval between a and x.

Formulas (9) and (10), taken together, are known as **Taylor's formula**. The number M appearing in the error estimate depends on several things—the particular function involved, the degree n of the Taylor polynomial, and the choice of x. Once these choices are made, we can find M by first computing the $(n+1)$st derivative of f, taking its absolute value, and determining the maximum over the interval between a and x.

EXAMPLE 10.1.5

Letting $f(x) = e^x$, use the 5th-degree Taylor polynomial about zero to approximate

$$\text{(i) } e^{-1}, \quad \text{(ii) } e^{-1/2}, \quad \text{(iii) } e^{1/2}.$$

Estimate the error in all cases.

SOLUTION We have already computed the Taylor polynomials of e^x about zero in Example 10.1.3. For $n = 5$, we have

$$P_5(x) = 1 + x + \frac{x^2}{2} + \frac{x^3}{6} + \frac{x^4}{24} + \frac{x^5}{120}. \tag{11}$$

(i) With $x = -1$, Taylor's formula gives

$$e^{-1} = P_5(-1) + R_5(-1).$$

To estimate $R_5(-1)$, we have to find M, the maximum of $|f^{(6)}(x)|$ in the interval $[-1, 0]$. In this case, $|f^{(6)}(x)| = |e^x| = e^x$, which is an increasing function. Therefore, the maximum occurs at the right-hand endpoint, and the maximum value equals 1. Applying (10) with $n = 5$, $a = 0$, $M = 1$, and $x = -1$ gives the estimate

$$\left| R_5(-1) \right| \le \frac{1}{6!} = \frac{1}{720} \approx 0.00139.$$

We conclude that

$$e^{-1} \approx P_5(-1) = 1 - 1 + \frac{1}{2} - \frac{1}{6} + \frac{1}{24} - \frac{1}{120} = \frac{11}{30} \approx 0.3667,$$

and since the error is less than 0.0015, we can conclude that the approximation is correct through at least the first decimal places.

(ii) In this case,

$$P_5\left(-\frac{1}{2}\right) = 1 - \frac{1}{2} + \frac{1}{2!4} - \frac{1}{3!8} + \frac{1}{4!16} - \frac{1}{5!32} = \frac{2,329}{3,840} \approx 0.60651$$

and

$$\left| R_5\left(-\frac{1}{2}\right) \right| \le \frac{M}{6!64}.$$

In this case, M is the maximum of $|f^{(6)}(x)| = e^x$ over the interval $[-\frac{1}{2}, 0]$, and, as before, the maximum occurs at the right-hand endpoint, so that $M = 1$. We conclude that $e^{-1/2} \approx 0.60651$, and the absolute value of the error satisfies

$$\left| R_5(-1) \right| \le \frac{1}{6!64} = \frac{1}{46,080} \approx 0.0000217.$$

(iii) Estimating the remainder for $x = \frac{1}{2}$ introduces a new difficulty. Once again, M is the maximum value of e^x, but this time on the interval $[0, \frac{1}{2}]$. Since e^x is increasing, the maximum occurs at $x = \frac{1}{2}$. So $M = e^{1/2}$, and

$$\left| R_n\left(\frac{1}{2}\right) \right| \le \frac{e^{1/2}\left(\frac{1}{2}\right)^{n+1}}{(n+1)!}. \tag{12}$$

But this estimate, although correct, is not very useful, because it involves the very quantity, namely, $e^{1/2}$, we are trying to approximate!

The way around this problem is to *overestimate* M in order to come up with a number we can use. For instance, we know that $e < 4$; so $e^{1/2} < 2$, and making that substitution in (12) gives us the workable estimate

$$\left| R_n\left(\frac{1}{2}\right) \right| < \frac{2 \cdot \left(\frac{1}{2}\right)^{n+1}}{(n+1)!} = \frac{1}{(n+1)!2^n}.$$

Of course, this estimate is less accurate than (12). Nevertheless, it is correct and easy to compute, and we sometimes have to settle for an estimate that is less than optimal to get one that is usable.

With $n = 5$, we obtain the approximation

$$e^{1/2} \approx P_5\left(\frac{1}{2}\right) = 1 + \frac{1}{2} + \frac{1}{2!4} + \frac{1}{3!8} + \frac{1}{4!16} + \frac{1}{5!32} = \frac{6,331}{3,840} \approx 1.6487,$$

with the error estimate

$$\left| R_5\left(\frac{1}{2}\right) \right| < \frac{1}{6!32} = \frac{1}{23,040} \approx 0.0000434.$$

As the last example shows, there is not always a single "correct" estimate for the remainder in these approximation problems. Formula (10) provides a bound on the size of the error, but it may not be the most accurate possible estimate, and any larger bound is even less optimal but still correct. We may have to forego the best possible estimate in favor of simplifying the calculations involved.

EXAMPLE 10.1.6

Use the 2nd-degree Taylor polynomial of $f(x) = \sqrt{x}$ around an appropriate number a to estimate

(i) $\sqrt{10}$ and   (ii) $\sqrt{8}$.

Estimate the size of the error in both cases and compare your results to the actual estimates given by a calculator.

SOLUTION   The first three derivatives are

$$f'(x) = \frac{1}{2\sqrt{x}}, \quad f^{(2)}(x) = -\frac{1}{4x^{3/2}}, \quad f^{(3)}(x) = \frac{3}{8x^{5/2}}.$$

Taking $a = 9$ gives us a point close to both 8 and 10 at which these derivatives are easy to compute. Using it, we get the Taylor series

$$\sqrt{x} = 3 + \frac{1}{6}(x - 9) - \frac{1}{216}(x - 9)^2 + R_2(x), \tag{13}$$

with

$$\left| R_2(x) \right| \leq \frac{M\,|x - 9|^3}{6}, \tag{14}$$

where M is the maximum of $3/(8x^{5/2})$ on the interval between 9 and x.

(i) If $x = 10$, (13) becomes

$$\sqrt{10} = 3 + \frac{1}{6} - \frac{1}{216} + R_2(10) = \frac{683}{216} + R_2(10).$$

In this case, M is the maximum of $3/(8x^{5/2})$ on the interval $[9, 10]$, which occurs at the left-hand endpoint because the function is decreasing. Therefore, $M = 3/(8 \cdot 9^{5/2}) = \frac{1}{648}$, and

$$\left| R_2(x) \right| \leq \frac{1}{(648) \cdot 6} \approx 0.0002572.$$

We conclude that $\sqrt{10} \approx \frac{683}{216} \approx 3.162037$, with an error no greater than 0.00026 in absolute value.

The actual value of $\sqrt{10}$ to six decimal places is 3.162278, which means our approximation is actually off by about 0.00024.

(ii) With $x = 8$, formula (13) becomes

$$\sqrt{8} = 3 - \frac{1}{6} - \frac{1}{216} + R_2(8) = \frac{611}{216} + R_2(8).$$

This time, M is the maximum of $3/(8x^{5/2})$ on $[8, 9]$, which occurs at 8. Since we cannot easily evaluate $8^{5/2}$, we use the fact that $\sqrt{8} > 2$ to overestimate M as follows:

$$M = \frac{3}{8 \cdot 8^{5/2}} = \frac{3}{8^3 \sqrt{8}} < \frac{3}{8^3 \cdot 2} = \frac{3}{1{,}024}.$$

Therefore,

$$\left| R_2(8) \right| \leq \frac{M}{6} < \frac{3}{6 \cdot 1{,}024} \approx 0.000488,$$

and we conclude that $\sqrt{8} \approx \frac{611}{216} \approx 2.828704$ with an error no greater than 0.00049 in absolute value.

The actual value of $\sqrt{8}$ to six decimal places is 2.828427, which means our approximation is actually off by about 0.000277.

Linear approximation

Taylor's formula for the case $n = 1$ is simply the linear approximation formula we first encountered in Chapter 3

$$f(x) = f(a) + f'(a)(x - a) + R_1(x),$$

with an estimate on the size of the error,

$$\left| R_1(x) \right| \leq \frac{M(x - a)^2}{2},$$

where M is the maximum of $f''(x)$ on the interval between a and x.

EXAMPLE 10.1.7

Using $f(x) = \ln x$ and $a = 1$, estimate the error involved in the linear approximation $\ln(1.1) \approx 0.1$.

SOLUTION   The first two derivatives are $f'(x) = 1/x$ and $f''(x) = -1/x^2$. Taylor's formula with $a = 1$ and $n = 1$ says that $\ln x = (x - 1) + R_1(x)$. Taking $x = 1.1$, we have $\ln(1.1) = 0.1 + R_1(1.1)$, and

$$\left| R_1(1.1) \right| \leq \frac{M(0.1)^2}{2},$$

where M is the maximum of $1/x^2$ on the interval [1, 1.1]. Since $1/x^2$ is decreasing over the interval, the maximum value occurs at $x = 1$ and equals 1. Therefore, we have the error estimate

$$\left| R_1(1.1) \right| \leq \frac{(0.1)^2}{2} = 0.005.$$

Estimating definite integrals

We can estimate $\int_a^b f(x)\,dx$ by replacing the given function by one of its Taylor polynomials about a. To illustrate the method, we first consider an example in which the integral is easy to compute.

EXAMPLE 10.1.8

Estimate $\int_0^{0.2} e^x \, dx$ by using the Taylor polynomial of degree 3 about zero. Compare the estimate to the actual value of the integral obtained by using the fundamental theorem of calculus.

SOLUTION   The 3rd-degree Taylor polynomial of e^x about zero is

$$1 + x + \frac{x^2}{2} + \frac{x^3}{6}$$

(see Example 10.1.3). By using it in place of e^x, we get the estimate

$$\int_0^{0.2} e^x \, dx \approx \int_0^{0.2} \left(1 + x + \frac{x^2}{2} + \frac{x^3}{6}\right) dx$$

$$= \left(x + \frac{x^2}{2} + \frac{x^3}{6} + \frac{x^4}{24}\right)\Bigg|_0^{0.2}$$

$$= 0.2 + \frac{(0.2)^2}{2} + \frac{(0.2)^3}{6} + \frac{(0.2)^4}{24} = 0.2214.$$

By comparison, a direct calculation yields

$$\int_0^{0.2} e^x \, dx = e^x \Bigg|_0^{0.2} = e^{0.2} - 1 \approx 0.221403.$$

The procedure is most useful in those cases when the antiderivative is hard, or even impossible, to compute, as in the next example. In general, the approximation is better if the integral is taken over a small interval.

EXAMPLE 10.1.9

Estimate $\int_0^{0.1} e^{-x^2} \, dx$ by using the Taylor polynomial of degree 2 around zero.

SOLUTION   Letting $f(x) = e^{-x^2}$, we obtain the following table:

k	$f^{(k)}(x)$	$f^{(k)}(0)$	$f^{(k)}(0)/k!$
0	e^{-x^2}	1	1
1	$-2xe^{-x^2}$	0	0
2	$(-2 + 4x^2)e^{-x^2}$	-2	-1

Therefore, the Taylor polynomial is $1 - x^2$, and we have the approximation

$$\int_0^{0.1} e^{-x^2}\, dx \approx \int_0^{0.1} (1 - x^2)\, dx = \left(x - \frac{x^3}{3} \right)\Bigg|_0^{0.1} = 0.1 - \frac{0.001}{3} \approx 0.099667.$$

By using the remainder estimate of Theorem 10.1.4, it is possible to find a bound on the approximation error in such approximations. In the last example, for instance, we can show that the error is less than 0.0002. However, the method is rather technical, and we will not go into it.

An application to differential equations

Euler's method for differential equations, which we studied in Section 9.1, uses linear approximation to estimate the solution of an initial value problem. There are other, more accurate methods based on higher-order Taylor polynomials. Without going into a full description of these methods, we will give a simple example of such an approximation.

EXAMPLE 10.1.10

Suppose $y(t)$ is the solution of the differential equation

$$y' = t - y \tag{15}$$

with initial condition

$$y(0) = 1. \tag{16}$$

Approximate $y(0.1)$ by using a 3rd-degree Taylor approximation.

SOLUTION   What we seek is the Taylor polynomial of $y(t)$ of degree 3 about zero:

$$P(t) = y(0) + y'(0)t + \frac{y''(0)}{2!}t^2 + \frac{y^{(3)}(0)}{3!}t^3.$$

To find the coefficients, we first observe that (16) gives $y(0) = 1$. Moreover, by substituting $t = 0$ into both sides of (15), we get

$$y'(0) = 0 - y(0) = -1. \tag{17}$$

To find $y''(0)$, we first differentiate both sides of (15) with respect to t, which gives

$$y'' = 1 - y'. \tag{18}$$

PROBLEM-SOLVING TACTIC

The key idea here is to use the differential equation $y' = f(t, y)$ itself to find the Taylor coefficients.

- $y(0)$ is given by the initial condition.
- $y'(0)$ is obtained by substituting $t = 0$ and $y = y(0)$ into $f(t, y)$.
- $y''(0)$ is obtained by differentiating f and then substituting the previous values.

For higher-order approximations, we continue in the same way.

Substituting $t = 0$ into both sides of this equation and using (17) give

$$y''(0) = 1 - y'(0) = 2. \tag{19}$$

Finally, to find $y^{(3)}(0)$, we differentiate both sides of (18) to get

$$y^{(3)} = -y'',$$

and evaluate both sides at $t = 0$, which gives

$$y^{(3)}(0) = -y''(0) = -2. \tag{20}$$

Combining (16–20) gives

$$P(t) = 1 - t + t^2 - \frac{2}{3!}t^3 = 1 - t + t^2 - \frac{1}{3}t^3.$$

Using this polynomial, we get the approximation to six decimal places:

$$y(0.1) \approx P(0.1) \approx 0.909667.$$

Remark   In this case, an explicit solution of the initial value problem is given by

$$y(t) = 2e^{-t} + t - 1,$$

from which we get $y(0.1) = 2e^{-0.1} + 0.1 - 1 \approx 0.909675$ to six decimal places.

How to derive Taylor's formula

We start with the **fundamental theorem of calculus**:

$$f(x) - f(a) = \int_a^x f'(t)\,dt.$$

In this formulation, x is a *constant*. For simplicity, we will assume that $x > a$. (The case of $x < a$ is similar, but slightly more technical.)

We apply integration by parts to the right-hand side, with

$$u = f'(t) \quad \text{and} \quad \frac{dv}{dt} = 1.$$

Then

$$\frac{du}{dt} = f''(t).$$

As for v, instead of taking $v = t$, we will use

$$v = t - x.$$

(Remember that in this integral x plays the role of a constant, and we can adjust any antiderivative by adding or subtracting a constant.) The integration by parts formula gives

$$
\begin{aligned}
f(x) - f(a) &= \int_a^x f'(t)\,dt \\
&= (t - x)f'(t)\Big|_a^x - \int_a^x (t - x)f''(t)\,dt \\
&= f'(a)(x - a) + \int_a^x (x - t)f''(t)\,dt.
\end{aligned}
$$

We will rewrite this in the form

$$
f(x) = f(a) + f'(a)(x - a) + R_1(x), \tag{21}
$$

where

$$
R_1(x) = \int_a^x (x - t)f''(t)\,dt. \tag{22}
$$

We now have an explicit form of the remainder $R_1(x)$, but an exact evaluation is generally impossible. However, we can estimate the integral in (22) by taking M to be the maximum of $|f''(t)|$ for t between a and x and using some standard properties of the integral (see Exercise 59):

$$
\begin{aligned}
|R_1(x)| &= \left| \int_a^x (x - t)f''(t)\,dt \right| \\
&\leq \int_a^x |(x - t)f''(t)|\,dt \\
&\leq \int_a^x M(x - t)\,dt = -M\frac{(x - t)^2}{2}\Big|_a^x = \frac{M(x - a)^2}{2}.
\end{aligned}
$$

This estimate, together with Formula (21), comprises Taylor's formula for the case $n = 1$. For higher-degree approximations, the proof proceeds in the same vein. Instead of stopping at Formulas (21) and (22), we continue integration by parts until we have written $f(x)$ as a sum of its nth-degree Taylor polynomial and an integral remainder involving $f^{(n+1)}(t)$, which we then can estimate. The details are not hard to work out, but we will not go into them here. (See Exercise 59.)

Practice Exercises 10.1

1. Compute the nth-degree Taylor polynomial of $f(x)$ about a in each of the following cases:

    (a) $f(x) = 2x^3 - \frac{3}{2}x^2 + 2x + \frac{1}{2}$, $a = 1$, $n = 3$

    (b) $f(x) = (1 + x)^{-3}$, $a = 0$, $n = 4$

2. Use the 2nd-degree Taylor polynomial of $x^{1/3}$ about 8 to approximate $(8.5)^{1/3}$. Estimate the size of the error.

Exercises 10.1

In Exercises 1–4, verify that the coefficients of the polynomial satisfy Formula (4).

1. $P(x) = 1 + \dfrac{x}{2} + \dfrac{x^2}{3} + \dfrac{x^3}{4} + \dfrac{x^4}{5}$

2. $P(x) = x + 2x^2 + 3x^3 + 4x^4 + 5x^5$

3. $P(x) = 1 - x + x^2 - x^3 + x^4 - x^5 + x^6$

4. $P(x) = 1 - \dfrac{x^2}{2} + \dfrac{x^4}{4} - \dfrac{x^6}{8}$

5. The function $f(x) = (1 + x)^n$ can be written as an nth-degree polynomial

$$f(x) = c_0 + c_1 x + c_2 x^2 + c_3 x^3 + \cdots + c_n x^n.$$

By computing derivatives and using Theorem 10.1.1, determine the coefficients in each of the following cases:
   (a) $n = 4$ (b) $n = 5$ (c) $n = 6$

6. (a) Compute the first four derivatives of $f(x) = (1+x)^n$ and use them to guess a formula in terms of n, k, and x for $f^{(k)}(x)$.
   (b) Check that your formula is correct by showing that if it holds for $f^{(k)}(x)$, it also holds for $f^{(k+1)}(x)$.
   (c) Use your formula and Theorem 10.1.1 to determine a formula for the coefficient c_k in the expansion

$$(1 + x)^n = c_0 + c_1 x + c_2 x^2 + \cdots + c_k x^k + \cdots + c_n x^n.$$

7. Find a formula for expanding $(u + v)^n$ for any u and v by writing

$$(u + v)^n = u^n \left(1 + \frac{v}{u}\right)^n$$

and applying the result of the previous exercise. This formula is known as the **binomial formula**.

In Exercises 8–11, compute the nth-degree Taylor polynomial of the given polynomial $f(x)$ about a. By expanding and collecting terms in your answer, verify that the two polynomials are the same.

8. $f(x) = x^2 + x + 1$, $a = 2$, $n = 2$

9. $f(x) = x^3 - 3x^2 + 2x + 1$, $a = 1$, $n = 3$

10. $f(x) = \frac{1}{2}x^3 - \frac{1}{3}x$, $a = 1$, $n = 3$

11. $f(x) = x^4$, $a = -1$, $n = 4$

In Exercises 12–21, compute the nth-degree Taylor polynomial of $f(x)$ about a. Simplify the coefficients as much as possible.

12. $f(x) = e^{-2x}$, $a = 0$, $n = 5$

13. $f(x) = \ln x$, $a = 1$, $n = 6$

14. $f(x) = \dfrac{1}{x^2}$, $a = -1$, $n = 5$

15. $f(x) = e^{-x^2}$, $a = 0$, $n = 4$

16. $f(x) = x^{-1/2}$, $a = 1$, $n = 4$

17. $f(x) = (1 + x)^{1/3}$, $a = 0$, $n = 3$

18. $f(x) = x \ln(1 + x)$, $a = 0$, $n = 5$

19. $f(x) = e^x + e^{-x}$, $a = 0$, $n = 6$

20. $f(x) = \sin x$, $a = 0$, $n = 5$

21. $f(x) = \cos x$, $a = \pi$, $n = 6$

In Exercises 22–29, compute the Taylor coefficients through degree 8 about the given point a. Then determine the pattern they follow and write a general formula for the kth coefficient.

22. $f(x) = e^{-x}$, $a = 0$ **23.** $f(x) = e^{2x}$, $a = 0$

24. $f(x) = x^{-1}$, $a = 1$ **25.** $f(x) = \ln x$, $a = 1$

26. $f(x) = \ln(1 - x)$, $a = 0$

27. $\dfrac{1}{(1 + x)^2}$, $a = 0$

28. $f(x) = \cos x$, $a = 0$

29. $f(x) = \sin x$, $a = 0$

30. Let $f(x) = xe^x$.
   (a) Compute the Taylor polynomials of degrees 1 through 5 about zero and compare them to those of e^x (see Example 10.1.3). What do you observe, and can you formulate a general rule that might explain it?
   (b) Try to find a pattern in the sequence of Taylor coefficients and use it to write a formula for the nth-degree Taylor polynomial. Compare it to that of e^x to see if it obeys the rule you formulated in (a).

31. Estimate $e^{-0.1}$ by using the 2nd-degree Taylor polynomial of e^x about zero. Estimate the size of the error.

32. Estimate the size of the error in approximating e by $P_5(1)$, where $P_5(x)$ is the 5th-degree Taylor polynomial of e^x about zero. You may use the fact that $e < 3$.

33. Find the approximations of $\sqrt{5}$ obtained by using the nth-degree Taylor polynomials of $\sqrt{x}$ about 4 for $n = 1, 2,$ and 3. In each case estimate the size of the error.

34. Use the 3rd-degree Taylor polynomial of $\ln(1 + x)$ about zero to approximate $\ln(1.1)$. Estimate the size of the error.

35. (a) Find a formula for the nth-degree Taylor polynomial of $\ln(1 + x)$ about zero.
   (b) In approximating $\ln(1.1)$, what choice of n will ensure you of an error with an absolute value no greater than 10^{-6}?

36. Use the 5th-degree Taylor polynomial of cos x about zero to approximate the cosine of two radians and estimate the size of the error. (*Hint:* In estimating the error, recall that $|\cos t| \le 1$ for every t.)

In Exercises 37–40, estimate the integral by using the Taylor polynomial of degree n around a. Then evaluate the integral by using the fundamental theorem of calculus and compare the result.

37. $\displaystyle\int_0^1 e^{x/2}\,dx, \;\; n = 3, \; a = 0$

38. $\displaystyle\int_0^{0.5} \cos x\,dx, \;\; n = 2, \; a = 0$

39. $\displaystyle\int_0^1 xe^x\,dx, \;\; n = 4, \; a = 0$

40. $\displaystyle\int_1^{1.5} \ln x\,dx, \;\; n = 4, \; a = 1$

In Exercises 41–42, estimate the integral by using the Taylor polynomial of degree n around a.

41. $\displaystyle\int_0^{0.1} \frac{1}{1+x^2}\,dx, \;\; n = 2, \; a = 0$

42. $\displaystyle\int_0^{0.5} e^{-x^2/2}\,dx, \;\; n = 4, \; a = 0$

43. A parcel of land has a river as its northern boundary. The other boundaries are straight, and the southern boundary is 100 feet long. For the width of the parcel, the river follows the curve

$$h(t) = \frac{40}{(t^2 + 5)^2}, \quad 0 \le t \le 1,$$

where t and h are both in hundreds of feet. (See Figure 10.1.5.) Approximate the area of the parcel by integrating the 2nd-degree Taylor polynomial of $h(t)$ around zero.

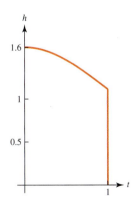

Figure 10.1.5

In Exercises 44–47, find the nth-degree Taylor polynomial of y(t) about zero, where y(t) is the solution of the given initial value problem, and use it to approximate y(0.1).

44. $y' = y + t + 1, \;\; y(0) = 1, \; n = 2$

45. $y' = y + e^t - t, \;\; y(0) = 0, \; n = 3$

46. $y' = 2ty + 1, \;\; y(0) = 1, \; n = 3$

47. $y' = t + y^2, \;\; y(0) = -1, \; n = 2$

48. If $f'(x) = e^{-x^2/2}$ and $f(0) = 1$, find the 5th-degree Taylor polynomial of $f(x)$ around zero.

49. A book is published, and the initial orders total 38,000 copies. The orders arrive often and change according to the formula

$$\frac{dB}{dy} = 0.4(y + 0.2)^{-0.6}, \tag{23}$$

where y is in years and B is in thousands of copies.
(a) Find the 4th-degree Taylor polynomial of $B(y)$ around zero. Use a calculator to find the coefficients to three decimal places.
(b) Use the Taylor polynomial to estimate sales for y from 0 to 0.5 in increments of 0.1.
(c) Find $B(y)$ by solving Eq. (23) with initial condition $B(0) = 38$ and compare the actual values of B with the previous estimates. How does the accuracy of the estimates change as y increases? What does that say about predicting the future?

50. Systematics is the study of animal classification. Quite often, measurements are taken of many members of a species—length of hind foot, width of carapace, body weight, etc. This gives an expected range for the species, which can help differentiate between two very similar species. In measuring the dorsal fin of one fish species, it was found to be 3 centimeters in 600 of the fish measured. The fin lengths follow a variation rate of

$$\frac{dF}{dc} = -2(c - 3)F,$$

where F is in thousands of fish measured and c is in centimeters.
(a) Use a third-order Taylor polynomial to estimate the number of fish in the sample having a dorsal fin length of 3.1 centimeters.

(b) Check your answer by solving the differential equation with initial condition $F(3) = 600$ and then computing $F(3.1)$.

51. Assume that the U.S. population $p(t)$, in billions, is modeled by the logistic equation

$$\frac{dp}{dt} = 0.02p(1-p),$$

where t is measured in years. The population is expected to be 0.31 billion in the year 2010. Find the 2nd-degree Taylor polynomial of $p(t)$ around zero and use it to estimate the population in 2011.

52. Graph $f(x) = e^x$ and its Taylor polynomials of degrees 1 through 4 on the same screen, using a $[0, 3] \times [0, 10]$ window. Which of the polynomials fits the graph best? For which x do the polynomials best approximate the function? Next, use a $[-2, 0] \times [0, 1]$ window and answer the same questions.

In Exercises 53–58, compute the nth-degree Taylor polynomial of $f(x)$ about a and use a graphing calculator or computer program to graph the function and Taylor polynomial in a suitable window.

53. $e^x - x$, $a = 0$, $n = 3$

54. x^5, $a = 1$, $n = 2$

55. xe^{-x}, $a = 0$, $n = 5$

56. $\ln\left(\dfrac{x}{x+1}\right)$, $a = 1$, $n = 2$

57. $\dfrac{x-1}{x+1}$, $a = 0$, $n = 3$

58. $\cos x - \sin x$, $a = 0$, $n = 4$

59. In this exercise, you will complete one more step in the derivation of Taylor's formula. Assume, for simplicity, that $x > a$.

(a) Show that applying integration by parts to formula (22), with $u = f''(t)$ and $dv/dt = (t - x)$, and substituting the result in (21) lead to

$$f(x) = f(a) + f'(a)(x-a) + \frac{f''(a)}{2}(x-a)^2 + R_2(x),$$

where $R_2(x) = \frac{1}{2}\int_a^x (x-t)^2 f^{(3)}(t)\,dt$.

(b) Show that $|R_2(x)| \le M|x-a|^3/3!$. You can use the following general properties of the integral over an interval $[a, b]$:

(i) For any function $G(t)$, $|\int_a^b G(t)\,dt| \le \int_a^b |G(t)|\,dt$.

(ii) If $G(t) \le H(t)$ for $a \le t \le b$, then $\int_a^b G(t)\,dt \le \int_a^b H(t)\,dt$.

Solutions to practice exercises 10.1

1. (a) The following table gives the derivatives and Taylor coefficients:

k	$f^{(k)}(x)$	$f^{(k)}(1)$	$f^{(k)}(1)/k!$
0	$2x^3 - \frac{3}{2}x^2 + 2x + \frac{1}{2}$	3	3
1	$6x^2 - 3x + 2$	5	5
2	$12x - 3$	9	9/2
3	12	12	2

The third-degree Taylor polynomial is $P(x) = 3 + 5(x-1) + \frac{9}{2}(x-1)^2 + 2(x-1)^3$.

(b) The following table gives the derivatives and Taylor coefficients:

k	$f^{(k)}(x)$	$f^{(k)}(0)$	$f^{(k)}(0)/k!$
0	$(1+x)^{-3}$	1	1
1	$-3(1+x)^{-4}$	-3	-3
2	$4 \cdot 3(1+x)^{-5}$	$4 \cdot 3$	6
3	$-5 \cdot 4 \cdot 3(1+x)^{-6}$	$-5 \cdot 4 \cdot 3$	-10
4	$6 \cdot 5 \cdot 4 \cdot 3(1+x)^{-7}$	$6 \cdot 5 \cdot 4 \cdot 3$	15

The Taylor polynomial is $P(x) = 1 - 3x + 6x^2 - 10x^3 + 15x^4$.

2. If $f(x) = x^{1/3}$, the first three derivatives are

$$f'(x) = \frac{1}{3}x^{-2/3}, \quad f^{(2)}(x) = -\frac{2}{9}x^{-5/3},$$

$$f^{(3)}(x) = \frac{10}{27}x^{-8/3}.$$

Taylor's formula gives $f(x) = P_2(x) + R_2(x)$, with

$$P_2(x) = 2 + \frac{1}{12}(x-8) - \frac{1}{288}(x-8)^2$$

and

$$|R_2(x)| \le \frac{M|x-8|^3}{6},$$

where M is the maximum of $\frac{10}{27}x^{-8/3}$ on the interval $[8, 8.5]$. The maximum is taken at the left-hand endpoint, so that $M = 10/(27 \cdot 256)$.

Thus, we get the approximation

$$(8.5)^{1/3} \approx P_2(8.5) = 2 + \frac{0.5}{12} - \frac{0.25}{288} \approx 2.0407986,$$

and we conclude that the absolute value of the error no larger than $(10 \times (0.5)^3)/(27 \cdot 256 \cdot 6) \approx 0.00003$.

10.2 Taylor Series

In a number of examples in the last section we used a Taylor polynomial to approximate the value of a given function $f(x)$. If the function has derivatives of all orders in an interval around a, then there is an nth-degree Taylor polynomial about a for *every* n, given by

$$P_n(x) = \frac{f(a)}{0!} + \frac{f'(a)}{1!}(x-a) + \frac{f^{(2)}(a)}{2!}(x-a)^2 + \cdots + \frac{f^{(n)}(a)}{n!}(x-a)^n. \quad (24)$$

It seems natural to ask whether we can make the approximation $f(x) \approx P_n(x)$ as good as we want by taking n large enough. To put it another way, writing $f(x) = P_n(x) + R_n(x)$,

- does $\lim_{n \to \infty} P_n(x) = f(x)$, or, what amounts to the same thing,
- does $\lim_{n \to \infty} R_n(x) = 0$?

The answer, as you might expect, depends on the function f, the number a, and the particular choice of x. Later in this section we will resolve a number of important cases.

If $\lim_{n \to \infty} P_n(x) = f(x)$, we can think of $f(x)$ as being represented by an infinite sum of powers of $(x-a)$. That is,

$$f(x) = \frac{f(a)}{0!} + \frac{f'(a)}{1!}(x-a) + \frac{f^{(2)}(a)}{2!}(x-a)^2 + \frac{f^{(3)}(a)}{3!}(x-a)^3 + \cdots, \quad (25)$$

where $\cdots$ indicates that the sum continues indefinitely. The right-hand side of this equation is called the **Taylor series** for $f(x)$ about a.

A more concise way to write Eq. (25) is as follows:

$$f(x) = \sum_{k=0}^{\infty} \frac{f^{(k)}(a)}{k!}(x-a)^k. \quad (26)$$

In this notation, the Greek letter Σ (capital sigma) is used to indicate summation—substituting successive integer values for k, beginning with the number displayed below the Σ, and adding. The ∞ symbol on top indicates that the sum continues indefinitely. More precisely, it indicates that we take a limit as $n \to \infty$ of the Taylor polynomial $P_n(x)$. By using the Σ notation, we can rewrite Eq. (24) in the form

$$P_n(x) = \sum_{k=0}^{n} \frac{f^{(k)}(a)}{k!}(x-a)^k,$$

and Eq. (26) is another way of writing

$$f(x) = \lim_{n \to \infty} \sum_{k=0}^{n} \frac{f^{(k)}(a)}{k!}(x-a)^k.$$

In terms of the Taylor series, our question can be rephrased as follows: *If a function has derivatives of all orders in an interval around a, for what x does it equal its Taylor series? Or, to put it another way, for what x is Eq. (26) valid?*

An example: the geometric series

A simple but instructive example is furnished by the function

$$f(x) = \frac{1}{1-x}, \quad x \neq 1.$$

For degrees 0 through 4, the derivatives and Taylor coefficients at $a = 0$ are displayed in the following table:

k	$f^{(k)}(x)$	$f^{(k)}(0)$	$f^{(k)}(0)/k!$
0	$(1-x)^{-1}$	1	1
1	$(1-x)^{-2}$	1	1
2	$2(1-x)^{-3}$	2	1
3	$3!(1-x)^{-4}$	3!	1
4	$4!(1-x)^{-5}$	4!	1

At this point it is not hard to spot a pattern—the coefficients are all equal to 1. Why is that so? Because each time we take the derivative, we bring down the exponent, thereby increasing the coefficient to the next higher factorial. More explicitly, if $f^{(k)}(x) = k!\,(1-x)^{-(k+1)}$, then $f^{(k)}(0)/k! = 1$, and taking the next derivative gives

$$f^{(k+1)}(x) = \frac{d}{dx}\,k!\,(1-x)^{-(k+1)} = (k+1)!\,(1-x)^{-(k+2)},$$

so that $f^{(k+1)}(0)/(k+1)! = 1$.

Thus, the nth-degree Taylor polynomial is

$$P_n(x) = 1 + x + x^2 + \cdots + x^n. \tag{27}$$

The right-hand side is called a **finite geometric series** (or, sometimes, a **geometric progression**). How well does it approximate the function, and for what x does $P_n(x) \to 1/(1-x)$ as $n \to \infty$? In this case, we have precise answers because there is a simple form for the remainder $R_n(x)$. To find it, we begin by multiplying both sides of (27) by x, which gives

$$x P_n(x) = x + x^2 + x^3 + \cdots + x^{n+1}. \tag{28}$$

Next, we subtract Eq. (28) from Eq. (27). On the right-hand side, most of the terms cancel—every term except the 1 in (27) and the x^{n+1} in (28) appear twice, once in each equation. Therefore, the subtraction results in the equation

$$(1-x) P_n(x) = 1 - x^{n+1}. \tag{29}$$

If $x \neq 1$, dividing both sides by $(1-x)$ gives

$$P_n(x) = \frac{1-x^{n+1}}{1-x} = \frac{1}{1-x} - \frac{x^{n+1}}{1-x}. \tag{30}$$

Finally, by rearranging terms, we get

$$\frac{1}{1-x} = P_n(x) + \frac{x^{n+1}}{1-x}. \tag{31}$$

The left-hand side is the given function, and the right-hand side consists of the Taylor polynomial plus a remainder. Therefore, we conclude that

$$R_n(x) = \frac{x^{n+1}}{1-x}. \tag{32}$$

For which values of x does $\lim_{n\to\infty} R_n(x) = 0$? To answer, we begin by observing the following.

Lemma 10.2.1

If x is fixed and $n \to \infty$, then

$$\lim_{n\to\infty} |x|^n = \begin{cases} 0 & \text{if } |x| < 1 \\ \infty & \text{if } |x| > 1. \end{cases}$$

PROOF Recall that $|x|^n = e^{n \ln |x|}$. If $|x| > 1$, then $\ln |x| > 0$ and $n \ln |x| \to \infty$ as $n \to \infty$, which means that $e^{n \ln |x|} \to \infty$. On the other hand, if $0 < |x| < 1$, then $\ln |x| < 0$ and $n \ln |x| \to -\infty$ as $n \to \infty$, which means that $e^{n \ln |x|} \to 0$.

With this preparation, we can now determine whether $\lim_{n\to\infty} P_n(x) = f(x)$.

Theorem 10.2.1

Let $P_n(x) = 1 + x + x^2 + \cdots + x^n$. Then,

$$P_n(x) = \frac{1 - x^{n+1}}{1-x}. \tag{33}$$

If $|x| < 1$, $\lim_{n\to\infty} P_n(x) = 1/(1-x)$, which we symbolize by writing

$$\sum_{k=0}^{\infty} x^k = \frac{1}{1-x} \quad \text{for } |x| < 1. \tag{34}$$

On the other hand, $P_n(x)$ does not have a finite limit if $|x| \geq 1$.

Remark The sum on the left-hand side of (34) is called an **infinite geometric series.** If $|x| < 1$, we say the series **converges** to $1/(1-x)$. If $|x| \geq 1$, we say the series **diverges**.

In general, we call an infinite sum of the form $\sum_{k=0}^{\infty} g_k$ an **infinite series**. The g_k's may either be numbers or expressions involving a variable such as x. In the case of the geometric series, for instance, $g_k = x^k$.

We say the series **converges** if $(g_0 + g_1 + \cdots + g_n)$ has a finite limit as $n \to \infty$; otherwise, we say the series **diverges**.

PROOF   Using (31), we can write

$$\left| \frac{1}{1-x} - P_n(x) \right| = \frac{|x|^{n+1}}{|1-x|}.$$

Keeping x fixed and letting $n \to \infty$, we get

$$\lim_{n \to \infty} \left| \frac{1}{1-x} - P_n(x) \right| = \lim_{n \to \infty} \frac{|x|^{n+1}}{|1-x|} = \begin{cases} 0 & \text{if } |x| < 1 \\ \infty & \text{if } |x| > 1. \end{cases}$$

We conclude that if $|x| < 1$,

$$\lim_{n \to \infty} \left[\frac{1}{1-x} - P_n(x) \right] = 0,$$

which is the same as $\lim_{n \to \infty} P_n(x) = 1/(1-x)$.

To see what happens if $|x| > 1$, we rewrite (30) in the form

$$P_n(x) = \frac{x^{n+1} - 1}{x - 1},$$

for $x \neq 1$. By analyzing the quotient on the right, we see that if $x > 1$, $P_n(x) \to \infty$ as $n \to \infty$. If $x < -1$, we have $\lim_{n \to \infty} |P_n(x)| = \infty$, but the sign of $P_n(x)$ will alternate between positive and negative as $n \to \infty$, and the limit does not exist. Finally, if $x = 1$, $P_n(x) = n + 1 \to \infty$, and if $x = -1$, $P_n(x)$ alternates between the values 1 and zero, and therefore does not have a well-determined limit.

Geometric series appear in many contexts in mathematics, among them decimal expansions of real numbers and present and future value problems.

Decimal expansions of real numbers

Every real number can be written in decimal form with either a finite or infinite number of digits, and conversely, every (finite or infinite) decimal expansion represents a real number. The representation is unique—that is, no two decimal expansions represent the same real number—if we rule out those decimals that end in an infinite string of 0's or 9's.

Decimal expansions with repeating patterns represent rational numbers.

EXAMPLE 10.2.1

Use the geometric series formula to represent each of the following infinite, repeating decimal expansions as a simple fraction:

(i) $0.\overline{7} = 0.777\ldots$ and   (ii) $1.34\overline{27} = 1.34272727\ldots.$

SOLUTION   (i) We can represent the decimal by an infinite sum, which we can then factor into the product of a fraction and an infinite geometric series.

$$0.777\ldots = \frac{7}{10} + \frac{7}{100} + \frac{7}{1,000} + \cdots$$

$$= \frac{7}{10}\left\{1 + \frac{1}{10} + \left(\frac{1}{10}\right)^2 + \left(\frac{1}{10}\right)^3 + \cdots\right\}$$

$$= \frac{7}{10}\cdot\frac{1}{1-\frac{1}{10}} = \frac{7}{10}\cdot\frac{10}{9} = \frac{7}{9}.$$

(ii) First, we rewrite the number so that the repeating pattern begins right after the decimal point:

$$1.34272727\ldots = 10^{-2}\cdot(134.27272727\ldots).$$

Next, write the repeating part as a series:

$$0.272727\ldots = \frac{27}{100} + \frac{27}{100^2} + \frac{27}{100^3} + \frac{27}{100^4} + \cdots$$

$$= \frac{27}{100}\sum_{k=0}^{\infty}\left(\frac{1}{100}\right)^k$$

$$= \frac{27}{100}\left\{\frac{1}{1-\frac{1}{100}}\right\} = \frac{27}{99} = \frac{3}{11}.$$

Therefore,

$$1.34272727\ldots = 10^{-2}\cdot\left(134 + \frac{3}{11}\right) = \frac{1,477}{1,100}.$$

In general, a decimal expansion of the form $0.d_1d_2d_3d_4,\ldots$, where the d_i's are integers between 0 and 9 inclusive, represents the real number that is the sum of the infinite series

$$\sum_{k=1}^{\infty}\frac{d_k}{10^k}. \tag{35}$$

This series converges to a real number for any set of digits $d_1, d_2, d_3, d_4,\ldots$. To see why, we observe that $0 \le d_k/10^k \le 9/10^k$ and

$$\sum_{k=1}^{\infty}\frac{9}{10^k} = 9\sum_{k=1}^{\infty}\frac{1}{10^k} = \frac{9}{10}\left(\frac{1}{1-\frac{1}{10}}\right) = 1.$$

Using a comparison test (which we will not prove, but which is similar to the comparison test for improper integrals given in Section 6.5), we can conclude that series (35) converges.

Present and future value

In Section 6.2 we used improper integrals to find the present and future values of a continuous income stream under the assumption that interest is continuously compounded. We can use an infinite geometric series in an analogous way in cases when the deposits are not continuously made and the interest is not continuously compounded.

> **EXAMPLE 10.2.2**

Suppose you want to establish a trust fund that will provide an annual scholarship in your name to a worthy undergraduate at your alma mater, and, since you expect your alma mater to be around for hundreds of years, you want the scholarship to last forever. As a hedge against inflation, you would like the size of the scholarship to increase by 3% every year, starting with $1,000 at the end of the first year. You endow the fund by investing a fixed amount that will earn an annual interest rate of 8%, compounded quarterly. How large must your endowment be?

SOLUTION To begin with, recall that if you invest A dollars at 8% annual interest, compounded quarterly, the amount in the fund at the end of k years will be $A(1 + \frac{0.08}{4})^{4k}$. Therefore, in order to have D dollars at the end of k years, you have to choose A to satisfy the equation

$$D = A(1.02)^{4k}.$$

This equation is easily solved for A:

$$A = D(1.02)^{-4k}. \tag{36}$$

(A is called the **present value** of D dollars payable after k years at 8% interest, compounded quarterly.)

According to formula (36), the amount you have to invest in order to award a scholarship of $1,000 at the end of 1 year is

$$a_1 = 1{,}000(1.02)^{-4}.$$

On top of that, you want to award a scholarship of $1,000(1.03)$ at the end of 2 years (the previous year's award plus 3%). To do that, you must increase the initial amount by

$$a_2 = 1{,}000(1.03)(1.02)^{-8}.$$

Similarly, at the end of 3 years you want the fund to pay out $1,000(1.03)^2$ (again, the previous year's award plus 3%). Therefore, you need to increase the initial investment by

$$a_3 = 1{,}000(1.03)^2(1.02)^{-12}.$$

Continuing in this way, we see that the total initial investment is given by the infinite series

$$a_1 + a_2 + a_3 + \cdots = 1{,}000(1.02)^{-4} + 1{,}000(1.03)(1.02)^{-8} + 1{,}000(1.03)^2(1.02)^{-12} + \cdots$$
$$= 1{,}000(1.02)^{-4}\left\{1 + (1.03)(1.02)^{-4} + (1.03)^2(1.02)^{-8} + \cdots\right\}.$$

Letting $r = (1.03)(1.02)^{-4} \approx 0.95$, we have

$$a_1 + a_2 + a_3 + \cdots = 1{,}000(1.02)^{-4} \sum_{k=0}^{\infty} r^k = 1{,}000(1.02)^{-4} \frac{1}{1-r}.$$

Replacing r by $(1.03)(1.02)^{-4}$, we obtain \$19,072.26.

The exponential and logarithm functions

As we already saw in Example 10.1.3, taking successive derivatives of the exponential function $f(x) = e^x$ simply reproduces the function over and over—that is, $f^{(k)}(x) = e^x$, and the kth Taylor coefficient about zero is given by $f^{(k)}(0)/k! = 1/k!$. Taylor's formula takes the form

$$e^x = \sum_{k=0}^{n} \frac{x^k}{k!} + R_n(x). \tag{37}$$

As we see in the following theorem, the Taylor series converges to e^x for every choice of x.

Theorem 10.2.2

For every x, $e^x = \lim_{n \to \infty} \sum_{k=0}^{n} x^k/k!$. In other words,

$$e^x = \sum_{k=0}^{\infty} \frac{x^k}{k!}. \tag{38}$$

To justify this result, we need the following fact: *For any positive number r,*

$$\lim_{m \to \infty} \frac{r^m}{m!} = 0. \tag{39}$$

Although we will not give a formal proof of this fact, we will explain the idea behind it. Given the sequence of terms,

$$\frac{r}{1!}, \quad \frac{r^2}{2!}, \quad \frac{r^3}{3!}, \quad \frac{r^4}{4!}, \quad \frac{r^5}{5!}, \quad \cdots,$$

we observe that each term comes from the previous one by multiplying it by r over the next consecutive integer. In other words, the second term is just the first term multiplied by $r/2$, the third term is just the second term multiplied by $r/3$, and so forth. In general, the $(m + 1)$st term is just the mth term multiplied by $r/(m + 1)$. As m gets larger and larger, it eventually reaches the point where

$$\frac{1}{2} > \frac{r}{m+1} > \frac{r}{m+2} > \frac{r}{m+3} > \cdots.$$

In fact, that happens as soon as $m \geq 2r$. From that point on, each term of the sequence is *less than half* its predecessor. Therefore, the terms must get arbitrarily small, and $r^m/m! \to 0$ as $m \to \infty$.

With the help of (39), we can show that $R_n(x) \to 0$ as $n \to \infty$ in Eq. (37). We begin by recalling the remainder estimate

$$|R_n(x)| \le \frac{M|x|^{n+1}}{(n+1)!},$$

where M is the maximum value of e^t on the interval between 0 and x. If we keep x fixed and let $n \to \infty$, M also stays fixed. Therefore,

$$\lim_{n \to \infty} \frac{M|x|^{n+1}}{(n+1)!} = M \lim_{n \to \infty} \frac{|x|^{n+1}}{(n+1)!} = M \cdot 0 = 0.$$

It follows that $|R_n(x)|$ can be made arbitrarily small for all n sufficiently large, which means that $R_n(x) \to 0$ as $n \to \infty$. This shows the validity of (38).

The **natural logarithm** has the following Taylor series representation:

Theorem 10.2.3

If $0 < x \le 2$,

$$\ln x = \sum_{k=1}^{\infty} (-1)^{k-1} \frac{(x-1)^k}{k} = (x-1) - \frac{(x-1)^2}{2} + \frac{(x-1)^3}{3} - \cdots. \quad (40)$$

By setting $t = x - 1$ in this formula, so that $x = t + 1$, we get an equivalent version, representing $\ln(1+t)$ as a Taylor series about zero: If $-1 < t \le 1$,

$$\ln(1+t) = \sum_{k=1}^{\infty} (-1)^{k-1} \frac{t^k}{k} = t - \frac{t^2}{2} + \frac{t^3}{3} + \cdots. \quad (41)$$

To derive formula (40), we begin by making a table of the first few derivatives and Taylor coefficients about 1, as follows:

k	$f^{(k)}(x)$	$f^{(k)}(1)$	$f^{(k)}(0)/k!$
0	$\ln x$	0	0
1	x^{-1}	1	1
2	$-x^{-2}$	-1	$-1/2$
3	$2x^{-3}$	2	$1/3$
4	$-3!x^{-4}$	$-3!$	$-1/4$
5	$4!x^{-5}$	$4!$	$1/5$

At this point, it is clear what is happening: $f^{(k)}(x) = (-1)^{k-1}(k-1)! x^{-k}$, so that $f^{(k)}(1) = (-1)^{k-1}(k-1)!$ and

$$\frac{f^{(k)}(1)}{k!} = (-1)^{k-1} \frac{(k-1)!}{k!} = (-1)^{k-1} \frac{1}{k}.$$

To check that this pattern continues, we take it one step further, from the kth to the $(k + 1)$st derivative:

$$f^{k+1}(x) = \frac{d}{dx} f^{(k)}(x) = (-1)^{k-1}(k - 1)! \frac{d}{dx} x^{-k} = (-1)^k k! \, x^{-(k+1)}, \qquad (42)$$

so that $f^{(k+1)}(1) = (-1)^k k!$ and $f^{(k+1)}(1)/(k + 1)! = (-1)^k(1/(k + 1))$.

Taylor's formula now gives

$$\ln x = \sum_{k=1}^{n}(-1)^{k-1}\frac{(x - 1)^k}{k} + R_n(x),$$

and we need to show that $\lim_{n\to\infty} R_n(x) = 0$ if $0 < x \leq 2$. That is a bit complicated if $0 < x < 1$, and we will restrict our attention to the case in which $1 \leq x \leq 2$. Once again, we use the remainder estimate:

$$|R_n(x)| \leq \frac{M|x - 1|^{n+1}}{(n + 1)!},$$

where M is the maximum of $|f^{(n+1)}|$ on $[1, x]$. In this case, using (42) with $k = n$, we have

$$|f^{(n+1)}(x)| = |(-1)^n n! \, x^{-(n+1)}| = \frac{n!}{x^{n+1}}.$$

The right-hand side is a decreasing function for positive x, so that its maximum occurs at the left-hand endpoint of the interval $[1, 2]$. Therefore, $M = n!$, and the error estimate takes the form

$$|R_n(x)| \leq \frac{|x - 1|^{n+1}}{n + 1} \leq \frac{1}{n + 1}$$

for $1 \leq x \leq 2$. It follows that $R_n(x)$ can be made arbitrarily small for all n sufficiently large, so that $\lim_{n\to\infty} R_n(x) = 0$.

We have now verified that the Taylor series representation given by (40) is valid for $1 \leq x \leq 2$, and we will accept without proof the validity for $0 < x < 1$. On the other hand, formula (40) is not true if either $x \leq 0$ or $x > 2$. In the former case, $\ln x$ is not defined, but the case of $x > 2$ is less obvious. We will examine that further in the next section when we take up the subject of power series.

Practice Exercises 10.2

1. Represent the infinite decimal $0.3\overline{14}$ as a simple fraction.

2. Find the Taylor series of $(3 - x)^{-2}$ about zero.

3. Use a calculator or spreadsheet program to estimate $\ln(1.2)$ by adding the terms of an appropriate Taylor series through degree 10.

Exercises 10.2

In Exercises 1–8, determine if the infinite series converges and, if so, determine the sum.

1. $\displaystyle\sum_{k=0}^{\infty}\left(\frac{2}{3}\right)^k$

2. $\displaystyle\sum_{k=0}^{\infty}\left(\frac{3}{2}\right)^k$

3. $1 - \dfrac{1}{5} + \left(\dfrac{1}{5}\right)^2 - \left(\dfrac{1}{5}\right)^3 + \left(\dfrac{1}{5}\right)^4 - \cdots$

4. $1 - 1 + 1 - 1 + 1 - 1 + \cdots$

5. $1 + e^{-1} + e^{-2} + e^{-3} + \cdots$

6. $\displaystyle\sum_{k=0}^{\infty} 2^{-3k}$

7. $\dfrac{3}{2} - \dfrac{3}{4} + \dfrac{3}{8} - \dfrac{3}{16} + \cdots$ **8.** $\displaystyle\sum_{k=2}^{\infty}(0.03)^k$

In Exercises 9–12, represent the given infinite decimal as a simple fraction.

9. $0.\overline{21}$

10. $0.1244444\ldots$

11. $2.\overline{13}$

12. $0.09999\ldots$

13. In order to establish a perpetual fund that pays a flat amount of $1,000 at the end of every year, you invest a certain lump sum in an account paying 5% annual interest, compounded monthly, and leave it there without adding anything to it. How much must you invest?

14. Suppose you want to establish a perpetual fund that pays out a certain amount at the end of every year, starting with $5,000 at the end of the first year and increasing the amount by 2% every year. How much must you endow the fund with if your investment pays 6% annual interest, compounded continuously?

15. Dudley took his brand new "high-bounce" ball up to his room and dropped it from the window to the large, flat driveway 27 feet below. His cousin measured each bounce and saw that it always rebounded to 82% of its previous height.
(a) Describe the behavior of the ball from the time it is dropped to the third time it touches the drive and sketch a diagram. What is the total vertical distance it travels during that time?
(b) If the ball is allowed to bounce indefinitely, what is the total vertical distance it will travel?

16. Zonko's, a joke shop in the town of Hogsmeade, sells a "high-bounce" ball with a magic spell on it that makes it rebound to 110% of its previous height. Ron purchased one and dropped it from a window to a courtyard 40 feet below.
(a) Describe the behavior of the ball from the time it is dropped to the third time it touches the courtyard and sketch a diagram. What is the total vertical distance it travels during that time?
(b) What can you say about the total vertical distance the ball will travel if it is allowed to bounce indefinitely?
(c) Unfortunately, the spell restricts the ball to 4,000 feet total distance traveled. How many bounces will Ron see before the spell stops the ball? (*Hint:* Use formula (30) in the text to get a formula in terms of n for the total distance traveled after n bounces.)

17. Suppose a patient takes 10 milligrams of a certain anti-inflammatory steroid at the beginning of each day. The drug is eliminated exponentially from the patient's system in such a way that a single 10 milligram dose will leave $10e^{-(0.8)t}$ milligrams remaining in the patient's system after t days.

At the end of the first day, the patient has taken one 10 milligram dose, of which $10e^{-(0.8)}$ remains in his system. At the end of the second day, the patient has $10e^{-1.6}$ of the first day's dose remaining, plus $10e^{-0.8}$ of the second day's dose, for a total of $10(e^{-0.8} + e^{-1.6})$.
(a) By the end of the third day, the patient has taken a total of three doses. How many milligrams remain in his system?
(b) How many milligrams remain in the patient's system at the end of n days? Express your answer in terms of a finite geometric series.
(c) Assuming the treatment is continued indefinitely, use an infinite geometric series to approximate the amount remaining in the patient's system after a very long period.

18. A patient is injected once a day with 50 units of a certain drug. Suppose this drug is eliminated exponentially, with any single injection leaving an amount of $50e^{ct}$ remaining after t days for some constant c.
(a) Suppose that for any single injection, exactly one-half the amount administered is eliminated after 36 hours. Use that information to determine c.
(b) If the patient is injected every day for n days, how many units remain in the patient's system at the end of n days? Express your answer in terms of a finite geometric series.
(c) Assuming the treatment is continued indefinitely, use an infinite geometric series to approximate the number of units of the drug remaining in the patient's system after a very long period.

19. The managers of a fish farm harvest their fish twice a year. The fish reproduce at the rate of 22% per year. How large must the initial stock be for them to harvest 1,000 each time without running out of fish? Model the fish population as money invested, treating the reproduction rate as continuously compounded interest.

20. A calculator with a 10-digit display uses the Taylor series of e^x to calculate e. The display shows e to be equal to 2.718281828. Up to what power of x does the calculator have to compute the series in order to come up with this decimal representation? Answer the question by calculating each partial sum to 10 digits until you find one that agrees with the calculator's version of e.

In Exercises 21–28, find the Taylor series about a. If possible, write a general formula using the Σ notation.

21. $f(x) = \dfrac{1}{2+x}$, $a = 0$ **22.** $f(x) = \dfrac{1}{x^2}$, $a = 1$

23. $f(x) = \ln(2 - x)$, $a = 1$

24. $f(x) = e^{-3x}$, $a = 0$

25. $f(x) = 2^x$, $a = 0$ **26.** $f(x) = \dfrac{x}{1-x}$, $a = 0$

27. $f(x) = (1 - 3x)^{-2}$, $a = 0$

28. $f(x) = x \ln x - x$, $a = 1$

29. Find the Taylor series for $\ln(1 - t)$ about zero and check that it is equal to the series you get by replacing t by $-t$ in formula (41).

30. Find the Taylor series for $t \ln(1 + t)$ about zero and check that it is equal to the series you get by multiplying the series in formula (41) by t.

31. Let $f(x) = (1 + x)^r$, where r is a constant.
  (a) Show that the Taylor series for $f(x)$ around zero is

$$1 + \sum_{k=1}^{\infty} \frac{r(r - 1) \cdots (r - k + 1)}{k!} x^k.$$

  (b) Show that if r is a positive integer, the series reduces to a polynomial of degree r. Compute the polynomial for $r = 2, 3, 4$, and 5.
  (c) Write the series in the simplest possible form for the case of $r = -1$. How does it relate to the geometric series?
  (d) Write the series in the simplest possible form for the case of $r = -2$. How does it relate to the series you obtained in (c)? Can you think of a reason?

32. Show that the Taylor series for $\cos x$ about zero is

$$1 - \frac{x^2}{2!} + \frac{x^4}{4!} - \frac{x^6}{6!} + \cdots.$$

(*Hint:* You should be able to determine a pattern for the derivatives and Taylor coefficients after the first four or five.)

33. Continuation of the previous problem Show that the Taylor series of $\cos x$ converges to $\cos x$ for every x. (*Hint:* Use the facts that $|\cos x| \le 1$ and $|\sin x| \le 1$ in estimating the remainder and also use formula (39).)

34. Use a graphing calculator or computer program to graph $\cos x$ and the finite sums

$$1 - \frac{x^2}{2!} + \frac{x^4}{4!} - \frac{x^6}{6!} + \cdots \pm \frac{x^{2m}}{(2m)!}$$

(called the **partial sums** of the Taylor series) for $m = 1, 2$, and 3 to see how well the polynomials fit the graph of $\cos x$ near zero.

35. Show that the Tayor series for $\sin x$ about zero is

$$x - \frac{x^3}{3!} + \frac{x^5}{5!} - \frac{x^7}{7!} + \cdots.$$

(*Hint:* You should be able to determine a pattern for the derivatives and Taylor coefficients after the first four or five.)

36. Continuation of the previous problem Show that the Taylor series of $\sin x$ converges to $\sin x$ for every x. (*Hint:* Use the facts that $|\cos x| \le 1$ and $|\sin x| \le 1$ in estimating the remainder and also use formula (39).)

37. Use a graphing calculator or computer program to graph $\sin x$ and the finite sums

$$x - \frac{x^3}{3!} + \frac{x^5}{5!} - \frac{x^7}{7!} + \cdots \pm \frac{x^{2m+1}}{(2m + 1)!}$$

(called the partial sums of the Taylor series) for $m = 1, 2$, and 3 to see how well the polynomials fit the graph of $\sin x$ near zero.

In Exercises 38–46, use a calculator or spreadsheet program to approximate the given number by adding the terms of an appropriate Taylor series through the degree specified.

38. e, $n = 5$

39. e, $n = 10$

40. $e^{-0.2}$, $n = 4$

41. $\ln(1.5)$, $n = 10$

42. $\ln(0.9)$, $n = 6$

43. $\ln 2$, $n = 50$

44. $\cos(\pi/5)$, $n = 6$

45. $\sin(\pi/9)$, $n = 5$

46. $\sin(\pi/180)$, $n = 3$

Solutions to practice exercises 10.2

1. First we write $0.3\overline{14} = (10)^{-1} 3.\overline{14}$. Next, we use the geometric series formula:

$$0.\overline{14} = \frac{14}{100} \sum_{k=0}^{\infty} \left(\frac{1}{100} \right)^k = \frac{14}{99}.$$

Therefore, $0.3\overline{14} = (10)^{-1}(3 + \frac{14}{99}) = \frac{311}{990}$.

2. We first make a table of the first few derivatives and Taylor coefficients:

k	$f^{(k)}(x)$	$f^{(k)}(0)$	$f^{(k)}(0)/k!$
0	$(3 - x)^{-2}$	3^{-2}	3^{-2}
1	$2(3 - x)^{-3}$	$2 \cdot 3^{-3}$	$2 \cdot 3^{-3}$
2	$3!(3 - x)^{-4}$	$3!3^{-4}$	$3 \cdot 3^{-4}$
3	$4!(3 - x)^{-5}$	$4!3^{-5}$	$4 \cdot 3^{-5}$
4	$5!(3 - x)^{-6}$	$5!3^{-6}$	$5 \cdot 3^{-6}$

From this table we see the pattern

$$f^{(k)}(x) = (k+1)!(3-x)^{-(k+2)}$$

and

$$\frac{f^{(k)}(0)}{k!} = (k+1) \cdot 3^{-(k+2)}.$$

To verify that this pattern continues, we check that if the kth derivative has this form, then

$$f^{(k+1)}(x) = \frac{d}{dx}(k+1)!(3-x)^{-(k+2)}$$
$$= (k+2)!(3-x)^{-(k+3)},$$

so that $f^{(k+1)}(0)/(k+1)! = (k+2) \cdot 3^{-(k+3)}$. Therefore,

the Taylor series is

$$\sum_{k=0}^{\infty} \frac{(k+1)}{3^{k+2}} x^k.$$

3. We can use either the Taylor series for $\ln(1+x)$ about zero or the series for $\ln x$ about 1. In both cases, we get

$$\ln(1.2) = \sum_{k=1}^{\infty} (-1)^{k+1} \frac{(0.2)^k}{k} = \sum_{k=1}^{\infty} (-1)^{k+1} \frac{1}{k \, 5^k}.$$

The sum of the first 10 terms is

$$\frac{1}{5} - \frac{1}{2 \cdot 5^2} + \frac{1}{3 \cdot 5^3} - \cdots + \frac{1}{9 \cdot 5^9} - \frac{1}{10 \cdot 5^{10}}.$$

A calculator or spreadsheet evaluates this sum to eight decimal places as 0.18232156.

10.3 Power Series

A **power series** about a is an infinite series of the form

$$\sum_{k=0}^{\infty} c_k (x-a)^k = c_0 + c_1(x-a) + c_2(x-a)^2 + c_3(x-a)^3 + \cdots, \qquad (43)$$

where the coefficients $c_0, c_1, c_2, \ldots$ are constants. The Taylor series of a given function $f(x)$ is an example of a power series, with $c_k = f^{(k)}(a)/k!$.

In addition to giving us a useful way of representing functions we already know, such as exponentials and logarithms, power series give us a way of defining new functions,

HISTORICAL PROFILE

Brook Taylor (1685–1731) was an accomplished musician and painter as well as a mathematician. He attended Cambridge University and had already written several important mathematical articles by the time he graduated in 1709. He was elected to the Royal Society in 1712 and served as its secretary from 1714 to 1718. His scientific work covered topics as diverse as capillary action, magnetism, and the oscillations of vibrating strings.

In addition to developing the series that bear his name, Taylor is credited with inventing the method of integration by parts and creating a new branch of mathematics, now known as the calculus of finite differences. His mathematical productivity was diminished by declining health and a series of personal tragedies. Both his first and second wives died in childbirth, and only one of those children survived. Nevertheless, his contributions to mathematics surpass the one theorem that bears his name.

enlarging the class of functions beyond those we already know and enabling us to deal with problems that cannot be solved using elementary formulas. Later in this section, for example, we will see how power series provide a powerful method for finding integrals and solving differential equations that are too complicated for the techniques we already know.

To keep things simple, we will restrict our attention in this section to the case in which $a = 0$, so that (43) takes the form

$$\sum_{k=0}^{\infty} c_k x^k = c_0 + c_1 x + c_2 x^2 + c_3 x^3 + \cdots. \tag{44}$$

Following the terminology for infinite series introduced in the last section, we say the power series

- **converges** if $\sum_{k=0}^{n} c_k x^k$ has a finite limit as $n \to \infty$, and
- **diverges** if it does not.

The radius of convergence

The following theorem, which we will not prove, shows that the set of x for which a given power series of the form (44) converges is an interval that is symmetric about zero.

Theorem 10.3.1

Given a power series $\sum_{k=0}^{\infty} c_k x^k$, either

- the series converges for every x, or
- the series diverges for every $x \neq 0$, or
- there is a positive number R so that the series converges for $|x| < R$ and diverges for $|x| > R$.

For instance, we saw in Theorems 10.2.1 and 10.2.2 that

- $\sum_{k=0}^{\infty} x^k / k!$ converges for all x, and
- $\sum_{k=0}^{\infty} x^k$ converges for $|x| < 1$ and diverges for $|x| > 1$.

An example of a power series that diverges for all $x \neq 0$ is given in Example 10.3.1 below.

The number R is called the **radius of convergence** of the power series. By allowing R to represent 0 and ∞ as well as a positive number, we can summarize all three cases of Theorem 10.3.1 by saying the following: There is an R so that the power series converges for $|x| < R$ and diverges for $|x| > R$.

If $R \neq 0$, we refer to the interval $(-R, R)$ as the **interval of convergence** of the power series.

There are a number of techniques for finding the radius of convergence of a given power series. One of the simplest and most effective is the following, called the **ratio test**.

> ## Theorem 10.3.2
>
> Given a power series $\sum_{k=0}^{\infty} c_k x^k$, suppose that
>
> $$\lim_{k \to \infty} \left| \frac{c_{k+1}}{c_k} \right| = A,$$
>
> where A is either zero, a positive number, or ∞.
>
> - If $A = 0$, then $R = \infty$.
> - If $A = \infty$, then $R = 0$.
> - If $0 < A < \infty$, then $R = 1/A$.

Although we will not prove this theorem, the following example illustrates how to use it.

EXAMPLE 10.3.1

Use the ratio test to find the radius and interval of convergence of each of the following power series:

$$\text{(i)} \quad \sum_{k=0}^{\infty} \frac{1}{3^k} x^k = 1 + \frac{1}{3} x + \frac{1}{3^2} x^2 + \frac{1}{3^3} x^3 + \cdots.$$

$$\text{(ii)} \quad \sum_{k=0}^{\infty} (k+1) x^k = 1 + 2x + 3x^2 + 4x^3 + \cdots.$$

$$\text{(iii)} \quad \sum_{k=0}^{\infty} (-1)^k \frac{x^k}{k!} = 1 - x + \frac{x^2}{2!} - \frac{x^3}{3!} + \cdots.$$

$$\text{(iv)} \quad \sum_{k=0}^{\infty} k! x^k = 1 + x + 2! x^2 + 3! x^3 + \cdots.$$

SOLUTION   (i) In this case, $c_k = 1/3^k$, and (replacing k by $k+1$) $c_{k+1} = 1/3^{k+1}$. Therefore,

$$\lim_{k \to \infty} \left| \frac{c_{k+1}}{c_k} \right| = \lim_{k \to \infty} \frac{1/3^{k+1}}{1/3^k} = \lim_{k \to \infty} \frac{3^k}{3^{k+1}} = \lim_{k \to \infty} \frac{1}{3} = \frac{1}{3}.$$

We conclude that $R = 3$, and the interval of convergence is $(-3, 3)$.

(ii) Here we have $c_k = (k+1)$, and (again replacing k by $k+1$) we have $c_{k+1} = (k+2)$. Therefore,

$$\lim_{k \to \infty} \left| \frac{c_{k+1}}{c_k} \right| = \lim_{k \to \infty} \frac{k+2}{k+1} = 1,$$

which means that $R = 1$ and the interval of convergence is $(-1, 1)$.

(iii) Here, $c_k = (-1)^k/k!$ and $c_{k+1} = (-1)^{k+1}/(k+1)!$, so that

$$\frac{c_{k+1}}{c_k} = -\frac{k!}{(k+1)!} = \frac{-1}{k+1}.$$

It follows that

$$\lim_{k\to\infty}\left|\frac{c_{k+1}}{c_k}\right| = \lim_{k\to\infty}\frac{1}{k+1} = 0.$$

Therefore, $R = \infty$, and the series converges for every x.

(iv) Since $c_k = k!$ and $c_{k+1} = (k+1)!$,

$$\lim_{k\to\infty}\left|\frac{c_{k+1}}{c_k}\right| = \lim_{k\to\infty}\frac{(k+1)!}{k!} = \lim_{k\to\infty}(k+1) = \infty.$$

Therefore, $R = 0$ and the series diverges for every $x \neq 0$.

Properties of power series

A power series defines a function on its interval of convergence—that is, for each input value x in the interval $(-R, R)$ there is a well-determined, finite output value. We can think of this class of functions as extending the class of polynomials by allowing *infinite* sums, provided the series converges. And, in fact, power series have many of the properties of polynomials.

Theorem 10.3.3

Suppose $f(x) = \sum_{k=0}^{\infty} a_k x^k$ and $g(x) = \sum_{k=0}^{\infty} b_k x^k$. Then

$$f(x) + g(x) = \sum_{k=0}^{\infty}(a_k + b_k)x^k, \tag{45}$$

and the formula is valid on the smaller of the intervals of convergence of f and g. Also, for any constant c and any integer m,

$$cx^m f(x) = \sum_{k=0}^{\infty} ca_k x^{k+m}, \tag{46}$$

and, if $c \neq 0$, the series on the right has the same radius of convergence as f.

We will omit the proof of this theorem, but it is not hard to derive it. Formula (45), for instance, follows from the fact that the limit of a sum is the sum of the limits, together with the usual method for adding polynomials term by term. Formula (46) follows from the same basic property of limits.

EXAMPLE 10.3.2

Find a power series representation for $\dfrac{x^2}{x-1}$ and determine its radius of convergence.

SOLUTION   From the geometric series formula (see Theorem 10.2.1), we have

$$\frac{1}{1-x} = \sum_{k=0}^{\infty} x^k,$$

with radius of convergence $R = 1$. Therefore,

$$\frac{x^2}{x-1} = -\frac{x^2}{1-x} = -\sum_{k=0}^{\infty} x^{k+2} = -x^2 - x^3 - x^4 - \cdots,$$

with a radius of convergence equal to 1.

EXAMPLE 10.3.3

Let $f(x) = \frac{1}{2}(e^x + e^{-x})$. Find a power series representation in two ways:
(i) by computing its Taylor series around zero, and
(ii) by adding power series representing e^x and e^{-x}.

SOLUTION   (i) If we compute the first two derivatives and Taylor coefficients, we get the following table:

k	$f^{(k)}(x)$	$f^{(k)}(0)$	$f^{(k)}(0)/k!$
0	$\frac{1}{2}(e^x + e^{-x})$	1	1
1	$\frac{1}{2}(e^x - e^{-x})$	0	0
2	$\frac{1}{2}(e^x + e^{-x})$	1	$1/2!$

Since $f^{(2)}(x) = f(x)$, the derivatives simply alternate between the same pair of functions, $\frac{1}{2}(e^x \pm e^{-x})$, and we conclude that

$$f^{(k)}(0) = \begin{cases} 0 & \text{if } x \text{ is odd} \\ 1 & \text{if } x \text{ is even.} \end{cases}$$

Therefore, we get the Taylor series

$$\sum_{j=0}^{\infty} \frac{x^{2j}}{(2j)!} = 1 + \frac{x^2}{2!} + \frac{x^4}{4!} + \cdots.$$

It is not hard to show, using Taylor's formula with remainder, that the series converges for all x and has the same values as the given function. However, the next method yields the same conclusion much more easily.
(ii) We know that for every x,

$$e^x = \sum_{k=0}^{\infty} \frac{x^k}{k!} = 1 + x + \frac{x^2}{2!} + \frac{x^3}{3!} + \frac{x^4}{4!} + \cdots. \tag{47}$$

If we replace x by $-x$, we get

$$e^{-x} = \sum_{k=0}^{\infty} \frac{(-x)^k}{k!} = 1 - x + \frac{x^2}{2!} - \frac{x^3}{3!} + \frac{x^4}{4!} - \cdots, \tag{48}$$

which is also valid for every x. Multiplying (47) and (48) each by $\frac{1}{2}$ and adding, we see that

$$\frac{1}{2}(e^x + e^{-x}) = 1 + \frac{x^2}{2!} + \frac{x^4}{4!} + \cdots$$

for all x.

Remark Formula (46) can be generalized to the product of a power series and a polynomial, and can be extended even further to the product of two power series. Briefly put, multiplying power series is similar to multiplying polynomials, with the following pattern for the coefficients:

$$(a_0 + a_1 x + a_2 x^2 + \cdots)(b_0 + b_1 x + b_2 x^2 + \cdots)$$
$$= a_0 b_0 + (a_1 b_0 + a_0 b_1)x + (a_2 b_0 + a_1 b_1 + a_0 b_2)x^2 + \cdots,$$

valid on the smaller of the two intervals of convergence of the given power series.

Derivatives and integrals of power series

We can also differentiate a power series term by term, just like a polynomial.

Theorem 10.3.4

Suppose $f(x) = \sum_{k=0}^{\infty} a_k x^k$ for $-R < x < R$. Then

$$f'(x) = \sum_{k=1}^{\infty} k a_k x^{k-1} = a_1 + 2a_2 x + 3a_3 x^2 + \cdots, \qquad (49)$$

valid for $-R < x < R$.

Once again, we omit the proof. However, we will provide a few examples.

EXAMPLE 10.3.4

Find a series representation for $(1 - x)^{-2}$ in two ways:
  (i) by computing its Taylor series around zero, and
  (ii) by differentiating both sides of the geometric series formula.

SOLUTION (i) As usual, we begin with a table of derivatives and Taylor coefficients:

k	$f^{(k)}(x)$	$f^{(k)}(0)$	$f^{(k)}(0)/k!$
0	$(1-x)^{-2}$	1	1
1	$2(1-x)^{-3}$	2	2
2	$3!(1-x)^{-4}$	$3!$	3
3	$4!(1-x)^{-5}$	$4!$	4
4	$5!(1-x)^{-6}$	$5!$	5

At this point the pattern is clear, and we obtain the Taylor series

$$\sum_{k=0}^{\infty}(k+1)x^k = 1 + 2x + 3x^2 + 4x^3 + \cdots.$$

By using the remainder estimate in Taylor's formula, one can show that the series equals the given function for $|x| < 1$, but the proof is somewhat technical.

(ii) The geometric series formula says that

$$(1-x)^{-1} = 1 + x + x^2 + x^3 + x^4 + \cdots,$$

valid for $|x| < 1$. Taking derivatives on both sides, we get

$$(1-x)^{-2} = 1 + 2x + 3x^2 + 4x^3 + \cdots = \sum_{k=0}^{\infty}(k+1)x^k,$$

also valid for $|x| < 1$.

By repeatedly applying Theorem 10.3.4, we get power series representations of the higher-order derivatives as well, such as

$$f^{(2)}(x) = \sum_{k=2}^{\infty} k(k-1)x^{k-2} = 2a_2 + 3\cdot 2a_3 x + 4\cdot 3a_4 x^2 + 5\cdot 4a_5 x^3 + \cdots, \qquad (50)$$

$$f^{(3)}(x) = \sum_{k=3}^{\infty} k(k-1)(k-2)x^{k-3} = 3\cdot 2a_3 + 4\cdot 3\cdot 2a_4 x + 5\cdot 4\cdot 3a_5 x^2 + \cdots, \qquad (51)$$

and so forth. An interesting and useful consequence of these formulas is that the Taylor series is the *only* power series representing a given function. More precisely, we can state the following.

Theorem 10.3.5

Suppose $f(x) = \sum_{k=0}^{\infty} a_k x^k$ for all x in some interval around zero. Then

$$a_k = \frac{f^{(k)}(0)}{k!}.$$

This theorem is similar to Theorem 10.1.1 for polynomials and is proved in exactly the same way. We first observe that $f(0) = a_0$. Next,

• setting $x = 0$ in (49) gives $f'(0) = a_1$,
• setting $x = 0$ in (50) gives $f^{(2)}(0) = 2a_2$,
• setting $x = 0$ in (51) gives $f^{(3)}(0) = 3!a_3$,

and so forth. At the kth stage, we have

$$f^{(k)}(x) = k(k-1)\cdots 3 \cdot 2a_k + \text{ terms involving positive powers of } x,$$

and setting $x = 0$ gives $f^{(k)}(0) = k!a_k$.

EXAMPLE 10.3.5

Find the Taylor series of the function $f(x) = e^{-x^2}$.

SOLUTION If we try to do this by the usual method of computing successive derivatives, the computations soon get out of hand. For instance,

$$f'(x) = -2xe^{-x^2}$$
$$f^{(2)}(x) = 2e^{-x^2}(2x^2 - 1)$$
$$f^{(3)}(x) = 4xe^{-x^2}(3 - 2x^2).$$

Instead, we start with the exponential series formula,

$$e^t = \sum_{k=0}^{\infty} \frac{t^k}{k!} = 1 + t + \frac{t^2}{2!} + \frac{t^3}{3!} + \frac{t^4}{4!} + \cdots \tag{52}$$

and substitute $t = -x^2$, which gives

$$e^{-x^2} = \sum_{k=0}^{\infty} (-1)^k \frac{x^{2k}}{k!} = 1 - x^2 + \frac{x^4}{2!} - \frac{x^6}{3!} + \frac{x^8}{4!} - \cdots. \tag{53}$$

Since (52) is valid for all t, Eq. (53) is valid for all x. By appealing to Theorem 10.3.5, we conclude that the series in (53) is, in fact, the Taylor series of $f(x)$.

As with polynomials, we can also find the antiderivative of a power series term by term.

Theorem 10.3.6

Suppose $f(x) = \sum_{k=0}^{\infty} a_k x^k$ for $-R < x < R$. Then the power series

$$\sum_{k=0}^{\infty} \frac{a_k}{k+1} x^{k+1} = a_0 x + \frac{a_1}{2} x^2 + \frac{a_2}{3} x^3 + \cdots$$

is an antiderivative of $f(x)$ for $-R < x < R$.

EXAMPLE 10.3.6

Use a power series to approximate $\displaystyle\int_0^1 e^{-x^2}\, dx$.

SOLUTION   The Taylor series expansion of e^{-x^2} was derived in Example 10.3.5 and is given by formula (53). Applying Theorem 10.3.6, we obtain an antiderivative:

$$F(x) = \sum_{k=0}^{\infty}(-1)^k \frac{x^{2k+1}}{k!\,(2k+1)} = x - \frac{x^3}{3} + \frac{x^5}{2!\,5} - \frac{x^7}{3!\,7} + \frac{x^9}{4!\,9} - \cdots.$$

Using the fundamental theorem of calculus and the fact that $F(0) = 0$, we get

$$\int_0^1 e^{-x^2}\,dx = F(1) = \sum_{k=0}^{\infty}\frac{(-1)^k}{k!(2k+1)} = 1 - \frac{1}{3} + \frac{1}{2!5} - \frac{1}{3!7} + \frac{1}{4!9} - \cdots.$$

The following table displays the approximations obtained by truncating the series at $k = n$ for various choices of n, rounded to nine decimal places:

n	3	5	10	20
$\displaystyle\sum_{k=0}^{n}\frac{(-1)^k}{k!\,(2k+1)}$	0.742857143	0.746729197	0.746824134	0.746824133

On the basis of this table, it seems reasonable to make the approximation

$$\int_0^1 e^{-x^2}\,dx \approx 0.7468241.$$

Remark   The integral of $f(x) = e^{-x^2}$ plays an important role in statistics, but its antiderivative cannot be expressed as a finite combination of known, simple functions. To approximate the integral, we must either use a numerical integration method, such as the trapezoid or Simpson's rules, or else use a power series approximation, as we have just done.

An application to differential equations

Power series are a useful tool in solving initial value problems for differential equations. As an example, we will consider a **second-order linear differential equation**, by which we mean an equation of the form

$$y'' + f(t)y' + g(t)y = h(t), \tag{54}$$

where $f(t)$, $g(t)$, and $h(t)$ are given functions. An initial value problem consists of Eq. (54) and a pair of initial values

$$y(0) = y_0, \quad y'(0) = y_1, \tag{55}$$

where y_0 and y_1 are given numbers. A solution is a function $y(t)$ satisfying both (54) and (55).

EXAMPLE 10.3.7

Verify that $y(t) = 2e^t + e^{-t} - t^2 - 3$ is a solution of the initial value problem

$$y'' - y = 1 + t^2, \quad y(0) = 0, \quad y'(0) = 1.$$

SOLUTION First, we display $y(t)$ and its first two derivatives:

$$y(t) = 2e^t + e^{-t} - t^2 - 3$$
$$y'(t) = 2e^t - e^{-t} - 2t$$
$$y''(t) = 2e^t + e^{-t} - 2.$$

Subtracting the first equation from the third gives

$$y''(t) - y(t) = (2e^t + e^{-t} - 2) - (2e^t + e^{-t} - t^2 - 3) = 1 + t^2,$$

which shows that $y(t)$ satisfies the differential equation. Next, setting $t = 0$ in the first and second equations gives

$$y(0) = 0 \quad \text{and} \quad y'(0) = 1.$$

Second-order linear differential equations play an important role in the study of mechanical vibrations and in modeling electric circuits. They have also been used to model the blood glucose regulatory system in studying diabetes.[1]

There are standard methods for solving an initial value problem of the type given by (54) and (55) in the case where $f(t)$ and $g(t)$ are constants. (We will not take them up here, but you can consult any book on differential equations.) However, if they are not constants, the equation may be difficult to solve. One technique is to look for a solution in power series form, say,

$$y(t) = \sum_{k=0}^{\infty} a_k t^k = a_0 + a_1 t + a_2 t^2 + \cdots.$$

We then take the first two derivatives and use Eqs. (54) and (55) to determine the coefficients.

EXAMPLE 10.3.8

Find a power series solution $y(t) = \sum_{k=0}^{\infty} a_k t^k$ of the differential equation

$$y'' - y' + ty = 1 + t^2 \tag{56}$$

with initial conditions

$$y(0) = 1, \quad y'(0) = -1. \tag{57}$$

SOLUTION We write out the power series for $y(t)$ and its first two derivatives as follows:

$$y(t) = a_0 + a_1 t + a_2 t^2 + a_3 t^3 + a_4 t^4 + a_5 t^5 + \cdots$$
$$y'(t) = a_1 + 2a_2 t + 3a_3 t^2 + 4a_4 t^3 + 5a_5 t^4 + \cdots$$
$$y''(t) = 2a_2 + 6a_3 t + 12a_4 t^2 + 20a_5 t^3 + \cdots.$$

Setting $t = 0$ in the first of these equations and using the initial condition (57), we have

$$1 = y(0) = a_0.$$

Similarly, setting $t = 0$ in the second equation gives

$$-1 = y'(0) = a_1.$$

[1] M. Braun, *Differential Equations and Their Applications*, New York: Springer-Verlag, 1975, Section 2.7.

Next, we substitute the three power series into Eq. (56) and collect terms:

$$(2a_2 - a_1) + (6a_3 - 2a_2 + a_0)t + (12a_4 - 3a_3 + a_1)t^2$$
$$+ (20a_5t^3 - 4a_4 + a_2)t^3 + \cdots = 1 + t^2.$$

Equating the coefficients of like powers of t on both sides of this equation leads to the following set of equations:

$$2a_2 - a_1 = 1$$
$$6a_3 - 2a_2 + a_0 = 0$$
$$12a_4 - 3a_3 + a_1 = 1$$
$$20a_5 - 4a_4 + a_2 = 0,$$

and so forth. One by one, we solve for each a_i in terms of the previous ones. Starting with $a_0 = 1$ and $a_1 = -1$, as given above, we get

$$2a_2 = 1 + a_1 = 1 - 1 = 0.$$

Therefore, $a_2 = 0$. Next,

$$6a_3 = 2a_2 - a_0 = 2 \cdot 0 - 1 = -1.$$

Therefore, $a_3 = -\frac{1}{6}$. Next,

$$12a_4 = 1 + 3a_3 - a_1 = 1 - \frac{1}{2} + 1 = \frac{3}{2}.$$

Therefore, $a_4 = \frac{1}{8}$. Next,

$$20a_5 = 4a_4 - a_2 = \frac{1}{2} - 0 = \frac{1}{2}.$$

Therefore, $a_5 = \frac{1}{40}$. In this way, we see that

$$y(t) = 1 - t - \frac{1}{6}t^3 + \frac{1}{8}t^4 + \frac{1}{40}t^5 + \cdots.$$

By continuing in this way, we can display the solution as a power series to any degree we choose.

The following theorem, which we state without proof, guarantees the validity of the power series method.

Theorem 10.3.7

Given an initial value problem of the form

$$y'' + f(t)y' + g(t)y = h(t), \quad \text{with } y(0) = y_0 \quad \text{and } y'(0) = y_1,$$

suppose that $f(t)$, $g(t)$, and $h(t)$ are equal to their Taylor series in an interval around zero. Then there is a unique solution $y(t)$, and it can be represented by a power series throughout the same interval.

Practice Exercises 10.3

1. Find the radius of convergence of the power series

$$\sum_{k=1}^{\infty} \frac{x^k}{k^2} = x + \frac{x^2}{2^2} + \frac{x^3}{3^2} + \cdots .$$

2. **(a)** Find the Taylor series of $f(x) = (1+x^2)^{-1}$ and determine its interval of convergence.
(b) Find a power series that represents an antiderivative of $f(x)$ on its interval of convergence.

Exercises 10.3

In Exercises 1–10, find the radius of convergence of the given power series.

1. $\displaystyle\sum_{k=1}^{\infty} \frac{x^k}{k}$

2. $\displaystyle\frac{x}{2} - \frac{x^2}{3} + \frac{x^3}{4} - \frac{x^4}{5}$

3. $\displaystyle\sum_{k=1}^{\infty} k^2 x^k$

4. $\displaystyle\sum_{k=1}^{\infty} 2^k x^k$

5. $\displaystyle\sum_{k=0}^{\infty} \frac{2k+1}{2^k} x^k$

6. $\displaystyle\frac{1}{2 \cdot 1} + \frac{1}{3 \cdot 2} x + \frac{1}{4 \cdot 3} x^2 + \frac{1}{5 \cdot 4} x^3 + \cdots$

7. $\displaystyle\sum_{k=0}^{\infty} \frac{k!}{10^k} x^k$

8. $\displaystyle\sum_{k=0}^{\infty} \frac{10^k}{k!} x^k$

9. $\displaystyle\sum_{k=0}^{\infty} \frac{k}{2k+1} x^k$

10. $\displaystyle x - 2\left(\frac{2}{3}\right) x^2 + 3\left(\frac{2}{3}\right)^2 x^3 - 4\left(\frac{2}{3}\right)^3 x^4 + \cdots$

In Exercises 11–19, find a power series representation of the given function and determine an interval on which it is valid.

11. $x^2 e^x$

12. $\ln(1 + x^2)$

13. $\displaystyle\frac{x}{1 - x^2}$

14. $\displaystyle\frac{1 + x}{1 - x}$

15. $\displaystyle\ln\left(\frac{1 + x}{1 - x}\right)$

16. $\displaystyle\frac{e^x - 1}{x}$, $x \neq 0$

17. $x \cos x$

18. $\sin(x^2)$

19. $\displaystyle\frac{\sin x}{x}$, $x \neq 0$

20. Verify that term-by-term differentiation of the power series $\sum_{k=0}^{\infty} x^k / k!$ yields the same series back again. Explain why that was to be expected. (*Hint:* What is the Taylor series of e^x?)

21. Verify that term-by-term differentiation of the Taylor series of $\sin x$ about zero yields the Taylor series of $\cos x$ about zero.

22. Find the Taylor series expansion of $\ln(1 + x^2)$, take its derivative term by term, and verify that the new series is the Taylor series of $2x/(1 + x^2)$.

23. Find the power series representation of $(1 - x^2)^{-2}$ by first finding the Taylor series of $(1 - u)^{-1}$, then taking its derivative, and finally substituting $u = x^2$.

24. Use a power series to approximate $\int_0^1 e^{-x^2/2} \, dx$.

25. Find a power series that is an antiderivative of $\ln(1 + x^2)$ and use it to approximate $\int_0^{1/2} \ln(1 + x^2) \, dx$.

26. Use a power series to approximate $\int_0^{0.1} 1/(1 + x^2) \, dx$.

In Exercises 27–30, find the power series solution through terms of degree 5.

27. $y'' + y = 0$, $y(0) = 1$, $y'(0) = 1$

28. $y'' - y = 0$, $y(0) = 1$, $y'(0) = -1$

29. $y'' - ty' + y = 1 + t$, $y(0) = 2$, $y'(0) = 0$

30. $y'' + ty = e^t$, $y(0) = 0$, $y'(0) = 1$

31. Cacti are very hardy plants, adapted well to the harsh environment of the desert. One species lives in an area with only tiny amounts of rainfall, so its seedlings grow very slowly. A study of seedling height during the first year of the plant's life showed that the mean height is modeled by the differential equation

$$h'' - 4h = 6t - 5t^3,$$

where h is in centimeters and t is the time in years, with $0 \leq t \leq 1$.
(a) At the very start of germination, the plant has no height and is not growing. What are the initial conditions?
(b) Writing $h(t)$ as a power series, find the coefficients through degree 5.
(c) Estimate the mean height at the end of 3 months.

Solutions to practice exercises 10.3

1. Using the ratio test with $c_k = 1/k^2$ for $k \geq 1$, we have

$$\lim_{k \to \infty} \left| \frac{c_{k+1}}{c_k} \right| = \lim_{k \to \infty} \frac{k^2}{(k+1)^2} = 1.$$

Therefore, $R = 1$.

2. (a) We start with the geometric series formula $(1 - t)^{-1} = \sum_{k=0}^{\infty} t^k$ and substitute $t = -x^2$, which gives

$$(1 + x^2)^{-1} = \sum_{k=0}^{\infty} (-1)^k x^{2k} = 1 - x^2 + x^4 - x^6 + \cdots.$$

Since the geometric series has radius of convergence $R = 1$, the series for $(1 + x^2)^{-1}$ converges if $x^2 < 1$ and diverges if $x^2 > 1$. Therefore, its interval of convergence is $(-1, 1)$.

(b) Taking the antiderivative term by term gives

$$F(x) = \sum_{k=0}^{\infty} (-1)^k \frac{x^{2k+1}}{2k+1} = x - \frac{x^3}{3} + \frac{x^5}{5} - \frac{x^7}{7} + \cdots,$$

with $F'(x) = f(x)$, valid for $-1 < x < 1$.

Chapter 10 Summary

- If a function $f(x)$ has derivatives through order n in an interval around a point a, then its nth-degree **Taylor polynomial** about a is given by

$$P_n(x) = f(a) + \frac{f'(a)}{1!}(x - a) + \frac{f^{(2)}(a)}{2!}(x - a)^2 + \cdots + \frac{f^{(n)}(a)}{n!}(x - a)^n.$$

If, in addition, $f(x)$ has continuous derivatives through order $(n + 1)$ in an interval around a, then

$$f(x) = P_n(x) + R_n(x),$$

where the remainder (error) $R_n(x)$ satisfies the inequality

$$|R_n(x)| \leq \frac{M |x - a|^{n+1}}{(n + 1)!},$$

with M being the maximum value of $|f^{(n+1)}|$ over the interval between a and x. This is known as **Taylor's theorem**, and it gives a way to approximate the value of a given function $f(x)$.

- If a function $f(x)$ has derivatives of all orders in an interval around a, then there is an nth-degree Taylor polynomial $P_n(x)$ about a for *every* n. If, in addition, $\lim_{n \to \infty} R_n(x) = 0$, then $\lim_{n \to \infty} P_n(x) = f(x)$. That is,

$$f(x) = \frac{f(a)}{0!} + \frac{f'(a)}{1!}(x - a) + \frac{f^{(2)}(a)}{2!}(x - a)^2 + \frac{f^{(3)}(a)}{3!}(x - a)^3 + \cdots,$$

where $\cdots$ indicates that the sum continues indefinitely. The right-hand side of this equation is called the **Taylor series** for $f(x)$ about a and is also written as

$$f(x) = \sum_{k=0}^{\infty} \frac{f^{(k)}(a)}{k!}(x - a)^k.$$

- Most functions we have encountered here can be represented as Taylor series. For example, we have

$$\frac{1}{1-x} = \sum_{k=0}^{\infty} x^k = 1 + x + x^2 + \cdots, \quad \text{if } |x| < 1$$

$$e^x = \sum_{k=0}^{\infty} \frac{x^k}{k!} = 1 + \frac{x}{1!} + \frac{x^2}{2!} + \cdots, \quad \text{for all } x$$

$$\ln(1+x) = \sum_{k=1}^{\infty} (-1)^{k-1} \frac{x^k}{k} = x - \frac{x^2}{2} + \frac{x^3}{3} + \cdots \quad \text{for } |x| < 1.$$

- A **power series** about a is an infinite series of the form

$$\sum_{k=0}^{\infty} c_k (x-a)^k = c_0 + c_1(x-a) + c_2(x-a)^2 + c_3(x-a)^3 + \cdots,$$

  where the coefficients $c_0, c_1, c_2, \ldots$ are constants. The Taylor series of a given function $f(x)$ is an example of a power series, with $c_k = f^{(k)}(a)/k!$. A power series may converge only for $x = a$, or for all x, or for all x such that $a - R < x < a + R$, for some positive number R, called the **radius of convergence**. It can be computed from the relation

$$\lim_{k \to \infty} \left| \frac{c_{k+1}}{c_k} \right| = \frac{1}{R},$$

  provided that the limit exists. Power series are differentiated and integrated term by term like polynomials. They expand our collection of functions for solving calculus problems, such as computing integrals and finding solutions to differential equations.

Chapter 10 Review Questions

- Write the 3rd-degree Taylor polynomial about zero for a function $f(x)$ with $f(0) = 5$, $f'(0) = 0$, $f''(0) = 7$, and $f'''(0) = 2$.

- Describe how you would use Taylor polynomials to find an approximate value for e^2 without using a calculator.

- Describe how you would use Taylor polynomials to find an approximate value for $\ln(1.5)$ without using a calculator.

- Describe how you would use Taylor polynomials to approximate the solution to an initial value problem. Give an example.

- Which number is bigger, 1 or $0.999\ldots$?

- What is the Taylor series of $f(x) = 1/(1-x)$ about zero? What is its radius of convergence?

- Use a power series to approximate $\int_0^1 e^{-x^4} \, dx$.

Chapter 10 Review Exercises

In Exercises 1–2, compute the nth-degree Taylor polynomial of the given polynomial about a. By expanding and collecting terms in your answer, verify that the two polynomials are the same.

1. $f(x) = x^3 + 2x^2 - 4x + 3$, $n = 3$, $a = 1$

2. $f(x) = x^4 + x^2 + 1$, $n = 4$, $a = -1$

In Exercises 3–8, compute the nth-degree Taylor polynomial of $f(x)$ about a. Simplify the coefficients as much as possible.

3. $f(x) = e^{x/2}$, $n = 3$, $a = 0$

4. $f(x) = \dfrac{1}{1 + x^2}$, $n = 2$, $a = 0$

5. $f(x) = \sqrt{1 + x}$, $n = 4$, $a = 0$

6. $f(x) = x \ln x$, $n = 5$, $a = 1$

7. $f(x) = \cos x$, $n = 4$, $a = \pi$

8. $f(x) = \sin^2 x$, $n = 4$, $a = 0$

In Exercises 9–12, compute enough Taylor coefficients to observe a pattern and use it to find a formula for the nth coefficient. Then, compute the next two coefficients and see if they fit your formula.

9. $f(x) = \dfrac{1}{1 - x}$, $a = 0$ **10.** $f(x) = \dfrac{1}{1 - x}$, $a = 2$

11. $f(x) = \sin x$, $a = \dfrac{\pi}{2}$

12. $f(x) = \dfrac{1}{2}(e^x - e^{-x})$, $a = 0$

13. Approximate $e^{-0.5}$ by using the 5th-degree Taylor polynomial of e^{-x} around zero. Find a bound on the error.

14. Use a 6th-degree Taylor polynomial to approximate $\sin(1)$ and find a bound on the error.

In Exercises 15–16, estimate the integral by using the Taylor polynomial of degree n around a.

15. $\displaystyle\int_0^{0.2} (1 - x^2)^{-1/2}\, dx$, $n = 2$, $a = 0$

16. $\displaystyle\int_1^{1.2} (\ln x)^2\, dx$, $n = 3$, $a = 1$

17. If $f(x)$ is an antiderivative of $(1 + x^2)^{-1}$ with $f(0) = 0$, find the 3rd-degree Taylor polynomial of $f(x)$ around zero.

18. If $y(t)$ is the solution of the initial value problem

$$\frac{dy}{dt} = t^2 + y^2, \quad y(0) = -1,$$

find the 3rd-degree Taylor polynomial of $y(t)$ about zero and use it to approximate $y(0.2)$.

19. If $y(t)$ is the solution of the initial value problem

$$\frac{dy}{dt} = e^y, \quad y(0) = 0,$$

find the 5th-degree Taylor polynomial of $y(t)$ about zero and use it to approximate $y(0.5)$.

20. You have been assigned the task of preparing a slide show about company revenue for the board of directors. Your data indicate that the marginal revenue is given by

$$\frac{dR}{dx} = \frac{30x}{(1 + x)^2},$$

where R is in millions of dollars and x is in thousands of units. You realize that your presentation would have more impact if you could discuss the actual revenue.
(a) Taking $R(0) = 0$, find the Taylor polynomial of R of degree 4 around zero.
(b) Use the Taylor polynomial to estimate the revenue at a production level of 100 units.

In Exercises 21–24, determine whether the given series converges and, if so, find its sum.

21. $1 - 2 + 4 - 8 + 16 - \cdots$

22. $1 - \dfrac{1}{2} + \dfrac{1}{4} - \dfrac{1}{8} + \dfrac{1}{16} - \cdots$

23. $\dfrac{4}{3} - \dfrac{4}{9} + \dfrac{4}{27} - \dfrac{4}{81} + \cdots$ **24.** $\displaystyle\sum_{k=0}^{\infty} (1.02)^{-k}$

In Exercises 25–27, represent the given infinite decimal as a simple fraction.

25. $2.2222\ldots$ **26.** $0.\overline{015}$ **27.** $1.0\overline{321}$

28. A mining company owns a parcel of land containing mineral deposits. Geologists estimate that there are 5.2×10^8 cubic meters of mineral in the parcel. The plan is to mine 1.3×10^6 cubic meters the first year and increase that amount by 2.7% each year after that.
(a) Complete the following table, in which $a = 1.3 \times 10^6$, the initial number of cubic meters removed:

Year	Amount removed this year	Total removed by end of year
1	a	a
2	$a + 0.027a = a(1.027)$	
3		
4		

(b) Write a formula using summation notation that shows how much material has been removed by the end of the kth year. Then, simplify it by means of formula (30) in the text.
(c) How many years will it take to remove all of the material?

29. Continuation of the previous problem Under pressure from conservationists, the company changes its plan. Starting with the same initial amount removed, it now plans to *decrease* the amount by 1% per year. Can the company expect the mining to go on forever?

In Exercises 30–35, find the Taylor series about a. If possible, write a general formula using the Σ notation.

30. $f(x) = e^{-0.1x}$, $a = 0$ **31.** $f(x) = \dfrac{2}{(2+x)^2}$, $a = 0$

32. $f(x) = (2-x)^{-1}$, $a = 1$

33. $f(x) = (x+1)e^x$, $a = 0$

34. $f(x) = \sin x$, $a = \dfrac{\pi}{2}$

35. $f(x) = \cos x + x \sin x$, $a = 0$

36. A calculator with a 10-digit display shows $\ln(1.1)$ to be equal to 0.0953101798. Using the Taylor series of $\ln(1+x)$ about zero, to what power of x do you have to go in order to come up with this decimal representation? Answer the question by calculating each partial sum to 10 digits until you find one that agrees with the calculator's version of $\ln(1.1)$.

37. Let $f(x) = (1+x)^{1/2}$.
(a) Find the Taylor series of $f(x)$ about zero. Try to write a general formula using the Σ notation.
(b) Use the partial sum of degree 4 to estimate $\sqrt{0.9}$. Compare your answer to that given by your calculator.

In Exercises 38–43, find the radius of convergence of the given power series.

38. $1 + 2x + 3x^2 + 4x^3 + \cdots$

39. $\displaystyle\sum_{k=1}^{\infty} \frac{x^k}{k(k+1)}$ **40.** $\displaystyle\sum_{k=1}^{\infty} k2^{k-1}x^k$

41. $1 - \dfrac{x}{3} + \dfrac{x^2}{5} - \dfrac{x^3}{7} + \dfrac{x^4}{9} + \cdots$

42. $\displaystyle\sum_{k=0}^{\infty} \frac{k!}{2^{k+1}} x^k$

43. $\dfrac{1}{3} + \dfrac{x}{3^2} + \dfrac{2x^2}{3^3} + \dfrac{2^2 x^3}{3^4} + \dfrac{2^3 x^4}{3^5} + \cdots$

In Exercises 44–47, find a power series representation of the given function and determine an interval on which it is valid.

44. $f(x) = \dfrac{1}{1+x^2}$ **45.** $f(x) = x^2 \ln(x+1)$

46. $f(x) = xe^{-x^2/2}$

47. $f(x) = \dfrac{1 - \cos x}{x}$, $x \neq 0$

48. Show that

$$\frac{1}{(1+x)^3} = \sum_{k=0}^{\infty} (-1)^k \frac{(k+1)(k+2)}{2} x^k$$

by starting with the series representation of $1/(1+x)$ and differentiating twice.

In Exercises 49–50, use a power series to approximate the given integral.

49. $\displaystyle\int_0^{1/2} \frac{x^2}{1-x^2}\, dx$ **50.** $\displaystyle\int_0^1 \frac{\sin x}{x}\, dx$

51. Find the terms through degree 5 of the power series solution of the second-order initial value problem

$$y'' - ty = 1, \quad y(0) = 1, \quad y'(0) = 0.$$

52. Verify that term-by-term differentiation of the Taylor series of $\cos x$ about zero yields the Taylor series of $-\sin x$ about zero.

Chapter **10** Practice Exam

1. Find the Taylor polynomials of degrees 1 through 3 about zero of the function $f(x) = xe^{2x}$.

2. Given a function $f(x)$ with $f(3) = 5$, $f'(3) = 2$, $f''(3) = -1$, $f^{(3)}(3) = 7$, and $f^{(4)}(3) = 6$, find its Taylor polynomials of degrees 1 through 4 about 3.

3. Find the Taylor polynomial of degree 3 of $f(x) = 1 + x + x^2 + x^3$ about the point $a = 1$.

4. Determine the Taylor polynomial of degree 3 of $f(x) = 1/(3-x)$ about the point $a = 2$.

5. Determine the Taylor polynomial of degree 2 of $f(r) = r \ln r$ about 1.

6. Use the 3rd-degree Taylor polynomial of $f(x) = \ln(1+x)$ about zero to approximate $\ln(1.2)$.

7. Estimate $\int_0^{0.2} e^{-t^4}\, dt$ by using the Taylor polynomial of degree 4 around zero.

8. Let $y(t)$ be the solution of the differential equation $y' = 8 + t - y^2$ satisfying $y(2) = 3$. Approximate $y(2.2)$ using its 2nd-degree Taylor polynomial about 2.

9. Find the Taylor series expansion of the function $f(x) = e^{-x/2}$ about zero. For what values of x does it converge to $e^{-x/2}$?

10. Find the Taylor series expansion of the function $f(x) = 1/(2-x)$ about the point $a = 1$. For what values of x does it converge to $1/(2-x)$?

11. Assume that the government puts \$50 billion into consumers' hands through a tax cut, and that consumers spend 80% of their annual income and save 20% of it. The tax cut will thus result in $(0.8) \cdot \$50$ billion additional spending during the first year, which realized as income will generate $(0.8) \cdot (0.8) \cdot \$50$ billion additional spending during the second year, and so on. Find the total additional spending the tax cut will generate in the long run.

12. A patient is injected once a day with eight units of a certain drug. Suppose this drug is eliminated exponentially, so that any amount A in the system reduces to $Ae^{-0.4t}$ remaining after t days.
(a) Write formulas giving the amount remaining in the patient's system at the end of 1 day and at the end of 2 days.
(b) Assuming the treatment is continued indefinitely, use an infinite geometric series to approximate the number of units of the drug remaining in the patient's system after a very long period.

13. The spruce budworm population $B(t)$ in a forest is modeled by the initial value problem

$$\frac{dB}{dt} = 0.3B\left(1 - \frac{B}{9}\right) - \frac{0.4B^2}{1+B^2}, \quad B(0) = 3,$$

where both t and B are in appropriate units. Use the 2nd-degree Taylor polynomial of $B(t)$ about zero to approximate $B(10)$.

Source: Such a model was introduced in 1978 by Ludwig, Jones, and Holling who studied the interaction between the spruce budworm and the balsam fir forest in eastern Canada. See "Qualitative Analysis of Insect Outbreak Systems: The Spruce Budworm and Forest," *Journal of Animal Ecology*, Vol. 47 pp. 315–332. Note that the population is assumed to grow according to a logistic model, and to decline at a rate $k^2 B^2 / (m^2 + B^2)$ due to predation.

14. Assume that there are about 26,500 gray whales in the North Pacific today and that their population $w(t)$ is governed by the differential equation

$$\frac{dw}{dt} = 0.014w \ln \frac{40}{w} - 0.174,$$

where w is measured in thousands and t in years. Use the 2nd-degree Taylor polynomial approximation of $w(t)$ to estimate the whales' population 20 years later.

...... **Chapter 10 Projects** •••••••••••••••••••••••••••••••••

1. **Geometric series models**   The geometric series can be used to model a number of phenomena in nature and social science.

(a) **A population growth model with immigration**   The population of a community can change in two ways:

- by changes in the birth and death rates (**intrinsic change**) and
- by immigration (or emigration).

If we divide a long period of time (such as a century) into equal units (such as years), we can write

$$P_{n+1} = r_n P_n, \quad n = 0, 1, 2, 3, \ldots,$$

where P_n is the size of the population during the nth period and r_n is the proportion by which it changes. If the reproduction rate r_n is a fixed number, say, r, then

$$P_1 = r P_0$$
$$P_2 = r P_1 = r^2 P_0$$
$$P_3 = r P_2 = r^3 P_0$$
$$\vdots$$
$$P_n = r P_{n-1} = r^n P_0.$$

With immigration (or emigration),

$$P_{n+1} = r P_n + b_n,$$

where r is the reproduction rate (same in every period) and b_n is the net increase in population due to immigration. Iterating gives

$$P_1 = r P_0 + b_0$$
$$P_2 = r P_1 + b_1 = r(r P_0 + b_0) + b_1 = r^2 P_0 + r b_0 + b_1$$
$$P_3 = r P_2 + b_2 = r(r^2 P_0 + r b_0 + r b_1) + b_2 = r^3 P_0 + r^2 b_0 + r b_1 + b_2$$
$$\vdots$$
$$P_n = r^n P_0 + r^{n-1} b_0 + r^{n-2} b_1 + \cdots + r b_{n-2} + b_{n-1}.$$

Question 1. Show that if $b_n = b$ is fixed, then

$$\lim_{n \to \infty} P_n = \begin{cases} \dfrac{b}{1-r} & \text{if } 0 < r < 1 \\ \infty & \text{if } r \geq 1. \end{cases}$$

(b) **The multiplier effect in economics** If the government puts money into the hands of consumers, say, by cutting taxes or by purchasing goods or services from the private sector, the people receiving it will spend a certain proportion and save the rest. People who receive the amount spent—merchants, landlords, and so forth—spend a proportion of that amount and save the rest. Then, at the third stage, the next set of receivers spends part and saves part, and so it goes on and on down the line. In economics this phenomenon is called the **multiplier effect**.

The average proportion spent throughout a community is called the **marginal propensity to consume**, and the average proportion saved is called the **marginal propensity to save**. If we denote them by c and s, respectively, then both are numbers in the interval $[0, 1]$, with $c + s = 1$.

Thus, if M is the amount originally put into circulation, the receivers spend cM. The next group—those receiving this amount—spend, on average, a proportion again equal to c. Therefore, they spend $c \cdot cM$, or $c^2 M$. The third group—those receiving this next amount—spend $c^3 M$, and the money continues to circulate this way through the community.

Let S_n be the total money that has been circulated in the first n stages of this process, beginning with the original amount furnished by the government. Write a formula for S_n and show that $\lim_{n \to \infty} S_n = M/s$.

(c) **A genetic disease model** Retinablastoma is a form of eye cancer afflicting children. The disease may occur either because of a mutation or because it has been inherited. Studies have shown that the proportion of the population that develops retinablastoma by mutation is approximately $m = 2 \times 10^{-5}$. Of those that contract the disease, a certain percentage recover and reproduce, and statistics indicate that the reproduction rate is $r = 0.35$. In other words, for every 100 persons who contract retinablastoma in one generation, there are 35 in the next generation inheriting the disease.

Consider successive generations, starting with some past generation (the 1st) and ending with one far in the future, and let p_k stand for the proportion of the kth generation afflicted by the disease. If we go far enough back in time, we can assume that virtually all the cases of retinablastoma in the 1st generation were caused by mutation—or, to put it another way, the number of inherited cases was small enough

to ignore. Therefore,

$$p_1 = m.$$

For any later generation, say, the kth, there are two distinct classes of afflicted people. First, there is a proportion m that developed the disease directly by mutation; second, there is a proportion $r p_{k-1}$ that inherited it from the $(k-1)$st generation. Thus,

$$p_k = m + r p_{k-1} \quad \text{for } k \geq 2.$$

Question 2.   By proceeding recursively, as we did in the population problem, show that the proportion of the nth generation suffering from retinablastoma is given by the formula

$$p_n = m + rm + r^2 m + \cdots + r^{n-1} m,$$

where $m = 2 \times 10^{-5}$ and $r = 0.35$.

Question 3.   By letting $n \to \infty$, find the long-range tendency—that is, the number toward which the proportion of the population afflicted with retinablastoma tends.

2. Higher-order Taylor methods for differential equations   In Example 10.1.10, we used a Taylor polynomial to approximate the solution $y(t)$ of an initial value problem

$$y' = f(t, y), \quad y(0) = y_0,$$

for t very close to zero. In general, to approximate the solution at a point, say, $t = b$, by using kth-order Taylor polynomials, we proceed as follows:

- Divide the interval $[0, b]$ into n equal segments, each of step size $\Delta t = b/n$.
- Use the kth-order Taylor polynomial $P(t)$ of $y(t)$ about zero, computed as in Example 10.1.10, to approximate $y(t)$. Let $y_1 = P(t_1)$, where $t_1 = t_0 + \Delta t$.
- Taking (t_1, y_1) as a new initial condition, use the kth-order Taylor polynomial of $y(t)$ about t_1 to approximate $y(t)$ at $t_2 = t_1 + \Delta t$. Let $y_2 = P(t_2)$.
- Continue in this way until you reach $b = t_n$. Then $y_n \approx y(b)$.

This procedure is the higher-order version of Euler's method, which we studied in Chapter 9, and it generally gives a more accurate approximation.

For example, let $y(t)$ be the solution of the differential equation

$$y' = y - 2t, \tag{58}$$

with initial condition $y(0) = 1$. We will approximate $y(0.5)$ by using 2nd-order Taylor polynomials, with $n = 5$ and $\Delta t = 0.1$. For the first step, we have $y(0) = 1$ and

$$y'(0) = 1 - 0 = 1.$$

Taking derivatives on both sides of (58) gives

$$y'' = y' - 2, \tag{59}$$

so that

$$y''(0) = 1 - 2 = -1.$$

Therefore, the 2nd-degree Taylor polynomial of $y(t)$ around zero is

$$P(t) = 1 + t - \frac{1}{2!}t^2 = 1 + t - \frac{t^2}{2},$$

and $y_1 = P(0.1) = 1.095$.

Next, we repeat the process, using (58) and (59) with the initial condition $(t_1, y_1) = (0.1, 1.095)$ to get

$$y(0.1) = 1.095, \quad y'(0.1) = y(0.1) - 2(0.1) = 0.895,$$

$$y''(0) = y'(0.1) - 2 = -1.105.$$

Therefore, the 2nd-degree Taylor polynomial at this step is

$$P(t) = 1.095 + 0.895(t - 0.1) - 0.5525(t - 0.1)^2,$$

and $y_2 = P(0.2) = 1.179$ to three decimal places. Continuing in this way, we generate the following table, rounded to three decimal places:

k	t_k	y_k	y_k'	y_k''
0	0	1	1	-1
1	0.1	1.095	0.895	-1.105
2	0.2	1.179	0.779	-1.221
3	0.3	1.251	0.651	-1.349
4	0.4	1.309	0.509	-1.491
5	0.5	1.353	0.353	-1.647

From this we get the approximation $y(0.5) \approx y_5 = 1.353$.

Question 1. Verify the data in this table.

Question 2. The solution to Eq. (58) with $y(0) = 1$ is $y(t) = 2t + 2 - e^t$. Compute $y(0.5)$ and calculate the approximation error $|y(0.5) - y_5|$.

Question 3. Use 2nd-degree Taylor polynomials to estimate $y(0.5)$ for Eq. (58) with $y(0) = 1$, but this time using $n = 10$ and $\Delta t = 0.05$. Calculate the approximation error $|y(0.5) - y_{10}|$.

Question 4. In using 2nd-order approximation, as we have just done, the error is approximately $C \cdot (\Delta t)^2$, where C is some constant and Δt is the step size. It follows that if the step size is halved, the error decreases by about one-fourth. Check your answer to questions 2 and 3 to verify that it did. Then repeat the procedure using step size $\Delta t = 0.025$ to see if the error is again reduced by about one-fourth.

Question 5. For the same initial value problem, estimate $y(0.5)$ using 3rd-degree Taylor polynomials and the following step sizes:
   (a) 0.1, **(b)** 0.05, **(c)** 0.025.
Compare these approximation errors with those of questions 2 and 3. By what factor did the error change when you reduced the step size by one-half?

Question 6. Given the initial value problem

$$y' = t^2 - y, \quad y(0) = 0,$$

estimate $y(1)$ by using 2nd-degree polynomials with
   (a) $\Delta t = 0.2$, **(b)** $\Delta t = 0.1$, **(c)** $\Delta t = 0.05$

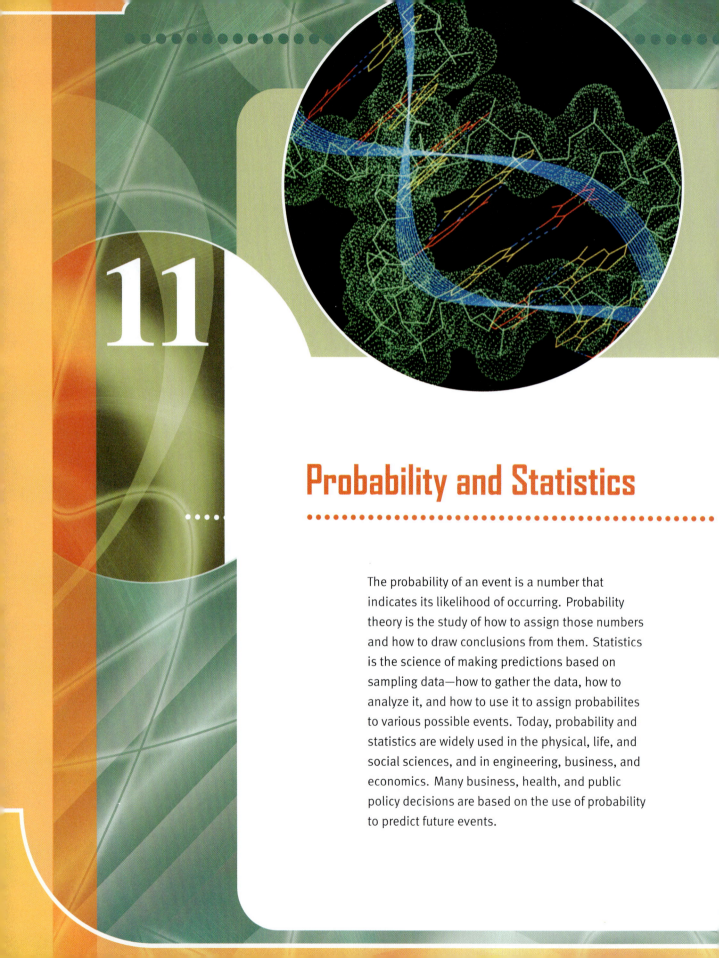

11

Probability and Statistics

The probability of an event is a number that indicates its likelihood of occurring. Probability theory is the study of how to assign those numbers and how to draw conclusions from them. Statistics is the science of making predictions based on sampling data—how to gather the data, how to analyze it, and how to use it to assign probabilites to various possible events. Today, probability and statistics are widely used in the physical, life, and social sciences, and in engineering, business, and economics. Many business, health, and public policy decisions are based on the use of probability to predict future events.

11.1   Sample Spaces and Assignment of Probabilities

The context in which probability is most naturally encountered is in analyzing the result of an **experiment**, by which we mean an activity that

- has an observable outcome and
- can, in principle, be repeated many times under the same conditions.

The experiment does not have to be done in a laboratory or involve scientific equipment. For example, tossing a coin and observing whether it lands heads or tails is an experiment. So is choosing a card from a deck and observing its value (such as king of hearts, seven of clubs, etc.). A different type of experiment—one that might be done as part of a quality-control procedure—is testing a certain type of lightbulb by turning it on, letting it burn continuously until it goes out, and recording the length of its lifetime.

Each of these experiments has a set of possible outcomes that cannot be predicted with certainty—as opposed, say, to dropping a ball out of a window, whose outcome is completely determined from the outset (assuming that friction and wind resistance are negligible). Not only will the ball move down rather than up until it reaches the earth's surface, but we can write a formula giving its height at any time. The types of experiments we will be concerned with are those in which we can identify a complete set of possible outcomes, but we do not know which of them will occur.

HISTORICAL PROFILE

Andrei Kolmogorov (1903–1987) had a very hard childhood. His mother died giving birth to him; his father lived in exile until after the Russian Revolution and later died in combat in 1919. After leaving school, young Andrei worked for a while as a railway conductor, but found time to write a treatise on Newtonian mechanics. At Moscow State University, which he entered in 1920, he studied metallurgy and Russian history in addition to mathematics, and wrote an article on property ownership in fifteenth- and sixteenth-century Russia.

By 1922 Kolmogorov's mathematical work was advanced enough to win international attention. By 1925 he had published eight articles, culminating in his first published work in probability. In 1931 he became a professor at Moscow University, and in 1933 he published a monograph on probability that laid out the axiomatic foundations of the subject much as Euclid had done for geometry over 2,000 years earlier.

Kolmogorov won many prizes for his mathematical work, and he was awarded honorary degrees in Paris, Stockholm, and Warsaw. Through it all he remained a dedicated teacher, gathering his students for long walks on Sundays at which current mathematical problems were discussed as well as other cultural activities, such as painting, literature, and architecture. Kolmogorov himself was very interested in these subjects and particularly in the poetry of the great Russian author Pushkin. He also devoted a large portion of his time to a high school for gifted students, writing texts and developing courses, as well as taking its pupils on hikes and excursions.

> **Definition 11.1.1**
>
> A complete set of possible outcomes of an experiment is called a **sample space** of the experiment. Each individual outcome is called a **sample point**.

EXAMPLE 11.1.1

Describe a sample space for each of the following experiments:
  (i) Flip a coin and observe the side that faces up when it lands.
  (ii) Flip two coins and observe how they land.
  (iii) Roll a die and observe the number on the side facing up.

SOLUTION (i) There are two possible outcomes—heads and tails, which we can represent by the letters H and T. These are the sample points of the experiment, and the sample space is represented by the set

$$S = \{H, T\}.$$

(ii) There are four possible outcomes, and together they make up the sample space. Once again writing H for heads and T for tails, we have

$$S = \{HH, HT, TH, TT\}.$$

Notice that we consider HT and TH to be different. That is based on the assumption that the two coins are distinguishable in some way—they may have different dates stamped on them, or one may have a nick in its edge. It is certainly reasonable to assume that there are some features distinguishing one from the other. If we call one of them *coin one* and the other *coin two*, the sample point HT represents the case of coin one landing heads and coin two landing tails, whereas TH is the opposite.

(iii) In this case, there are six possible outcomes, represented by the sample points 1, 2, 3, 4, 5, and 6. The sample space is

$$S = \{1, 2, 3, 4, 5, 6\}.$$

In each of these examples the sample space was a finite set, but infinite sets may also occur.

EXAMPLE 11.1.2

Suppose a quality-control department is conducting tests to determine the average lifetime of a certain type of transistor. A transistor, chosen at random from a particular lot, is run until it wears out, and its lifetime is measured in hours. Describe a sample space for this experiment.

SOLUTION In theory, the result of this experiment can be *any* positive real number, and we take the sample space to be

$$S = [0, \infty) = \{x : 0 \leq x < \infty\}.$$

In practice, of course, neither extremely large nor extremely small numbers occur, but we do not know in advance what bounds to set. For that and other reasons, it is more convenient to allow *all* nonnegative real numbers as sample points.

In analyzing an experiment and its sample space, we are usually interested in knowing the probability that a certain event occurs. In tossing two coins, for example, we may be interested in computing the probability that at least one of the coins lands heads, or we may want to know the probability that both coins land the same way. Looking at the sample space

$$S = \{HH, HT, TH, TT\},$$

we see that each of these events can also be described by listing a certain subset of S. Saying that "at least one coin lands heads" is the same as specifying the set of outcomes $\{HH, HT, TH\}$, and saying that the two coins land the same way amounts to specifying the subset $\{HH, TT\}$.

In fact, identifying an event with a particular subset of the sample space—namely, the set of outcomes satisfying the conditions describing the event—is very useful, for it gives us a precise language for discussing events. With that in mind, we will use the word "event" in a limited and specific way, as follows.

Definition 11.1.2

A subset E of a given sample space S is called an **event**.

If the outcome of performing the experiment is a sample point in E, we say that the event E occurs. In rolling a die, for example, the subset

$$E = \{1, 3, 5\}$$

is an event. In words, we can describe it by saying "the die comes up odd." If, in a trial, the outcome is 5, then the event E occurs. If the outcome is 2, the event E does not occur.

Another event is the subset $F = \{1, 5, 6\}$. This event is more awkward to describe in words, but it is an event nevertheless. (We could say "the die comes up less than 2 or more than 4," but that does not give us any more information than listing the subset.) In this case, an outcome of 6 in a trial means the event occurs, but an outcome of 3 means it does not.

Since events are subsets of the sample space, we can apply the usual set operations to them.

Union of Two Events

If E and F are two events in the same sample space S, their **union** $E \cup F$ is the event consisting of all outcomes that are either in E or F, or both.

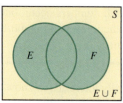

Figure 11.1.1

The event $E \cup F$ occurs when either E or F, or both occur. The diagram in Figure 11.1.1, called a Venn diagram, depicts the union of E and F.

> **Intersection of Two Events**
>
> If E and F are two events in the same sample space S, their **intersection** $E \cap F$ is the event consisting of all outcomes that are common to both E and F.

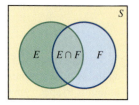

Figure 11.1.2

The event $E \cap F$ occurs if both E and F occur. The Venn diagram in Figure 11.1.2 depicts the intersection $E \cap F$.

> **Complement of an Event**
>
> If E is an event in the sample space S, its **complement** E' is the event consisting of all outcomes in S that are not in E.

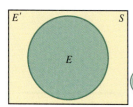

Figure 11.1.3

The event E' occurs if and only if the event E does not occur. The Venn diagram in Figure 11.1.3 depicts the complement of E'.

EXAMPLE 11.1.3

In the experiment of rolling a single die, with sample space $S = \{1, 2, 3, 4, 5, 6\}$, consider the events

$$E = \{1, 3, 5\} \quad \text{and} \quad F = \{1, 2, 3, 4\}.$$

Find $E \cup F$, $E \cap F$, E', and F'.

SOLUTION

$$E \cup F = \{1, 2, 3, 4, 5\} \quad \text{and} \quad E \cap F = \{1, 3\}.$$

The complements are

$$E' = \{2, 4, 6\} \quad \text{and} \quad F' = \{5, 6\}.$$

Any sample space S is a subset of itself, and we refer to it in probability terms as the **certain event**. In the case of rolling a die, the event described by saying "the die is either even or odd" or by saying "the die is between 1 and 6 inclusive" is a certain event. In both cases the subset in question is the full sample space S. At the other extreme is the empty set, which is also a subset of S. We refer to it as an **impossible event** and denote it by $\emptyset$.

If two events have no outcome in common, we say they are **mutually exclusive** or **disjoint**. More formally, we have the following:

Definition 11.1.3

Two events E and F are **mutually exclusive** (or **disjoint**) if $E \cap F = \emptyset$.

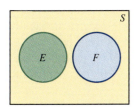

Figure 11.1.4

A Venn diagram illustrating this situation is shown in Figure 11.1.4.

For example, in the sample space $S = \{1, 2, 3, 4, 5, 6\}$ that models the roll of a single die, we see that the events $\{1, 3, 5\}$ (the die is odd) and $\{2, 4, 6\}$ (the die is even) are mutually exclusive. So are the events

$$E = \{1, 2\} \quad \text{and} \quad F = \{3, 5, 6\}.$$

It follows from the definition that any event and its complement are mutually exclusive. In fact, for any event E we have

$$E \cup E' = S \quad \text{and} \quad E \cap E' = \emptyset.$$

Assigning probabilities

Given an experiment and its sample space, we assign a number to every event E, called the **probability** of E, and denoted by $P(E)$. There are a number of rules that must be observed in assigning probabilities. Behind these rules is the following guiding principle: *The probability of an event is an estimate of the proportion of times that event will occur if the experiment is repeated over and over.* That proportion is called the **relative frequency** of the event. In other words,

$$\text{The relative frequency of } E = \frac{\text{number of occurrences of } E}{\text{number of repetitions of the experiment}}.$$

The probability $P(E)$ is intended as an "ideal" estimate of the relative frequency, a number toward which the relative frequency tends as the number of repetitions grows larger and larger. In tossing a coin, for instance, we usually assign a probability of $\frac{1}{2}$ to the event "heads" in the belief that it represents the approximate proportion of heads in a very large number of tosses and that, in fact, the proportion will tend toward $\frac{1}{2}$ as the coin is tossed over and over.

This interpretation leads to the following rules for assigning probabilities:

Basic Probability Axioms

1. For any event $E, 0 \le P(E) \le 1$.
2. $P(S) = 1$ and $P(\emptyset) = 0$.
3. If E and F are mutually exclusive, $P(E \cup F) = P(E) + P(F)$.

These are not the *only* rules governing the assignment of probabilities, but all the others can be derived from this basic set. Toward the end of this section we will state and prove a number of them. First, however, we will discuss some practical methods for assigning probabilities.

If the sample space is finite, a simple and intuitive method is to assign a probability between 0 and 1 to each individual outcome in such a way that all these probabilities add up to 1. Then, for any event E, we take $P(E)$ to be the sum of the probabilities of each of its outcomes, with the further understanding that $P(\emptyset) = 0$. It is not hard to check that this assignment satisfies the three rules listed above.

For example, suppose you toss a coin twice in a row. A sample space for this experiment is $S = \{HH, HT, TH, TT\}$, with each pair of letters designating a possible sequence of heads and tails. If we assume the coin is a fair one, each of these outcomes should occur with relative frequency $\frac{1}{4}$ over many repetitions of the experiment. Therefore, it is natural to assign the following probabilities:

outcome	*HH*	*HT*	*TH*	*TT*
probability	$\frac{1}{4}$	$\frac{1}{4}$	$\frac{1}{4}$	$\frac{1}{4}$

This chart is called the **probability table** of the experiment. In this case,

$$P(S) = P(HH) + P(HT) + P(TH) + P(TT) = \frac{1}{4} + \frac{1}{4} + \frac{1}{4} + \frac{1}{4} = 1.$$

EXAMPLE 11.1.4

By following the relative frequency principle, determine the probability table for the experiment of rolling a single fair die. Then, use the table to find the probabilities of the following events:

   (i) The top face is odd.
   (ii) The top face is less than 5.

SOLUTION A sample space for this experiment is

$$S = \{1, 2, 3, 4, 5, 6\}.$$

Since the die is fair, we can assume that no outcome is more likely to occur than any of the others. Therefore, each should have a relative frequency about $\frac{1}{6}$ in a large number of repetitions of the experiment. On that basis, we assign the probabilities according to the following table:

outcome	1	2	3	4	5	6
probability	$\frac{1}{6}$	$\frac{1}{6}$	$\frac{1}{6}$	$\frac{1}{6}$	$\frac{1}{6}$	$\frac{1}{6}$

(i) Saying that the top face is odd describes the event $E = \{1, 3, 5\}$, and

$$P(E) = P(1) + P(3) + P(5) = \frac{1}{6} + \frac{1}{6} + \frac{1}{6} = \frac{1}{2}.$$

(ii) Saying that the top face is less than 5 describes the event $F = \{1, 2, 3, 4\}$, and we have

$$P(F) = P(1) + P(2) + P(3) + P(4) = 4 \cdot \frac{1}{6} = \frac{2}{3}.$$

Equiprobable sample spaces

The preceding example illustrates a situation that often arises—a finite sample space in which every outcome is equally likely to occur. We call such a sample space **equiprobable**.

If an equiprobable sample space contains a total of N outcomes, then each outcome is assigned probability $1/N$. Therefore, the probability of any event E is simply the number of outcomes in E times $1/N$. We are thus led to the following formula for determining the probability of any event E in a sample space S with equally likely outcomes:

$$P(E) = \frac{\text{number of outcomes in } E}{\text{number of outcomes in } S}. \tag{1}$$

EXAMPLE 11.1.5

Suppose you toss a coin three times in a row. Determine the probabilities of each of the following events:

(i) The first two tosses are heads.
(ii) At least two of the three tosses are heads.
(iii) Exactly two of the three tosses are heads.

SOLUTION The sample space consists of eight outcomes. We can describe it as follows:

$$S = \{HHH, HHT, HTH, HTT, THH, THT, TTH, TTT\}.$$

If we assume that the coin is fair, each of these outcomes is equally likely, and we can use formula (1).

(i) Let E be the event that the first two tosses are both heads. As a subset, $E = \{HHH, HHT\}$, so that

$$P(E) = \frac{2}{8} = \frac{1}{4}.$$

(ii) Let F be the event that at least two of the tosses are heads. As a subset, $F = \{HHH, HHT, HTH, THH\}$, so that

$$P(F) = \frac{4}{8} = \frac{1}{2}.$$

(iii) Let G be the event that exactly two of the tosses are heads. Then $G = \{HHT, HTH, THH\}$, so that

$$P(G) = \frac{3}{8}.$$

EXAMPLE 11.1.6

Suppose you roll a pair of fair dice. Construct a sample space for this experiment so that all outcomes are equally likely and find the following probabilities:
  (i) The top faces add up to 10.
  (ii) The top faces are equal.

SOLUTION   The top face of each die will be an integer between 1 and 6. In order to get a sample space with equally likely outcomes, we treat the dice as *distinguishable* from one another, just as we did with the two coins in Example 11.1.1. Therefore, we can represent each outcome by *ordered* pair of integers of the form (i, j), with i and j taking values from 1 to 6. There are 36 such ordered pairs, as displayed below:

$$
\begin{array}{cccccc}
(1,1) & (1,2) & (1,3) & (1,4) & (1,5) & (1,6) \\
(2,1) & (2,2) & (2,3) & (2,4) & (2,5) & (2,6) \\
(3,1) & (3,2) & (3,3) & (3,4) & (3,5) & (3,6) \\
(4,1) & (4,2) & (4,3) & (4,4) & (4,5) & (4,6) \\
(5,1) & (5,2) & (5,3) & (5,4) & (5,5) & (5,6) \\
(6,1) & (6,2) & (6,3) & (6,4) & (6,5) & (6,6).
\end{array}
$$

  (i) There are exactly three outcomes for which the top faces add up to 10, namely, $(4, 6)$, $(5, 5)$, and $(6, 4)$. Therefore, the probability of that event is $\frac{3}{36}$, or $\frac{1}{12}$.
  (ii) There are exactly six outcomes with the top faces equal. Therefore, the probability of that event is $\frac{6}{36}$, or $\frac{1}{6}$.

Some nonequiprobable sample spaces

As the last two examples show, assigning probabilities in the equiprobable case amounts to counting the number of outcomes in a given event. However, not all sample spaces are equiprobable, as the next two examples show.

EXAMPLE 11.1.7

Suppose the starting center on your school's basketball team is a 60% free-throw shooter. If she comes to the line for two shots, what is the probability she will make at least one?

SOLUTION   If we write H for "hit" and M for "miss," we can describe a sample space for this experiment as follows:

$$S = \{HH, HM, MH, MM\}.$$

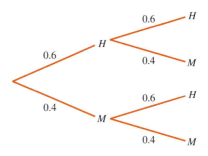

Figure 11.1.5

In this case, however, the four outcomes are not equally likely. To assign probabilities to them, we reason as follows. The center has a 60% chance of making her first free-throw. This means that if she comes to the line 100 times for a pair of free-throws, she will make the first shot approximately 60 of those times. Out of those 60 times she has a 60% chance of making the second shot as well. Since 60% of 60 is 36, we conclude that she will make *both* free-throws about 36 out of 100 times. In other words, the relative frequency of making both free-throws is best approximated by 0.36, and we accordingly assign $P(HH) = 0.36$.

To determine $P(HM)$, we argue in a similar way. Out of a 100 pairs of free-throws, she will make the first shot approximately 60 times, and out of those 60 she will miss the second shot 40% of the time. Since 40% of 60 is 24, we conclude that she will hit the first and miss the second approximately 24 times out of 100. Therefore, we assign $P(HM) = 0.24$.

In a similar way, we conclude that $P(MH) = 0.24$ and $P(MM) = 0.16$. (You should work these out for yourself as an exercise.) This method of computing probabilities can be illustrated by the diagram in Figure 11.1.5, called a **tree diagram**. The two branches coming out of the leftmost vertex represent the possible results of the first free-throw, and each is labeled with its probability.

Each of the other four branches represents the possible results of the second shot, and they are also labeled with their probabilities. Each path through the tree from left to right represents an outcome of the experiment of shooting a pair, and its probability is obtained by multiplying the probabilities of its individual branches. For example,

$$P(HM) = (0.6)(0.4) = 0.24.$$

Notice that

$$P(S) = P(HH) + P(HM) + P(MH) + P(MM)$$
$$= 0.36 + 0.24 + 0.24 + 0.16 = 1.$$

Now, suppose E is the event that she makes at least one of the free-throws. Then $E = \{HH, HM, MH\}$, so that

$$P(E) = P(HH) + P(HM) + P(MH) = 0.36 + 0.24 + 0.24 = 0.84.$$

EXAMPLE 11.1.8

Suppose a consumer testing service wants to measure the lifetime (in hours) of a certain kind of lightbulb. They test 1,000 of them and come up with the following data, with the time rounded to the nearest hour:

lifetime in hours	<450	450–649	650–849	850–1,049	≥1,050
number of lightbulbs	50	180	470	210	90

Based on this table, determine a sample space for this experiment and assign a probability to each outcome.

SOLUTION   The outcomes of this experiment are

$$s_1 = \text{lifetime in hours is smaller than 450}$$
$$s_2 = \text{lifetime in hours is from 450 to 649}$$
$$s_3 = \text{lifetime in hours is from 650 to 849}$$
$$s_4 = \text{lifetime in hours is from 850 to 1,049}$$
$$s_5 = \text{lifetime in hours is at least 1,050,}$$

and the sample space is

$$S = \{s_1, s_2, s_3, s_4, s_5\}.$$

From the data table we assign probabilities to the outcomes by computing their relative frequencies. We have

$$P(s_1) = \frac{50}{1,000} = 0.05$$

$$P(s_2) = \frac{180}{1,000} = 0.18$$

$$P(s_3) = \frac{470}{1,000} = 0.47$$

$$P(s_4) = \frac{210}{1,000} = 0.21$$

$$P(s_5) = \frac{90}{1,000} = 0.09.$$

We can summarize this information in the following table:

outcome	s_1	s_2	s_3	s_4	s_5
probability	0.05	0.18	0.47	0.21	0.09

Notice here that the outcomes are *not* equally likely to occur. Once again we observe that

$$P(S) = P(s_1) + P(s_2) + P(s_3) + P(s_4) + P(s_5)$$
$$= 0.05 + 0.18 + 0.47 + 0.21 + 0.09$$
$$= 1.$$

Further rules for probabilities

The following theorems give us useful formulas for computing probabilities. They can all be derived from the basic probability axioms.

Theorem 11.1.1

For any event E,

$$P(E') = 1 - P(E).$$

PROOF Since $S = E \cup E'$ and $E \cap E' = \emptyset$, we get

$$1 = P(S) = P(E) + P(E').$$

Subtracting $P(E)$ from both sides gives $P(E') = 1 - P(E)$.

EXAMPLE 11.1.9

If you toss a fair coin four times, what is the probability of getting at least one head?

SOLUTION A sample space for this experiment consists of strings of H's and T's of length four, such as $HHHH$, $HHHT$, $HHTT$, and so forth. There are 16 such strings. One way to see that is by making a complete list of them. A shorter way is to observe that there are eight strings of length 3 made up of H's and T's (as we saw in Example 11.1.5), and we can get every string of length 4 by adding either an H or a T to one of those. In other words, every string of length 3 gives us *two* strings of length 4, and we therefore have 16 of the latter.

If we let E denote the event "at least one head," its complement E' can be described as "all tails." Thus, $E' = \{TTTT\}$, and its probability is given by $P(E') = \frac{1}{16}$. Using Theorem 11.1.1, we conclude that $P(E) = \frac{15}{16}$.

If every outcome of event E is also in event F, we say that E is a **subset** of F and write $E \subset F$. For instance, in tossing two coins, the event $\{HH\}$ ("both are heads") is a subset of $\{HH, HT, TH\}$ ("at least one is heads.")

Theorem 11.1.2

Suppose E and F are events in the same sample space.

If $E \subset F$, then $P(E) \leq P(F)$.

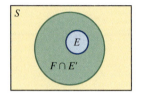

Figure 11.1.6

PROOF   As illustrated by the diagram in Figure 11.1.6, we can decompose F into two mutually exclusive sets, as follows:

$$F = E \cup (F \cap E').$$

Using axiom 3, we get

$$P(F) = P(E) + P(F \cap E'),$$

and since $P(F \cap E') \geq 0$ by axiom 1, we conclude that $P(E) \leq P(F)$.

Recall that the third basic axiom said that the probability of the union of two events is the sum of their probabilities, provided that the events are mutually exclusive. The next two theorems extend that property to three or more mutually exclusive events.

Theorem 11.1.3

Suppose E, F, and G are in events in the same sample space satisfying the conditions

$$E \cap F = \emptyset, \quad E \cap G = \emptyset, \quad F \cap G = \emptyset.$$

Then

$$P(E \cup F \cup G) = P(E) + P(F) + P(G). \tag{2}$$

PROOF   Since $E \cap F = \emptyset$ and $E \cap G = \emptyset$, we have $E \cap (F \cup G) = \emptyset$. Therefore, by applying axiom 3 to E and $F \cup G$, we get

$$P[E \cup (F \cup G)] = P(E) + P(F \cup G).$$

Next, applying the same axiom to F and G gives

$$P(F \cup G) = P(F) + P(G).$$

Combining the last two equations gives formula (2).

In a similar way, we can prove the following result for any number of mutually exclusive events.

Theorem 11.1.4

Suppose $E_1, E_2, \ldots, E_n$ are events in the same sample space such that $E_i \cap E_j = \emptyset$ whenever $i \neq j$. Then

$$P(E_1 \cup E_2 \cup \cdots \cup E_n) = P(E_1) + P(E_2) + \cdots + P(E_n).$$

EXAMPLE 11.1.10

A bowl contains 15 slips of paper, of which three are red, two are white, four are blue, one is yellow, and five are black. If one is drawn at random, what is the probability it is neither white nor black?

SOLUTION   The event "neither white nor black" is the same as "red or blue or yellow," which is the union of three events—"red," "blue," and "yellow"—whose respective probabilities are $\frac{3}{15}$, $\frac{4}{15}$, and $\frac{1}{15}$. Since every pair of these is mutually exclusive, the probability of their union is the sum:

$$\frac{3}{15} + \frac{4}{15} + \frac{1}{15} = \frac{8}{15}.$$

Of course, we can reach the same conclusion by observing that the number of slips that are either red, blue, or yellow equals eight out of a total of 15 in the box.

The next theorem gives us a formula for computing the probability of a union of nonmutually exclusive events.

Theorem 11.1.5

Let E and F be two events in the same sample space. Then

$$P(E \cup F) = P(E) + P(F) - P(E \cap F).$$

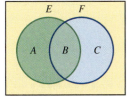

Figure 11.1.7

PROOF   The proof of this theorem is illustrated by Figure 11.1.7, in which the surrounding rectangle represents the sample space S, the circle on the left represents the subset E, and the circle on the right represents F.

The union $E \cup F$ is made up of three disjoint pieces:

- the intersection $E \cap F$, which has been labeled B in Figure 11.1.7,
- the part of E that does *not* meet F, labeled A in the diagram, and
- the part of F that does *not* meet E, labeled C in the diagram.

Notice that A, B, and C are mutually exclusive, and that

$$E = A \cup B, \quad B = E \cap F, \quad F = B \cup C.$$

Since $E \cup F = E \cup C$ and $E \cap C = \emptyset$, we have

$$P(E \cup F) = P(E) + P(C). \tag{3}$$

Moreover, $F = B \cup C$ and $B \cap C = \emptyset$, so that

$$P(F) = P(B) + P(C).$$

Subtracting $P(B)$ from both sides gives

$$P(C) = P(F) - P(B) = P(F) - P(E \cap F). \tag{4}$$

Substituting (4) into (3) gives

$$P(E \cup F) = P(E) + P(F) - P(E \cap F).$$

EXAMPLE 11.1.11

In a study of two commonly used antiinflammatory medicines, it was found that 6% of the patients treated had an allergic reaction to the first type, 4% to the second type, and 1% to both. Based on these data, what is the probability that a patient will have no allergic reaction to either drug?

SOLUTION   Let A_i represent the event that the patient has an allergic reaction to the ith medicine, and let E denote the event that there is no allergic reaction to either. Then $A_1 \cup A_2 = E'$. (The complement of "no allergic reaction to either" is the event "an allergic reaction to at least one.")
From the given data we get

$$P(A_1) = 0.06, \quad P(A_2) = 0.04, \quad P(A_1 \cap A_2) = 0.01.$$

Therefore, $P(A_1 \cup A_2) = 0.06 + 0.04 - 0.01 = 0.09$, and

$$P(E) = 1 - 0.09 = 0.91.$$

Practice Exercises 11.1

1. Suppose a bowl contains six red chips and three white ones. You remove three chips in succession from the bowl.
 (a) Describe a sample space for this experiment. Do you think the outcomes are equally likely? Explain.

 (b) Consider the events
$$A = \text{at least one is white}$$
$$B = \text{at most one is white}$$
$$C = \text{all white}$$
$$D = \text{all red.}$$

Which pairs of events are mutually exclusive? Which are complements of one another? Which are subsets of others?

2. Among the patients of a medical clinic, 13% have high blood pressure, 9% have chronic severe headaches, and 5% have both.
 (a) What is the probability that a patient of that clinic suffers from neither?
 (b) What is the probability a patient suffers from chronic severe headaches but does not have high blood pressure?

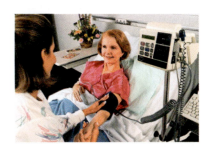

Exercises 11.1

In Exercises 1–9, describe a sample space for the given experiment. If the sample space is finite, specify the number of outcomes.

1. A die is rolled and a coin is tossed simultaneously, and the top face of each is recorded.

2. A bowl contains four chips, numbered 1 through 4. Two chips are drawn *with replacement*. That is, a chip is drawn at random and returned to the bowl after its number is recorded. Then, the bowl is shaken to mix up the chips, and a second chip is drawn at random.

3. A bowl contains four chips, numbered 1 through 4. Two chips are drawn *without replacement*. That is, the second is drawn without replacing the first.

4. A three-letter string is made up of the letters a, b, and c, with each letter appearing exactly once.

5. A three-letter string is made up of the letters a, b, c, and d, with each letter appearing no more than once.

6. An on-line shopping company records the number of hits on its Web site during a particular hour.

7. You keep tossing a coin until it lands heads.

8. You observe the length of the line at a fast-food restaurant.

9. You observe the time it takes a customer to be served at a fast-food restaurant.

10. Suppose you toss a fair coin three times in a row.
 (a) List all the outcomes in each of the following events:

$$A = \text{exactly one tail}$$
$$B = \text{at least one tail}$$
$$C = \text{at most one tail.}$$

 (b) Write a formula expressing A in terms of B and C.
 (c) What is $A \cup B$? Describe it in words and list all its outcomes.
 (d) What is B'? Describe it in words and list all its outcomes.
 (e) What is $B \cap A'$? Describe it in words and list all its outcomes.

11. A corporation efficiency expert records the time it takes an assembly line worker to perform a particular task. Let E be the event "more than 5 minutes." F the event "less than 8 minutes," and G the event "more than 4.5 minutes."
 (a) Describe a sample space for this experiment, treating time as a continuous variable.
 (b) Describe the following events in words in the simplest possible way:

$$E \cap F, \quad F', \quad E' \cap G, \quad F \cap G', \quad E \cup G'.$$

 (c) Simplify the expression of each of the following events:

$$G' \cap E, \quad E \cup G, \quad F \cup G, \quad E \cap G.$$

12. Consider the events

$$E = \{1, 2\}, \quad F = \{2, 3\}, \quad G = \{1, 5, 6\},$$

connected with the experiment of rolling a single die.
 (a) Are E and F mutually exclusive?
 (b) Are F and G mutually exclusive?

13. Suppose you toss a coin three times and consider the events:

$$E = \text{more heads than tails}$$
$$F = \text{at least one tail}$$
$$G = \text{at most one head.}$$

Which pairs of events are mutually exclusive?

14. Which of the following probabilities are feasible for an experiment having sample space $\{s_1, s_2, s_3\}$?
 (a) $P(s_1) = 0.4, \quad P(s_2) = 0.4, \quad P(s_3) = 0.4$
 (b) $P(s_1) = 0.3, \quad P(s_2) = 0.2, \quad P(s_3) = 0.5$
 (c) $P(s_1) = 0.5, \quad P(s_2) = 0.7, \quad P(s_3) = -0.2$
 (d) $P(s_1) = 2, \quad P(s_2) = 1, \quad P(s_3) = \frac{1}{2}$
 (e) $P(s_1) = \frac{1}{4}, \quad P(s_2) = \frac{1}{2}, \quad P(s_3) = \frac{1}{4}$

15. Suppose that the sample space of an experiment is $S = \{s_1, s_2, s_3, s_4, s_5, s_6\}$, and that $P(\{s_1, s_3, s_4\}) = 0.4$, and $P(\{s_2\}) = 0.1$. Find $P(\{s_5, s_6\})$.

16. An experiment with outcomes $s_1, s_2, s_3, s_4, s_5, s_6$ is described by the probability table:

outcome	s_1	s_2	s_3	s_4	s_5	s_6
probability	0.05	0.25	0.05	0.01	0.63	0.01

Let $E = \{s_1, s_2\}$ and $F = \{s_3, s_5, s_6\}$.
(a) Determine $P(E)$ and $P(F)$.
(b) Determine $P(E')$.
(c) Determine $P(E \cap F)$.
(d) Determine $P(E \cup F)$.

17. Jane is throwing a party and has invited her two parents, her mom's two brothers, her three sisters, and her son and his girlfriend. The guests arrive one by one in random order.
(a) What is the probability that the first person Jane greets is one of her relatives?
(b) What is the probability that the first person to arrive is one of her siblings?
(c) What is the probability that the first guest to arrive is a blood relative of Jane's father?

18. Harry bought a package of Every-Flavor Beans and sorted them by color. Of the 72 beans in the package, he counted the colors he disliked: pale green (12 beans), goldenrod (seven beans), black-and-white (three beans), and orange spot (two beans).
(a) If Harry mixed them all together and then chose one at random, what is the chance he would pick a color he disliked?
(b) What is the probability he would pick pale green or goldenrod?
(c) Ron also dislikes those colors, except for the orange spot. Find the probability that Harry would pick a color that Ron dislikes.

19. Suppose you roll a pair of fair dice. Find the probabilities of each of the following events (see Example 11.1.6):
(a) A = the top faces add up to 7.
(b) B = the top faces add up to 11.
(c) C = the sum of the top faces is even.
(d) D = the top faces are both odd.
(e) E = at least one of the top faces is odd.

20. Suppose you roll a pair of fair dice. Express each of the following events in terms of the events of the previous problem and find its probability without counting the number of outcomes.
(a) Both of the top faces are even.
(b) At least one of the top faces is even.
(c) The sum of the top faces is odd.
(d) The sum of the top faces is either 7 or 11.
(e) The sum of the top faces is neither 7 nor 11.
(f) The sum of the top faces is either 3, 5, or 9.

21. A tetrahedron is a four-sided solid, each face of which is an equilateral triangle. Suppose you are given a pair of tetrahedral dice, each with its faces labeled 1 through 4. You roll the dice and record the number on the bottom face. Describe a sample space for that experiment. Next, describe each of the following events as a subset of the sample space and determine its probability, assuming that all outcomes are equally likely.
(a) Both numbers are even.
(b) At least one number is even.
(c) The sum of the numbers is 6.
(d) The numbers are the same.
(e) A 2 or 3 occurs, but not both.

22. Tay-Sachs disease is a hereditary neurological condition that is fatal at an early age to anyone born with two Tay-Sachs genes. Carriers have one Tay-Sachs gene t and one normal gene N. If both parents are carriers, there are four equally likely gene pairings, given by the following table:

		Mother	
		N	t
Father	N	NN	Nt
	t	tN	tt

(a) What is the probability an offspring will be a carrier?
(b) What is the probability an offspring will have the disease?

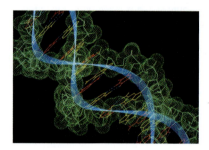

23. Continuation of the previous exercise Make a table of possible gene pairs for the offspring in case the mother is a carrier and the father is not (i.e., has two normal genes) and answer the same two questions as in the previous exercise.

24. Suppose a coin is weighted so that it comes down heads 60% of the time, and you toss it twice in row. Consider the events "both tosses are heads" and "both tosses are tails."
(a) Do you think these events are equally likely?
(b) Determine the probability that both tosses come down heads. (*Hint:* See Example 11.1.7.)
(c) Determine the probability that both tosses come down tails.

25. Suppose a coin is weighted so that it comes down heads 55% of the time. If you toss it twice in a row, what is the probability of getting at least one head?

26. A hospital hired a management firm to study the efficiency of its admission procedures. After several weeks of observation, the firm came up with the following table displaying the time spent waiting to be admitted, rounded to the nearest minute:

number of minutes at admitting desk	percentage of patients
≤10	8
11–15	23
16–20	41
21–30	24
>30	4

Based on this table, determine the probabilities of each of the following events:
(a) A patient waits no more than 30 minutes.
(b) A patient waits more than 10 but not more than 20 minutes.
(c) It takes more than 20 minutes to admit a patient.

27. Suppose E and F are events satisfying $P(E) = 0.6$, $P(F) = 0.5$, and $P(E \cap F) = 0.4$. Find
(a) $P(E \cup F)$
(b) $P(E \cap F')$ (*Hint:* $E = (E \cap F) \cup (E \cap F')$.)

28. Suppose E and F are events satisfying $P(E) = \frac{1}{4}$, $P(F) = \frac{2}{3}$, and $P(E \cup F) = \frac{3}{4}$. Find
(a) $P(E \cap F)$ **(b)** $P(E' \cap F)$ **(c)** $P(E' \cup F')$

29. An educational survey queried 100 people about a particular math class. The results were as follows:

38 enjoyed the math class.

26 were math majors.

20 were math majors who enjoyed the class.

Find the probability that a person

(a) is a math major who did not enjoy the class.
(b) is a math major or enjoyed the class (or both).
(c) is not a math major.
(d) is not a math major but enjoyed the class.

30. It was shown in Example 11.1.9 that there are 16 outcomes to the experiment of tossing a coin four times.
(a) Use a similar argument to show that there are 32 outcomes to the experiment of tossing a coin five times.
(b) How many outcomes are there to the experiment of tossing a coin six times? Justify your answer.
(c) If n is a positive integer, how many outcomes are there to the experiment of tossing a coin n times? Explain.

31. If you toss a fair coin six times, what are the probabilities of the following events?
(a) All tosses are heads.
(b) At least five tosses are heads.
(c) At most one toss is heads.

In the next two problems, assume that an individual child is equally likely to be a boy as a girl.

32. In a family of five children, what is the probability that all the children are of the same sex?

33. In a family of four children, which of the following events has greater probability?
(a) Two are boys and two are girls.
(b) Three children are of the same sex, and the other is of a different sex.

Solutions to practice exercises 11.1

1. (a) The sample space S consists of all three-letter strings made up of R's and W's. More precisely, listing the outcomes in alphabetical order, we get

$S = \{RRR, RRW, RWR, RWW, WRR, WRW, WWR, WWW\}$.

The outcomes are not equally likely. Because of the greater number of red to start with, the probability of WWW, for instance, is smaller than RWW.
(b) By listing the events,

$A = \{RRW, RWR, RWW, WRR, WRW, WWR, WWW\}$,

$B = \{RRR, RRW, RWR, WRR\}$,

$C = \{WWW\}$, $D = \{RRR\}$,

we see that

- $A \cap D = \emptyset$, $B \cap C = \emptyset$, and $C \cap D = \emptyset$,
- A and D are complements of one another,
- C is a subset of A, and D is a subset of B.

2. Letting B be the event "the patient has high blood pressure" and C be the event "the patient has chronic severe headaches," we have the given information $P(B) = 0.13$, $P(C) = 0.09$, and $P(B \cap C) = 0.05$.
(a) The event "the patient has neither" is the complement of $B \cup C$. We have $P(B \cup C) = 0.13 + 0.09 - 0.05 = 0.17$, and, therefore, the probability of a patient's having neither ailment is 0.83.
(b) Since $P(C) = P(B \cap C) + P(B' \cap C)$, we get

$$P(B' \cap C) = P(C) - P(B \cap C) = 0.09 - 0.05 = 0.04.$$

11.2 Conditional Probability

In many situations we are interested in computing the probability of a particular event *given the knowledge that another event has already occurred.* For instance, suppose that 0.2% of the population is infected with a certain disease. In the absence of any further information, we would conclude that the probability of a randomly chosen person being infected is 0.002 or $\frac{1}{500}$. However, knowing that the person has tested positive for the disease would certainly change our estimate of the probability. We might not be able to say with certainty that the person has the disease, because the test might have a small probability of giving a "false positive" result—that is, of indicating that a healthy person has the disease—but we would surely assign a much higher value than 0.002 to the probability.

In Example 11.2.7 below we will study just such a situation, one that involves the well-known Pap test for cervical cancer. Before getting to that, however, we need to study a few simpler cases.

EXAMPLE 11.2.1

In a family with two children, what is the probability that both children are of the same sex? How does the probability change if you know in advance that at least one child is a girl?

SOLUTION   The sample space of this experiment is

$$S = \{gg, gb, bg, bb\},$$

where the genders of the children are written in order of their ages—that is, g stands for girl, b stands for boy, and the first letter represents the older child. Assuming that boys and girls are born with the same frequency, we conclude that all four outcomes of the experiment are equally likely. The event that the children are of the same sex is

$$E = \{gg, bb\}.$$

Therefore, by applying formula (1) for equiprobable sample spaces, we obtain

$$P(E) = \frac{2}{4} = \frac{1}{2}.$$

To answer the second question, we need to compute the probability of E, given the information that the event

$$G = \{gg, gb, bg\}$$

has occurred. In that case, we should ignore the outcome bb, since we know it has not occurred, and use the event G as the full sample space. By the same token, we should ignore those outcomes of E that are not in G, which means replacing the event E by $E \cap G$. Since all outcomes in G are equally likely, we can conclude that the probability of E *given that G has occurred* is the ratio

$$\frac{\text{number of outcomes in } E \cap G}{\text{number of outcomes in } G}.$$

Since $E \cap G$ consists of the single outcome gg, we conclude that the probability of both children being of the same sex *given that at least one is a girl* is $\frac{1}{3}$.

The same reasoning applies to any events E and G in a finite sample space with equally likely outcomes. The probability of E, *given the information that G has occurred*, is called the **conditional probability** of E given G and denoted by $P(E|G)$. To compute it, we ignore any outcomes not in G and compare the size of $E \cap G$ to that of G. In other words, in a finite sample space with equally likely outcomes,

$$P(E|G) = \frac{\text{number of outcomes in } E \cap G}{\text{number of outcomes in } G}.$$

If we let $N(A)$ denote the number of outcomes in an event A, we can rewrite this formula as follows:

$$P(E|G) = \frac{N(E \cap G)}{N(G)},$$

provided the sample space is equiprobable.

By dividing both the numerator and denominator of this quotient by $N(S)$ (the number of elements in S), we get

$$P(E|G) = \frac{N(E \cap G)/N(S)}{N(G)/N(S)}.$$

Now, the numerator and denominator are the (unconditional) probabilities of $E \cap G$ and G, respectively. Therefore, we get another formula for the conditional probability:

$$P(E|G) = \frac{P(E \cap G)}{P(G)}. \tag{5}$$

This version of the conditional probability formula omits any reference to the number of outcomes. It can be interpreted as saying the conditional probability of E given G is the ratio of the (ideal) relative frequencies of $E \cap G$ and G, and it can be applied to more general situations—those in which S is not finite or the outcomes are not equally likely. With that in mind, we use formula (5) to *define* the meaning of conditional probability, as follows.

Definition 11.2.1

If G is a given event of a sample space with $P(G) > 0$, then the conditional probability of any other event E given G, denoted by $P(E|G)$, is the number defined by

$$P(E|G) = \frac{P(E \cap G)}{P(G)}. \tag{6}$$

EXAMPLE 11.2.2

Suppose that you toss a coin three times. Find the probability that at least two of the tosses are heads. Then find the probability that at least two are heads, *given that the first is heads*. Assume that all outcomes are equally likely.

SOLUTION   Here the sample space is

$$S = \{HHH, HHT, HTH, THH, HTT, THT, TTH, TTT\}.$$

It consists of eight outcomes. The event that at least two are heads is

$$E = \{HHH, HHT, HTH, THH\},$$

and it contains four outcomes. Therefore, $P(E) = \frac{4}{8} = \frac{1}{2}$.

We next want to compute the conditional probability $P(E|G)$, where G is the event that the first is heads, that is,

$$G = \{HHH, HHT, HTH, HTT\}.$$

Therefore, $P(G) = \frac{4}{8} = \frac{1}{2}$.

To use formula (6), we need to compute $P(E \cap G)$. We have

$$P(E \cap G) = P(\{HHH, HHT, HTH\}) = \frac{3}{8}.$$

Thus,

$$P(E|G) = \frac{P(E \cap G)}{P(G)} = \frac{\frac{3}{8}}{\frac{4}{8}} = \frac{3}{4}.$$

EXAMPLE 11.2.3

In a preelection poll, 40% of the voters support increased funding for the U.S. Environmental Protection Agency (EPA), 30% support increased funding for the Education Department (ED), and 10% support increased funding for both.

(i) If a voter supports increased funding for ED, what is the probability that this voter also supports increased funding for the EPA?

(ii) If a voter supports increased funding for EPA, what is the probability that this voter also supports increased funding for the ED?

SOLUTION   In this example the sample space S consists of all the voters.

(i) Here we let

$$E = \text{the voter is in favor of EPA}$$
$$G = \text{the voter is in favor of ED}$$
$$E \cap G = \text{the voter is in favor of both}$$

and the desired probability is

$$P(E|G) = \frac{P(E \cap G)}{P(G)} = \frac{0.1}{0.3} = \frac{1}{3}$$

(ii) Here E and G interchange roles:

$$E = \text{the voter is in favor of ED}$$
$$G = \text{the voter is in favor of EPA}$$

and

$$P(E|G) = \frac{P(E \cap G)}{P(G)} = \frac{0.1}{0.4} = \frac{1}{4}.$$

The multiplication rule

Equation (6) leads to a very useful way of computing $P(E \cap G)$. Indeed, multiplying both sides of Eq. (6) by $P(G)$ gives the following rule, known as the **multiplication rule**:

$$P(E \cap G) = P(G)P(E|G). \tag{7}$$

EXAMPLE 11.2.4

Consider the experiment of dealing two cards in succession from a standard deck. What is the probability that both cards are aces?

SOLUTION If we define the events

$$G = \text{the first card is an ace}$$
$$E = \text{the second card is an ace},$$

then the desired probability is $P(E \cap G)$. Since there are four aces in a deck of 52 cards, we have $P(G) = \frac{4}{52}$, or $\frac{1}{13}$. To compute $P(E|G)$, we reason as follows. If the first card is an ace, there are 51 cards left, of which three are aces. Therefore, $P(E|G) = \frac{3}{51}$, or $\frac{1}{17}$. Using (7) gives

$$P(E \cap G) = P(G)\, P(E|G) = \frac{1}{13} \cdot \frac{1}{17} = \frac{1}{221}.$$

EXAMPLE 11.2.5

A certain industrial product is tested by two quality-control inspectors acting in tandem. The first one rejects the items that he finds defective and passes the others to the second person, who then tests them a second time. Statistical data have shown that the first expert passes 8% of the defective units he tests, and the second passes 30% of the defective units passed on to him. What is the probability that a defective item is passed by both experts?

SOLUTION We define the events

$$G = \text{a defective unit is passed by the first inspector}$$
$$E = \text{a defective item is passed by the second inspector}.$$

The given information can then be translated into the formulas

$$P(G) = 0.08 \quad \text{and} \quad P(E|G) = 0.3.$$

The desired probability is $P(E \cap G)$, which can be found by the multiplication rule, as follows:

$$P(E \cap G) = P(G)\, P(E|G) = (0.08) \cdot (0.3) = 0.024.$$

Independent events

As we have observed in the previous examples, the conditional probability $P(E|G)$ is not, in general, equal to $P(E)$. However, that does occur in special cases—those in which the occurrence of the event G has no effect on the probability of E. In tossing a coin twice, for instance, the result of the first toss does not change the probability that the second toss will be heads. We say that such events are **independent**.

If we examine the formula for conditional probability,

$$P(E|G) = \frac{P(E \cap G)}{P(G)},$$

we see that $P(E|G) = P(E)$ if $P(E \cap G) = P(E)\,P(G)$. That leads to the following definition of independence.

Definition 11.2.2

Two events E and G are *independent* if

$$P(E \cap G) = P(E)\,P(G). \tag{8}$$

Two events E and G are said to be **dependent** if they are not independent.

EXAMPLE 11.2.6

Suppose that statistical evidence indicates that the probability of a newborn child being a boy is 0.51. If two children are born on the same day, what is the probability that both are girls?

SOLUTION According to the statistical data, the probability of a newborn child being a girl is 0.49. We define the events

$$E = \text{the second child is a girl}$$
$$G = \text{the first child is a girl.}$$

Then,

$$E \cap G = \text{both children are girls.}$$

Using the reasonable assumption that E and G are independent, we obtain

$$P(E \cap G) = P(E)\,P(G) = (0.49) \cdot (0.49) \approx 0.24.$$

Bayes's formula

There are a number of instances in which we know a certain conditional probability, say, $P(E|G)$, but we really want to know the conditional probability of the same events in the opposite order—that is, $P(G|E)$. Such situations occur, for example, in interpreting medical tests, as illustrated below in Example 11.2.7. We begin with the conditional formula

$$P(G|E) = \frac{P(E \cap G)}{P(E)}. \tag{9}$$

The multiplication rule says that

$$P(E \cap G) = P(E|G)\,P(G),$$

so we can rewrite (9) as

$$P(G|E) = \frac{P(E|G)\,P(G)}{P(E)}. \tag{10}$$

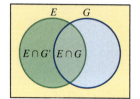

We next observe that the event E is the union of two disjoint events, $E \cap G$ and $E \cap G'$. That is, $E = (E \cap G) \cup (E \cap G')$, as shown in Figure 11.2.1

Using the additive property of probability for disjoint events, we obtain

$$P(E) = P(E \cap G) + P(E \cap G'). \tag{11}$$

Figure 11.2.1

By applying the multiplication rule to each of the terms on right, we get

$$P(E) = P(E|G)\,P(G) + P(E|G')\,P(G'). \tag{12}$$

Finally, substituting this into (10) gives

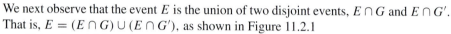

Bayes's formula

$$P(G|E) = \frac{P(E|G)\,P(G)}{P(E|G)\,P(G) + P(E|G')\,P(G')}. \tag{13}$$

EXAMPLE 11.2.7

The Papanicolaou or Pap test is an important screening test for cervical cancer. It is known from statistical data that it is 95% positive when the disease is present, and it is 5% (false) positive when the disease is not present. If 0.1% of the female population has the disease, then what is the probability that a woman, chosen at random, who tests positive does indeed have the disease?

SOLUTION   Let us consider the following events:

$$G = \text{the tested woman has the disease}$$
$$E = \text{the test result is positive.}$$

The desired probability is $P(G|E)$. To compute this quantity using Bayes's formula, we use the given data to compute the following probabilities:

$$P(E|G) = 0.95, \quad P(G) = 0.001, \quad P(E|G') = 0.05,$$
$$P(G') = 1 - P(G) = 1 - 0.001 = 0.999.$$

Then Bayes's formula gives

$$P(G|E) = \frac{P(E|G)\,P(G)}{P(E|G)\,P(G) + P(E|G')\,P(G')}$$
$$= \frac{(0.95)\cdot(0.001)}{(0.95)\cdot(0.001) + (0.05)\cdot(0.999)} \approx 0.019.$$

Practice Exercises 11.2

1. A bowl contains six slips of paper, numbered 1 through 6. You take two of them. What is the probability both are odd?

2. There are two coins in a box. Coin A is a fair coin, and the probability of its landing heads when flipped is $\frac{1}{2}$. Coin B, on the other hand, is weighted so that the probability of heads is $\frac{1}{3}$. Suppose you choose a coin at random and flip it twice.
   (a) What is the probability that the coin chosen lands heads both times, given that it is coin B?
   (b) What is the probability that it is coin B, given that it lands heads both times?

Exercises 11.2

1. A coin is tossed twice
   (a) What is the probability that at least one is heads?
   (b) What is the probability that at least one is heads, given that the second is tails?

   In Exercises 2–3, assume that the probabilities of a given child being a boy or a girl are equal.

2. Assume that a family has three children.
   (a) What is the probability that exactly two of the children are boys?
   (b) What is the probability that at least two of them are boys?
   (c) What is the probability that exactly two of them are boys, given that at least two are boys?
   (d) What is the probability that at least two of them are boys, given that at least one is a boy?

3. A family with four children is chosen at random.
   (a) Find the probability that at least three of the chidren are girls.
   (b) Find the probability that at least three are girls, given that the youngest is a girl.

   (c) Find the probability that exactly three are girls, given that the youngest is a girl.

4. Find the following conditional probabilities involved in tossing a pair of fair dice:
   (a) the probability that at least one of the dice is a 5, given that they add up to 7.
   (b) the probability that the dice add up to 7, given that at least one is a 5.

5. Find the following conditional probabilities involved in tossing a pair of fair dice:
   (a) the probability that both dice are even, given that they add up to 6
   (b) the probability that the dice add up to 6, given that they are both even

6. A marble is chosen at random from an urn that contains 20 red, 15 white, and 25 blue marbles.
   (a) What is the probability the marble is white, given that it is not red?
   (b) What is the probability the marble is blue, given that it is not red?

(c) What is the probability the marble is red, given that it is not white?

(d) What is the probability the marble is red, given that it is not blue?

7. A college English class consists of 16 boys and 12 girls. Four of the girls and six of the boys are seniors.

(a) What is the probability that a student selected at random is not a senior?

(b) What is the probability that a student selected at random is not a senior, given that the selected student is a boy?

(c) What is the probability that a student selected at random is a senior, given that the selected student is a boy?

(d) What is the probability that a student selected at random is a senior, given that the selected student is a girl?

8. A study of alcoholics showed that 38% had an alcoholic father, 7% had an alcoholic mother, and 4% had alcoholic parents.

(a) What is the probability that both parents were alcoholic, given that the mother was?

(b) What is the probability that the father was alcoholic, given that at least one of the parents was?

9. A bowl contains seven slips of paper, of which three are red and four are white. Two slips are drawn without replacement from the bowl. Use the multiplication principle to find the probabilities that

(a) both are red.

(b) neither is red.

(c) at least one is red.

10. Suppose you draw a card at random from a standard deck, then return it, shuffle the deck, and draw again.

(a) What is the probability that both cards are clubs?

(b) What is the probability that neither card is a club?

(c) What is the probability that one is a club and the other is not?

11. Suppose you draw a card at random from a standard deck, do *not* return it, and draw a second. Answer the same three questions posed in the last exercise.

12. A class consists of 8 girls and 12 boys. If two of the children are selected at random in succession, find the probability that

(a) the first is a boy and the second is a girl.

(b) one is a boy and one is a girl.

(c) both are boys.

13. The board game Monopoly® has been popular for decades, and its probabilities have been studied. For example, if your piece is on either Go or Oriental Avenue, the probability that it will land on one of the first 10 squares (Go = #1) in one turn is given by the following tables (under the assumptions that you roll again on doubles and get out of jail immediately):

		at	
		Go	Oriental
Land	Go	0.012	0.003
	Mediterranean Avenue	0.0	0.0
	Community Chest	0.024	0.0
	Baltic Avenue	0.056	0.0
	Income Tax	0.093	0.0

		at	
		Go	Oriental
Land	Reading Railroad	0.122	0.0
	Oriental Avenue	0.138	0.0
	Chance	0.063	0.0
	Vermont Avenue	0.138	0.027
	Connecticut Avenue	0.111	0.056

(a) What is the probability you land on Oriental Avenue in one turn, given that you are currently on Go?

(b) What is the probability you land on Connecticut Avenue in one turn starting from Oriental Avenue?

(c) Starting at Go and taking two turns, what is the probability you land first on Oriental and second on Connecticut?

Source: Taken from http://www.tkcs-collins.com/truman/monopoly/monopoly.shtml. Copyright © 1997 by Truman Collins.

14. Statistics show your favorite football team defeats its opponent 60% of the time under good weather conditions (say, no rain). The weather service predicts that the chances of good weather the day of the game are 75%. What is the probability that the weather will be good on the day of the game and your favorite team will win?

15. A company produces a certain auto component, and 3% of the components produced are defective. A quality-control test detects only 90% of the defective components. Find the probability that a component you buy will be defective.

16. Of the members of a certain credit union 25% have mortgages, 35% have car loans, and 5% have both a mortgage and a car loan. A member is selected at random from the credit union's membership.

(a) Find the probability that the member has a car loan, given that the member has a mortgage.

(b) Find the probability that the member has a mortgage, given that the member has a car loan.

(c) Are the events "the member has a car loan" and "the member has a mortgage" independent?

17. Suppose you toss a coin three times and consider the following events:

$$A = \text{the first is heads}$$
$$B = \text{at least two are heads}$$
$$C = \text{all three are the same.}$$

Which pairs are independent?

18. A family of four children is chosen at random.

(a) Are the events "exactly two of the children are girls" and "the youngest child is a girl" independent?

(b) Are the events "exactly three of the children are girls" and "the youngest child is a girl" independent?

19. Suppose A and B are events in the same sample space and both have nonzero probability. Can they be both independent and mutually exclusive?

20. Suppose A and B are independent, and $P(A) = \frac{1}{3}$ and $P(B) = \frac{1}{4}$. Find $P(A \cup B)$.

21. Suppose that the probability that your computer will fail to work during the next month is 0.15 and the probability that your printer will fail to work during the same period is 0.20. Find the probability that both your computer and your printer fail to work during the next month. Assume that these events are independent.

22. Assume that each engine of a two-engine airplane has 0.0001 probability of failing during an 8-hour flight, and that the failure of one is independent of the other.

(a) Find the probability that both engines will fail.

(b) Find the probability that neither engine will fail.

(c) Find the probability that at least one of the engines will not fail.

23. If a pair of dice are tossed, compute the probability that both are even in two ways:

(a) by taking the ratio of the number of outcomes in the event to the size of the total sample space, and

(b) by treating the two dice as independent and computing the probability of each one's being even.

24. The distribution of blood types within the population is roughly as follows: 38% have type A, 12% have type B, 4% have type AB, and 46% have type O. Suppose two unrelated blood donors arrive at a blood bank.

(a) What is the probability that both have type A?

(b) What is the probability that at least one has type A?

25. In addition to a specific blood type, each person has an Rh blood factor that is either positive or negative. The Rh factor is controlled by a pair of genes, each inherited from one parent, and the positive gene is dominant. In other words, a person will have Rh positive blood if either gene is positive. According to genetic data, 61% of the genes that control the Rh factor are positive.

(a) What is the probability that a randomly chosen person has Rh negative blood?

(b) Genetic evidence indicates that blood type and Rh factor are independent. Under that assumption, what is the probability that a randomly chosen person has O negative blood? (Use the data of Exercise 24 for blood types.)

26. During World War II each soldier's blood was tested and the type was stamped on his ID tag. It has been estimated that 88% of those with type A were correctly tagged, and 3% of those who did not have type A were tagged as type A. What is the probability that a soldier who was tagged as type A had type A blood? (Use the data of Exercise 24 for blood types.)

27. A high-school chemistry teacher has observed that 72% of his students pass the first test. Moreover, 91% of those who pass the first test also pass the second, and 12% of those who fail the first test rally to pass the second.

(a) What is the probability that a randomly chosen student will pass both tests?

(b) What is the probability that a student who passed the second test also passed the first?

28. It is known from statistical data that a blood exam for detecting a certain disease is positive in 92% of the tests when the disease is present, but is (false) positive in 4% of the tests when the disease is not present. If 6% of a given population has this disease, then what is the probability that a person who tests positive actually has the disease?

29. A car company produces the same car in two different plants. Plant A produces 600 cars during a period, and plant B produces 400 cars during the same period. Statistical data show that 3% of the cars produced by plant A are defective, while the corresponding number for plant B is 4%. Suppose that a car is selected at random from the total pool of 1,000 and is found to be defective.

(a) What is the probability that it comes from plant A?

(b) What is the probability that it comes from plant B?

30. Based on its statistical data, an auto insurance company considers that 25% of the drivers are accident-prone, and there is a 0.5 probability that an accident-prone driver will have an accident within a 2-year period. For drivers that are not prone to accidents, the corresponding probability is 0.2. Find the probability that a policy holder is accident-prone, given that he has an accident within the first 2 years.

31. Assume that in a city of 5 million residents a crime has been committed and that DNA samples from the scene of the murder match those of six residents. Assume that one of these is the guilty party.

(a) What is the probability that a randomly selected resident matches the DNA samples, given that this person is innocent?

(b) What is the probability that a resident who matches the DNA fragments is innocent?

32. Show that if A and B are independent, then A' and B are independent. Then show that A' and B' are independent. (*Hint:* Start with the equation $P(B) = P(A \cap B) + P(A' \cap B)$.)

33. Show that the conditional probability $P(E|G)$, where G is kept fixed, satisfies the three rules for assigning probabilities discussed in Section 11.1.

Solutions to practice exercises 11.2

1. Let O_1 be the event that the first slip is odd and O_2 the event that the second is odd. Since there are initially six slips, of which three are odd, $P(O_1) = \frac{3}{6} = \frac{1}{2}$. If the first is odd, then there remain five slips, of which two are odd, which means that $P(O_2|O_1) = \frac{2}{5}$. Therefore, the probability that both are odd is given by the multiplication rule, as follows:

$$P(O_1 \cap O_2) = P(O_1) \cdot P(O_2|O_1) = \frac{1}{2} \cdot \frac{2}{5} = \frac{1}{5}.$$

2. Let A be the event that coin A is chosen, let B be the event that coin B is chosen, and let HH be the event the coin comes down heads both times.

(a) If coin B is chosen, the probability of landing heads on any toss is $\frac{1}{3}$. Since both tosses are independent, the probability that it lands heads twice in succession is the product $\frac{1}{3} \cdot \frac{1}{3}$, or $\frac{1}{9}$.

(b) In the previous part we concluded that $P(HH|B) = \frac{1}{9}$. By similar reasoning, $P(HH|A) = \frac{1}{4}$. Moreover, $P(A) = P(B) = \frac{1}{2}$. Therefore, by Bayes's theorem,

$$P(B|HH) = \frac{\frac{1}{9} \cdot \frac{1}{2}}{\left(\frac{1}{9} \cdot \frac{1}{2}\right) + \left(\frac{1}{4} \cdot \frac{1}{2}\right)} = \frac{4}{13}.$$

11.3 Discrete Random Variables

In many applications of probability there is a numerical value attached to each outcome of an experiment. In tossing a pair of dice, for instance, we are usually interested in the sum of the top faces. As another example, imagine a coin-tossing game in which you toss a coin twice and count the number of heads that occur. The sample space for this experiment is

$$S = \{HH, HT, TH, TT\},$$

and to each outcome we assign the number of H's that appear. That is, we make the assignments:

$$HH \to 2, \quad HT \to 1, \quad TH \to 1, \quad TT \to 0.$$

What we have, then, is a rule that assigns a number from the set $\{0, 1, 2\}$ to each outcome of the sample space S. Such a rule is called a **function**. Up until now, in our study of calculus, we have used the word *function* to mean a rule that assigns a unique numerical value to every number in some interval of the real axis or to every point in some domain of n-space. In general, the word function is used to describe a correspondence between elements of two sets S and T, in which to every element of S is assigned a unique element of T. In the case in question—tossing a coin twice and counting the number of heads—the set S is the sample space and T is the set $\{0, 1, 2\}$. In probability language, this type of function is called a **random variable**.

> ### Definition 11.3.1
>
> A random variable is a function that assigns a real number to every outcome of a sample space.

The set of real numbers that can occur as values of a random variable is called its **range**. Random variables are often denoted by capital letters, such as X, Y, or Z.

Of course, there may be many different random variables associated with a given experiment. In tossing a coin three times, for example, we might let X be the number of heads, Y be the number of tails, and Z be the number of heads minus the number of tails. If we consider the outcome HTT, for instance, then $X(HTT) = 1$, $Y(HTT) = 2$, and $Z(HTT) = -1$. In this case, the range of both X and Y is the set $\{0, 1, 2, 3\}$, whereas the range of Z is the set $\{-3, -1, 1, 3\}$.

In each of our examples so far, the range of the random variable was a finite set of numbers. Finite sets and sets that can be written in the form $\{x_1, x_2, \ldots\}$ are called discrete. A random variable whose range is discrete is called a **discrete random variable**. More formally, we can define it as follows:

> ### Definition 11.3.2
>
> A random variable is called **discrete** if it takes only a finite number of values or if its values can be placed in one-to-one correspondence with the positive integers.

Every random variable with a finite range is discrete, but there are also discrete random variables whose ranges are infinite. As an example, suppose you toss a coin until it comes up heads, then stop. The sample space for this experiment is the infinite set given by

$$S = \{H, TH, TTH, TTTH, TTTTH, \ldots\},$$

where each outcome is a string of T's followed by a single H. If we let X be the number of tosses up to and including the first H, so that $X(H) = 1$, $X(TH) = 2$, $X(TTH) = 3$, and so forth, then X is a discrete random variable whose range is the set of positive integers.

Probability distributions and histograms

The key thing to know about a discrete random variable, if possible, is its probability of taking on any particular value. Suppose the range of X is either a finite set of the form $\{x_1, x_2, \ldots, x_n\}$ or an infinite discrete set of the form $\{x_1, x_2, x_3, \ldots\}$. For each number x_j in the range, the equation $X = x_j$ defines an event in the sample space—namely, the set of all outcomes that take the value x_j. Letting p_j represent the probability of that event, we can write

$$p_j = P(X = x_j).$$

For example, if we toss two coins and let X be the number of heads, we have

$$P(X = 0) = P(TT) = \frac{1}{4}$$

$$P(X = 1) = P(HT) + P(TH) = \frac{1}{4} + \frac{1}{4} = \frac{1}{2}$$

$$P(X = 2) = P(HH) = \frac{1}{4}.$$

We can summarize this information in a table, as follows:

x	0	1	2
$P(X = x)$	$\frac{1}{4}$	$\frac{1}{2}$	$\frac{1}{4}$

The assignment of a probability p_j to each number x_j in the range of X is called the **probability distribution** of X.

Notice that the entries in the second row of the distribution table add up to 1. That is no coincidence, as the following theorem shows:

Theorem 11.3.1

If X is a random variable with a finite range $\{x_1, x_2, \ldots, x_n\}$ and $p_i = P(X = x_i)$, then

$$p_1 + p_2 + \cdots + p_n = 1.$$

PROOF   Let E_1 be the set of outcomes for which $X = x_1$, and let F_1 be the outcomes for which $X \neq x_1$. The events E_1 and F_1 are mutually exclusive, and together they comprise the entire sample space of X. Therefore,

$$P(E_1) + P(F_1) = 1,$$

and, since $p_1 = P(E_1)$, we have

$$p_1 + P(F_1) = 1. \tag{14}$$

Now, let E_2 be the set of those outcomes in F_1 for which $X = x_2$ and let F_2 be the set of outcomes for which X equals neither x_1 nor x_2. The events E_2 and F_2 are mutually exclusive, and $F_1 = E_2 \cup F_2$. Therefore,

$$P(F_1) = P(E_2) + P(F_2) = p_2 + P(F_2).$$

Substituting this into Eq. (14) gives

$$p_1 + p_2 + P(F_2).$$

Next, divide F_2 into two mutually exclusive events—the set E_3 of those outcomes for which $X = x_3$ and the set F_3 of those outcomes for which $X \neq x_3, x_2,$ or x_1.

Proceeding in this way for n steps leads to the equation

$$p_1 + p_2 + \cdots + p_n = 1.$$

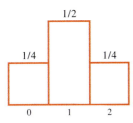

A visual way of displaying the information in a probability distribution table is by drawing a bar graph. In the case of tossing two coins, we mark the numbers 0, 1, and 2 along a horizontal axis in increasing order, and then over each of those numbers we draw a rectangle whose height is the probability of that number and whose base has length 1. Thus, *the area of the rectangle over each number is the probability that X equals that number.* This diagram is called the **probability histogram** of X, and it is shown to the left. Observe that each rectangle is drawn so that its base has the value of X at the center.

EXAMPLE 11.3.1

In the experiment of tossing two dice, let X be the random variable that gives the sum of the top faces. Find the probability distribution of X and draw its probability histogram.

SOLUTION The sample space for this experiment, which has 36 outcomes, was displayed in Example 11.1.6. For convenience, we repeat it here:

$$\begin{aligned}
S = \{ & (1, 1), (1, 2), (1, 3), (1, 4), (1, 5), (1, 6), \\
& (2, 1), (2, 2), (2, 3), (2, 4), (2, 5), (2, 6), \\
& (3, 1), (3, 2), (3, 3), (3, 4), (3, 5), (3, 6), \\
& (4, 1), (4, 2), (4, 3), (4, 4), (4, 5), (4, 6), \\
& (5, 1), (5, 2), (5, 3), (5, 4), (5, 5), (5, 6), \\
& (6, 1), (6, 2), (6, 3), (6, 4), (6, 5), (6, 6) \}.
\end{aligned}$$

The range of X is the set of integers between 2 and 12, inclusive. If x is any one of those numbers, we can find $P(X = x)$ by simply listing all the pairs (i, j) with $i + j = x$, then counting them, and dividing by 36. For instance, to find $P(X = 4)$, we list all pairs that add up to 4, namely, $(1, 3), (2, 2),$ and $(3, 1)$. There are three of them, and therefore

$$P(X = 4) = \frac{3}{36} = \frac{1}{12}.$$

By doing the same for each x from 2 to 12, we arrive at the following probability distribution:

x	2	3	4	5	6	7	8	9	10	11	12
$P(X = x)$	$\frac{1}{36}$	$\frac{2}{36}$	$\frac{3}{36}$	$\frac{4}{36}$	$\frac{5}{36}$	$\frac{6}{36}$	$\frac{5}{36}$	$\frac{4}{36}$	$\frac{3}{36}$	$\frac{2}{36}$	$\frac{1}{36}$

The probability histogram of X looks like this:

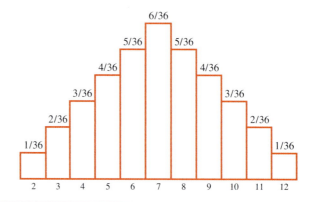

Histograms are also used to display a table of data. The guiding principle is to let the area of each rectangle represent the relative frequency of each data group.

EXAMPLE 11.3.2

In a study of the side effects of a certain medicine, the systolic blood pressure of each of 20 patients was measured, and the data were organized in the following table:

systolic blood pressure	115–119	120–124	125–129	130–135
number of patients	6	8	4	2

Draw a histogram displaying these data.

SOLUTION   First, we measure the relative frequency of each blood pressure group, or, in other words, the proportion of patients in each group. To do that, we simply divide the number in each group by 20, the total number of patients. The results are given in the third row of the following expanded table:

systolic blood pressure	115–119	120–124	125–129	130–135
number of patients	6	8	4	2
relative frequency	0.3	0.4	0.2	0.1
height of rectangle	0.06	0.08	0.04	0.02

To construct a histogram, we mark off the points 115, 120, 125, 130, and 135 along a horizontal axis and construct a rectangle over each interval formed by successive pairs of these numbers. Each of these rectangles will have a base of length 5 and its area will equal the relative frequency for that interval. In order to achieve that, we need to make the height equal to the relative frequency divided by 5 (the width of the rectangle). The resulting heights are shown in the fourth row of the expanded table. The histogram is drawn below:

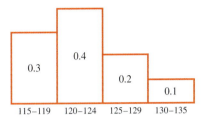

The expected value or mean

The probability distribution and histogram associated with a random variable are important tools, but they sometimes contain more information than we need and may be cumbersome to work with. A much more succinct piece of information is the **expected value**, denoted by $E(X)$, which is a number that measures the "average" value of a random variable.

Definition 11.3.3

If X is a random variable whose range is $\{x_1, x_2, \ldots, x_n\}$, the expected value is defined by

$$E(X) = x_1 p_1 + x_2 p_2 + \cdots + x_n p_n,$$

where $p_i = P(X = x_i)$. The expected value of a random variable is also called its **mean** and is often denoted by the Greek letter μ.

EXAMPLE 11.3.3

Suppose you toss a pair of dice and let X be the sum of the top faces. Find the expected value $E(X)$.

SOLUTION   We already determined the probability distribution for X in Example 11.3.1. For easy reference, let's repeat it here.

x	2	3	4	5	6	7	8	9	10	11	12
$P(X = x)$	$\frac{1}{36}$	$\frac{2}{36}$	$\frac{3}{36}$	$\frac{4}{36}$	$\frac{5}{36}$	$\frac{6}{36}$	$\frac{5}{36}$	$\frac{4}{36}$	$\frac{3}{36}$	$\frac{2}{36}$	$\frac{1}{36}$

According to the definition, we compute $E(X)$ by multiplying each entry on the top line by the number under it and then adding the results. That gives

$$E(X) = 2 \cdot \frac{1}{36} + 3 \cdot \frac{2}{36} + 4 \cdot \frac{3}{36} + 5 \cdot \frac{4}{36} + 6 \cdot \frac{5}{36} + 7 \cdot \frac{6}{36}$$

$$+ 8 \cdot \frac{5}{36} + 9 \cdot \frac{4}{36} + 10 \cdot \frac{3}{36} + 11 \cdot \frac{2}{36} + 12 \cdot \frac{1}{36},$$

which equals 7.

To understand the significance of the expected value, imagine that you repeated the experiment of tossing the dice N times, observing and recording the sum X of the top faces each time. Let n_2 be the number of times that X takes the value 2, let n_3 be the number of times it takes the value 3, and so forth, up to n_{12} (the number of times that X takes the value 12). Then we must have

$$n_2 + n_3 + \cdots + n_{12} = N.$$

Next, we will average all the values of X that were taken in the N trials of the experiment. We have

$$\text{Average} = \frac{\overbrace{2 + \cdots + 2}^{n_2 \text{times}} + \overbrace{3 + \cdots + 3}^{n_3 \text{times}} + \cdots + \overbrace{12 + \cdots + 12}^{n_{12} \text{times}}}{N}$$

$$= \frac{2n_2 + 3n_3 + \cdots + 12n_{12}}{N}$$

$$= 2\frac{n_2}{N} + 3\frac{n_3}{N} + \cdots + 12\frac{n_{12}}{N}.$$

The fraction n_i / N is the relative frequency of the occurrence of $X = i$, so that for very large N we have

$$\frac{n_2}{N} \approx P(X = 2) = \frac{1}{36}$$

$$\frac{n_3}{N} \approx P(X = 3) = \frac{2}{36},$$

and so forth, from which we see that the average is approximately equal to

$$2 \cdot \frac{1}{36} + 3 \cdot \frac{2}{36} + 4 \cdot \frac{3}{36} + 5 \cdot \frac{4}{36} + 6 \cdot \frac{5}{36} + 7 \cdot \frac{6}{36}$$

$$+ 8 \cdot \frac{5}{36} + 9 \cdot \frac{4}{36} + 10 \cdot \frac{3}{36} + 11 \cdot \frac{2}{36} + 12 \cdot \frac{1}{36},$$

which is precisely $E(X)$.

The same interpretation applies to any random variable X whose range is finite, say, $\{x_1, x_2, \ldots, x_m\}$. Suppose we repeat the experiment associated with X a large number of times, say, N, and let n_i denote the number of times the value x_i occurs. We can compute the average of all the values that occur just as we did for the dice—that is

$$\text{Average} = \frac{x_1 n_1 + x_2 n_2 + \cdots + x_m n_m}{N}$$

$$= x_1 \frac{n_1}{N} + x_2 \frac{n_2}{N} + \cdots + x_m \frac{n_m}{N}.$$

As before, replacing the relative frequency n_i/N by the probability p_i gives us the expected value.

EXAMPLE 11.3.4

Let X be the number of computers sold by a computer store in a given day. Assume that the sales for the last 200 days are given by the following table:

number of computers sold	6	7	8	9	10
number of days	20	30	80	40	30

Estimate the probability distribution of the random variable X and use it to compute the mean μ.

SOLUTION   The range of X consists of the integers from 6 to 10, inclusive. If x is any of those numbers, we use its relative frequency, given in the table above, to estimate its true probability of occurring. In other words, $P(X = x)$ is the number of days on which x computers were sold divided by 200 (the total number of days in question). Thus,

$$P(X = 6) = \frac{20}{200} = 0.10,$$

and so forth. In that way we arrive at the following probability distribution:

x	6	7	8	9	10
$P(X = x)$	0.10	0.15	0.40	0.20	0.15

The mean (which is the same as the expected value) of X is

$$\mu = 6 \cdot (0.10) + 7 \cdot (0.15) + 8 \cdot (0.40) + 9 \cdot (0.20) + 10 \cdot (0.15) = 8.15.$$

We can interpret $\mu = 8.15$ as saying that, in the long run, the computer store will be selling an average of 8.15 computers a day. Notice that 8.15 is not one of the numbers in the range of X. In general, the mean represents an *average* value of X and need not be one of the possible values taken by the random variable.

EXAMPLE 11.3.5

Roll a single die and let X be the number on top. Find $E(X)$.

SOLUTION   The possible values of X are 1, 2, 3, 4, 5, 6, each with probability $\frac{1}{6}$. Therefore,

$$E(X) = \frac{1}{6}(1 + 2 + 3 + 4 + 5 + 6) = 3.5.$$

The variance

As we have seen, the expected value (or mean) is an average and averages provide a certain amount of information, but they can be deceiving. For instance, the set of numbers 48, 49, 50, 51, 52 have 50 as their average, but so do the numbers 10, 30, 50, 70, 90. In order to use the average intelligently, it helps to have some measure of how closely the numbers are spread around the average—are they clustered close to it or are they widely dispersed? For a random variable X with mean μ, there are two numbers that measure the dispersion. The first is called the **variance** of X and denoted by Var(X).

Definition 11.3.4

If X is a random variable with range $\{x_1, x_2, \ldots, x_n\}$, the variance is defined by

$$\text{Var}(X) = (x_1 - \mu)^2 p_1 + (x_2 - \mu)^2 p_2 + \cdots + (x_n - \mu)^2 p_n,$$

where $p_i = P(X = x_i)$.

Notice that the variance is never negative, since every term in the sum is a square multiplied by a probability, both of which are greater than or equal to zero. Roughly speaking, a small variance indicates that the values of X are closely clustered about the mean—or, at least that the values clustered there have a greater probability of occurring. The larger the variance, the less likely the values are to lie close to the mean. As you can see in the formula above, the differences $(x_j - \mu)$ are squared in computing the variance. That is to avoid cancellation of terms with opposite signs; otherwise, two values that were both far away from μ but on opposite sides might nullify one another.

EXAMPLE 11.3.6

Suppose W, X, and Y are discrete random variables, with the following probability tables:

values of W	10	20	30	40
probability	0.25	0.25	0.25	0.25

values of X	23	24	26	27
probability	0.25	0.25	0.25	0.25

values of Y	10	20	30	40
probability	0.1	0.4	0.4	0.1

(i) Verify that all three random variables have a mean of 25.

(ii) Without computing, guess which of the two random variables W and X has the smaller variance and give a reason for your guess. Check your guess by computing the two variances.

(iii) Do the same with the random variables W and Y.

SOLUTION

(i) $E(W) = (0.25)(10 + 20 + 30 + 40) = 25.$

$E(X) = (0.25)(23 + 24 + 26 + 27) = 25.$

$E(Y) = (0.1)(10) + (0.4)(20) + (0.4)(30) + (0.1)(40) = 25.$

(ii) We would expect X to have a smaller variance, because the variables cluster closer to the mean. And, in fact,

$$\text{Var}(W) = (0.25) \cdot \left[(-15)^2 + (-5)^2 + 5^2 + (15)^2 \right] = 125$$

and

$$\text{Var}(X) = (0.25) \cdot \left[(-2)^2 + (-1)^2 + 1^2 + 2^2 \right] = 2.5.$$

(iii) In this case, both variables have the same range. However, the values of W are all equally probable, whereas the values of Y that are closer to the mean have a higher probability than those that are further. This suggests that Y has a smaller variance. In fact, $\text{Var}(W) = 125$, as we have just seen, and

$$\text{Var}(Y) = (0.1)(-15)^2 + (0.4)(-5)^2 + (0.4)5^2 + (0.1)(15)^2 = 65.$$

EXAMPLE 11.3.7

Compute the mean and variance of the random variable in each of the following experiments:

(i) You roll a single die and let X be the number on the top face.

(ii) A bowl contains six slips of paper, three marked with the number 3 and the others with a 4. You draw one of the slips at random and let Y be the number on the slip.

SOLUTION   (i) In this case, the range is the set of numbers $\{1, 2, 3, 4, 5, 6\}$, and X has the following probability distribution:

x	1	2	3	4	5	6
$P(X = x)$	$\frac{1}{6}$	$\frac{1}{6}$	$\frac{1}{6}$	$\frac{1}{6}$	$\frac{1}{6}$	$\frac{1}{6}$

Using this table, we get

$$\mu = 1 \cdot \frac{1}{6} + 2 \cdot \frac{1}{6} + 3 \cdot \frac{1}{6} + 4 \cdot \frac{1}{6} + 5 \cdot \frac{1}{6} + 6 \cdot \frac{1}{6}$$

$$= \frac{1}{6} \cdot (1 + 2 + 3 + 4 + 5 + 6) = \frac{7}{2}$$

and

$$\text{Var}(X) = \frac{1}{6} \cdot \left\{ \left(1 - \frac{7}{2}\right)^2 + \left(2 - \frac{7}{2}\right)^2 + \left(3 - \frac{7}{2}\right)^2 \right.$$

$$\left. + \left(4 - \frac{7}{2}\right)^2 + \left(5 - \frac{7}{2}\right)^2 + \left(6 - \frac{7}{2}\right)^2 \right\}$$

$$= \frac{35}{12}.$$

(ii) The range of Y is the set of numbers $\{3, 4\}$. Its probability distribution is

x	3	4
$P(X = x)$	$\frac{1}{2}$	$\frac{1}{2}$

Using this table, we get

$$\mu = 3 \cdot \frac{1}{2} + 4 \cdot \frac{1}{2} = \frac{7}{2},$$

and

$$\text{Var}(Y) = \left(3 - \frac{7}{2}\right)^2 \cdot \frac{1}{2} + \left(4 - \frac{7}{2}\right)^2 \cdot \frac{1}{2} = \frac{1}{4}.$$

Notice that both of these random variables have the same mean, but X has a much larger variance than Y, reflecting the fact that the values of X are more widely distributed about the mean.

There is an alternate way of computing the variance that is sometimes easier, especially when working with a hand calculator.

Theorem 11.3.2

Suppose X is a random variable with a finite range of values $\{x_1, \ldots, x_n\}$. If $p_i = P(X = x_i)$ and μ is the mean of X, then

$$\text{Var}(X) = x_1^2 p_1 + \cdots + x_n^2 p_n - \mu^2.$$

PROOF By definition, $\text{Var}(X) = (x_1 - \mu)^2 p_1 + (x_2 - \mu)^2 p_2 + \cdots + (x_n - \mu)^2 p_n$. If we expand a typical term on the right-hand side, it looks like this:

$$(x_i - \mu)^2 p_i = x_i^2 p_i - 2\mu x_i p_i + \mu^2 p_i.$$

Adding these for i from 1 to n and collecting terms, we get

$$\text{Var}(X) = x_1^2 p_1 + \cdots + x_n^2 p_n - 2\mu(x_1 p_1 + \cdots + x_n p_n) + \mu^2(p_1 + \cdots + p_n). \quad (15)$$

But we know from Theorem 11.3.1 that $p_1 + \cdots + p_n = 1$, and the definition of the mean tells us that $x_1 p_1 + \cdots + x_n p_n = \mu$. Substituting these into Eq. (15) gives

$$\begin{aligned}
\text{Var}(X) &= x_1^2 p_1 + \cdots + x_n^2 p_n - 2\mu^2 + \mu^2 \\
&= x_1^2 p_1 + \cdots + x_n^2 p_n - \mu^2.
\end{aligned}$$

The standard deviation

In defining the variance, we squared each of the numbers $(x_j - \mu)$ in order to avoid cancellation. However, squaring may cause some distortion of the units involved. For that reason, we use the square root of the variance to measure how closely the values are clustered around the mean. It is called the **standard deviation** and is symbolized by $\sigma(X)$.

Definition 11.3.5

The standard deviation $\sigma(X)$ of a random variable X is defined by

$$\sigma(X) = \sqrt{\text{Var}(X)}.$$

EXAMPLE 11.3.8

Compute the variance and standard deviation of the random variable X in Example 11.3.4.

SOLUTION   We already found that $\mu = 8.15$ in Example 11.3.4. To compute the variance, let's use the formula of Theorem 11.3.2.

$$\text{Var}(X) = 36(0.10) + 49(0.15) + 64(0.40) + 81(0.20) + 100(0.15) - (8.15)^2$$
$$= 1.3275.$$

Thus,

$$\sigma(X) = \sqrt{1.3275} \approx 1.152.$$

APPLYING TECHNOLOGY

Given a data table such as the one in Example 11.3.4, a graphing calculator can be used to compute the mean, variance, and standard deviation of the corresponding probability distribution. With a TI-83 Plus, for example, that can be done either with the STAT or LIST editor.

EXAMPLE 11.3.9

Referring back to Example 11.3.4, consider the following data for computer sales over a 200-day period:

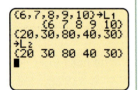

Figure 11.3.1

number of computers sold	6	7	8	9	10
number of days	20	30	80	40	30

Letting X be the number of computers sold in a given day, use a TI-83 Plus calculator to find the mean, variance, and standard variance of X.

Figure 11.3.2

SOLUTION   The top row of the table lists the possible values of X, and the second row gives the frequency with which the particular value occurs (not to be confused with the **relative frequency**, which in this case would be the frequency divided by 200, the total number of days). We enter each row into the calculator as a list, as shown in Figure 11.3.1, in which we have used the { and } keys to mark each list and saved the values as L_1 and the frequencies as L_2. Next, using the LIST MATH menu displayed in Figure 11.3.2, we obtain the one-variable statistics for X, displayed in Figure 11.3.3. In particular, we obtain the mean (denoted by $\overline{x}$) and the standard deviation (denoted by σ_x). To find the variance, we simply square the standard deviation. Thus, we get

$$E(X) = \overline{x} = 8.15, \quad \sigma(X) = \sigma_x = 1.152171862, \quad \text{Var}(X) = \sigma(X)^2 = 1.3275.$$

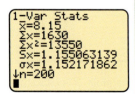

Figure 11.3.3

A few remarks on the other statistics listed on the calculator screen: n is the number of samples (in this case, the number of days), and Σx is the sum of the values, each multiplied by the frequency of its occurrence (in this case, the total number of computers sold during the 200 days), and S_x is known as the **sample standard deviation**. It is used in many sampling situations in place of the actual standard deviation, to which it is related by the formula

$$S_x = \sigma_x \sqrt{\frac{n}{n-1}}.$$

Practice Exercise 11.3

1. Let X be the number of heads in four tosses of a fair coin.
   (a) Construct the table giving the probability distribution of X and draw its histogram.
   (b) Compute the mean, variance, and standard deviation of X.

Exercises 11.3

1. Let X be the number of heads in three tosses of a fair coin.
   (a) Construct the table giving the probability distribution of X and draw its histogram.
   (b) Compute the mean, variance, and standard deviation of X.

2. Consider the experiment of tossing two tetrahedral (four-sided) fair dice, each with faces numbered 1 through 4. Let X be the random variable that gives the sum of the bottom faces.
   (a) Find the probability distribution of X and its probability histogram.
   (b) Compute the probability that $3 \leq X \leq 7$.

3. Consider the experiment of tossing two six-sided fair dice and let X be the absolute value of the difference of the top faces. Find the probability distribution of X and its probability histogram. Then compute its expected value and variance.

4. It is hyposthesized that when a flock of birds perch on a wire, they will position themselves so that the space between any two is about twice the pecking distance—that is, the maximum distance at which one bird can peck another. That way, no two will be close enough to peck each other. To estimate the pecking distance, you could take a picture of a crowded wire, measure the space between the birds, and compute the mean distance. How would you then determine the pecking distance? If the hypothesis is true, what would you expect the variance to look like? In what way would the variance be different if the spacing were random?

5. Let X be the random variable that denotes the number of bikes sold by a bike store in a given day. Assume that the sales for the last 400 days are given by the following table:

number of bikes sold	6	7	8	9	10
number of days	40	60	160	80	60

Find the probability distribution of the random variable X and draw its histogram. Then compute the expected value, variance, and standard deviation of X.

6. An ornithologist studying cliff swallows (*Petrochelidon pyrrhonota*) has found large colonies of nests that are accessible by ladder. He counted the number of eggs in 187 nests and displayed his data in the following table:

number of eggs	4	5	6	7
number of nests	37	56	51	43

(a) Find the expected number of eggs per nest.
(b) Estimate the probability that the next nest will have four eggs.

7. Archaeologists were excavating an ancient village. They identified the central gathering structure, saw the individual homes spread out around it, and measured the distance from the center of the gathering place to the center of each home. The following table shows their findings:

distance in meters	10–19	20–29	30–39	40–49	50–59
number of homes	15	12	14	7	5

Compute the relative frequency of each distance interval and display the data in histogram form. Using the midpoint of each interval, estimate the average distance of a home from the gathering place.

8. Compute the expected value $E(X)$ and the variance $\text{Var}(X)$ for the random variable X with the following probability distribution:

value of X	10	12	16	18	20
probability	0.1	0.1	0.2	0.4	0.2

9. Suppose you draw one card from an ordinary, well-shuffled deck. If the card is an ace, you win $3. If it is a king, queen, or jack, you win $2. If it is a ten, you win $1. If it is any other card, you lose $1. Let X be your net gain on a single play (positive if you win, negative if you lose).
(a) What is $E(X)$?
(b) If you play the game 100 times, how much would you expect to win or lose? Round your answer to the nearest cent.

10. Suppose a random variable X has range $\{-2, -1, 0, 1, 2\}$, and each value is equally likely to occur. Find $E(X)$ and $\text{Var}(X)$.

11. A random variable X has only three possible values: 1, 2, and 4. The expected value (mean) is 3 and the variance is $\frac{3}{2}$. Find the probability distribution of X.

12. Suppose you roll a single die and let X be the number on the top face. Now let $Y = 2X + 3$. Find $E(Y)$ and $\text{Var}(Y)$.

13. Suppose X is a random variable whose range is a finite set $\{x_1, x_2, \ldots, x_n\}$, with $p_i = P(X = x_i)$. If $Y = aX + b$

for some numbers a and b, show that $E(Y) = aE(X) + b$ and $\text{Var}(Y) = a^2\text{Var}(X)$.

14. Consider the experiment of counting the passengers of a particular United Airlines flight from Chicago to San Francisco. Assume that the capacity of the plane is 300 seats and that you are given the following data for the last 1,000 days of flight service:

passengers	days
0–100	10
101–160	30
161–200	90
201–220	120
221–250	250
251–270	350
271–290	100
291–300	50

(a) Estimate the probability of this event: "In the next flight the number of passengers will be more than 100 and less than or equal to 250."
(b) Estimate the probability of this event: "There will be more than 250 passengers in the next flight."
(c) Assume that every passenger pays $328 for this flight, and let X be the revenue of the flight in hundreds of dollars. What is the economic meaning of the expectation of X?

Solutions to practice exercise 11.3

1. (a) The range of X is the set of numbers $\{0, 1, 2, 3, 4\}$. The sample space consists of 16 outcomes (see Example 11.1.9). Of these, there is only one outcome, $TTTT$, for which $X = 0$, and one outcome, $HHHH$, for which $X = 4$. There are four outcomes for which $X = 1$, namely, $HTTT, THTT, TTHT$, and $TTTH$. Similarly, there are four outcomes for which $X = 3$. The remaining six outcomes give $X = 2$.

Using this information, we construct the following probability table:

x	0	1	2	3	4
$P(X = x)$	$\frac{1}{16}$	$\frac{1}{4}$	$\frac{3}{8}$	$\frac{1}{4}$	$\frac{1}{16}$

To construct the histogram, we plot the points 0, 1, 2, 3, 4 on a horizontal axis and raise a rectangle over each of them,

with the given point at the center of the base. Each rectangle has base length 1, so that its area and height are the same. Therefore, we take the rectangle over zero to have height $\frac{1}{16}$, the rectangle over 1 to have height $\frac{1}{4}$, and so forth.

The histogram is shown below.

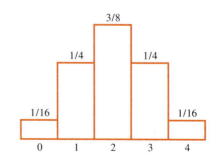

(b) The mean is given by

$$\mu = E(X) = 0 \cdot \frac{1}{16} + 1 \cdot \frac{1}{4} + 2 \cdot \frac{3}{8} + 3 \cdot \frac{1}{4} + 4 \cdot \frac{1}{16} = 2.$$

The variance is

$$\mathrm{Var}(X) = 0 \cdot \frac{1}{16} + 1 \cdot \frac{1}{4} + 4 \cdot \frac{3}{8} + 9 \cdot \frac{1}{4} + 16 \cdot \frac{1}{16} - 4 = 1,$$

and the standard deviation is $\sigma(X) = \sqrt{\mathrm{Var}(X)} = 1$.

11.4 The Binomial Distribution

An important application of probability is to polling, in which a random collection of people from a given population are asked the same "yes-no" question, such as: *Are you voting for or against proposition 16 in next year's election* or *do you prefer brand A toothpaste over brand B*? Suppose we ask n people, and let X be the number who answer "yes." Then X is a discrete random variable, and we say it has a **binomial distribution**, where the word *binomial* come from the Latin for "having two names"—in this case, the two answers, "yes" and "no."

The same type of random variable arises in other situations in which we repeatedly pose a question with two answers, such as "true-false" or "success-failure." For instance, in testing a commercial product for quality control, we may try n samples and label each as *good* or *defective*; or we may test a new medical procedure on n patients and label each as *patient recovered* or *patient did not recover*.

Definition 11.4.1

Suppose an experiment has exactly two outcomes, one (call it "success") occurs with probability p, the other (call it "failure"), occurring with probability $(1 - p)$. If the experiment is repeated n times, all under identical conditions but independent of one another, and we let X be the total number of successes, then we say that X has a **binomial distribution** with parameters n and p.

The simplest example of a binomial distribution occurs in tossing a coin n times and counting the number of heads. In fact, if we assume the coin is weighted so that the probability of heads is p and the probability of tails is $(1 - p)$, we can use a coin toss to model any binomial situation.

EXAMPLE 11.4.1

Suppose a coin is weighted so that it has a $\frac{1}{3}$ probability of landing heads on any toss and a $\frac{2}{3}$ probability of landing tails. If you toss it three times, what is the probability of getting exactly k heads for $k = 0, 1, 2, 3$?

SOLUTION The sample space for this experiment consists of the following 3-letter strings:

$$HHH, HHT, HTH, HTT, THH, THT, TTH, TTT.$$

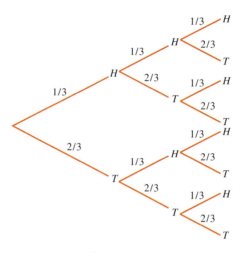

Figure 11.4.1

Each of these outcomes is represented by a path through the tree diagram of Figure 11.4.1, where each branch represents a single toss. The only one of these outcomes with exactly three heads is HHH. Since the probability of H on a single toss is $\frac{1}{3}$, the probability of getting three H's in three independent tosses is $\frac{1}{3} \cdot \frac{1}{3} \cdot \frac{1}{3} = \frac{1}{27}$.

What about the probability of getting exactly two heads? There are three strings satisfying that condition, HHT, HTH, and THH, and the probability of each of those is $\left(\frac{1}{3}\right)^2 \left(\frac{2}{3}\right) = \frac{2}{27}$. For instance, the probability of HHT is $\frac{1}{3} \cdot \frac{1}{3} \cdot \frac{2}{3}$, as shown in Figure 11.4.1, and the probability of HTH is $\frac{1}{3} \cdot \frac{2}{3} \cdot \frac{1}{3}$. In a similar way, we see that the probability of THH is $\frac{2}{3} \cdot \frac{1}{3} \cdot \frac{1}{3}$. Adding the probabilities of each of the three outcomes gives $\frac{6}{27}$ as the probability of two heads.

The other probabilities are obtained in the same way. The following table summarizes the results. You can check that the entries in the right-hand column add up to 1.

k (number of heads)	Outcomes	Probability
0	TTT	8/27
1	HTT, THT, TTH	12/27
2	HHT, HTH, THH	6/27
3	HHH	1/27

We can generalize Example 11.4.1 by tossing a weighted coin n times and letting p be the probability of *heads* on a single toss. It follows that $(1 - p)$ is the probability of *tails*. To specify the probability of getting k heads, we will use the following standard notation:

Definition 11.4.2

The symbol $\binom{n}{k}$ represents the number of strings consisting of k H's and $(n - k)$ T's. The numbers $\binom{n}{k}$ are called **binomial coefficients**.

For instance, in Example 11.4.1 we saw that

$$\binom{3}{0} = 1, \quad \binom{3}{1} = 3, \quad \binom{3}{2} = 3, \quad \binom{3}{3} = 1.$$

That is, there are exactly one string with no H's, three strings with one H, three with two H's, and one with three H's.

EXAMPLE 11.4.2

Find the binomial coefficients $\binom{4}{k}$ for k from zero to 4.

SOLUTION There are 16 strings consisting of H's and T's. A tree diagram will help you to find them all. The following table lists them according to the number of H's:

k (number of H's)	Strings	$\binom{4}{k}$
0	$TTTT$	1
1	$HTTT, THTT, TTHT, TTTH$	4
2	$HHTT, HTHT, HTTH, THHT, THTH, TTHH$	6
3	$HHHT, HHTH, HTHH, THHH$	4
4	$HHHH$	1

We will shortly describe a method for computing $\binom{n}{k}$ without listing all possible strings. First, however, we show how these numbers are used in computing binomial probabilities.

Theorem 11.4.1

Suppose a coin is weighted so that on any single toss the probability of heads is p and the probability of tails is $(1 - p)$. Then the probability of getting exactly k heads in n tosses is

$$\binom{n}{k} p^k (1 - p)^{n-k}. \tag{16}$$

The rationale behind this theorem is similar to that of Example 11.4.1. Each outcome with k heads is represented by a string of k H's and $(n - k)$ T's, and the probability of any single one of those strings is $p^k (1 - p)^{n-k}$. Therefore, the total probability of getting k heads and $(n - k)$ tails is the number of such strings times $p^k (1 - p)^{n-k}$.

We can apply Theorem 11.4.1 to many situations by interpreting them as weighted coin tosses.

EXAMPLE 11.4.3

In a forthcoming election in a large city, 60% of the residents favor the Democratic candidate for mayor and 40% favor the Republican. If you choose four residents at random, what is the probability they all favor the Democrat? What is the probability that two favor the Democrat and the other two favor the Republican?

SOLUTION   We can think of this as a weighted coin toss, with *heads* corresponding to *Democrat* and *tails* to *Republican*. In that case, $p = 0.6$ is the probability of heads and 0.4 the probability of tails. The probability of all four residents favoring the Democrat is the same as all four tosses being heads, which is

$$\binom{4}{4}(0.6)^4 = 1 \cdot (0.6)^4 = 0.1296.$$

The probability of two favoring the Democrat and two favoring the Republican is

$$\binom{4}{2}(0.6)^2(0.4)^2 = 6 \cdot (0.6)^2(0.4)^2 = 0.3456.$$

Pascal's triangle

To compute the numbers $\binom{n}{k}$, we begin by observing that

$$\binom{n}{0} = 1 \quad \text{and} \quad \binom{n}{n} = 1.$$

That is simply because there is only one string, $TT \cdots T$, with zero H's and n T's, and there is only one string, $HH \cdots H$, with n H's and zero T's.

The key to computing the other binomial coefficients is the following formula:

Theorem 11.4.2

If k and n are integers with $1 \leq k \leq n - 1$, then

$$\binom{n}{k} = \binom{n-1}{k-1} + \binom{n-1}{k}.$$

PROOF   By definition, $\binom{n}{k}$ is the number of strings of k H's and $(n-k)$ T's. We can divide those strings into two nonoverlapping groups:

- those that begin with H and
- those that begin with T.

For the first kind, we have to fill in the remaining $(n-1)$ letters of the string with $(k-1)$ H's and $(n-k)$ T's. There are $\binom{n-1}{k-1}$ ways to do that. For the second kind, we have to fill in the remaining $(n-1)$ letters with k H's and $(n-1-k)$ T's. There are $\binom{n-1}{k}$ ways to do that. Adding these together gives all the possible strings with k H's and $(n-k)$ T's.

We can use Theorem 11.4.2 to build a tableau of binomial coefficients. The first row consists of two 1's. Each other row starts and ends with 1, and its other numbers are obtained by adding two consecutive entries from the preceding row. Thus, the first two rows are

$$
\begin{array}{ccc}
1 & 1 & \\
1 & 2 & 1.
\end{array}
$$

To get the third row, we begin and end with 1 and fill in with sums of adjacent numbers from the second row:

$$
\begin{array}{cccc}
& 1 & 1 & \\
1 & 2 & 1 & \\
1 & 3 & 3 & 1.
\end{array}
$$

As you can see, these are precisely the numbers $\binom{n}{k}$ for $n = 1, 2, 3$ and $0 \le k \le n$. Continuing in this way, we get the binomial coefficients up to any order we want. Here is the tableau for n from 1 to 9:

$$
\begin{array}{ccccccccccc}
 & & & & & 1 & & 1 & & & \\
 & & & & 1 & & 2 & & 1 & & \\
 & & & 1 & & 3 & & 3 & & 1 & \\
 & & 1 & & 4 & & 6 & & 4 & & 1 \\
 & 1 & & 5 & & 10 & & 10 & & 5 & & 1 \\
1 & & 6 & & 15 & & 20 & & 15 & & 6 & & 1 \\
1 & 7 & & 21 & & 35 & & 35 & & 21 & & 7 & & 1 \\
1 & 8 & 28 & & 56 & & 70 & & 56 & & 28 & & 8 & 1 \\
1 & 9 & 36 & 84 & & 126 & & 126 & & 84 & & 36 & 9 & 1.
\end{array}
$$

This tableau for the binomial coefficients is known as **Pascal's triangle** in honor of the French mathematician and philosopher Blaise Pascal (1623–1662), although it was known for centuries before he rediscovered it. (It has been found, for example, in Chinese manuscripts dating back to the fourth century.)

EXAMPLE 11.4.4

Find the binomial coefficients $\binom{8}{3}$ and $\binom{10}{4}$.

SOLUTION   Looking at the 8th row of Pascal's triangle and counting from the left (starting with $k = 0$), we see that

$$
\binom{8}{0} = 1 \quad \binom{8}{1} = 8 \quad \binom{8}{2} = 28 \quad \binom{8}{3} = 56.
$$

To find $\binom{10}{4}$, we must add another row to the triangle, as follows:

$$1 \quad 9 \quad 36 \quad 84 \quad 126 \quad 126 \quad 84 \quad 36 \quad 9 \quad 1$$
$$1 \quad 10 \quad 45 \quad 120 \quad 210 \quad 252 \quad 210 \quad 120 \quad 45 \quad 10 \quad 1.$$

Again counting from the left, starting with $k = 0$, we see that $\binom{10}{4} = 210$.

With Pascal's triangle, we can answer all binomial probability questions.

EXAMPLE 11.4.5

Suppose you draw a card from an ordinary, well-shuffled deck. Then, after noting its value, you return it to the deck, reshuffle, and repeat the process four more times for a total of five draws. What is the probability that

(i) exactly three are spades? (ii) at least three are spades?

SOLUTION (i) The situation described is binomial with $n = 5$ and $p = (\frac{1}{4})$ (the probability of a *spade* on a single draw), analogous to the case of tossing a weighted coin five times, with $\frac{1}{4}$ probability of heads on each toss. Therefore, the probability of exactly three spades is

$$\binom{5}{3}\left(\frac{1}{4}\right)^3 \left(\frac{3}{4}\right)^2 = 10 \cdot \frac{3^2}{4^5} \approx 0.0879.$$

(ii) To find the probability of at least three spades, we add the probabilities of getting exactly three spades, exactly four spades, and exactly five spades:

$$\binom{5}{3}\left(\frac{1}{4}\right)^3 \left(\frac{3}{4}\right)^2 + \binom{5}{4}\left(\frac{1}{4}\right)^4 \left(\frac{3}{4}\right) + \binom{5}{5}\left(\frac{1}{4}\right)^5 = 10 \cdot \frac{3^2}{4^5} + 5 \cdot \frac{3}{4^5} + 1 \cdot \frac{1}{4^5}$$

$$= \frac{90}{1,024} + \frac{15}{1,024} + \frac{1}{1,024} = \frac{106}{1,024}$$

$$\approx 0.1035.$$

There is another method for computing binomial coefficients, which we state without proof (but see the exercises). Recall that $m!$ denotes the product of all the integers from 1 to m, so that $3! = 1 \cdot 2 \cdot 3 = 6$, $4! = 1 \cdot 2 \cdot 3 \cdot 4 = 24$, and so forth. We also define $0! = 1$.

Theorem 11.4.3

If k and n are integers with $0 \le k \le n$, then

$$\binom{n}{k} = \frac{n!}{k!(n-k)!} = \frac{n(n-1)\cdots(n-k+1)}{k!}$$

The mean and the variance

We say that a random variable X has a **binomial distribution** with parameters n and p if

$$P(X = k) = \binom{n}{k} p^k (1 - p)^{n-k} \quad \text{for } k = 0, 1, \ldots, n.$$

We can interpret X as the number of heads in n tosses of a weighted coin, with p being the probability of heads on each toss.

Theorem 11.4.4

If X has a binomial distribution with parameters n and p, then

$$E(X) = np \quad \text{and} \quad \text{Var}(X) = np(1 - p).$$

We will not prove this theorem. However, the formula for the expected value is very plausible on intuitive grounds. For instance, if a coin is weighted so that the probability of heads on any toss is 0.3, then you would expect about 30% of the tosses to land heads. Thus, in tossing 100 times, you would expect about 30 heads. The same holds for any p and n. Since p represents the proportion of heads, you would expect about np heads in n tosses.

EXAMPLE 11.4.6

If you roll a pair of dice 30 times, what is the expected number of 6's? If X is the number of 6's, what is $\text{Var}(X)$?

SOLUTION We saw earlier that the probability of getting a 6 on a single roll of a pair of dice is $\frac{5}{36}$. Therefore,

$$E(X) = \frac{150}{36} = \frac{25}{6} \quad \text{and} \quad \text{Var}(X) = 30 \cdot \left(\frac{5}{36}\right) \cdot \left(\frac{31}{36}\right) \approx 3.588.$$

APPLYING TECHNOLOGY

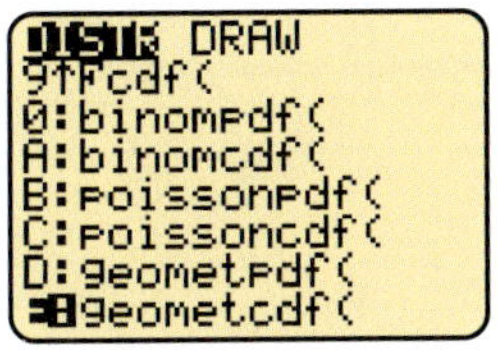

Graphing calculators are useful in avoiding the complicated computations that often arise in computing binomial probabilities. The TI-83 Plus, for example, has a DISTR menu (part of which is shown in Figure 11.4.2), which contains the most important probability distributions. For a random variable X having a binomial distribution with parameters n and p, menu item 0:binompdf calculates the probability $P(X = k)$ for a given k, as defined by formula (16), and item A:binomcdf computes the cumulative probability $P(X \leq k)$.

Figure 11.4.2

EXAMPLE 11.4.7

An automobile parts dealer claims that 12% of the spark plugs it receives from a certain manufacturer are defective, a statement that the manufacturer denies. To resolve the dispute, it tests 100 spark plugs chosen at random from the latest shipment. Assuming the dealer's claim is true, find the probabilities of each of the following events. Use a graphing calculator.

  (i) None of the plugs tested are defective.
 (ii) At most 12 are defective.
(iii) More than 15 are defective.
(iv) Between 8 and 16, inclusive, are defective.

SOLUTION   Let X be the number of defective plugs among the 100 tested.
  (i) We want to find $P(X = 0)$. Using binompdf(100, .12, 0) we find that it is approximately 2.8×10^{-6}, or 0.0000028.

 (ii) Using binomcdf(100, .12, 12), we find that $P(X \leq 12) = 0.576$ (to three decimal places).

(iii) We first observe that $P(X > 15) = 1 - P(X \leq 15)$. Next, using binomcdf(100, .12, 15), we get $P(X \leq 15) = 0.85855$ (to five decimal places). Therefore, $P(X > 15) = 1 - 0.85855 = 0.14145$.

(iv) Since $P(8 \leq X \leq 16) = P(7 < X \leq 16) = P(X \leq 16) - P(X \leq 7)$, the cumulative probabilities shown in Figure 11.4.3 (rounded to five decimal places) give $P(8 \leq X \leq 16) = 0.91256 - 0.07614 = 0.83642$.

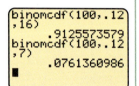

```
binomcdf(100,.12
,16)
        .9125573579
binomcdf(100,.12
,7)
        .0761360986
■
```

Figure 11.4.3

Practice Exercises 11.4

1. Find the binomial coefficients $\binom{7}{5}$ and $\binom{11}{3}$.

2. Suppose you roll a single die eight times. What is the probability of getting exactly two 3's?

Exercises 11.4

1. Suppose a coin is weighted so that the probability of heads on a single toss is 0.2. Find the probability of getting k heads in three tosses for $k = 0, 1, 2, 3$.

2. Suppose a coin is weighted so that the probability of heads on a single toss is $\frac{1}{6}$. Find the probability of getting k heads in four tosses for k from zero through 4.

3. Suppose you roll a single die four times. What is the probability of getting
   (a) exactly one 5?   **(b)** at most one 5?

4. Suppose a box contains three red chips and five white ones. You choose a chip, note its color, and return it to the box. Then you shake the box to mix up the chips and repeat the procedure. What is the probability of getting exactly two white chips if you draw
   (a) three times?   **(b)** four times?

5. A dog-food manufacturer is testing a new flavor of dog food. A researcher fills one bowl with the new flavor and another with the leading brand, and then she brings in a hungry dog. The experiment is repeated with 10 dogs, and nine of them choose the new flavor. What is the probability of that happening if there were no preference between the two foods?

6. Suppose the dog-food manufacturer of the previous exercise decides to test 100 dogs. Assuming there is no preference between the two foods, find the following probabilities:
   (a) More than 60 of the dogs surveyed choose the new flavor.
   (b) Less than 30 of the dogs surveyed choose the new flavor.

7. Use Pascal's triangle to find the following binomial coefficients:
   (a) $\binom{5}{3}$ **(b)** $\binom{6}{4}$ **(c)** $\binom{9}{5}$

8. Extend Pascal's triangle through row 14.

9. Suppose the probability of a new-born baby being male is 0.5. In a family of five children, what is the probability that exactly four are boys? What is the probability that at least four are boys?

10. Suppose an eight-digit number is made up by choosing a digit from 1 to 9 at random to fill each place. What is the probability the number contains exactly three 9's?

11. Suppose a 50-digit number is made up by choosing a digit from 1 to 9 at random to fill each place. What is the probability the number contains
   (a) no more than four 9's?   **(b)** more than 10 9's?

12. Suppose a batter gets a hit 30.6% of his times at bat. In 10 times at bat, what is the probability he gets
   (a) no hits?   **(b)** three hits?
   (c) more than eight hits?

13. In a state election, 58% of the voters favor the Republican candidate for governor, while 42% favor the Democrat.
   (a) If you choose 12 voters at random, what is the probability that seven favor the Republican?
   (b) What is the expected number of voters favoring the Republican in a random sample of 50 voters?

14. In the election of the previous problem, suppose you sample 1,000 voters. What is the probability that
   (a) less than 55% favor the Republican?
   (b) more than 60% favor the Republican?

15. Suppose the Notre Dame football team has a 0.8 probability of winning any game it plays next season. If it plays 10 games, what is the probability it will
   (a) go undefeated?
   (b) lose exactly one game?
   (c) lose less than three games?

16. Suppose that 12% of the population has type B blood. If six blood donors are chosen at random, what is the probability that two will have type B blood? What is the probability that less than two will have type B blood?

17. Use Theorem 11.4.3 to compute the following binomial coefficients:
   (a) $\binom{14}{2}$ **(b)** $\binom{15}{3}$ **(c)** $\binom{20}{4}$ **(d)** $\binom{100}{3}$

18. Suppose that 2% of the computer chips manufactured by a certain company are defective. What is the probability of finding one defective chip in a batch of 50? What is the probability of finding more than one? (*Hint:* "More than one" is the complement of "one or less.")

19. If X has a binomial distribution with $n = 100$ and $p = 0.52$, find $E(X)$ and $\text{Var}(X)$.

20. Suppose you roll a pair of dice 12 times.
   (a) How many times do you expect to roll a 7?
   (b) If X is the number of 7's, what is $\text{Var}(X)$?

21. Every New Year's Eve the city hosts a family-friendly street fair with music, food, crafts, and games. To raise money, it sells commemorative T-shirts at the fair. Past experience shows the probability of an adult buying a shirt is 0.69. The city anticipates that 6,000 adults will attend the fair this year.

  (a) How many T-shirts should the city expect to sell?

  (b) The profit is $8 per shirt. What is the expected profit?

  (c) Experience suggests that the actual number of T-shirts sold will not vary from the expected value by more than twice the standard deviation (in either direction). If that is correct, find a lower and upper bound on the number they will sell and the profit they will make.

22. An advertisement claims that 78% of the people in a town approve of the new manufacturing plant scheduled to be built soon. The town's newspaper hires a polling expert to survey people at random about their opinions on the matter.

  (a) If the advertisement is correct, what is the probability that none of the first five people surveyed approves of the plant?

  (b) The polling expert surveys 200 people and she finds that 47 approve. How does that compare to the number of people expected to approve the plant?

  (c) What should she conclude about the advertisement's claim?

23. Under the circumstances described in the previous exercise, what is the probability that less than 48 people out of 200 surveyed approve of the plant, assuming the advertisement is correct?

24. If you look at Pascal's triangle, you observe a symmetry in each row that suggests the following:

$$\binom{n}{k} = \binom{n}{n-k}.$$

Can you explain that in terms of strings of H's and T's?

25. If you look at Pascal's triangle, you observe that the second entry of the nth row is always n. That is,

$$\binom{n}{1} = n.$$

Can you explain that in terms of strings of H's and T's?

26. To prove Theorem 11.4.3, let $F(n, k) = n!/k!(n-k)!$, for any integers n and k with $0 \leq k \leq n$.

  (a) Show that $F(n, 0) = 1$ and $F(n, n) = 1$.

  (b) Show that $F(n, k)$ satisfies the formula stated in Theorem 11.4.2. That is, if $1 \leq k \leq (n-1)$, $F(n, k) = F(n-1, k-1) + F(n-1, k)$.

  (c) Explain why steps (a) and (b) lead to the conclusion that

$$F(n, k) = \binom{n}{k}.$$

Solutions to practice exercises 11.4

1. By using the 7th row of Pascal's triangle and counting from the left, starting with $k = 0$, we get $\binom{7}{5} = 21$. To find $\binom{11}{3}$, we must extend the triangle to 11 rows. We have already computed the 10th row in Example 11.4.4, and using it, we get the 11th, which starts as follows: 1 11 55 165. Therefore, $\binom{11}{3} = 165$.

2. The probability of getting a 3 on a single roll is $\frac{1}{6}$. Using the binomial distribution with $n = 8$ and $p = \frac{1}{6}$ gives the probability

$$\binom{8}{2}\left(\frac{1}{6}\right)^2\left(\frac{5}{6}\right)^6 = 28\frac{5^6}{6^8} \approx 0.26.$$

11.5　Continuous Random Variables

In the previous section we considered discrete random variables. Now, we shall consider random variables whose range consists of an entire interval. This type of variable arises in problems in which we measure some quantity whose values may vary continuously. Some examples are the waiting time for treatment at a hospital emergency room, the weight of a laboratory animal in an experiment, and the lifetime of a transistor in a quality-control test.

  In such cases it does not make sense to assign a probability to each value in the range. There are an infinite number of them, spread out over an entire interval. Instead, we use a function called a **probability density function** to indicate how the probability is distributed throughout the range.

Definition 11.5.1

A continuous function $f(x)$, $a < x < b$, is called a probability density function on the interval (a, b) if

$$f(x) \geq 0 \quad \text{for } a < x < b \quad \text{and} \quad \int_a^b f(x)\, dx = 1.$$

In this definition, the interval (a, b) may be an interval of finite length, a half-line, or the entire x-axis. In other words, we allow the possibility that $a = -\infty$ or $b = \infty$, or both. If either a or b is a finite number, it may also be included in the domain of f. For example, $a < x < b$ may be replaced by $a \leq x \leq b$.

EXAMPLE 11.5.1

Verify that the function $f(x) = e^{-x}$, $x \geq 0$ is a probability density function on the interval $[0, \infty)$.

SOLUTION First, $e^{-x} > 0$ for all x. Second,

$$\int_0^\infty e^{-x}\, dx = \lim_{h \to \infty} \int_0^h e^{-x}\, dx = \lim_{h \to \infty} \left. (-e^{-x}) \right|_0^h$$

$$= \lim_{h \to \infty} (1 - e^{-h}) = 1.$$

The probability density function enables us to find the probabilities associated with a continuous random variable. In fact, we define a **continuous random variable** as being one that has a probability density function.

Definition 11.5.2

A random variable X whose range is an interval (a, b) is called continuous if there is a probability density function $f(x)$ on (a, b) so that for any c and d with $a \leq c < d \leq b$,

$$P(c < X < d) = \int_c^d f(x)\, dx.$$

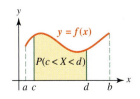

Figure 11.5.1

From a geometric point of view, the probability that $c < X < d$ is the area under the graph of the probability density function $f(x)$ from $x = c$ to $x = d$, as shown in Figure 11.5.1. In this sense, the graph plays a role similar to that of the histogram in the case of a discrete random variable.

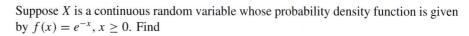

EXAMPLE 11.5.2

Suppose X is a continuous random variable whose probability density function is given by $f(x) = e^{-x}$, $x \geq 0$. Find

(i) $P(0 < X < 1)$ and   (ii) $P(X > 2)$.

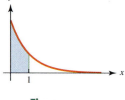

Figure 11.5.2

SOLUTION   (i) $P(0 < X < 1) = \int_0^1 e^{-x} \, dx = 1 - e^{-1} \approx 0.632$.

In this case, the probability is the area under the graph of $y = e^{-x}$ from $x = 0$ to $x = 1$, shaded in Figure 11.5.2.

(ii) $P(X > 2) = \int_2^\infty e^{-x} \, dx = \lim_{h \to \infty} (e^{-2} - e^{-h}) = e^{-2} \approx 0.135$.

In this case, the probability is the area under the graph of $y = e^{-x}$ to the right of $x = 2$, shaded in Figure 11.5.3.

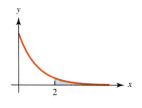

Figure 11.5.3

The simplest example of a continuous random variable is one whose range is an interval $[a, b]$ of finite length and whose probability density function is constant. It models the case in which the probability is evenly distributed throughout the interval— as would be the case, for instance, if a point of the interval were chosen in a completely random way.

EXAMPLE 11.5.3

Given a 4-inch length of ribbon, you cut it into two pieces at a point chosen completely at random. Find the probability that the length of the left-hand piece is between 2 and 2.25 inches.

SOLUTION   The length of the left-hand piece is equal to the coordinate of the cutting point. If we let X denote that coordinate, then X is a random variable whose range is the interval $[0, 4]$. Since we are choosing the point completely at random, there is no reason to consider any location inside the interval more probable than any other. Therefore, we take the density function to be constant, say,

$$f(x) = \lambda, \quad 0 \leq x \leq 4,$$

where the Greek letter λ stands for some fixed real number. To determine its value, we use the fact that the integral of $f(x)$ over $[0, 4]$ must equal 1, which gives

$$1 = \int_0^4 \lambda \, dx = 4\lambda,$$

so that $\lambda = \frac{1}{4}$. The probability that the length of the left-hand piece is between 2 and 2.25 is given by

$$P(2 < X < 2.25) = \int_2^{2.25} \frac{1}{4} \, dx = \frac{0.25}{4} = \frac{1}{16}.$$

A feature of continuous random variables that is quite different from the discrete case is that there is zero probability of a particular value being taken on. In the previous example, for instance, the probability that X is *exactly* 2.215307 inches long is zero. Intuitively, that is because there are so many possible values, covering an entire interval, that no single value can have positive probability. Here is a formal statement of that property.

Theorem 11.5.1

If X is a continuous random variable, then $P(X = c) = 0$ for any number c.

To see why, suppose c is any number in the range of X. Then for any small, positive number h,

$$0 \leq P(X = c) \leq P(c - h < X < c + h). \tag{17}$$

The probability that X is between $c - h$ and $c + h$ is the area under the graph of the probability density function from $c - h$ to $c + h$, as shown in Figure 11.5.4. That area is approximately $2hf(c)$, the width of the interval times the height of the graph over c, which means it goes to zero as $h \to 0$. By combining that observation with Eq. (17), we conclude that $P(c) = 0$.

As a corollary, we get the following:

Theorem 11.5.2

If X is a continuous random variable and c is any number in its range,

$$P(X \leq c) = P(X < c).$$

This is true simply because $P(X = c) = 0$, so that

$$P(X \leq c) = P(X < c) + P(X = c) = P(X < c).$$

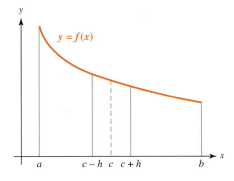

Figure 11.5.4

An important application of continuous random variables is in modeling the lifetime of a piece of equipment, such as an electronic component.

EXAMPLE 11.5.4

Assume that the lifetime in hours of a certain type of transistor is a continuous random variable X whose probability density function has the form

$$f(x) = \lambda e^{-0.01x}, \quad x \geq 0,$$

where λ is a constant.
   (i) Find the value of λ.
   (ii) Find the probability that the lifetime of a given transistor of that type is no more than 150 hours.

SOLUTION   (i) Since $f(x)$ is a probability density function on $[0, \infty)$, we must have

$$1 = \lambda \int_0^\infty e^{-(0.01)x} \, dx = -100\lambda e^{-(0.01)x} \Big|_0^\infty = 100\lambda.$$

Therefore, $\lambda = 0.01$.
   (ii) Letting X be the lifetime of the transistor, we get

$$P(X \leq 150) = (0.01) \int_0^{150} e^{-(0.01)x} \, dx$$

$$= -e^{(0.01)x} \Big|_0^{150} = 1 - e^{-1.5} \approx 0.777.$$

PROBLEM-SOLVING TACTIC

It is a good idea to always check the two properties of a probability density function $f(x)$, $a < x < b$. They are

(1) $f(x) \geq 0$

and

(2) $\displaystyle\int_a^b f(x) \, dx = 1.$

Here the first is true and the second reads

$$\lambda \int_0^\infty e^{-0.01x} \, dx = 1,$$

which will give you λ.

The cumulative distribution function

It is important to realize that the probability density function $f(x)$ of a continuous random variable X is *not* a probability. Rather, it measures how the probability is distributed or "spread out" over the range of X. There is, however, another useful function associated with X that *is* a probability, called its **cumulative distribution function**.

Definition 11.5.3

If X is a continuous random variable whose range is the interval (a, b), its cumulative distribution function $F(x)$ is defined by

$$F(x) = P(X \leq x), \quad \text{for } a < x < b. \tag{18}$$

Using the probability density function, we can also write this in the form

$$F(x) = \int_a^x f(t) \, dt, \quad \text{for } a < x < b. \tag{19}$$

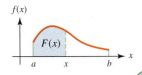

$f(x)$

$F(x)$

a x b x

Figure 11.5.5

Geometrically, the cumulative distribution function is the area under the graph of the probability density function from a to x. In Figure 11.5.5, $F(x)$ is the shaded area.

EXAMPLE 11.5.5

Find the cumulative distribution function of a random variable X whose probability density function is

$$f(x) = 0.01e^{-0.01x}, \quad x \geq 0.$$

SOLUTION Using formula (19), we have

$$F(x) = \int_0^x 0.01e^{-0.01t}\, dt$$

$$= -e^{-0.01t}\, \Big|_0^x = 1 - e^{-0.01x}.$$

The probability density function of a continuous random variable determines its cumulative distribution function, according to formula (19). The converse is also true: If we know the cumulative distribution function, we can recapture the probability density. The following theorem tells us how:

Theorem 11.5.3

If $f(x)$ is the probability density function of a continuous random variable and $F(x)$ is its cumulative distribution function, then

$$F'(x) = f(x).$$

PROOF This is really a special case of the fundamental theorem of calculus. If $f(x)$ is continuous, then

$$F'(x) = \frac{d}{dx} \int_a^x f(t)\, dt = f(x).$$

In Example 11.5.5, for instance,

$$F'(x) = \frac{d}{dx} \left[1 - e^{-0.01x}\right] = 0.01e^{-0.01x} = f(x).$$

Knowing the cumulative distribution function gives us a simple way of computing probabilities, as the next theorem shows.

Theorem 11.5.4

If X is a continuous random variable with cumulative distribution function $F(x)$, then

$$P(c < X < d) = F(d) - F(c)$$

for any c and d in the range of X with $c < d$.

PROOF Using the properties of the definite integral, we have

$$P(c < X < d) = \int_c^d f(t)\,dt$$

$$= \int_a^d f(t)\,dt - \int_a^c f(t)\,dt = F(d) - F(c).$$

Since $P(X = c)$ and $P(X = d)$ are both zero, we can also conclude that

$$P(c \le X < d) = P(c < X \le d) = P(c \le X \le d) = F(d) - F(c).$$

EXAMPLE 11.5.6

Let $f(x) = \frac{3}{2}x - \frac{3}{4}x^2$, $0 \le x \le 2$.
   (i) Verify that $f(x)$ is a probability density function.
   (ii) Determine the cumulative distribution function $F(x)$ and verify that $F'(x) = f(x)$.
   (iii) Use $F(x)$ to compute $P(\frac{1}{2} \le X \le 1)$, where X is a random variable with $f(x)$ as its probability density function.

SOLUTION   (i) We first need to check that $f(x) \ge 0$ for $0 \le x \le 2$. To do that, we factor $f(x)$ as follows:

$$f(x) = \frac{3}{4}x(2 - x).$$

In this form, it is clear that both factors, x and $2 - x$, are positive if x is between zero and 2.
   Next we have to check that $\int_0^2 f(x)\,dx = 1$. That is a straightforward calculation:

$$\int_0^2 \left(\frac{3}{2}x - \frac{3}{4}x^2\right) dx = \left(\frac{3}{4}x^2 - \frac{1}{4}x^3\right)\Bigg|_0^2 = 3 - 2 = 1.$$

(ii) We have, for $0 < x < 2$,

$$F(x) = P(0 \le X \le x) = \int_0^x \left(\frac{3}{2}t - \frac{3}{4}t^2\right) dt$$

$$= \left(\frac{3}{4}t^2 - \frac{1}{4}t^3\right)\Bigg|_0^x = \frac{3}{4}x^2 - \frac{1}{4}x^3.$$

Taking the derivative gives

$$F'(x) = \frac{d}{dx}\left(\frac{3}{4}x^2 - \frac{1}{4}x^3\right) = \frac{3}{2}x - \frac{3}{4}x^2.$$

(iii) Applying Theorem 11.5.4, we get

$$P\left(\frac{1}{2} \le X \le 1\right) = F(1) - F\left(\frac{1}{2}\right)$$

$$= \left(\frac{3}{4} - \frac{1}{4}\right) - \left(\frac{3}{4}\cdot\frac{1}{4} - \frac{1}{4}\cdot\frac{1}{8}\right) = \frac{11}{32}.$$

The expected value

Just as in the discrete case, the **expected value** of a continuous random variable is an average of its values, weighted according to their probabilities. It is defined as follows:

> ### Definition 11.5.4
>
> If X is a continuous random variable with probability density function $f(x)$, $a < x < b$, its **expected value** $E(X)$ is defined by
>
> $$E(X) = \int_a^b xf(x)\,dx. \tag{20}$$

To see the rationale behind this definition, let's assume that both a and b are finite and partition the interval $[a, b]$ into n subintervals, each of length

$$\Delta x = \frac{b - a}{n}$$

and choose a point x_j in the jth subinterval for $j = 1, \ldots, n$. If we let

$$p_j = f(x_j) \cdot \Delta x,$$

then p_j is approximately the probability that the value of X falls in the jth subinterval. Motivated by the discrete case, we may reasonably consider the number

$$x_1 p_1 + \cdots + x_n p_n$$

as an approximate average of the values of X, so that

$$E(X) \approx x_1 p_1 + \cdots + x_n p_n = x_1 f(x_1)\Delta x + \cdots + x_n f(x_n)\Delta x.$$

The right-hand side is a Riemann sum for the definite integral of the function $xf(x)$ over the interval $[a, b]$. As n gets larger and larger, $\Delta x \to 0$, and the Riemann sum approaches the integral.

As in the discrete case, the expected value is also called the **mean** and denoted by the Greek letter μ.

EXAMPLE 11.5.7

Assume that the lifetime in hours of a certain type of lightbulb is a continuous random variable X with probability density function

$$f(x) = \frac{1}{750} e^{-x/750} \, dx, \quad x \geq 0.$$

Find the average lifetime of this type of bulb.

SOLUTION   The average lifetime is the same as the expected value,

$$E(X) = \int_0^\infty \frac{x}{750} e^{-x/750} \, dx.$$

To compute it, we use integration by parts with $u = x$ and $dv/dx = \frac{1}{750} e^{-x/750}$, as follows:

$$E(X) = -xe^{-x/750} \Big|_0^\infty + \int_0^\infty e^{-x/750} \, dx$$

$$= 0 - 750 e^{-x/750} \Big|_0^\infty = 750.$$

The variance and standard deviation

The **variance** of a continuous random variable is defined in a similar way as in the discrete case, using the integral in place of the sum and the probability density function in place of the probability distribution table.

Definition 11.5.5

If X is a continuous random variable with mean μ and probability density function $f(x)$, $a < x < b$, its variance $\text{Var}(X)$ is defined by

$$\text{Var}(X) = \int_a^b (x - \mu)^2 f(x) \, dx. \tag{21}$$

As in the discrete case, the variance indicates how narrowly or widely the values of X are concentrated around the mean $E(X)$, with a smaller variance indicating a narrower spread. And, just as in the discrete case, the **standard deviation**, denoted by

$\sigma(X)$, is defined by

$$\boxed{\sigma(X) = \sqrt{\text{Var}(X)}.}\qquad(22)$$

Another similarity with the discrete case is the following alternate formula for computing the variance:

Theorem 11.5.5

If X is a random variable with mean μ and probability density function $f(x)$, $a < x < b$, then

$$\text{Var}(X) = \int_a^b x^2 f(x)\,dx - \mu^2.\qquad(23)$$

The proof is similar to that of Theorem 11.3.2, with the sum replaced by an integral and the probability distribution table by the probability density function. The details are left as an exercise for you.

EXAMPLE 11.5.8

For the lightbulb of Example 11.5.7, find the variance and standard deviation of X.

SOLUTION   We already know from example 11.5.7 that $\mu = 750$. To compute the variance, we use formula (23) and apply integration by parts twice:

$$
\begin{aligned}
V(X) &= \int_0^\infty \frac{x^2}{750} e^{-x/750}\,dx - 750^2\\[4pt]
&= \int_0^\infty x^2 \frac{d}{dx}(-e^{-x/750})\,dx - 750^2\\[4pt]
&= -x^2 e^{-x/750}\Big|_0^\infty + \int_0^\infty 2x e^{-x/750}\,dx - 750^2\\[4pt]
&= 0 + 2\int_0^\infty x \frac{d}{dx}(-750 e^{-x/750})\,dx - 750^2\\[4pt]
&= -1{,}500 x e^{-x/750}\Big|_0^\infty + 2\cdot 750 \int_0^\infty e^{-x/750}\,dx - 750^2\\[4pt]
&= 0 - 2\cdot 750^2 e^{-x/750}\Big|_0^\infty - 750^2\\[4pt]
&= 2\cdot 750^2 - 750^2 = 750^2.
\end{aligned}
$$

Therefore, we have

$$\text{Var}(X) = 750^2,$$

and the standard deviation is equal to

$$\sigma(X) = \sqrt{750^2} = 750.$$

The median

Another type of average is called the **median**. For a continuous random variable X, it is a number m that divides the range of X into two equally probable intervals. More precisely, we have the following:

Definition 11.5.6

If X is a continuous random variable with probability density function $f(x)$, $a < x < b$, the median is the number m such that

$$\int_a^m f(x)\,dx = \frac{1}{2} = \int_m^b f(x)\,dx. \tag{24}$$

In principle, there may be more than one such number, but for all commonly used continuous random variables there is only one. In particular, if $f(x) > 0$ for $a < x < b$, then there is a unique m satisfying (24).

Geometrically, we can describe the median m by saying the line $x = m$ divides the region under the graph of $f(x)$ into equal pieces, each of area $\frac{1}{2}$.

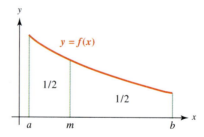

Note   In general, the median is *different* from the mean.

EXAMPLE 11.5.9

Compute the mean and median of a random variable X with probability density function

$$f(x) = \frac{1}{750} e^{-x/750}, \quad x \geq 0.$$

SOLUTION   We already computed the mean of X in Example 11.5.7 and found that $\mu = 750$. To compute the median, we look for a positive number m such that

$$\int_0^m \frac{1}{750} e^{-x/750}\,dx = \frac{1}{2}.$$

Computing the integral gives the equation

$$e^{-x/750}\Big|_0^m = -e^{-m/750} + 1 = \frac{1}{2},$$

and solving for m gives $m = 750 \ln 2 \approx 519.86$.

Practice Exercises 11.5

1. A security alarm is controlled by a computer chip, with a back-up chip that takes over if the first fails. The combined lifetime (in years) of the two chips is a random variable X, with probability density function

$$f(x) = \frac{1}{9}xe^{-x/3}, \quad 0 \le x < \infty.$$

(a) Find the cumulative distribution function.

(b) Find the probability that the combined lifetime is greater than 2 years.

2. Suppose a random variable X has a probability density function of the form $f(x) = 1 - cx$, $0 \le x \le 2$, where c is a constant.

(a) Find the value of c.

(b) Find $E(X)$ and $\text{Var}(X)$.

Exercises 11.5

In Exercises 1–6, show that the given function is a probability density function.

1. $f(x) = \dfrac{x^2}{9}, \ 0 \le x \le 3$

2. $\dfrac{3}{4}(1 - x^2), \ -1 \le x \le 1$

3. $f(x) = \dfrac{1}{x^2}, \ x \ge 1$

4. $f(x) = 0.2e^{-0.2x}, \ x \ge 0$

5. $f(x) = xe^{-x}, \ x \ge 0$

6. $f(x) = xe^{-x^2/2}, \ x \ge 0$

7. A test having many problems and a short time limit is intended to challenge a student's ability to solve problems quickly. The designers are interested in estimating the number of students who will finish the test, and they believe the probability density function to be given by

$$f(x) = 2 - 2x, \quad 0 \le x \le 1,$$

where x is the proportion who finish.

(a) Graph this function and verify that $f(x) \ge 0$ for $0 \le x \le 1$.

(b) Use elementary geometry to confirm that it is indeed a probability density function.

(c) Use elementary geometry to determine the probability that at least 60% of the students will finish the test.

8. Biologists want to know if a particular mammal prefers certain places to sleep at night, or if it merely beds down near where it happens to be at sunset. They mark a location near the center of its territory and measure the distance X from the location to the sleeping place.

(a) If the mammal does not favor any particular sleeping place within its territory, what does the probability distribution function of X look like?

(b) Suppose the mammal shows a preference for two distinct places. Which of the following graphs best represents the probability distribution function?

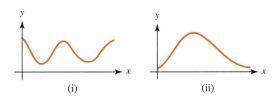

(i) (ii)

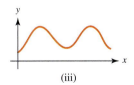

(iii)

In Exercises 9–16, determine whether the given function is a probability density function.

9. $f(x) = -\frac{1}{2}x + 1, \ 0 \le x \le 2$

10. $f(x) = \frac{2}{3}(x - 1), \ 0 \le x \le 3$

11. $f(x) = \frac{3}{2}x^2, \ -1 \le x \le 1$

12. $f(x) = \frac{4}{15}x^3, \ -1 \le x \le 2$

13. $f(x) = x(1 - x)$, $0 \le x \le 1$

14. $f(x) = 2(1 - x)$, $0 \le x \le 1$

15. $f(x) = e^{-0.1x}$, $x \ge 0$

16. $f(x) = \dfrac{2x}{(1 + x^2)^2}$, $x \ge 0$

17. Find the value of c that makes the function $f(x) = cx^2(1 - x)$, $0 \le x \le 1$, a probability density function.

18. Find the value of c that makes the function $f(x) = cx^{-3}$, $x \ge 1$, a probability density function.

19. Assume that the life in hours of a certain type of light-bulb is a random variable with probability density function $f(x) = cx^{-2}$, $1{,}000 \le x \le 4{,}000$.
(a) Find the constant c.
(b) Compute the probability that the component's life is at least 3,000 hours.

20. Assume that the daily demand for a certain product in thousands of units has probability density function $f(x) = \frac{1}{18}(9 - x^2)$, $0 \le x \le 3$.
(a) Find the probability that the demand is at least 1,000 units
(b) Find the probability that the demand is at most 2,000 units
(c) Find the probability that the demand is between 1,000 and 2,000 units.

21. In a psychology experiment, a mouse is required to find its way through a maze in order to reach a piece of cheese. Suppose the number of minutes it takes a mouse to reach the cheese on its first try is a random variable X with probability density function $f(x) = xe^{-x}$, $x \ge 0$. Find the probability that
(a) the mouse takes at least 1 minute.
(b) the mouse reaches the cheese in less than 5 minutes.

In Exercises 22–25, verify that the given function is a probability density function and find the corresponding cumulative distribution function.

22. $f(x) = \dfrac{1}{2}$, $-1 \le x \le 1$

23. $f(x) = \dfrac{1}{2\sqrt{x}}$, $1 \le x \le 4$

24. $f(x) = \dfrac{3}{x^4}$, $x \ge 1$

25. $f(x) = \begin{cases} 1 + x, & -1 \le x \le 0 \\ 1 - x, & 0 < x \le 1 \end{cases}$

26. A random variable X whose range is the interval $[0, 2]$ has the cumulative distribution function $F(x) = x - (x^2/4)$, $0 \le x \le 2$.
(a) Find $P(1 < X < 2)$.
(b) Find the probability density function.

27. A random variable X whose range is the interval $(2, 6)$ has the cumulative distribution function $F(x) = \frac{1}{2}\sqrt{x - 2}$, $2 \le x \le 6$.
(a) Compute $P(X \le 3)$.
(b) Compute $P(4 \le X \le 5)$.
(c) Find the probability density function.

28. If $F(x)$ is the cumulative distribution function of a continuous random variable whose range is the interval (a, b), what are $\lim_{x \to a^+} F(x)$ and $\lim_{x \to b^-} F(x)$? Explain.

In Exercises 29–34, find the expected value and variance of a continuous random variable having the given probability density function.

29. $f(x) = 2 - 2x$, $0 \le x \le 1$

30. $f(x) = \dfrac{1}{2}$, $1 \le x \le 3$

31. $f(x) = \dfrac{3}{4}(2x - x^2)$, $0 \le x \le 2$

32. $f(x) = \dfrac{3}{16}\sqrt{x}$, $0 \le x \le 4$

33. $f(x) = 2e^{-2x}$, $x \ge 0$

34. $f(x) = \dfrac{3}{x^4}$, $x \ge 1$

35. For the mouse of Exercise 21, what is the mean time it takes the mouse to reach the cheese?

36. Assume that the daily demand for a certain product in thousands of units is a random variable X with probability density $f(x) = \frac{1}{18}(9 - x^2)$, $0 \le x \le 3$. Find the expected demand.

37. Assume that the lifetime in hours of a certain electronic component is a random variable X with probability density function $f(x) = \frac{1}{10}e^{-x/10}$, $x \ge 0$. Find the expected value, variance, and standard deviation of X.

38. If the range of X is an unbounded interval (i.e., a half-line or the entire axis), then $E(X)$ or $\text{Var}(X)$ may not exist because the improper integral may not converge to a finite number.
(a) Let $f(x) = x^{-2}$, $x \ge 1$. Show that $f(x)$ is a probability density function, but the expected value does not exist.
(b) Let $f(x) = 2x^{-3}$, $x \ge 1$. Show that $f(x)$ is a probability density function and the expected value exists, but the variance does not.

In each of the following exercises, compute the median and the expected value of the continuous random variable with the given probability density.

39. $f(x) = \dfrac{1}{3}$, $0 \le x \le 3$

40. $f(x) = 1 - \dfrac{x}{2}$, $0 \le x \le 2$

41. $f(x) = 3x^2,\ 0 \le x \le 1$

42. $f(x) = \dfrac{3}{4}(1 - x^2),\ -1 \le x \le 1$

43. $f(x) = e^{-x},\ x \ge 0$

44. $f(x) = 2(x + 1)^{-3},\ x \ge 0$

45. In some of the Exercises 39–44, the mean and the median were the same. Can you identify a common property of the graphs for which that was true? Can you state a general condition under which the mean and median will be the same?

46. The number of minutes it takes to be served at a certain bank window is a random variable whose density function is $f(x) = \frac{1}{3}e^{-x/3},\ x \ge 0$.

(a) What is the mean waiting time?

(b) What is the median waiting time?

(c) Based on knowledge of the mean and median, what would you roughly expect if you were to record the waiting times of 1,000 customers?

47. Suppose the score a student makes on a standardized test is a random variable. A random sample of 21 test scores was as follows:

$$\{100, 94, 91, 88, 88, 85, 82, 80, 78, 76, 73,$$
$$70, 70, 70, 69, 66, 62, 54, 42, 36, 20\}.$$

If you had to guess the mean and median from these data, what would you guess?

Solutions to practice exercises 11.5

1. (a) Using integration by parts, we obtain for $x \ge 0$

$$F(x) = \frac{1}{9} \int_0^x t e^{-t/3}\, dt = \frac{1}{3} \int_0^x t \frac{d}{dt}(-e^{-t/3})\, dt$$

$$= -\frac{1}{3} t e^{-t/3} \Big|_0^x + \frac{1}{3} \int_0^x e^{-t/3}\, dt = -\frac{x}{3} e^{-x/3} - e^{-t/3} \Big|_0^x$$

$$= 1 - e^{-x/3} - \frac{x}{3} e^{-x/3}.$$

(b) $P(X > 2) = 1 - P(X \le 2) = 1 - F(2)$

$$= e^{-2/3}\left(1 + \frac{2}{3}\right) \approx 0.8557.$$

2. (a) Since

$$1 = \int_0^2 (1 - cx)\, dx = \left(x - \frac{cx^2}{2}\right) \Big|_0^2 = 2 - 2c,$$

we must have $c = \frac{1}{2}$.

(b) $E(X) = \displaystyle\int_0^2 \left(x - \frac{x^2}{2}\right) dx = \left(\frac{x^2}{2} - \frac{x^3}{6}\right) \Big|_0^2 = \frac{2}{3}.$

$$\text{Var}(X) = \int_0^2 \left(x^2 - \frac{x^3}{2}\right) dx - \left(\frac{2}{3}\right)^2$$

$$= \left(\frac{x^3}{3} - \frac{x^4}{8}\right) \Big|_0^2 - \frac{4}{9} = \frac{2}{9}.$$

11.6 Some Important Continuous Densities

In this section we will describe three of the most important continuous probability densities and some of their applications.

The uniform distribution

A random variable X is said to have a **uniform distribution** over an interval $[a, b]$ if its probability density function is given by

$$f(x) = \frac{1}{b - a}, \quad a \le x \le b.$$

Its graph is a horizontal straight line, as shown in Figure 11.6.1.

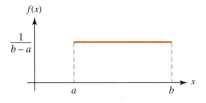

Figure 11.6.1

To check that $f(x)$ is a probability density, we observe that $f(x) \geq 0$, and

$$\int_a^b \frac{1}{b-a}\, dx = \frac{x}{b-a}\Big|_a^b = \frac{b-a}{b-a} = 1.$$

Next, we compute the mean and variance:

$$E(X) = \int_a^b \frac{x}{b-a}\, dx = \frac{1}{2}\frac{x^2}{b-a}\Big|_a^b = \frac{1}{2}\frac{b^2-a^2}{b-a} = \frac{a+b}{2}.$$

$$\begin{aligned}
\mathrm{Var}(X) &= \int_a^b \frac{x^2}{b-a}\, dx - \left(\frac{a+b}{2}\right)^2 \\
&= \frac{1}{3}\frac{x^3}{b-a}\Big|_a^b - \left(\frac{a+b}{2}\right)^2 \\
&= \frac{1}{3}\frac{b^3-a^3}{b-a} - \frac{1}{4}(a+b)^2 \\
&= \frac{1}{3}\frac{(b-a)(b^2+ab+a^2)}{b-a} - \frac{1}{4}(a+b)^2 \\
&= \frac{1}{3}(b^2+ab+a^2) - \frac{1}{4}(a^2+2ab+b^2) \\
&= \frac{1}{12}(a^2-2ab+b^2) = \frac{1}{12}(b-a)^2.
\end{aligned}$$

In summary,

$$E(X) = \frac{1}{2}(a+b), \quad \mathrm{Var}(X) = \frac{1}{12}(b-a)^2, \quad \sigma(X) = \frac{1}{2\sqrt{3}}(b-a).$$

The uniform distribution is used to model an experiment in which a number is selected in a completely random way from a given interval.

EXAMPLE 11.6.1

Buses on a certain route arrive every 45 minutes.
  (i) What is the probability that a person arriving at a random time will have to wait for at least 15 minutes for a bus?
  (ii) What is the average waiting time?

SOLUTION The waiting time is a random variable X uniformly distributed over $[0, 45]$.

(i) The probability of waiting at least 15 minutes is equal to

$$P(X \geq 15) = \int_{15}^{45} \frac{1}{45} \, dx = \frac{x}{45} \Big|_{15}^{45} = \frac{45 - 15}{45} = \frac{30}{45} = \frac{2}{3}.$$

(ii) The average waiting time is given (in minutes) by

$$E(X) = \frac{0 + 45}{2} = 22.5.$$

The exponential distribution

We say that a random variable X has an **exponential distribution** if its probability density function has the form

$$\boxed{f(x) = \lambda e^{-\lambda x}, \quad x \geq 0,}$$

where λ is a given positive number (called a **parameter**).

To check that $f(x)$ is indeed a density function, we observe that it is positive for all x, and

$$\int_0^\infty \lambda e^{-\lambda x} \, dx = -e^{-\lambda x} \Big|_0^\infty = 1.$$

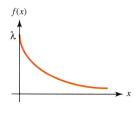

Figure 11.6.2

Figure 11.6.2 shows the graph of $f(x)$.

The exponential distribution is used to model the waiting time until the next occurrence of a particular event. For example, the lifetime of a certain type of lightbulb—in other words, the waiting time until the bulb stops functioning—is a random variable with an exponential distribution. We already encountered this situation in Examples 11.5.7 and 11.5.8. Other examples of exponential distributions are the time until the next earthquake occurs, the time until your next phone call arrives, and the time a customer waits in a checkout lane.

The mean, variance, and standard deviation are as follows:

$$\boxed{E(X) = \frac{1}{\lambda}, \quad \text{Var}(X) = \frac{1}{\lambda^2}, \quad \sigma(X) = \frac{1}{\lambda}.}$$

These formulas were derived for the special case of $\lambda = \frac{1}{750}$ in Examples 11.5.7 and 11.5.8, and the same method works for any parameter λ. The details are left to you (see the exercises).

EXAMPLE 11.6.2

Suppose that the length of time in minutes it takes for a person to finish her money transactions at a bank machine is an exponential random variable with $\lambda = \frac{1}{10}$.

(i) If someone arrives immediately ahead of you, find the probability that you have to wait more than 15 minutes.

(ii) What is the average waiting time?

SOLUTION   (i) Let X denote the length of time that the person ahead of you takes to complete her transaction. Then the desired probability is given by

$$P(X > 15) = \int_{15}^{\infty} \frac{1}{10} e^{-x/10} \, dx$$

$$= -e^{-x/10}\Big|_{15}^{\infty} = e^{-1.5} \approx 0.223.$$

(ii) The average waiting time is the expected value. Since $\lambda = \frac{1}{10}$, the expected value equals 10 minutes.

The normal distribution

We say that X is a **normal random variable**, or that X is **normally distributed**, with parameters μ and σ, if the probability density function of X is given by

$$f(x) = \frac{1}{\sigma\sqrt{2\pi}} e^{-(x-\mu)^2/2\sigma^2}, \qquad -\infty < x < \infty. \tag{25}$$

In this formula μ can be any fixed number, but we require that $\sigma > 0$. By using more advanced techniques, it can be shown that

$$\frac{1}{\sigma\sqrt{2\pi}} \int_{-\infty}^{\infty} e^{-(x-\mu)^2/2\sigma^2} \, dx = 1, \tag{26}$$

which means that $f(x)$ is indeed a probability density. And by using this formula, substitution, and integration by parts, it can be shown that

$$E(X) = \mu, \quad \text{Var}(X) = \sigma^2, \quad \sigma(X) = \sigma.$$

An important special case is that in which $\mu = 0$ and $\sigma = 1$. It is called the **standard normal distribution**. Its density is

$$f(x) = \frac{1}{\sqrt{2\pi}} e^{-x^2/2}, \qquad -\infty < x < \infty, \tag{27}$$

whose graph is drawn in Figure 11.6.3.

As you can see, the graph is a bell-shaped curve. It is not hard to verify, using the standard graphing techniques of calculus, that the graph is symmetric about $x = 0$. It has both a relative and absolute maximum there, and it has inflection points at $x = \pm 1$.

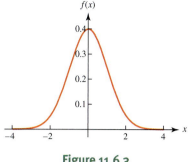

Figure 11.6.3

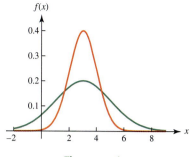

Figure 11.6.4

In general, the graph of a normal density function with parameters μ and σ is bell-shaped, symmetric about $x = \mu$, and has inflection points at $x = \mu - \sigma$ and $x = \mu + \sigma$. In Figure 11.6.4 we show the graphs of two normal density functions drawn on the same set of axes—the taller with $(\mu, \sigma) = (3, 1)$ and the shorter with $(\mu, \sigma) = (3, 2)$.

Normal random variables are frequently used to model many real-life phenomena, such as the distribution of SAT scores and the errors occurring in physical measurements.

Standardizing

In principle, we can compute the probabilities associated with a normal random variable by integrating the probability density function, that is,

$$P(a < X < b) = \frac{1}{\sigma \sqrt{2\pi}} \int_a^b e^{-(x-\mu)^2/2\sigma^2} \, dx,$$

but it is impossible to evaluate this integral by any simple method, and we have to resort to approximation techniques. It is sufficient, however, to construct a table of approximate values for the *standard* normal, because any normal random variable can be brought into standard form by a linear change. The key fact is the following:

Theorem 11.6.1

Suppose X is normally distributed with parameters μ and σ, and Z is defined by setting

$$Z = \frac{X - \mu}{\sigma}. \tag{28}$$

Then Z has a standard normal distribution.

The proof of this theorem is deferred to the exercises at the end of this section.

We can use formula (28) to convert any normal random variable into a standard one and then use a table of probabilities for the standard normal distribution, such as the one provided at the end of this chapter. What the table actually shows are the values (accurate to four decimal places) of the cumulative distribution function $\Phi(x)$ of a standard normal random variable. That is,

$$\Phi(x) = \frac{1}{\sqrt{2\pi}} \int_{-\infty}^{x} e^{-t^2/2}\, dt, \tag{29}$$

and the table at the end of this chapter lists the values of $\Phi(x)$ for $x > 0$.

To find the values of $\Phi(x)$ for $x < 0$, we use the equation

$$\Phi(-x) = 1 - \Phi(x). \tag{30}$$

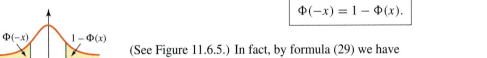

Figure 11.6.5

(See Figure 11.6.5.) In fact, by formula (29) we have

$$\Phi(-x) = \frac{1}{\sqrt{2\pi}} \int_{-\infty}^{-x} e^{-t^2/2}\, dt.$$

Now we let $u = -t$ and obtain

$$\Phi(-x) = \frac{1}{\sqrt{2\pi}} \int_{\infty}^{x} e^{-u^2/2}\, (-du)$$

$$= \frac{1}{\sqrt{2\pi}} \int_{x}^{\infty} e^{-u^2/2}\, du$$

$$= \frac{1}{\sqrt{2\pi}} \int_{-\infty}^{\infty} e^{-u^2/2}\, du - \frac{1}{\sqrt{2\pi}} \int_{-\infty}^{x} e^{-u^2/2}\, du$$

$$= 1 - \Phi(x).$$

If Z is a standard normal variable, then Eq. (30) implies that

$$P(Z \le -x) = P(Z \ge x),$$

since $P(Z \ge x) = 1 - P(Z \le x)$.

EXAMPLE 11.6.3

If X is a normal random variable with parameters $\mu = 3$ and $\sigma = 2$, then find the probability that X takes values between 1 and 6.

SOLUTION   If we set $Z = (X - 3)/2$, then Z has a standard normal distribution. Therefore,

$$P(1 < X < 6) = P\left(\frac{1-3}{2} < \frac{X-3}{2} < \frac{6-3}{2} \right)$$

$$= P\left(-1 < Z < \frac{3}{2} \right)$$

$$= \Phi\left(\frac{3}{2}\right) - \Phi(-1)$$

$$= \Phi\left(\frac{3}{2}\right) - [1 - \Phi(1)]$$

$$= \Phi\left(\frac{3}{2}\right) + \Phi(1) - 1.$$

By using the standard normal table (or by using a computer program, such as Mathematica), we find that

$$\Phi\left(\frac{3}{2}\right) \approx 0.9332 \quad \text{and} \quad \Phi(1) \approx 0.8413.$$

Therefore, we get $P(1 < X < 6) \approx 0.9332 + 0.8413 - 1 = 0.7745$.

EXAMPLE 11.6.4

The weekly sales total for a particular product is normally distributed with mean 800 and standard deviation 200. What is the probability that a given week's sales will
- (i) exceed 1,100?
- (ii) be less than 700?
- (iii) be between 600 and 1,000?

SOLUTION   If X denotes the number of units of the item sold in a week, then $Z = (X - 800)/200$ has a standard normal distribution. For part (i), we have

$$P(X > 1,100) = P\left(\frac{X - 800}{200} > \frac{1,100 - 800}{200}\right)$$

$$= P\left(Z > \frac{3}{2}\right)$$

$$= 1 - \Phi\left(\frac{3}{2}\right) \approx 0.0668.$$

For (ii) we have

$$P(X < 700) = P\left(\frac{X - 800}{200} < \frac{700 - 800}{200}\right)$$

$$= P\left(Z < -\frac{1}{2}\right)$$

$$= \Phi\left(-\frac{1}{2}\right) \approx 0.3085.$$

PROBLEM-SOLVING TACTIC

The key step in computing probabilities for a normal random variable X is to *subtract the mean* and *divide by the standard deviation* to reduce the computations to a standard case. Here, this is done by going from X to $(X - 800)/200$.

And for (iii) we have

$$P(600 < X < 1{,}000) = P\left(\frac{600-800}{200} < \frac{X-800}{200} < \frac{1{,}000-800}{200}\right)$$

$$= P(-1 < Z < 1)$$

$$= \Phi(1) - \Phi(-1) \approx 0.6826.$$

Grading on the curve

An examination is said to be graded on the curve if the test scores of students are used for determining the normal parameters μ and σ. Then the test scores are viewed as normally distributed with parameters μ and σ. The typical assignment of letter grades is as follows: A to those whose score is greater than $\mu + \sigma$, B to those whose score is between μ and $\mu + \sigma$, C to those whose score is between $\mu - \sigma$ and μ, D to those whose score is between $\mu - 2\sigma$ and $\mu - \sigma$, and F to those getting a score below $\mu - 2\sigma$. Of course, this can be modified in many different ways. However, according to the above assignment we have the following percentages of letter grades:

$$A:\ P\{X > \mu + \sigma\} = P\left\{\frac{X-\mu}{\sigma} > 1\right\} = 1 - \Phi(1) = 0.1587.$$

$$B:\ P\{\mu < X < \mu + \sigma\} = P\left\{0 < \frac{X-\mu}{\sigma} < 1\right\} = \Phi(1) - \Phi(0) = 0.3413.$$

$$C:\ P\{\mu - \sigma < X < \mu\} = P\left\{-1 < \frac{X-\mu}{\sigma} < 0\right\} = \Phi(0) - \Phi(-1) = 0.3413.$$

$$D:\ P\{\mu - 2\sigma < X < \mu - \sigma\} = P\left\{-2 < \frac{X-\mu}{\sigma} < -1\right\} = \Phi(2) - \Phi(1) = 0.1359.$$

$$F:\ P\{X < \mu - 2\sigma\} = P\left\{\frac{X-\mu}{\sigma} < -2\right\} = \Phi(-2) = 0.0228.$$

Figure 11.6.6

APPLYING TECHNOLOGY

Normal probabilities can be computed with a graphing calculator. For example, we can use a TI-83 Plus to find the probability that $a \le X \le b$ for a normal random variable with mean μ and standard deviation. We use the item 2:normalcdf on the DISTR menu (shown in Figure 11.6.6), entering the numbers a, b, μ, and σ in that order.

EXAMPLE 11.6.5

```
normalcdf(90,95,
100,7.5)
    .1612811851
normalcdf(100,11
0,100,7.5)
    .4087887176
■
```

Figure 11.6.7

If X is normally distributed with $\mu = 100$ and $\sigma = 7.5$, find

(i) $P(90 \le X \le 95)$ and   (ii) $P(X \le 110)$.

SOLUTION   (i) Using normalcdf(90, 95, 100, 7.5), as shown in Figure 11.6.7, we obtain $P(90 \le X \le 95) = 0.16128$ to five decimal places.

> (ii) The problem here is that we cannot enter $-\infty$ into the calculator as the lower bound. To get around that, we use $a = 100$ (the mean) as the lower bound. Figure 11.6.6 shows that $P(100 \leq X \leq 110) = 0.4088$ (rounded to four decimal places). Because $\mu = 100$, we know that $P(X < 100) = 0.5$. Therefore, we conclude that $P(X \leq 110) = 0.9088$, to four decimal places.

Practice Exercises 11.6

●●●

1. The time (in days) between accidents at a factory is an exponential random variable with $\lambda = 0.016$.
   (a) If an accident has just occurred, what is the probability another will occur within 30 days?
   (b) What is the average time between accidents?

2. The sodium level in a healthy person is a normal random variable with $\mu = 141.5$ and $\sigma = 3.25$.
   (a) What is the probability that a healthy person has sodium level above 145?
   (b) What proportion of healthy people have sodium levels between 140 and 143?

Exercises 11.6

●●●

1. An automated machine produces a certain part every 2 minutes. What is the probability that an inspector randomly arriving will have to wait 30 seconds or less for the part?

2. Suppose a computer program rounds off numbers to the nearest integer, so that the round-off error at each step is a random number in the interval $(-1, 1)$.
   (a) What is the probability that the round-off error at a given step is less than $\frac{1}{3}$ in absolute value?
   (b) Over a sequence of many round-off steps, what do you expect the average error to be?
   (c) What is the standard deviation of the round-off error?

3. A 30-mile section of Interstate 80 through Indiana is a completely straight east-west road. Accidents happen from time to time along that highway, and the points at which they occur are uniformly distributed over the 30-mile stretch.
   (a) If an accident occurs, what is the probability that it happens within 3 miles of the eastern end of the highway?
   (b) If four accidents occur independently, what is the probability that all four occurred in that same 3-mile stretch?

4. The time in months until the breakdown of a computer is a random variable with exponential density with $\lambda = 0.3$.
   (a) Determine the probability that no breakdown occurs for 9 months.
   (b) What is the average length of time the computer will operate before breaking down?

5. During a certain time of day, the time in hours between two aircraft arrivals at an airport has been observed to be a random variable with an exponential density function, where $\lambda = 20$.
   (a) What is the expected length of time between arrivals?
   (b) Determine the probability that the time between two successive arrivals will be less than the expected amount.

6. Some mammals are solitary, living their lives separately from other members of their species. Occasionally, though, their paths cross and they interact with one another. For a member of a certain species, the average number of days between contacts is 75. Let X denote the number of days that pass before such a mammal encounters another member of its species.
   (a) Assume X has an exponential distribution. What is λ?
   (b) What is the probability there will be no contact for at least 100 days?
   (c) If the population of that species were greatly reduced due to poaching, would λ decrease or increase?

7. A statistician eats dinner in the same crowded restaurant every Saturday night and records the length of time he must wait for a table. Over 10 weeks, the waiting times in minutes are 12, 20, 14, 10, 23, 18, 5, 9, 15, 18. He suspects that the waiting time has an exponential distribution. If so, what would be his best guess for λ?

8. A nuclear energy plant emits a detectable amount of radioactive gas on the average of once every 10 days. Assume that the time between emissions is an exponentially distributed random variable.
   (a) What is the probability that 2 weeks go by during which the plant does not emit a detectable amount of radioactive gas?

(b) If an omission has just been detected, what is the probability of another being detected within 24 hours?

9. Suppose that X has an exponential distribution with parameter λ, and let a and b be two positive numbers.
(a) Compute the conditional probability that X is greater than $a + b$ given that it is greater than b.
(b) Compare that to the probability that X is greater than a.
(c) Statisticians sometimes say that the exponential distribution is "memoryless." Can you see why?

10. A statistician has been studying the time (in days) between on-the-job accidents at two factories, and she has concluded that in both cases it has an exponential distribution with $\lambda = 0.01$. Suppose that factory A has just had an accident, whereas factory B has not had an accident in 80 days. Which factory has a greater chance of having an accident within the next 50 days?

11. Show that if X is a random variable with probability density function $f(x) = \lambda e^{-\lambda x}$, then $E(X) = 1/\lambda$ and $\text{Var}(X) = 1/\lambda^2$.

In Exercises 12–17, use the table at the end of the chapter to determine the probability, assuming Z is a standard normal random variable.

12. $P(0 < Z < 1.34)$

13. $P(1.2 \leq Z \leq 1.6)$

14. $P(Z > 2.01)$

15. $P(Z < -0.48)$

16. $P(Z \geq -1.07)$

17. $P(-1.64 \leq Z \leq 0.85)$

18. If X is a normal random variable with $\mu = 100$ and $\sigma = 4$, find
(a) $P(X \leq 110)$ **(b)** $P(98 < X < 102)$
(c) $P(X > 94)$

19. If X is a normal random variable with $\mu = 5$ and $\sigma = 0.2$, find
(a) $P(X \geq 5.3)$ **(b)** $P(X < 4.56)$
(c) $P(4.88 \leq X \leq 5.26)$

20. A machine produces cylindrical rods for use in auto parts. The diameter of the rods may vary slightly because of machine vibrations and other small perturbations. Suppose the diameter is a normally distributed random variable with a mean of 1 centimeter and a standard deviation of 0.004 centimeters.
(a) What is the probability that a rod produced by the machine will have a diameter that is more than 1.01 centimeters?
(b) Approximately what proportion of the rods have diameters between 0.997 and 1.003 centimeters?

21. A brand of orange juice is sold in 2-quart containers, but the actual amount may vary slightly. Suppose that the amount is a normally distributed random variable with $\mu = 64$ ounces and $\sigma = 0.1$ ounce. Find the probability that a container of orange juice chosen at random contains less than 63.9 ounces.

22. For a certain product, the number of items sold in a week is normally distributed with mean 900 and standard deviation 200. What percentage of weeks will sales
(a) exceed 1,200?
(b) be less than 800?
(c) be between 700 and 1,100?

23. Doctors studying the blood-clotting time of a healthy person not taking any medication conclude that in their test group it has a mean of 8.1 seconds and a variance of 4.8 seconds. Assuming that the clotting time is normally distributed with that mean and variance, find the probability that
(a) the blood will clot in less than 2 seconds.
(b) it will take more than 10 seconds for the blood to clot.

24. A gasoline company has 2 million credit card holders. During the preceding month the average amount billed to a card holder was $22, and the standard deviation was $6. Assume that billings are normally distributed.
(a) What is the probability that a card holder had a bill that exceeded $26?
(b) Estimate the number of customers who received bills exceeding $26.

25. Assume that the scores in a given exam are normally distributed with an expected value of 75 and standard deviation of 7. Also assume that the cutoff for an A grade is 89. Estimate the percentage of students who will get an A on the exam.

26. The weight of melons, in pounds, in a farm is normally distributed, with mean $\mu = 4.5$ and standard deviation $\sigma = 0.5$.
(a) Compute the probability that a randomly chosen melon from the farm weights between 4 and 5.5 pounds.
(b) Compute the probability that in a sample of two melons each one weighs between 4 and 5.5 pounds.

27. Assume that in a sunflower field the number of seeds in a sunflower head is normally distributed with expected value $\mu = 500$ and standard deviation $\sigma = 40$.
(a) Find the probability that a randomly selected sunflower head has more than 600 seeds.
(b) Find the probability that in a sample of two sunflower heads each one has more than 600 seeds.

28. By evaluating the integrals $\int_{-\infty}^{\infty} x\,e^{-x^2/2}\,dx$ and $\int_{-\infty}^{\infty} x^2\,e^{-x^2/2}\,dx$, show that a standard normal random variable Z has $E(Z) = 0$ and $\text{Var}(Z) = 1$. (*Hint:* The first integral can be done by substitution, and the second by integration by parts, using formula (26).)

29. In this exercise we consider the proof of Theorem 11.6.1. Suppose X is a normal random variable with mean μ and standard deviation σ, and $Z = (X - \mu)/\sigma$. We wish to show that Z has a standard normal distribution.
(a) Let $F(z)$ be the cumulative distribution function of Z. Verify that

$$F(z) = P(Z \leq z) = P(X \leq \mu + \sigma z)$$

$$= \frac{1}{\sigma\sqrt{2\pi}} \int_{-\infty}^{\mu+\sigma z} e^{-(x-\mu)^2/2\sigma^2}\,dx.$$

(b) According to Theorem 11.5.3, the probability density function $f(z)$ is the derivative of $F(z)$. Use the fundamental theorem of calculus and the chain rule to show that

$$F'(z) = \frac{1}{\sqrt{2\pi}} e^{-z^2/2},$$

which is the probability density function of the standard normal.

Solutions to practice exercises 11.6

1. The time between successive accidents is a random variable X with probability density function $f(x) = 0.016 e^{-0.016}$.
(a) $P(X \leq 30) = 0.016 \int_0^{30} e^{-0.016x}\,dx =$
$(-e^{-0.016x})\big|_0^{30} = 1 - e^{-0.48} \approx 0.38.$
(b) Using integration by parts, we obtain

$$E(X) = \int_0^{\infty} (0.016)x e^{-0.016x}\,dx = \int_0^{\infty} x\,\frac{d}{dx}\,(e^{-0.016x})\,dx$$

$$= -xe^{-0.016x}\Big|_0^{\infty} + \int_0^{\infty} e^{-0.016x}\,dx$$

$$= 0 - \frac{1}{0.016} e^{-0.016x}\Big|_0^{\infty} = \frac{1}{0.016} = 62.5.$$

2. Let X be the sodium level and $Z = (X - 141.5)/3.25$. Z is a standard normal random variable.

(a) $P(X > 145)$
$$= P\left(\frac{X - 141.5}{3.25} > \frac{145 - 141.5}{3.25}\right) = P\left(Z > \frac{3.5}{3.25}\right).$$

Rounding to two decimal places and using the standard normal table, we get

$$P\left(Z > \frac{3.5}{3.25}\right) \approx P(Z > 1.08) = 1 - P(Z \leq 1.08)$$

$$= 1 - 0.8599 = 0.1401.$$

(b) Again, we standardize, round to two decimal places, and use the standard normal table:

$$P(140 < X < 143)$$

$$= P\left(\frac{140 - 141.5}{3.25} < \frac{X - 141.5}{3.25} < \frac{143 - 141.5}{3.25}\right)$$

$$= P(-0.46 < Z < 0.46)$$

$$= P(Z < 0.46) - P(Z \leq -0.46)$$

$$= \Phi(0.46) - [1 - \Phi(0.46)] = 0.3544.$$

Chapter 11 Summary

- A **sample space** of an experiment is a set listing all possible outcomes. Each individual outcome is called a **sample point**, and a subset of a sample space is called an **event**.

- The assignment of probabilities must obey the following three **basic probability axioms:**

  1. For any event E, $0 \leq P(E) \leq 1$.
  2. $P(S) = 1$ and $P(\emptyset) = 0$.

3. If E and F are mutually exclusive, $P(E \cup F) = P(E) + P(F)$.

- If G is a given event of a sample space with $P(G) > 0$, then the **conditional pro-bability** of any other event E given G, denoted by $P(E|G)$, is defined by

$$P(E|G) = \frac{P(E \cap G)}{P(G)}.$$

- Two events E and G are **independent** if $P(E \cap G) = P(E)\,P(G)$.

- **Bayes's formula**: $P(G|E) = \dfrac{P(E|G)\,P(G)}{P(E|G)\,P(G) + P(E|G')\,P(G')}$.

- A **random variable** is a function that assigns a real number to every outcome of a sample space. It is called **discrete** if every number in its range can be surrounded by an interval that excludes all the other numbers in the range.

- A random variable X is said to have a **binomial distribution** with parameters n and p if

$$P(X = k) = \binom{n}{k} p^k (1 - p)^{n-k},$$

where n is a positive integer, $0 < p < 1$, and k is an integer from zero to n. One example is the number of heads in n tosses of a weighted coin, where p is the probability of heads on a single toss.

- If X is a random variable whose range is $\{x_1, x_2, \ldots, x_n\}$, then its **expected value** is defined by

$$E(X) = x_1 p_1 + x_2 p_2 + \cdots + x_n p_n,$$

where $p_i = P(X = x_i)$. Its **variance** is defined by

$$\text{Var}(X) = (x_1 - \mu)^2 p_1 + (x_2 - \mu)^2 p_2 + \cdots + (x_n - \mu)^2 p_n.$$

- A continuous function $f(x)$, $a < x < b$, is called a **probability density function** on the interval (a, b) if

$$f(x) \geq 0, \quad \text{for } a < x < b, \quad \text{and} \quad \int_a^b f(x)\,dx = 1.$$

- A random variable X whose range is an interval (a, b) is called **continuous** if there is a probability density function $f(x)$ on (a, b) so that for any c and d with $a \leq c < d \leq b$,

$$P(c < X < d) = \int_c^d f(x)\,dx.$$

- If X is a continuous random variable with probability density function $f(x)$, $a < x < b$, then its **expected value** $E(X)$ is defined by

$$E(X) = \int_a^b x f(x)\,dx,$$

and its **variance** is defined by

$$\text{Var}(X) = \int_a^b (x - \mu)^2 f(x)\,dx = \int_a^b x^2 f(x)\,dx - \mu^2.$$

- A random variable X is said to have a **uniform distribution** over an interval (a, b) if its probability density function is given by

$$f(x) = \frac{1}{b - a}, \quad a \le x \le b.$$

- A random variable X has an **exponential distribution** if its probability density function has the form

$$f(x) = \lambda e^{-\lambda x}, \quad x \ge 0,$$

where λ is a given positive number.

- We say that X is a **normal random variable**, or that X is **normally distributed**, with parameters μ and $\sigma > 0$, if the probability density function of X is given by

$$f(x) = \frac{1}{\sigma \sqrt{2\pi}} e^{-(x - \mu)^2 / 2\sigma^2}, \quad -\infty < x < \infty.$$

If $\mu = 0$ and $\sigma = 1$, the distribution is called the **standard normal distribution**.

Chapter 11 Review Questions

- What is the relative frequency of an event?

- What is an equiprobable sample space? Give an example.

- Using the axioms, write a formula for computing the probability of the union of two events that are not mutually exclusive.

- State the multiplication rule.

- Give an example of a discrete random variable.

- Explain what Bayes's formula says and how it is used.

- Explain why the expected value of a random variable with a finite range can be interpreted as an average.

- What is the standard deviation of a random variable, and what does it measure?

- Define the cumulative distribution function for a continuous random variable and relate it to the probability density function using the fundamental theorem of calculus.

- Write the expected value, variance, and standard deviation of the binomial, uniform, exponential, and normal distributions.

Chapter 11 Review Exercises

In Exercises 1–4, describe a sample space for the given experiment. If the sample space is finite, specify the number of outcomes.

1. A two-letter string is made up of the letters a, b, c, and d, with each letter appearing no more than once.

2. You count the number of misprints in a particular issue of *The New York Times*.

3. You toss a coin until you have either obtained two heads or made four tosses, whichever comes first.

4. You measure the total amount of snow that fell in Buffalo, New York, during the month of December.

5. Suppose you toss a coin four times and consider the events:

$$A = \text{the first is heads}$$
$$B = \text{the last is tails.}$$

List the outcomes in $A \cup B$, $A \cap B$, and A'. Are A and B' mutually exclusive?

6. Suppose you toss a pair of dice and consider the events:

$$A = \text{both of the top faces are even}$$
$$B = \text{at least one of the top faces is even}$$
$$C = \text{at most one of the top faces is even.}$$

Which pairs of events are mutually exclusive?

7. If $\{s_1, s_2, s_3, s_4\}$ is a sample space with

$$P(s_2) = P(s_1), \quad P(s_3) = \frac{1}{3}P(s_2), \quad P(s_4) = 2P(s_3),$$

find $P(s_i)$ for $i = 1, 2, 3, 4$.

8. If $\{s_1, s_2, s_3\}$ is a sample space with

$$P(\{s_1, s_2\}) = \frac{1}{2} \quad \text{and} \quad P(\{s_2, s_3\}) = \frac{2}{3},$$

find $P(s_i)$ for $i = 1, 2, 3$.

9. Suppose you toss a fair coin three times in a row. Find the probabilities of the following events:
   (a) All tosses are heads.
   (b) At most one toss is heads.
   (c) At least one toss is heads.

10. Suppose you roll a pair of fair dice. Find the probabilities of the following events:
    (a) The sum of the top faces is either 6 or 7.
    (b) The sum of the top faces is neither 6 nor 7.

11. Suppose a bowl contains three red chips and five white ones, and you choose two chips in succession without replacement. Find the probability that both chips are white. (*Hint:* Use a tree diagram as in Example 11.1.7.)

12. An experiment with outcomes s_1, s_2, s_3, s_4, s_5 is described by the probability table

outcome	s_1	s_2	s_3	s_4	s_5
probability	0.1	0.3	0.2	0.015	0.25

Let $A = \{s_1, s_2, s_3\}$, $B = \{s_3, s_4, s_5\}$, and $C = \{s_1, s_3, s_5\}$. Find
(a) $P(A \cup B)$ (b) $P(B \cap C)$ (c) $P(B' \cap C)$

13. Suppose E and F are events satisfying $P(E) = 0.3$, $P(F) = 0.5$, and $P(E \cap F') = 0.1$. Find $P(E \cap F)$ and $P(E \cup F)$.

14. A survey of 50 junior-high-school students found that 32 had read *Harry Potter and the Sorcerer's Stone*, 44 had seen the movie, and 28 had done both. Based on these data, what is the probability a student has neither read the book nor seen the movie?

15. In a family with six children, what is the probability that
    (a) all are girls?
    (b) there are exactly five girls and one boy?

16. A female mallard duck (*Anas platyrhynchos*) has eight drakes (males) competing for her attention.
    (a) If she is willing to choose any one of them, what is the probability of a particular drake being chosen?
    (b) She will reject the three drakes whose feathers are not in prime condition. What is the probability of a particular drake being chosen, given that its feathers are in prime condition?

17. Suppose you roll a pair of fair dice.
    (a) What is the probability that the sum of the top faces is 8, given that they are both odd?

(b) What is the probability that the sum of the top faces is 8, given that at least one is even?

(c) What is the probability that both of the top faces are odd, given that their sum is 8?

18. If $P(A) = 0.3$, $P(B) = 0.5$, and $P(A \cup B) = 0.6$, find $P(A|B)$ and $P(B|A)$.

19. Suppose you draw two cards in succession, without replacement, from a well-shuffled deck.

(a) What is the probability the second is an ace, given that the first is an ace.

(b) What is the probabililty the second is an ace, given that the first is not an ace.

20. Is $P(B|A) = 1 - P(B'|A)$? Justify your answer.

21. Is $P(B|A) = 1 - P(B|A')$? Justify your answer.

22. At a large midwestern university, 68% of the freshmen take mathematics, 46% take economics, and 22% take both.

(a) What is the probability that a student who is taking economics is also taking mathematics?

(b) What is the probability that a student who is taking at least one of the two subjects is taking mathematics?

(c) What is the probability that a student who is taking at least one of the two subjects is taking both?

23. A chef is planning the dishes for tomorrow's lunch. Based on the ingredients available, her assistants estimate there is a 90% chance she will make veal piccante. They also know that 85% percent of the times she makes that dish she also prepares sweet corn tamales as a vegetarian alternative. The assistants will start preparing these dishes if there is at least a $\frac{3}{4}$ probability the chef will make both of them. Advise them on what to do.

24. A bowl contains 10 slips of colored paper, of which two are red, three are white, and five are blue. A blindfolded person draws two of them out of the bowl, without replacement. Find the following probabilities:

(a) Both are red.

(b) Neither is red.

(c) Exactly one is red.

25. You draw two cards from a deck, without replacement.

(a) What is the probability the second is a spade, given that the first is a heart? a club? a diamond? a spade?

(b) Use the multiplication principle to find the probability that the first is a heart and the second is a spade. Do the same with each of the other suits as the first card.

(c) Find the probability that the second card is a spade.

26. In drawing two cards from a deck, consider the events "the first is a king" and "the second is a queen." Are they independent if the drawing is done

(a) with replacement (i.e., you return the first card to the deck and reshuffle before choosing the second)?

(b) without replacement?

27. Suppose you toss a coin three times. Which of the following pairs of events are independent?

(a) "The first is heads" and "the third is heads"

(b) "The third is heads" and "there is at least one head"

(c) "The third is heads" and "there is at least one head and one tail"

(d) "All three are the same" and " the first is heads"

(e) "All three are the same" and "at least one is heads"

28. Suppose that A and B are independent events with $P(A) = 0.6$ and $P(B) = 0.3$. Find

(a) $P(A \cap B')$ **(b)** $P(A \cap B|B)$

(c) $P(A|A \cup B)$

29. Suppose that 64% of all American households watch the broadcast of the Super Bowl. If you call two of them at random during the broadcast, what is the probability that both will be watching it? That neither will be watching it? If you call three, what is the probability that none will be watching the game?

30. You are given two silver dollars, one of which is fair and the other is weighted so that it lands heads 80% of the time. You choose one of the coins at random and toss it twice in succession. If it comes down heads both times, what is the probability you chose the weighted coin?

31. Suppose a random variable X has the following probability distribution:

value of X	-3	-2	-1	0	1	2	3
probability	0.08	0.1	0.12	0.15	0.25	0.2	0.1

Find $E(X)$ and $\text{Var}(X)$.

32. A metal-stamping machine is cutting out metal plates in a special shape. Samples indicate that the thickness of the plate is governed by the following probabilities:

thickness in millimeters	probability
1.15	0.01
1.16	0.02
1.17	0.12
1.18	0.37
1.19	0.36
1.20	0.1
1.21	0.01
1.22	0.01

(a) The customer wants the average thickness of the plate to be between 1.18 and 1.20 millimeters. Will his demand be satisfied?

(b) Approximately what percentage of the plates produced will be less than 1.18 millimeters thick? Greater than 1.20?

33. A hospital hired a management firm to reorganize its emergency-room admission procedures. As part of its initial survey, the firm counted the number of patients coming in for treatment between midnight and 1 A.M. on 30 successive weeknights, resulting in the following data table:

number of patients	0	1	2	3	4	5	6	7	8
number of days	2	3	6	9	7	6	4	2	1

(a) Using these data, construct a table giving the probability distribution for X, the number of patients arriving between midnight and 1 A.M. on a random weeknight.
(b) Find the mean, variance, and standard deviation of X.
(c) What is the probability that more than six patients will arrive between midnight and 1 A.M. on any given weeknight

34. Suppose a coin is weighted so that it lands heads 75% of the time. If you toss it 4 times, what is the probability of getting
(a) exactly two heads?   **(b)** more than two heads?

35. Suppose you make up a six-letter code word, with each letter chosen randomly from a to z and repeats allowed. Write a formula that gives the probability the word will contain exactly k a's, for k from zero to 6.

36. A traffic engineering company has determined that a car driving on the turnpike has a 13% chance of taking the next offramp.
(a) What is the probability that more than one of the next five cars will take the off-ramp?
(b) In the next hour 500 cars will pass that spot on the turnpike. How many are expected to take the off-ramp?

37. Find
(a) $\binom{9}{4}$ **(b)** $\binom{14}{6}$ **(c)** $\binom{3,124}{0}$ **(d)** $\binom{2,714}{1}$

38. Given that $\binom{21}{11} = 352{,}716$ and $\binom{21}{12} = 293{,}930$, find
(a) $\binom{21}{10}$ **(b)** $\binom{22}{12}$

39. An 18-digit number is made up by choosing a digit from 1 to 9 at random to fill each place.
(a) What is the probability there are no 5's?
(b) What is the expected number of 5's?
(c) If the actual number of 5's is within one standard deviation of the mean in either direction, find a lower and upper bound for the number of 5's.

40. Find the constant c for which $f(x) = cx, 0 \le x \le 4$, is a probability density function.

41. Let $f(x) = 2x^{-3}, 1 \le x$. Verify that $f(x)$ is a probability density function. Then find the cumulative distribution function $F(x)$ and verify that $F'(x) = f(x)$.

42. Let
$$f(x) = \begin{cases} x & \text{for } 0 \le x \le 1 \\ 2 - x & \text{for } 1 < x \le 2. \end{cases}$$
(a) Verify that $f(x)$ is a probability density function.
(b) Find the cumulative distribution function $F(x)$.
(c) Verify that $F'(x) = f(x)$.

43. Suppose X is a continuous random variable whose cumulative distribution function is
$$F(x) = 1 - \sqrt{1 - x}, \quad 0 \le x \le 1.$$
(a) Find $P(\frac{1}{4} \le X \le \frac{3}{4})$.
(b) Find the probability density function of X.

In Exercises 44–47, find the mean and variance of a continuous random variable with the given probability density function.

44. $f(x) = \dfrac{1}{4}, \ -2 \le x \le 2$

45. $f(x) = \dfrac{4}{x^5}, \ x \ge 1$

46. $f(x) = \dfrac{1}{3}e^{-x/3}, \ x \ge 0$

47. $f(x) = x + \dfrac{1}{2}, \ 0 \le x \le 1$

48. It can be shown that $f(x) = 1/\pi(1 + x^2), \ -\infty < x < \infty$ is a probability density function.
(a) Show that the expected value does not exist.
(b) Find the median.

In Exercises 49–51, find the mean and median of a continuous random variable with the probability density function.

49. $f(x) = \dfrac{1}{4\sqrt{x}}, \ 0 < x \le 4$

50. $f(x) = 3e^{-3x}, \ x \ge 0$

51. $f(x) = xe^{-x}, \ x \ge 0$

52. Harry and Sally are supposed to meet at noon on the steps of the Metropolitan Museum of Art. Sally is always exactly on time, but Harry is known to arrive anywhere between 10 minutes early and 20 minutes late with a uniform probability distribution.
(a) What is the probability that Harry will arrive before Sally?

(b) What is the probability that Sally will have to wait more than 15 minutes?

(c) On average, how long should Sally expect to wait?

53. If X is a random variable that is uniformly distributed over an interval (a, b), what is the probability that X will be within one standard deviation of the mean in either direction?

54. After studying the operation of its emergency room, a local hospital concluded that the waiting time had an exponential distribution with a mean of 20 minutes.

(a) What is the probability density function of the waiting time?

(b) What is the probability a person arriving at the waiting room will wait more than 30 minutes?

55. A study showed that the time (in minutes) between flight departures at a busy airport has an exponential distribution with $\lambda = 2$.

(a) A plane has just taken off. What is the probability there will be another departure in less than a minute?

(b) What is the average waiting time between departures?

56. If X is a random variable having an exponential distribution with parameter λ, what is the probability that X will be within one standard deviation of the mean in either direction?

57. If X is a normal random variable with $\mu = 10$ and $\sigma = 2.8$, find

(a) $P(X \le 12)$ **(b)** $P(8 < X < 11)$

(c) $P(X > 9)$

58. A tree farmer knows that one variety's height ranges from 5 to 9 feet after 4 years of growth. Her records show that the average tree height is 7 feet. If the heights are normally distributed, it is reasonable to assume that the upper and lower bounds on the height are three standard deviations away from the mean.

(a) Under that assumption, what are μ and σ?

(b) Out of 100 trees planted, about how many should she expect to be under 6 feet?

59. A survey at grocery store A shows that the average total purchase is \$41.25 with a variance of \$101. A similar survey at grocery store B gives an average of \$27.93 with a variance of \$52.70. Assuming the purchase amounts are normally distributed, compare the probabilities of a customer making a purchase of more than \$30 at each of the stores.

60. It can be shown that if X_1 and X_2 are independent random variables, both normally distributed with $E(X_i) = \mu_i$ and $\text{Var}(X_i) = \sigma_i^2$ for $i = 1, 2$, then the difference $X_1 - X_2$ is normally distributed with mean $\mu_1 - \mu_2$ and standard deviation $\sqrt{\sigma_1^2 + \sigma_2^2}$. Suppose that the SAT scores of seniors at Adams High School are normally distributed with a mean of 920 and a standard deviation of 70, while those at Taft High School are normally distributed with a mean of 840 and a standard deviation of 80.

(a) What is the probability that a randomly chosen senior at Adams will achieve a higher SAT score than a randomly chosen senior at Taft?

(b) What is the probability their scores will be within 50 points of one another?

Chapter 11 Practice Exam

1. Suppose A and B are events in some sample space and

$$P(A) = 0.5, \quad P(B) = 0.7, \quad P(A \cup B) = 0.8.$$

(a) Find $P(A \cap B)$ **(b)** $P(A \cup B')$

2. Suppose A and B are events in some sample space with $P(A) = \frac{1}{2}$ and $P(B) = \frac{2}{5}$.

(a) If A and B are mutually exclusive, what is $P(A \cup B)$?

(b) If A and B are independent, what is $P(A \cup B)$?

3. According to a survey, 78% of college students saw the movie *Toy Story*, 64% saw *Toy Story II*, and 56% saw both. Based on these data, what is the probability that a randomly chosen college student did not see either movie?

4. Suppose you toss a coin five times in succession and consider the events:

A = at least one is heads

B = at most one is heads

C = all are heads

D = none are heads.

Are A and B mutually exclusive? A and C? A and D?

5. In rolling a pair of honest dice, let A be the event "dice total four or less" and B the event "both dice are odd."

(a) Find $P(A|B)$.   **(b)** Find $P(B|A)$.

6. An urn contains two red and three white chips. Two are chosen, without replacement.

(a) What is the probability that both are red?

(b) What is the probability that at least one is white?

(c) What is the probability both are white, given that at least one is white?

7. Suppose you roll a pair of dice, one of which is red and the other green. Consider the events

$$A = \text{the red die is a } 2$$
$$B = \text{the dice add up to } 6$$
$$C = \text{the dice add up to } 7$$

Answer the following questions and justify your answer:
(a) Are A and B independent?
(b) Are A and C independent?

8. Professors Softie and Ratso teach different sections of a probability course every fall. Professor Softie invariably fails exactly 2% of his students, whereas Professor Ratso invariably fails 25% of his. Naturally, Professor Softie is more popular, and every year 70% of the students taking the course enroll in his section. The unlucky 30% who register too late wind up in Professor Ratso's section.
(a) If you choose a probability student at random, what is the probability he will fail the course?
(b) If the student does fail the course, what is the probability he was in Professor Ratso's section?

9. Daily sales records for a manufacturer of jet engines show that it will sell zero, one, or two engines per day with the following probabilities: 0.6 probability of selling zero, 0.3 of selling one, and 0.1 probability of selling two. Let X be the number sold on a given day. Find
(a) $E(X)$ **(b)** $\text{Var}(X)$ **(c)** $\sigma(X)$

10. One million lottery tickets were sold for $1 each. There is a first prize of $100,000 and three second prizes of $50,000 each. If you buy a ticket, what are your expected earnings? A negative number represents a loss or expense.

11. Suppose an urn contains two white and three red chips. You draw a chip from the urn 20 times with replacement.
(a) Write a formula giving the probability of choosing k white for k from 0 to 20.
(b) How many times do you expect to draw a white?

12. Suppose that in a tray containing 20 silicon wafers, four are defective. If two of these wafers are chosen at random for inspection,

(a) what is the probability that both are defective?
(b) what is the probability that both are defective, given that at least one is defective?

13. (a) What choice of the constant c makes the function

$$f(x) = c(x + 2)^{-2}, \quad x \geq 0$$

a probability density function?
(b) If X is a random variable with that probability density function, find $P(1 \leq X \leq 3)$.

14. Suppose that a random variable X has the probability density function $f(x) = 2e^{-2x}$, $x \geq 0$. Find the cumulative distribution function.

15. Find the mean and median of a random variable whose cumulative distribution function is $F(x) = x^2/4$, $0 \leq x \leq 2$.

16. Find the mean and variance of a random variable having probability density function $f(x) = x/8$, $0 \leq x \leq 4$.

17. Suppose X is the coordinate of a point randomly chosen from the interval $(1, 3)$. Find $E(X)$ and $\text{Var}(X)$.

18. Suppose that the time between accidents on a busy stretch of interstate highway is governed by an exponential probability distribution, with an average of 7.5 days between accidents. An accident had just occurred.
(a) What is the probability there will be another accident within the next 24 hours?
(b) What is the probability there will not be another accident during the entire next week?

19. Suppose X has a normal distribution with mean 80 and variance 25. Find
(a) $P(X > 88)$ **(b)** $P(X < 78)$
(c) $P(X = 80)$.

20. Assume that the birth weights (in grams) of babies born in the United States are normally distributed with $\mu = 3,315$ and $\sigma = 575$. Approximately what percentage of babies born will weigh
(a) less than 2,500 grams?
(b) more than 4,000 grams?

Chapter 11 Projects

1. Life insurance valuation   Assume that you are working for a life insurance company and you want to find the present value of a term life insurance policy of $100,000 bought by customers at the age of 25 and kept until they reach the age of 65. You estimate that long-term interest rates are about 6%. In this project you will compute the present value using a spreadsheet program. For this, let the random variable X denote the age at which a person living in the United States dies.

(a) In the first column of your spreadsheet list the values of X, which are the integers from 25 to 65.

(b) In the second column list the number p_j, where p_j is the probability of a 25-year-old person dying at the age of j years. (You can find these data in your library or on the Internet.)

(c) In the third column you compute $100{,}000 \cdot p_j$.

(d) In the fourth column you compute the present value of each entry of the third column.

(e) Finally, add up the entries of the fourth column to find the present value of this policy.

2. **The Poisson distribution**[1] In many instances, we are interested in counting the number of times a certain phenomenon occurs within a given time interval. For example, we may be interested in

- the number of customers arriving at a check-out line within a given 5-minute period,
- the number of hits at a certain Web site within a particular minute, or
- the number of accidents in a certain factory within a given year.

If we let X be the number of occurrences within a specified time period, then X is a discrete random variable. Under certain reasonable assumptions (which we will not go into here), the probabilities associated with X are given by the following formula:

$$p_k = P(X = k) = e^{-\lambda}\frac{\lambda^k}{k!}, \quad \text{for } k = 0, 1, 2, 3, \ldots,$$

where λ is equal to the *average* number of occurrences over a time period of that length. In that case, we say that X has a Poisson distribution with parameter λ.

Question 1. Show that $p_0 + p_1 + p_2 + \cdots = 1$.

Question 2. Suppose that the number of telephone calls arriving at a university switchboard averages three per minute and has a Poisson distribution.

(a) What is the probability that exactly two calls will arrive in the next minute?

(b) What is the probability that less than three calls will arrive in the next minute?

Question 3. Although the range of a Poisson distribution covers all nonnegative integers, in practice the probabilities p_k eventually become so small as to be negligible. For $\lambda = 1$, compute the probabilities $p_0, p_1, p_2, \ldots$, until they become less than 10^{-4}. What is the first k for which that happens? Answer the same questions for $\lambda = 0.5$ and $\lambda = 1.5$.

Question 4. One of the early tests of how well a Poisson distribution fits a given set of data was made by a German scientist named Bortkiewicz. He studied the number of fatalities due to horse kicks that occurred in 10 Prussian cavalry corps over a 20-year period. Instead of simply using a time period, the unit Bortkiewicz used was a corps-year, so that his data involved the number of fatalities per corps per year. His data are summarized in the following table:[2]

---

[1]This project assumes knowledge of Chapter 10.

[2]L. Bortkiewicz, *Das Gesetz der Kleinen Zahlen*, Leipzig: Teubner, 1898.

k (number of deaths)	0	1	2	3	4
number of corps-years with k deaths	109	65	22	3	1

On the basis of these data, answer the following:

(a) Find the average number of deaths per corps-year.

(b) Using that average as λ, find $P(X = k)$ for k from zero to 4, as given by the Poisson distribution.

(c) For each k, compute the proportion of corps-years for which k deaths occurred and compare it to the probabilities in part (b).

Question 5. One of the earliest studies involving radioactive particles was performed in 1910 by Ernest Rutherford and Hans Geiger. By using what came to be known as a Geiger counter, they counted the number of α-particles emitted from a radioactive source in $\frac{1}{8}$ of a minute. After repeating the procedure 2,608 times, they came up with an average number of 3.87 particles emitted per $\frac{1}{8}$ of a minute. The proportion of times that k particles were detected, for k from 0 to 9, is given in the following table:[3]

k (number of particles)	0	1	2	3	4	5	6	7	8	9
proportion of times	0.02	0.08	0.15	0.2	0.2	0.16	0.10	0.05	0.02	0.01

Use a Poisson distribution with $\lambda = 3.87$ to compute the probability of k particles being emitted in $\frac{1}{8}$ of a minute for k from zero to 9. Compare these probabilities with the observed proportions given in the table.

3. **The central limit theorem**   This is one of the most striking theorems in all of mathematics. The simplest version is also called the **DeMoivre–Laplace limit theorem** in honor of the two mathematicians who discovered it, independently of one another, around 1800. Roughly speaking, the theorem says that if X has a binomial distribution with parameters n and p, and if n is very large, then X can be approximated by a normal distribution. More precisely, we first standardize X by subtracting its mean np and dividing by its standard deviation $\sqrt{np(1 - p)}$. The new random variable we get that way,

$$\frac{X - np}{\sqrt{np(1 - p)}},$$

has a probability distribution that is approximately standard normal. Specifically, if n is large, then for any numbers a and b,

$$P\left(a \leq \frac{X - np}{\sqrt{np(1 - p)}} \leq b\right) \approx P(a \leq Z \leq b), \tag{31}$$

[3]E. Rutherford, J. Chadwick, and C.D. Ellis, *Radiation from Radioactive Substances*, London: Cambridge University Press, 1951.

where Z has a standard normal distribution. Another way to express it is by writing

$$\lim_{n\to\infty} P\left(a \le \frac{X - np}{\sqrt{np(1 - p)}} \le b\right) = P(a \le Z \le b).$$

Question 1. Suppose X has a binomial distribution with $p = \frac{1}{2}$ and $n = 36$. Find $P(13 \le X \le 21)$ in two ways:

(a) by an exact calculation (with a calculator) using the binomial distribution, and

(b) by applying formula (31) and using the standard normal table at the end of this chapter.

For (b), you have to change the inequality $13 \le X \le 21$ by subtracting $np = 18$ and dividing by $\sqrt{np(1 - p)} = 3$ on all three sides of the inequality:

$$-1.67 \approx \frac{13 - 18}{3} \le \frac{X - 18}{3} \le \frac{21 - 18}{3} = 1.$$

Then you replace the expression in the middle by a standard normal random variable and use the table for that distribution.

Question 2. Because the binomial is discrete and the standard normal is continuous, we can improve the estimate obtained from the Central Limit Theorem by replacing the numbers a and b by fractions that extend the size of the interval. In question 1, for instance, we can replace $13 \le X \le 21$ by the inequality $12.5 \le X \le 21.5$. This does not change the probability for the binomial, but it does for the standard normal, and it results in a better estimate.

This method is known as the **continuity correction**. Redo part (b) of question 1 by using the continuity correction.

Question 3. Assume that 38% of the population has type A blood. Use the Central Limit Theorem to estimate the probability that in a population of 200 people, there are at least 70 with type A blood. Do the calculation two ways: without the continuity correction and with it.

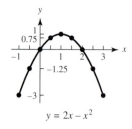

$$y = 2x - x^2$$

Area $\Phi(x)$ Under the Standard Normal Curve to the Left of x

x	.00	.01	.02	.03	.04	.05	.06	.07	.08	.09
.0	.5000	.5040	.5080	.5120	.5160	.5199	.5239	.5279	.5319	.5359
.1	.5398	.5438	.5478	.5517	.5557	.5596	.5636	.5675	.5714	.5753
.2	.5793	.5832	.5871	.5910	.5948	.5987	.6026	.6064	.6103	.6141
.3	.6179	.6217	.6255	.6293	.6331	.6368	.6406	.6443	.6480	.6517
.4	.6554	.6591	.6628	.6664	.6700	.6736	.6772	.6808	.6844	.6879
.5	.6915	.6950	.6985	.7019	.7054	.7088	.7123	.7157	.7190	.7224
.6	.7257	.7291	.7324	.7357	.7389	.7422	.7454	.7486	.7517	.7549
.7	.7580	.7611	.7642	.7673	.7704	.7734	.7764	.7794	.7823	.7852
.8	.7881	.7910	.7939	.7967	.7995	.8023	.8051	.8078	.8106	.8133
.9	.8159	.8186	.8212	.8238	.8264	.8289	.8315	.8340	.8365	.8389
1.0	.8413	.8438	.8461	.8485	.8508	.8531	.8554	.8557	.8599	.8621
1.1	.8643	.8665	.8686	.8708	.8729	.8749	.8770	.8790	.8810	.8830
1.2	.8849	.8869	.8888	.8907	.8925	.8944	.8962	.8980	.8997	.9015
1.3	.9032	.9049	.9066	.9082	.9099	.9115	.9131	.9147	.9162	.9177
1.4	.9192	.9207	.9222	.9236	.9251	.9265	.9279	.9292	.9306	.9319
1.5	.9332	.9345	.9357	.9370	.9382	.9394	.9406	.9418	.9429	.9441
1.6	.9452	.9463	.9474	.9484	.9495	.9505	.9515	.9525	.9535	.9545
1.7	.9554	.9564	.9573	.9582	.9591	.9599	.9608	.9616	.9625	.9633
1.8	.9641	.9649	.9656	.9664	.9671	.9678	.9686	.9693	.9699	.9706
1.9	.9713	.9719	.9726	.9732	.9738	.9744	.9750	.9756	.9761	9767
2.0	.9772	.9778	.9783	.9788	.9793	.9798	.9803	.9808	.9812	.9817
2.1	.9821	.9826	.9830	.9834	.9838	.9842	.9846	.9850	.9854	.9857
2.2	.9861	.9864	.9868	.9871	.9875	.9878	.9881	.9884	.9887	.9890
2.3	.9893	.9896	.9898	.9901	.9904	.9906	.9909	.9911	.9913	.9916
2.4	.9918	.9920	.9922	.9925	.9927	.9929	.9931	.9932	.9934	.9936
2.5	.9938	.9940	.9941	.9943	.9945	.9946	.9948	.9949	.9951	.9952
2.6	.9953	.9955	.9956	.9957	.9959	.9960	.9961	.9962	.9963	.9964
2.7	.9965	.9966	.9967	9968	.9969	.9970	.9971	.9972	.9973	.9974
2.8	.9974	.9975	.9976	.9977	.9977	.9978	.9979	.9979	.9980	.9981
2.9	.9981	.9982	.9982	.9983	.9984	.9984	.9985	.9985	.9986	.9986

Answers to Odd-Numbered Exercises

CHAPTER 0

Exercises 0.2

1. $8, -1, 1/27, a^3, -a^3, a^3 + 3a^2 + 3a + 1, 3x^2 + 3xh + h^2$

3. $1/2, 2, -1, 3/a, a/3, \dfrac{-1}{a(a+h)}$

5. $0, -3, -c^3 + 2c + 1, a^3 - 3a^2 + a + 2,$
$3a^2h + 3ah^2 + h^3 - 2h$

7. $h(1-a) = (1-a)^2 - (1-a) + 1 =$
$1 - 2a + a^2 - 1 + a + 1 = a^2 - a + 1 = h(a)$

9. $x \le 1$ **11.** $x > 1$ **13.** $t \ne \pm 1$ **15.** $u \ne 0$

17. $x = \sqrt{\dfrac{1-y}{y}}, 0 < y \le 1$ **19.** $1 \le y < \infty$

21. $y \ne 0$ **23.** $0 < y \le 2/3$

25. See Table 0.2.25 below.

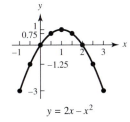

$y = 2x - x^2$

27. See Table 0.0.27 below.

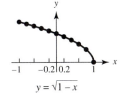

$y = \sqrt{1-x}$

29. No **31.** No **33.** No **35.** No **37.** No **39.** Yes

41. Yes, $g(x) = \dfrac{1}{\sqrt{x}}$. domain: $x > 0$, range: $y > 0$

43. Yes, $g(x) = \dfrac{1}{\sqrt[3]{x}}$. domain: $x \ne 0$, range: $y \ne 0$

45. Yes, $g(x) = x^2 + 2$. domain: $x \ge 0$, range: $y \ge 2$

47. $g(x) = -\dfrac{1}{x}$,
domain: $x < 0$,
range: $y > 0$

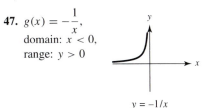

$y = -1/x$

Exercises 0.3

1. Increasing on $(-2, -1)$ and $(1, 2)$, decreasing on $(-1, 1)$.

3. Increasing on $(0, 6)$, decreasing on $(-6, 0)$.

5. Decreasing on $(-4, 4)$.

7. Decreasing for all x.

9. Decreasing on $(-\infty, -1)$, increasing on $(-1, \infty)$.

11. Decreasing on $(-\infty, 0)$ and $(0, \infty)$.

13. Increasing on $(-\infty, 1)$, decreasing on $(1, \infty)$.

15. Odd, symmetric about the origin.

17. Even, symmetric about the y-axis.

19. Odd, symmetric about the origin.

21. Even, symmetric about the y-axis.

x	-1	-0.5	0	0.5	1	1.5	2	2.5	3
$f(x)$	-3	-1.25	0	0.75	1	0.75	0	-1.25	-3

Table 0.2.25

x	-1	-0.8	-0.6	-0.4	-0.2	0	0.2	0.4	0.6	0.8	1
$f(x)$	1.41	1.34	1.26	1.18	1.10	1	0.89	0.77	0.63	0.45	0

Table 0.2.27

23. Odd, symmetric about the origin.

25.

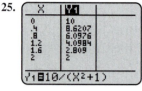

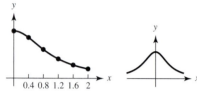

27. y-intercept at $y = 4$, x-intercept at $x = -\frac{4}{3}$

29. y-intercept at $y = 0$, x-intercepts at $x = 0$ and $x = 1$

31. y-intercept at $y = 1$, x-intercept at $x = -1$

33. No intercepts

35. (i) and (a), (ii) and (c), (iii) and (b)

37.

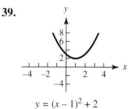

$y = x^2 - 2$

39.

$y = (x-1)^2 + 2$

41.

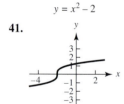

$y = \sqrt[3]{x} + 2$

43.

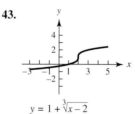

$y = 1 + \sqrt[3]{x - 2}$

45.

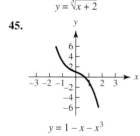

$y = 1 - x - x^3$

47.

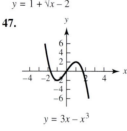

$y = 3x - x^3$

Exercises 0.4

1. slope $\frac{1}{2}$, y-intercept -1 **3.** slope 0, y-intercept 4

5. slope $\frac{1}{2}$, y-intercept $-\frac{5}{2}$ **7.** slope 0, y-intercept -3

9. $y = -3x + 2$ **11.** $y = -\frac{1}{2}x$ **13.** $y = -\frac{5}{4}$

15. $y = -2x + 3$ **17.** $y = -\frac{4}{3}x + \frac{11}{3}$

19. $y = 1$ **21.** $y = \frac{5}{3}x - 3$

23. $C = \frac{5}{9}F - \frac{160}{9}$; $140/9$, $190/9$, $240/9$, $290/9$; rate is $5/9$

25. $I = 400 + 0.08R$, Rate$= 0.08$

27. (a) $x = 200$ (b) 700 (c) 1.5

29. $s = 12 + 1.2t$: 36 miles east of the Skyway; 84 miles east of the Skyway.

31.

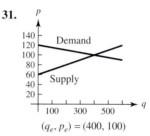

$(q_e, p_e) = (400, 100)$

33. (a) $q = -10p + 1300$ (b) $(q_e, p_e) = \left(\frac{10{,}350}{11}, \frac{395}{11}\right)$

35. $y = 2.55x + 0.7$ **37.** $y = 0.043x - 0.77$

39. $y = 49.95x + 674.72$, 874.5 **43.** Yes

45. (a) Neither   (b) Perpendicular   (c) Parallel   (d) Neither

Exercises 0.5

1. $y = -(x - 3)^2 + 20$
Opens downward
Vertex: $(3, 20)$
Axis of symmetry: $x = 3$

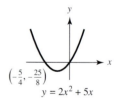

$y = -x^2 + 6x + 11$

3. $y = 2\left(x + \frac{5}{4}\right)^2 - \frac{25}{8}$
Opens upward
Vertex: $\left(-\frac{5}{4}, -\frac{25}{8}\right)$
Axis of symmetry: $x = -\frac{5}{4}$

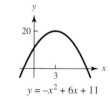

$\left(-\frac{5}{4}, -\frac{25}{8}\right)$

$y = 2x^2 + 5x$

5. $y = -0.1(x + 6)^2 + 7.2$
Opens downward
Vertex: $(-6, 7.2)$
Axis of symmetry: $x = -6$

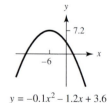

$y = -0.1x^2 - 1.2x + 3.6$

7. $y = 3\left(x - \frac{1}{3}\right)^2 + \frac{2}{3}$
Opens upward
Vertex: $\left(\frac{1}{3}, \frac{2}{3}\right)$
Axis of symmetry: $x = \frac{1}{3}$

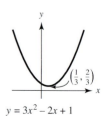

$\left(\frac{1}{3}, \frac{2}{3}\right)$

$y = 3x^2 - 2x + 1$

9. $x = -2, 7$; positive on $(-\infty, -2)$ and $(7, \infty)$, negative on $(-2, 7)$

11. $x = \frac{1}{3}, \frac{1}{2}$; positive on $(-\infty, \frac{1}{3})$ and $(\frac{1}{2}, \infty)$, negative on $(\frac{1}{3}, \frac{1}{2})$

13. $x = 2, 6$; positive on $(-\infty, 2)$ and $(6, \infty)$, negative on $(2, 6)$

15. $x = 1 + \sqrt{2}, 1 - \sqrt{2}$; positive on $(-\infty, 1 - \sqrt{2})$ and $(1 + \sqrt{2}, \infty)$, negative on $(1 - \sqrt{2}, 1 + \sqrt{2})$

17. $x = -\frac{1}{2}, \frac{2}{5}$; positive on $(-\infty, -\frac{1}{2})$ and $(\frac{2}{5}, \infty)$, negative on $(-\frac{1}{2}, \frac{2}{5})$

19. Break-even points at $x = 10, 50$; profit if $10 < x < 50$.

21. (a) $q = -30p + 1380$ (b) $R = -30p^2 + 1380p$

23. (a) $p = \$30$ (b) $p = \$38$

25. $p = \$80$ **27.** $q_e = 10, \ p_e = 30$

Exercises 0.6

1. Increases without bound to the right, decreases without bound to the left.

3. Decreases without bound to the right, increases without bound to the left.

5. n odd, $a_n > 0$ **7.** n odd, $a_n < 0$

9. Vertical asymptote at $x = -1$. The graph climbs toward $x = -1$ from both directions.

11. Vertical asymptote at $x = -4$. The graph climbs as it approaches $x = -4$ from the left, falls as it approaches from the right.

13. Vertical asymptotes at $x = \pm 2$. The graph falls as it approaches $x = 2$ from the left, climbs as it approaches from the right. The same is true for $x = -2$.

15. No vertical asymptote.

17. Vertical asymptote at $x = -1$. The graph falls as it approaches $x = -1$ from the left, climbs as it approaches from the right.

19. Vertical asymptotes at $x = 0$ and $x = 2$. The graph climbs as it approaches $x = 0$ from the left, falls as it approaches from the right. The graph falls as it approaches $x = 2$ from the left, climbs as it approaches from the right.

21. 16   **23.** 0.027   **25.** $\frac{1}{16}$ **27.** 27   **29.** 4

31. 4   **33.** 12   **35.** $\frac{1}{27}$

37. Domain: all real numbers.
No vertical asymptote.
Increasing for all x.
Symmetric about the origin.
$f(x) > 0$ for $x > 0$
$f(x) < 0$ for $x < 0$

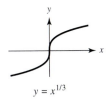
$y = x^{1/3}$

39. Domain: all real numbers.
No vertical asymptote.
Decreases on $(-\infty, 0)$.
Increases on $(0, \infty)$.
Symmetric about y-axis.
$f(x) > 0$ for all $x \neq 0$

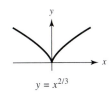
$y = x^{2/3}$

41. Domain: $x \geq 0$.
No vertical asymptote
Increasing on $(0, \infty)$.
No symmetry.
$f(x) > 0$ for $x > 0$

$y = x^{3/2}$

43. $2^{7/2} = 8\sqrt{2} \approx 11.3$

45. (a) Natural domain is $r \neq 0$. In this context, the domain is all $r > 0$. (b) Decreasing for $r > 0$, increasing for $r < 0$. There is a vertical asymptote at $r = 0$. The graph climbs toward the asymptote as it approaches from either side. (c) $(1.002)^{-2} \approx 0.996$. The force decreases by about 0.4%.

Exercises 0.7

1. (a) 2   (b) $0.2222\ldots$ or $2/9$ (c) 2

3. (a) 3.5   (b) 0.2

5. (a) 12   (b) 16   (c) 4   (d) 36

7. 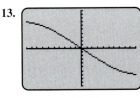   **9.**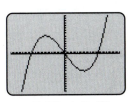

11.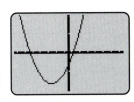

Default $[-40, 40] \times [-30, 30]$

13.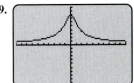

Default $[-30, 30] \times [-15, 15]$

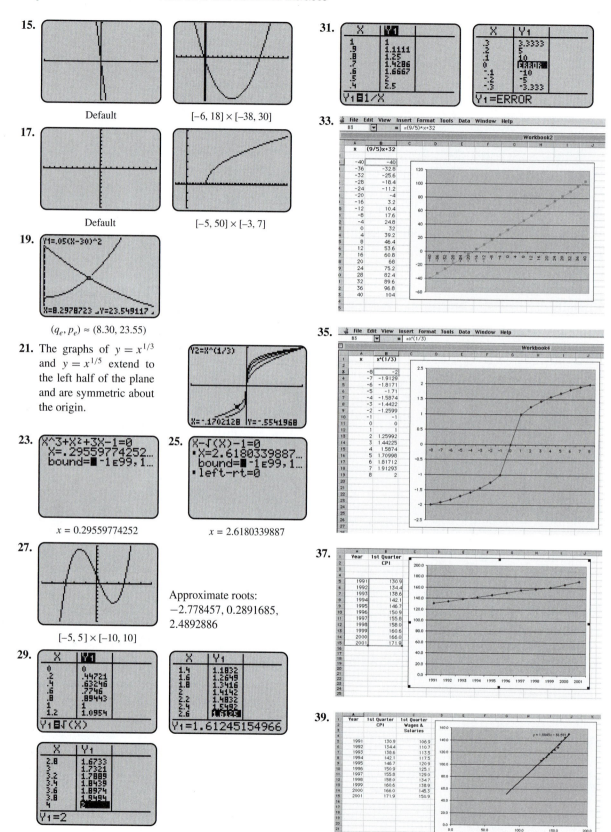

15.

Default   [−6, 18] × [−38, 30]

17.

Default   [−5, 50] × [−3, 7]

19.

Y1=.05(X−30)^2

X=8.2978723 Y=23.549117

$(q_e, p_e) \approx (8.30, 23.55)$

21. The graphs of $y = x^{1/3}$ and $y = x^{1/5}$ extend to the left half of the plane and are symmetric about the origin.

Y2=X^(1/3)

X=−.1702128 Y=−.5541968

23.
X^3+X²+3X−1=0
X=.29559774252…
bound=■ −1ᴇ99, 1…

$x = 0.29559774252$

25.
X−√(X)−1=0
▪X=2.6180339887…
bound=■ −1ᴇ99, 1…
▪left−rt=0

$x = 2.6180339887$

27.

[−5, 5] × [−10, 10]

Approximate roots:
−2.778457, 0.2891685, 2.4892886

29.

X	Y1
0	0
.2	.44721
.4	.63246
.6	.7746
.8	.89443
1	1
1.2	1.0954

Y1◼√(X)

X	Y1
1.4	1.1832
1.6	1.2649
1.8	1.3416
2	1.4142
2.2	1.4832
2.4	1.5492
2.6	1.6125

Y1=1.61245154966

X	Y1
2.8	1.6733
3	1.7321
3.2	1.7889
3.4	1.8439
3.6	1.8974
3.8	1.9494
4	2

Y1=2

31.

X	Y1
1	1
.9	1.1111
.8	1.25
.7	1.4286
.6	1.6667
.5	2
.4	2.5

Y1◼1/X

X	Y1
.3	3.3333
.2	5
.1	10
0	ERROR
−.1	−10
−.2	−5
−.3	−3.333

Y1=ERROR

33.

File Edit View Insert Format Tools Data Window Help
B3 =(9/5)*x+32

x	(9/5)x+32
−40	−40
−36	−32.8
−32	−25.6
−28	−18.4
−24	−11.2
−20	−4
−16	3.2
−12	10.4
−8	17.6
−4	24.8
0	32
4	39.2
8	46.4
12	53.6
16	60.8
20	68
24	75.2
28	82.4
32	89.6
36	96.8
40	104

35.

File Edit View Insert Format Tools Data Window Help
B3 =x^(1/3)

x	x^(1/3)
−8	−2
−7	−1.9129
−6	−1.8171
−5	−1.71
−4	−1.5874
−3	−1.4422
−2	−1.2599
−1	−1
0	0
1	1
2	1.25992
3	1.44225
4	1.5874
5	1.70998
6	1.81712
7	1.91293
8	2

37.

Year	1st Quarter CPI
1991	130.9
1992	134.4
1993	138.6
1994	142.1
1995	146.7
1996	150.9
1997	155.8
1998	158.0
1999	160.6
2000	166.0
2001	171.9

39.

Year	1st Quarter CPI	1st Quarter Wages & Salaries
1991	130.9	106.9
1992	134.4	110.7
1993	138.6	113.5
1994	142.1	117.5
1995	146.7	120.9
1996	150.9	125.1
1997	155.8	129.0
1998	158.0	134.7
1999	160.6	138.9
2000	166.0	145.3
2001	171.9	150.9

y = 1.0645x − 16.591

41.

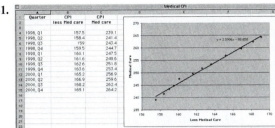

Chapter 0: Review Exercises

1. $25, 2, 1, 9/25, 0, 45, 12t, 2x + h$

3. Natural domain: all $x \neq 4$; $-1/10, -1/5, -1/4, -1/2,$
    $-2, -1/(1 + h)$

5. **(a)** $-3/4, 5/4, 3/2, 11/8$ **(b)** $1/2, 2$ **(c)** $-3/8$
    (d) $[-\frac{9}{16}, 2]$ **(e)** $(-\frac{9}{16}, \frac{1}{2})$ and $(\frac{3}{2}, 2)$ **(f)** No
    (g) Yes   **(h)** Yes

7. **(a)**

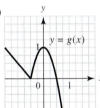

(b)

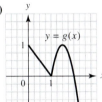

(c)

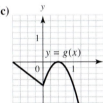

(d)

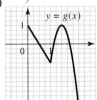

(e)

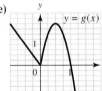

(f)

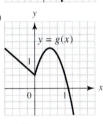

9. **(a)** even
    (b) odd
    (c) even
    (d) odd

11. **(a)** All $x \geq 0$ and $\neq 2$ **(b)** All $x \neq 1$ or 2

13. Inverse: $g(x) = (1 - x)^{1/4}$
    Domain: $(-\infty, 1]$
    Range: $[0, \infty)$

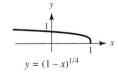

15. $-2/3$ **17.** $y = -3x - 1$ **19.** $5x - 2y = 22$

21. Positive on $(\frac{2-\sqrt{5}}{2}, \frac{2+\sqrt{5}}{2})$,
    negative on $(-\infty, \frac{2-\sqrt{5}}{2})$ and $(\frac{2+\sqrt{5}}{2}, \infty)$

23. $f(x) = -3\left(x - \frac{5}{3}\right)^2 + \frac{1}{3}$
    max. value: $1/3$;
    min. value: none
    y-intercept: -8
    x-intercepts: 2 and $4/3$

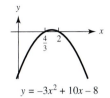

25. No vertical asymptote
    Even function

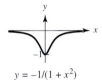

27.

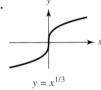

    Natural domain:
    all real numbers

29.

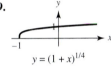

    Natural domain:
    $[-1, \infty)$

31. **(a)** $P(q) = 20q - 5000$ **(b)** 250

33. **(a)** $R = 3,000p - 20p^2$
    (b) $P = -20p^2 + 3,800p - 140,000$
    (c) $\$95$

Chapter 0: Exam

1. 9.4 **2.** $(0, 3]$ **3.** 3 **4.** (i) b   (ii) c   (iii) c   (iv) e

5. **(a)** -1 **(b)** $t^2 - 4t + 2$ **(c)** $h + 4$

6.

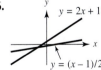

    Inverse: $g(x) = (x - 1)/2$

7.

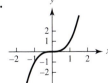

8.

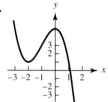

9. $y = -\frac{1}{2}x - \frac{1}{2}$

10. (a)

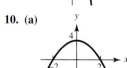

Even

(b)

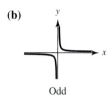

Odd

(c)

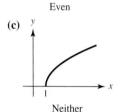

Neither

11. 400

12. $y = -\left(x - \frac{3}{2}\right)^2 - \frac{11}{4}$
opens downward
vertex: $\left(\frac{3}{2}, -\frac{11}{4}\right)$
axis of symmetry: $x = \frac{3}{2}$

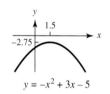

$y = -x^2 + 3x - 5$

13. (a) $q = -8p + 120$ **(b)** $R = -8p^2 + 120p$ **(c)** \$7.50

14. (a) $h = -16t^2 + 32t + 48$ **(b)** At $t = 3$ seconds
(c) 64 feet

CHAPTER 1

Exercises 1.1

1. (a) -2 **(b)** 3 **(c)** 2

3. $x = 1$ and $x = 3$

5. (a) 2 **(b)** 1 **(c)** 2 **(d)** $g(-1) = 2$, $g(1) = 2$

7. $x = 2$ and $x = 3$

9. (a) 8.001 **(b)** No

11. (a) $f(t) = \begin{cases} 25, & 0 < x \le 4 \\ 30, & 4 < t \le 5 \\ 35, & 5 < t \le 6 \\ 40, & 6 < t \le 7 \end{cases}$

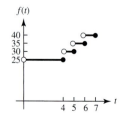

(b) The function has no limit at $t = 4, 5, 6$.
$\lim_{t \to 4^-} f(t) = 25, \ \lim_{t \to 4^+} f(t) = 30;$
$\lim_{t \to 5^-} f(t) = 30, \ \lim_{t \to 5^+} f(t) = 35;$
$\lim_{t \to 6^-} f(t) = 35, \ \lim_{t \to 6^+} f(t) = 40$

13. -7 **15.** $1/2$ **17.** 1 **19.** 0 **21.** $2/5$ **23.** 0 **25.** 3

27. 0 **29.** 6 **31.** $3/2$ **33.** $1/2$ **35.** $-1/6$ **37.** 1
39. 12

41. -1 **43.** $2/5$ **45.** -1 **47.** 0 **49.** 0 **51.** 0

53. (a) 0 **(b)** 1 **55.** $5/6$ **57.** 1 **59.** $1/2$

61. 0

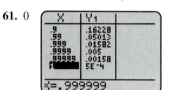

63. limit does not exist

65. $1.\overline{3}$

Exercises 1.2

1. (a) -1 **(b)** 0 **(c)** ∞ **(d)** -1 **(e)** 0

3. At $x = -1, 0, 2$

5. (a) $-\infty$ **(b)** ∞ **(c)** 0 **(d)** -1 **(e)** 0

7. ∞ **9.** ∞ **11.** ∞ **13.** $-\infty$ **15.** $-\infty$ **17.** ∞

19. $2/5$ **21.** 1 **23.** $-\infty$ **25.** 0 **27.** 0 **29.** $1/2$

31. (a) $\lim_{x \to \infty} C(x) = \infty$ **(b)** $\lim_{x \to \infty} \frac{C(x)}{x} = 0.17$

33.

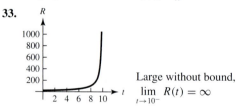

Large without bound,
$\lim_{t \to 10^-} R(t) = \infty$

35. (a) $R(q) = \dfrac{2400q}{q + 2}$

(b) $\lim_{q \to 0^+} R(q) = 0, \ \lim_{q \to 10} R(q) = 2000,$
$\lim_{q \to \infty} R(q) = 2400$
(c) \$2400

37. vertical asymptote: $x = 2$
horizontal asymptote: $y = 2$
positive on $(-\infty, -1/2)$
and $(2, \infty)$
negative on $(-1/2, 2)$

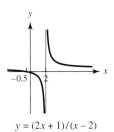

$y = (2x + 1)/(x - 2)$

39. vertical asymptotes: $x = 2$
and $x = -2$
horizontal asymptote: $y = 0$
positive on $(-2, 0)$ and $(2, \infty)$
negative on $(-\infty, -2)$ and $(0, 2)$

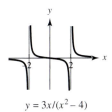

$y = 3x/(x^2 - 4)$

41. no vertical asymptote
horizontal asymptote: $y = 1$
positive for all $x \neq 0$

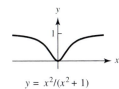

$y = x^2/(x^2 + 1)$

43. vertical asymptotes: $x = 1$
and $x = -1$
horizontal asymptote: $y = 1$
positive on $(-\infty, -3)$, $(-1, 1)$
and $(3, \infty)$
negative on $(-3, -1)$ and $(1, 3)$

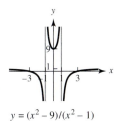

$y = (x^2 - 9)/(x^2 - 1)$

45. vertical asymptotes: $x = 3$
and $x = -1$
horizontal asymptote: $y = 0$
positive on $(-\infty, -1)$
and $(3, \infty)$
negative on $(-1, 3)$

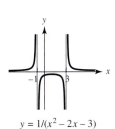

$y = 1/(x^2 - 2x - 3)$

47. vertical asymptote: $x = 1$
horizontal asymptote: $y = 1$
positive on $(-\infty, 0)$ and $(1, \infty)$
negative on $(0, 1)$

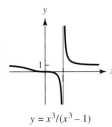

$y = x^3/(x^3 - 1)$

49. $y = \frac{1}{3}x - \frac{5}{3}$ **51.** $y = 2x$

53.

X	Y1
10	.8031
100	.97131
1000	.99701
10000	.9997
100000	.99997
500000	.99999

Y1=.999994000054

$$\lim_{x \to \infty} \frac{\sqrt{x^2 + 9}}{x + 3} = 1$$

55.

X	Y1
100	90
1000	968.38
10000	9900
100000	99684
500000	499293
1E6	999000
1E7	1E7

X=10000000

$$\lim_{x \to \infty} t - \sqrt{t} = \infty$$

57.

X	Y1
100000	315.23
1E6	999
1E9	31622
1E12	999999
1E15	3.16E7
1E18	1E9

Y1=999999999

$$\lim_{x \to \infty} \frac{x\sqrt{x} + 1}{x + \sqrt{x} + 1} = \infty$$

59.

X	Y1
100000	-1E5
1E6	-1E6
1E9	-1E9
1E12	-1E12
1E15	-1E15
1E18	-1E18

Y1=-1000000001

$$\lim_{x \to \infty} \frac{\sqrt{x^4 + 1}}{1 - x} = -\infty$$

61.

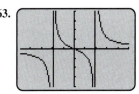

63.

65.

67. $\displaystyle \lim_{h \to \infty} v = \lim_{h \to \infty} \sqrt{2gR \frac{h}{h + R}} = \sqrt{2gR} \approx 11{,}182.5 \text{ m/sec}$

Exercises 1.3

1. At $x = -1, 1, 4, 6, 8$ **3.** None   **5.** At $x = 4$

7. At $x = 6$ **9.**

11.

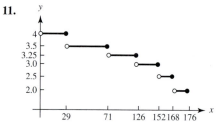

Discontinuous at $x = 29$, 71, 126, 152, 168

13. Discontinuous at $x = 1$, continuous for all other x.

15. Continuous for all x.   **17.** Continuous for all x.

19. Discontinuous at $x = 0$, continuous for all other x.

21. Continuous for all x.

23. (a) $f(0) = 2$, $f(1) = 3$, $f(2) = 1/2$, $f(4) = 2$
    (b) At $x = 1$, 2

25. (a) $f(1) = 1$, $f(4) = 2$, $f(5) = 3$, $f(8) = 6$,
    $f(9) = 27$ **(b)** At $x = 8$

27. At $x = 2$ **29.** 3 **31.** -3 **33.** 4/3 **35.** $f(0) = -2$
    37. $f(0) = 0$

39.

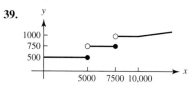

Discontinuous at $x = 5000$, 7500

41. Yes, twice.   **43.** Two

45. No, because $f(x) > -2$ on $[-3, 5]$.

47. No. The theorem does not apply because g is not coninuous throughout the interval.

49. At least 2 roots, one in $(-1, 0)$ and one in $(2, 3)$.

51. (a), (c), (d), (e)

53. $p = 1.98$, $q = 4.95$

55. 1.703125 with $|\text{error}| \leq 0.015625$

57. 2.671875 with $|\text{error}| \leq 0.046875$

59. 0.79836 **61.** -1.39824

63. -1.699628, 0.239123, 2.460505

Exercises 1.4

1. $\delta = \varepsilon/3$ **3.** $\delta = 2\varepsilon$ **5.** $\delta = \min(1, \varepsilon/5)$

7. $\delta = \min(1, \varepsilon/19)$

Chapter 1: Review Exercises

1. (a) -1 **(b)** 1 **(c)** -1 **(d)** 4

3. (a) -1 **(b)** 3 **(c)** 2

5. At $x = -2$, 0, 1, 5 **7.** At $x = -5$ **9.** 6 **11.** $-1/5$

13. -8 **15.** 0 **17.** $-\infty$ **19.** $-\infty$ **21.** 4 **23.** 0

25.

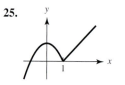

Continuous for all x

27.

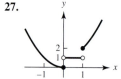

Discontinuous at $x = 0$, 1

29. (d) **31.** (a)

33. vertical asymptote: none
    horizontal asymptote: $y = 1$

35. vertical asymptotes: $x = 10$ and $x = -10$
    horizontal asymptote: none

37. vertical asymptote: none
    horizontal asymptote: $y = 3$

39. vertical asymptote: none
    horizontal asymptote: $y = 3$
    $0 \leq y < 3$ for all x

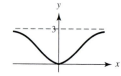

Chapter 1: Exam

1. 2 **2.** does not exist **3.** -1 **4.** $-1/3$

5. $x = 2$ and $x = -3$ **6. (a)** 4/3 **(b)** -8

7. 4 **8. (a)** 1 **(b)** -1

9. (a) $-\infty$ **(b)** ∞ **(c)** 2

10. (a) $\lim\limits_{x \to -2^-} f(x) = 1$, $\lim\limits_{x \to -2^+} f(x) = 0$,
    $\lim\limits_{x \to 2^-} f(x) = 2$, $\lim\limits_{x \to 2^+} f(x) = 2$,
    $\lim\limits_{x \to 5^-} f(x) = 4$, $\lim\limits_{x \to 5^+} f(x) = 2$,
    $\lim\limits_{x \to 7^-} f(x) = 0$, $\lim\limits_{x \to 7^+} f(x) = 0$
    (b) $x = -2$, 5 **(c)** $x = -2$, 2, 5

11. vertical asymptotes: $x = 3$ and $x = -3$
    horizontal asymptotes: $y = 5$

12. vertical asymptotes: none
    horizontal asymptotes: $y = 0$

13. vertical asymptote: $x = -1$
    horizontal asymptotes: none

14. natural domain: all $x \neq \pm 1$
    range: $(-\infty, \infty)$
    vertical asymptotes: $x = \pm 1$
    horizontal asymptote: $y = 0$
    negative on $(-\infty, -1)$
    and $(0, 1)$
    positive on $(-1, 0)$ and $(1, \infty)$

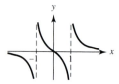

15. At $x = 0$ **16.** 1/4

17. Three roots, one in $(-2, -1.5)$, one in $(-0.5, 0)$, one in $(0.5, 1)$

18. 1

CHAPTER 2

Exercises 2.1

1.

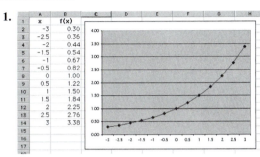

3.

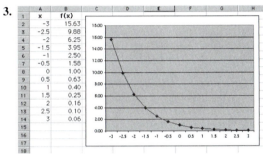

5. Apply $\left(\frac{2}{3}\right)^x = \left(\frac{3}{2}\right)^{-x}$ to the table of exercise 1 to get Table 2.1.5 (see below).

7. Apply $\left(\frac{5}{2}\right)^x = (0.4)^{-x}$ to the table of exercise 3 to get Table 2.1.7 (see below).

9. Same table as exercise 1. **11.** Same table as exercise 2.

13.

r	2^r
1	2
1.7	3.249009585
1.73	3.317278183
1.732	3.321880096
1.73205	3.321995226
1.7320508	3.321997068

$2^{\sqrt{3}} = 3.321997085\ldots$

15. 9 **17.** 4/3 **19.** 1 **21.** 27

23. (a) $500(1.07)^t$ (b) \$983.58

25. $(1.06)^7$, $(1.06)^{12}$ **27.** \$22,819.35

29. $y = 500\left(\frac{8}{5}\right)^t$, with y in thousands.

31. 344, 382 **33.** 52.77%

35. (a) $y_0 = 50$, $b = \left(\frac{3}{5}\right)^{1/10} \approx 0.95$ (b) 18 mg
(c) $80\left(\frac{3}{5}\right)^{3/2} \approx 37.18$ mg

37.

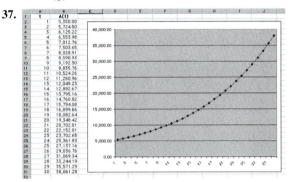

39.

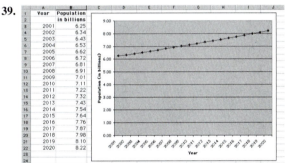

Exercises 2.2

1. (a) \$540 (b) \$541.50 (c) \$541.64

3. (a) \$405.47 (b) \$405.52

5. 6% compounded 4 times a year.

7. (a) 1,038.21 (b) 1,051.27 (c) 1,183.82

9. (a) $10,000e^{0.045/2} = 10,227.55$ dollars
(b) $10000e^{3(0.045)/2} = 10,698.30$ dollars
(c) $10,000e^{3(0.045)} = 11,445.37$ dollars

x	-3	-2.5	-2	-1.5	-1	-0.5	0	0.5	1	1.5	2	2.5	3
$f(x)$	3.38	2.76	2.25	1.84	1.5	1.22	1.0	0.82	0.67	0.54	0.44	0.36	0.30

Table 2.1.5

x	-3	-2.5	-2	-1.5	-1	-0.5	0	0.5	1	1.5	2	2.5	3
$f(x)$	0.06	0.10	0.16	0.25	0.40	0.63	1.00	1.58	2.50	3.95	6.25	9.88	15.63

Table 2.1.7

11. $1,883.53

13. $e^3 \approx 20.0855$ **15.** $e^{-2} \approx 0.1353$

17. $e^{3/4} \approx 2.1170$ **19.** $10e^{-2} \approx 1.3534$

21. $1000e^{1/2} \approx 1648.7213$

23. 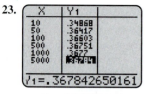   **25.**

$1/e \approx 0.3678794412$ $e^{-3/2} \approx 0.2231301601$

27. See Figure 2.2.27 below.   **29.**

Solution $x \approx 0.56714329$

Exercises 2.3

1. 3   **3.** 0   **5.** 3/2   **7.** -1 **9.** -3

11. 1   **13.** 3/2   **15.** $-1/2$

17. (a) 1.602   (b) -0.398 (c) -0.602 (d) 0.301

19. 29.0402265   **21.** 261.8183008   **23.** 1.155346656

25. 3.170   **27.** -0.848 **29.** 2.585   **31.** 1.661

33. 4.907   **35.** 1.9535   **37.** $2 + b$ **39.** $(1 + b)/2$

41. $b/2$ **43.** $2b$ **45.** $2/b$ **47.** 6.802

49. $10^{5.5} \approx 316, 228$ times greater   **51.** 99.77

53. 100 times more acidic   **55.**

Solution $y \approx 8.420$

Exercises 2.4

1. $e^{7\ln 3}$ **3.** $e^{5\ln x}$ **5.** $e^{-3\ln a}$ **7.** $e^{(\ln a)/3}$ **9.** -0.5 **11.** 5

13. -0.4 **15.** 9   **17.** 1.693   **19.** -1.386 **21.** 2.386

23. 2   **25.** $3t^3$ **27.** $-3x$ **29.** $1/x$ **31.** $(x - 1)^2$

33. 4   **35.** $\frac{1}{2}(\ln 7 - 1)$ **37.** 2   **39.** $e - 1$ **41.** $e^{5/3}$

43. 2   **45.** 3/2   **47.** 2.274149   **49.** 25.7786519

51. 1.0986123   **53.** $\dfrac{\ln 12}{\ln 5} \approx 1.5439593$

55. $\dfrac{\ln 10}{\ln 2} \approx 3.3219281$

59.   **61.**

$y = 1 + \ln x$ $y = \ln x - 1$

63. (d)   **65.** (b)   **67.** $\dfrac{\ln 3}{0.075} \approx 14.65$ years

69. (a) $k = \ln \frac{105}{43} \approx 0.89276$,
$A = \frac{22,050}{43} \approx 512.79$ (b) 0.166 seconds

71. (a) $y = 4000e^{0.0811t}$ (b) 7653   (c) $\dfrac{\ln 3}{0.0811} \approx 13.55$

73. $a = 8$, $b = 3$, $r = \frac{1}{10}\ln(\frac{3}{2}) \approx 0.04$

75.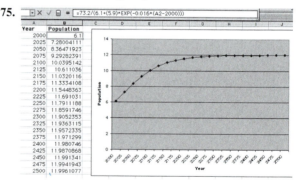

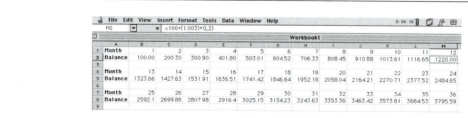

Figure 2.2.27

77.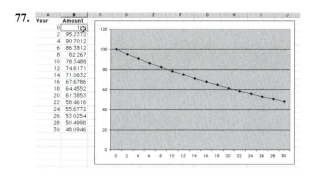

79. 57.08 mg

Chapter 2: Review Exercises

1. (a) $1/3$ (b) 81 (c) $1/1000$ (d) 1000

3. (a) $y = 600 \left(\frac{5}{3}\right)^{t/3}$ (b) $600 \left(\frac{5}{3}\right)^2 \approx 1667$

5. (a) \$5308.39   (b) \$5978.09

7. $100 \left(1 + \dfrac{0.06}{n}\right)^n$ dollars

9. (a) \$907.03   (b) \$905.21   (c) \$904.84

11. e^{-2} **13.** e^{-2}

15. (a) -1.262 (b) 1.201 (c) 2.631
 (d) 0.3155 (e) -1.771

17. b   **19.** b   **21.** 1   **23.** 2   **25.** 3

27. $1/2$ **29.** 2   **31.** 80,000   **33.** $\dfrac{\ln 4}{0.08} \approx 17.33$

35. (a) ∞ (b) 0 (c) 0 (d) ∞ (e) ∞ (f) $-\infty$

37. 1.95420   **39.** $0, 8e^{-0.1}$ **41.** $\dfrac{\ln(2.5)}{0.04} \approx 22.9$ years

43. 12.5 grams   **45.** $\dfrac{\ln 3}{14} \approx 0.0785$

Chapter 2: Exam

1. \$1127.27   **2.** \$3731.08

3. \$3669.30   **4.** $e \approx 2.72$ dollars

5. (a) $e^{1/4}$ (b) e^{-6} **6.** (a) $1/2$ (b) -2

7. (a) -2.930 (b) 3.465 **8.** 3

9. (a) $(2x + 1)^3$ (b) $3x^2 + 4 - \ln 2$

10. (a)
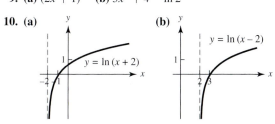
(b)

(c)
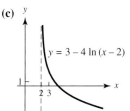

11. (a)
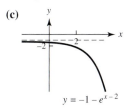
(b)

(c)
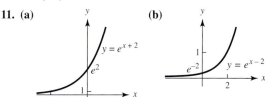

12. $1 + \dfrac{\ln 23}{\ln 5} \approx 2.94819$ **13.** $\ln 4 \approx 1.3863$

14. $-1 + \sqrt{2} \approx 0.41421$

15. (a) Vertical asymptote: none, horizontal asymptotes:
  $y = 0$ (as $x \to -\infty$) and $y = 2$ (as $x \to \infty$).
 (b) Vertical asymptote: $x = \ln 2$, horizontal asymptotes:
  $y = 0$ (as $x \to -\infty$) and $y = -2$ (as $x \to \infty$).

16. 125 grams   **17.** $\dfrac{\ln 4}{0.07} \approx 19.8$ years

18. $300 \left(\frac{5}{3}\right)^3 \approx 1388.89$ dollars   **19.** 6.838 days

20. $10^{0.4} \approx 2.51$ times more severe

CHAPTER 3

Exercises 3.1

1. 2   **3.** C, D   **5.** $0, -\frac{3}{4}, 0$ **7.** C, B, A

9. (a) $y = x - \frac{1}{4}$ (b) $y = 0$ (c) $y = -6x - 9$

11. $y = 12x + 16$ **13.** $y = \frac{3}{4}x - \frac{1}{4}$ **15.** $(1, 1)$ and $(-1, -1)$

17. $(0, 0)$ **19.** $y = \frac{1}{6}x + \frac{3}{2}$ **21.** $\left(\frac{1}{16}, \frac{1}{4}\right)$ **23.** None

25. (a) $3 + h$ (b) 3   (c) $2x + 1$
 (d) (i) $y = 3x - 1$ (ii) $y = -3x - 4$ (iii) $y = -x - 1$

27. (a) $2x + 3$ (b) 5   (c) $y = -x - 5$

29. (a) $y = \dfrac{1}{2\sqrt{x + 4}}$
 (b) (i) $y = \frac{1}{4}x + 2$ (ii) $y = \frac{1}{2}x + \frac{5}{2}$ (iii) $y = \frac{1}{6}x + \frac{13}{6}$

31. (a) $3x^2 - 2$ (b) $y = -2x + 1$

33. (a) $0.02t$ (b) $t = 4$

35. (a) $36t^2$ **(b)** 900
   (c) The rate of grape production is 12 times greater.

37. (a) Slope of AB is 1, slope of AC is 2/3. Average rate over [2, 3] is 1, average rate over [2, 5] is 2/3. Instantaneous rate of change at 2 equals slope at A is $\approx 9/7$.
   (b) Slope of ED is $-3/2$, slope of EF is -1. Average rate over [6, 7] is $-3/2$, average rate over [7, 8] is -1. Instantaneous rate of change at 7 equals slope at E $\approx -3/2$.
   (c) Sales are increasing fastest when $t = 2$, decreasing fastest when $t = 7$. At $t = 5$ the rate changes from increasing to decreasing and equals zero there.

39. (a) $R = \begin{cases} 0.05i & \text{if } 0 \le i \le 6000 \\ 300 & \text{if } i > 6000 \end{cases}$

   (b)

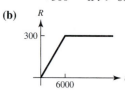

   Slope is 0.05 at $i = 4000$ and 0 at 8000.

   (c) 0.05 and 0

41. (a)

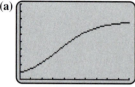

   (b)

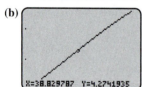

   (c) $\lim_{t\to\infty} p = \frac{4.8}{0.62} \approx 7.742$. The population asymptotically approaches an upper bound of about 7,742. The slope approaches zero.

Exercises 3.2

1. $\lim_{h\to 0} \frac{1}{h}\left(\frac{1}{3(x+h)} - \frac{1}{3x}\right) =$

$\lim_{h\to 0}\left(\frac{-1}{3x(x+h)}\right) = -\frac{1}{3x^2}$

3. $\lim_{h\to 0} \frac{1}{h}\left(\frac{1}{2-(x+h)} - \frac{1}{2-x}\right) =$

$\lim_{h\to 0}\left(\frac{1}{(2-x-h)(2-x)}\right) = \frac{1}{(2-x)^2}$

5. $\frac{5}{3}x^{2/3}$ **7.** $(1.4)x^{0.4}$

9. $-\dfrac{1}{2x^{3/2}}$ **11.** $-4x^{-5}$

13. $-\frac{1}{4}x^{-5/4}$ **15.** -1

17. $-0.4,\ -1.1,\ -1.4$

19. -48 **21.** $y = 12x - 16$ **23.** $(64, 16)$

25. (a) $\dfrac{0.46}{\sqrt{t}}$ **(b)** $m'(5) \approx 0.2057$ and $m'(50) \approx 0.06505$. The rate decreases over time.

27. all -1 **29.** all -2 **31.** $15t^4 + 6t^2 - 5$

33. $\frac{3}{2}(x^2 - x^{-4})$ **35.** $2s + 6s^{-3} + \frac{5}{2}s^{-1/2}$

37. $\dfrac{1}{4\sqrt{u}} - \dfrac{1}{u^{3/2}}$ **39.** $z^2 - \dfrac{z}{2} + \dfrac{1}{2}$

41. $y = -4x - 6$ **43.** $y = 7x - 11$

45. $(2, -1)$ **47.** $(1/2, 5/8)$ and $(-1/2, -5/8)$

49. $D'(1) = 4.476$, $D'(3) = 1.244$, $D'(7) = 46.044$. Fastest at $m = 7$, slowest at $m = 3$.

51. $q'(k) = 5.4k^{-0.7}$, $q'(8) = (5.4)8^{-0.7} \approx 1.26$. The output is increasing at the rate of $1.26 per worker per dollar invested.

53. 14   **55.** See Table 3.2.55 below.

57.

$f'(8) = \frac{1}{3}8^{-2/3} = \frac{1}{12} \approx 0.08333$

59. (a)

   (b) $y = 3.296x - 0.296$

   (c)

Age	2	4	6	8	10	12	14	16
Rate	-0.27	-0.125	-0.08	-0.06	-0.055	-0.045	-0.014	-0.0005

Table 3.2.55

61.

(a) $t = 0$ **(b)** It decreases.   **(c)** 8.27

63. See Figure 3.2.63 below.

Exercises 3.3

1. (a) height = 240 ft., velocity = −32 ft/sec,
speed = 32 ft/sec   **(b)** $t = 1/2$ sec   **(c)** $t = 8$ sec

3. (a) −33.6, −32.16, −32.016 (all ft/sec)   **(b)** −32 ft/sec

5. (a) −3.32 meters/sec   **(b)** $\sqrt{100/0.83} \approx 11$ seconds
(c) $-1.66\sqrt{100/0.83} \approx -18.22$, 18.22 (both meters/sec)

7. (a) $H'(1) = -9.28$, $H'(3) = -145.92$ (both ft/sec)
(b) 99.49 mi/hr. Yes, it is within the guidelines.

9. (a) $A(30) = B(30) + 10$ **(b)** $B'(10) = A'(10) + 5$
(c) $A'(20) = 2B'(20)$ **(d)** $A''(1) = B''(1)$

11. (a) −32 ft/sec² **(b)** −32 ft/sec²

13. (a) At $t = 1$ velocity and speed are both decreasing.
(b) At $t = 2$ velocity is decreasing and speed is increasing.

15. (a) $dx/dt > 0$, $dx^2/dt^2 > 0$
(b) $dx/dt > 0$, $dx^2/dt^2 < 0$
(c) $dx/dt < 0$, $dx^2/dt^2 < 0$
(d) $dx/dt < 0$, $dx^2/dt^2 > 0$

17. a   **19.** b   **21.**

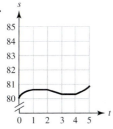

23. 2   **25.** $2t^{-3}$ **27.** 0   **29.** 44   **31.** $-\frac{2}{9}x^{-5/3}$

33. $60x^2 - 72x + 12$ **35.** $-\frac{15}{8}u^{-7/2}$

37. $MC = 4.5$, $MR = 10 - 0.02x$, $MP = 5.5 - 0.02x$

39. (a) $q = -2p + 5200$
(b) $R = -2p^2 + 5200p$, $MR = -4p + 5200$;
(c) $C = -800p + 2,080,000$, $MC = -800$;
(d) $MP = -4p + 6000$;
(e) Increase because $MP(1050) > 0$.
(f) Decrease because $MP(1550) < 0$

41. $A'(2) = -0.06$. Calcium is being eliminated at a rate of
0.06 mg per hour.

43. (a) $V'_A(1) = -2.4$
(b) $V'_A(2) = 3$
(c) $V'_B(3) = 1.1 + V'_A(3)$
(d) $V'_B(4) = 2V'_A(4)$

45. (a) $C(2) = 50$
(b) $C'(2) = -3.5$
(c) $C'(t) = -kC(t)$

47.

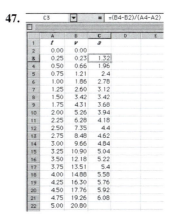

49. See Figure 3.3.49 below.

C3			=	=(D2-B2)/(D1-B1)														
							3.2.63											
	A	B	C	D	E	F	G	H	I	J	K	L	M	N	O	P	Q	R
1	**x**	1.92	1.93	1.94	1.95	1.96	1.97	1.98	1.99	2	2.01	2.02	2.03	2.04	2.05	2.06	2.07	2.08
2	**f(x)**	3.337	3.3722	3.4075	3.4432	3.4792	3.5156	3.5523	3.5894	3.6269	3.6647	3.7028	3.7414	3.7803	3.8196	3.85926	3.89932	3.9398
3	**f'(x)**		3.515	3.55	3.585	3.62	3.655	3.69	3.73	3.765	3.795	3.835	3.875	3.91	3.948	3.986	4.0255	
4																		

Figure 3.2.63

C3			=	=(D2-B2)/(D1-B1)														
								3.3.49										
A	B	C	D	E	F	G	H	I	J	K	L	M	N	O	P	Q	R
x	−1.6	−1.4	−1.2	−1	−0.8	−0.6	−0.4	−0.2	0	0.2	0.4	0.6	0.8	1	1.2	1.4	1.6
y	0.1109	0.1497	0.1942	0.2420	0.2897	0.3332	0.3683	0.3910	0.3989	0.3910	0.3683	0.3332	0.2897	0.2420	0.1942	0.1497	0.1109
		0.2082	0.2306	0.2388	0.2281	0.1964	0.1445	0.0767	0.0000	−0.0767	−0.1445	−0.1964	−0.2281	−0.2388	−0.2306	−0.2082	

Figure 3.3.49

51. (a) 0.1758, −0.0324

(b) Sales are increasing after 3 months, decreasing after 17 months

(c) At $t \approx 6.87$

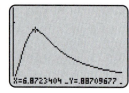

Exercises 3.4

1. (a) $\frac{37}{6}$ **(b)** $\frac{71}{12}$

3. (a) 20.125 **(b)** 1.9875 **(c)** 5.003

5. (a) 25.031468 **(b)** 0.28125 **(c)** 0.999946

7. 9.99 **9.** $\frac{619}{1200}$ **11.** 0.0098 **13.** $\frac{323}{108} \approx 2.99$ **15.** 0.997

17. $f(3) = 2.5$ and $f'(3) \approx \frac{4-2.5}{4-3} = 1.5$. Therefore $f(2.9) \approx 2.5 + 1.5 \cdot (-0.1) = 2.35$, and $f(2.99) \approx 2.5 + 1.5 \cdot (-0.01) = 2.485$. The second approximation should be more accurate because x is closer to a in that case. We would guess the approximations are smaller than the true values because the tangent line lies below the graph.

19. $f(5) = 6$ and $f'(5) \approx \frac{6-4}{5-2} = \frac{2}{3}$. Therefore $f(4.9) = 6 + \frac{2}{3} \cdot (-0.1) \approx 5.933$, and $f(4.95) = 6 + \frac{2}{3} \cdot (-0.05) \approx 5.967$. The second approximation should be more accurate because x is closer to a in that case. We would guess the approximations are greater than the true values because the tangent line lies above the graph.

21. 8622

23. (a) 437.5 **(b)** 437.50 **(c)** Actual difference is 438.45

25. 39.4 ft/sec **27.** 0.01 in **29.** 6.1 billion

31. (a) 0.0001 **(b)** 0.000098 **(c)** 0.000099 **(d)** 0.05849

33. 0.0016

35.

	A	B	C
1	t	v	s
2	0.00	0.00	0
3	0.25	0.23	0
4	0.50	0.66	0.0575
5	0.75	1.21	0.2225
6	1.00	1.86	0.525
7	1.25	2.60	0.99
8	1.50	3.42	1.64
9	1.75	4.31	2.495
10	2.00	5.26	3.5725
11	2.25	6.28	4.8875
12	2.50	7.35	6.4575
13	2.75	8.48	8.295
14	3.00	9.66	10.415
15	3.25	10.90	12.83
16	3.50	12.18	15.555
17	3.75	13.51	18.6
18	4.00	14.88	21.9775
19	4.25	16.30	25.6975
20	4.50	17.76	29.7725
21	4.75	19.26	34.2125
22	5.00	20.80	39.0275
23			

37. $3.6 < x < 14$ **39.** Yes, $f(x)$ is continuous at $x = 0$ but not differentiable.

41. $f(x)$ is not continuous at $x = 0$, therefore not differentiable.

43. (a)

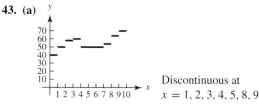

Discontinuous at $x = 1, 2, 3, 4, 5, 8, 9$

(b) zero

45. (a) $f(x) = \begin{cases} 0.07x, & \text{if } 0 < x \le 100{,}000 \\ 3000 + 0.04x & \text{if } x > 100{,}000 \end{cases}$

(b) Continuous for all $x > 0$.

(c) Not differentiable at $x = 100{,}000$

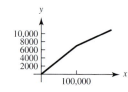

47. Not continuous, and therefore not differentiable, at $x = 2$, 6, 7. Continuous but not differentiable at $x = 0, 1, 3$, 4, 5. Differentiable, and therefore continuous, at $x = -2, -1, 8$.

Exercises 3.5

1. $2x + 4 + \dfrac{1}{x}$ **3.** $-\dfrac{1}{x}$ **5.** $\dfrac{x-1}{x^2}$ **7.** 2

9. $y = 3x - 2$ **11.** $(\frac{1}{2}, -\ln 2)$ **13.** $(\frac{1}{\sqrt{2}}, \frac{1}{2}(1 + \ln 2))$

15. $-\dfrac{1}{x}$ **17.** $\dfrac{1}{2x}$ **19.** $-\dfrac{3}{t}$

21. (a) $y = x - 1$

(b) See Table 3.5.21b on top of page A15.

(c)

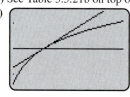

Tangent line is always above graph.

23. (a) With $\ln 2 \approx 0.69315$, linear approximation gives $\ln(2.01) \approx 0.69815$ and $\ln(1.9) \approx 0.64315$.

(b) $y = \frac{1}{2}x - 1 + \ln 2$

(c) Linear approximations are larger because the tangent line is always above the graph.

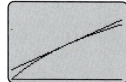

25. $2e^x + \dfrac{1}{2x^{3/2}}$ **27.** $\dfrac{e^t}{2}$ **29.** $2 + e^x$ **31.** $y = e(x - 1)$

33. $(\ln 3, 3)$ **35.** $(-\ln 2, 1 + \ln 2)$ **39.** $\frac{1}{3}e^{x/3}$ **41.** $3e^{3x+1}$

	$\ln(1.1)$	$\ln(1.2)$	$\ln(0.9)$	$\ln(0.8)$
Linear approximation	0.1	0.2	−0.1	−0.2
Calculator	0.09531	0.18232	−0.10536	−0.22314

Table 3.5.21b

43. $-e^{-t}$ **47.** $-\dfrac{\ln 2}{3}$ **49.** $\dfrac{2^x \ln 2}{3} - \dfrac{2}{3x^2}$ **51.** $-\dfrac{\ln 2}{2^t}$

53. $2^{1+t} \ln 2$

55. $P(10) = 1000(1.023)^{10} \approx 1255$,
$P'(10) = 1000(1.023)^{10} \ln(1.023) \approx 29$

57. **(a)** $-20e^{-0.2t}$ percent per day **(b)** 10 days

59. **(a)** Positive. As time increases the habit becomes stronger.
(b) $H'(t) = 0.1e^{-0.1t}$, which is always positive. **(c)** Use
$H = 1 - e^{-kt}$, where $0 < k < 0.1$ if the index increases
more slowly, and $k > 0.1$ if it increases faster.

61. **(a)** $H(t) = 40 + 25e^{-t \ln(5/3)/20}$
(b) $H(30) = 40 + 25e^{-3(\ln 5/3)/2} \approx 51.62$,
$H'(30) = \frac{5}{4} \ln \left(\frac{3}{5}\right) \left(\frac{3}{5}\right)^{3/2} \approx -0.3$ degrees/minute
(c) 63 minutes

63. **(a)** $H = 400 + Ae^{kt}$ with $A = -80e^{3/2}$ and $k = -3/40$
(b) $H(0) = 41.465$, $H'(0) = 26.89$
(c) The rate decreases.

65. **(a)** Approx. 1.364 degrees/hour.
(b) Approx. 0.03 degrees/hour

67. Approx. 0.564

69. **(a)** 0.8% **(b)** $dy/dt = -0.008y$, $y(0) = 2000$,
with t in years after 1999 and y is millions of hectares.
(c) $y = 2000e^{-0.008t}$

Exercises 3.6

1. $e^{-x}(1 - x)$ **3.** $\dfrac{1 - \ln x}{x^2}$ **5.** $e^{-3x} \left(\dfrac{1}{x} - 3 \ln x\right)$

7. $\dfrac{12x}{(x^2 + 4)^2}$ **9.** $\dfrac{e^x + 1 - xe^x}{(e^x + 1)^2}$ **11.** $\dfrac{2e^x}{(e^x + 1)^2}$

13. $x + 2x \ln x$ **15.** $\dfrac{x + 1 + \ln x}{(x + 1)^2}$ **17.** $5/18$

19. **(a)** $37/5$ **(b)** $5/27$ **21.** $y = 5x + 3$

23. $y = \dfrac{3}{4}x - \dfrac{e}{4}$

25. $y = 3ex - 2e$

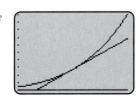

27. $x = -1$

29. **(a)** $R = 5000pe^{-0.01p}$ **(b)** $R'(p) = 50e^{-0.01p}(100 - p)$
(c) $R(200) = 1,000,000e^{-2}$, $R'(200) = -5000e^{-2}$
(d) decrease

31. $\dfrac{9w + 0.15w^2}{(9 + 0.3w)^2}$

33. **(a)** $e^{0.0035t}(0.94t + 0.001645t^2)$ **(b)** Yes

Exercises 3.7

1. $-\dfrac{e^{1/x}}{x^2}$ **3.** $2(e^x + e^{-x})(e^x - e^{-x})$ **5.** $\dfrac{e^x}{1 + e^x}$

7. $10(x^2 + x + 1)^9(2x + 1)$ **9.** $\dfrac{4x^3}{x^4 + 1}$

11. $(1 + x^4)^{1/2} + 2x^4(1 + x^4)^{-1/2}$ **13.** $-\dfrac{e^{-\sqrt{x}}}{2\sqrt{x}}$

15. $\ln(2x + 1) + \dfrac{2x}{2x + 1}$ **17.** $-\dfrac{2e^{2x} + e^x}{(e^{2x} + e^x + 1)^2}$

19. $\dfrac{4e^{2x}}{(e^{2x} + 1)^2}$ **21.** $\dfrac{2x + 3}{x^2 + 3x + 5}$ **23.** $\dfrac{1}{2(x + \sqrt{x})}$

25. $\dfrac{5x^2 + 2x}{\sqrt{2x + 1}}$

27. $y = \frac{2}{3}x + \frac{5}{3}$ **29.** $y = 2x - \frac{1}{2}$

31. $y = \dfrac{2}{\sqrt{15}}(x - 3) + \sqrt{15}$

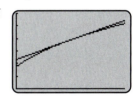

33. $x = \pm 1$ **35.** -3 **37.** $-3/4$, $y = -\frac{3}{4}x + 2$ **39.** -6

41. $3/2$ **43.** 4 **45.** $-3/10$ **47.** $-2/49$ **49.** 281.25

51. $v = 1 - \dfrac{2t}{\sqrt{2t^2 + 100}}$

Exercises 3.8

1. $y' = \dfrac{1}{3y^2}$ and $y' = \frac{1}{3}x^{-2/3}$

3. $y' = \dfrac{x}{y}$ and $\dfrac{x}{\sqrt{x^2 - 1}}$

5. $y' = 2xe^{-y}$ and $y' = \dfrac{2x}{x^2+1}$

7. $y' = -\dfrac{2x+y}{x+2y}$ **9.** $y' = \dfrac{2}{e^y + e^{-y}}$

11. $y' = \dfrac{2xy}{y+1}$

13. (a) $= -1/(2\sqrt{3})$ **(b)** $(-\sqrt{2}, 1/\sqrt{2}), (\sqrt{2}, -1/\sqrt{2})$

17. 0   **21.** $200e^{-0.8} \approx 90$ barrels per day

23. $2/5\pi \approx 0.13$ ft/min   **25.** $0.04\pi \approx 0.126$ cm/sec

27. $2\sqrt{10}/3 \approx 2.1$ ft/min

29. (a) $\dfrac{dA}{dt} = 2\pi r \dfrac{dr}{dt}$

    (b) $A = 9\pi$ cm^2, $\dfrac{dA}{dt} = 1.2\pi$ cm^2/sec

31. (a) 50 miles   **(b)** 50 mph

Chapter 3: Review Exercises

1. $\displaystyle\lim_{h\to 0}\dfrac{(x+h)^2 - 3(x+h) + 2 - (x^2 - 3x + 2)}{h} =$

    $\displaystyle\lim_{h\to 0}\dfrac{2xh + h^2 - 3h}{h} = 2x - 3$

3. $\displaystyle\lim_{h\to 0}\dfrac{1}{h}\left(\dfrac{2}{3(x+h)-1} - \dfrac{2}{3x-1}\right) =$

    $\displaystyle\lim_{h\to 0}\dfrac{1}{h}\left(\dfrac{-6h}{(3x+3h-1)(3x-1)}\right) = \dfrac{-6}{(3x-1)^2}$

5. $1/6$, $y = \frac{1}{6}x + \frac{3}{2}$ **7.** 3, $y = 3x - 1$

9. $1/4$, $y = \frac{1}{4}x + \frac{1}{4}$ **11.** 2, $y = 2x - 1$

13. $5x^4 - 12x^3 + 6x^2 + 2x - 5$ **15.** $(2x-3)e^{x^2-3x+1}$

17. $(3x+1)^{-2/3}$ **19.** $\dfrac{4(\ln x)^3}{x}$

21. $-e^{-x}$ **23.** $x^2(1 + 3\ln x)$ **25.** $\dfrac{3e^{3x}}{2\sqrt{e^{3x}+1}}$

27. 3   **29.** None   **31.** $4/9$ **33.** $-\dfrac{2}{x^2}$ **35.** $-(2x+1)^{-3/2}$

37. $4e^{2x}(1+x)$ **39. (a)** $11/2$ ft/sec   **(b)** $27/14$ ft/sec
    (c) 20 ft/sec

41. (a) $-400e^{-0.8}$ **(b)** $1000e^{-0.8}$ **43.** 22.5, no

45. (a) $\frac{99}{14} \approx 7.07$ **(b)** $\frac{82}{9} \approx 9.11$ **(c)** 9.9   **47.** 0.473

49. 0.005   **51.** About 6500 years ago

53. $3/2$ **55.** -30 **57.** $1/8$

59. (a) $-3, 3, 6$ **(b)** $-3, 0, 3, 5, 6$

61. $\dfrac{dy}{dx} = \dfrac{x(1+y^2)}{y}$ **63.** $\dfrac{dy}{dx} = \dfrac{2x + 2xy - e^y}{xe^y - x^2}$

73.

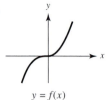

$y = f(x)$ Differentiable for all x

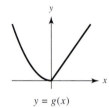

$y = g(x)$ Not differentiable at $x = 0$

75. (a) $\dfrac{1}{16\pi}$ inches per second   **(b)** 1 square inch per second

Chapter 3: Exam

1. c

2. $\displaystyle\lim_{h\to 0}\dfrac{1}{h}\left(\dfrac{1}{2(x+h)+5} - \dfrac{1}{2x+5}\right) =$

    $\displaystyle\lim_{h\to 0}\dfrac{-2}{(2x+2h+5)(2x+5)} = -\dfrac{2}{(2x+5)^2}$

3. $\dfrac{1}{\sqrt{x}} + \dfrac{9}{x^4}$ **4.** $y = -\frac{1}{16}x + \frac{3}{4}$

5. $(2, 23)$ and $(-2, -21)$

6. (a) $s'(0) = -1$, $s'(2) = -1/9$ **(b)** $-1/3$ **(c)** $1/4$

7. $-400e^{-4} \approx -7.3$

8. $100 + 300\ln(10) \approx 790.78$

9. $10 + \dfrac{h}{300}$ **10.** $\$30$ **11.** a

12. Not continuous at $x = 2, 4$. Not differentiable at $x = 1$,
    $2, 3, 4$.

13. Differentiable at $x = 0$.   **14.** $f'(x) = 2xe^{x^2+1}$

15. $y = x - 1$ **19.** $202.6°$C   **20.** -2

21. $\dfrac{32\pi}{3}$ cubic feet   **22.** -5

24. (b) 5 billion   **(c)** $\dfrac{10}{1 + e^{-3.3}} \approx 9.644$ billion
    (d) 10 billion

CHAPTER 4

Exercises 4.1

1. Critical point at $x = \frac{7}{4}$, increasing on $\left(\frac{7}{4}, \infty\right)$, decreasing
    on $\left(-\infty, \frac{7}{4}\right)$, local minimum at $x = \frac{7}{4}$.

3. Critical points at $x = -3$ and $x = 2$, increasing on $(-\infty, -3)$ and $(2, \infty)$, decreasing on $(-3, 2)$, local maximum at $x = -3$, local minimum at $x = 2$.

5. Critical points at $x = -2, 0, 2$, increasing on $(0, \infty)$, decreasing on $(-\infty, 0)$, local minimum at $x = 0$.

7. Critical point at $x = -1$, increasing on $(-1, \infty)$, decreasing on $(-\infty, -1)$, local minimum of at $x = -1$.

9. Critical point at $x = 0$, increasing on $(0, \infty)$, decreasing on $(-\infty, 0)$, local minimum of $f(x)$ at $x = 0$.

11. Critical point of $f(x)$ at $x = 1$, increasing on $(-\infty, 1)$, decreasing on $(1, \infty)$, local maximum of $f(x)$ at $x = 1$.

13. Critical point of $f(x)$ at $x = 1$, increasing on $(1, \infty)$, decreasing on $(-\infty, 1)$, local minimum $x = 1$.

15. No critical point, increasing on $(-\infty, \infty)$.

17. Critical point of $f(x)$ at $x = 1$, increasing on $(0, 1)$, decreasing on $(1, \infty)$, local maximum at $x = 1$.

19. Critical point at $x = 0$, increasing on $(0, \infty)$, decreasing on $(-\infty, 0)$, local minimum at $x = 0$.

21. (a) $y = 10$ (b) Positive in $(0, 10)$, negative in $(10, 20)$. (c) Most amount of rain when $y = 10$, least when $y = 0$ or $y = 20$.

23. The purchases will not reach a maximum because $S'(t) = \frac{1}{3}t^{-2/3} > 0$ for $t > 0$, which means that $S(t)$ is always increasing.

25. Decrease

27. (a)

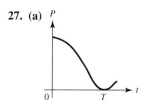

(b) Negative in $(0, T)$, positive in (T, ∞).
(c) Maximum at $t = 0$, minimum at $t = T$.

29. b **31.** c

33. (a) $(-3, -1)$ and $(1, 3)$ (b) $(-1, 1)$
(c) Two points $x = 1$ and $x = -1$
(d) $x = -1$ (e) $x = 1$

35.

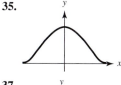

37.

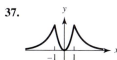

39.

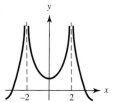

41. c **43.** d

45.

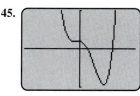

Local maximum at $x \approx 0.5387$, local minimum at $x \approx 2.361$ (using equation solver).

47. Global maximum at $x = \frac{1}{3}$ with value $f\left(\frac{1}{3}\right) = \frac{16}{3}$.

49. Global maximum at $x = 1$ with value $f(1) = 1/e$.

51. Global minimum at $x = 0$ with $f(0) = 0$.

53. Global maximum at $x = 0$ with value $f(0) = 1$.

55. Global minimum at $x = 1$ with value $f(1) = 0$.

57. Global minimum at $x = 0$ with value $f(0) = -1$.

59. (b) Since $f'(x) = 5x^4 + 1 \geq 1$, $f(x)$ has no critical points and $f(x)$ is always increasing.

61. Global minimum of $f(x)$ at $x = 0$ with value $f(0) = 1$.

63. $\dfrac{dp}{dt} = 0.03p\left(1 - \dfrac{p}{12}\right) > 0$ if $0 < p < 12$. No maximum value.

Exercises 4.2

1. B **3.** A **5.** (a), (c), (f), (g), (k) **7.** (b), (f), (i) **9.** (f)

11. (i) **13.** (a), (g) **15.** (c) **17.** (h)

19. Concave down on $(-\infty, -1)$, concave up on $(-1, \infty)$, inflection point at $(-1, 10)$.

21. Concave up on $(-\infty, -1)$ and $(4, \infty)$, concave down on $(-1, 4)$, inflection points at $\left(-1, -\frac{65}{6}\right)$ and $\left(4, -\frac{235}{3}\right)$.

23. Concave up on $(-\infty, 2)$ and $(2, \infty)$, no inflection point.

25. Concave down on $(-\infty, 1)$, concave up on $(1, \infty)$, no inflection point.

27. Concave up on $(0, e)$, concave down on (e, ∞), inflection point at $(e, 1)$.

29. Concave down on $(-\infty, -1)$ and $(1, \infty)$, concave up on $(-1, 1)$, inflection points at $\left(-1, \frac{1}{4}\right)$ and $\left(1, \frac{1}{4}\right)$.

31. Concave up on $(-\infty, -\frac{2}{\sqrt{3}})$ and $(\frac{2}{\sqrt{3}}, \infty)$, concave down in $(-\frac{2}{\sqrt{3}}, \frac{2}{\sqrt{3}})$, inflection points at $(-\frac{2}{\sqrt{3}}, \frac{3}{16})$ and $(\frac{2}{\sqrt{3}}, \frac{3}{16})$.

33. Local maximum at $x = -\sqrt{2}$ with value $f(-\sqrt{2}) = 1 + 4\sqrt{2}$ and local minimum at $x = \sqrt{2}$ with value $f(\sqrt{2}) = 1 - 4\sqrt{2}$.

35. Local minimum at $x = -\frac{1}{3}$ with value $f\left(-\frac{1}{3}\right) = -\frac{1}{3}e^{-1}$.

37. Local minimum at $x = 2$ with value $f(2) = 4$ and local maximum at $x = -2$ with value $f(-2) = -4$.

39. Local maximum at $x = 1$ with value $f(1) = e$.

41. Local minimum at $x = -1$ with value $f(-1) = 2$, local maximum at $x = 0$ with value $f(0) = 3$, and local minimum at $x = 1$ with value $f(1) = 2$.

43. Local minimum at $x = \frac{1}{4}$ with value $f\left(\frac{1}{4}\right) = -\frac{1}{4}$.

45. On the tenth day.

47. Critical points at $x = 0$ and $x = -1$. Local minimum at $x = 0$ with value $f(0) = -1$. No local extremum at $x = -1$.

49. Critical points $x = -1$, $x = 0$, and $x = 1$. Local maximum at $x = 0$, with value $f(0) = 1$. Local minimum of $f(x)$ at $x = -1$ with value $f(-1) = 0$. Local minimum of $f(x)$ at $x = 1$ with value $f(1) = 0$.

51. **(a)** The graph of the total enrollment is concave down. **(b)** Change *smaller* to *larger*.

53. **(a)** Employment

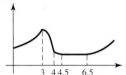

(b) Local minimum at $t = 0$ and throughout the interval $[4.5, 6.5]$. Local maximum at $t = 3$. Derivative is positive for $t < 3$ and $t > 6.5$, negative for $3 < t < 4$ and $4 < t < 4.5$, zero for $4.5 < t < 6.5$.
(c) Concave up for $t < 3$ and $4 < t < 4.5$ and $t > 6.5$, with second derivative positive. Concave down for $3 < t < 4$, with second derivative negative.
(d) Inflection points at $t = 3$, $t = 4$.
(e) *ever-increasing rate* and *more and more* and *slowed down the layoff rate*.

55.
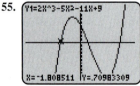
Concave down on $(-\infty, -1.83)$ and $(0.67, 3.67)$, concave up on $(-1.83, 0.67)$ and $(3.67, \infty)$. Inflection points at $x = -1.83$, 0.67, and 3.67. (All numbers rounded to 2 decimal places.)

57.
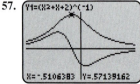
(a) $x = -1/2$, where $f(x)$ has a local (and global) maximum. **(b)** $x \approx -1.26$, where $f(x)$ has an inflection

point and changes from concave up to concave down.
(c) $x \approx 0.26$, where $f(x)$ has an inflection point and changes from concave down to concave up.

59.
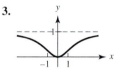

Exercises 4.3

1.
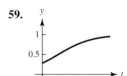

3.

5.
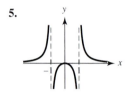

7.
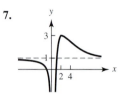

9. Symmetry: none.
Asymptotes: none.
Local maximum at $(-1, 7)$.
Local minimum at $(1, 3)$.
Inflection point at $(0, 5)$
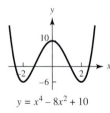
$y = x^3 - 3x + 5$

11. Symmetry: about y axis.
Asymptotes: none.
Local maximum at $(0, 10)$.
Global minimum at $(-2, -6)$ and $(2, -6)$.
Inflection points at $\left(-\frac{2}{\sqrt{3}}, \frac{10}{9}\right)$ and $\left(\frac{2}{\sqrt{3}}, \frac{10}{9}\right)$.
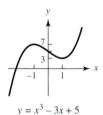
$y = x^4 - 8x^2 + 10$

13. Symmetry: about $x = 2$.
Vertical asymptote: $x = 2$.
Horizontal asymptote: $y = 0$.
No local extrema.
No inflection points.
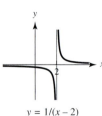
$y = 1/(x - 2)$

15. Vertical asymptote: $x = -1/2$.
Horizontal asymptote: $y = 1/2$.
No local extrema.
No inflection points.
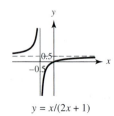
$y = x/(2x + 1)$

17. Symmetry: about $(0, 0)$.
Vertical asymptote: $x = 0$.
Slant asymptote: $y = x$.
Local minimum at $(-2, -4)$.
Local maximum at $(2, 4)$.
No inflection points.

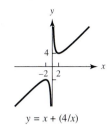

$y = x + (4/x)$

(c) $\lim\limits_{x \to \infty} f(x) = \infty$, $\lim\limits_{x \to -\infty} f(x) = -\infty$

(d)

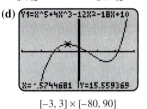

$[-3, 3] \times [-80, 90]$

19. Symmetry: about $(0, 0)$.
Vertical asymptote: none.
Horizontal asymptote: $y = 0$.
Global maximum at $(2, \frac{1}{4})$.
Global minimum at $(-2, -\frac{1}{4})$.
Inflection points at
$(-2\sqrt{3}, -\frac{\sqrt{3}}{8})$, $(2\sqrt{3}, \frac{\sqrt{3}}{8})$, $(0, 0)$.

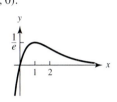

$y = x/(x^2 + 4)$

21. Symmetry: none.
Vertical asymptote: none.
Horizontal asymptote: $y = 0$.
Global maximum at $(1, e^{-1})$.
No local minimum.
Inflection point at $(2, 2e^{-2})$.

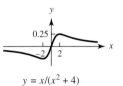

$y = xe^{-x}$

23. Symmetry: about y axis.
Asymptotes: none.
Global minimum at $(0, 2)$.
No local maximum.
No inflection points.

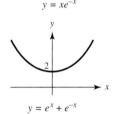

$y = e^x + e^{-x}$

25. Symmetry: about y axis.
Asymptotes: none.
Global minimum at $(0, 2\ln 2)$.
No local maximum.
Inflection points at $(-2, 3\ln 2)$
and $(2, 3\ln 2)$.

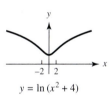

$y = \ln(x^2 + 4)$

27. $f'(x) = 5x^4 + 12x^2 - 24x - 18$ and
$f''(x) = 20x^3 + 24x - 24$
(a) Since $f''(0) < 0$ and $f''(1) > 0$, there is an inflection point between $x = 0$ and $x = 1$. And since $d/dx\, f''(x) = 60x^2 + 24 > 0$ for all x, $f''(x)$ is increasing and cannot cross the x-axis more than once. Therefore, there is exactly one inflection point. Using the equation solver with initial guess 0.5 gives the approximate solution $x_1 \approx 0.7063$ to $f''(x) = 0$.
(b) Since $f''(x) < 0$ on $(-\infty, x_1)$, with x_1 as in (a), $f'(x)$ is decreasing throughout that interval and can have at most one root there. Similarly, $f'(x)$ is increasing on (x_1, ∞) and can have at most one root there. Moreover, $f'(x_1) \neq 0$. Therefore, $f'(x) = 0$ has at most two solutions. Using the equation solver with initial guess 0 gives the approximate solution $x \approx -0.567$, and using it with initial guess 1 gives $x \approx 1.52$.

29. The infection is spreading fastest when 80 trees are infected, since $R(x)$ takes a maximum value of $R(80) \approx 164$ at $x = 80$.

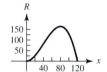

31. (a)

P

(b) No, the greatest depth at which the plant can survive is about 6.9 feet.

33. (a)

N

1827

$-10 \quad 10$

s

(b) The score achieved by the largest number of students was the same as the average score.

Exercises 4.4

1. (a) Maximum value of 1 at $x = 0$, minimum value of -3 at $x = -1$.
(b) Maximum value of 1 at $x = 3$, minimum value of -3 at $x = 2$.
(c) Maximum value of 1 at $x = 0$ and $x = 3$, minimum value of -19 at $x = -2$.

3. Maximum value of 9 at $x = 2$, minimum value of 3 at $x = 0$.

5. Maximum value of $1/4$ at $x = 2$, minimum value of $1/5$ at $x = 1$.

7. Maximum value of 6561 at $x = 0$, minimum value of 0 at $x = 3/2$.

9. Maximum value of $\sqrt{2}$ at $x = 0$ and $x = 2$, minimum value of 1 at $x = 1$.

11. Global maximum at $x = 3$ with value 0, global minimum at $x = 1$ with value -4.

13. Global minimum at $x = 2$ with value 4, no global maximum.

15. Global maximum at $x = 1/2$ with value $1/(2e)$, no global minimum.

17. Global minimum at $x = -1$ with value -3, global maximum at $x = 1$ with value 1.

19. Global minimum at $x = e^{-1/2}$ with value $-1/(2e)$, no global maximum.

21. Global minimum at $x = 2$ with value $e^2/4$, no global maximum.

23. Global minimum at $x = \sqrt{3}$ and $x = -\sqrt{3}$ with value -9, global maximum at $x = 0$ with value 0.

25. Global minimum at $x = \ln 3$ with value -9, no global maximum.

27. Global minimum at $x = 0$ with value -1, no global maximum.

29. 5 seconds, 600 feet   **31.** 19, minimum value 690

Exercises 4.5

1. $400   **3.** 40

5. (a) $q = -2p + 66$ (b) $16.50   (c) $18.50

7.

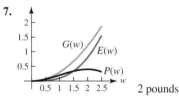

2 pounds

9. (a) $1 - e^{-0.032w} - 0.01w$
(b) $31.25 \ln (3.2) \approx 36.35$ inches

11. Minimum: 4 million, maximum: 164 million.

13. 10 inches wide and 15 inches long.

15. 80 by 32 feet.

17. (a) $r = ky(M - y)$, $0 \le y \le M$
(b) $r'(y) > 0$ if $y < M/2$ and $r'(y) < 0$ if $y > M/2$ and $r'(M/2) = 0$. The disease is spreading most rapidly when exactly half the community is infected.

19. $15\sqrt{14} \approx 56$ mph

21. 5 orders per year of 54,000 boxes each.

23. $(\frac{1}{2}, \frac{1}{\sqrt{2}})$ **25.** $60   **27.** 1

29. (a) $x(30 - 2x)^2$, $0 \le x \le 15$
(b) $x = 5$ inches, maximum volume equals 2000 cubic inches.

Exercises 4.6

Most answers in this section are approximations.

1. Local minima at $(-1.25723, 10.92465)$ and $(1.38725, -19.59187)$, local maximum at $(-0.43002, 15.13373)$. Inflection points at $(0.68103, -4.85644)$ and $(-0.88102, 12.888)$. The x-intercepts are at $x = 0.519303$ and 1.93300.

3. Local minimum at $(0.96242, -0.65120)$, local maximum at $(-0.96242, 0.65120)$. The x-intercepts are at $x = -1.62213$, $x = 0$, and $x = 1.62213$.

5. Local maximum: $(0.12808, -0.87297)$
Local minimum: $(5.20526, -66.31221)$
Inflection point: $(2.66667, -33.59359)$
x-intercept: $x = 7.758841$

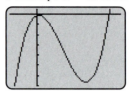

$[-3, 9] \times [-70, 5]$

7. Local maximum: $(0.57590, 3.94630)$
Local minimum: $(-2.13805, -17.28567)$ and $(0.812150, 3.92140)$.
Inflection points: $(-1.19648, -8.60727)$ and $(0.69648, 3.93366)$
x-intercepts: $x = -3$ (exact) and $x = -0.46557$

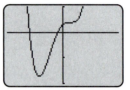

$[-5, 5] \times [-20, 10]$

9. Local maximum: $(-1.93045, -19.86606)$ and $(1.03603, 1.01796)$
Local minimum: $(-1.03603, -25.01796)$ and $(1.93045, -4.13394)$.
Inflection points: $(0, -12)$ (exact), $(-1.54919, -22.16007)$ and $(1.54919, -1.83993)$
x-intercepts: $x = 0.76756485$, $x = 1.32116$ and $x = 2.26633$

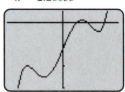

$[-3, 3] \times [-30, 5]$

11. Local minimum: $(1.44364, -1.30241)$
No local maximum, no inflection point
x-intercepts: $x = 0.58939$ and $x = 2.12680$

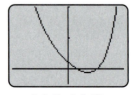

$[-4, 4] \times [-5, 20]$

13. Local minimum: $(0, 0.5)$ (exact)
Local maxima: $(\pm 0.91018, 0.60355)$
Inflection points: $(\pm 0.52, 0.56)$ and $(\pm 1.55, 0.45)$
No x-intercept.

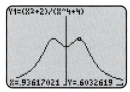

$[-4, 4] \times [0, 1]$

15. Local minimum: $(0, 0)$ (exact)
x-intercept: $x = 0$
Local maxima: $(\pm \sqrt{e - 1}, 0.36788) \approx$
$(\pm 1.3108325, 0.36788)$
Inflection points: $(\pm 2.13986, 0.30812)$ and
$(\pm 0.42999, 0.14318)$

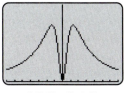

$[-7, 7] \times [0, 0.5]$

17. Local minimum: $(-0.453156, -0.58561)$
Local maximum: $(4.0822, 4.5906)$
Inflection points: $(0, -1/2)$ (exact), $(2.08851, 1.98680)$,
$(6.04255, 3.15703)$, and $(0.02120, -0.49470)$
x-intercepts: $x = \pm 1$ (exact)

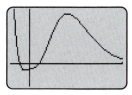

$[-2, 10] \times [-2, 5]$

19. Vertical asymptote: $x = 1.79632$
Horizontal asymptote: $y = 0$ (exact)
Local maximum: $(-1.2599, 0.53160)$
Inflection points: $(0.40324, -0.18592)$,
$(-0.41368, 0.22623)$, and $(-1.91275, 0.42107)$
x-intercept: $x = 0$ (exact)

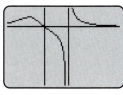

$[-3, 6] \times [-3, 1]$

21. Vertical asymptotes: $x = \pm 2$ (exact)
Horizontal asymptote: $y = 0$ (exact)
No local maximum or minimum.

Inflection point: $(0, 0)$ (exact)
x-intercept: $x = 0$ (exact)

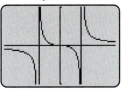

$[-5, 5] \times [-10, 10]$

23. (a)

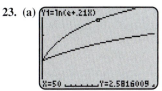

The efforts appear to be successful.

(b)

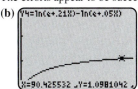

$$\lim_{t \to \infty} |D_p(t) - D_a(t)| = \lim_{t \to \infty} \ln \left(\frac{e + 0.21t}{e + 0.05t} \right) = \ln 4.2 \approx 1.435$$

Chapter 4: Review Exercises

1. Decreasing on $(-\infty, 1)$, increasing on $(1, \infty)$. Local minimum at $x = 1$ with value -12.

3. Increasing on $(-\infty, -\frac{1}{3})$ and $(3, \infty)$, decreasing on $(-\frac{1}{3}, 3)$. Local maximum at $x = -\frac{1}{3}$ with value $-121/27$, local minimum at $x = 3$ with value -23.

5. Decreasing on $(-\infty, -2)$ and $(0, \infty)$, increasing on $(-2, 0)$. Local minimum at $x = -2$ with value $-1/4$.

7. Decreasing on $(-\infty, 0)$ and $(\frac{2}{3}, \infty)$, increasing on $(0, \frac{2}{3})$. Local minimum at $x = 0$ with value 0, local maximum at $x = 2/3$ with value $\frac{4}{9}e^{-2}$.

9. Decreasing on $(-\infty, \frac{1}{3} \ln 2)$, increasing on $(\frac{1}{3} \ln 2, \infty)$. Local minimum at $x = \frac{1}{3} \ln 2$ with value $3/2^{2/3}$.

11. Increasing on $(-\infty, -\frac{\sqrt{15}}{5})$ and $(\frac{\sqrt{15}}{5}, \infty)$, decreasing on $(-\frac{\sqrt{15}}{5}, \frac{\sqrt{15}}{5})$. Local maximum at $-\frac{\sqrt{15}}{5}$ with value $\frac{\sqrt{15}}{5}(\frac{2}{5})^{1/3}$, local minimum at $\frac{\sqrt{15}}{5}$ with value $-\frac{\sqrt{15}}{5}(\frac{2}{5})^{1/3}$.

13. Concave up on $(-\infty, -1)$ and $(1, \infty)$, concave down on $(-1, 1)$. Inflection points at $(-1, 7)$ and $(1, -13)$.

15. Concave up for all x.

17. Concave down on $(0, \frac{1}{2})$, concave up on $(\frac{1}{2}, \infty)$. Inflection point at $(\frac{1}{2}, \frac{1}{4} + \frac{1}{2} \ln 2)$.

19. Concave down on $(-\infty, -\sqrt{1.5})$ and $(0, \sqrt{1.5})$, concave up on $(-\sqrt{1.5}, 0)$ and $(\sqrt{1.5}, \infty)$. Inflection points at $(-\sqrt{1.5}, -\sqrt{1.5}\,e^{-1.5})$, $(0, 0)$, and $(\sqrt{1.5}, \sqrt{1.5}\,e^{-1.5})$.

21. Concave up on $(-\infty, -1)$ and $(0, \infty)$, concave down on $(-1, 0)$. Inflection points at $(-1, 1)$ and $(0, 1)$.

23. d

25. (a) Decreasing on $(-\infty, 0)$, increasing on $(0, \infty)$.
(b) Concave up on $(-\infty, \infty)$.

27. d **29.** f

31. Critical point at $x = 1$, which is a local minimum because $f''(1) = 1 > 0$.

33. Critical point at $x = \sqrt{e}$, which is a local maximum because $f''(\sqrt{e}) = -2/e < 0$.

35.

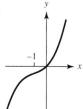

37. Domain: $(-\infty, \infty)$
Asymptotes: none
Increasing on $(-\infty, -\frac{2}{3})$ and $(\frac{2}{3}, \infty)$
Decreasing on $(-\frac{2}{3}, \frac{2}{3})$
Local max at $x = -2/3$, local min at $x = 2/3$
Concave down on $(-\infty, 0)$, up on $(0, \infty)$
Inflection point: $(0, 4)$

$y = 9x^3 - 12x + 4$

39. Domain: $x \neq \pm 2$, symmetric about y-axis
Asymptotes: vertical $x = \pm 2$, horizontal $y = 1$
Increasing on $(-\infty, -2)$ and $(-2, 0)$
Decreasing on $(0, 2)$ and $(2, \infty)$
Local max at $x = 0$
Concave down on $(-2, 2)$, up on $(-\infty, -2)$ and $(2, \infty)$
No inflection point

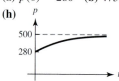

$y = x^2/(x^2 - 4)$

41. Domain: $(-\infty, \infty)$, symmetric about $(0, 0)$
Asymptotes: none
Increasing on $(-\infty, -27)$ and $(27, \infty)$
Decreasing on $(-27, 27)$
Local max at $x = -27$, local min at $x = 27$
Concave down on $(-\infty, 0)$, up on $(0, \infty)$
Inflection point: $(0, 0)$

$y = x^{5/3} - 15x$

43. No global max or min.

45. Global min at $x = e^{-1}$ with value $-e^{-1}$.

47. (a) Global max at $x = 1$, global min at $x = 0$.
(b) Global max at $x = 1$, global min at $x = -1$.
(c) Global max at $x = 4$, global min at $x = 3$.

49. \$9.50 **51.** $x = 60$, $y = 40$ **53.** $\left(\frac{1}{10}, \frac{3}{10}\right)$

Chapter 4: Exam

1. $x = -2$ (local min), $x = 0$ (local max), $x = 2$ (local min)

2. (a) Increasing on $(-\infty, \infty)$.
(b) Increasing on $(-\infty, -\frac{2}{3})$ and $(0, \infty)$, decreasing on $(-\frac{2}{3}, 0)$.

3. Global maximum at $x = 3$ with value $27e^{-3}$.

4. Concave down on $(0, e^{-3/2})$, up on $(e^{-3/2}, \infty)$

5. e **6.** $(\sqrt{2}, -13 + 2\sqrt{2})$ and $(-\sqrt{2}, -13 - 2\sqrt{2})$

7.

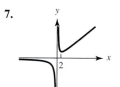

8.

9. (a) $2e \approx 5.44$ (b) $e^2 \approx 7.39$

10. One. By the intermediate value theorem, there is at least one solution, and, since $f(x)$ is decreasing, there cannot be more than one.

11. 400 **12.** 30 by 20 yards

17. (a) $p(0) = 280$ (b) 475.3 million (c) 800 (g) 4
(h)

18. $\frac{4 + 2\sqrt{229}}{3} \approx 11.42$ hundred units

CHAPTER 5

Exercises 5.1

1. $P(x) = -0.4x^2 + 2000x - 44{,}640$

3. $\frac{2}{3}x^{3/2}$ **5.** $\frac{1}{2}s^2 - \ln|s|$ **7.** $\frac{1}{6}x^6 - \frac{3}{4}x^{4/3} - \frac{1}{3}x^{-3}$

9. $\frac{1}{3}\pi r^3 - \pi r^2 + c$ **15.** $A = 1/5$ **17.** $A = \frac{1}{2}$ **19.** $A = \frac{1}{12}$

21. $A = -1$ **23.** $A = \frac{1}{5}$, $B = -\frac{1}{5}$

25. $2x - 2$, $2x - 1$, $2x$, $2x + 1$, $2x + 2$

27. $\frac{1}{4}x^2 - 2$, $\frac{1}{4}x^2 - 1$, $\frac{1}{4}x^2$, $\frac{1}{4}x^2 + 1$, $\frac{1}{4}x^2 + 2$

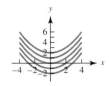

29. $-2e^{-x}$, $-2e^{-x} + 1$, $-2e^{-x} + 2$, $-2e^{-x} + 3$, $-2e^{-x} + 4$

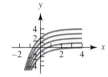

31. $F_1(1) = 0$, $F_2(1) = -1$, and $F_1(3) - F_2(3) = 1$

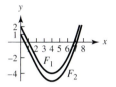

33. (a) dw/dt

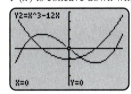

(b) dw/dt is not continuous, while $w(t)$ is.

(c), (d)

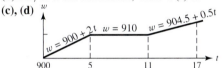

35. $(-1/x^2) + c$ **37.** $\frac{1}{3}e^{3t} + c$ **39.** $\frac{1}{2}(e^x - e^{-x}) + c$

41. $\frac{5}{6}y^6 + \frac{1}{2}y^2 + c$

43. $F(x) = x^3 - 12x$

$F(x)$ is decreasing when $f(x)$ is negative.
$F(x)$ is increasing when $f(x)$ is positive.
$F(x)$ is concave up when $f(x)$ is increasing.
$F(x)$ is concave down when $f(x)$ is decreasing.

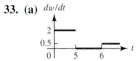

45. $F(x) = \frac{3}{4}x^{4/3} - \frac{1}{4}x^4$

$F(x)$ is decreasing when $f(x)$ is negative.
$F(x)$ is increasing when $f(x)$ is positive.
$F(x)$ is concave up when $f(x)$ is increasing.
$F(x)$ is concave down when $f(x)$ is decreasing.

47. $\frac{1}{5}x^5 - \frac{12}{5}x^{5/3} + c$ **49.** $-\frac{100}{0.06}e^{-0.06t} + 350t + c$

51. $x^2 + x - 1$ **53.** $\frac{1}{2}x^2 - e^{-x} + 2$

55. $12.5e^{0.08t} + 79987.5$

57. $y = 2x^3 - 2x^2 + x + c$, $c = -2, -1, 0, 1, 2$

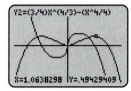

59. $y = \frac{1}{2}x^2 + \ln(x^2) + c$, $c = -\frac{1}{2}, \frac{1}{2}, \frac{3}{2}, \frac{5}{2}, \frac{7}{2}$

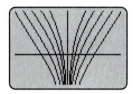

61. (a) $P(x) = 15x - 0.005x^2 - 1000$ (b) 9000   (c) 10,250

63. $(1/0.03)\ln(5/4) \approx 7.438$ **65.** 7/4

67. (a) $y(t) = \int -0.005t\, dt$,   (b) $y(0) = 1$, (c) 49%   (d) 20 years

69. $J(t) = e^t - 0.9t + 7.5t^2 + 35$ **71.** 3,060   **73.** 40 ft/sec

75. (a) $dV/dt = -20e^{-0.2t}$, $V(0) = 100$, $V(t) = 100e^{-0.2t}$

79. See Figure 5.1.79 below.

81. See Figure 5.1.81 below.

03	▼	= =C3+(0.25)*C2											
								acceleration					
A	B	C	D	E	F	G	H	I	J	K	L	M	N
t	0	0.25	0.5	0.75	1	1.25	1.5	1.75	2	2.25	2.5	2.75	3
a	0.00	0.34	0.76	1.30	2.04	2.92	4.15	5.80	8.12	10.80	14.26	16.82	18.05
v	0.00	0.00	0.09	0.28	0.60	1.11	1.84	2.88	4.33	6.36	9.06	12.62	15.83

Figure 5.1.79

A	B	C	D	E	F	G	H	I	J	K	L
x	0.000	0.100	0.200	0.300	0.400	0.500	0.600	0.700	0.800	0.900	1.000
Estimated	0.000	0.100	0.197	0.290	0.381	0.468	0.553	0.635	0.714	0.791	0.865
Actual to 3 decimal places	0.000	0.098	0.193	0.285	0.374	0.461	0.544	0.624	0.702	0.778	0.850

Figure 5.1.81

Exercises 5.2

1. $\ln(x^2 + 1) + c$ **3.** $-e^{-t} + c$ **5.** $\frac{2}{9}(3x + 4)^{3/2} + c$

7. $\frac{1}{4}[\ln(x^2 + 1)]^2 + c$ **9.** $4\sqrt{x} - x - 4\ln(1 + \sqrt{x}) + c$

11. (a) $2\sqrt{x} - 2\ln(\sqrt{x} + 1) + c$
    (b) The answers can differ by a constant.

13. No, since $du = e^t dt$ **15.** $2(x^2 + 4)^{1/2} + c$

17. $\frac{1}{12}(x^2 - 3)^6 + c$ **19.** $\dfrac{-1}{2(x^2 + x + 1)^2} + c$

21. $-2.5e^{-0.1t^4} + c$ **23.** $\frac{3}{2}\ln|2x + 1| + c$

25. $(\ln|t|)^2 + c$ **27.** $\ln|\ln|t|| + c$

29. $x - 2\sqrt{x} + 2\ln|1 + \sqrt{x}| + c$

31. $y(t) = \frac{1}{6}(4t + 1)^{3/2} + \frac{17}{6}$ **33.** $r(x) = \ln\dfrac{e^x + e^{-x}}{2}$

35. $5 - \frac{1}{2}\ln 2$ **37.** 3828

39. (a) $R(h) = 37\ln(0.4h^2 + h + 1) + 168$ **(b)** $R(1) \approx 200$

41. $E(y) = 4.2(0.1y^2 + 1)^{-3} - 0.6$, erosion will reach the fence in about three years.

43. (a) $t = \dfrac{1}{k}\ln\left(\dfrac{r}{r - kx}\right)$,   **(b)** $x = \dfrac{r}{k}(1 - e^{-kt})$,   **(c)** $\dfrac{r}{k}$

45. $F(x) = \frac{1}{2} - (e^x + e^{-x})^{-1}$
$F(x)$ is decreasing when $f(x)$ is negative.
$F(x)$ is increasing when $f(x)$ is positive.
$F(x)$ is concave up when $f(x)$ is increasing.
$F(x)$ is concave down when $f(x)$ is decreasing.

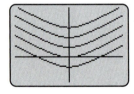

47. $y = \sqrt{x^2 + 1} + c$ for $c = -2, -1, 0, 1, 2$

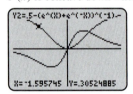

49. $y = (\ln x)^2 + c$ for $c = -2, -1, 0, 1, 2$

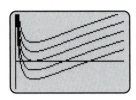

Exercises 5.3

1. $\frac{1}{2}e^{2x}(x - \frac{1}{2}) + c$ **3.** $-2e^{-t/2}(t + 2) + c$

5. $-e^{-t}(t^2 + 2t + 2) + c$ **7.** $\frac{1}{3}x^3(\ln x - \frac{1}{3}) + c$

9. $\frac{2}{9}x(3x + 2)^{3/2} - \frac{4}{135}(3x + 2)^{5/2} + c$

11. $\frac{1}{3}x^2(x^2 + 1)^{3/2} - \frac{2}{15}(x^2 + 1)^{5/2} + c$

13. $\dfrac{1}{(x - 1)(x + 5)} = \dfrac{1}{6}\left(\dfrac{1}{x - 1} - \dfrac{1}{x + 5}\right)$.

15. $\ln\left|\dfrac{x - 5}{x - 4}\right| + c$ **17.** $\ln\left|\dfrac{t}{t + 1}\right| + c$

19. $\dfrac{1}{a - b}\ln\left|\dfrac{x - a}{x - b}\right| + c$

21. $\frac{1}{3}\ln|x - 1| + \frac{2}{3}\ln|x + 2| + c$

23. $\frac{1}{5}\ln|x - 4| + \frac{4}{5}\ln|x + 1| + c$

25. $\frac{68}{3}$ feet

27. See Figure 5.3.27 below. The treatment worked.

29. 8.17 days

31. (a) The range is expanding since $y'(t) > 0$ for all t,
    (b) $y(t) = \ln\dfrac{10t}{t + 9} + 17.7$,   **(c)** $\ln 10 + 17.7$

33. $F(x) = \frac{1}{2}\ln(1 + x) - \frac{1}{2}\ln(1 - x)$
$F(x)$ is decreasing when $f(x)$ is negative.
$F(x)$ is increasing when $f(x)$ is positive.
$F(x)$ is concave up when $f(x)$ is increasing.
$F(x)$ is concave down when $f(x)$ is decreasing.

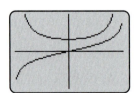

Figure 5.3.27

35. $F(x) = x \ln x$

$F(x)$ is decreasing when $f(x)$ is negative.
$F(x)$ is increasing when $f(x)$ is positive.
$F(x)$ is concave up when $f(x)$ is increasing.
$F(x)$ is concave down when $f(x)$ is decreasing.

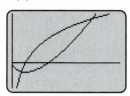

37. $y = x^2 + x \ln x - x + c$ for $c = -3, -2, -1, 0, 1, 2, 3$

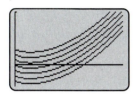

Supplementary exercises for sections 5.2 and 5.3

39. $\frac{1}{30}(3x + 5)^{10} + c$

41. $2 \ln |x + 4| - 3 \ln |x - 1| + c$

43. $\frac{1}{3}(t^2 + 9)^{3/2} + c$ **45.** $\frac{1}{2} \ln \left| \frac{t-1}{t+1} \right| + c$

47. $-\dfrac{1}{t - 1} + c$ **49.** $\frac{1}{24}(x^4 - 4)^6 + c$

51. $\frac{1}{2}(t^2 + \frac{1}{2}e^{4t}) + c$ **53.** $\frac{1}{2}x^2(\frac{1}{2} + \ln |x|) + c$

55. $e^{0.5t^2} + c$ **57.** $\ln \left| \frac{x-2}{x-1} \right| + c$

59. $x(\ln x)^3 - 3x(\ln x)^2 + 6x \ln x - 6x + c$

61. $2 \ln |t - 3| + c$ **63.** $x - 2\ln(1 + e^x) + c$

65. $\frac{1}{2} \ln \left| \frac{e^x - 1}{e^x + 1} \right| + c$

67. $\frac{2}{3}x^{3/2} - x + 2\sqrt{x} - 2 \ln(\sqrt{x} + 1) + c$

69. $\frac{2}{3} \ln \left| \frac{\sqrt{x} - 1}{\sqrt{x} + 2} \right| + c$ **71.** $\frac{1}{3}(2t + 1)^{3/2} + \frac{2}{3}$

73. $y = t(\ln |t|)^2 - 2t \ln |t| + 2t - \frac{1}{2}t^2 + \frac{1}{2}$

75. $y(t) = \frac{2}{9}t(3t + 1)^{3/2} - \frac{4}{135}(3t + 1)^{5/2} + \frac{4}{135}$

Exercises 5.4

1. (a) 0.760 **(b)** 0.635 **(c)** 0.691 **3.** 23.4

5. counting rectangles gives: a) 1.125 b) 2.875 c) −4.225 d) −1.35

7. (a) 39,760 m^2, **(b)** 37,820 m^2

9. −3 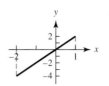 **11.** −6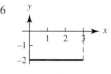

13. $-\frac{3}{2}$  **15. (a)** 1.155, **(b)** 1.43

17. (a) $\int_1^7 r(t)\, dt$, **(b)** $\Delta t = 1$, **(c)** 390

19. $\int_0^2(-32t - 50)\, dt$ **21.** $\int_0^{100} 0.05e^{-0.01t}\, dt$

23. 8.7 **25.** −20

27. (a) 0.692835, error $\approx 312 \times 10^{-6}$
(b) 0.693069, error $\approx 78 \times 10^{-6}$
(c) 0.693127652, error $\approx 20 \times 10^{-6}$

29. −3.3438 **31.** 1.3537 **33.** 1.9820

35. 3.995 **37.** 1.2328

Exercises 5.5

1. 2 **3.** 4 **5.** 3

7. 4 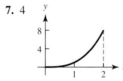 **9.** $\frac{4}{3}$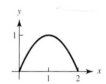

11. $\frac{3}{2} + \ln 2$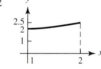

13. $\frac{2}{3}$

15. $e - e^{-1}$ 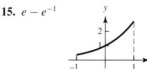 **17.** $\frac{1}{4}$ **19.** −2

21. −2/3 **23.** −256/5 **25.** 0

27. $(e^2 - e^{-2})/2$ **29.** 26/3 **31.** $2 - e^{-1} - e$

33. $2 \ln 2 - \frac{3}{4}$ **35.** integral 2/3, area 3

37. integral $= \frac{1}{2} \ln 2 =$ area

39. integral $= 2 \ln(2) - 1 =$ area **41.** −2

43. $A(t) = \int_0^t x\, dx = t^2/2$ **45.** $x^2 + 16/x$

47. $\sqrt{t^2 - 3t + 5}$ **49.** $1 + t^2 e^{t^4}$ **51.** $2t|t|e^{t^2}$

53. $\frac{552}{7}(e^{-0.35} - 1)$ **55.** $22,400/3$

57. $P(x) = -0.1x^2 + 50x - 2{,}500$

59. max at \$162,000 when $x = 700$ **61.** 9,024

63. Yes, since production increased by 300, which is more than 66 claimed.

65. (a)

year	growth rate of A	growth rate of B
5	1.8	5.51571
12	10.368	9.3267

(b)

by year	growth of A	growth of B
5	3	17.2366
12	41.472	69.9503
20	192	158.397

67. 6.435 using left endpoints

Exercises 5.6

1. $10\left(1 - \dfrac{1}{e}\right)$ **3.** 9   **5.** 1/3   **7.** 0   **9.** 20/3

11. $\frac{1}{4}\ln 31$ **13.** 9   **15.** $\ln\frac{4}{3}$

17. $\int_{-1}^{2}[f(x) - g(x)]\,dx + \int_{2}^{4}[g(x) - f(x)]\,dx + \int_{4}^{6}[f(x) - g(x)]\,dx$

19. 2/3   **21.** 21   **23.** 45/4

25. $(2, 1)$, $\ln(16/9) + 1$

27. $(0, 1)$, $2e + 2e^{-1} - 4$ **29.** $\frac{17\sqrt{17}-45}{48} \approx 0.522767$

31. 2/3   **33.** 27/2   **35.** 3   **37.** $\frac{20}{3}\ln 2$ **39.** 10

41. $(2\ln 4 - \ln 7)/9$ **43.** 6   **45.** yes   **47.** 4, No

49. 22.4226   **51.** 4,834

53. (a) maximum at $y = 1$,
(b) growth of $10.736e^{-1}$ hundred over $[0, 1]$, decay of 3.79 hundred over $[1, 6]$

55. The solution for $e^{-x} = x$ is $x \approx 0.5671$ and the area ≈ 0.272

57. 1.5734

Exercises 5.7

1. (a) $|E_M| \le \frac{1}{300}$, $|E_T| \le \frac{1}{150}$
(b) Midpoint: 0.33, with error $\frac{1}{300}$;
Trapezoidal: 0.34, with error $\frac{1}{150}$

3. (a) 1.09286   **(b)** 26; 1.0984   **(c)** 0.0002

5. 3.56134   **7.** 1.7072, $|E_T| \le \dfrac{8}{1{,}200} \approx 0.0067$

9. Midpoint: 1.89938, Trap: 1.87885

11. Midpoint: 1.44241, Trap: 1.44327

13. Midpoint: 0.379863, Trap: 0.37993

15. Midpoint: 0.147391, Trap: 0.146879

17. 530 feet   **19.** 8.09375   **21.** 2,813

23. 12, 0.693149   **25.** 41.2 milligrams

27. 0.7468241328

29.
```
fnInt((1+X²)^(-1
),X,0,1)
        .7853981634
π/4
        .7853981634
■
```

Chapter 5: Review Exercises

1. g **3.** $\frac{1}{3}\ln(e^{3x} + e^{-3x})$ **5.** $\frac{1}{2}e^{x^2+1} + c$

7. $(-4t - 16)e^{-0.25t} + c$ **9.** $\dfrac{1}{2}\ln\left|\dfrac{x-5}{x-3}\right| + c$

11. $\frac{2}{3}(t-4)^{3/2} + 8\sqrt{t-4} + c$

13. $-\frac{1}{2}e^{-2x} - x^4 + x + 12.5$ **15.** $\frac{1}{6}(e^6 + 11)$

17. $C(x) = \frac{1}{8}x^2 + 5{,}000x - 50{,}050$

19. ≈ 2.325 **21.** ≈ 2.55 **23.** 6.0

25. -2 **27.** $-4\ln 2$ **29.** $2e^2 - 2e$

31. 0.3093   **33.** 9   **35.** 3   **37.** $\dfrac{1}{3 + t^4}$

39. $\sqrt{t^2 - 2t + 8}$ **41.** $12 + 3e^2$

43. $\approx 3{,}515$ **45. (a)** 1.10967   **(b)** 1.11499

47. 671.83 million   **49.** 6,000

51. ≈ 0.533 mile   **53.** $\approx 7{,}396$

Chapter 5: Exam

1. 17/4   **2.** $1 + \ln 2 + e$ **3.** $e^{x^2-5} + c$ **4.** $y(1) = \frac{56}{15}$

5. $\frac{1}{16}(1 + 3e^4)$ **6.** $\frac{23}{2}$ **7.** 8.4   **8.** 1.1796   **9.** -0.3

10. $g'(1) = 2$ **11.** $\frac{1}{2}(\ln 22 - \ln 4)$ **12.** $-1{,}680$

13. $\frac{64}{3}$ **14.** 28,500   **15.** $R(t) = 100e^{0.2t} + 300$

16. 2/9   **17.** $\dfrac{1}{6}\ln\left|\dfrac{x-7}{x-1}\right| + c$ **18.** $\frac{11}{18}$ kilometers

19. (a) ≈ 1.114 square kilometers
(b) ≈ 1.208 square kilometers

20. $\frac{1}{2}\sqrt{x} - \frac{1}{8}\ln(4\sqrt{x} + 1) + c$ **21.** 3/4

CHAPTER 6

Exercises 6.1

1. $CS = 1$, $PS = 2.5$

3. $q_e = 8$, $p_e = 6.8$, $CS = 12.8$, $PS = 19.2$

5. $q_e = 3000$, $p_e = 40$, $CS = 90,000$, $PS = 45,000$

7. $q_e = 25$, $p_e = 15$, $CS = \$187.50$, $PS = 125$

9. $q_e = 20$, $p_e = 15$, $CS = 80$, $PS = 160$

11. $q_e = 10$, $p_e = 30$, $CS = 400$, $PS = 100$

13. $q_e = 3$, $p_e = 5$, $CS = 7.91$, $PS = 4.50$

15. $q_e = 2$, $p_e = 9$, $CS = 14.67$, $PS = 7.33$

17. $CS \approx 1,000$, $PS = 800$

19. (a) supply function $c = S(b)$, where b is the number of babies, and c is calories per baby. (c) $b_0 c_0$ (b)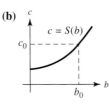

21. (a) $PS = 13.72$ (b) $CS = 72.2867$

23.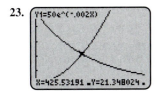
$q_e = 429.608$,
$p_e = 21.175$,
$CS = 5315.83$,
$PS = 5655.1$

25.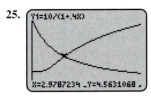
$q_e = 3.01722$,
$p_e = 4.53127$,
$CS = 6.11776$,
$PS = 4.6303$

Exercises 6.2

1. $\$6352.45$

3. (a) $n = 1$ (b) $r = 0.23$, $t = 5$
(c) $1,000(1.23)^5 \approx 2,815$

5. 8902.16 7. $(\ln 2)/7$ 9. $62,542.14$ 11. $18,917$

13. The present value of the income stream is $\$35,739,057$. That plus the initial $\$6,000,000$ is more than $\$40,000,000$. So, he should choose the second option.

15. $\$269,655.10$ 17. 10.2806 million

19. $\$2,926.85$ 21. 17.2562 years

23. (b) (i) $1,038.07$ (ii) $2,139.83$

25. $74,062.86$ 27. $2,549.27$

29. (c) $A(3, 650) = 15,570.09$, $\dfrac{S}{n}(e^{rT} - 1) = 15,569.31$

Exercises 6.3

1. Yes 3. Yes 5. $y = Ce^{-t}$ 7. $M = Ce^{0.1t} - 100$

9. $y = Ce^{e^t} + 1$ 11. $y = \dfrac{Ce^t - 2}{Ce^t - 1}$ 13. $y = Ce^{rt} + a$

15. $y = -\dfrac{4}{4t - 1}$ 17. $y = 3e^{-t} + 9$

19. $M = 1050e^{0.1t} - 1000$ 21. $y = \dfrac{e^{x^2} + 3}{3 - e^{x^2}}$

23. $y = cx$,
$c = -3, -2, -1, 0, 1, 2, 3$
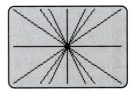

25. $y = \ln(t + e^c)$,
$c = 0, 0.5, 1, 1.5, 2, 2.5, 3$

27. $y^2 = 2e^{-x^2} + c$, (a) $y = \sqrt{2e^{-x^2} - 1}$
(b) $y = -\sqrt{2e^{-x^2} + 2}$

29. $\$170,430$ 31. $\$3,300,000$ per year

33. (a) $dy/dt = 100 - 0.12y$ (b) 597.4 mg

35. (a) $dV/dt = 200 - 0.01V$, $V(0) = 50$
(b) $V(t) = 20,000 - 19,950e^{-0.01t}$ (c) 640 words

37. (a) $dP/dt = 1000 - 0.02P$, $P(0) = 40,000$,
$P(t) = 50,000 - 10,000e^{-0.02t}$
(b) Not successful since $P(1) \approx 40,198$, $P(2) \approx 40,392$,
$P(3) \approx 40,582$

39. (a) $dM/dt = 0.09M - 20 \cdot 25,000 = 0.09M - 500,000$,
$M(0) = M_0$ (b) $\$5,555,555.56$

41. (a) $dM/dt = 0.12M - 12A$, $M(0) = 10,000$,
$A = 221.637$ (b) $A = 222.444$

43. $dw/dt = 0.12w$, $w(0) = 2500$.

45. $dP/dt = 0.025P + 0.05$, $P(0) = 30$.

Exercises 6.4

1. $p_0 = 327$, $r = 0.015$, $k = r/K$

3. (a) $dp/dt = 0.7p - 0.01p^2$, $p(0) = 35$,
$p(t) = \dfrac{70}{1 + e^{-0.7t}}$ (b) $K = 70$ (c) $p(8) = 69.7$

5. $p(t) = 5e^{(0.005 \ln 56)t}$,   **(a)** 37.4166
     (b) model not good over long time

7. $r = 0.04$, $k = 0.001$; $p = 0$ and $p = 40$

9. $r = 0.25$, $k = 0.05$; $p = 0$ and $p = 5$

11. $dP/dt = 0.03p(1 - p/9)$

13. **(a)** $y = \dfrac{3000}{1 + ce^{-3000at}}$,   **(b)** $c = 149$, $a \approx 0.000318$,
     (c) $y(5) \approx 1325$

15. $p(t) = \dfrac{17,556}{76 + 155e^{-0.03t}}$

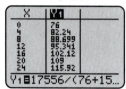

     Part of table (with X in years after 1900)

17. $p(t) = \dfrac{1}{1 + 0.25e^{-0.05t}}$ **19.** $p(t) = \dfrac{20}{4 + e^{-0.03t}}$

21. **(a)** $y = \dfrac{36,000}{1 + 719e^{-36,000at}}$, $a \approx 0.00002243$
     (b) $y(10) \approx 29,416$

23. 2100 deer

25. **(b)** $p(t) = \dfrac{30}{3 + 2(t + 1)^{0.2}e^{-0.2t}}$
     (c) $p(5) = 7.4022$, $\lim_{t \to \infty} p(t) = 10$

27. **(a)** $p = k$ **(b)–(c)** $p = 0$ and $p = K$
     (d) $p = 0$ and $p = K\left(1 - \ln\dfrac{d}{r}\right)$

Exercises 6.5

1. 1/64   **3.** 10   **5.** 20   **7.** 1/3   **9.** 1

11. $f(x) \ge 0$ and $\int_0^\infty \lambda e^{-\lambda x} = 1$ **13.** 1/2   **15.** -1

17. 3/2   **19.** diverges   **21.** converges to $-1/4$

23. diverges   **25.** diverges   **27.** converges to ln 2

29. diverges   **33.** 133,333.33   **35.** \$1,000,000

37. 2,000,000   **39.** **(a)** $\int_0^h \dfrac{180}{(t+1)^3} \, dt$ **(b)** 90

41. 2,500   **43.** converges   **45.** diverges

47. diverges   **49.** converges   **51.** converges

53.
```
fnInt(1/(1+X²),X
,0,1000)
        1.569796327
■
```
55.
```
fnInt(e^(-X²/2),
X,-4,4)
        2.506469499
√(2π)
        2.506628275
■
```

57. 1.74, $0 < \text{Error} < \int_0^{0.0001} x^{-1/2} \, dx = 0.02$

Chapter 6: Review Exercises

1. $CS \approx 1,910$

3. $q_e = 2$, $p_e = 9$, $CS = \frac{44}{3}$

5. $q_e = 100$, $p_e = \$30$, $CS = \$500 = PS$

7. $FV = \frac{3600}{0.06}(e^{0.48} - 1)$

9. **(a)** Option II is better ($PV \approx \$432,332$ for II)
     (b) Option I is better ($PV \approx \$632,121$ for II)

11. **(a)** $PV = \int_0^5 1000\left(\dfrac{5 - t}{t + 1}\right)e^{-0.04t} dt$
     (b) Trapezoidal rule with $\Delta t = 1$ gives 5,958.60. Better estimate (using Simpson's rule or calculator) is 5,491.45

13. $k = 4$ **15.** $y = 3e^{x^2/2} - 1$ **17.** $y = -\ln\left(\dfrac{1}{2} - \dfrac{x^2}{2}\right)$

19. $dM/dt = 9000 + 0.09M$, $M(0) = 2000$;
     $M(t) = 102,000e^{0.09t} - 100,000$,
     $M(25) = 102,000e^{2.25} - 100,000 \approx 867,749$

21. **(a)** $135(1 - e^{-2.4}) \approx \$122,753$
     (b) total interest $= 324,000 - 122,753 = \$201,247$

23. $p = \dfrac{5}{5 - 4e^{-t}}$, $\lim_{t \to \infty} p(t) = 1$ **25.** $K = 50$

27. $y = 0$ and $y = 5$ **29.** $\int_0^\infty \dfrac{x^2}{(x^3 + 1)^4} \, dx = \dfrac{1}{9}$

31. $\int_0^\infty 0.5e^{-0.5x} = 1$ **33.** $10,000/0.1 = 100,000$

35. It converges: $\int_0^1 \dfrac{e^t}{\sqrt{e^t - 1}} \, dt = 2\sqrt{e - 1}$

37. *Hint:* first replace $\dfrac{e^x}{e^x + e^{-x}}$ by $\dfrac{e^{2x}}{e^{2x} + 1}$. Answer: It diverges.

39. It converges. $\int_0^1 (\ln x)^3 \, dx = -6$

41. Converges.   **43.** Converges.   **45.** Diverges.

Chapter 6: Exam

1. 5   **2.** $q_e = 2$, $p_e = 25$

3. $q_e = 2$, $p_e = 4$, $CS = 16 \ln 2 - 8$

4. $q_e = 4$, $p_e = 7$, $PS = 8$ **5.** 247,651.62

6. 1,719,880.78

7. **(a)** 1,449,487.39
     (b) A lump sum of 1,949,487.39 is a better option.

8. 33,861.5   **9.** $dM/dt = 0.07M - 36,000$, $M(35) = 0$

10. $k = 4$ **11.** $y(x) = \sqrt{x^2 + 2x + 8}$

12. $Q(t) = 8e^{2t} - 3$ **13.** $K = 20$

14. $dp/dt = 0.02p(1 - p/K)$, $p(0) = 25$, $K \approx 50.56$,
     $p(40) = 34.64$ million

15. 800,000   **16.** 2   **17.** 4/3

18. $\displaystyle\int_1^\infty \frac{dx}{2+x^4} \le \int_1^\infty \frac{dx}{x^4} < \infty$

CHAPTER 7

Exercises 7.1

1.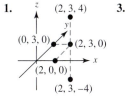

3.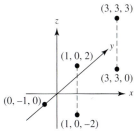

5. $(3, 1, 2)$ **7.**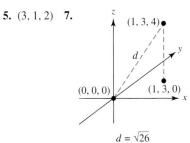

$d = \sqrt{26}$

9.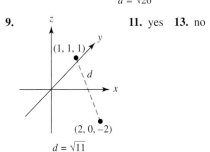

11. yes   **13.** no

$d = \sqrt{11}$

15.

day	1	2	3	4	5
distance	3.06	6.15	11.33	21.10	39.62

z represents the change in depth

17. 6, 12, 22, 8   **19.** 0, 5, 73, 69

21. 0, $-24/25$, 1, 1, 1   **23.** 0, 0, 1, $\frac{1}{2}\ln 13$

25. All (x, y) with $y > x$ **27.** All (x, y) with $x + y \ne 1$

29. All (x, y) with either $x > 0$, $y > 1$ or $x < 0$, $y < 1$

31.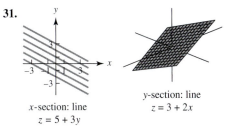

x-section: line
$z = 5 + 3y$

y-section: line
$z = 3 + 2x$

33.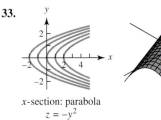

x-section: parabola
$z = -y^2$

y-section: line
$z = x - 4$

35.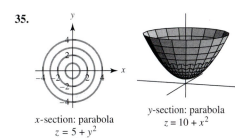

x-section: parabola
$z = 5 + y^2$

y-section: parabola
$z = 10 + x^2$

37.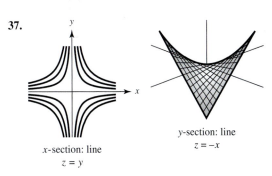

x-section: line
$z = y$

y-section: line
$z = -x$

39. (a)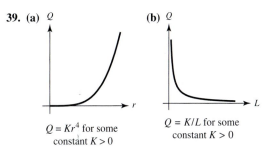

$Q = Kr^4$ for some
constant $K > 0$

(b)

$Q = K/L$ for some
constant $K > 0$

41. (a) $R(x, y) = 2{,}000x + 3{,}000y$

(b)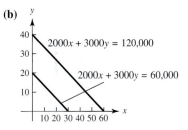

$2000x + 3000y = 120{,}000$

$2000x + 3000y = 60{,}000$

43. (a) $1{,}000e^{rt} = 5{,}000$ $(\ln 5)/(0.08) \approx 20.12$ years

45.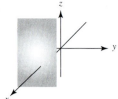

47.

49.

51.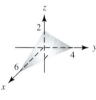

53.

55.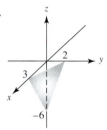

57.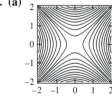

59. $z - 3 = 2(x - 4) + 5(y + 3)$

61. $z - \frac{9}{2} = -\frac{5}{2}(x + 4) + \frac{7}{2}(y + 2)$ **63.** $z = -3x + y + 2$

65. $z = -\frac{3}{2}x + y + 5$

67. (a) negative   **(b)** positive   **(c)** $(0, 0, 2{,}000)$

69. No intersection   **71.** The point $(3, 2, 1)$

73. $z = \frac{1}{2}x + 2y + 500$

75. (a)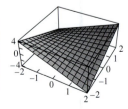

(b)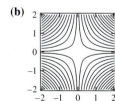

(c)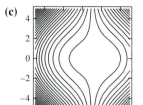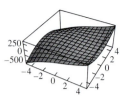

(d)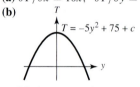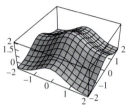

Exercises 7.2

1. $\partial f / \partial x = -8$, $\partial f / \partial y = 5$

3. $\partial f / \partial x = 3x^2 y^2 - 4xy$, $\partial f / \partial y = 2x^3 y - 2x^2 + 1$

5. $\partial M / \partial r = 1000 t e^{rt}$, $\partial M / \partial t = 1000 r e^{rt}$

7. $\dfrac{\partial R}{\partial s} = -\dfrac{5t^2}{s^2}$, $\dfrac{\partial R}{\partial t} = \dfrac{10t}{s}$

9. $\dfrac{\partial l}{\partial x} = \dfrac{2x}{x^2 - 5y + 3}$, $\dfrac{\partial l}{\partial y} = -\dfrac{5}{x^2 - 5y + 3}$

11. $\partial Q / \partial K = 4K^{-0.6} L^{0.6}$, $\partial Q / \partial L = 6K^{0.4} L^{-0.4}$

13. $\dfrac{\partial M}{\partial t} = 360{,}000 \left(1 + \dfrac{r}{360}\right)^{360t} \ln\left(1 + \dfrac{r}{360}\right)$,
$\dfrac{\partial M}{\partial r} = 1{,}000t \left(1 + \dfrac{r}{360}\right)^{360t-1}$

15. 5   **17.** 40   **19.** No, $\partial F / \partial t$ should be negative.

21. (a) $\partial T / \partial x = 10x$, $\partial T / \partial y = -10y$

(b)

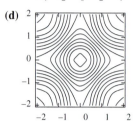

$T = -5y^2 + 75 + c$

Maximum at $y = 0$

(c)

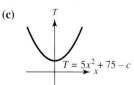

$T = 5x^2 + 75 - c$

Minimum at $x = 0$

(d) If there were a local minimum (or maximum), *both* the x- and y-sections through that point would have a local minimum (or maximum), which cannot be the case.

23. $L(x, y) = 23 + 6(x - 1) + 20(y - 2)$

25. $L(x, y) = 1 + 2(x - 1) + 4(y + 1)$

27. $L(x, y) = 16 + 2(x - 4) + 32(y - 1)$

29. $L(x, y) = 8 + 5x - 3y$

31. $L(x, y) = -2 + 2(x + 3) - \frac{8}{7}(y - 5)$

33. (a) 96 cubic inches   (b) 0.8 cubic inches

35. (a) $2\pi r_0 h_0$ (b) $2\pi r_0 h_0 + \pi h_0$

37. (a) $\dfrac{\partial V}{\partial T} = \dfrac{0.5}{P}$, $\dfrac{\partial V}{\partial P} = -\dfrac{0.5T}{P^2}$ (b) -1.1

39. $MPK \approx 15.83$, $MPL \approx 9.237$

41. $\dfrac{\partial^2 f}{\partial x^2} = \dfrac{2}{y}$, $\dfrac{\partial^2 f}{\partial y^2} = \dfrac{2x^2}{y^3}$, $\dfrac{\partial^2 f}{\partial x \partial y} = \dfrac{-2x}{y^2} = \dfrac{\partial^2 f}{\partial y \partial x}$

43. $\dfrac{\partial^2 f}{\partial x^2} = 0$, $\dfrac{\partial^2 f}{\partial y^2} = -\dfrac{x}{y^2}$, $\dfrac{\partial^2 f}{\partial x \partial y} = \dfrac{1}{y} = \dfrac{\partial^2 f}{\partial y \partial x}$

45. $\partial f/\partial x\,(1, 2) \approx 4$, $\partial f/\partial y\,(1, 2) \approx 0.5$

47. $\partial f/\partial x\,(5, 6) \approx 2$, $\partial f/\partial y\,(5, 6) \approx -20/3$

49. $\partial f/\partial x\,(0, 0) \approx -2$, $\partial f/\partial y\,(0, 0) \approx 10$

51. (a) -0.75 minute per degree   (b) -24 minutes per tsp
(c) 53.25 minutes

53. $\dfrac{\partial f}{\partial x} = -\dfrac{z}{(x + y^2)^2}$, $\dfrac{\partial f}{\partial y} = \dfrac{-2yz}{(x + y^2)^2}$, $\dfrac{\partial f}{\partial z} = \dfrac{1}{x + y^2}$

55. $\dfrac{\partial f}{\partial x} = y^2 \ln z$, $\dfrac{\partial f}{\partial y} = 2xy \ln z$, $\dfrac{\partial f}{\partial z} = \dfrac{xy^2}{z}$

57. $\dfrac{\partial f}{\partial x} = \dfrac{x}{\sqrt{x^2 + y^2 + z^2}}$, $\dfrac{\partial f}{\partial y} = \dfrac{y}{\sqrt{x^2 + y^2 + z^2}}$,
$\dfrac{\partial f}{\partial z} = \dfrac{z}{\sqrt{x^2 + y^2 + z^2}}$

59. (a) 660 cubic inches   (b) 620 cubic inches

Exercises 7.3

1. $(1, -2)$ **3.** $(0, 0)$ **5.** $(0, 0)$ and $(1, 1)$

7. $(0, 1)$ **9.** $(0, 0)$ **11.** Local minimum at $(-1, -2)$

13. Saddle point at $(2, 5)$

15. Local minimum at $(1, 1)$, saddle point at $(0, 0)$

17. Local maximum at $(2, 1)$

19. There is a local (and global) minimum at $(0, 0)$, but second derivative test is inconclusive.

21. There is a local (and global) maximum at $(0, 0)$, but second derivative test is inconclusive.

23. Saddle point at $(0, 1)$ **25.** Local minimum at $(0, 0)$

27. It is a critical point but not a local maximum or minimum.

29. Global minimum at $(3, 1)$ **31.** Global maximum at $(1, \frac{1}{2})$

33. Global maximum at $(3, -2)$

35. Maximum of 1,400 at $(300, 200)$

37. Maximum of 120,000 at $(2,000, 1,000)$

39. (a) $R(x, y) = 50x + 6y + \frac{5}{2}xy - 3x^2 - y^2$
(b) $x = 20$, $y = 28$ maximize revenue because $D > 0$ and $\partial^2 f/\partial x^2 < 0$.
(c) $x = 25.22$, $y = 32.53$

41. (a) $(\frac{3}{2}, \frac{5}{2})$

(b)
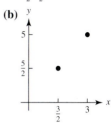

The stove is at the midpoint of the line segment connecting the sink and refrigerator.

43. local maximum   **45.** saddle point   **47.** $a = 1$, $b = 2/3$

Exercises 7.4

1. $y = \frac{3}{2}x + \frac{2}{3}$ **3.** $y = \frac{3}{2}x + \frac{5}{6}$ **5.** $\frac{3}{4}x + \frac{13}{12}$ **7.** $y = \frac{1}{5}x + \frac{6}{5}$

9. $y = 0.27x + 2.05$, $y(6) = 3.67$ billion dollars

11. $y = 234.465x - 669.202$ (with x in years after 1960, y in billions of dollars), $y(50) = 11,054$

13. $y = 2.02x + 64.32$ (with x in years after 1900, y in millions), $y(110) \approx 286.63$

15. (a) The slope is much too large in absolute value to fit the given data.
(b) $y = -0.914x + 236.619$

19. (a)
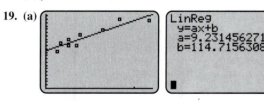

(b) ≈ 9.23 (c) ≈ 114.7

21.
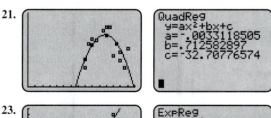

23.
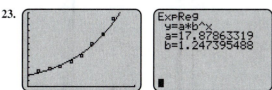

Exercises 7.5

1. min at $(100, 300)$ **3.** max at $(500, 500)$

5. min at $(4, 7)$

7. max at $(1/\sqrt{2}, 1/\sqrt{2})$ and $(-1/\sqrt{2}, -1/\sqrt{2})$
min at $(1/\sqrt{2}, -1/\sqrt{2})$ and $(-1/\sqrt{2}, 1/\sqrt{2})$

9. min at $(1, 1)$ and $(-1, -1)$

11. (a) C, $2{,}100$ **(b)** tangentially
(c) $\dfrac{\partial f}{\partial x}(x_0, y_0) = \lambda \dfrac{\partial g}{\partial x}(x_0, y_0)$ and
$\dfrac{\partial f}{\partial y}(x_0, y_0) = \lambda \dfrac{\partial g}{\partial y}(x_0, y_0)$ for some $\lambda \neq 0$

13. $(3/10, 1/10)$ **15.** Minimum value of 9 at $(2, -2, 1)$

17. $(2, 2, 2)$ **19.** $(5/6, 5/6, -5/3)$

21. Find the extreme points of $f(x, y, z) = z$, subject to $x^2 + y^2 = 8$ and $x + y + z = 5$. Solution: maximum at $(-2, -2, 9)$, minimum at $(2, 2, 1)$.

23. $x = 25$, $y = 75$ **25.** $K = 20{,}000$, $L = 40{,}000$

27. $x = y = 18$, $z = 36$

29. (a) $r = 10$, $h = 5$ **(b)** $500\pi/3 \approx 523.6$ cubic feet
(c) If $|t|$ is small, taking $r = 10 + t$ results in $h = 5 - t$
and $V = \dfrac{500\pi}{3} - t^2(15 - t) < \dfrac{500\pi}{3}$.

31. length $=$ width $= 40$ cm, height $= 20$ cm

33. $(7/6, -2/3, 7/6)$

35. All are equal.

Chapter 7: Review Exercises

1. A, B **3.** $z = 3 - 5(x - 1) + \frac{2}{7}(y - 2)$

5. (a) yes   **(b)** no   **(c)** no   **7.** $f(x, y) = 5x + 2y + 3$

9. $z = -9$: circle of radius 3 about $(0, 0)$
$z = -4$: circle of radius 2 about $(0, 0)$
$z = 0$: single point $(0, 0)$
$z = -9$ and $z = -4$: empty

11. $\partial f/\partial x = 2xy - 2y^3 + 1$, $\partial f/\partial y = x^2 + 6y - 6xy^2 - 5$

13. $\dfrac{\partial f}{\partial x} = \dfrac{2x + y}{x^2 + xy}$, $\dfrac{\partial f}{\partial y} = \dfrac{x}{x^2 + xy}$

15. $\dfrac{\partial f}{\partial x} = 2x \ln(1 + y^2)$, $\dfrac{\partial f}{\partial y} = \dfrac{2x^2 y}{1 + y^2}$

17. $\partial f/\partial x(1, -2) = 5$, $\partial f/\partial y(1, -2) = 2$

19. $\dfrac{\partial f}{\partial y} e^{x/y} = -\dfrac{xe^{x/y}}{y^2}$

21. $z = 14 + 4(x - 1) + 12(y - 2)$ **23.** 3.4

25. $L(x, y) = 5 + 2(x - 10) - (y - 20)$, 4.6

27. $\dfrac{\partial^2 f}{\partial x^2} = \dfrac{8 - 2x^2 + 2y^2}{(4 + x^2 + y^2)^2}$, $\dfrac{\partial^2 f}{\partial y^2} = \dfrac{8 + 2x^2 - 2y^2}{(4 + x^2 + y^2)^2}$,
$\dfrac{\partial^2 f}{\partial x \partial y} = \dfrac{\partial^2 f}{\partial y \partial x} = -\dfrac{4xy}{(4 + x^2 + y^2)^2}$

29. $(4, 1)$

31. $(0, 0)$, saddle point; $(1, 0)$ and $(-1, 0)$, both local maxima

33. local maximum

35. Saddle point at $(0, 0)$, local max at $(-1, -1)$

37. $x = 5$, $y = 7$ **39.** $y = \frac{3}{7}x + \frac{13}{7}$

41. $(b - 2)^2 + (a + b - 1)^2 + (2a + b - 1)^2 + (4a + b - 2)^2 +$
$(5a + b - 5)^2 + (6a + b - 4)^2$

43. 7.5 thousand gallons

45. Minimum of 8 at $(1, 2)$ and $(-1, -2)$, no maximum

47. $(-1, 3/2, 1/2)$

Chapter 7: Exam

1. 3   **2.**

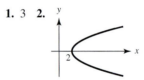

3. $z = 4 + (x - 2) - 2(y - 6)$ **4.** -2 **5.** $24 + 3e^6$

6. $z = x - y - 1$ **7.** $L(x, y) = 9 + 8(x - 1) + 4(y - 2)$

8. $(2, -8/3)$ **9.** Saddle point   **10.** Local minimum

11. (a) $y = -3x + \frac{62}{3}$ **(b)** 8.67

12. 14   **13.** $K = 1800$, $L = 7200$

CHAPTER 8

Exercises 8.1

1. 2π, $5\pi/2$, 3π **3.** $3\pi/4$, $5\pi/4$ **5.** 225, 315

7. 150, 210, 330   **9.** $\pi/3$ each

11.   **13.**

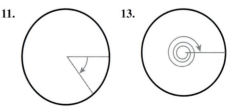

15. 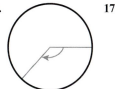   **17.**

19. $-\pi$ **21.** $3\pi/2$ **23.** -2π **25.** $5\pi/4$ **27.** $13\pi/4$

29. $20\pi/3$ **31.**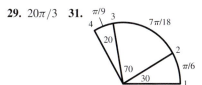

Exercises 8.2

1. $\sin\theta = 1/\sqrt{5},\ \cos\theta = 2/\sqrt{5},\ \tan\theta = 1/2$

3. $\sin\theta = 1/3,\ \cos\theta = \sqrt{8}/3,\ \tan\theta = 1/\sqrt{8}$

5. approx 47.12 feet   **7.** **(a)** 1   **(b)** both equal $1/2$

9. **(a)** $\tan\theta = \dfrac{\text{length of shadow}}{\text{length of stick}}$ **(b)** $r = \dfrac{\text{distance AS}}{\theta}$

11. $\sin(5\pi/4) = -1/\sqrt{2},\ \cos(5\pi/4) = -1/\sqrt{2},$
$\tan(5\pi/4) = 1$

13. $\sin(-\pi/3) = -\sqrt{3}/2,\ \cos(-\pi/3) = 1/2,$
$\tan(-\pi/3) = -\sqrt{3}$

15. -1 **17.** $-1/2$ **19.** $-1/2$ **21.** -1 **23.** $2/\sqrt{13}$

25. $\sin\theta = -4/5,\ \cos\theta = 3/5$

27. $\sin(u-v) = \sin u\ \cos v - \cos u\ \sin v$
$\cos(u-v) = \cos u\ \cos v + \sin u\ \sin v$

29. $T(t) = 70 + 5\sin\left(\dfrac{\pi}{12}t - \dfrac{\pi}{2}\right)$

31. $p(t) = 100 + 20\sin\left(\dfrac{7\pi}{3}t + \dfrac{\pi}{2}\right)$

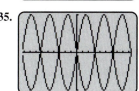

$p = 100 + 20\sin\left(\frac{7\pi}{3}t + \frac{\pi}{2}\right)$

33.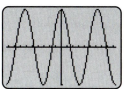

The graphs are identical.

35.

Each is the negative of the other.

39. $\sin(3x) = 3\sin x\ \cos^2 x - \sin^3 x,$
$\cos(3x) = \cos^3 x - 3\cos x\ \sin^2 x$

41.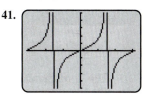

The graphs are identical.

43. $\dfrac{1-\sqrt{3}}{2\sqrt{2}}$ **45.** $\dfrac{1+\sqrt{3}}{1-\sqrt{3}}$ **47.** $\left(\dfrac{2-\sqrt{2}}{4}\right)^{1/2}$

49. $\left(\dfrac{4+\sqrt{2}+\sqrt{6}}{8}\right)^{1/2}$ **51.** 1.2490

53. $0.6435,\ 0.9273,\ \pi/2$

55. $\tan^{-1}(1/4) \approx 0.245$ **(b)** $\cos^{-1}(3/4) \approx 0.723$

57. See Table 8.2.57 below. Guess: the limit as $t \to 0$ equals zero.

Exercises 8.3

1. $2\cos x + 3\sin x$ **3.** $\dfrac{\sin x}{(1+\cos x)^2}$

5. $3\sin x\ \cos x\ (\sin x + \cos x)$ **7.** $4x^3\sec^2(1+x^4)$

9. $-18\cos(3t)\sin(3t)$ **11.** $\dfrac{-2}{\sin^2(2x+1)}$

13. $e^{2t}\ (5\cos(3t) - \sin(3t))$ **15.** $\csc x\ (1 - x\cot x)$

17. $e^{\tan x}\ \sec^2 x$

19. **(a)** $D'(t) = 10\cos^2 t\ \sin t - 5\sin^3 t$ **(b)** 0

t	-0.01	-0.001	-0.0001	0.0001	0.001	0.01
$g(t)$	0.005	0.0005	0.00005	-0.00005	-0.0005	-0.005

Table 8.2.57

21. $y = -x + \frac{1}{2}$

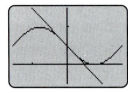

25. $\pi/2$ and $-\pi/2$ (local max at both) and 0 (local min)

27. Increasing on $(0, 3\pi/4)$ and $(7\pi/4, 2\pi)$, decreasing on $(3\pi/4, 7\pi/4)$

29. Increasing on $(\pi/3, 5\pi/3)$, decreasing on $(0, \pi/3)$ and $(5\pi/3, 2\pi)$

31. At $x = \pi/2$ and $x = 3\pi/2$

33. (a) Least happy at $t = 0, 14, 28, \ldots$.
Most happy at $t = 7, 21, 35, \ldots$
(b) -0.25 at $t = 0, 14, 28, \ldots$, and
0.95 at $t = 7, 21, 35, \ldots$.
(c) Frequency had increased, range has decreased.

35. Maximum is 121.63, minimum is 78.37.

37. and 39. $f(x)$ increasing (decreasing) if $f'(x)$ positive (negative), and $f(x)$ concave up (down) if $f'(x)$ increasing (decreasing)

37.

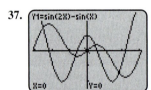

39.

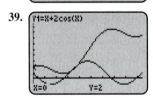

41. $90/\pi$ **43.** $88/375$ radians/sec

45. Increasing on $(\pi/2, \pi)$ and $(3\pi/2, 2\pi)$, decreasing on $(0, \pi/2)$ and $(\pi, 3\pi/2)$.
Local minimum at $x = \pi/2$ and $3\pi/2$, local maximum at $x = 0, \pi$ and 2π.
Concave up on $(\pi/4, 3\pi/4)$ and $(5\pi/4, 7\pi/4)$.
Concave down on $(0, \pi/4)$, $(3\pi/4, 5\pi/4)$, and $(7\pi/4, 2\pi)$.
Inflection points at $x = \pi/4, 3\pi/4, 5\pi/4, 7\pi/4$.
No asymptotes.

$y = \cos^2 x, 0 \le x \le 2\pi$

47. Decreasing on $(-\pi, 0)$ and $(0, \pi)$. No local extrema.
Concave up on $(-\pi, -\pi/2)$ and $(0, \pi/2)$.
Concave down on $(-\pi/2, 0)$ and $(\pi/2, \pi)$.
Inflection points at $x = \pi/2$ and $-\pi/2$.
Vertical asymptotes at $x = -\pi, 0, \pi$.

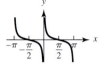

$y = \cot x, -\pi < x < \pi$

49. Increasing on $(0, 2\pi)$.
No local extrema.
Concave up on $(\pi, 2\pi)$, down on $(0, \pi)$.
Inflection point at $x = \pi$
No asymptotes.

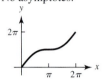

$y = x + \sin x, 0 \le x \le 2\pi$

51. (a) $x \ge x \sin x$ for $x \ge 0$
(b) Intersect at $x = \pi/2$ and $5\pi/2$.
Slope of $x \sin x$ equals 1 at those points.
Same behavior at $x = \dfrac{\pi}{2} + 2k\pi$,
for $k = 0, 1, 2, 3, \ldots$

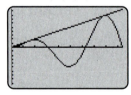

(c) Local maxima at $x \approx 2.028758$ and $x \approx 7.9786657$
(d) $x \sin x \ge -x$ for $x \ge 0$. The graphs intersect at $x = \dfrac{3\pi}{2} + 2k\pi$ for $k = 0, 1, 2, 3, \ldots$, where $y = x \sin x$ has slope -1. Local minimum at $x \approx 4.91318$.

53. (b) The graphs intersect and have equal slope at $x = \dfrac{\pi}{2} + 2k\pi$ for $k = 0, 1, 2, 3, \ldots$.
(c) The graphs intersect and have equal slope at $x = \dfrac{3\pi}{2} + 2k\pi$ for $k = 0, 1, 2, 3, \ldots$

55. $u = 2\sin(2t) - 2\cos(2t)$, $v = \cos(2t) + \sin(2t)$

Exercises 8.4

1. $-\frac{1}{3}\cos(3x) + c$ **3.** $\tan(x + 1) + c$

5. $\sin t - t\cos t + c$ **7.** $t^2 \sin t + 2t\cos t - 2\sin t + c$

9. $-\frac{2}{3}\cos^3 x + c$ **11.** $-\frac{1}{3}\cot^3 x + c$ **13.** 2   **15.** π **17.** 0

19. 0   **21.** 2   **23.** $\frac{1}{2}\ln 2$ **25.** 7/3   **27.** 1.764 thousand

29. (a) At $y = 1$ (b) $\ln\left(\frac{\sec 1 + \tan 1}{\sec 1 - \tan 1}\right) \approx 2.45$

31. $\ln 2 \approx 0.693147$ **33.** $\pi \approx 3.14159$

35. (a) $\sin(e^x) + c$
(b) $y = \sin(e^x) + 4 - \sin 1 \approx \sin(e^x) + 3.16$
Level oscillates faster and faster.

37. Graphs intersect at $x \approx 0.739$, area ≈ 1.035

39. Approx 3.86 **47.** $c_1 = 2/\pi$, $c_2 = -2/\pi$, $c_3 = 4/\pi$

49. $y = \sqrt{2 \sin t + 4}$ **51.** $y = t + 2\pi$

55. (c) $y = \tan^{-1} x + c$

Chapter 8: Review Exercises

1. (a) $\pi/30$ (b) $5\pi/6$ (c) $23\pi/12$

3. (a) 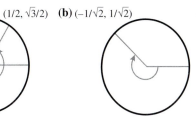 $(1/2, \sqrt{3}/2)$ (b) $(-1/\sqrt{2}, 1/\sqrt{2})$

5. $\sin \theta = 3/5$, $\cos \theta = 4/5$, $\tan \theta = 3/4$

7. (a) $2/\sqrt{3}$ (b) $-\sqrt{2}$ (c) 0 (d) -2

13. $\tan x + x \sec^2 x$ **15.** $\dfrac{2 \sin x \cos x}{(1 + \cos^2 x)^2}$

17. $2e^x (\cos x - \sin x)$ **19.** At $x = \pi/3$

21. Increasing on $(-\pi/3, \pi/3)$.
Decreasing on $(-\pi, -\pi/3)$ and $(\pi/3, \pi)$.
Local maximum at $x = \pi/3$.
Local minimum at $x = -\pi/3$.
Concave up on $(-\pi, 0)$, down on $(0, \pi)$.
Inflection point at $x = 0$.
No vertical asymptote.

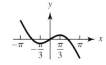

$y = 2 \sin x - x, \, -\pi < x < \pi$

25. $\frac{1}{6} \sin^2(3x) + c$ **27.** $x \tan x - \ln|\sec x| + c$

29. $\ln 2$ **31.** $4/3$ **33.** $\dfrac{4\pi}{3} + \ln\left(\dfrac{2 - \sqrt{3}}{2 + \sqrt{3}}\right) \approx 1.555$

35. $\pi/2$ **37.** 2 **39.** 0

Chapter 8: Exam

1. $\pi/10$ **2.** $10\pi/3$

3. $\sin \theta = -\sqrt{5}/3$, $\tan \theta = -\sqrt{5}/2$, $\sec \theta = 3/2$

4. $\frac{1+\sqrt{3}}{2\sqrt{2}}$ **5.** $5\pi/6$

6. (a) $y = \tan x$ (b) $y = \sin x$ (c) $y = \sec x$
(d) $y = \cos x$

7. $-\sin t \, \sec^2(\cos t)$

8. $y = -\sqrt{3}\left(x - \dfrac{\pi}{6}\right) + \dfrac{3}{2}$

9. Absolute maximum at $x = 11\pi/6$, absolute minimum at $x = \pi/6$.

10. Decreasing on $(-\pi/4, \pi/4)$.
Increasing on $(-\pi/2, \pi/4)$ and $(\pi/4, \pi/2)$.
Local maximum at $x = -\pi/4$.
Local minimum at $x = \pi/4$.
Concave down on $(-\pi/2, 0)$.
Concave up on $(0, \pi/2)$.
Inflection point at $x = 0$.
Vertical asymptotes at $x = \pm\pi/2$.

$y = \tan x - 2x, \, -\pi/2 < x < \pi/2$

11. $u = \cos(t/2) - 4\sin(t/2)$, $v = -2\cos(t/2) - \frac{1}{2}\sin(t/2)$

12. $\ln 2$ **13.** 211.46

14. $\frac{1}{5}e^{2x}(2\cos x + \sin x) + c$ **15.** 0

CHAPTER 9

Exercises 9.1

1. (a)

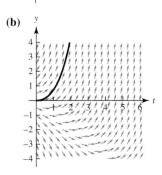

(b)

(c)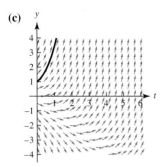

(d)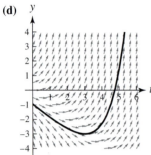

Solution (a) is a linear function.

3. (d) Slope is negative for $x > 0$, positive for $x < 0$, zero for $x = 0$.

5. (a) Slope is negative for $y > 1$, positive for $y < 1$, zero for $y = 1$.

7.

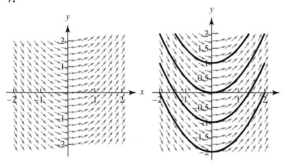

9.

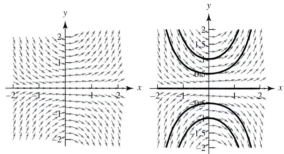

11. (a) Increasing because $y'(t) > 0$.

(b) 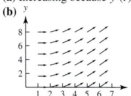   **(c)**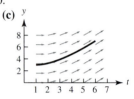

13. (a) 1.44141   **(b)** $y(t) = e^t - 1$, $y(1) = e - 1 \approx 1.71828$

15. (a) 2.18238   **(b)** $y(x) = \sqrt{x^2 + 4}$, $y(1) = \sqrt{5} \approx 2.236$

17. $87,795.86   **19.** 0.366 billion

21. (a) With t in years and r in millions of dollars per year, see Table 9.1.21 below.

(b) $(0.25)(0.25 + 0.787 + 1.243 + 1.485 + 1.493 + 1.344 + 1.149 + 0.983) = 2.1835$ million dollars

23. 491

Exercises 9.2

1. $y = 0$, asymptotically stable; $y(t) = ce^{-0.1t}$

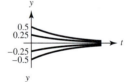

3. $y = 1$, asymptotically stable; $y(t) = 1 - ce^{-t}$

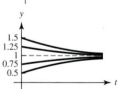

n	1	2	3	4	5	6	7	8
t_n	0.25	0.5	0.75	1	1.25	1.5	1.75	2
r_n	0.25	0.787	1.243	1.485	1.493	1.344	1.149	0.983

Table 9.1.21

5. $y = 3$, unstable;
$y(t) = 3 + ce^t$

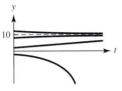

7. $p = 0$, unstable,
$p = 10$, asymptotically stable;
$$p(t) = \frac{10c}{c + (10 - c)e^{-0.1t}}$$

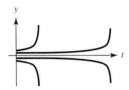

9. $y = 0$, unstable;
$$y(t) = \begin{cases} (c - 2t)^{-1/2} & \text{if } y(0) > 0 \\ -(c - 2t)^{-1/2} & \text{if } y(0) < 0 \end{cases}$$

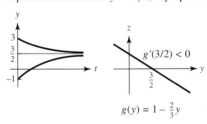

11. $y = 2$, unstable

13. $y = 0$, unstable; $y = 4$, asymptotically stable

15. $y = -1$, unstable; $y = 2$, asymptotically stable

17. $y = 0$ and $y = 4$, asymptotically stable; $y = 2$, unstable

19. $k = 0$, unstable; $k = 16$, asymptotically stable

21. $p = 0$ and $p = K$, asymptotically stable; $p = T$, unstable

23. Equilibrium solution: $y = 3/2$, asymptotically stable.

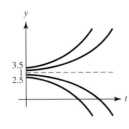

$g(y) = 1 - \frac{2}{3}y$

25. Equilibrium solution: $y = 1/2$, unstable.

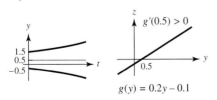

$g(y) = 0.2y - 0.1$

27. Equilibrium solutions: $y = 0$, unstable, and $y = 6$, asymptotically stable.

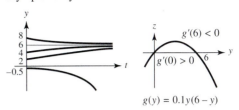

$g(y) = 0.1y(6 - y)$

29. Equilibrium solutions: $y = 0$, asymptotically stable, and $y = 5$, unstable.

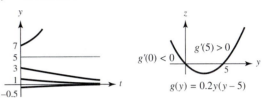

$g(y) = 0.2y(y - 5)$

31. Equilibrium solutions: $y = 4$, asymptotically stable, and $y = -1$, unstable.

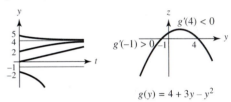

$g(y) = 4 + 3y - y^2$

33. Equilibrium solutions: $y = 2$, asymptotically stable, and $y = 6$, unstable.

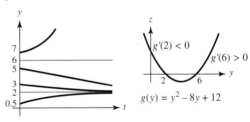

$g(y) = y^2 - 8y + 12$

35. Equilibrium solutions: $y = 0$ and $y = 2$, asymptotically stable, and $y = 0.5$, unstable.

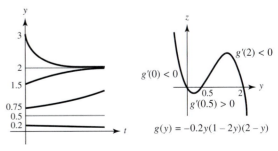

$g(y) = -0.2y(1 - 2y)(2 - y)$

37. Equilibrium solutions: $y = 4$, asymptotically stable, $y = -4$, unstable, and $y = 0$, unstable.

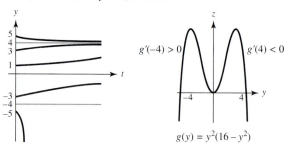

$$g(y) = y^2(16 - y^2)$$

39. The graph intersects the t-axis. It has 3 inflection points, at $y = -1$, 1, and 3

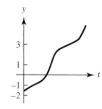

41. (a) The variables involved are the time t, measured in suitable units (such as years), and the density $y(t)$, measured in number of plants per unit of area.
(b) Three. The first is $y = 0$, the second is $y = m$, where m is the minimum density required for cross-pollination, and the third is $y = M$, where M is the optimum density for the plants.
(c) $\dfrac{dy}{dt} = -ry\left(1 - \dfrac{y}{m}\right)\left(1 - \dfrac{y}{M}\right)$, where $0 < r < 1$ and $0 < m < M$.
(d) $y = 0$ and $y = M$ are asymptotically stable, and $y = m$ is unstable. The threshold is m and carrying capacity is M.

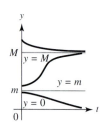

43. (a) 123 birds **(b)** 31 birds

45. (a)

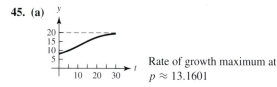

Rate of growth maximum at $p \approx 13.1601$

(b) 20 **(c)** threshold level: 1, carrying capacity: 20

47. $0.9e^{-1} \approx 0.331$

Exercises 9.3

1. (a) $dk/dt = 0.22k^{0.3} - 0.1k$
(b) $k_s = \left(\dfrac{0.22}{0.1}\right)^{1/(1-0.3)} \approx 3.08443$
(c) $k_i = \left(\dfrac{\alpha s}{\delta}\right)^{1/(1-\alpha)} = (0.66)^{1/(1-0.3)} \approx 0.552339$,
(d) $k = k_s$ is asymptotically stable since $g'(k_s) = -0.07 < 0$.

(e)

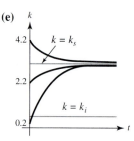

(f)

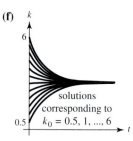

3. (a) $dk/dt = 0.34k^{0.3} - 0.1k$ **(b)** $k_s \approx 5.74454$
(c) $k_i \approx 1.02869$ **(d)** $k = k_s$ is asymptotically stable since $g'(k_s) = -0.07 < 0$. **(e)** and **(f)** as in exercise 1.

5. (a) $k(5) \approx 2.1789$ **(b)** $k(t) = [5/3 + (2^{0.55} - 5/3)e^{-0.0825t}]^{1/0.55}$ and $k(5) \approx 2.1733$ **(c)** $k_s \approx 2.5314$, asymptotically stable since $g'(k_s) = -0.0825 < 0$

7. $k(t) = (3 + 2e^{-0.05t})^2$

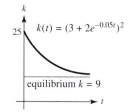

9. (a) $dk/dt = 0.2(0.6\sqrt{k} + 1)^2 - 0.1k$
(b) $k_s = \left(\dfrac{1}{\sqrt{2}} - 0.6\right)^{-2} \approx 87.17$, asymptotically stable since $g'(k_s) = \dfrac{0.12}{\sqrt{2}} - 0.1 < 0$.
(c) $k(15) \approx 30.546$
(d) Using Mathematica's NDSolve and Plot (with $\Delta t = 2$, and $n = 100$): $k(200) \approx 83.1266$

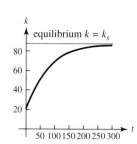

Chapter 9: Review Exercises

1. 3 **3.** $e - 1 \approx 1.718$

5. (a) $y'(0) = 3$, $L(t) = 1 + 3t$ **(b)** $y(0.1) \approx L(0.1) = 1.3$

7.

n	1	2	3	4	5
t_n	0.2	0.4	0.6	0.8	1.0
y_n	4.2	5.48	6.872	8.421	10.189
$y(t_n)$	4.246	5.613	7.160	8.977	11.195
error	0.046	0.133	0.288	0.556	1.006

Error is largest at $t = 1$.

9. 1.085 million

11. $y = 2$, unstable, and $y = 8$, asymptotically stable

13. $y = 0$, unstable, and $y = 8(1 - \ln 3 + \ln 2)$, asymptotically stable

15. Because $y'(t) = -1 - y(t)^2$ is negative for all t.

17.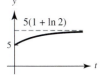

19.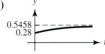

21. (a) $dp/dt = 0.02p(1 - 2p) + 0.001$, $p(0) = 0.28$, with p in billions. (b) $p = 0.25 + 0.05\sqrt{35} \approx 0.5458$, asymptotically stable (c)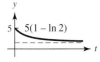

23. (a) $dy/dt = 0.2ye^{1-(y/5)} - 0.1y$, $y(0) = 5$. Equilibrium solution: $y = 5(1 + \ln 2)$. Asymptotically stable

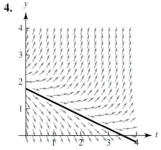

(b) $dy/dt = 0.2ye^{1-(y/5)} - 0.4y$, $y(0) = 5$. Equilibrium solution: $y = 5(1 - \ln 2)$. Asymptotically stable

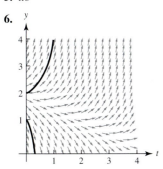

(c) $k = 0.1(2e - 1) \approx 0.44$

Chapter 9: Exam

1. positive **2.** negative **3.** no

4.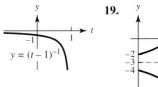

$y = -\frac{1}{2}t + \frac{7}{4}$

5. no

6.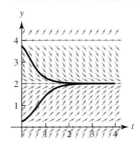

7. $y = 4$, unstable, $y = 2$, asymptotically stable, $y = 0$, unstable

8.

$y(0)$	-0.25	1	3	4.25
$y(t)$	increasing	increasing	decreasing	increasing

9. Concave up to concave down: $y(0) = 0.25$
Concave down to concave up: $y(0) = 3.75$

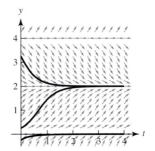

10.

$y(0)$	-0.25	0.25	3.25
$\lim_{t\to\infty} y(t)$	0	2	2

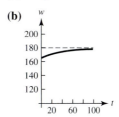

11. 1.45

12. (a) 180 pounds

(b)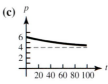

(c) 170.6 pounds

13. (a) $\dfrac{dp}{dt} = 0.02p\left(1 - \dfrac{p}{8}\right) - 0.01p$, $p(0) = 6$, with t in years and p is millions. (b) $p = 0$, unstable, and $p = 4$, asymptotically stable

(c)

14. (a) $\dfrac{dp}{dt} = 0.02p\left(1 - \dfrac{p}{8}\right) - 0.0144$, $p(0) = 6$

(b) $p = 0.8$, unstable, and $p = 7.2$, stable
(c) The graph is increasing and concave down, and $\lim_{t \to \infty} p(t) = 7.2$.

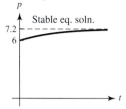

CHAPTER 10

Exercises 10.1

5. (a) 1, 4, 6, 4, 1   **(b)** 1, 5, 10, 10, 5, 1
    (c) 1, 6, 15, 20, 15, 6, 1

7. $(u+v)^n = c_0 u^n + c_1 u^{n-1} v + c_2 u^{n-2} v^2 + \cdots + c_{n-1} u v^{n-1} + c_n v^n$, where $c_k = \dfrac{n(n-1)\cdots(n-k+1)}{k!}$.

9. $1 - (x-1) + (x-1)^3$

11. $1 - 4(x+1) + 6(x+1)^2 - 4(x+1)^3 + (x+1)^4$

13. $(x-1) - \frac{1}{2}(x-1)^2 + \frac{1}{3}(x-1)^3 - \frac{1}{4}(x-1)^4 + \frac{1}{5}(x-1)^5 - \frac{1}{6}(x-1)^6$

15. $1 - x^2 + \frac{1}{2}x^4$ **17.** $1 + \frac{1}{3}x - \frac{1}{9}x^2 + \frac{5}{81}x^3$

19. $2 + x^2 + \frac{1}{12}x^4 + \frac{1}{360}x^6$

21. $-1 + \frac{1}{2}(x-\pi)^2 - \frac{1}{24}(x-\pi)^4 + \frac{1}{720}(x-\pi)^6$

23. $2^k/k!$ **25.** $c_k = \dfrac{(-1)^{k+1}}{k}$, $k = 1, 2, 3, \ldots$, and $c_0 = 0$.

27. $(-1)^k (k+1)$

29. $c_{2k} = 0, \; c_{2k+1} = (-1)^k/(2k+1)!$

31. 0.905, $|\text{error}| \le 0.000167$

33. For $n = 1$, $\sqrt{5} \approx 2.25$ with $|\text{error}| \le 1/64$.
For $n = 2$, $\sqrt{5} \approx 2.234375$ with $|\text{error}| \le 1/512$.
For $n = 3$, $\sqrt{5} \approx 2.236328$,
$|\text{error}| \le \frac{5}{16,384} \approx 0.000305$

35. (a) $x - \dfrac{x^2}{2} + \dfrac{x^3}{3} - \cdots + (-1)^{n+1}\dfrac{x^n}{n}$ **(b)** $n \ge 5$

37. Taylor polynomial: 1.296875. Fundamental Theorem: $2(\sqrt{e} - 1) \approx 1.29744$.

39. Taylor polynomial: 0.99167. Fundamental Theorem: 1

41. $\int_0^{0.1} (1 - x^2)\, dx \approx 0.09967$.   **43.** 13,867 square feet

45. $t + \frac{1}{2}t^2 + \frac{1}{3}t^3$; 0.10533   **47.** $-1 + t - \frac{1}{2}t^2$; -0.905

49. (a) $P_4(y) = 38 + 1.051y - 1.576y^2 + 4.202y^3 - 13.658y^4$
   (b) See Table 10.1.49b below.   **(c)** $B(y) = (y+0.2)^{0.4} + 38 - 0.2^{0.4} \approx (y+0.2)^{0.4} + 37.475$. See Table 10.1.49c below.

51. $0.31 + 0.00428t + 0.00002t^2$; 0.3143

53. $P_3(x) = 1 + \dfrac{x^2}{2} + \dfrac{x^3}{6}$

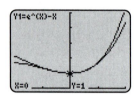

55. $P_5(x) = x - x^2 + \dfrac{x^3}{2} - \dfrac{x^4}{6} + \dfrac{x^5}{24}$

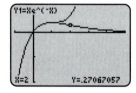

57. $P_3(x) = -1 + 2x - 2x^2 + 2x^3$

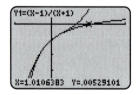

Exercises 10.2

1. Converges to 3   **3.** Converges to 5/6

5. Converges to $e/(e-1)$ **7.** Converges to 1

9. 7/33   **11.** 211/99   **13.** \$19,545.80

y	0	0.1	0.2	0.3	0.4	0.5
$P_4(y)$	38	38.092	38.159	38.176	38.088	37.803

Table 10.1.49b

y	0	0.1	0.2	0.3	0.4	0.5
$B(y)$	38	38.093	38.168	38.233	38.290	38.342

Table 10.1.49c

15. (a) 107. 5896 feet   **(b)** 273 feet

17. (a) $10(e^{-0.8} + e^{-1.6} + e^{-2.4})$

   (b) $10\sum_{k=1}^{n} e^{-0.8k}$ **(c)** $10\sum_{k=1}^{\infty} e^{-0.8k} \approx 8.16$ mg

19. $\dfrac{1000}{e^{0.11} - 1} \approx 8{,}600$ **21.** $\sum_{k=0}^{\infty}(-1)^k \dfrac{x^k}{2^{k+1}}$

23. $-\sum_{k=1}^{\infty} \dfrac{(x-1)^k}{k}$ **25.** $\sum_{k=0}^{\infty} \dfrac{(\ln 2)^k}{k!} x^k$

27. $\sum_{k=0}^{\infty}(k+1)3^k x^k$ **29.** $-\sum_{k=1}^{\infty} \dfrac{t^k}{k}$

31. (b) $1 + 2x + x^2$; $1 + 3x + 3x^2 + x^3$;
   $1 + 4x + 6x^2 + 4x^3 + x^4$;
   $1 + 5x + 10x^2 + 10x^3 + 5x^4 + x^5$

   (c) $\sum_{k=0}^{\infty}(-1)^k x^k$ **(d)** $\sum_{k=0}^{\infty}(-1)^k (k+1)x^k$

37.

39. $\sum_{k=0}^{10} \dfrac{1}{k!} \approx 2.71828180$

41. $\sum_{k=1}^{10} \dfrac{(-1)^{k+1}(0.5)^k}{k} \approx 0.405435$

43. $\sum_{k=1}^{50} \dfrac{(-1)^{k+1}}{k} \approx 0.683247$

45. $\dfrac{\pi}{9} - \dfrac{(\pi/9)^3}{3!} + \dfrac{(\pi/9)^5}{5!} \approx 0.34202$

Exercises 10.3

1. 1   **3.** 1   **5.** 2   **7.** 0   **9.** 1

11. $\sum_{k=0}^{\infty} \dfrac{x^{k+2}}{k!}$; $(-\infty, \infty)$ **13.** $\sum_{k=0}^{\infty} x^{2k+1}$; $(-1, 1)$

15. $\sum_{k=0}^{\infty} \dfrac{2x^{2k+1}}{2k+1}$; $(-1, 1)$ **17.** $\sum_{k=0}^{\infty}(-1)^k \dfrac{x^{2k+1}}{(2k)!}$; $(-\infty, \infty)$

19. $\sum_{k=0}^{\infty}(-1)^k \dfrac{x^{2k}}{(2k+1)!}$; $(-1, 1)$

23. $\sum_{k=0}^{\infty}(k+1)x^{2k}$; $(-\infty, \infty)$

25. $\sum_{k=1}^{\infty} \dfrac{(-1)^{k+1}}{k(2k+1)2^{2k+1}} \approx 0.0389$ (summing to $n = 5$)

27. $1 + t - \dfrac{t^2}{2!} - \dfrac{t^3}{3!} + \dfrac{t^4}{4!} + \dfrac{t^5}{5!}$

29. $2 - \dfrac{t^2}{2} + \dfrac{t^3}{6} - \dfrac{t^4}{24} + \dfrac{t^5}{60}$

31. (a) $h(0) = 0,\ h'(0) = 0$ **(b)** $t^3 - \dfrac{t^5}{20}$ **(c)** 0.0156 cm

Chapter 10: Review Exercises

1. $(x-1)^3 + 5(x-1)^2 + 3(x-1) + 2$

3. $1 + \dfrac{x}{2} + \dfrac{x^2}{8} + \dfrac{x^3}{48}$

5. $1 + \frac{1}{2}x - \frac{1}{8}x^2 + \frac{1}{16}x^3 - \frac{5}{128}x^4$

7. $-1 + \dfrac{(x-\pi)^2}{2} - \dfrac{(x-\pi)^4}{24}$ **9.** $c_n = 1$

11. $c_{2k} = \dfrac{(-1)^k}{(2k)!},\ c_{2k+1} = 0,\ k = 0, 1, 2, \ldots$

13. $0.60651,\ |\text{error}| \leq \dfrac{1}{6!2^5} \approx 0.000043403$

15. $\displaystyle\int_0^{0.2} \left(1 + \dfrac{x^2}{2}\right) dx \approx 0.2013$ **17.** $x - \dfrac{x^3}{3}$

19. $P_5(t) = t + \dfrac{t^2}{2} + \dfrac{t^3}{3} + \dfrac{t^4}{4} + \dfrac{t^5}{5}$; 0.68854

21. Diverges   **23.** Converges to 1   **25.** $\dfrac{20}{9}$ **27.** $\dfrac{3437}{3330}$

29. Yes. If it mines forever it will take out 1.3×10^8 cubic meters.

31. $\sum_{k=0}^{\infty} \dfrac{(-1)^k (k+1)x^k}{2^{k+1}}$ **33.** $1 + \sum_{k=1}^{\infty} \left(\dfrac{1}{k!} + \dfrac{1}{(k-1)!}\right) x^k$

35. $1 + \sum_{k=1}^{\infty}(-1)^k \left(\dfrac{1}{(2k)!} - \dfrac{1}{(2k-1)!}\right) x^{2k}$

37. (a) $1 + \dfrac{1}{2}x + \sum_{k=2}^{\infty}(-1)^{k+1} \dfrac{1 \cdot 3 \cdot 5 \cdots (2k-3)}{k! 2^k} x^k$

   (b) 0.9486836

39. 1   **41.** 1   **43.** 3/2

45. $\sum_{k=1}^{\infty}(-1)^{k+1} \dfrac{x^{k+2}}{k}$, $(-1, 1)$

47. $\sum_{k=0}^{\infty}(-1)^k \dfrac{x^{2k+1}}{(2k+2)!}$, $(-\infty, \infty)$

49. $\sum_{k=1}^{\infty} \dfrac{1}{(2k+1)\, 2^{2k+1}} \approx 0.0493$ (summing to $n = 5$)

51. $1 + \frac{1}{2}t^2 + \frac{1}{6}t^3 + \frac{1}{40}t^5$

Chapter 10: Exam

1. $P_1(x) = x$, $P_2(x) = x + 2x^2$, $P_3(x) = x + 2x^2 + 2x^3$

2. $P_1(x) = 5 + 2(x - 3)$, $P_2(x) = 5 + 2(x - 3) - \frac{1}{2}(x - 3)^2$,
$P_3(x) = 5 + 2(x - 3) - \frac{1}{2}(x - 3)^2 + \frac{7}{6}(x - 3)^3$
$P_4(x) = 5 + 2(x - 3) - \frac{1}{2}(x - 3)^2 + \frac{7}{6}(x - 3)^3 + \frac{1}{4}(x - 3)^4$

3. $4 + 6(x - 1) + 4(x - 1)^2 + (x - 1)^3$

4. $1 + (x - 2) + (x - 2)^2 + (x - 2)^3$

5. $(r - 1) + \frac{1}{2}(r - 1)^2$ **6.** 0.18267

7. $\int_0^{0.2} e^{-t^4}\,dt \approx \int_0^{0.2}(1 - t^4)\,dt = 0.199936$

8. $P_2(2.2) = 3.1$, where $P_2(t) = 3 + (t - 2) - \frac{5}{2}(t - 2)^2$

9. $\displaystyle\sum_{k=0}^{\infty} (-1)^k \frac{x^k}{k!\,2^k}$;   all x.

10. $\displaystyle\sum_{k=0}^{\infty} (x - 1)^k$; $0 < x < 2$ **11.** 200 billion dollars

12. **(a)** $8e^{-0.4}$ at end of 1 day, $8e^{-0.4} + 8e^{-0.8}$ at end of 2 days
   (b) $\displaystyle\sum_{k=1}^{\infty} 8e^{-k(0.4)} = \frac{8e^{-0.4}}{1 - e^{-0.4}} \approx 16.266$

13. $B(10) \approx P_2(10) = 6.3$, where
$P_2(t) = 3 + 0.24t + 0.00912t^2$.

14. $w(20) \approx P_2(20) = 26.116$, where
$P_2(t) = 26.5 - 0.02125t + 0.0000875t^2$.

CHAPTER 11

Exercises 11.1

1. There are 12 outcomes.
$S = \{(H, 1), (H, 2), (H, 3), (H, 4), (H, 5), (H, 6),$
$(T, 1), (T, 2), (T, 3), (T, 4), (T, 5), (T, 6)\}$

3. There are 12 outcomes.
$S = \{(1, 2), (1, 3), (1, 4), (2, 1), (2, 3), (2, 4),$
$(3, 1), (3, 2), (3, 4), (4, 1), (4, 2), (4, 3)\}$

5. There are 24 outcomes.
$$S = \left\{\begin{array}{llllll} abc, & abd, & acb, & acd, & adb, & adc, \\ bac, & bad, & bca, & bcd, & bda, & bdc, \\ cab, & cad, & cba, & cbd, & cda, & cdb, \\ dab, & dac, & dba, & dbc, & dca, & dcb \end{array}\right\}$$

7. There are infinitely many outcomes.
$S = \{H, TH, TTH, TTTH, TTTTH, \ldots\}$

9. There are infinitely many outcomes. $S = (0, \infty)$, the set of all positive real numbers.

11. **(a)** $S = (0, \infty)$, the set of all positive real numbers.
   (b) $E \cap F$: "more than 5 and less than 8 minutes."
   F': "at least 8 minutes."
   $E' \cap G$: "more than 4.5 but not more than 5 minutes."
   $F \cap G'$: "not more than 4.5 minutes."

$E \cup G'$: "either more than 5 minutes or else not more than 4.5."
   (c) $G' \cap E = \emptyset$ (the empty set)
$E \cup G = G$
$F \cup G = S$ (the entire sample space)
$E \cap G = E$

13. E and G **15.** 0.5 **17.** **(a)** $8/9$ **(b)** $1/3$ **(c)** $4/9$

19. **(a)** $1/6$ **(b)** $1/18$ **(c)** $1/2$ **(d)** $1/4$ **(e)** $3/4$

21. The sample space consists of the 16 outcomes (i, j), where $i = 1, 2, 3, 4$ and $j = 1, 2, 3, 4$.
   (a) $\{(2, 2), (2, 4), (4, 2), (4, 4)\}$, with probability $1/4$
   (b) $\{(1, 2), (1, 4), (2, 1), (2, 2), (2, 3), (2, 4), (3, 2), (3, 4),$
$(4, 1), (4, 2), (4, 3), (4, 4)\}$, with probability $3/4$
   (c) $\{(2, 4), (3, 3), (4, 2)\}$, with probability $3/16$
   (d) $\{(1, 1), (2, 2), (3, 3), (4, 4)\}$, with probability $1/4$
   (e) $\{(1, 2), (1, 3), (2, 1), (2, 2), (2, 4), (3, 1), (3, 3), (3, 4),$
$(4, 2), (4, 3)\}$, with probability $5/8$

23. **(a)** $1/2$ **(b)** 0

<table>
<tr><td></td><td></td><td colspan="2" align="center">Mother</td></tr>
<tr><td></td><td></td><td align="center">N</td><td align="center">t</td></tr>
<tr><td rowspan="2">Father</td><td align="center">N</td><td align="center">NN</td><td align="center">Nt</td></tr>
<tr><td align="center">N</td><td align="center">NN</td><td align="center">Nt</td></tr>
</table>

25. 0.7975 **27.** **(a)** 0.7 **(b)** 0.2

29. **(a)** 0.06 **(b)** 0.44 **(c)** 0.74 **(d)** 0.18

31. **(a)** $1/64$ **(b)** $7/64$ **(c)** $7/64$ **33.** **(b)**

Exercises 11.2

1. **(a)** $3/4$ **(b)** $1/2$

3. **(a)** $5/16$ **(b)** $1/2$ **(c)** $3/8$

5. **(a)** $2/5$ **(b)** $2/9$

7. **(a)** $9/14$ **(b)** $5/8$ **(c)** $3/8$ **(d)** $1/3$

9. **(a)** $1/7$ **(b)** $2/7$ **(c)** $5/7$

11. **(a)** $1/17$ **(b)** $19/34$ **(c)** $13/34$

13. **(a)** 0.138 **(b)** 0.056 **(c)** 0.007728

15. 0.003 **17.** A and C, B and C, but not A and B

19. No   **21.** 0.03 **23.** **(a)** $1/4$ **(b)** $1/4$

25. **(a)** 0.1521 **(b)** 0.069966

27. **(a)** 0.6552 **(b)** 0.9512

29. **(a)** $9/17$ **(b)** $8/17$

31. **(a)** $\frac{5}{4,999,999}$ **(b)** $5/6$

Exercises 11.3

1. **(a)**

x	0	1	2	3
$P(X = x)$	$1/8$	$3/8$	$3/8$	$1/8$

(b) $E(X) = 3/2$, $\text{Var}(X) = 3/4$, $\sigma(X) = \sqrt{3}/2$

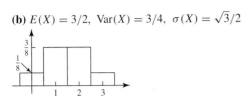

3.

x	0	1	2	3	4	5
$P(X = x)$	1/6	5/18	2/9	1/6	1/9	1/18

$E(X) = 35/18$, $\text{Var}(X) = 2.05$

5.

x	6	7	8	9	10
$P(X = x)$	1/10	3/20	2/5	1/5	3/20

$E(X) = 8.15$,
$\text{Var}(X) = 1.3275$, $\sigma(X) = \sqrt{1.3275} \approx 1.15217$

7.

Distance	10–19	20–29	30–39	40–49	50–59
Frequency	15/53	12/53	14/53	7/53	5/53

Average distance is 29.783

9. (a) $2/13 \approx 0.15$ **(b)** \$15.38

11.

x	1	2	4
$P(X = x)$	1/6	1/4	7/12

Exercises 11.4

1. $P(k \text{ heads}) = \binom{3}{k}(0.2)^k(0.8)^{3-k}$

k	0	1	2	3
$P(k \text{ heads})$	0.512	0.384	0.096	0.008

3. (a) $\binom{4}{1}\left(\frac{1}{6}\right)\left(\frac{5}{6}\right)^3 = \frac{125}{324} \approx 0.386$

(b) $\binom{4}{0}\left(\frac{5}{6}\right)^4 + \binom{4}{1}\left(\frac{1}{6}\right)\left(\frac{5}{6}\right)^3 = \frac{125}{144} \approx 0.868$

5. $10\left(\frac{1}{2}\right)^{10} = \frac{5}{512} \approx 0.01$

7. (a) 10 **(b)** 15 **(c)** 126

9. $\binom{5}{4}(0.5)^5 \approx 0.156$, $(0.5)^5\left(1 + \binom{5}{4}\right) \approx 0.1875$

11. (a) $\sum_{k=0}^{4} \binom{50}{k}\left(\frac{1}{9}\right)^k\left(\frac{8}{9}\right)^{50-k} \approx 0.3348$

(b) $1 - \sum_{k=0}^{10} \binom{50}{k}\left(\frac{1}{9}\right)^k\left(\frac{8}{9}\right)^{50-k} \approx 0.0196$

13. (a) $\binom{12}{7}(0.58)^7(0.42)^5 \approx 0.2285$ **(b)** 29

15. (a) $(0.8)^{10} \approx 0.1074$ **(b)** $\binom{10}{9}(0.8)^9(0.2) \approx 0.2684$
(c) $\sum_{k=0}^{2} \binom{10}{k}(0.2)^k(0.8)^{10-k} \approx 0.6778$

17. (a) 91 **(b)** 455 **(c)** 4845 **(d)** 161,700

19. $E(X) = 52$, $\text{Var}(X) = 24.96$

21. (a) 4140 **(b)** \$33,120 **(c)** Number sold between 4068 and 4212, profit between \$32,544 and \$33,696.

23. 3.46672×10^{-60}

Exercises 11.5

7. (a)

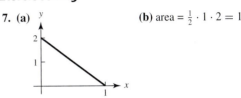

(b) area $= \frac{1}{2} \cdot 1 \cdot 2 = 1$

(c)

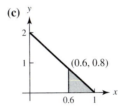

probabiliy $=$ shaded area $=$ $\frac{1}{2}(0.4)(0.8) = 0.16$

9. Yes **11.** Yes **13.** No **15.** No **17.** 12

19. (a) $4000/3$ **(b)** $1/9$

21. (a) $2e^{-1} \approx 0.736$ **(b)** $1 - 6e^{-5} \approx 0.96$

23. $F(x) = x^{1/2} - 1$, $1 < x < 4$

25. $F(x) = \begin{cases} \dfrac{1}{2} + x + \dfrac{x^2}{2} & \text{if } -1 < x \le 0 \\ \dfrac{1}{2} + x - \dfrac{x^2}{2} & \text{if } 0 < x < 1 \end{cases}$

27. (a) $1/2$ **(b)** $\frac{1}{2}(\sqrt{3} - \sqrt{2}) \approx 0.159$ **(c)** $f(x) = \dfrac{1}{4\sqrt{x-2}}$

29. $E(X) = 1/3$, $\text{Var}(X) = 1/18$

31. $E(X) = 1$, $\text{Var}(X) = 1/5$

33. $E(X) = 1/2$, $\text{Var}(X) = 1/4$ **35.** 2 minutes

37. $E(X) = 10$, $\text{var}(X) = 100$, $\sigma(X) = 10$

39. $E(X) = 1.5$, median $= 1.5$

41. $E(X) = 3/4$, median $= \left(\frac{1}{2}\right)^{1/3} \approx 0.7937$

43. $E(X) = 1$, median $= \ln 2$

47. mean $= 71.143$, median $= 73$

Exercises 11.6

1. $1/4$ **3. (a)** $1/10$ **(b)** $\left(\frac{1}{10}\right)^4 = 0.0001$

5. (a) 3 minutes **(b)** $1 - e^{-1} \approx 0.632$

7. $\frac{1}{14.4} \approx 0.0694$ **9. (a)** $e^{-\lambda a}$ **(b)** $e^{-\lambda a}$ **13.** 0.0603

15. 0.3156 **17.** 0.7158

19. (a) 0.0668 **(b)** 0.0139 **(c)** 0.6289

21. 0.1587 **23. (a)** 0.0027 **(b)** 0.1922

25. About 2.3 %

Chapter 11: Review Exercises

1. There are 12 outcomes.
$S = \{ab, ac, ad, ba, bc, bd, ca, cb, cd, da, db, dc\}$

3. There are 11 outcomes.
$S = \{HH, HTH, THH, HTTH, HTTT, THTH,$
$THTT, TTHH, TTHT, TTTH, TTTT\}$

5. $A \cup B = \{HHHH, HHHT, HHTH, HHTT,$
$HTHH, HTHT, HTTH, HTTT,$
$THHT, THTT, TTHT, TTTT\}$
$A \cap B = \{HHHT, HHTT, HTHT, HTTT\}$
Not mutually exclusive.

7. $P(s_1) = 1/3$, $P(s_2) = 1/3$, $P(s_3) = 1/9$, $P(s_4) = 2/9$

9. (a) $1/8$ **(b)** $1/2$ **(c)** $7/8$ **11.** $5/14$

13. $P(E \cap F) = 0.2$, $P(E \cup F) = 0.6$

15. (a) $1/64$ **(b)** $3/32$ **17. (a)** $2/9$ **(b)** $1/9$ **(c)** $2/5$

19. (a) $1/17$ **(b)** $4/51$

21. In general, no. For example, draw a card and let A be "the card is a spade" and B be "the card is a heart."

23. Start preparing the dishes. The probability the chef will prepare both is 0.765.

25. (a) $13/51$, $13/51$, $13/51$, $12/51$ **(b)** All are $13/204$, except the probability that the first and second are both spades, which is $1/17$. **(c)** $1/4$

27. (a) independent **(b)** not independent **(c)** independent **(d)** independent **(e)** not independent

29. 0.4096, 0.1296, 0.046656

31. $E(X) = 0.39$, $\mathrm{Var}(X) = 3.0379$

33. (a) See Table 11.33 below.
(b) $E(X) = 3.65$, $\mathrm{Var}(X) = 3.5775$,
$\sigma(X) = \sqrt{3.5775} \approx 1.89$ **(c)** $3/40$

35. $\binom{6}{k}\left(\frac{1}{26}\right)^k\left(\frac{25}{26}\right)^{6-k}$, $k = 0, 1, \ldots, 6$

37. (a) 126 **(b)** 3003 **(c)** 1 **(d)** 2714

39. (a) $\left(\frac{8}{9}\right)^{18}$ **(b)** 2 **(c)** Between 1 and 3 inclusive

41. $F(x) = 1 - x^{-2}$, $x > 1$

43. (a) $\sqrt{3/4} - \sqrt{1/4} \approx 0.366$ **(b)** $\dfrac{1}{2\sqrt{1-x}}$, $0 < x < 1$

45. $\mu = 4/3$, $\sigma^2 = 2/9$ **47.** $\mu = 7/12$, $\sigma^2 = 11/144$

49. Mean $= 4/3$, median $= 1$

51. Mean $= 2$, median ≈ 1.678

53. $1/\sqrt{3}$ **55. (a)** $1 - e^{-2} \approx 0.8647$ **(b)** $1/2$ minute

57. (a) 0.76 **(b)** 0.40 **(c)** 0.64

59. Store A: 0.87, Store B: 0.39

Chapter 11: Exam

1. (a) 0.4 **(b)** 0.7 **2. (a)** $9/10$ **(b)** $7/10$ **3.** 0.14

4. No, no, yes **5. (a)** $1/3$ **(b)** $1/2$

6. (a) $1/10$ **(b)** $9/10$ **(c)** $1/3$

7. (a) No. $P(A) = 1/6$, $P(B) = 5/36$,
$P(A \cap B) = 1/36 \neq P(A) \cdot P(B)$
(b) Yes. $P(A) = 1/6$, $P(C) = 1/6$,
$P(A \cap C) = 1/36 = P(A) \cdot P(C)$

8. (a) 0.089 **(b)** $\frac{0.075}{0.089} \approx 0.84$

9. (a) 0.5 **(b)** 0.45 **(c)** $\sqrt{0.45} \approx 0.67$ **10.** $-\$0.75$

11. (a) $\binom{20}{k}\left(\frac{2}{5}\right)^k\left(\frac{3}{5}\right)^{20-k}$, $k = 0, 1, 2, \ldots, 20$. **(b)** 8

12. (a) $3/95$ **(b)** $3/35$ **13. (a)** 2 **(b)** $4/15$

14. $F(x) = 1 - e^{-2x}$, $x > 0$

15. (a) $4/3$ **(b)** $\sqrt{2}$ **16. (a)** $8/3$ **(b)** $8/9$

17. $E(X) = 2$, $\mathrm{Var}(X) = 1/3$

18. (a) $1 - e^{-2/15} \approx 0.125$ **(b)** $e^{-14/15} \approx 0.393$

19. (a) 0.0548 **(b)** 0.3446 **(c)** 0 **20. (a)** 7.8% **(b)** 11.7%

x	0	1	2	3	4	5	6	7	8
$P(X = x)$	1/20	3/40	3/20	9/40	7/40	3/20	1/10	1/20	1/40

Table 11.33

Photo Credits

Index